Zoophysiology Volume 37

Editors:
S.D. Bradshaw   W. Burggren
H.C. Heller   S. Ishii   H. Langer
G. Neuweiler   D.J. Randall

Springer-Verlag Berlin Heidelberg GmbH

Zoophysiology

J.N. Maina

The Gas Exchangers

Structure, Function, and Evolution
of the Respiratory Processes

With 108 Figures and 33 Tables

 Springer

Prof. John N. Maina
Faculty of Veterinary Medicine
Dept. of Veterinary Anatomy
University of Nairobi
P.O. Box 30197
Nairobi, Kenya

Faculty of Health Sciences
Dept. of Anatomical Sciences
The University of Witwatersrand
7 York Road, Parktown
Johannesburg 2193
South Africa

Cover illustration: A plastron of a crane fly larva *Dicranomyia* (Fig. 107)

ISBN 978-3-642-63756-8
ISSN 0720-1842

Library of Congress Cataloging-in-Publication Data

Maina, J.N.
 The gas exchangers: structure, function, and evolution of the respiratory processes / J.N. Maina.
  p.   cm. – (Zoophysiology; v. 37)
 Includes bibliographical references and index.
 ISBN 978-3-642-63756-8   ISBN 978-3-642-58843-3 (eBook)
 DOI 10.1007/978-3-642-58843-3
 1. Respiration. 2. Physiology, Comparative. I. Title. II. Series.
 QP121.M276 1998
 573.2 — dc21

Cover design: Design & Production GmbH, Heidelberg
Typesetting: Best-set Typesetter Ltd., Hong Kong

SPIN: 10521202   31/3137 – 5 4 3 2 1 0 – Printed on acid-free paper

"Biologists are seeking to integrate studies of morphology, development, physiology, ecology, systematics, and behaviour in order to understand how species and lineages deal with their environments and how they have diversified." Wake (1990)

Dedication

This book would not have been written without the patience, understanding, and support of my wife, Wanjuku, and our children, Ndegwa, Wanjiru, and Kireru. I am particularly indebted to them for their forbearance when the work was a major preoccupation. They kept urging me on especially when the "spirit was willing but the body weak" by firmly and persistently inquiring when "the book" would be ready. I dedicate the work to them.

It has been both a challenge and pleasure writing this book. I hope that the reader will partake in some of it.

Acknowledgement

I am immensely grateful to the numerous colleagues who have collaborated with me over the years and given me their ideas and time unreservedly. All cannot be mentioned here. The particularly noteworthy ones include Prof. A.S. King (Emeritus Prof.), University of Liverpool; Dr. M.A. Abdalla, King Saud University; Prof. G.M.O. Maloiy, formerly of the University of Nairobi, Prof. S.P. Thomas, Duquesne University, and Prof. C.M. Wood, McMaster University. I am indebted to Prof. D.J. Randall, Department of Zoology, University of British Columbia, the editor responsible for this Volume, for reading the manuscript and making most helpful suggestions. My most sincere thanks go to Dr. Dallas M. Hyde, Chair, Department of VM: Anatomy, Physiology and Cell Biology, University of California (Davis) for having me in his laboratory and putting all necessary facilities at my disposal during my 1996 year Fulbright sabbatical leave, when substantial organization of this work was carried out. I wish to record my gratitude to Messrs. M. Mwasela-Tangai and J. Gachoka for assistance with the preparation of the illustrations and Ms L. Mburu and J. Muhia for initial word processing of some parts of the manuscript. My past and present graduate students have been integral in shaping the thoughts expressed in this exposition. To all of them, I express my appreciation. Special thanks go to Profs. R.D. Farley (University of California – Berkeley) and D.O. Okello (University of Nairobi) for providing me with original copies of Figures 15 and 54, respectively, for reproduction and the following publishers for their generous permission to reproduce materials from their publications:

Academic Press Ltd.
Annual Review Inc.
Birkhahauser-Verlag
Cambridge University Press
Elsevier/North Holland Publishing Company
Harvard University Press
Kluwer Publications Ltd.
Longman Group (UK) Ltd.
McGraw-Hill Inc.
McMillan Magazines Ltd.
Oxford University Press
Pergamon Press
Springer-Verlag
The Company of Biologists Ltd.
Wiley/Liss Inc.

I am indebted to the editorial staff at the Springer-Verlag office (Heidelberg) for their numerous courtesies.

Johannesburg, March 1998 _J.N. Maina_

Preface

"Amongst animals, diversity of form and of environmental circumstances have given rise to a multitude of different adaptations subserving the relatively unified patterns of cellular metabolism. Nowhere else is this state of affairs better exemplified than in the realm of respiration". Jones (1972).

The field of comparative respiratory biology is expanding almost exponentially. With the ever-improving analytical tools and methods of experimentation, its scope is blossoming to fascinating horizons. The innovativeness and productivity in the area continue to confound students as well as specialists. The increasing wealth of data makes it possible to broaden the information base and meaningfully synthesize, rationalize, reconcile, redefine, consolidate, and offer empirical validation of some of the earlier anecdotal views and interpretations, helping resolve the issues into adequately realistic and easily perceptible models. Occasional reflections on the advances made, as well as on the yet unresolved problems, helps chart out new grounds, formulate new concepts, and stimulate inquiry. Moreover, timely assessments help minimize isolation among investigators, averting costly duplication of effort. This exposition focuses on the diversity of the design of the gas exchangers and gives a critical appraisal of the plausible factors that have motivated or constrained the evolvement of respiration. The cause-and-effect relationship between the phylogenetic, developmental, and environmental factors, conditions, and states which at various thresholds and under certain backgrounds conspired in molding the gas exchangers is argued. Convergence as well as divergence, retrogression and progression, parallel as well as serial developments have occurred in this stochastic process manifest with recurrent catastrophic crises. Gould (1994) asserts that "life's pathways are more contingent and chancy than predictable and directional". Such a caveat notwithstanding, with judicious use of data, an integrative multidisplinary approach into the evolvement of the gas exchangers should help develop a conceptual framework of the appropriation and synthesis of the myriad parameters and components involved in the assembly of the gas exchangers. The repertoire of the functional mechanisms conceived, the investments made to support such developments, the losses incurred, and the gains reaped in the course of the development of respiration can be grasped. The adaptive attributes common to or unique to broadly defined groups of organisms can be adduced. On that basis, the connectivity, the diversity, and the chronology of the changes that occurred during and after the incorporation of molecular O_2 into the respiratory process can be gauged. To place the historical developments of respiration in their proper perspective, an ecopaleobiological approach has been adopted. Borrowing heavily from works of others, pertinent past and present states, events, and circumstances have been associated. I have been consciously eclectic in order to remain both brief and focused on the subject matter. The embracing approach adopted provides instructive parametric insights into the permutations and spatial and temporal vectorial shifts of the causative factors during the genesis of the respiratory processes. This should explain how and why particular respiratory traits were acquired, some lost, and the means by which solutions to attendant challenges were found as the respiratory machines were forged. In evolutionary terms, contemporary animals are living edifices of past events and developments. They give us a narrow, nebulous window through which to espy and presage the assaults that life has endured and the changes which may have occurred. Kardong (1995) observes that "the architectural design of an organism expresses something about the processes that produced, it, the history out of which it came, and the functions its parts perform."

To present a truly comparative account, detailed considerations of individual animal species have been avoided. Instead, broad taxonomic groups are considered. The shared and dissimilar strategies, principles, mechanisms, and themes are instead rationalized. Data on individual species are given only to explain a feature, a process, a concept, or a theory. Effort has been made to balance out physiology with morphology, the latter having been largely relegated to the background in practically all earlier publications on the subject. This approach is adopted in the firm belief that morphology is not simply a synonym of anatomy, i.e., a description of the structure, topography, and composition of inert biological entities for their own sake. The discipline delves into the logical basis of form, casting back on the true spirit of the etymology of the word. Perceptively, Thompson (1959, p.14) expressed this notion as follows: "morphology is not only a study of material things and of the forms of material things, but has its dynamical aspect, under which we deal with the interpretation, in terms of force, of the operations of Energy."

This book was written with a broad readership in mind. Graduate students as well as established biologists, be they zoologists, physiologists, morphologists, paleontologists, or ecologists who work or may contemplate working on materials and aspects of respiration in whole organisms will find it useful. Scientists in Earth Sciences with an interest in the interactive developmental processes between the physical and biological realms will find the book of certain interest and perhaps stimulating. No apology is made that a rather lengthy section of the work has been devoted to the biophysical aspects of the respiratory media, habitats, and the accretion of molecular O_2. The "history" of the evolution of the respiratory processes and that of gas exchangers in particular would be in great default without a clear grasp of the setting in which the changes occurred, the intractable challenges faced, and the successes and failures encountered as animals strived not only to survive but also to establish themselves and flourish in alien habitats. Some of the aspects covered here have taken me outside the limits of my personal experience and expertise. In such cases, I well realize that I may not have articulated the issues with the competence required by the specialists in those fields. Should any infelicities of judgment, inference, or omission have occurred, I beg the reader's indulgence, and would be most grateful if such aspects were pointed out to me. While consciously avoiding teleology, a rather difficult thing for one to do especially when rationalizing design, I have not hesitated to use explicative interpretations and suppositions where firm data are lacking. Fishman (1983) observed that "the understanding of biological systems rests on a combination of mechanisms, which deal with beginnings, and teleology, which deals with the concept of purpose and end", while Jones (1972) asserts that "frank and self-controlled (teleological) speculation is an essential part of the comparative approach". In such a long process, and one which has to be studied indirectly (due to lack of fossilized materials of soft tissues like the gas exchangers), it has to be conceded that "background noise", discernible as anomaly between the expected outcome and the reality, is to be anticipated. Evolutionary reconstruction of the history of a natural process is one of the most challenging problems that biologists have been called upon to solve. The skepticism with which such accounts are received was voiced by Rancour-Laferriere (1985), who stated that "anyone who attempts to narrate the evolution of something over geologic time is telling a story". Although, given the limitations imposed by the wide gaps in the fossil records, such skepticism is warranted, I consider such sentiments extreme and perhaps somewhat misplaced. When presenting a chronological account of a classical natural event, a measure of subjectivity should be allowed. More often than not, the best that can be accomplished is to build as complete a case as possible – based on intuitive sense, logical inferences, and known facts. A historically realistic model or a simulation of what could have happened should be fabricated and tested. This should show that, given specified conditions and circumstances, particular events were not only inevitable but most likely occurred. In the general purview of biology, Stebbins (1984) realistically observes that "the only law that holds without exception in biology is that exceptions exist for every law". I have explicitly stated where experimental facts are lacking and personal view has been proffered. I hope that such thoughts will provide areas for consideration and debate.

Due to limitation of space, it is not possible to give an encyclopedic review of the utterly immense aspects of the evolution and comparative respiratory morphology and physiology of the gas exchangers in a single work. There are excellent accounts, monographs, treatises, and perspectives on different areas of these aspects, especially on the vertebrates. The key publications are included in the reference list; they should be consulted where particular details are desired. While offering a synthesis of more recent findings, this account is meant to complement such comprehensive publications and not supersede them. I have purposely placed particular emphasis on the respiratory processes of the invertebrates, a group on which relatively less is known. In doing so, I hoped to avoid the conventional "prejudice" (against the group) and the parochial but somewhat legitimate interest and popular emphasis on the vertebrate aspects. In most people, mention of the word animal conjures up a vertebrate in their mind. With real setbacks, mammals in particular (a highly specialized taxon in most respects) have been used as the focal point (model group) in discussing comparative pulmonary structure and function. A prospective look at the design of the gas exchangers, i.e., from the lowest to the highest forms rather than retrospectively (i.e., back in time) has been made here in order to hopefully better fit the developments into the natural polarity of evolutionary change. This approach perhaps offers a more satisfactory causal explanation of the early developments in respiration. Moreover, reading evolution backwards has its intrinsic bias in that one inevitably sees an outcomes as the only possible result from a series of events. Nature is extremely complex and too unpredictable to be read easily. I hope that a satisfactory balance in the presentation has been achieved here. The designs of the gas exchangers in the so-called primitive animals (a very unfortunate term which has come to be equated with crude, mediocre, or imperfect, just as complex, superior, and advanced have come to be taken to mean better or efficient) are most instructive in understanding the structure and the operations of the more sophisticated processes of respiration. Tenney and Boggs (1985) observed that "the great events in respiratory system evolution can be appreciated best in lower classes". The diversity of the stratagems adopted for procuring molecular O_2 and eliminating CO_2 among animals having different lifestyles, occupying different habitats, and which have acquired remarkable morphological eccentricities are most evident when the underlying cost-benefit analysis and the compromises, the trade-offs (and even payoffs?), and the alternative solutions to the respiratory imperatives are comparatively examined. The protracted transactions and the transformations that went into the design and adoption of particular respiratory schemes and the perceivable limitations or capacities inherent in different gas exchangers can be recognized. A heuristic approach in biology reveals subtle areas of convergence, homology, analogy, and homeostasis, providing an explanation of the mechanisms which yielded particular states and phenomena. Nature, however, does not yield its secrets willingly. They have to be patiently gleaned and teased through protracted, well-planned inquiries and meticulous attention to details. A multidisplinary approach in such studies allows a broad understanding of a problem and offers robust answers.

The chapters in this account are sequenced to survey the respiratory events as they are conceived to have more or less chronologically unfolded. It is hoped that this approach offers a more logical understanding of the developments. Chapter 1 examines the fundamental attributes of life and its evolution on Earth, the accretion and fluctuations of O_2 and CO_2 in the biosphere, and the enigmatic nature of molecular O_2 in respiration and evolution of life itself and the intricate role it has played in the ecological adaptations of the animal life. Chapter 2 surveys the forms and designs of the gas exchangers, the underlying engineering principles which form the basis of their manifold plans and constructions, and argues out the factors which have imposed the different architectural contrivances. Chapter 3 considers the biophysical dictates of water and air as respiratory media and livable habitats, the adaptive respiratory needs for survival in unique environments and circumstances, the effects of the different respiratory milieus and habitats on the fabrication of the gas exchangers, the essence of the diverse respiratory stratagems adopted by different animal taxa, and the effect of gravity on the design and function of gas exchangers. Chapter 4 looks at the structure and function of the archetype aquatic gas exchanger – the gills – and considers the placenta as an ephemeral liquid-to-liquid gas exchanger. Chapter 5 examines the limitations and constraints which confronted animals during the switch from water- to air breathing and the pivotal import of transitional (bimodal) breathing in the traumatic process of terrestrial invasion. Chapter 6 outlines a cost-benefit analysis of what went into the utilization of the atmospheric O_2 and the structural and functional prescriptions in the evolved air-breathing organs and structures. I must regardless concede that in this treatise, not all aspects and issues pertaining to the evolution of respiration and the form and function of the gas exchangers have been covered and those which are may not have been exhausitively discussed and satisfactorily resolved. Based on personal preferences on the subject matter and keeping in perspective the restriction of space, many elisions were made. Only the weighed imperatives were included. My objective will have been more than adequately met if the discourse provides areas for reflection, points for discussion, or, better still, aspects for further investigations on the many yet gray areas.

Contents

Chapter 1

Perspectives on Life and Respiration: How, When, and Wherefore

"A reconstruction of the remote past must necessarily be based upon inference, rarely from systematically collected data, more often from an inadequate number of facts which chance has placed in the way of competent investigators who can recognize their significance. As time goes on some gaps are filled, others remain forever empty, but the picture as a whole becomes progressively clearer." Beadle (1974).

1.1 Life: Diversity, Complexity, and Uniformity Fabricated on Simplicity

Humankind has always been fascinated by the spectacle of extreme states and phenomena. The *Guinness Book of Records*, which after the *Holy Bible* is alleged to be the second most widely read book, is according to the publishers compiled "in hope of providing a means for peaceful setting of arguments about record performances". Though not given much attention outside the professional realms, the elegance and constellation of life on Earth is enchanting, bewildering, and intellectually intriguing. More than 284 000 species of plants, 750 000 species of insects, and 280 000 species of other animals have been catalogued (e.g., Dixon 1994; Service 1997). Of this plethora, vertebrates represent only one phylum and a mere 50 000 species or so (e.g., Pough et al. 1989). Nature's fortitude for survival is remarkable. For example, albeit the tumultuous crises which preceded the Tertiary period when colossal population clashes occurred and many species were wiped out, by the end of the period, there were as many as 2500 families of animals (Benton 1995). From molecular sequence studies of different microcosms (e.g., Pace et al. 1985; Ward et al. 1990; Winker and Woese 1991; Olsen and Woese 1993), it is becoming unequivocally evident that, compared with the Metazoa and the Metaphyta (i.e., the visible world), the microbial domain (i.e., the microworld) presents a more complex biodiversity than was hitherto thought, and quantitatively remains largely unknown to us (e.g., Embley et al. 1994; Lovejoy 1994). The existing numerical data on the taxonomic diversity of animal life differ remarkably. It is envisaged that life's copious tree comprises between 5 and 50 million species of animals (e.g., May 1988, 1990, 1992; Hammond 1992). This hopelessly wide range owns up to the fact that we know very little of the actual richness and geographical patterning of contemporary animal life (e.g., Wilson 1992; Colwell and Coddington 1994). Since most of the crucial environmental processes are driven by the microbial activities which involve biochemical nutrient recycling operations such as nitrogen fixation, sulfur oxidation and reduction, ammonification, methanogenesis, and methane oxidation (e.g., Capone et al. 1997), it can legitimately be anticipated that the microorganismal biomass (which must form the base of life) should overwhelm that of plants and animals. The methanogenic bacteria which occur in the hindguts of many arthropods contribute substantially to the loading of the atmosphere with methane (e.g., Hackstein and Stumm 1994). Fenchel (1992) estimated that a 1 cm^3 core of coastal marine sediment contains 4×10^{10} bacterial cells, 10^4 heterotrophic flagellates and amoebae, 10^8 chlorophyll a-containing microorganisms, and about 2000 ciliates. Expressing a personal view on the ambitious program named Systematics Agenda 2000, which envisages that all the Earth's species will be discovered, described, and classified in the next 25 years (Anonymous 1993), Wicksten (1994) doubts that "the world's organisms will be described within the next century – if ever"! The same may be said on the Fifty-Year Plan (e.g., Raven and Wilson 1992). Applying the lengthy traditional methods of identifying and cataloguing species, May (1990), Hawksworth (1991), and Hammond (1992) envisage that it will take a couple of centuries to gain an adequate understanding of species diversity on Earth. With speciation being a continuous process (e.g., Butlin and Tregenza 1997; Smith et al. 1997), there may not be an actual end to the task. From the microbes to the whales, living animals differ in mass by at least 21 orders of magnitude (McMahon and Bonner 1983; Schmidt-Nielsen 1984; Brown 1995). The staggering numerical density, specific diversity, and allometric disparity is a clear testimony of the tenacity, resiliency, richness, and innovativeness of nature's designs for survival. It remains a great challenge to the biologists to fully explain the factors which drive and determine specific patterning (Hutchison 1959; Brown 1981; Cracraft 1994; Butlin and Tregenza 1997).

Until the enunciation of the theory of evolution through natural selection (Darwin and Wallace 1858; Darwin 1859), the complexity of life on Earth was scientifically inexplicable. Since then, the kaleidoscopy has been perceived as integral to the dynamic process of evolution (e.g., Ruthen 1993), with the degree of profuseness and variability a mark of success and suitability to an environment. The spatial and temporal distribution of species is set by definite physical, biological, and environmental controls. Through the about 4 billion years (billion = 1 thousand million years = 10^9) of existence on Earth, living forms have adopted behavioral strategies, developed biochemical and functional capacities, and appropriated certain devices which enable them to occupy species-specific niches. Such stable states have been attained amidst profound ambient changes that have included variations in the rotation of the Earth, changes in the Earth's orbit, fluctuations in the average surface temperature, physical displacements of the continental plates, pulses in the O_2 and CO_2 levels, and fluctuations in the availability of important nutrients (e.g., Hayes et al. 1976; Hunt 1979; Ben-Avraham 1981; Boucot and Gray 1982; Walker et al. 1983; Bray 1985; Raymo and Ruddiman 1992). In their different ecological settings, the nature, dynamics, and heterogeneity of environments detail the dissimilitude of animal life: form and function are molded by the physical and biotic factors in the ecotopes. Based on the separate traits they have acquired through interfacing with environments, animal life can be grouped into aquatic, terrestrial, and aerial assemblages. Finer divisions place them into, e.g., nocturnal, diurnal, fossorial, pelagic, arboreal, benthonic types, etc. Since environments are dynamic spaces (e.g., Schaffer and Kot 1986; Doebeli 1993; Rand and Wilson 1993), organisms must equally be phenotypically fluid to continually adapt to the external cues (e.g., Stearns 1982;

Prosser 1986). While freely interacting with the environment, organisms must not compromise their physical and biological integrity as self-sustaining, self-regulating dynamical entities. Energy must be obtained from the environment to regulate their internal states and defend homeostasis. To steady the environmental oscillations, strictly, organisms are a process in a nonsteady state (e.g., Levins and Lewontin 1985; Wainright 1988). As urged by Simpson (1953), when considering adaptation, "it is equally or more useful to focus neither on environment nor on organisms but on the complex interrelationship in which they are not really separable". The term comprehensive selective regime was used by Baum and Larson (1991) to define the combination of all the environmental and organismic factors which prescribe how natural selection acts on organismal variations. Chronobiology (circadian rhythmicity), a process which has evolved to sense and counterbalance the nuances of the environment, pervades all levels of biological organization from molecular, cellular, to organismal (e.g., Aschoff and Pohl 1970; Bünning 1973; Sweeney 1987; Ishii et al. 1989; Prinzinger and Hinninger 1992; Lloyd and Rossi 1993; Martin and Palumbi 1993; Page 1994). Through a mechanism termed endogenous clock or pacemaker (e.g., Sassone-Corsi 1996), circadian rhythmic activities occur even under constant environmental conditions (e.g., Aldrich and McMullan 1979; Prinzinger and Hinninger 1992). In such cases, it is thought that the property helps organisms anticipate the exigencies of life through programmed cyclic regulation of specific target genes. The common inhibitory neurotransmitter in the central nervous system, GABA (γ-aminobutyric acid), has been implicated in circadian rhythmicity (e.g., Wagner et al. 1997).

Life occurs in backgrounds which after cursory glance can resolutely be dismissed as being implicitly inhospitable. Nature, however, appears to abhor a vacuum. Practically every nook and cranny on Earth is filled with some kind of life. Animals which have adapted to extreme environmental circumstances have particularly intrigued biologists (e.g., Madigan and Marrs 1997). Habitats such as the subzero temperature glaciers (e.g., Arrigo et al. 1997), the fiercely hot bubbling hydrothermal (volcanic) vents in the deep seas (e.g., Meredith 1985; Jonnston et al. 1986), the highly desiccating tropical deserts, the fresh (nearly ion-free) water, the hypersaline lakes, the remote virtually anoxic reducing muds, and submarine environments where the hydrostatic pressures may exceed 1000 atmospheres have all been variably conquered and occupied. Life has been found at a depth of about 4 km below the surface of the Earth where the temperature is 75 °C (Ehrlich 1996; Frederickson and Onstott 1996) and in the ice-free, cold, dry valleys of the Antarctica, regions considered to be the closest terrestrial analogs of the Martian and other extraterrestrial planetary environments (Friedmann 1982; Friedmann and Ocampo 1976). While few, if any, multicellular organisms can tolerate a temperature above 50 °C (Huey and Kingsolver 1993), some microbes, i.e., the so-called hyperthermophiles, thrive at and above 100 °C (Madigan and Marrs 1997), the boiling point of water at sea level. In their quest to secure new habitats, organisms have devised design-specific solutions to the vast threats which have resolutely besieged them. These have ranged from momentous changes such as variations in solar insolation (e.g., Newman and Rood 1977; Frils-Christensen and Lassen 1991) and volcanic activity to minor spatial and temporal shifts in temperature and levels of respiratory gases in their immediate habitats. Such changes have had dramatic effects on the form, distribution, and lifestyles of animal life. Hippopotamuses are reported to have roamed the Yorkshire Dales some 125 000 years before present (Shackleton 1993) and in northern Africa, the Sahara-Sahel boundary crept northwards by 10° of latitude between 18 and 8 thousand years before present (Petit-Maire 1991). The overall fitness of an organism in a particular environmental setting is an aggregate effect of the different adaptive strategies an animal has requisitioned during its evolutionary existence (Kozlowski 1993). This arsenal confers the adaptive capacity to withstand adverse changes in the environment and, if necessary, to actively carve out and exploit new, more hospitable and resourceful habitats.

From the smallest known entities, the fundamental subatomic particles such as quarks, to the observable universe (the largest entity of which humankind has knowledge), a scale which ranges from under 10^{-15} to 10^{27} (Ronan 1991), nature is governed by four fundamental forces. These are gravity, electromagnetism, the weak force (the force responsible for radioactive decay), and the strong nuclear force (the force that holds the nucleus of an atom together). Gravity, of which the carrier particle (the graviton) is still undiscovered, is the weakest of the forces (strength 10^{-38}) but has an infinite range. The weak nuclear force, of which the carriers are the electrically charged W^+ and the W^- and the neutral Z° particles, has a range of 10^{-15} and a strength of 10^{-13}, the electromagnetic force, of which the carrier particles are photons, has an infinite range, a strength of 10^{-2}, and operates between electrically charged particles, while the strong nuclear force which acts on the quarks is powerful (strength $= 1$) but has a range of only 10^{-12}. By studying the universal properties of matter, science strives to understand and test how these characteristics contrived matter culminating in the phenomenon of life. It is hoped that the principles which govern organismal existence, design[1], behavior, function, and life-style can be better grasped as a part of such a broad fundamental and integrative approach. The quest for the elusive grand unified theory (e.g., Hawking 1993; Weinberg 1994), an attempt to integrate a number of independent mathematical equations which seek to demonstrate that three of the fundamental forces of nature (electromagnetism, and the weak and strong nuclear forces) are essentially performances of the same superforce, continues with zeal and zest. After unification with gravity, the theoretical physicists contemplate advancing the "theory of everything", an encompassing principium which should explain the formation of the Universe at the Big Bang, the subsequent existence of the material world, and perhaps the development of life (Moore 1990a; Maynard-Smith and Szathmary 1996; Ronan 1991) and the end of it all at the Big Crunch. Though some measure of success has been achieved, especially in harmonizing the weak and electromagnetic forces, from what is

---

[1] The term design, which is borrowed from engineering, is used in this book in a sense to mean "creative natural arrangement of parts (= components) in a device (= gas exchanger) for a particular purpose (respiration)". Vogel (1988) defined biological design as "functionally competent arrangement of parts resulting from natural selection".

4

known presently, the cosmos abounds with mystery and deliberately shuns order and conformity. While the Newtonian laws apply in the intermediate scales of biology, at subatomic (the so-called nanoworld) level, quantum mechanics reigns, and at the cosmological level, relativity holds true. Recently, a concept dubbed new physics was professed to be highly rewarding in analyzing the dynamics of the life processes (Davies 1989; Stonier 1990). The approach is based on arguments that: (1) life is transmitted through what are bits of information (algorithms) inscribed in the genome and (2) the frequency and amplitude of biological events are regulated through ultrafast integrated information processing. Applying an electronic analogy, Lloyd and Rossi (1993) asserted that "the living state is an ensemble of oscillators" and that information processing may be the "common denominator" (the missing link?) in the formulation of a "unifying theory". Inasmuch as our understanding of biological processes (especially neurobiology) continues to accumulate with the advances in computer technology, this supposition appears to be credible (e.g., Adams 1979; Kawasaki 1993). Other lines of inquiries attempting to simplify and explain biological order and diversity include the so-called life-history theory (e.g., Charnov 1993; Ruthen 1993) and the synthetic theory of biological organization (e.g., Eigen and Schuster 1979; Fontana and Buss 1993). According to these concepts, it is argued that for a given organism, if the fundamental parameters which govern and regulate resource procurement, utilization, regeneration, and self-perpetuation in a particular environment are known, it would be possible to model the life patterns which optimize fitness (Charnov 1997; Godfray 1997). Reductionism and all mechanistic approaches to biology endeavor to explain natural phenomena by manipulating fewer and simpler components that are responsive to exact simple physical laws (e.g., Popper 1968, 1969). However, when dealing with complex natural dynamical entities like organisms (e.g., Mann 1982; Brown 1993), it is not only difficult to correctly identify such components but also practically impossible to predict the outcomes of their myriad nonlinear interactions. Due to the hierarchical organization of biological systems from atomic, molecular, cellular, tissues, organs, organisms, through populations and communities to ecosystems, boundary conditions exist (Brown 1994): it is practically impossible to predict the outcomes from one level of organization to another (Polanyi 1968). Despite the caveats, the rationale behind the rather esoteric reductionistic investigations in biology includes among others the warranted recognition that although biological systems may evolve through means different from those that accrete the physical ones, their forms and states are fundamentally governed by the universal, permeative properties of matter and energy (e.g., Nagel 1961; Brooks 1994). The structure and the mechanistic chemical juxtaposition of organic molecules is a programmed process which generates exact structural configurations and arrangements which yield stable states. Life cannot violate the immutable laws of physics and chemistry during its development. Thompson (1959, p. 8) points out that "the forces which operate in the body are of the same character as are the inorganic forces". He envisages that mathematics and physics will greatly contribute in explaining biological phenomena even though they may not fully account for certain aspects that he consigns to the "soul". As cautioned by Giebisch et al. (1990) and Nurse (1997), when dealing with an intricate process such as life, reductionistic approaches may fail simply due to the fact that the underlying assumption that encompassing properties of an organism can be understood by studying its individual parts could be fatally flawed. The aggregate expression of the functional processes in an organism is not necessarily equal to that expressed by the intact animal. Thompson (1959, p. 41) observed that "the life of the body is more than the sum of the properties of the cells of which it is composed" and Hoagland and Dodson (1995) note that "an organism is greater than the sum of its parts". The total of the O_2 consumption of the individual tissues of the body, for example, may be lower (e.g., Weymouth et al. 1944; Itazawa and Oikawa 1983; Oikawa and Itazawa 1993) or higher (e.g., Terroine and Roche 1925; Crandall and Smith 1952; Vernberg 1954; Lilja 1997) than that of the whole animal. Based on a similar observation, Von Bertalanffy and Estwick (1953) proposed that the decrease in the mass-specific metabolic rate of an intact animal is regulated by "factors lying in the organism as a whole which do not appear in tissues excised from the intact animal". Brown (1994) contends that "while physical scientists seek precise answers to relatively simple problems, biologists on the whole seek approximate answers to very complex problems".

In what may be deemed convergence between natural and human engineering designs, modern research into natural configurations has led to the fascinating discipline of bionics (= *biomechanics*) (e.g., Nachtigall 1991; Witt and Lieckfeld 1991). Astonishing parallels in the "plans" and "constructions" of living organisms with technical principles abound. Though nature uses essentially the same structural materials as those used by human beings (i.e., materials found on Earth), while it is possible to mimic them, it is virtually impossible to exactly duplicate natural designs. During the long period of evolution, through progressive and yet recursive trial-and-error processes (e.g., Schaeffer 1965a), nature has honed and produced perfect or near-perfect innovations. Natural solutions for complex problems are often strikingly simple and fascinating (e.g., Hayes 1994): minimal resources are committed in configuring highly cost-effective structures. Amidst the remarkable diversity of form at the organismal level, however, the differences that distinguish the various kinds of life in the macroscopic and microscopic worlds disappear down the organizational cascade as similarities preponderate. From the perspective of structural chemistry, at the molecular and atomic levels, there are no differences between the living and nonliving worlds: organic molecules are made up of essentially the same elements (atoms) which comprise the inorganic ones but are arranged into complexes with unique properties. Life's diversity can be attributed to differences in the characters and arrangements of protein molecules which constitute more than half of the nonwater mass in a cell (Hoagland and Dodson 1995). Though the number of molecules which formed with the accrual of the Earth (inorganic evolution) and development of life (organic evolution) (Kirschner 1994; Weinberg 1994) is immense and to this bounty chemists keep on throwing in new ones, from this vast collection, life has been extremely selective on what it has appropriated. Of the about 8 million now known chemical compounds (e.g., Morgan 1995), only a very small number has been incorporated in the evolved biochemistry. Proteins, some of the largest and most complex molecules known and perhaps the most important organic factors, as they form enzymes which catalyze chemical reactions, are configured around only 20 different amino acids. From the infinite three-dimensional possible dispositions, the forms and behaviors of proteins are limitless (e.g., Ronan 1991). Amazingly, the amino acids are produced through a code which is written in only four molecules (the nucleotide bases) which present 64 possible arrangements in triplets, the codons: 61 codons are distributed among 20 amino acids, the other 3 serving as stop codons. Proteins are intrisically dynamic molecules. Flauenfelder et al. (1991) pointed out that evolution occurs through changes in the primary sequence of proteins, a process which leads to changes in the structure and the conformational energy landscapes. From the well over 100 known elements, about 99% of the living matter is fundamentally made up of four elements, i.e., C, O_2, N_2, and H_2. Of the 28 selected elements in the human body, H_2 and O_2 are the most abundant, respectively comprising 63 and 25.5% of it. Carbon is central to life. It forms chains and rings that can be elaborated into an immense number of complex compounds and makes up about 1% of the mass of the Earth. Carbon-based fuels contribute about 75% of the energy that is currently used on Earth. Methane is the most abundant organic molecule in the Universe (Ancilotto et al. 1997).

Succinctly put, biology uses a characteristic set of elements and compounds to carry out an infinite array of processes. It is bewildering that the most complex state of organization that matter has consummated has been fabricated through remarkable simplicity, essentially during a chemical circumstance which entailed microscopic architecture around the carbon atom. Though the chemical constituents of living organisms have been recognized and the biochemical processes which support life are now reasonably well understood, the actual origin of life remains a mystery (Szathmàry 1997). It is now known that organic molecules abound in the cosmos. Such molecules could have been seeded on Earth from space (e.g., Ronan 1991; Cohen 1995), landing at the right place at the right time. The recent report by NASA scientists on chemical fingerprints of extremely primitive life in a 4.5-billion-year-old piece of Martian rock (Kerr 1996a; McKay et al. 1996) may in future totally change our concept and the very definition of life.

1.2 The Earth: a Highly Dynamic Planet

Though from space it looks serene and motionless, the Earth is a perpetually metamorphosing planet. Life, the most complex organization of matter, has astonishingly evolved against a highly dynamic setting. Geometrically about a sphere (but strictly an ellipse) some 12700 km in diameter, a mass of about 6×10^{21} tonnes and a volume of about $1.1 \times 10^{12} km^3$, it spins round on an inclined axis (23.5° to the perpendicular of the plane of its orbit) at a speed of about 28 km/s (at the equator), moving around the sun (from a distance of 150 million km) along a slightly elliptical orbit about 300 million km in length at an incredible speed in excess of 1700 km/s. The rotation is completed in 23 h and 56 min and the revolution (around the sun) takes 365.25 days: the rotation causes day and night while the revolution occasions seasons. Different places on the Earth's surface move at different speeds, the speed at the equator being the greatest. The spin greatly influences the shape (e.g., Dixon 1987), distorting the spherical figure to a slightly flattened shape and creating many irregularities on the surface: the polar diameter is about 43 km less than the equatorial one. In what has been called the Colioris Effect, due to the rotation, in the Northern Hemisphere a mass of air around a high pressure area is deflected in a clockwise direction and counterclockwise around a low pressure one: in the Southern Hemisphere there is an opposite effect. A centrifugal acceleration which tends to oppose gravity makes the value of gravity at the equator ($9.780 \, \mathrm{m \, s^{-2}}$) 0.35% less than that of $9.832 \, \mathrm{m \, s^{-2}}$ at the poles: the maximum gravity ($10.5 \, \mathrm{m \, s^{-2}}$) is reached at the boundary of the liquid core some 2900 km below the Earth's surface. Near the surface, gravity decreases by about $0.003 \, \mathrm{m \, s^{-2} \, km^{-1}}$ distance above sea level. While the angular momentum in an elliptical orbit remains constant, according to Kepler's Second Law of Planetary Motion, at periastron (when the radius of the orbit is small) the speed is faster and at apastron (when the radius of the orbit is greater), the speed is slower. The giant outer gas planets, namely Jupiter, Saturn, Uranus, and Neptune (the Jovian planets), have solid cores surrounded by cold atmospheres of light gases such as methane, ammonia, helium, and hydrogen. The Earth, like the other three inner planets, i.e., Mercury, Venus, and Mars (terrestrial planets), is made up of a dense iron-nickel core (some 3400 km in radius at a temperature of 3700 °C), a rocky mantle 2900 km thick, and an outer shell, the crust (lithosphere) some 50 km thick. The light gases were lost from the terrestrial planets due to the fact that: (1) they are closer to the sun and hence received more heat that highly excited the gas molecules and (2) their smaller gravitational forces were not adequate to retain the fast-moving molecules. Compelling an escape velocity of 11.8 km/s, the Earth has been able to hold onto most of its gases except for the very light ones like hydrogen, neon, krypton, and argon. Hydrogen, produced by photodissociation of water vapor in the stratosphere, is presently estimated to be escaping from the Earth's atmosphere at a rate of 3×10^8 atoms $\mathrm{cm^{-2} \, s^{-1}}$ (Donahue 1966; Joseph 1967; Hunten 1973; Hunten and Strobel 1974; Hunten and Donahue 1976). A planet or a satellite has to be more than 10% of the mass of the Earth to be able to hold its atmosphere. About two thirds of the Earth's surface is covered with water. The gravitational pulls of the Moon and the Sun cause tides and the movement of the air masses weather. Water (hydrosphere), air (atmosphere), and the superficial layer of the lithosphere (the soil) constitute the biosphere. They are the theaters in which life has experimented and developed.

1.3 Factors that Encouraged the Evolution of Life on Earth

Whether by default or design, the Earth is peculiarly well conditioned for habitation, at least by the kind of life which we know. As prescribed by nature, organisms adapt and evolve into environments: they do not carve them out to suit themselves. The fundamental factors which "allowed" life to develop on Earth included: (1) an O_2-rich atmosphere with the moderate level being just appropriate for the respiratory needs for life (Urey 1959), (2) location at a right distance from the sun (orbiting between the terribly hot Venus and the hard frozen Mars) for the temperature to support presence of water in both liquid and vapor form, (3) the accretion of the giant planets like Jupiter, Saturn, Uranus, and Neptune drastically reduced the number of comets and other debris in orbit, minimizing the devastating effects of collision with the Earth (Weidenschilling and Marzari 1996), (4) the presence of atmospheric gases like CO_2 provided a mild greenhouse effect which kept the planet warm, (5) the mass of the Earth (and hence its gravity) is just right to hold and prevent loss of most atmospheric gases to the outer space without undue pressure on life, and (6) the Earth's magnetosphere is adequately strong to prevent sputtering of the atmospheric gases by the constant bombardment of energized ions. Jupiter's largest moon, Ganymede, with a magnetic field about one tenth that of Earth but greater than that of Mercury, Venus, and Earth's moon (Gurnett et al. 1996), has a thin atmosphere (Stevenson 1996) with O_2 in a frozen state (Vidal et al. 1997). Though Mars, with its distant orbit which is 50% farther from the Sun than Earth, now presents a desolate, cold, and dry surface with a small ice cap especially at its north pole, in the past the planet appears to have experienced episodes during which an atmosphere may have existed to create a greenhouse effect adequate to generate ample liquid water on the surface (Kargel and Strom 1996): over time, the planet has lost large quantities of CO_2 as well as O_2 and H_2, gases derived from breakdown of atmospheric water by sunlight, leaving a thin 7-km-thick gaseous envelope. The planet's atmosphere has been worn out (sputtered) by energetic O^+ ions created from escaping O_2 and hurled back to the atmosphere by the solar wind fields (Johnson and Liu 1996). The highly rarefied Martian atmosphere compares with that of Earth at an altitude of 50 km, the atmospheric pressure being less than 1% of that on Earth. It cannot be completely ruled out that under similar or different circumstances a kind of life could have evolved elsewhere in the Universe (e.g., Powell 1993). Possible past occurrence of primitive life on Mars has been advanced (e.g., Kerr 1996b; McKay et al. 1996) and existence of life in other celestial bodies argued (e.g., Chyba 1997).

The Earth's atmosphere weighs about 500 million million tonnes. About 50% of it is in the lower layers about 5 km from the Earth's surface. The troposphere, the part of the atmosphere where the air is well mixed, extends up to an altitude of about 10 km above sea level and constitutes 80% of the total mass of the atmosphere. Barometric pressure is a consequence of the Earth's gravitational pull on the air which envelopes it. From the surface of the Earth, barometric pressure drops exponentially. However, at a given altitude, the actual pressure depends on factors such as latitude, season, and the prevailing weather conditions (e.g., Bouverot 1985). For every 5.5-km ascent from sea level (where the barometric pressure is 1013 mbar), the barometric pressure drops by a half and the temperature drops at a rate of 1 °C for every 150 m. Beyond 500 km, the atmosphere (exosphere) is highly rarefied and contains free atoms of O_2, H_2, and helium. The most important respiratory gases in air are O_2 and CO_2. At normal pressures, N_2 is considered to be physiologically inert, but at high pressures it is harmful to life.

1.4 Oxygen: a Vital Molecular Resource for Life

The Lord God formed the man from the dust of the ground and breathed into his nostrils the breath of life, and the man became a living being. (Genesis 2:7)

There are few, if any, processes in biology which are as encompassing and critical for life as respiration. For most animals, procuring O_2 from outside and delivering it to the tissues and voiding CO_2 produced from tissue oxidative metabolism are some of the main tasks of respiration. Laitman et al. (1996) assert that "the acquisition and processing of O_2 and its by-products is the primary mission of any air-breathing vertebrate". Just now (as you read this line) you are breathing O_2 and would die in a couple of minutes without it. Beyond about 3 min of cessation of breathing, irreparable damage, especially of the central nervous system, could occur even after successful resuscitation, and after about 6 min one would be declared brain dead. From a practical standpoint, this bespeaks the importance and urgency of procuring molecular O_2 at the right time and in the right quantities. In the course of evolution, preference would have been given to the organs and systems which support such crucial process. Though other cells such as the neurons (e.g., Schömig et al. 1987), endothelial cells (e.g., Mertens et al. 1990), and smooth muscle cells (e.g., Paul 1989) can cover energy deficits by anaerobic glycolysis, except for the hearts of the freeze-tolerant animals which stop at subzero temperatures (e.g., Storey and Storey 1986, 1988), the continuous mechanical performance of the myocardial cells is totally dependent on transient changes in cytosolic Ca^{2+} (interacting with contractile proteins) and sustained oxidative production of energy in the mitochondria (e.g., Driedzic and Gesser 1994; Piper et al. 1994). While there are assertions that life can exist without O_2, such states can only exist in the simplest forms of life (Hochachka et al. 1973; Herreid 1980; Fenchel and Finlay 1990a, 1991a, 1994). Intestinal parasites are alleged to live without molecular O_2 and intertidal molluskan facultative anaerobes remain for days without it (Ghiretti 1966). In adverse conditions, adaptively, some animals enter latent (ametabolic) states where in cryptobiosis (Hochachka and Guppy 1987), the most extreme of such conditions, life virtually stops. However, even in such states, an infinitesimal amount of energy must be produced by the cells to sustain the crucial molecular processes of life such as protein turnover and ion flux. Even before the discovery of O_2 was made by Priestley and the composition of air demonstrated by Lavoisier (see Perkins 1964 for an account of these elegant discoveries), it was recognized that breathing, the mechanical pumping of air in and out of the body, a process which occurred in the majority of animals, was essential for life. For a long time and until recently, a test for death was the failure of breathing and the common method of killing was by strangulation. Presently, the phrase "the breath of life" is commonly used to indicate the need for continuous movement of air in and out of the body to support life. The comprehensive need for O_2 for life was termed the call for oxygen by August Krogh (Krogh 1941).

Generation, storage, and utilization of energy are processes central to the activities and the very existence of living cells, just as they are relevant to the proper economic management of the Earth's resources. Excess energy is largely conserved in form of carbohydrate and lipid molecules to be utilized in event of deficit. The acquisition and utilization of energy in life occurs according to Maxwell's Laws of Thermodynamics. According to the first law, the amount of energy in the Universe is fixed: no more of it can be created nor can the existing amount be destroyed but can be converted from one form to another. In face of the steadily decreasing amount of freely usable (accessible) energy in the Universe (according to the Second Law of Thermodynamics), the natural state of matter is chaos (e.g., Prigogine and Stengers 1984). Since living systems constitute highly organized complex states of matter, from a casual glance, it would seem that life runs uphill in a downhill Universe, i.e., it proceeds counter to the natural dissipation of energy. In such a case, life would appear to negate the Second Law of Thermodynamics. This, however, is not the case. In fact, instead of threatening life, the Second Law of Thermodynamics actually guarantees it. Unlike the closed thermostatic state of the ordinary (controlled) chemical reactions, living systems are open thermodynamic processes which access energy from outside (especially from the sun) to steady entropy (chaos) through effervescent repair and rebuilding at the molecular level. Generally, organisms are efficient conduits of energy in the vastness of the Universe. Evolution by natural selection is thought to be driven by competition for the dwindling amount of energy in the Universe (e.g., Blum 1955). Much of the energy on Earth is conserved in the covalent bonds, especially of the large organic molecules such as glucose, fatty acids, starch, and glycogen. Intricate interdependency exists in nature where, directly or indirectly, living things rely on each other in appropriating raw materials and harnessing energy. Over time, evolution has modified morphological design and physiological processes to eliminate or reduce unnecessary expenditure of energy. More optimal states are established to better manage the finite resources. Superfluous structures are eliminated and hence support of unused or underutilized capacities is avoided.

No molecule has been as pervasive in its influence on life and paradoxical in its roles as O_2. In all evolved complex animal life, O_2 is the most important molecular factor contracted from the ambient milieu. The metabolic rate of an organism correlates with the efficiency of procuring it. Nature has been particularly inventive in the development of gas exchangers and the respiratory processes. The many examples of convergence show that permeative forces have been involved in programming the design of the gas exchangers. Regarding the procurement and utilization of energy, living things are essentially open thermodynamic systems in a self-regulating steady state. A continuous influx and efflux of matter and energy occur as the necessary physiological and behavioral adjustments are made to maintain homeostasis. Such a dynamic state ensures that, though intimately relating to its immediate environment, an organism remains a viable, discrete entity. While life can be defined as a process of capturing and utilizing energy and raw materials, empirically, death is the cessation of all such activities, i.e., when energy production falls to zero. In such a state, the energy in a body is at equilibrium with that in the environment. For individual animals, the continuity of respiration is terminated at death, and for a species at extinction. Animals will generally live for weeks without food, days without water, but only minutes without O_2. Activities such as feeding, thermoregulation, locomotion, and even reproduction (e.g., Hurst and McVean 1996) can be adjusted, delayed, or abandoned altogether, depending on species and circumstances (McNamara and Houston 1996).

Energy is decisive in all biological events from molecular, biochemical, ecological to evolutionary levels (e.g., Bennett 1988). It is required for building, servicing, and maintaining the general infrastructural integrity of an organism as well as driving the physiological processes and fortifying homeostasis against external perturbations. The rate of respiration indicates the speed at which an animal uses its resources to meet the demands placed on it by the environment and the lifestyle it leads. Those species capable of maintaining a high rate of O_2 to CO_2 exchange ratio in relation to the volume and the complexity of the protoplasmic mass are able to establish stable tissue fluid gas concentrations under different environmental circumstances and metabolic states. It is unequivocally evident from the design of the gas exchangers that such specialized taxa are the most successful. From the earliest recorded fossils, which are 3.8 to 3.5 million years old (Schopf 1978, 1993; Mojzsis et al. 1996), to the first well-documented composite organisms, the ediacaran Metazoa which occurred more than 600 million years ago (e.g., Gould 1989; Knoll 1991; Levinton 1992; Runnegar 1992), for over 80% of its tenure on Earth, life remained exclusively unicellular (Gould 1994) and anaerobic. It is thought that lack of O_2 in most of the Precambrian may have been the main factor which repressed further progress (Knoll 1991): the so-called Cambrian explosion, when the biota underwent remarkable diversity, has been associated with the presence of and the increasing levels of molecular O_2.

1.5 Anaerobic Metabolism and Adaptive Success in Animals

The capacity to procure, transport, and utilize large amounts of O_2 has bestowed a monumental selective advantage on the evolution and adaptive radiation of the terrestrial vertebrate fauna. Predator avoidance correlates with the level of energy expenditure and the kind of food eaten (e.g., McNab 1966). Terrestrial species with low metabolic rates rely heavily on burrows or passive integumental structures such as shells, plates, and spines for protection. The evolution of efficiently ventilated and perfused gas exchangers and carrier-mediated O_2 transport systems appear to have been fundamental for supporting energetically demanding life-styles. Metabolic rate expresses the integral speed at which energy is mobilized, transformed, and utilized by an organism for biological activities (e.g., Kleiber 1965; Calder 1987; Brown et al. 1993; Lundberg and Persson 1993) and hence expresses the vitality of life (e.g., Zeuthen 1970; Calder 1984). In mammals, factors such as enzymatic activities (Emmett and Hochachka 1981), enzyme contents of tissues (Drabkin 1950), O_2 consumption (R.E. Smith 1956), and protein turnover (Munro and Downie 1964) reflect the effect of body size on metabolism. An inverse correlation between the specific metabolic rate (amount of O_2 consumed per gram body mass per unit time) of animal species and their life spans has been established (e.g., Adelman et al. 1988; Shigenaga et al. 1989). Dwarf mice live much longer than normal ones by as much as 350 days for males and 470 days for females (Brown-Borg et al. 1996).

Thompson (1959, p. 42) forthrightly stated that "size of body is no mere accident". Metabolic rate determines vital aspects such as life patterns, population fluctuations, behavioral ecology, and reproductive efficiency (e.g., Prothero 1986; Calder 1987; DeAngelis et al. 1991; Dunham 1993; McNamara and Houston 1996). In the modern ecosystems, the chance of extinction is directly proportional to body size (e.g., Carroll 1988). Diverse factors such as phylogeny, habitat, ambient temperature, O_2 consumption, food intake, latitude, climate, season, body size, shape, level of development, degree of activity, sex, and age to varying extents determine the metabolic rate (e.g., Zeuthen 1953; Else and Hubert 1985; Crews et al. 1987; Labra and Rosenmann 1994). Unlike metabolic substrates, e.g., carbohydrates and fats, which can be stored in large quantities in the body, except in a few heterothermic and anaerobic parasites (e.g., Ghiretti 1966), O_2 has to be derived from the external environment in the necessary measures. In the human being, about 12000 l of air are filtered by the lung everyday (Burri 1985). The amount of O_2 dissolved in blood or plasma is insufficient for tissue requirements even at rest. The quantity falls far short of the amount which would be required to service physical activity when the uptake may increase by as much as 30 times during vigorous exercise, e.g., flight (Thomas 1987). A human being at rest requires 200 to 250 ml O_2 min^{-1} but during maximal exercise the amount increases to about 5.5 l min^{-1} (Comroe 1974; Weibel et al. 1987a). A 70-kg human being has only 1.55 l of O_2 in the body at any one moment, 370 ml being in the alveolar gas, about 280 in the arterial blood, about 600 in the capillary and venous blood, 60 ml dissolved in body tissues, and 240 ml bound to the muscle myoglobin (Farhi 1964): the total amount is adequate to support life for only 6 minutes, but irreparable damage starts to occur within about 3 min of cessation of breathing. Snyder (1983) observed that the amount of O_2 dissolved in tissue (about 0.8 ml/kg) is sufficient to support aerobic metabolism for only a few seconds. However, in the champion divers, e.g., the Weddell seal, *Leptonychotes weddelli* (Kooyman 1985), in a 450-kg animal, the O_2 stored in the muscles can support aerobic metabolism at a rate of 4.2 ml O_2 kg^{-1} min^{-1} for about 15 min. At an estimated O_2 consumption rate of 1.6 ml O_2 per kg, a 20-tonne sperm whale, *Physeter catodon* can dive for 50 min while maintaining aerobic metabolism (Butler 1991a). The body stores of CO_2 in solution and in form of HCO_3^- ions exceed those of O_2 (Farhi and Rahn 1955). It is perhaps owing to its intrinsically great toxicity (e.g., Fenchel and Finlay 1990b) (Sect. 1.16.1) that animals have not evolved capacities of storing appreciable amounts of O_2 in the body tissues and cavities: the rate of O_2 uptake from the environment is approximately equal to its utilization. The infinitesimal amounts held in the bodies of most organisms, either chemically bound or in solution, are able to support aerobic requirements for only a short period of time. In some fish with physoclistic swim bladders, however, O_2 is held in the swim bladder at high pressures and concentrations (Saunders 1953). Such stores can be utilized during hypoxia (Randall and Daxboeck 1984) with adequate tissue oxygenation being sustained for several hours.

1.6 Evolved Mechanisms and Strategies of Procuring Molecular O_2

Respiration has been pivotal in all the evolutionary and adaptive changes which have occurred in animal life. This is evinced by the fact that to a fair measure, the functional competencies of the gas exchangers correspond with the general phylogenetic statuses of animals. Respiration encompasses an impressive arsenal of biomechanical, physiological, and behavioral strategies and mechanisms that are involved in making available to an organism a sample of the external respiratory milieu from which molecular O_2 is extracted and into which CO_2 is voided. External respiration involves movement of two vectorial quantities in opposite directions, namely, influx of O_2 from the environment and efflux of CO_2 from the organism. Oxygen is delivered to the tissue cells across a panoply of structural compartments through self-regulating convective and diffusive processes. The operation starts with convective delivery of O_2 by a respiratory medium (water and/or air) to the gas-exchanging site, diffusion across the tissue barrier, binding to carrier pigments, convective transport by blood circulation, and ultimately diffusion from the blood into the cells (Figs. 1,2,3). In a steady nonlimiting state, servomechanically, the flow of O_2 from the environment across the steps to the mitochondria is constant (Weibel 1982; Wagner 1993). Mitochondria contain all enzymes associated with the processes of oxidative phosphorylation in their inner and cristae membranes and the enzymes of the TCA cycle in the mitochondrial matrix. The influx of O_2 into the mitochondria, the terminal O_2 sink which determines the flow of O_2 across the lung through the cardiovascular system (e.g.,

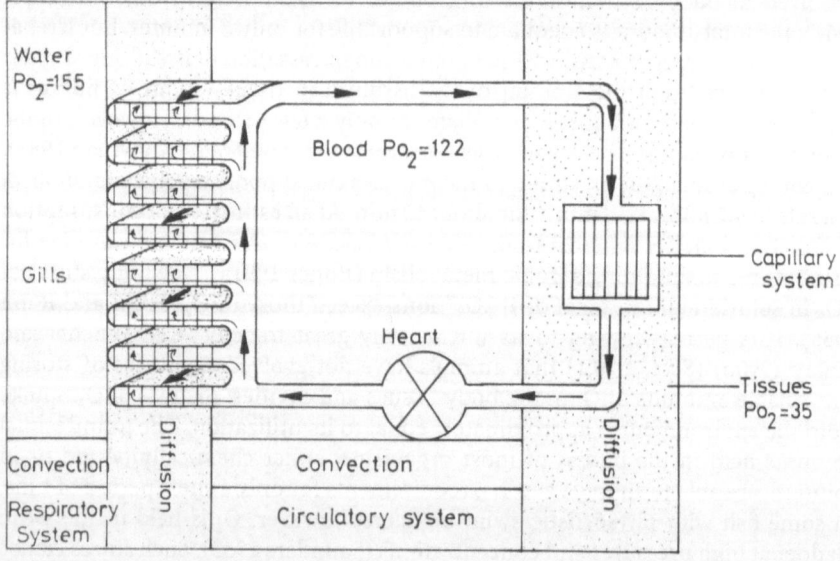

Fig. 1. Sites of convective and diffusive gas transfer in an aquatic breather. Water and blood spatially relate in a countercurrent manner at the gills. (After Satchell 1971)

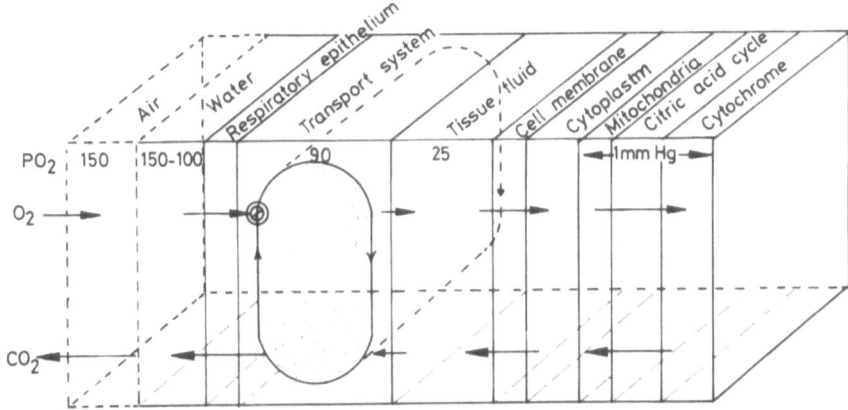

Fig. 2. The intricate stratified arrangement of the gas exchange components in the diffusional pathway of O_2 from the external milieu to the mitochondria. The partial pressure of O_2 gradually decreases towards the tissue cells. CO_2 is greatest in the tissues and is eliminated in the opposite direction to that of O_2. (After Hughes 1978)

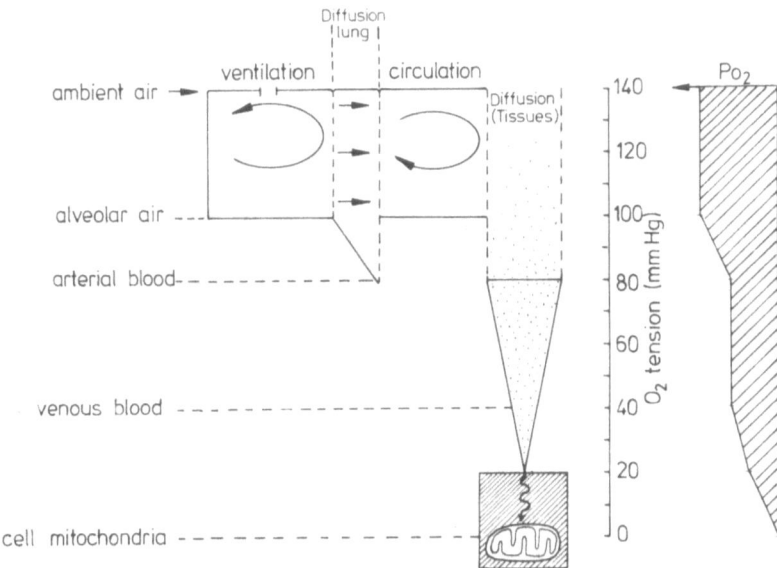

Fig. 3. Cascade process of delivery of O_2 from the ambient milieu in an air breather with a convective lung. The ventilatory and circulatory systems maintain a partial pressure gradient across the air/blood interface. The PO_2 decreases towards the tissue cells and the mitochondria. The utilization of O_2 in the mitochondria maintains the flow. The components of the pathway are quantitatively sized to optimize flow. (Wood and Lenfant 1976)

Suarez 1992), is set by the phosphorylation potential of the tissue cells (Folkow and Neil 1971; Taylor et al. 1987a, 1989) especially in the skeletal muscle mass, which constitutes about 50% of the body mass in mammals and as much as 80% in fish (Goldspink 1985): the PO_2 drops from about 20 kPa in water or air to almost zero in the vicinity of the mitochondria of the outlying tissues (Wittenberg and Wittenberg 1987, 1989; Graiger et al. 1995). Between the capillaries of the heart muscle and the mitochondria, the PO_2 drop is 2.7 kPa (Tamura et al. 1989) and that between the cytosol and the mitochondria is less than 0.03 kPa (Wittenberg and Wittenberg 1987, 1989; Clark et al. 1987). In a rat (mean body mass 266 g), the total mitochondrial surface area in the liver, kidney, heart, brain, and lung is 460 m^2 and in a 1.8-kg bandicoot the value is 5520 m^2 for the same organs (Else and Hubert 1985). The drop in the PO_2 between the capillaries and the surrounding tissues is inversely proportional to the permeability of the tissue to O_2 (Meng et al. 1992). In exercising human gastrocnemius muscle, the O_2 tension decreases with the intensity of contraction, indicating that O_2 may be a limiting factor for mitochondrial respiration (Fellenius et al. 1984): under such conditions, the PO_2 in the extracellular fluid may decrease by 70% and in the cells by 30% of the resting values. To increase the flow of O_2 from water to the tissues, the heart muscle of the hemoglobinless Antarctic icefish is profusely supplied with blood (Fitch et al. 1984; Johnston and Harrison 1985): the muscle mitochondrial volume density rivals that of the insect flight muscles (Londraville and Sidell 1990) but myoglobin is lacking in significant quantities (Douglas et al. 1985; Feller and Gerdy 1987; Sidell et al. 1987). The complexity of the O_2 conduction line differs between animals. It depends on factors such as body size, environment occupied, and lifestyle. In the Protozoa, O_2 diffuses across the cell membrane into the protoplasm. The simple invertebrates lack a circulatory system while the more complex ones and the vertebrates have a circulatory system which convectively transports O_2 from the respiratory site(s) to the body tissue cells. In insects (Sect. 6.6.1), O_2 is delivered directly to the tissue cells by the trachea. Internal respiration entails utilization of O_2 at cellular level to generate the high energy molecule adenosine triphosphate (ATP), with CO_2 and water as secondary products. Expulsion of CO_2 occurs in the opposite direction to that of O_2, i.e., from the tissue cells to the gas exchanger, driven by the same mechanisms (Hills 1996) but carried through somewhat different processes (e.g., Davenport 1974; Heisler 1989): CO_2 excretion occurs through a passive process along an electrochemical gradient from the site of production (Bidani and Crandall 1988; Nikinmaa 1990).

Depending on the function(s) they carry, organ systems have different needs for O_2 (Else and Hubert 1983). By default, the gas exchangers are the only organs in which a conflict of interests can occur. For efficient uptake of O_2, the designs must effect transfer O_2 with minimal utilization of it. Without compromising their functional integrity, as little tissue as possible must be committed in the construction of the gas exchangers. In the avian lung, the bodies of the extremely thin epithelial cells which line the air capillaries are confined to the atria and to a lesser extent in the infundibulae, which are non gas exchange sites (Figs. 88,90). In its thinnest sections, the blood-gas barrier of the avian lung consists of an endothelial, an epithelial, and a common basement membrane (Maina and King 1982a): interstitial spaces are largely lacking (Figs. 29,40). On average, the vertebrate lung is estimated to utilize as much as 10% of the total body O_2 consumption (Slonim and Hamilton 1971).

While O_2 transfer in simple organisms occurs by simple diffusion across the entire body surface, in the more complex ones this takes place at specialized sites where tissue barriers are crossed. In spite of the intrinsic differences in the structural complexities, after about 2 billion years of evolution of aerobic metabolism, the transfer of molecular O_2 in all evolved gas exchangers still occurs by passive diffusion – only ways have changed but means have remained essentially the same. Aerobic metabolism must have evolved at a critical point when the ambient PO_2 was just adequate to drive the gas across the cell membranes of the amphiaerobes. The gradual increase of the PO_2 in the bioshere lead to reduction in the respiratory effort, supporting greater metabolic capacity. The delivery of O_2 to the cells/tissues appears to have been machanistically optimized right from the point of incorporation of molecular O_2 into the aerobic processes. For a substance that is needed continuously throughout life, the alternative method of acquisition, i.e., by active means, as was envisaged to occur in the vertebrate lungs at the turn of this century (see, e.g., Haldane 1922), is improbable. To support respiration driven through active acquisition of O_2 would obligate an enormous investment in energy, rendering the entire process uneconomical and perhaps untenable within the present designs of the gas exchangers and the activities which animals perform. Gas exchangers are largely multifunctional organs. Many play nonrespiratory functions which are in some cases equal to, if not more important than, respiration (Sect. 6.10). Fish and crustacean gills, for example, serve indispensable osmoregulatory roles, are the principal pathway for ammonia and urea excretion, and are involved in the regulation of levels of some blood chemical factors, e.g., hormones (e.g., Zadunaisky 1984; Gilles and Pequeux 1985; C.M. Wood et al. 1989, 1994). The human lungs are important sites for elaboration, metabolism, and regulation of the concentrations of various active pharmacological agents in the body (e.g., Slonim and Hamilton 1971). The lung removes from circulation or destroys such chemical factors as prostaglandins, serotonin, and bradykinin, converts angiotensin I to angiotensin II, and synthesizes lipids such as the pulmonary surfactant (Sect. 6.10): the high O_2 consumption of the vertebrate lung can be attributed to these metabolic processes. While the design requirements for gas exchange demand minimal tissue infrastructure, there must be a critical mass necessary to carry out the nonrespiratory roles. The definitive organization of the gas exchangers must integrate these rather constraining needs. The gas exchangers are unique in that there are no tissue cells which are ubiquitous to them as hepatocytes are to the liver, osteocytes to bone, or neurons to nervous tissue. A cell membrane with no distinct specializations (Figs. 4,5) as in the Protozoa and in many simple Metazoa is the most elementary gas exchanger (e.g., Mangum 1994). The unconventionalized designs of the gas exchangers can be ascribed to the fact that the simple passive process of diffusion of the respiratory gases does not oblige a specific structural plan. Based on the fundamental feature that at a respiratory site all that need exist is an O_2 concen-

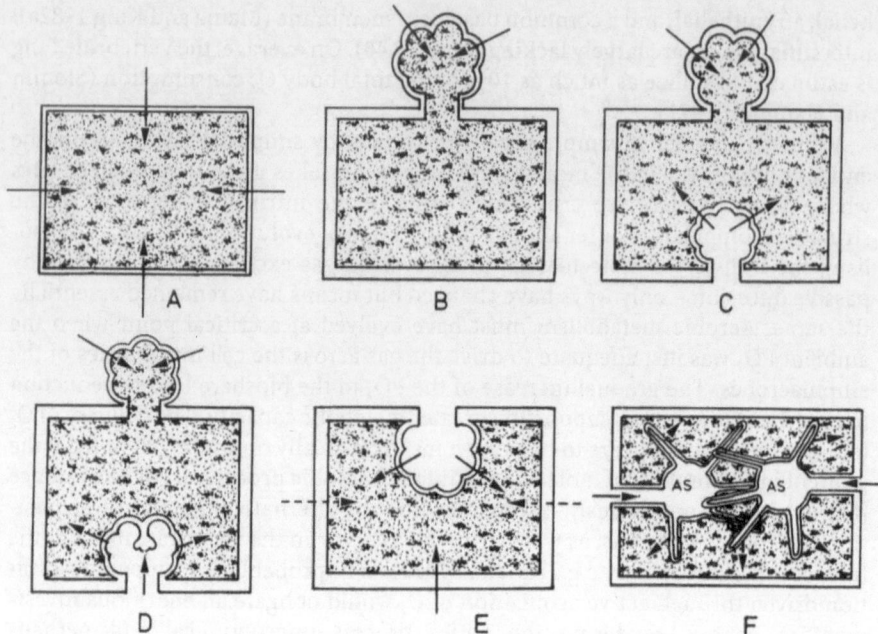

Fig. 4A–F. Basic designs of gas exchangers. The least specialized gas exchanger (**A**) is found in the unicells where the transfer (shown with →) occurs by diffusion across a plain cell membrane. In the more complex animals, gas exchangers have formed either as evaginations (**B**) generally called gills and specialized for aquatic breathing, or invaginations (**C–F**) generally called lungs for aerial breathing. **B** An example of a unimodal aquatic breather. **C** A bimodal breather with an "unspecialized" accessory respiratory organ. **D** A bimodal breather with a "specialized" accessory respiratory organ. **E** A terrestrial air breather with the lung (an invagination) as the exclusive gas exchanger [in amphibians diffusion across the skin (shown with →) is an important respiratory pathway in appropriate environments]. **F** Insect tracheal system where air is delivered directly to the tissue cells across the air sacs, *AS*, especially in the larger species. (Maina 1994)

tration differential (a partial pressure gradient), exquisite experiments have been used to identify morphologically obscure respiratory sites by use of O_2-sensitive bioindicators. Suitable flagellates such as *Polytoma* (Thorpe 1932) and a protozoa such as *Bodo* (Fox 1921) have been utilized as markers of areas of rapid O_2 influx in the *Simulium* and larval *Omorgus* (Ichneumonidae), respectively, and in *Cryptochaetum* (Agromyzidae). Terebellid worms (e.g., Weber 1978) and sea anamones, e.g., *Metridium senile* (Sassaman and Mangum 1972; Schick 1991) present some localized areas of the body where the thickness of the skin is drastically reduced. In the large scyphozoan, *Cyanea capillata*, and the tube anemone, *Ceriantheopsis americanus*, it is still debatable whether the skin over the body column plays any significant role in respiration (Sassaman and Mangum 1974; Mangum 1994). Up to now, gas exchangers still remain to be discovered!

18

animal body ambient medium	examples and remarks
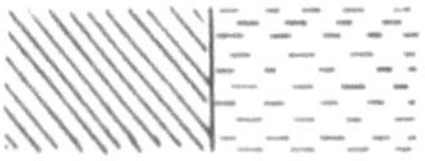 integuments — water	eventually the surrounding medium may be air
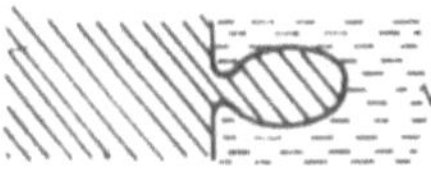 gill — water	crustaceans fishes molluscs tadpoles also – book gill of limulus –podia of urchins
 water lung — water	holothurians (sea cucumber) also – buccopharyngeal respiration – cloacal respiration
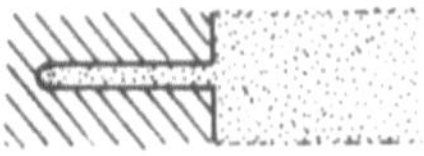 trachea — air	spiders insects also – book lung of scorpions tracheal lung of spiders
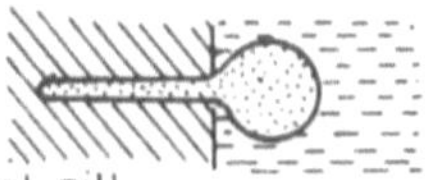 tracheal gill — water	aquatic larvae of insects e g mayflies
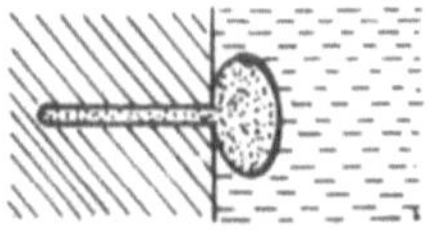 compressible gas gill — water	aquatic insects e. g. Notonecta

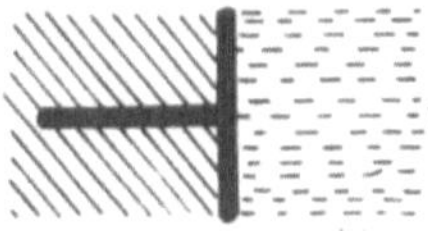 incompressible gas gill — water	aquatic insects
	   e g Aphelocheirus
	       syn.: plastron

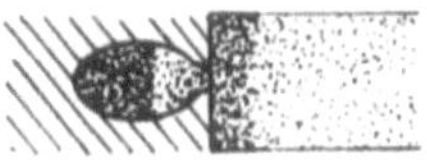 air lung — air	– vascularized cavities (land snail) – air chamber of some air breathing fish – simple air sac (sphenodon) – alveolar lungs (reptiles and mammals) – parabronchial lung of birds

Fig. 5. Scheme showing the different anatomical designs of gas exchangers in water and air breathers and specialized modes of gas exchange. The organization of gas exchangers is mainly determined by the kind of respiratory medium utilized, habitat occupied, lifestyle, and the evolutionary level of development of a particular organism. (Dejours 1988)

1.7 Explicating the Process of Evolution of Respiration: Limitations

Reconstructing the pathways and the stages through which gas exchangers have evolved is an undertaking beset with many difficulties. The main obstacle lies in the uncertainty of our grasping the nature, the severity, and the direction of the changes which occurred in what was a highly dynamic biosphere of the ancient Earth. There is a particularly glaring lack of fossilized materials of the gas exchangers except on one Devonian species, *Bothriolepis* (e.g., Denison 1941; Wells and Dorr 1985). Soft tissues are very seldom adequately geologically preserved. When it occurs, however, by their very nature of being delicate, the materials are reduced from their three-dimensional form to two-dimensional films through intense heat and compression (e.g., Behrensmeyer and Kidwell 1985), making their recognition and interpretation very difficult and unreliable.

Air breathing has only evolved in the lineages of the two osteichthyan (bony) fishes, the actinopterygian (ray-finned fishes), e.g., bichirs, gar, and bowfin and sarcopterygian (lobe-finned fishes), e.g., lungfishes (Romer 1972; Pough et al. 1989; Cloutier and Forey 1991). Though a very distant point in evolution, this, nevertheless, provides a focal point for seeking convergence and divergence of the animal groups from which air breathers evolved. The near total extinction of the crossopterygian fishes (the sole survivor being the coelacanth, *Latimeria chalumnae*), a group thought to be a direct progenitor of the tetrapods (e.g., Pough et al. 1989; Gorr et al. 1991), makes the discernment of the evolution of the respiratory processes much harder. More often than not, one has to more or less rely on circumstantial evidence. From molecular genetic studies, it has now been proved that the popular phrase that "ontogeny recapitulates phylogeny" is too simplistic for developmental biology (on its own) to be reliably meaningful in reconstructing phylogenies (e.g., Humburger 1980; Alberch 1985; De Queiroz 1985; Gans 1985; Northcutt 1990; Marshall and Schultze 1992). Neoteny and/or pedomorphy (e.g., Semlitsch and Wilbur 1989; Wake and Roth 1989) are but two characteristics which manifest nonconforming developments. Morphological, experimental embryological, physiological, biomechanical, paleontological, molecular, and biochemical investigations need to be integrated to effectively elucidate evolutionary transformations, mechanisms, and pathways. In amphibians in particular, dramatic changes in form, location, and function of the gas exchangers and the circulatory system occur during metamorphosis (e.g., Infantino et al. 1988; Hou and Burggren 1991; Burggren and Bemis 1992; Newman 1992; Burggren and Infantino 1994; Fig. 47). These entail radical modifications of the ventilatory mechanisms and transformations of the respiratory organs, changes which must be accompanied by appropriate neurophysiological reorganizations for proper motor functional control and coordination (Burggren et al. 1990).

Albeit the glaring lack of data, the evolutionary and adaptive developments of the gas exchangers can be patched together by collateral evidence gathered by studies of: (1) animals which have adapted to atypical habitats, (2) those that possess transitional respiratory features, (3) those at different phylogenetic levels of development, (4) those which have unique behavioral lifestyles, and (5) those which show peculiar developmental changes. Study of the ancient air-breathing fish such as the Dipnoi, Holostei, and Polypterida should provide a fertile ground for such inferential studies. For biologists, there is always some unique kind of satisfaction when a previously unknown phenomenon or state of natural history fits into a theoretical prediction. The excitement of discovering the so-called living fossils, rare animals which fill the evolutionary gaps, is great. The capture of the Carboniferous actinistian, the coelacanth *Latimeria chalumnae* in 1938 (see R.E. Smith 1956; Thomson 1986; Fricke 1988; Cloutier and Forey 1991; Bruton et al. 1992), a group thought to be long extinct, is a classical instance. It is now, however, profitable to recognize that since the first successful engineering of transgenic mice by Gordon et al. (1980), it is no longer necessary to formulate a scientific question to suit a biological system. These days, it is possible to specifically design an organism (e.g., a transgenic organism) to best answer a particular question (e.g., Cory and Adams 1988; Adams and Cory 1991; Taketo et al. 1991; Ho 1994). Such new experimental paradigms, which hitherto were not possible to configure, provide useful tools for analytical manipulations at molecular, organismal, and ecological levels. When adopted, they should contribute greatly in the advancement of comparative respiratory biology.

The diversity of the organization of the gas exchangers was highlighted in Maina (1994). In this account, the essence of the contrast is discussed with particular focus on: (1) the conditions under which the respiratory processes evolved, (2) the physical characteristics of the media in which these changes occurred, and (3) the different strategies which animals adopted to extract O_2 from the external milieus. The simplest respiratory organs are generally found in the aquatic animals or those organisms which subsist in cryptozoic (humidic) habitats. In their rudimentary form, they occur in the form of permeable, moist, well-vascularized surface membranes, e.g., the integument of the invertebrates such as the earthworms and planaria and in vertebrates such as the eel and the frog, e.g., *Rana*. At the more advanced stages, cardiorespiratory coupling developed to enhance the transfer of O_2 to the complex highly aerobic tissues. In those aquatic animals where the integument is the main respiratory pathway and in fish which have lost the bucco-pharyngeal ventilatory capacity of the gills (e.g., mackerel), locomotion provides an important respiratory activity.

1.8 Plans and Performance Measures of the Gas Exchangers

Over and above the simple diffusive respiration of the unicellular organisms and the lower invertebrates, the gills and lungs are distinctively suited for respiration in water and air, respectively. Owing mainly to the remarkable differences between the two fluid respiratory media (Sect. 3.2), on very rare occasions the two organs are contrived to operate in both respiratory media with equal efficiency. With the progressive organizational complexity of animals, forms, and processes such as closed circulation, double circulation, convective movement of the respiratory media, and presence of respiratory pigments in the body fluids and erythrocytes evolved to match the intensifying demands. Such transformations invite interpretations and speculations. Compared with the more recent enhancements, the ancient elements of the respiratory improvements would have been optimized and conserved in due course. Both the physical characteristics of the respiratory media and the respiratory needs have determined the definitive functional designs of the gas exchangers. The various schemes of the respiratory systems, however, are not congruous with the classic concept of Darwinian radiative animal evolution, which is artistically presented as a branching tree with birds and mammals sitting somewhere at the top. Unlike the brain, which shows progressive development reaching the pinnacle in the human being, the gas exchangers of mammals do not present the ultimate pulmonary design. The structural and functional attributes of a gas exchanger cannot be easily predicted based on any single phylogenetic factor in a simple and direct way.

The importance of O_2 in the survival of organisms is reflected in the dramatic effect that hypoxia and hyperoxia have on the structure and function of the gas exchangers. Parameters such as blood O_2 carrying capacity, O_2 affinity, and myoglobin concentration in tissues can change within a matter of hours in response to aspects such as sojourn at high altitude or after being subjected to severe exercise. In the respiratory system, the working capacities at all steps, be they convective or diffusive, must be appropriately sized and regulated for optimal function. Decrease in size and increase in activity calls for more elaborate gas exchangers and more efficient means of O_2 uptake and transport. Amidist these permutations, certain conflicts, compromises, and tradeoffs occur. For example, whereas intucking of the gas exchangers was essential for avoidance of desiccation on land, affording better protection against trauma, and achieving a more extensive respiratory surface area (Figs. 4,5), such organs could only be ventilated tidally, a pattern functionally inferior to the continuous unidirectional process which is possible in the evaginated gas exchangers (Figs. 6,18). In the erythrocytes, the main organic phosphate 2,3-diphosphoglycerate (2,3-DPG) and CO_2 combine with the same basic groups of the hemoglobin competing with each other (Davenport 1974). The effects of 2,3-DPG and CO_2 on the hemoglobin dissociation curve are not additive: the shift brought about by the two factors together is less than the sum of each separately. The avian trachea presents a good example illustrating the nature and extents of the compromises and structural adjustments effected to enhance the efficiency of the gas exchangers. To attain flight, birds totally committed the forelimbs for this singular function. Ipso facto (i.e., to substitute for the roles which the forelimbs played), birds evolved a long neck (and with it a long trachea) for defense, procuring food, construction of the nests, and preening. For animals of the same body mass, the avian trachea is three times longer than that of a mammal (Hinds and Calder 1971). In order to compensate for what may have led to a greater resistance to air flow (as resistance to air flow in a tube is directly proportional to the length but inversely proportional to the radius to the fourth power in laminar flow – and to the 4.75 power if the flow is turbulent), the avian trachea acquired a diameter 1.3 times greater than that of mammals (Hinds and Calder 1971): air flow in the trachea of the ostrich has been shown to be laminar (Schmidt-Nielsen et al. 1969). The net effect of these adjustments, i.e., increase in the diameter and the length of the trachea in birds, ensued in an overall resistance similar to that of the trachea of a mammal of the same size. In gaining a large tracheal volume and hence a large tracheal dead space (TDS) which is 4.5 times greater than in a mammal (Hinds and Calder 1971), it could

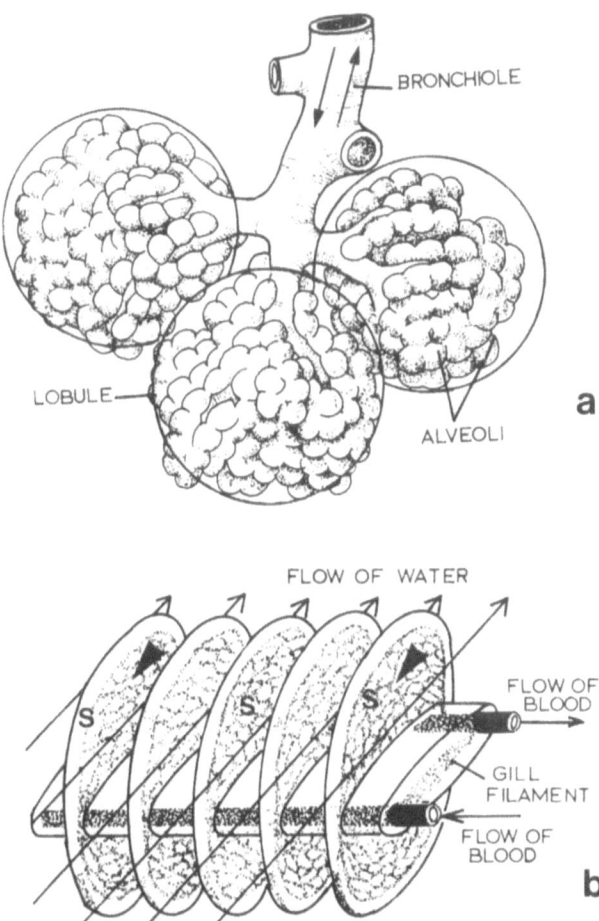

BRONCHIOLE

LOBULE

ALVEOLI

a

FLOW OF WATER

S   S   S

FLOW OF
BLOOD

GILL
FILAMENT

FLOW OF
BLOOD

b

Fig. 6a,b. Spatial arrangement of the respiratory media in an invaginated gas exchanger (a) and an evaginated one (b). Except for the highly specialized bird lung, the gas exchange zone of the invaginated gas exchangers is ventilated tidally (⇆, a) while the evaginated ones are ventilated unidirectionally (→, b). s Secondary lamellae. In the fish gills (b), the direction of the flow of the blood in the secondary lamellae (➤) runs counter to that of the water in the interlamellar space (→). (Kylstra 1968)

conceivably be concluded that natural selection imparted an impediment on the function of the avian respiratory system. This, however, is not the case. The lower respiratory frequency (RF) of birds, which is 0.32 to 0.42 times that of a mammal of equal size, countered the limitations caused by the large TDS. Moreover, the lower RF has afforded room for remarkable configurations of the trachea (e.g., Forbes 1882). Extreme trachea lengths occur in birds such as the trumpet bird, *Phonygammus keraudrenii*, the magpie goose, *Anseranas semipalmata*, and the whooping crane, *Grus americana* (Clench 1978; McLelland 1989). Although

23

P. keraudrenii has the same body mass as a common flicker (*Colaptes auratus*), which has a trachea only 38 mm long, its trachea may be over 800 mm in length and compares with that of the much larger ostrich, *Struthio camelus* (Clench 1978). In *G. americana*, the overall tracheal length is 1.5 m (Welty 1979). Tracheal coiling has been taken to be an acoustic adaptation for lowering the pitch or amplifying sound (e.g., Greenewalt 1968; Gaunt et al. 1987). Furthermore, tracheal loops have been said to increase the tracheal respiratory surface area enhancing evaporative water loss during panting (e.g., Prange et al. 1985) without running a risk of respiratory alkalosis (Schmidt-Nielsen et al. 1969; Bech and Johansen 1980). Compared with birds with straight trachea, those with tracheal convolutions or other tracheal prolongations adaptively have relatively wider tracheal diameters (Hinds and Calder 1971). Perhaps to play similar roles, trachea diverticula have been reported in snakes (Young 1992).

Miscellaneous tissues and organs such as the cell membrane, skin, buccal cavity, gastrointestinal tract, gills, and lungs variably serve as respiratory sites. Because they characterize the more phylogenetically advanced animals, the air-breathing organs (lungs) are assumed to be the better (more efficient) gas exchangers. Except in the bimodal breathers (Chap. 5), gas exchangers are refined to operate best in only one respiratory medium. The human being at the epitome of evolution soon succumbs when the lungs are flooded with liquid (Sect. 6.11). In all respiratory organs, be they water or air breathing, O_2 dissolves in a thin film of water as it traverses the tissue barrier (Sect. 6.1). The flux of the respiratory gases occurs under the prevailing partial pressure gradients across the water or air-blood barrier and is maintained by utilization (O_2) and production (CO_2) in the tissue cells and the physical movements of the external and internal respiratory media. There has been protracted debate as to whether the diffusion of O_2 across the cell membrane is entirely passive or is facilitated (e.g., Longmuir and Bourke 1959; Scholander 1960; Burns and Gurtner 1973; Wittenberg 1976). Hemoglobin, myoglobin, and a specific carrier (cytochrome P_{450}) have been implicated in facilitated diffusion of O_2 in tissues such as the lung, placenta, and the liver (Kreuzer 1970; Wittenberg and Wittenberg 1989). The significance of facilitated diffusion of O_2 in tissues is not well known. The process may, however, be consequential in states of reduced O_2 flow across the blood-gas barrier, e.g., in cases of interstitial edema (Burns et al. 1975, 1976), and in hypoxic conditions (Longmuir 1976).

In both aquatic and terrestrial animals, the complexity of the gas exchangers correlates with the mode of life, habitat occupied, environment, and the general metabolic capacities (e.g., Hughes and Morgan 1973; Gehr et al. 1981; Maina et al. 1989a; Hughes 1995). In nature, the high metabolic needs of the endotherms have not been satisfied except by air breathing. The diffusion of O_2 occurs at a rate of $2.3 \times 10^{-5} \, \text{cm}^2 \text{s}^{-1}$ (Grote 1967) across an extremely thin, expansive blood-gas barrier. The process is completed within 250 to 500 ms (West 1974). For the typical O_2 uptake of 200 ml per min, a concentration gradient of only 0.057 kPa (a value which is negligible compared with the prevailing inspired air-arterial blood PO_2 gradient of about 6.7 kPa in the mammalian lung) is all that is necessary. The requisite structural and functional attributes of an efficient respiratory organ are an extensive surface area, a thin partitioning between the respiratory media, and efficient ventilatory and perfusive mechanisms to maintain the highest possible pressure gradient across the barrier. In simple animals, the extensive surface area (per unit body volume) is more than adequate for gas transfer, while in the larger ones respiration is restricted to specialized sites. Such areas are brought about by outfolding (evagination) or intucking (invagination = cavitation = sacculation) of a part of the body surface (Figs. 4,5). The organs in the first category have been termed gills and those in the second lungs. The gills (Chap. 4) are the archetype aquatic respiratory organs while the lungs (Chap. 6) are the model ones for air breathing: the bimodal (transitional) breathers (Chap. 5) have evolved organs which are used to extract molecular O_2 from both water and air. In the multicellular organisms, the consequential features which must be presented either singly or in combination for an organ to be designated a gas exchanger include: (1) movement of the external respiratory medium, (2) the PO_2 must be lower and the CO_2 higher in the effluent respiratory medium than in the influent one (e.g., Qasim et al. 1960), and (3) perceptible structural modifications such as infolding or outfolding and internal subdivision of the respiratory surface. Vestiges of lungs which are used for water breathing and a number of gills modified for air breathing exist, but these are rare.

1.9 The Early Anoxic Earth and the Evolution of Life

Of all concepts which have been enunciated in biology, that of evolution is probably the most important and encompassing. Dobzhansky (1973) declares that "nothing in biology makes sense, except in right of evolution". Wainright (1988) asserts that "evolution is the single most important and inclusive concept in biology". Its practical utility in biology is summed up by Nelson (1978) as follows: "the concept of evolution is an extrapolation, or an interpretation, of the order liness of ontogeny". Although debate still continues even on the validity of the concept itself and the mechanisms through which it occurs, no other plausible principle can: (1) satisfactorily organize and explain the diversity of the existing life forms, (2) account for and align the preserved fossils and the extinct forms with the extant species, and (3) explain in the context of the contemporary species the paintings and sculptures made within recorded history by the early human beings. Fossils bespeak terminated (failed?) experiments in evolution. Living things have a shared biology. The theory of evolution is grounded on the fundamental belief that life has a common origin (e.g., Brown and Doolittle 1995): through natural selection, animals and plants have progressively developed and genomically diverged in the continuum of time. However unpalatable it may sound, from congruent evidence derived from multiple proteins (Baldauf and Palmer 1993), animals and fungi are sister groups with plants constituting a more distant evolutionary lineage! Cladistic classifications attempt to reconstruct the evolutionary histories and establish relationships between different taxa from study of states of shared derived characters (e.g., Benton 1995; Huelsenbeck and Rannala 1997). Molecular genetic sequences form the basis of many modern phylogenetic reconstructions (e.g., Stewart et al. 1987; Dean and Golding 1997). To delve into the origin of life and understand the subsequent inputs and changes which culminated in the formation of the modern ecosystems for which utilization of molecular O_2 was central, different scientific disciplines like biology, astronomy, atmospheric physics, geophysics, astrophysics, geochemistry, inorganic and organic chemistry, oceanography, and geology should be integrated. An interdisplinary approach better illumines the convergence of experimental and analytical data, connecting events across temporal and spatial scales. Scientific disciplines gradually diffuse into each other. The often aggressively defended boundaries are more apparent than real. They are often created for self-interests and preservation and have profoundly hindered advances in ratiocinative thought.

The age of the Universe is estimated at between 10 to 20 billion years (e.g., Schopf 1980; Peebles et al. 1994). For a long time after its accretion, some 4.5 billion years ago, the Earth was in a state of perpetual physical and chemical turbulence (Schopf 1978, 1993; Mojzsis et al. 1996). The surface temperature was in excess of 1500 K and the high pressure primary atmosphere consisted of water vapor ($\sim 8 \times 10^{22}$ mol), CO_2 ($\sim 5 \times 10^{21}$ mol), N_2 ($\sim 3 \times 10^{20}$ mol), H_2S ($\sim 9 \times 10^{20}$ mol), and SO_2 ($\sim 7 \times 10^{19}$ mol) (Matsui and Abe 1986). With the cooling to below 650 K, the water vapor condensed, forming highly acidic primitive oceans. The minerals in the lithosphere soon neutralized the acids and the dissolved SO_2 formed sulfates and sulfides. Through outgassing (Allegre and Schneider 1994), the secondary atmosphere came to comprise mainly CO_2, N_2, water vapor, and traces of CH_4, NH_3, and SO_2. This composed the incipient neutral atmosphere which was essentially similar to the present one of Venus and Mars. Subsequently, the H_2O vapor was photochemically dissociated into H_2 and H^+, converting the secondary atmosphere into a reducing one. Some of the other most important changes to have occurred during the evolution of life on Earth have been: (1) variations in temperature and solar insolation (e.g., Foley et al. 1994; D'Hondt and Arthur 1997), (2) changes in the orbit (Imbrie et al. 1989), (3) plate tectonics (e.g., Raymo and Ruddiman 1992), (4) variations in the gaseous composition of the atmosphere (from a neutral, i.e., one where CO_2 and N_2 predominated to a highly reducing one where H_2 was the principal gas and finally an oxidizing one – with accretion of O_2) (e.g., Tappan 1974; Chappellaz et al. 1992), (5) decrease in the rate of rotation (e.g., Scrutton 1978), and (6) small fluctuations of the atmospheric pressure (e.g., Hinton 1971) and gravitational forces (e.g., Carey 1976; McElhinny et al. 1978). It is widely postulated that the chemical evolution of life occurred by combination and transformation of a vast range of simple inorganic molecules such as carbon monoxide, CO_2, N_2, H_2, and H_2O into complex biologically relevant organic compounds. This process is envisaged as having been induced by enormous energy influx probably from solar radiation, heat, meteorite impact events, radioactive decay, electrical discharges, and thunder shock waves (Calvin 1956). Though organic molecules themselves may have been extraterrestrial in origin (e.g., Cohen 1995), life is thought to have been fabricated in a chemically reducing atmosphere (e.g., Chang et al. 1983; Cloud 1983a,b; Jenkins 1991), probably around geothermal springs (e.g., Stong 1979) or on the surface of catalytic iron sulfide crystals (Russell and Daniel 1992; Kaschke and Russell 1994; Russell et al. 1994). Through long intricate condensation, polymerization and oxidation-reduction reactions of organic molecules such as amino acids, sugars, and other suitable molecules (the so-called primor-dial broth, or organic soup), the high-energy phosphate bonds (for intracellular energy transfer), specificity of protein molecules as organic catalysts, genetic coding of the nucleotides, and membrane ionic transfer processes developed (e.g., Bar-Nun and Shaviv 1975). Biogenesis of self-repairing, self-constructing, highly dynamic molecules resulted in the first living entity called protobiont by Oparin (1953) and concept organism by Chapman and Ragan (1980). This micro-scopic unit is the simplest ancestral prokaryote which possessed the most basic requisites for life. Organic evolution had to await the development of genetic and protein-synthesizing pathways. Interestingly, Lee et al. (1996) have described a self-replicating peptide. Organic molecules like amino acids, protein-like poly-mers, and nucleic acid polymers have been synthesized in the laboratory by passing an electric are through a mixture of gases such as CH_4, NH_3, H_2, and H_2O vapor, i.e., by simulating what are thought to have been the atmospheric condi-tions and circumstances which existed in the primeval past (see e.g., Sagan 1994).

1.10 Abundance of Molecular O_2 in the Earth's Biosphere

Of the nine solar system planets (eight – should Pluto finally turn out to be merely a piece space junk as recently suggested!), only the Earth has a veritable atmo-sphere suitable for life. While the atmosphere of Earth contains only tiny amounts of CO_2, those of Venus and Mars contain 96.5 and 98% CO_2, respectively. The atmospheres of Jupiter and Saturn are composed essentially of H_2 and helium. Mercury strictly lacks an atmosphere. The present atmosphere of Earth compares with that of Mars some 300 to 400 million years ago (Kargel and Strom 1996). Saturn's giant moon, Titan, has an atmosphere ten times larger than that of Earth and a surface pressure of 1.5 atm (Samuelson et al. 1981; Lorenz et al. 1997). The atmosphere is made up predominantly of molecular N_2 (82.2%), Ar (11.6%), 6% CH_4, and 0.2% H_2. Like Earth, Titan has a greenhouse effect (McKay et al. 1991; Lorenz et al. 1997). The greenhouse effect on Venus, which is caused by CO_2, generates surface temperatures of around 455 °C. Life evolved on Earth nearly 4 billion years ago (Schopf 1980; Balter 1996). The first obligatory aerobic eukary-otic cells appeared between 2.0 to 1.5 billion years ago (e.g., Reader 1986; Schopf and Walter 1983) and the first multicellular organisms about 600 million years ago, i.e., at the beginning of the Cambrian period (Nursall 1959; Cloud 1983a). The tenure of life on Earth constitutes only about 15% of biogeologic history. The complex organisms have existed for an even shorter period, i.e., about 5% of it (Figs. 7,8).

1.11 Shift from Anaerobiotic to Aerobiotic State in the Early Earth

The oxidative state of the Earth's biosphere has corresponded with the measure of the sources of O_2 and the abundance of inorganic and organic reduced com-

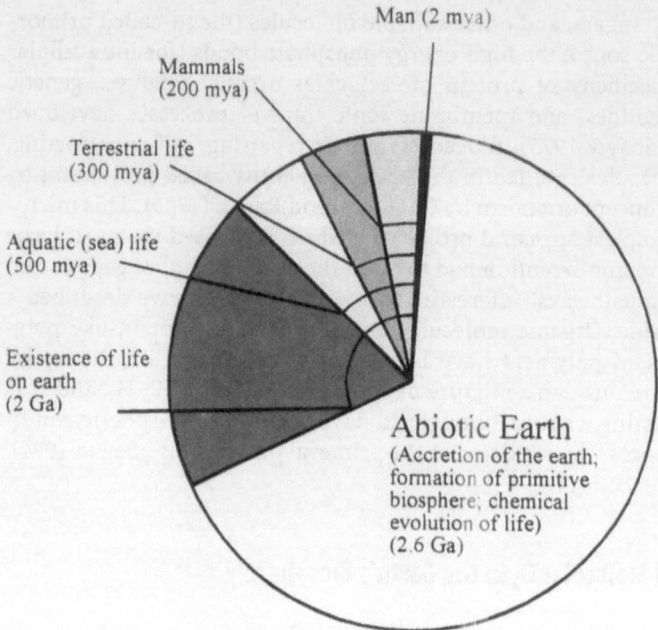

Fig. 7. Scheme showing the relatively very short period over which terrestrial and in general air-breathing organisms have lived on Earth. The dates (in *parentheses*) are averages from different publications

pounds. The appearance of an oxygenic environment entailed a change from an inefficient to a more advanced energetically O_2-dependent and metabolically highly versatile oxidizing ecosystems. Anaerobic fermentation is a highly inefficient source of energy as much of it is left locked in molecules such as alcohols and organic acids, end-products which must be removed before they accumulate to toxic levels. Fermentation of a molecule of glucose results in production of only 2 molecules of adenosine triphosphate (ATP) which contain only about 15 kcal of available energy, while in aerobic respiration 36 molecules of ATP equivalent to 263 kilocalories of utilizable energy are produced. Stated differently, to obtain the same amount of energy through fermentative respiration, a greater quantity of carbohydrate molecules must be utilized. The Pasteur effect, named in honor of Louis Pasteur (1822 to 1895), who first described the fermentation of yeast independent of O_2, has been defined in the broadest terms as "the stimulation of carbohydrate consumption by reduced O_2 tension" (Schmidt and Kamp 1996). The products of aerobic respiration are water and CO_2, two innocuous molecules which are easily eliminated into any environment at minimal risk and cost. While the energy derived from fermentative processes is just adequate to support life, aerobic respiration provided excess energy, which organisms invested towards attaining greater structural and functional complexity, resulting in greater success compared with their predecessors. Without O_2, life would probably not have developed above that of the unicellular fermentative prokaryotes.

Evolution of the earth's earliest biosphere and biota

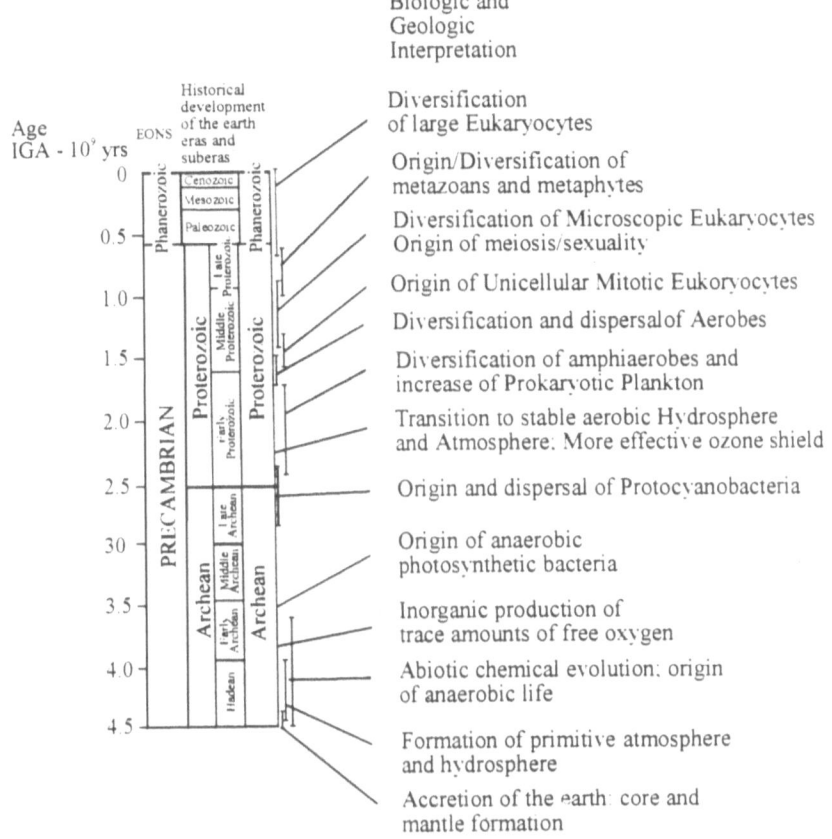

Fig. 8. Summary of the major biotic developments during the geological periods. (Schopf et al. 1983)

Oxygenic biochemistry has evolved regularly in the past in response to the changing levels of molecular O_2 in the biosphere (e.g., Fox et al. 1980). These conditions have changed from neutral to reducing state and varied from anoxia, hypoxia, and hyperoxia (relative to the present) by no means chronologically in that order (e.g., Frakes 1979; Hendry 1993; Fig. 9). Changes in atmospheric O_2 have paralleled biotic developments (Figs. 8,9). Until about 1 to 2 billion years ago, the atmosphere consisted essentially of carbon monoxide, NH_3, CH_4, H_2, H_2O vapor, and other simple hydrocarbons (Schopf 1978, 1983; Owens et al. 1979; Chapman and Schopf 1983; Grieshaber et al. 1994). Chemical evolution of life occurred within such a reducing atmosphere after the surface of the Earth had cooled to a level compatible with synthesis of the labile organic molecules (Chang et al. 1983). It has been argued by, among others, Oparin (1938) that the present conditions on Earth are no longer suitable for compounding life from inorganic

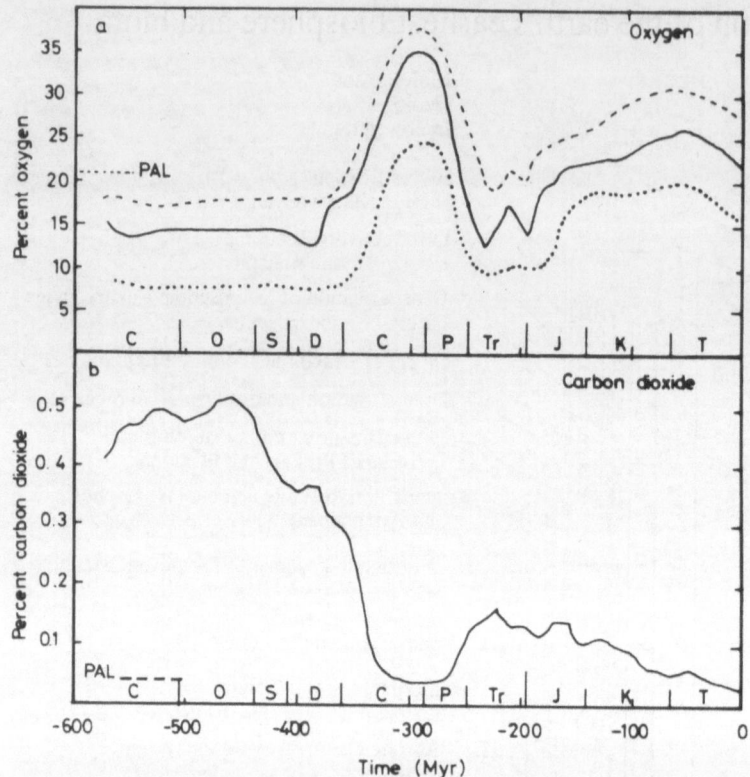

Fig. 9. Changes in the levels of O_2 and CO_2 in the late Paleozoic. Oxygen levels have fluctuated from a low of about 15% (reached towards the end of the Paleozoic, i.e., 250 million years ago) peaking at 35% by the late Carboniferous (268 mya). The present atmospheric level (PAL) of 21% is shown with a *dotted line*. CO_2 level was highest in the Ordovician-Silurian, dropped remarkably during the Devonian-Carboniferous, and increased in the late Permian. The PAL of CO_2 of about 0.036% is shown with a *dotted line*. Note that the relative levels of the two gases fluctuate in an inverse manner. *C* Cambrian; *O* Ordovician; *S* Silurian; *D* Devonian; *C* Carboniferous; *P* Permian; *Tr* Triassic; *J* Jurassic; *K* Cretaceous; *T* Tertiary. (After Graham et al. 1995; reprinted by permission from *Nature*, Vol. 375, pp. 117–120, copyright 1995 Macmillan Magazines Ltd.)

matter because the atmosphere is too highly oxidizing. Anaerobic microorganisms flourished in water in excess of 500 million years before O_2 production started (Fenchel and Finlay 1994). At the middle of the Precambrian era, a group of prokaryotes, the cyanobacteria (blue-green algae), evolved chlorophyll a, acquiring means of utilizing solar energy for the process of photosynthesis (e.g., Owens et al. 1979; DiMagno et al. 1995; Nisbet et al. 1995; Boussaad et al. 1997). Photosynthesis is the one large-scale process that abundantly converts simple inorganic compounds (CO_2, H_2O, and tiny amounts of minerals) into complex energy-rich organic carbohydrate (CH_2O) molecules. It is the source of all living matter on Earth and in that case all biological energy (Rabinowitch and Govindjee

1965). It is interesting to note that the discovery of flourishing hydrothermal vent communities in the 1970s (e.g., Meredith 1985) demonstrated that life could exist on Earth totally independent of solar radiation. The accumulation of O_2 (a product of the photosynthetic process) in the atmosphere resulted in the transformation of the Earth's nascent biotic ecosystems from an anaerobic to an aerobic state. This initiated decisive biological changes (Table 1). For successful progression of life from water to land, due to the harmful effects of the UV light, the presence of O_2, which generated a protective ozone layer in the atmosphere, was necessary. Compared with the present state, the solar UV light flux was more fierce and perhaps invariably lethal during the first 500 million years of the Earth's evolution (Gaustad and Vogel 1982).

Oxygen enrichment of the atmosphere resulted in an increase from 3 to 10% up to 100% of the present atmospheric level in the late Proterozoic and early Cambrian epochs (about 0.54 billion years ago) (Des Marais et al. 1992; Canfield and Teske 1996). It is believed that this led to the so-called Cambrian explosion, an event that was characterized by dramatic biotic developments which included: (1)

Table 1. Comparison of physical properties of the present O_2 atmosphere (21% O_2) with those of the relatively hyperoxic late Carboniferous (35% O_2) and relatively hypoxic end-Permian (15% O_2). (Graham et al. 1995)

	21% O_2 present	35% O_2 285 mya	15% O_2 250 mya	
Respiratory gases				Biological significance
Oxygen				
O_2 partial pressure (kPa)	21.2	35.3	15.1	Respiration, lignin biosynthesis
Krogh's maximum radius (cm)	0.11	0.14	0.09	Size limit for diffusion dependence
Water O_2 content (ml^{-1})	6.9	7.4	4.9	Aquatic respiration
Carbon dioxide				
Carbon dioxide partial pressure (kPa)	0.03	0.03	0.09	Effects on photosynthesis, moisture content and global energy balance
Water CO_2 content (ml^{-1})	0.31	0.31	0.31	Aquatic pH effects, acid-base balance and ion regulation
Air properties				
Density ($kg\,m^{-3}$)	1.29	1.56	1.12	Flight and respiratory mechanics, wind shear
Dynamic viscosity ($kg\,m^{-1}s^{-1}$)	18.2×10^{-6}	+	−	Boundary layer thickness
Specific heat ($js^{-1}deg^{-1}$)	1.006	+	−	Heat capacity and relative humidity
Thermal conductivity ($js^{-1}m^{-1}deg^{-1}$)	2.4×10^{-2}	+	−	Earth thermal budget, climate

+ and − indicate increase and decrease, respectively, relative to the present 21% O_2 atmosphere; mya, million years ago.

the evolution of the multicellular life (Conway-Morris 1993), (2) synthesis of the structural protein collagen which is widely distributed in the metazoans (Towe 1970), and (3) remarkable adaptive radiation and ecological diversification of the animal life (Conway-Morris 1993; Canfield and Teske 1996; Knoll 1996). In general, episodes of rapid evolutionary change correspond with occurrences of speciation (e.g., Gould and Eldridge 1977; Stanley 1979). In water, the surge in the O_2 level accelerated the biodegradation of the dissolved iron and the organic (bacterial and algal) matter. Precipitation of the resultant complexes to the bottom increased the level of oxygenation of the surface waters (Logan et al. 1995), making them more habitable. Without the ancient cyanobacteria, the Earth would still be having little, if any, reactive molecular O_2: like the atmospheres of Mars and Venus, CO_2 would still be the predominant atmospheric gas. In *Rhodobacter sphaeroides*, a metabolically versatile photosynthetic bacterium able to operate under a wide variety of environmental states, a decrease in O_2 availability leads to induction of the membranous photosynthetic apparatus (Yeliseev et al. 1997): the expression of gene-encoding components of the photosynthetic complexes, e.g., structural polypeptides, bacteriochlorophyll, and carotenoids, is closely directed by O_2 tension and light intensity. The momentous point at the end of the Early Proterozoic (some 2.0 to 1.5 billion years ago) (e.g., Kasting and Walker 1981), when the Earth changed from a mainly anoxic hydrosphere and atmosphere to an oxic one is marked by the time at which: (1) the production of banded iron abruptly stopped, (2) deposition of the highly oxidizable uraninite stopped, and (3) the first occurrence of blue-green algae of which the cells included thick-walled heterocysts which may have shielded the O_2-sensitive nitrogenase enzymes, as the modern ones do. It was not until after the oxidization of the reducing gases and mineralogic factors, when photosynthetic O_2 discharge into the biosphere finally exceeded the turnover rate of the reduced matter, that O_2 became a vital and permanent factor in a stable aerobic atmosphere. The transition from reducing to oxidizing oceans and atmosphere may have been accelerated by a declining discharge of reducing gases and oxidizable substrates through less tectonic activity (Walker 1978). Depending on ecological settings, the transitional point may have differed profoundly in different parts of the Earth.

The extensive invasion of land by plants during the Devonian enhanced the rate of production of O_2. This shifted the base of photosynthesis from water to land (McLean 1978; Knoll 1979, 1991) with the productivity of O_2 on land exceeding that of the oceans by a factor of 2 (Holland 1978). Practically all the molecular O_2 which was produced during the Earth's history, much of which is now held in diverse organic sinks, arose from green plant type and blue-green algal cyanobacterial photosynthesis (van Valen 1971; Walker 1974). The O_2 we respire today was de facto "excreted" by the cyanobacteria some 2 billion years ago during what is often called the age of the blue-green algae. As the level of O_2 in water rose and by diffusion the gas was transferred to the atmosphere, the terrestrial obligate anaerobes of the time perished with only a few, e.g., tetanus bacteria, surviving until today. Some molecular O_2 could, however, have been produced inorganically through UV light-induced photodissociation of water vapor in the primitive atmosphere after which H_2 was lost into the interplanetary space. Such a small quantity of O_2 would not have been of any biochemical consequence, as much of it would have been rapidly taken up by the unoxidized volcanically produced gases and mineralogic factors. However, a modicum level of photolytic (nonbiological) molecular O_2 may have nurtured the evolution of biological aerotolerance to O_2 in the elementary biota (Fay 1965; Holm-Hansen 1968) through development of specific biochemical pathways of mopping up and detoxifying intracellular O_2. This would have imparted a selective advantage to such moderately adapted microorganisms (e.g., Schopf and Walter 1983). Based on 16S ribosomal RNA sequencing of the prokaryotes (Fox et al. 1980), it has been shown that under modicum level of O_2, these fledgling life forms gave rise to the aerobic eukaryotes. The threshold for this transition (about 0.2% or 0.002 atm) has been incorrectly termed the Pasteur point as it resembles the Pasteur effect (e.g., Dejours 1975), the level of O_2 at which amphiaerobes change from anaerobic (fermentation) to aerobic metabolism. At that critical point, an organism converts from low efficiency fermentation to high efficiency aerobic energy-yielding catabolism. The actual point in time when this process occurred varied between organisms and environmental circumstances.

1.12 Accretion of Molecular O_2

As a general rule, the ecological resources available to organisms in any environment are finite (Hutchison 1959; MacArthur and Levine 1967; Levine 1976; Brown 1981). Environments cannot endure if they are continuously depleted of resources. Life will last only if and as long as there is frugal utilization and coherent cycling and revitalization of materials and energy (Smil 1997). Hydrologic and atmospheric O_2-CO_2 recycling are but two of the many global natural rotations of resources (e.g., McLean 1978). The changes in the partial pressures of O_2 and CO_2 in the biosphere are some of the most fascinating parameters which have influenced the direction and pace of development of life on Earth (Berkner and Marshall 1965; Rutten 1970). To stabilize and set optimal tolerable limits, at least since the Devonian, environmental PO_2 regulation was brought under direct biological control. This occurred through transfer between the biological sources (photosynthesis) and sinks (aerobic respiration) (Fig. 10). In effect, but not in mechanism, the processes of respiration and photosynthesis are diametrical: the former yields CO_2 and the latter O_2. Plants (photosynthetic autotrophs) and animals (heterotrophs) are involved in a continuous, intricate process of resource recycling, maintaining constant levels of CO_2 and O_2 in the atmosphere (Fig. 10). Using sunlight, atmospheric O_2 is continuously replenished by aquatic and terrestrial plant life and CO_2 and H_2O are produced by the respiratory processes. The end products of the way of life of each group become food for the other. The sun is the decisive source of the energy which drives this global animal-plant continuum of resource recycling. Averaged globally, the Earth receives $343\,W\,m^{-2}$ of energy from the sun in form of short wavelength radiation. About one third of it is reflected back into space by the atmosphere and the remainder ($240\,W/m^2$) is absorbed by the Earth's surface and the atmosphere. About one third ($103\,W/m^2$)

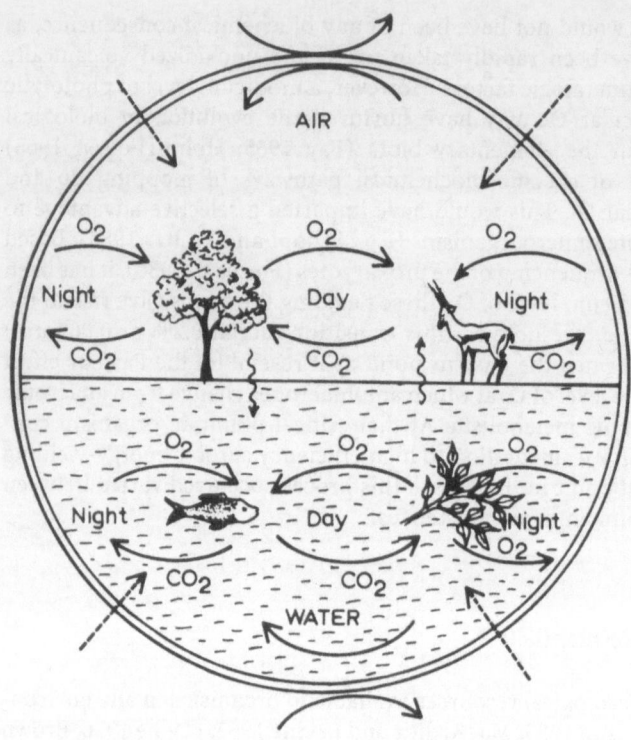

Fig. 10. Scheme showing the processes which regulate the O_2 levels in a closed habitat and the dynamics of gas transfer between water and air. The rather constant atmospheric O_2 and CO_2 levels are maintained by the cyclic balance between the photosynthetic and respiratory processes and supported by the fast diffusion rates of gases in air

of the net incoming solar radiation is reflected by the Earth's surface to the atmosphere (the Earth's albedo) in form of long wave-length radiation where it is absorbed by the greenhouse gases (e.g., water vapor, CO_2, ozone, methane, and nitrous oxide) and the clouds maintaining the surface temperature about 33 °C warmer than it would otherwise be without them (= the natural greenhouse effect). Oxygen and CO_2 are exchanged in air and to an extent with water by diffusion. Without a self-regulating O_2 and CO_2 recycling mechanism, life on Earth would have been short-lived. In the modern atmosphere, nitrogen constitutes 78.09%, O_2 20.95%, and CO_2 0.03%, the rest being composed of rare gases such as argon, hydrogen, krypton, xenon, etc. The present so-called normoxic atmospheric level of O_2 (21% by volume = about 0.2 atm) is strictly hyperoxic and far in excess of the optimum respiratory needs for life. Strictly, the modern terrestrial animals are exposed to an oxidative stress. According to the available data on amphiaerobic eukaryotic yeasts, systemic aerobic biochemistry can occur at values of 0.04% O_2 by volume (about 0.0004 atm) and organismic aerobiosis can occur at O_2 concentrations as low as 0.2 to 0.4% (i.e., 0.002 to 0.004 atm) (e.g., Rogers and Stewart 1973; Jahnke and Klein 1979). It is probably not coincidental that the Pasteur point is about 0.2% O_2 by volume, a value equal to the minimum O_2 level able to support organismal aerobiosis (Chapman and Schopf 1983). Adaptations to withstanding the harmful effects of the reactive factors of molecular O_2 was of particular importance in those life forms which produced the gas itself or were immobile and hence unable to escape from microhabitats with high concentrations of it. Unlike animals which normally operate under rather constant and somehow manageable O_2 tensions, green plants which produce the molecule itself had to evolve a complex range of molecular factors for protection against oxidative attack. Chloroplasts are the main source of antioxidants which include vitamins C and E (Crawford et al. 1994). Furthermore, while the land plants are exposed to air (21% O_2 by volume), the roots are located in the soil which, depending on type and firmness, may be virtually anoxic at depth (Currie 1962, 1984). In the marine angiosperms (Teal and Kanwisher 1966; Armstrong 1970), O_2 is known to diffuse from the roots, creating aerobic zones in the immediate area. This provides a unique microhabitat for some marine creatures such as the eulamellibranch bivalve, *Lucina floridana* (Britton 1970).

By the start of the Paleozoic era (about 600 million years ago), the PO_2 in the water and air had risen to the modest level of 0.2 kPa, i.e., one hundredth (= 0.2% O_2 by volume) of the modern sea level value. When the first vertebrates (ostracoderms) appeared some 550 million years ago (e.g., Forey and Janvier 1994), the PO_2 was only 0.9 kPa and by the Silurian-Permian periods, some 450 to 250 million years ago, the PO_2 had risen to 4.7 kPa when the amphibians ventured onto land (McClanahan et al. 1994). The terrestrial arthropods and amphibians were well entrenched on land by the Devonian period, by which time the PO_2 had risen to 10.7 kPa. The critical environmental threshold (10% of the present-day O_2 level) was crossed late in the Proterozoic era (Canfield and Teske 1996). The present level of 21 kPa was not reached until the Carboniferous period (350 million years ago) when the first reptiles appeared on land (e.g., Carroll 1988). The level of atmospheric O_2 has fluctuated greatly in the Phanerozoic (e.g., Tappan 1974; Cloud 1983a; Berner and Canfield 1989; Graham et al. 1995; Fig. 9). During the late Paleozoic, over a period of about 120 million years, O_2 rose to a hyperoxic level of 35% (compared to the present atmospheric level of 21%) and then dropped precipitously to a hypoxic low of 15% (Berner and Canfield 1989; Landis and Snee 1991). These changes were duplicated in the water (e.g., Hosler 1977; Solem 1985; Dejours 1994) and had a dramatic influence on the aquatic life (Table 1), inducing relocation to land. The greater availability of O_2 during the Mid-Devonian to Carboniferous hyperoxic episode would have made it possible for organisms, e.g., the arthropods, to attain larger body sizes (Graham 1994). Furthermore, the abundance of O_2 resulted in higher metabolic capacities and greater accessibility to resources instigating vast radiation of the animal life.

1.13 CO_2 Pulses in the Biosphere

The atmospheric and aquatic levels of CO_2 have undergone remarkable fluctuations in the past (e.g., Bender 1984; Shackleton and Pisias 1985; Walker 1985;

Barnola et al. 1987; Berger and Spitzy 1988; Jasper and Hayes 1990). The partial pressure of CO_2 is presumed to have been 100 to 1000 times more in the antediluvian Earth than now (Walker 1983). It is envisaged that CO_2, CH_4, and NH_3 produced a greenhouse effect which sustained liquid water (e.g., Owens et al. 1979; Kasting 1997; Sagan and Chyba 1997). The solar luminosity during that time was 25 to 30% lower than at present (e.g., Newman and Rood 1977). Models of the early Earth after the end of the heavy bombardment suggest that the PCO_2 may have been as high as 10 bar (Walker 1977, 1983). From analysis of air trapped in the ice cores, Raynaud et al. (1993) observed that the atmospheric CO_2 decreased from about 290 to 190 μmol mol^{-1} over a period of about 10 000 years during the last interglacial-glacial maxima. Whereas the atmospheric partial pressures of nitrogen and helium remained fairly constant across the Phanerozoic, those of neon, krypton, and argon may have increased through mantle and crustal degassing (Holland 1984; Warneck 1988). Carbon dioxide and O_2 oscillations have occurred in reverse manner (e.g., Delmas et al. 1980; Neftel et al. 1982; Graham et al. 1995; Fig. 9). This is due to the fact that over geological time, photosynthetic carbon fixation in the oceans has surpassed the respiratory oxidation of carbon (Holland 1984; Berner 1991; Walker 1987; Falkowski 1997): the difference between the two values has reflected the net increase in O_2 and reduction of CO_2 from the Earth's atmosphere. During the Devonian period (400 to 360 million years ago), the spread of rooted vascular plants to the elevated areas of the dry land may have enhanced chemical weathering leading to removal of CO_2 from the atmosphere (Berner 1997; Fig. 9). Mechanisms of phosphorus-mediated redox stabilization of the atmospheric and marine O_2 levels (e.g., Redfield 1958; Broecker 1982; Cappellen and Ingall 1996) and nitrogen fixation and denitrification in sequestration of CO_2 in the oceans over geological time scales have been described by McElroy (1983), Codispoti and Christensen (1985), Shaffer (1990), and Falkowski (1997). During the past 2 million years, reduction in the atmospheric CO_2 level has correlated with increases in the deposition of organic carbon from the surface waters to the marine sediments (e.g., Sarnthein et al. 1988; Mix 1989; Hansell et al. 1997): carbon is traded between the atmosphere, the oceans, and the terrestrial biosphere, and in geological time scales between the sediments and the sedimentary rocks. The equatorial Pacific Ocean is the greatest oceanic source of CO_2 to the atmosphere and is also the main site of organic carbon discharge to the deep sea (Murray et al. 1994).

The highest levels of CO_2 in the biosphere occurred in the Ordovician and Silurian, mainly owing to massive tectonic activities (Holland 1984). Between the late Miocene, Pliocene, and early Pleistocene (i.e., between 10 and 2 million years ago), the concentration of CO_2 fluctuated within 280 and 370 ppm by volume (van der Burgh et al. 1993). By the Carboniferous, the level had dropped almost to the present one of 0.036% subsequently rising threefold by the end of the Permian (Graham et al. 1995). In the Archean, CO_2 level may have been 100 times greater than it is today (Walker et al. 1983). In the geological recent past, the atmospheric concentration of O_2 has been fairly constant but that of CO_2 is estimated to have increased from 0.029 to 0.033% within a period of 50 years, i.e., between 1900 and 1950 (Callender 1940) owing to combustion of fossil fuels. From the start of this century, anthropogenic emission of CO_2 (mainly from activities such as burning fossil fuel, cement production, and changes in land use) has led to its increase in mole fraction from 0.00030 to 0.00034 (Revelle 1982). Between the years 1980 and 1989, the average annual anthropogenic production of CO_2 is estimated to have been 7.1 billion tonnes (Bolin et al. 1994). Since the industrial revolution, the concentration of CO_2 has risen from 280 to 350 ppm by volume (ppmv), the highest value reached in the last 160 000 years (Bazzaz and Fajer 1992; Bolin et al. 1994). Estimations at the Mauna Loa observatory in Hawaii indicated about 20% real rise in CO_2 levels between the years 1957 and 1987 (Barnola et al. 1987). Owing to the greenhouse effect, the present global temperature change correlates with the logarithm of the atmospheric CO_2 concentration (Thomson 1995). It is projected that through anthropogenic emissions, global concentrations of CO_2 will double by the end of the 21st century, a process which may cause a temperature rise of about 3 °C, resulting in serious ecological consequences (e.g., Bazzaz and Fajer 1992; Azar and Rodh 1997). If CO_2 emission were held at the present level, this would lead to a nearly constant rate of increase in the atmospheric concentrations for at least two centuries, stabilizing at 500 ppmv by the end of the 21st century, i.e., about twice the level of the time before the industrial revolution, which was about 280 ppmv (Bolin et al. 1994). Photosynthesis can be stimulated by increased level of CO_2: an inverse relationship between stomatal frequency in the leaves of C_3 plants and anthropogenic increase in atmospheric CO_2 concentration has been demonstrated (e.g., Wagner et al. 1996). Under optimal conditions of water and nutrient supply, there is a potential increase in photosynthesis by 20 to 40% when the level of CO_2 rises (Youvan and Marrs 1987). Undisturbed forests are important terrestrial sinks of CO_2 (Grace et al. 1995). The view that an atmosphere enriched with CO_2 will accelerate photosynthesis resulting in a "greener planet" and that the greenhouse effect will be brought under control by increased withdrawal of CO_2 from the atmosphere by the luxuriant plant growth has been deemed highly simplistic and short-sighted (e.g., Bazzaz and Fajer 1992). Presently, CO_2 is the most important gas in the causation of global warming by the greenhouse effect, a state which may lead to irreversible changes in plant physiology and pattern of the vegetation cover (Betts et al. 1997).

Although CO_2 is well mixed in the atmosphere, variations in its concentration in air over land and that over the oceans and between the Northern and Southern Hemispheres are well recognized. In the more industrial Northern Hemisphere, the concentration of CO_2 rises in winter and declines in summer mainly in response to seasonal growth in land vegetation (e.g., Chapin et al. 1996; Keeling et al. 1996): the seasonal cycle (peak to trough) is 15 to 20 ppmv in the far north. The equilibrium PCO_2 between air and seawater increases three times when temperature rises from 0 to 30 °C. The surface temperature of the sea modifies the CO_2 content of the oceanic biosphere, making the cold polar air contain as much as 20 ppm less CO_2 than the warmer continental and tropical air. In spite of such fluctuations, the composition of the atmosphere is considered to be reasonably homogenous. This is mainly attributable to the greater turbulence of air resulting from temperature differentials in various parts of the Earth and the high diffusivity of atmospheric gases at these temperatures. This accounts for the simplicity of the composition of the atmosphere and, for that matter, of any gas mixture. With as little information as temperature and barometric pressure, it is possible to accurately predict and estimate the changes in the tensions and concentrations which occur in a gas phase when O_2 is consumed and CO_2 released during respiration. Except for special microhabitats such as burrows and dens of mammals, reptiles, and birds, caves of freely roosting bats, within colonies of insects, and in pouches of marsupials (e.g., Mitchell 1964; Boggs et al. 1984; White et al. 1984), where relatively high levels of CO_2 and NH_3 and low concentrations of O_2 may occur, deviations in the basic composition of atmospheric gases are only naturally encountered at high altitude due to changes in barometric pressure. The rise in the atmospheric O_2 and the drop in that of CO_2 were major factors in the development of the modern respiratory organs. In the derelict aquatic habitats, high concentrations of CO_2 in water constituted a decisive driving force for transition to air breathing (Chap. 5). High atmospheric levels of O_2 led to a greater emphasis of the lungs for gas exchange, influenced the transition from buccal pumping to suctional breathing (Liem 1985; Brainerd 1994), and perhaps occasioned the change from O_2- to a CO_2-regulated respiratory control mechanism. The reduced water loss to O_2 extraction ratio was an important benefit derived from occupying a normoxic atmosphere. The early Paleozoic aquatic animals and subsequently the early amphibians subsisted in hypoxic conditions similar to the inimical ones which presently occur in habitats such as in burrows, tropical swamps, ocean sediments, and high altitudes. The present-day bimodal breathers and the developing amphibians face challenges similar to those which confronted the pioneers of transition from water- to air breathing.

1.14 The Overt and Covert Roles of O_2 in Colonization and Extinctions of Biota

Except for the recent past when the Earth presented much the same kind of climate as it does now (e.g., Boucot and Gray 1982), the planet is replete with recurrent catastrophic crises of varying magnitudes, a number of which have more or less directly or indirectly corresponded with the levels of O_2 in the biosphere. Five major episodes, in addition to numerous minor ones, have occurred. They took place at the end of the Ordovician, late Devonian, end of Permian, end of Triassic, and end of Cretaceous (e.g., Benton 1993; Weinberg 1994). Fairly fortuitously, life has navigated through these hazards, but at an enormous cost. About 99.99% of all animal species which have ever evolved on Earth are now extinct (Pough et al. 1989). During the mid-Paleozoic Mass Extinction, dubbed "the mother of all extinction" by Erwin (1993), an event which occurred towards the end of the Permian period (e.g., Robinson 1991; Allegre and Schneider 1994), adverse tectonic activity which lasted through the Triassic to the early Jurassic culminated in an abrupt temperature decline. Concurrently, a sudden drop in sea level occurred. The land submerged under water decreased from 30 to 5%, leading to exposure of expansive organic sediment (Erwin 1996). This resulted in increased utilization of O_2 in oxidative processes with an attendant upsurge in the discharge of CO_2 into the atmosphere. The consequence was an extremely severe hypoxia which was exacerbated by hypercarbia (Benton 1995; Wignall and Twitchett 1996): a worldwide deep sea anoxic outcome which occurred across the Permo-Triassic (or Paleozoic and Mezozoic) boundary (~250 million years ago) which lasted for 20 million years has been described by Isozaki (1997). The mid-Paleozoic crisis led to a near-annihilation of the marine biota (e.g., Tappan 1974; Weinberg 1994). Even under such dire circumstances, the level of O_2 in air was greater than that in water, making air breathing an evolutionary advantage. Nearly 90% of the aquatic animals succumbed (e.g., McGhee 1989) and on land, more than two thirds of the reptilian and amphibian species perished (Erwin 1993, 1994). In the only mass extinction which insects have had to endure, a sign of the severity of the prevailing conditions, 30% of the orders died (Erwin 1996). In all taxa, the particularly vulnerable groups were those which had preadapted to the earlier hyperoxic milieu of the Carboniferous. It has, however, been argued, e.g., Graham et al. (1995), that lack of O_2 per se was not the primary factor which precipitated this immense demise, since the drop in O_2 from the peak levels is envisaged to have been very gradual, occurring over a long period of time during which the animals should have adapted to the change. However, Erwin (1996) pointed out that the actual active period of extinction may have been as short 1 million years, if not less. In the late Triassic extinction (about 200 million years ago), 20% of the families of animals died out, eliminating some 50% of the species (Benton 1993). Animals such as the ammonoids and bivalves were severely decimated and the cocodonts disappeared. Contrary to expectation, no frequency or periodicity has been evident in the seven mass extinctions (i.e., in the Early Cambrian, Late Ordovician, Late Devonian, Late Permian, Early Triassic, Late Triassic, and End Cretaceous) which have occurred in the last 250 million years (Benton 1995): the episodes are separated by between 20 and 60 million years.

The ecological disaster which has caught most of the attention of scientists and the public, even though far less severe, is that of the dinosaurs and their contemporaries, the plesiosaurs and pterosaurs. This occurred between the Cretaceous and Tertiary (65 million years ago). The global faunal diversity was reduced by 60 to 80% (Raup and Jablonski 1993). Geological (terrestrial) as well as cosmic (extraterrestrial) events have been associated with the demise (e.g., Hallam 1987; Stanley 1987; Kerr 1988; Powell 1993). Based on a high concentration of iridium in some rock deposits at the end of the Cretaceous, a large asteroid or a comet about 10 km in diameter is thought to have impacted on Earth off the Yucatan Peninsula (Mexico) (e.g., Sheehan et al. 1991). The mass of dust thrown into the atmosphere blocked out the sun's rays, suppressed photosynthesis (perhaps causing a drop in the level of O_2 in both air and water), and occasioned severe changes in the Earth's climate. As in the "mid-Paleozoic Mass Extinction", sea level fluctuated by about 50 to 200 m over a period of 0.2 to 1 million years (Haq et al. 1987; Kerr 1996b; Stoll and Schrag 1996). The definite cause of the drop in sea level during the Cretaceous, a period during which the climate is predicted to have been fairly stable, where the equatorial temperatures were equal to the present ones (Herman and Spicer 1996), and the poles were ice-free (Barron et al. 1981), is still debatable.

1.15 Oxygen: a Paradoxical Molecule

Most profoundly, molecular O_2 has influenced the geology of the Earth and pervasively directed the trajectory and forward momentum of evolution of life. Since the first appearance of the gas in the biosphere in appreciable quantity some 2 billion years ago (e.g., Owens et al. 1979), the history of life is literally inscribed on this single molecule. The buildup of O_2 and evolution of oxygenic respiration (following the radical conversion of the incipient high CO_2-low O_2 atmosphere to low CO_2-high O_2 atmosphere) led to the transformation of the early simple anaerobic cells to the versatile eukaryotic ones which subsequently accreted into the aerobic, multicellular organisms (Siever 1979; Allegre and Schneider 1994; Orgel 1994). The transformation enhanced the efficiency of carbon and nutrient recycling, leading to a climactic increase in the organic biomass. Margulis (1979) envisages that the toxicity of the O_2 molecule enforced symbiotic associations on anaerobic bacteria, leading to development of eukaryotic cells. Symbiotic relationships are known to evolve under extreme circumstances (e.g., Childress et al. 1989; Rennie 1992). The presence and the resolutely increasing levels of O_2 may explain the short period (in evolutionary terms) of 700 million years which it took for the growth and change of the eukaryotic cells to multicellular organisms compared with the over 2 billion years which passed before the aerobic eukaryotic cells developed from the anaerobic prokaryotes (Gould 1994; Fig. 8).

In virtually all the solar system planets and their satellites, free atmospheric O_2 for which there are no known primary (geochemical) sources, is found in high concentration only on Earth. As a molecular factor, however, O_2 is not unique to our biosphere. Jupiter's moons Europa and Io, for example, have surface and atmospheric water, gaseous sodium, and small quantities of O_2 (Brown and Hill 1996; Kerr 1997), a state which is similar to that of Earth some 4 billion years ago. Due to its high reactivity with other elements at both the temperatures of the formation of the magma (500 to 1200 °C) and at ordinary surface temperatures, O_2 is the most abundant element of the average crustal rocks, followed by silicon, aluminum, and iron (Chapman and Schopf 1983). This notwithstanding, O_2 is a somewhat alien factor to life. A reducing (nonoxidizing) environment was a prerequisite for the chemical evolution of life from the first organic molecules in the primitive atmosphere (e.g., Cloud 1974, 1988; Miller and Orgel 1974; Tappan 1974; Schidlowski 1975; Chang et al. 1983). It is an intellectually intriguing contradiction that life, which evolved in absence of O_2, is now tractably dependent on it. Oxygen is a necessary resource for body growth and development (e.g., Adelman and Smith 1970; Priede 1977; Armstrong et al. 1992). Extended exposure to hypoxia in newborn and young mammals causes a decrease in body growth (e.g., Timiras et al. 1957; LaManna et al. 1992). While animals that are well adapted to high altitude, like the llama and many small rodents and birds, are known to reproduce successfully up to an altitude of 5 km, human fetal growth is retarded at altitudes beyond 3 km above sea level (e.g., Haas et al. 1980; Mayhew et al. 1990).

1.16 The Rise of the Level of Molecular O_2: a Curse or a Blessing?

1.16.1 The Deleterious Reactive Radicals of Molecular O_2

Although the aerobic life-style confers great advantages, as metabolism using O_2 yields 20 times more free metabolic energy than an anaerobic one, utilization of O_2 is accompanied by great danger (Pryor 1986; Joenje 1989; Sies 1991). Even under ordinary conditions, due to continuous formation of free radicals (chemically highly reactive molecules with an unpaired electron) which are intermediates of a number of biochemical reactions, O_2 is a highly toxic substance. Its utilization by aerobes is harmful in both the short and long terms (e.g., Comroe et al. 1945; Clark and Lambertsen 1971; Fridovich 1978; Halliwell 1978; Hill 1978; Slater 1984; Sies and Cadenas 1985). Physical exercise increases formation of reactive O_2 species but in the long term, endurance training improves antioxidant handling (Sen 1995). Strictly, molecular O_2 is not the toxic agent but rather its reactive derivatives (Cochrane 1991). Inflammatory cells (neutrophils, eosinophils, and macrophages), catalase negative bacteria, inhaled environmental pollutants (e.g., ozone and nitrous oxide), and even epithelial cells (e.g., intracellular production from mitochondrial respiration and xenobiotic drug metabolism) are potential sources of reactive O_2 species (Bast et al. 1991; Kinnula et al. 1992; Cohn et al. 1994). The oxidants have been associated with pathogenesis of respiratory disorders such as adult respiratory distress syndrome, emphysema, asthma, and pollutant-precipitated diseases (Adler et al. 1990; Barnes 1990). During ordinary aerobic respiration, complete reduction of a molecule of O_2 to H_2O requires four electrons which are sequentially utilized in the process:

$$O_2 \quad \xrightarrow{e^-} \quad O_2^- \quad \xrightarrow{e^-} \quad H_2O_2 \quad \xrightarrow{e^-} \quad OH^- \quad \xrightarrow{e^-} \quad H_2O.$$

Various active intermediates, highly reactive chemical species with one or more unpaired electrons, are produced. These include the superoxide anion radical (O_2^-), hydrogen peroxide (H_2O_2), hydroxyl radical (OH^-), and singlet oxygen ($^1O_2$). The oxidant by products of metabolism cause extensive damage to DNA, proteins, and other macromolecules (Borteux 1993; Epe 1995). Degenerative diseases associated with aging such as arteriosclerosis, Parkinson's disease, diabetes, cancer, decline in the immune system, and senility have been associated with the oxidative damages caused by molecular O_2 (e.g., Fraga et al. 1990; Wagner et al. 1992; Ames et al. 1993; Gutteridge 1993; Halliwell 1994). The assault by the reactive O_2 radicals on cell functional and structural integrity is intense. It is estimated that about 2 to 3% of the O_2 taken up by aerobic cells results in production of O_2^- radical and H_2O_2 (Chance et al. 1979). Approximately 10^{12} O_2 molecules are handled by a rat cell daily (Chance et al. 1979). This results in about 2×10^{10} (i.e., 2%) O_2^- and H_2O_2 active (partially reduced) species. Frage et al. (1990) estimated that there are about 9×10^4 attacks on the DNA per day per cell in a rat. Under a steady state, about 10% of protein molecules may undergo carbonyl modifications (Fridovich 1978; Ames et al. 1993; Orr and Sohal 1994). Ames et al. (1993) envisaged that O_2 free radicals are responsible for 10 000 or so DNA base modifications per cell per day. Such sustained attacks can easily overwhelm the cell's repair mechanisms. All cells have evolved a number of repair endonucleases which specifically recognize and repair the damages caused by the reactive species (Lindal 1990; Demple and Harrison 1994). When the biodegradative processes surpass the biosynthetic ones, the cumulative damages may result in significant loss in the cell functional capacity (Stadtman 1992). Oxygen free radical-mediated lipid peroxidation, for example, could easily lead to loss of membrane integrity and hence compromise the normal cellular activities. A substantial memory recovery was achieved by chronically treating old gerbils (15 to 18 months) with a free radical spin-trapping compound N-tert-butyl-β-phenylnitrone (Carney et al. 1991). The process resulted in a substantial decrease in the amount of oxidized protein in the brain and an increase in the amounts of glutamine synthetase and neutral protease activities. Oxygen, and particularly the resultant OH^- radical, increase the lethal effects of ionizing radiation (Von Sonntag 1987; Ames et al. 1993). Lipid peroxidation results in mutagenic factors (e.g., epoxides and alkoxyls) and $^1O_2$.

The formation of reactive O_2 species inside cells constitutes a serious threat to the functional and structural integrity of the cellular genome (Lindal 1990; Epe 1995). It is envisaged that control of O_2 toxicity could have necessitated the evolution of the nucleus and the nuclear membrane in the eukaryotic cells to minimize external affronts by molecular O_2 (e.g., Margulis 1981). The nucleus constitutes an anoxic and fairly safe location for the deoxyribonucleic acid (DNA). The DNA nearer to the nuclear membrane and that in the O_2-rich cytoplasm are more susceptible to damage. The mitochondrial DNA (mtDNA) is particularly more exposed to O_2 toxicity compared with the nuclear DNA (nDNA) (Gupta et al. 1990; Dyer and Ober 1994). The mtDNA from the rat liver causes more than ten times the level of oxidative DNA damage than does nDNA from the same tissue (Richter et al. 1988). This may be due to factors such as lack of mtDNA repair enzymes, lack of histones protecting mtDNA, and the proximity of mtDNA to oxidants produced during oxidative phosphorylation. A high turnover of mitochondria ensures removal of the damaged organelles (which generate greater quantities of oxidants) but overall, oxidative lesions accumulate in the mtDNA at a greater rate than in the nDNA (Ames et al. 1993). Generally, mitochondria are protected from injury by: (1) the complex extracellular O_2 diffusion pathway where at the cellular level the PO_2 will have dropped to almost zero and (2) by clustering, a process which reduces the area available for influx of O_2 into individual mitochondria (Gnaiger 1991). The PO_2 in blood capillaries of the cardiac muscle is ten times greater than in isolated mitochondria (Tamura et al. 1989). Clustering accounts for the difference in the O_2 uptake capacities between isolated liver mitochondria and the intact ones in the hepatocytes (Jones 1986). Connett et al. (1985) estimated that the PO_2 in the red muscle cell is about 0.07 kPa. Mitochondria in living tissues are estimated to operate at low O_2 levels, frequently below 2% of air saturation of a PO_2 of 0.5 kPa (Wittenberg and Wittenberg 1987, 1989; Graiger et al. 1995). Respiration in the mitochondria is not affected until the PO_2 drops to below 0.01 to 0.1 kPa (Oshino et al. 1974; Sugano et al. 1974). Just as in high quantities, lack of O_2 is equally injurious. In those tissues that are intolerant to hypoxia, e.g., the brain and the heart muscle, mitochondria undergo irreversible structural failures where the membrane potentials decline, and the degree and capacity of molecular coupling is lowered (Zimmer et al. 1985).

1.16.2 Senescence: the Effects of Molecular O_2

Senescence is a progressive and irrevocable loss of functional capacity due to degeneration of somatic cells in the last part of life. It has been associated with the use of O_2 by the cells even under normal physiological conditions (e.g., Floyd 1991; Ames et al. 1993; Sohal and Weindruch 1996). As part of the metabolic processes, aerobic organisms generate potentially destructive O_2 species which cause serious oxidative damage to biological macromolecules (e.g., Fridovich 1978). The damage occurs in form of peroxidation of membrane polysaturated fatty acid chains, alteration in DNA protein bases configuration, and carbonylation and loss of sulfhydryls in proteins. Changes due to aging directly or indirectly affect the O_2 uptake process itself (Horvath and Borgia 1984). A senile lung is defined as one which presents dilation of the air spaces without tissue destruction (e.g., Hyde et al. 1977; Pinkerton et al. 1982; Snider et al. 1985; Dios-Escolar et al. 1994). Studies on the effects of age on lung structure have, however, been inconlusive (e.g., Thurlbeck 1980). The impedement has been due to the fact that the effects of environmental pollution and lifestyle have been difficult to fully assess and exclude from the various studies. No aging structural changes in the lung were observed in the growing specific-pathogen-free inbred male BALB/cNNia mice between 38 days and 28 months by Masahiko et al. (1984). In adult humans, arterial O_2 tension decreases by an average of 0.28 to 0.54 kPa each decade while the alveolar PO_2 remains constant or increases slightly (Dill et al 1963; Sorbini et al. 1968). Pulmonary diffusing capacity decreases by 5 to 8% in each decade of life (Cohen 1964). With aging, the surface area of the erythrocytes decreases and the cells become more susceptible to osmotic changes and mechanical disruption while there is reduction in enzymatic activity, e.g., hexokinase, glucose-6-phosphate dehydrogenase, and lactate dehydrogenase (Prankerd 1961). Compared with young rats, old rats have a lower hind limb muscle respiratory capacity and whole body maximal O_2 consumption (Cartee and Farrar 1987). In both animals and plants, paraquat (methyl viologen), a commonly used herbicide, and antibiotics such as streptonigrin which enhance the rate of O_2 radical (O_2^-) production, are more toxic under aerobic than anaerobic conditions. Longevity correlates inversely with the rate of mitochondrial production of the O_2^-, H_2O_2, and OH^- species. In a number of mammals, the severity of oxidative DNA damage appears to correlate with the metabolic rate (Adelman et al. 1988; Shigenaga et al. 1989). Restriction of calorific intake, a process which lowers metabolic stress, reduces age-related changes and prolongs the life span of mammals (Sohal and Weindruch 1996). Overexpression of superoxide dismutase and catalase, antioxidative enzymes which respectively remove O_2^- and H_2O_2, extends the life span of *Drosophila melanogaster* (Orr and Sohal 1994). Reduction in production of active O_2 species in transgenic *Drosophila melanogaster* leads to a 30% increase in the metabolic potential (Orr and Sohal 1994). In two human diseases associated with premature aging, Werner syndrome and progeria, oxidized protein residues increase at a faster rate than normally (Stadtman 1992). The gradual decay of the cell integrity with age is an indication that the antioxidant defenses of the aerobic cells are inefficient in face of sustained assault by the reactive O_2 metabolites. Oxidative lesions on DNA increase with age (Ames et al. 1993): at the age of 2 years, a rat cell has about 2 million lesions per cell. This is about twice that of a young rat. Exogenous oxidant loading from activities such as smoking may deplete the endogenous antioxidant levels in the cells, such stress compromising the cell defenses (Csillag and Aldhous 1992).

Notwithstanding the deleterious effects of the endogenous reactive species, not all effects of the active species are harmful. Leukocytes and other phagocytic cells destroy bacteria or virus-infected cells by subjecting them to a lethal discharge of nitric oxide (NO), O_2^-, H_2O_2, OH^- and OCl^-, a potent oxidant mixture (Baldridge and Gerard 1933; Seifert and Schultz 1991; Stamler et al. 1992). Like O_2, among gases, NO (which, unlike O_2, is elaborated in animal tissues but in very small quantities; De Belder et al. 1993; Ballingand et al. 1995) is an enigmatic molecule. In the atmosphere, NO is a toxic chemical but in small regulated quantities in the body, it plays vital physiological, pharmacological, and immunological roles. It participates in processes such as blood pressure control (through smooth muscle relaxation), platelet inhibition, neurotransmission, destruction of pathogens, penile erection, and has even been associated with learning and long-term memory (Culotta and Koshland 1992; Koshland 1992; Stamler et al. 1992; De Belder et al. 1993; Ballingand et al. 1995). Nitric oxide and, amazingly, carbon monoxide (one of the most feared gases) are the first gases known to physiologically regulate levels of guanosine 3,5-monophosphate and thus act as biological messengers and signaling molecules in mammals (Moncada et al. 1991; Toda and Okamura 1991; Galla 1993; Verma et al. 1993; Katušic and Cosentino 1994). Above certain thresholds, active radicals of NO such as nitrosonium cation (NO^+), nitric oxide ($NO^·$), and nitroxyl anion (NO^-), similar to the redox states of O_2 (i.e., O_2^-, H_2O_2, and OH^-), form and oxidize biological molecules.

Evolution does not appear to have found any direct, enduring solution for neutralizing the harmful active species of molecular O_2. However, to ameliorate the effects, cellular biochemistry has been configured such that the most important processes are the reductive ones: biological oxidations entail removal of H_2 rather than addition of O_2. The increasing susceptibility of the cells to oxidative effects of the reactive O_2 radicals with age may be caused by several factors. These include: (1) an increase in the O_2 delivery rate to the cells as may occur with changes in the amounts of allosteric effectors which determine the O_2 binding capacity by the hemoglobin, (2) increased availability of the Fe(II) or Cu(II), metals which are catalytically involved in the production of the highly damaging O_2 free species by changes in the efficacy of metal binding proteins and chelating agents, (3) an age-dependent increase in the production of the reactive O_2 metabolites, (4) an intrinsic decline in the production of the endogenous antioxidant scavenger enzymes and metabolite defenses, (5) loss of capacity to biodegrade the products of cell oxidation, and (6) decrease in the capacity to mobilize and repair the damages caused by the active radicals. In the life cycles of practically all complex organisms, the integrity of the tissue cells appears to be guaranteed only up to the useful reproductive period. It is possible that reproduction takes such a heavy toll of an animal's resources that the necessary amount needed for maintenance of the integrity of the somatic cells is irrevocably compromised.

1.16.3 Biological Defenses Against O_2 Toxicity

Of the four fundamental elements of life, the so-called biogenetic elements, carbon, hydrogen, oxygen, and nitrogen, O_2 is a geochemical and biochemical anomaly. Although a structural component of biological tissues and an integral part of biochemical reactions, in combined as well as free diatomic state, depending on the organism and level of concentration, O_2 is a toxic factor. The accretion of molecular O_2 changed the entire global environment and dramatically influenced the nature and tempo of all subsequent evolutionary developments in life. Most animals die after extended exposure to 100% O_2 (1 atm), the endothelial lining of the lung being the main site of injury (e.g., Matalon and Egan 1981; Crapo et al. 1984; Block et al. 1986; Crapo 1986). Rats die between 60 to 72 h after exposure to 100% O_2 (Crapo 1987). Exposure of 21-day-old rats to >95% O_2 for 8 days induces cholinergic hyperresponsiveness as well as hypertrophy of the airway epithelial and smooth muscle layers (Hershenson et al. 1994). More than 50% of 72-h-old chick embryos die on exposure to 3 h of 5 atm of O_2 and 20 to 30% of those which hatch have deformities of the brain, eyes, upper jaw, legs, feet, and heart (Pizarello and Shircliffe 1967). In mammals, on exposure to hyperbaric O_2 (partial pressure of O_2 in the inspired air greater than 1 atm), nervous acute O_2 poisoning occurs. This is expressed in form of epileptiform convulsions (Barthelemy 1987). Nitrogenases, the enzymes necessary for the fixation of nitrogen (e.g., Leigh 1997), are inhibited by as little as 0.1% free O_2 (Postgate 1987). Many nitrogenous bacteria occupy anaerobic habitats, e.g., cells of leguminous plants under the soil. The reactivity of O_2 makes it a rather biologically enigmatic molecule (Cochrane 1991). It readily reacts with the reduced (H-rich) biochemical factors to produce energy (in aerobic respiration) and its highly reactive radicals oxidize and destroy enzymes. The reactive side products such as superoxides, peroxides, hydroxyl radicals, and $^1O_2$ are biologically toxic to strict anaerobes and at even moderate concentrations are highly toxic to aerobes (Fridovich 1976, 1978; Halliwell 1978). Practically all organic compounds tend to be easily oxidized and are hence potentially unstable in the presence of molecular O_2 (Miller and Orgel 1974). Adaptively, the epithelial cells which line the respiratory organs contain antioxidant enzymes which prevent initiation of pulmonary pathology after contact with oxidants such as ozone, nitrous oxide, and oxidants produced by local inflammatory reactions (Kinnula et al. 1992; Cohn et al. 1994).

It is overtly anomalous that life started in an O_2-free environment and subsequently became so heavily dependent on it. The stubbornly increasing levels of O_2 left the cells with no alternative but to somehow accommodate it or face certain annihilation. Its rather small molecular size and hence high intracellular diffusivity and a correct redox potential endeared O_2 to utilization as an electron acceptor in the biochemical energy production of the tricarboxylic chain processes. Interestingly, in myocardial cells, when energy depletion in a cell, e.g., in the case of ischaemia or hypoxia, is extended beyond the stage when reoxygenation leads to spontaneous recovery, in what has been termed the oxygen paradox (e.g., Piper et al. 1994), reoxygenation may intensify tissue damage (Hearse et al. 1973; Ganote 1983; Gorge et al. 1991) presumably due to excessive production of active free radical (McCord 1988; Turrens et al. 1991): life's biochemistry has not achieved a capacity of totally taming molecular O_2.

As a part of antioxidant defense system, gas exchangers (e.g., Crapo and McCord 1976) and some organisms, e.g., bacterial aerobes like *Escherichia coli* (Gregory and Fridovich 1973) possess a battery of simple nonenzymatic molecules and complex enzymes which scavenge the oxidative O_2 radicals (Forman and Fisher 1981; Freeman and Crapo 1982; Halliwell and Gutteridge 1985; Sies 1991). The former include glutathione, ascorbate, urate, bilirubin, ubiquinol, β-carotene, and tacopherol, while the latter comprise superoxide dismutase (SD), catalase, and glutathione peroxidase. Superoxide dismutase converts the superoxide radical (O_2^-) to H_2O_2 plus O_2 (e.g., Crapo and Tierney 1974; Fridovich 1975; Cassini et al. 1993) and catalases and peroxidases convert H_2O_2 to H_2O and O_2. Experimentally, rats are protected from O_2 toxicity by intravenous injection of liposome-entrapped catalase and superoxide dismutase (Turrens et al. 1984). Hyperoxia increases release of reactive O_2 species in the mitochondia (Turrens et al. 1982a,b). Adaptively, in old rats, even though glutathione synthesis is decreased, the tissue regeneration capacity appears to be increased to cope with oxidative stress (Ohkuwa et al. 1997). In rat lungs, on hyperoxic exposure, gene expression plays an important role in controlling manganese containing superoxide dismutase activity (Ho et al. 1996). Interestingly, molecular H_2, has been reported to destroy active O_2 radicals (Jones 1996). Aerobiosis could not have arisen directly from anaerobiosis. It was imperative that preadaptations for tolerating O_2 toxicity would have required time to evolve. It is speculated that photochemically produced O_2 at low manageable levels may have nurtured some degree of aerotolerance (Walker et al. 1983), a feature which would have imparted a selective advantage when a stable oxygenic atmosphere formed. The envisaged chronological order through which molecular O_2 was incorporated into the biochemical processes is: (1) development of defensive mechanisms (though it is difficult to explain how these could have been configured in absence of O_2 itself), (2) production of O_2 by the cyanobacteria, and (3) incorporation and utilization of O_2 at the cellular level. The transition to an aerobic environment brought about and required evolution of higher redox potential biochemistry (e.g., Williams and Da Silva 1978). This necessitated evolution of new catalysts which were based on transitional metals. It has been estimated that the befitting redox potential of the primitive oceans may have been about $-350\,mV$ (Osterberg 1974). Though O_2 is mainly utilized for energy production, many other nonrespiratory processes like collagen synthesis, oxidation of amino acids, and tanning of cuticle (in many insects) require molecular O_2. In the facultative invertebrate anaerobes, growth is arrested with reduction in the level of O_2 (e.g., Hammen 1969, 1976). To avoid oxidation of nitrogenase enzymes, the nitrogen-fixing cyanobacteria have evolved a novel method of coexistence ensuring intracellular anoxia. Special cells called heterocytes which lack the full complement of photosynthetic pigments and are hence incapable of producing O_2 have developed thick cell walls covered by a mucinous coat to limit O_2 inflow: these cells are endowed with the appropriate enzymes for destroying the harmful O_2 radicals.

1.17 The Evolution of Complex Metabolic Processes

The evolution of the eukaryotes from prokaryotes (some 2 billion years ago), the attainment of capacity for sexual reproduction (some 1 billion years ago) and, subsequently, development of complex multicellular life about 600 million to 1 billion years ago (Schopf and Oehler 1976; Schopf 1978; Wray et al. 1996) were pivotal points in the development and proliferation of animal life (Romer 1967). It is conjectured that mitochondria, the power houses of cells, evolved from free-living eubacteria-like endosymbionts which more than a billion years ago (e.g., Palmer 1997) invaded the eukaryotic cells (e.g., Margulis 1970, 1979; Vogel 1997). This is vindicated by the fact that organelles such as chloroplasts and mitochondria contain RNA and DNA which are different from those in the nuclei of the eukaryotic cells and in most ways resemble those of certain bacteria. The closest contemporary "relatives" of the mitochondria are the rickettsial group of the α-proteobacteria (Yang et al. 1985; Lang et al. 1997). The rat mitochondrial outer membrane localizes benzodiazepine receptor (MBR) which is expressed in wild-type and $TspO^-$ (tryptophan-rich sensory protein) strains of the facultative photoheterotroph, *Rhodobacter sphaeroides* (Yeliseev et al. 1997): functionally, MBR substitutes for $TspO^-$ and negatively regulates the expression of photosynthesis genes in response to O_2. This provides further evolutionary support for the origin of mammalian mitochondrion from a photosynthetic precursor. Since it was fully elucidated in the 1960s, the genetic code was thought to be universally identical as any mutations were deemed fatal – the frozen accident hypothesis of Crick (1966). Intellectually, this was a satisfying expectation as it supported the concept of a common origin of life and the parsimony of life in conservation of highly important factors. It is now known that the coding system in mitochondria in various mammalian phyla, certain bacteria, ciliated protozoa, mycoplasma, algae, and yeasts differs from the "universal code", in use of certain codons (e.g., Barrell et al. 1979; Anderson et al. 1981; Jukes 1985; Yamao et al. 1985; Jukes and Osawa 1993). It is envisaged that the universal code may be an evolutionary descendent of the mitochondrial-type code (Osawa et al. 1992) which shows extreme features of genomic economization perhaps to save space (Kurland 1992). By incorporating the endosymbionts, a process called secondary endosymbiosis, the eukaryotic cells did not, so to speak, have to reinvent respiration and photosynthesis through the long and costly trial-and-error process of genetic evolution. The origin of aerobic metabolism involved only extension and refinement of the preexisting anaerobic processes (the O_2-independent primitive glycolytic fermentation) by development of catalysts (enzymes) (Keevil and Mason 1978; Gunsalus and Sligar 1978; White and Coon 1980) which acted on the new addition to the energetically more efficient citric acid (Krebs) cycle (Williams and Da Silva 1978; Chapman and Schopf 1983; Schopf 1989). In advanced eukaryotes (e.g., vertebrates and plants), O_2 is involved in the sequence of chemical reactions only at the terminal stages, i.e., the most recently evolved sections of the pathway. The anaerobic bacteria are capable of effecting only the glycolytic pathway. With time, molecular O_2 has been incorporated in the biochemical synthetic pathways of compounds such as phenols, polyunsaturated fatty acids, amino acids, cyto-chromes, and bile pigments (in vertebrates) and elsewhere only in the relatively advanced organisms of particular lineages. The O_2-dependent metabolites of the aerobes have amply been used only to refine in a biochemical, structural, and functional manner the earlier established anaerobic systems: no totally new mo-lecular complexes have evolved (e.g., Rohmer et al. 1979; Chapman and Ragan 1980).

1.18 Oxygen and CO_2 as Biochemical Factors in Respiration

The development of aerobic biochemistry is of fundamental interest to respira-tory and evolutionary biologists. Compared with CO_2, O_2 is the more important factor in respiration. No permanent change especially of structural nature has been reported on exposure of living tissue to extended hypercapnia. In total anoxia, e.g., when breathing pure N_2, a human being loses consciousness within 30 to 40 s and permanent brain damage may occur. On breathing 30% CO_2 in O_2, a period of dyspnea occurs and loss of consciousness ensues. Complete resuscita-tion without any permanent damage can occur. Carbon dioxide-induced acidotic narcosis (Sieker and Hickam 1956) has been applied to treat some psychiatric disorders (e.g., Meduna 1950). Unequivocally, O_2 procurement is the primary goal of respiration and CO_2 elimination and hence acid-base regulation are sec-ondary roles. A complex neurohormonal system has evolved especially in verte-brates (e.g., Fedde and Kuhlmann 1978; Ballintijn 1982) to monitor O_2 levels in blood. A reduction of the PO_2 in the ambient air results in a drop of that in blood. Below a certain threshold level, the arterial PO_2 chemoreceptors send impulses to the respiratory center in the brain stem from where ventilatory rate is adjusted. Although the brain constitutes one of the so-called noble organs (the others being the heart and the lung), which are structurally and functionally highly protected against O_2 fluctuations (e.g., Zapol et al. 1979; Freedman et al. 1980), the blood-brain barrier is highly permeable to respiratory gases. The actual mechanism through which molecular CO_2 acts on the respiratory centers, if it does at all, has been highly debated since H^+ and HCO_3^- ions exist in equilibrium with carbonic acid (H_2CO_3) (e.g., Crone and Lassen 1970; Bradbury 1979). Hyperoxia-induced hypercapnia, which may result in total cessation of ventilation in the trout (e.g., Dejours 1973), in the crayfish (Massabuau et al. 1984), and in the green crab (Jouve-Duhamel and Truchot 1983), shows that at least in water breathers, animals are insensitive to or are incapable of responding to elevated CO_2 in face of O_2 assault. Hypercapnea does not appear to affect ventilation in hyperoxic water breathers (Dejours 1988). However, in the only discordant view, Thomas et al. (1983) reported data to the contrary on the rainbow trout (*Oncorhynchus mykiss*): marked increase in ventilation was observed in specimens made hypercapnic even under hyperoxia. Except in the fossorial animals which adaptively tolerate high CO_2 levels, whether in moderate hypoxia, normoxia, or hyperoxia, air breathers increase ventilatory rate after inhalation of CO_2-enriched air: the hypercapnia-induced hyperventilation enhances CO_2 clearance, reduces arterial hypercapnia, and establishes normal pH. In a hypoxic environment, increased ventilation and heart rate, processes which are energetically costly to maintain, are not only unproductive but harmful. Faced with such a crises, reduction of metabolism is a beneficial process to overcome hypoxia (e.g., Hochachka 1988). Hypothermic hypometabolism in a squirrel results in an 88% saving on energy (Wang 1978).

1.19 Homeostasis: the Role of Respiration

Respiration plays an integral part in regulation of the blood pH. To a large extent, respiratory activity is driven to meet the need. The PCO_2 of the arterial blood is the most important factor which governs respiration (Davenport 1974). The goal of regulating breathing is to minimize respiratory work while maintaining stable and optimal levels of respiratory gases and pH. The arterial blood O_2 concentrations are appropriately adjusted and the levels of CO_2 and H^+ ions in blood kept within the physiological range. At the normal steady-state respiratory rate, the alveolar and arterial PCO_2 are adjusted around a value of 5.3 kPa. The respiratory system presents a metabolic servomechanism designed to match pulmonary and metabolic gas exchange rates without altering the internal chemical concentrations of the body fluids. The error-correcting feedback signals are provided by the concentrations of O_2, CO_2, and H^+ in the arterial blood. In the higher vertebrates, birds and mammals and to an extent terrestrial reptiles, pulmonary ventilation is used to regulate the rate of CO_2 elimination, a role played by the gills in aquatic breathers. The gas exchanger and the circulatory system effect these processes at rates corresponding with the prevailing metabolic demands. Uptake of O_2 affects body fluid homeostasis especially the acid-base status to a very small extent through release of H^+ ions (after conversion of HCO_3^- ions to CO_2 on the binding of O_2 to hemoglobin – the Haldane effect) in the gas exchanger (e.g., Davenport 1974; Heisler 1989). In contrast, CO_2 is the main product of aerobic metabolism. It is involved in chemical reactions which affect the acid-base status of the body fluids. In a watery solution, on accumulation, CO_2 is a weak acid and has to be buffered to keep the body fluid pH relatively constant. Through carbonic anhydrase catalysis and chloride shift, the largest fraction of total CO_2 is carried in form of HCO_3^- ions in both blood plasma and the erythrocytes (Davenport 1974). This roughly constitutes about 90 to 95% of the total CO_2 in blood. At the gas exchanger, the HCO_3^- ions are converted back into CO_2, which diffuses out into the external medium (e.g., Perry and Laurent 1990). To ascertain CO_2 diffusion equilibrium, the amounts of carbonic anhydrase in the erythrocytes are much higher in the small than in the larger mammals up to an order of magnitude (Larimer and Schmidt-Nielsen 1960; Lindstedt 1984). Although hypoxia may be brought about by deficiency of O_2 in the environment or excessive utilization of it

Table 2. Comparison of blood respiratory features of the trout (*Salmo gairdneri*), the tadpole and adult bullfrog (*Rana catesbeiana*), and the snapping turtle (*Chelydra serpentina*) to show the differences in water and air breathers

	Trout[a]	Bullfrog[b]		Bullfrog		Turtle[c]
		Tadpole	Adult	Tadpole	Adult	
T (°C)	20	20	20	23	23	20
PCO_2 (mmHg)[d]	2.42	1.95	13.4	4.36	18.9	25.2
pH	7.80	7.83	7.90	7.80	7.70	7.76
HCO_3^- (mEq l^{-1})	4.63	4.0	32	8.0	27.5	49.0

[a] Trout – Randall and Cameron (1973).
[b] Tadpole and adult bullfrog – Erasmus et al. (1970/71) at 20 °C and Just et al. (1973) at 23 °C.
[c] Snapping turtle – Howell et al. (1970).
[d] To convert to kPa multiply by 0.133.

Table 3. Some physiological characteristics of typical aquatic and terrestrial animals. (Dejours 1988)

Parameter	Aquatic	Terrestrial
Respiration	Skin and/or gills	Tracheae or lungs
PCO_2/PO_2[a]	Low ratio	High ratio
(HCO_3^-)	Low	High
N end products	Mainly ammonia	Mainly urea and/or uric acid and other purine derivatives
Water turnover	High or very high	Low or very low
Temperature	Poikilothermy (most of them)	Homeothermy (some of them)
Locomotion	Swimming	Running, flying

[a] PO_2 and PCO_2 designate the difference in PO_2 and PCO_2 values between body fluids and ambient milieu.

by an organism, O_2 has no direct effect on the acid-base status. However, the increased ventilatory rate which befalls an animal in an attempt to supply the required amounts of O_2 may lead to respiratory acidosis due to increased metabolic CO_2 production or respiratory alkalosis due to excessive flushing out of CO_2 from the gas exchanger. There is no systemic difference in pH among the water-, bimodal-, and air breathers. Since the PCO_2 is higher in air breathers than in water breathers, that of the bimodal breathers falling in between, the adjustments in the pH are made by body fluid bicarbonate concentrations which increase with the degree of air breathing (Tables 2,3).

A truly comparative account on respiration must not only endeavor to examine how different organs and organ systems have been refined and integrated for the purpose of exchange of O_2 and CO_2 but must also explore these states and phenomena outside the purview of the so-called model animals. In the class Agnatha, the most frequently studied species is the Atlantic hagfish, *Myxine glutinosa* (e.g., Strathmann 1963), in the elamobranchs the commonly studied species are a variety of dogfish, e.g., *Scyliorhinus canicula*, *Squalus suckleyi*, *Squalus acanthias*, and skates; among the bony fish (class: Pisces) studies have been made largely on the subclass Teleosti, the most highly studied species being the cod (*Gadus morhua*), eel (*Anguilla anguilla*), goldfish (*Carassius auratus*), trout (*Onchorhynchus mykiss*, formerly *Salmo gairdneri*), and the sea raven (*Hemitripterus americanus*). In amphibians, the common grass frog (*Rana pipiens*), European frog (*Rana temporaria*), and the marine toad (*Bufo marinus*), all of which are anurans, are taken to be representative of the diverse class Amphibia. Within the class Reptilia, particular interest has been shown in the painted turtles, which fall either into the genus *Pseudemys* or *Chrysemys*. The laboratory white rat (*Rattus rattus*) and the guinea pig (*Carvia porcellus*) have been used widely among mammals, while in birds, the domestic fowl (*Gallus gallus* variant *domesticus*), muscovy duck (*Cairina moschata*), and the guinea fowl (*Numida meleagris*) have been used extensively. These few animals, most of which have been selected more for convenience and availability than for any concrete morphological or physiological merit, are far from being genuinely representative of the many taxa in the Animal Kingdom. Most of the discrepancies in the conclusions and observations that abound in comparative biology have arisen from unwarranted extrapolations of observations based on a handful of unrepresentative animals.

Chapter 2

Essence of the Designs of Gas Exchangers – the Imperative Concepts

"To understand completely respiratory adaptations to the environment, it is an implicit but fundamental requirement that we understand how such adaptations evolved, not just how they operate in living animals." Burggren (1991)

2.1 Innovations and Maximization of Respiratory Efficiency

Gas exchangers have developed and tractably adapted with the respiratory requirements of whole organisms in different states and habitats. The environmental factors that have profoundly influenced the general phenotype have simultaneously shaped the designs of the gas exchangers. On that account, the functional constructs of the gas exchangers cannot be understood without recognizing both these drives as well as the underlying physical principles that govern organismal biology. Form is a gestalt of structure. The importance of morphology and physiology as investigative approaches towards conceptual understanding of comparative evolution by natural selection cannot be overstated (Cracraft 1983; Duncker 1985; Greenberg 1985; Huey 1987). It should, nonetheless, be cautioned that it is oftentimes possible to mislead these aspects (especially morphology) in accurate reconstruction of phylogeny. For example, the so-called cryptic species (e.g., Bruna et al. 1996) or sibling species (e.g., Mayr 1942) are morphologically identical but genetically different. Such mismatched animals can be utilized to investigate the ecological and evolutionary events and mechanisms which enforce congruent morphologies. Until recently, morphological characteristics were the primary and practically the only means of organizing and classifying animals (e.g., Eldredge 1993; Rieppel 1993). Molecular genetics now offers a powerful means of supplementing morphological observations and validating phylogenetic relationships between different animals (e.g., Sibley and Ahlquish 1990; Graur 1993; Larson and Chippindale 1993; Luckett and Hartenberger 1993; Blair 1994; Hedges and Sibley 1994; Janke et al. 1994; Averof and Akam 1995; Penny and Hasegwa 1997). Considering the remarkable diversity of animal life, the different habitats occupied, lifestyles led, and the disparate metabolic potentials, unless autamorphic features of respiration are emphasized, the uniformity of the gas exchange and transport mechanisms between species is astonishing. A reductionistic (mechanistic) perception of a gas exchanger is that of a construction where an external medium and an internal one are separated by a barrier and a concentration gradient of O_2 and CO_2 occurs between the two compartments. This overly simplistic concept provides a useful conceptual framework for understanding the fundamental comparative principles of respiratory biology. The ultimate design of the gas exchanger must present those useful features which natural selection has selected, rigorously tested, and genomically conserved.

Energy is integral for building, servicing, and supporting the tissue infrastructure of organisms. To maximize the finite quantity available to them, cost-effective designs are essential. This calls for a logical plan of the constituent parts of the body. The most economic designs are those which demand least cost to construct, operate, and maintain while yielding the best possible results. Rosen (1967) deems optimization to be synonymous with "quest for minimum cost", Howell (1983) considers minimization of cost as "a pragmatic replacement for maximization of fitness", and Cannon (1939) termed optimal design simply "the wisdom of the body". Integrated arrangement occurs in the cardiopulmonary system of the fox (Longworth et al. 1989): though having among the highest mass-specific O_2 consumption in mammals (3.05 ml O_2 per second per kg), the animal has only an ordinary mass-specific morphometric pulmonary diffusing capacity (Weibel et al. 1983). The high unit O_2 flux across the blood-gas barrier is achieved by a large PO_2 gradient which drives O_2 from the alveolus to the capillary blood. At VO_{2max} (3.6 ml $O_2 s^{-1}$ per kg), the fox raises the alveolar PO_2 to 16 kPa by hyperventilating and maintains a low mean capillary PO_2 (12 kPa) by having a short capillary transit time (0.31 of a second) due to a high mass specific cardiac output (25 ml $s^{-1} kg^{-1}$). In the horse lung, at VO_{2max}, the capillary transit time is 0.4 to 0.5 of a second and the capillary blood is equilibrated with alveolar air after 75% of the transit time (Constantinopol et al. 1989). Among the evolved respiratory steps, the shape of the mammalian erythrocytes gives a good example of the multidimensionality in the enhancement of functional efficiency through morphological and biochemical refinements (Edsall 1972). With few exceptions, e.g., in the camel (Cohen 1978), where the cells are ellipsoidal, in mammals, the resting or minimum energy configuration is that of a biconcave disk (Nikinmaa 1990). The shape affords a high surface-to-volume ratio, with the definitive erythrocyte surface area of about 163 μm^2 being 70% larger than the surface of a spherical cell of equal volume. The camel's erythrocytes are subjected to considerable changes in the osmotic pressure of the plasma when the animal goes for 6 to 8 days without drinking water (Schmidt-Nielsen et al. 1957; Schmidt-Nielsen 1990).

Energy production by oxidative phosphorylation has been an irrevocable and continuous process since the evolution of the O_2-utilizing pathways in the primordial facultatively aerobic prokaryotic and eukaryotic unicells (Fig. 8). By way of binary division, unicellular organisms and sex cells of higher animals have been transmitted for millennia of generations. Though in finite amounts, these cells require O_2 for subdivision. The O_2 consumption of a single unfertilized mouse egg cell is 0.37 μl $O_2 mg^{-1} h^{-1}$, a value which increases to 0.38 after fertilization (Mills and Brinster 1967). In the pea, *Pisum sativum*, mitosis is only completed above an O_2 level of 0.004 of an atmosphere (Amoore 1961). Since the transition from anaerobiosis to aerobiosis, the function of the gas exchangers has remained essentially the same, i.e., taking up O_2 from the external milieu and discharging metabolically produced CO_2 into the same. The survival and adaptability of an organism are dependent on availability of the necessary resources and the capacity of the genotype to manipulate them (Phillipson 1981). In the Metazoa, molecular O_2 is a critical factor in energy production. For such an important procss, the strategies of procuring O_2 should have been differently optimized very early after the inauguration of aerobic metabolism. The past history of an organism sets the boundaries and the scope of its future development and the present constraints define its prevailing operational latitudes. Working on genetic variation, natural selection shapes and hones biological structures, increasing their fitness. Excessive design and redundancy are expunged, eliminating superfluousity and hence avoiding the unnecessary cost of supporting underutilized capacities. Astute designs are particularly necessary in those organs like the respiratory ones which must remain stable in face of changing needs and circumstances. The cost effectiveness of a gas exchanger can be gauged from the difference between the energy expended to secure molecular O_2 and that required by an organism for sustenance: optimization of the respiratory process endeavors to increase the net balance of O_2.

The modern gas exchangers are recent products of long-standing evolutionary developments in the ancestral animals. They give us an opening to conceive what may have evolved in the past. The environment has directed and regulated the amplitude and frequency of the adaptive changes which have occurred in the respiratory organs. Environments are highly dynamical systems (Peitgen and Richter 1986; Wainright 1988) which affect the development of organisms in complex and multidimensional ways. Even clonal populations show phenotypic variability. Dubbed Dollo's Law of irreversibility of Evolution (see Meyer 1988), it is considered statistically improbably that an organism can follow exactly the same evolutionary pathway in either direction, i.e., during progressive and retrogressive transformations. This is because environments "evolve" moment by moment. As soon as they are vacated, they are immediately taken over by other animals (e.g., Harvey 1993). The hypothesis frequently called Gause's Principle (see Moore 1990a) asserts that no two species can occupy the same niche. The pneumonate gastropods, a group which displays remarkable subtlety in respiratory strategies (Sect. 5.6.1), constitute an excellent prototype showing the effect environment has on the design of the gas exchangers. With realization or air breathing, the mantle cavity was transformed into a lung (Fig. 59) as the gills (ctenidia) gradually regressed: some species in the group have readopted water breathing through regrowth of similar but not identical structures (Cheatum 1934; Yonge 1952).

2.2 Safety Factors and Margins of Operation of Gas Exchangers

While biological systems largely function at a steady (unstressed) state, occasionally momentary severe conditions call for large adjustments in the level of operation. Highly trained human athletes and some elite animal species such as the horse and the dog (Snow 1985) can increase their O_2 consumption above rest 20- to 30-fold (e.g., Seeherman et al. 1981; Jones et al. 1989). To accommodate such adjustments, the relevant biological systems are intrinsically malleably designed and constructed (e.g., di Prampero 1985). In the case of the gas exchangers, maximal gas transfer is effected at the maximum O_2 consumption (VO_{2max}) when further increase in exercise does not result in a corresponding increase in O_2 uptake across the gas exchanger (e.g., Taylor et al. 1981, 1987a; Weibel et al. 1987b): supplementary energy is supplied anaerobically, with the accumulating lactic acid eventually inhibiting exercise (Margaria 1976; Taylor et al. 1981, 1989). Vo_{2max} increases with increasing PO_2 in the inspired air (e.g., Margaria et al. 1972; Welch and Pedersen 1981), transfusion of erythrocytes (e.g., Ekblom et al. 1975; Buick et al. 1984), and endurance exercise training (Saltin and Gollnick 1983; Saltin 1985). The scale of adjustment which enables a biological system to cope with functional loads is said to constitute a reserve capacity (=safety factor). Such capacities could strictly be viewed as "excessive constructions" over and above those necessary for minimum operation (Gans 1979). Biological designs appear to be configured for the worst-case scenarios and when these extremes are exceeded, death or irreparable damage occurs. In the Australian agamid lizard, *Amphibolurus nuchalis*, intense endurance exercise results in a decrease (rather than an increase) in the maximal O_2 consumption by a factor of 18% and pathological changes in the muscles and joints (Garland et al. 1987). Muscle fiber necrosis has been reported in human marathon runners (Hikida et al. 1983).

In engineering schemes (e.g., Gordon 1978; Petroski 1985), a safety factor is defined as the ratio between the load that just causes failure of a device (i.e., the component's maximal capacity (=strength = performance) to the maximum load that the device is anticipated to bear during operation. Within certain extents, biological systems change harmonically with the fluctuating strains and stresses to which they are subjected (e.g., Gilbert 1988). In composite systems (as are biological tissues), theoretically there should be room for infinite design creativity. Physical (constructional) and biological (phylogenetic, developmental, functional, and ecological) constraints, however, limit the number of possible outcomes (phenotypes) (Thompson 1959; Alberch 1980; Alexander 1985) delimiting the most optimal ultimate configuration(s). Increasing the operational safety margin calls for commitment of greater resources for construction and maintenance. From a perspective of cost-benefit analysis, through refinement brought about by natural selection acting on the phenotype, the measure of the safety factors contrived into different biological systems is aligned with specified needs (Diamond and Hammond 1992; Weibel et al. 1998). Excess functional capacities and extravagant structures precipitate unnecessary costs in form of energy required for maintenance of the infrastructure and operation. For an organ such as the brain, which consumes as much as 25% of the total resting O_2 consumption (Dejours 1990), optimal structure and function with inbuilt malleability is critical. Three-week exposure of rats to hypobaric hypoxia (0.5 of an atmosphere) causes an increase in the brain blood flow by 71% and microvessel density in the frontopolar cerebral cortex by 76% (LaManna et al. 1992). Lack of space in the body may be an important factor in determining the location and the definitive sizes of different organs (e.g., Diamond 1998). The unilateral development of normally paired organs such as the lung in, e.g., snakes and caecilians (e.g., Renous and Gasc 1989), animals with long cylindrical bodies, may be a consequence of a 'crowding-out effect. In bats (Maina et al. 1982a), the capacity to procure the large amounts of O_2 needed for flight has mainly been attained by development of remarkably large lungs (Maina and King 1984; Maina et al. 1991): compared with the thoracic cavity, the abdominal cavity is remarkably small. The gastrointestinal (GIT) system is simple (Makanya and Maina 1994; Makanya et al. 1995), and the transit time of the ingesta through the GIT is short (Klite 1965; Morrison 1980). Animals evolve just enough of what they need to overcome natural loads (Gans 1988): perfection and elegance are not pursued for their own sake. In what Taylor et al. (1987b) termed shared adaptive effort, unless there are certain underlying constraints, animals prefer to use compensatory combinations of multiple factors to achieve a greater broad-based safety margin of operation rather than few thinly stretched ones (e.g., Maina 1998). Gans (1985) envisaged that "it is the rule rather than the exception that each role utilizes multiple structures and that each structure inevitably supports multiple roles". In different species, biological functions are carried out in a variety of different ways (Brown 1994), indicating that "successful adaptations to particular conditions differ from species to species and from one group to another". This phenomenon has been called multiple realisability by Kitcher (1984) and Brooks (1994). Unless the "coupling", the "working", and the environment in which a gas exchanger operates are well understood (e.g., Connett et al. 1990; Wasserman 1994), respiratory safety margins may be perceived as overadaptations or even redundancies (e.g., Cannon 1939). By way of illustration, in mammals, the morphometric (DLo_2m) and the physiological (DLo_2p) diffusing capacities of the lung for O_2 differ by a factor of 2 (Gehr et al. 1978; Weibel et al. 1983). While the difference between the two values may be attributed to intrinsic deficiencies and assumptions in the model used to estimate the DLo_2m, an aspect discussed by Crapo et al. (1986, 1988) and recently critically reviewed by Weibel et al. (1993), it is thought that the disparity may register a fabricated functional reserve which is utilized during extreme circumstances (Weibel 1984a, 1990; Weibel and Taylor 1986). In a similar study which suggested the existence of a safety factor in the design of gas exchangers, Weibel and Taylor (1986) and Karas et al. (1987a) observed that the athletic animals (dog and pony) had a 2.5-fold greater maximum O_2 consumption (Vo_{2max}) compared with the less athletic ones (goat and calf): the latter group utilized one third of their pulmonary diffusing capacity while the former used three quarters of it at Vo_{2max}. The estimations made by Wagner and West (1972) and Hill et al. (1973) indicated that the blood-gas barrier of the human lung, which is normally $0.62\,\mu m$ thick (Gehr et al. 1978), would have to be increased four to ten times before it became a limiting factor for end capillary PO_2 equilibration.

Among animals, different structures, organs, and organ systems possess different functional reserves and redundancies (Alexander 1981, 1982b, 1996; Gans 1985, 1988; Karasov and Diamond 1985; Diamond et al. 1986; Karasov et al. 1986; Toloza et al. 1991; Diamond 1998). The human intestine has a factor of 2 to 2.5, leg bones of mammals 3, breast of most mammals 2, shell of a squid 1.3 to 1.4, and lung of a small dog 1.25. Calculations indicate that at VO_{2max}, in contrast to other animals like the pony, dog, calf, and goat (Karas et al. 1987b; Taylor et al. 1987b), the fox uses almost all the capillary transit time for O_2 equilibration, i.e., under extreme effort the fox virtually exhausts the capillary length in O_2 transfer. In the normal placenta at full term, for example, about 21% of the total volume comprises nonparenchymal tissue which is not involved in either gas exchange or metabolite transfer (Aherne and Dunnill 1966). The lung of the shrew, the smallest extant and most highly metabolically active mammal (Morrison et al. 1959; Fons and Sicart 1976; Gehr et al. 1980; Sparti 1992), is elegantly refined for gas exchange. The alveoli are as small as $30\,\mu m$ in diameter (Tenney and Remmers 1963), the alveolar surface area density is $2800\,cm^2\,cm^{-3}$, and the harmonic mean thickness of the blood-gas barrier is only $0.25\,\mu m$. This gives a mass pulmonary morphometric diffusing capacity of $0.143\,ml\,O_2\,s^{-1}\,mbar^{-1}\,g^{-1}$ compared with the values of 0.08 and 0.05 in the rat and man, respectively. Based on the geometry and profusity of the pulmonary blood capillaries in the alveolar septa of the lung of the shrew, Weibel (1979, 1984a) observed a resemblance between the honeycomb arrangement of the alveoli of the shrew with the intimately intertwined air- and blood capillaries of the bird lung (Maina 1982a, 1988a; Figs. 88, 89). The physiological diffusing capacities of the chorioallatois estimated by Piiper et al. (1980) of $8.7 \times 10^{-5}\,ml\,O_2\,s^{-1}\,mbar^{-1}$and that of $7.5 \times 10^{-5}\,ml\,O_2\,s^{-1}\,mbar^{-1}$ by Tazawa and Mochizuki (1976) compare with the morphometric diffusing capacities of O_2 of a 16-day-old chicken egg of $8.5 \times 10^{-5}\,ml\,O_2\,s^{-1}\,mbar^{-1}$ estimated by Wangensteen and Weibel (1982). Wangensteen and Weibel (1982) and Weibel (1984a) interpreted this similarity to bespeak the underlying optimization of gas transfer capacity across the chicken egg shell. While this may apply to individual eggs, egg shell conductance differs remarkably between and within species (Tazawa 1987). In the African parrot, *Enicognathus ferrugineus*, in a single clutch, the shell conductances may differ by a factor of 7 (Bucher and Barnhart 1984). Similar variations have been reported in the turkey, *Melleagris gallopavo* (Rahn et al. 1981; Tullet 1981), and in the chicken, *Gallus domesticus* (Tullet and Deeming 1982; Tazawa et al. 1983a; Visschedijk et al. 1985), eggs. Although large differences in the air cell PO_2 and PCO_2 must occur between the low and high conductance eggs, the O_2 consumption during the last stages of incubation are the same in both kinds of eggs (Tazawa 1987). This must allude to occurrence of either underutilized capacity (functional reserve) in the high conductance eggs or possible compensatory adjustments in the blood-gas uptake and transport variables to promote O_2 availability in the low conductance eggs. High conductance eggs should withstand environmental changes in O_2 levels better than the low conductance ones, improving chances of survival in hypoxic environments. The morphometric diffusing capacity of the human placenta (DPo_2m) was estimated to range between 0.05 to $0.08\,ml\,O_2$ per s per mbar (Mayhew et al. 1984). Based on estimations of O_2 consumption in the pregnant human uterus and the PO_2 in the maternal and fetal blood streams, Metcalfe et al. (1967) estimated the physiological diffusing capacity of the placenta (DPo_2p) to lie between 0.014 to $0.018\,mlO_2\,s^{-1}\,mbar^{-1}$. However, calculations based on diffusing capacities of CO_2 in pregnant women (Forster 1973) yielded a higher value of $0.025\,ml$ $O_2\,s^{-1}\,mbar^{-1}$. Based on a mathematical model which simulated the effects of uterine contractions on placental O_2 exchange, Longo et al. (1969) estimated a DPo_2p of $0.038\,ml\,O_2\,s^{-1}\,mbar^{-1}$. The present data indicate that the DPo_2p of the human placenta lies between 0.025 to $0.038\,ml\,O_2\,s^{-1}\,mbar^{-1}$, a value which is lower than the DPo_2m by a factor of about 2. As may apply to the mammalian lung, the difference between the DPo_2m and the DPO_2p may constitute a reserve which is exploited by the placenta during extreme circumstances (Mayhew et al. 1984). The reserve is brought about by factors such as vascular shunts, placental O_2 consumption, and regional inequalities of perfusion (Metcalfe et al. 1967; Mayhew et al. 1984, 1990). Lack of uniformity in the thickness of the villous membrane may lead to local inhomogeneities of diffusional resistances across the sporadically attenuated barrier (Mayhew et al. 1984; Jackson et al. 1985). Like other gas exchangers, the placenta is a multifunctional organ which must present a compromise design (Sect. 4.7). Though fundamentally constructed for gas exchange, the organ plays important endocrine roles and constitutes an important barrier which protects the fetus from harmful agents and factors in the maternal blood. In some situations, fetotropic viruses such as rubella virus, cytomegalovirus, hepatitis B virus, human immunodeficiency virus, enterovirus, and Theiler's murine encephalomyelities virus are prevented from affecting the fetus (e.g., Alford et al. 1964; Hayes and Gibas 1971; Maury et al. 1989; Garcia et al. 1991; Abzug 1994). It is material to note that the maximum morphometric pulmonary diffusing capacity of the lung for O_2 (DLo_2m) of 2.38 ml $O_2 s^{-1} mbar^{-1}$ (Gehr et al. 1978) is about 30 times greater than the morphometric diffusing capacity of the placenta (DPo_2m) of 0.08 ml $O_2 s^{-1} mbar^{-1}$. This should constitute an enormous functional reserve which guarantees O_2 supply to the fetus under different respiratory conditions and circumstances.

The trade-offs necessary for the development of optimal designs and functional reserves are manifest in many gas exchanges. For example, the completely aquatic lobster, *Homarus vulgaris*, which shows neither evidence for terrestrial adaptations nor propensity for air breathing when stranded on the beach (a real hazard for intertidal animals) will breath air, maintaining its O_2 uptake at near-aquatic levels (Thomas 1954). Facultative air breathing appears to be an acquired adaptive feature which allows solely aquatic animals to withstand a transient, stressful physiological condition. In eggs, high porosity compromises fitness by increasing water loss, but the same process improves it by enhancing the conductance of the egg shell to respiratory gases. In the mammalian lung, while intense subdivision of the pulmonary parenchyma provides a greater respiratory surface area, according to the Young-Laplace relationship ($P = T r^{-1}$), where P is the recoil pressure, T surface tension, and r the radius of curvature at the air-liquid interface, the decrease in the diameter (with the resultant increase in the surface radius of curvature of the alveolus) engenders greater surface tension. At an invariable recoil pressure, this increases the disposition of the alveoli to collapse as well as the energy needed to inflate them with air during inspiration (e.g., Wilson 1981; Wilson and Bachofen 1982). On expiration, the alveolar surface tension reaches a value close to zero (Schürch et al. 1985). The design of the lung of the shrew, e.g., *Microsorex hoyi* which weighs as little as 2.3 g (Lasiewski 1963a,b), has been greatly improved and probably driven to the very limit of a functionally operable mammalian lung. On the lung of the minute shrew, Weibel (1979) pointed out that "it may well be that a limit of bioengineering feasibility has been reached at all levels of the respiratory system". The same may apply for the lungs of the 2-g Cuban bee humming bird, and the Thai bumblebee bat (Suarez 1992). To maintain extremely small sizes especially during the larval stages of development, endothermy, a high-cost approach to life, is out of reach of the ectothermic air breathing vertebrates such as amphibians and reptiles. The smallest amphibian is the arrow-poison frog, *Sminthilus limbatus*, which in adulthood measures only 11 mm from nose to anus. In biological systems, through compromises, concessionary states are established by harmonizing the composite fitness components with the limiting ones. In the fish gills, the two main parameters that can be adjusted to increase transbranchial O_2 diffusion are respiratory surface area and the partial pressure gradient of O_2: the former can be increased through lamellar recruitment (e.g., Booth 1978) and the latter by increasing ventilatory and perfusion rates. Adjustments of these parameters leads to what has been called osmorespiratory compromise. While improving gas exchange, increasing the gill surface area results in osmoregulatory problems engendered by increased ionic loss and influx of water in the freshwater teleosts or water loss and ionic loading in the marine teleosts (Perry and Laurent 1993). It has been demonstrated by, e.g., Randall et al. (1972), Wood and Randall (1973), and Gonzalez and McDonald (1992), that increased O_2 consumption is accompanied by increase in Na^+ flux, a process that calls for increased energy expenditure. It was envisaged by, e.g., Satchell (1984) and Nilsson (1986), that fish oblige the osmorespiratory compromise by limiting the surface area and increasing the partial pressure gradient of O_2, a process which increases O_2 flux without provoking problems of ionic transfer. Part et al. (1984) demonstrated nonrespiratory areas in perfused gills of the rainbow trout, *Oncorhynchus mykiss*, supporting the long-held proposition that fish are able to regulate the surface area of the gills for the purposes of gaseous and osmotic exchange (e.g., Randall 1982; Butler and Metcalfe 1983). To enhance gas exchange, the more energetic species of fish appear to rely more on increasing the respiratory surface area while the less energetic ones utilize hemodynamic adjustment (Perry and McDonald 1993). In the rainbow trout, at rest, only 60% of the gill lamellae are perfused (Booth 1978). During exercise, the blood perfusing the gills is shunted from the less well-ventilated basal channels to the more central ones (Nilsson 1986). The complex anatomy of the circulatory system of the gills where the vascular arrangement has been differentiated into respiratory and nonrespiratory pathways (e.g., Gannon et al. 1973; Dunel-Erb and Laurent 1980a; Butler and Metcalfe 1983; De Vries and De Jager 1984) has been associated with the plasticity of the gills for varying the exposure of blood to water. In the European eel, *Anguilla anguilla*, the volume of the blood in the nonrespiratory vasculature of the gills comprises 5% of the total blood volume in the body (Bennett 1988). Interestingly, though the blood of the dogfish, *Scyliorhinus canicula*, like that of the other elasmobranchs, is almost isosmotic to seawater (Burger and Bradley 1951) and hence gill perfusion does not affect ionic flux across the water-blood barrier significantly, both kinds of vascular circuits were reported (Metcalfe and Butler 1986). Unlike the teleosts (Pettersson and Nilsson 1979), the gills of the dogfish, *S. canicula*, lack nervous control of blood flow across the gills (Metcalfe and Butler 1984).

The existence of safety margins in biology is well shown in some physiological processes. Oxygen regulators, for example, operate between two levels, a critical and a limiting one. In the lugworm, *Arenicola marina*, the critical PO_2 is near air saturation at 16 kPa (Toulmond 1975) but the anaerobic processes do not start until the PO$_2$ drops to 6 kPa (Schöttler et al. 1983). Between the two levels, an organism can maintain aerobic metabolism by rearranging and/or mobilizing different physiological factors. In extreme circumstances, the metabolic level of activity is reduced (Hochachka 1988). In principle, the function of a biological system is facilitated by a structure that is correct for the settings where function occurs. As animals establish themselves in relatively more stable environments, adopt more successful designs and acquire more efficient behavioral, physiological, and biochemical responses to external perturbations, flexibility for genetic change and transformation is gradually blunted. While prokaryotic organisms exhibit remarkable capacity to change metabolism in response to changes in substrate and environmental conditions, cells and tissues of eukaryotic organisms, particularly the higher animals, are considered less responsive since they live in more stable environments (e.g., Golspink 1985). Whether a catastrophic event occurs or not, in all species, the capacity to survive inevitably decays with geological time. Schopf (1984) estimated that the average species' longevity may be as short as 200 000 years. No vertebrate species has avoided extinction for more than a few million years (Carroll 1988). Effective adaptive changes demand selection and refinement of only those features which are favorable in a particular milieu. Apparently, not all features presented by an organism are intrinsically adaptive (e.g, Futuyuma 1986): most organisms present anachronistic features. These features may arise when a particular selective pressure drive stops midstream (e.g., if an environmental stress factor abates) or if a certain structure is diverted to configure a totally different one from that initiated. The common perception that "biological innovations tend to appear soon after environmental conditions become favorable to them" (e.g., Cloud 1974) is a gross oversimplification of the evolutionary process. Strictly, organisms are not passive participants totally subservient to the drifts of the environment: they identify the evolutionary pathways they wish to follow, set their own selective pressures, and actively engage the environment in determining the direction, rate of progress, and nature of change (e.g., Gillis 1991). Only the traits that require a driving force (e.g., natural selection) to establish and impart a performance advantage are adaptations (e.g., Baum and Larson 1991). For example, at the various stages of their evolution, invagination, compartmentalization, and ventilation of the gas exchangers were necessary improvements for respiratory efficiency (e.g., Brainerd et al. 1993). In organisms that were not adequately inventive, these requirements constituted limitations which in some cases stopped any further evolutionary progress. Retaining the buccal force pump and particularly the skin as a gas exchanger apparently consigned the amphibians to water or humidic habitats. The evolution of homologous structures in biology illustrates convergence in solutions to common problems (Gould 1966). In the reptilian lung, to increase the respiratory surface area, the subdivision of the lung was achieved through an inward growth of trabeculae towards the central air space, and in the mammalian lung, the process took place by outward projection of the alveolar septae from the bronchial system. In the latter, while an extensive respiratory surface was achieved, an efficient costodiaphragmatic ventilatory system was necessary to ventilate the more profusely compartmentalized lung.

2.3 Engineering Principles in the Design of the Gas Exchangers

Biological processes are intrinsically finite in magnitude and frequency. The boundaries within which operations occur are set by determinate physical constraints and regulated by evolved biological feedback mechanisms. For about 2 billion years, limitations set by diffusion firmly confined animal life to the protozoan domain. This grip was loosened only when O_2 rose above nascent levels and convective processes of transporting respiratory gases from the outside to the proximity of the gas exchangers onward to the tissue cells developed. A wide spectrum of respiratory plans, structures, and strategies based reasonably on common engineering plans has evolved in animals. With certain modifications, the sheet-flow (e.g., Fung and Sobin 1969; Tenney 1979; Farrell et al. 1980; Fung 1993) or tubular design (e.g., Guntheroth et al. 1982), depending on how one visualizes the thin blood conduits, occurs in practically all evolved gas exchangers, e.g., in fish (Figs. 11,49) and crustacean gills (Fig. 13) and in the reptilian

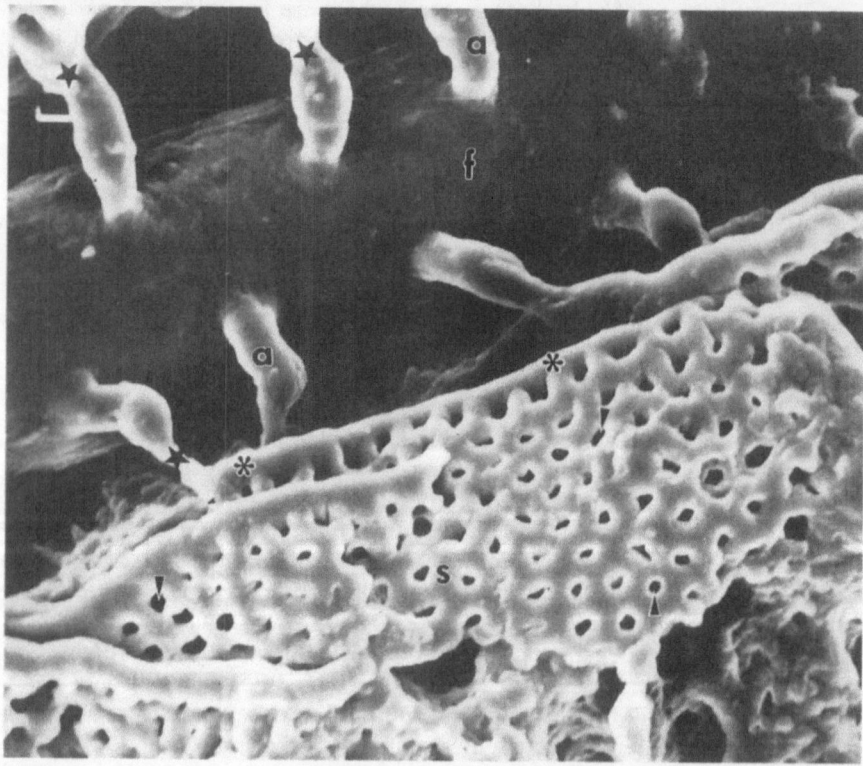

Fig. 11. Cast of the gills of a tilapiine fish, *Oreochromis alcalicus grahami*, showing a gill filament artery, *f*, giving rise to afferent lamellar arterioles, *a*, which supply blood to secondary lamellae, *s*. The constriction of the afferent lamellar arterioles, ★, are thought to be valves which regulate lamellar perfusion. ✳ marginal channel; ➤ pillar cells which subdivide the lamellar plate into spaces through which blood percolates. *Bar* 5 μm. (Maina 1994)

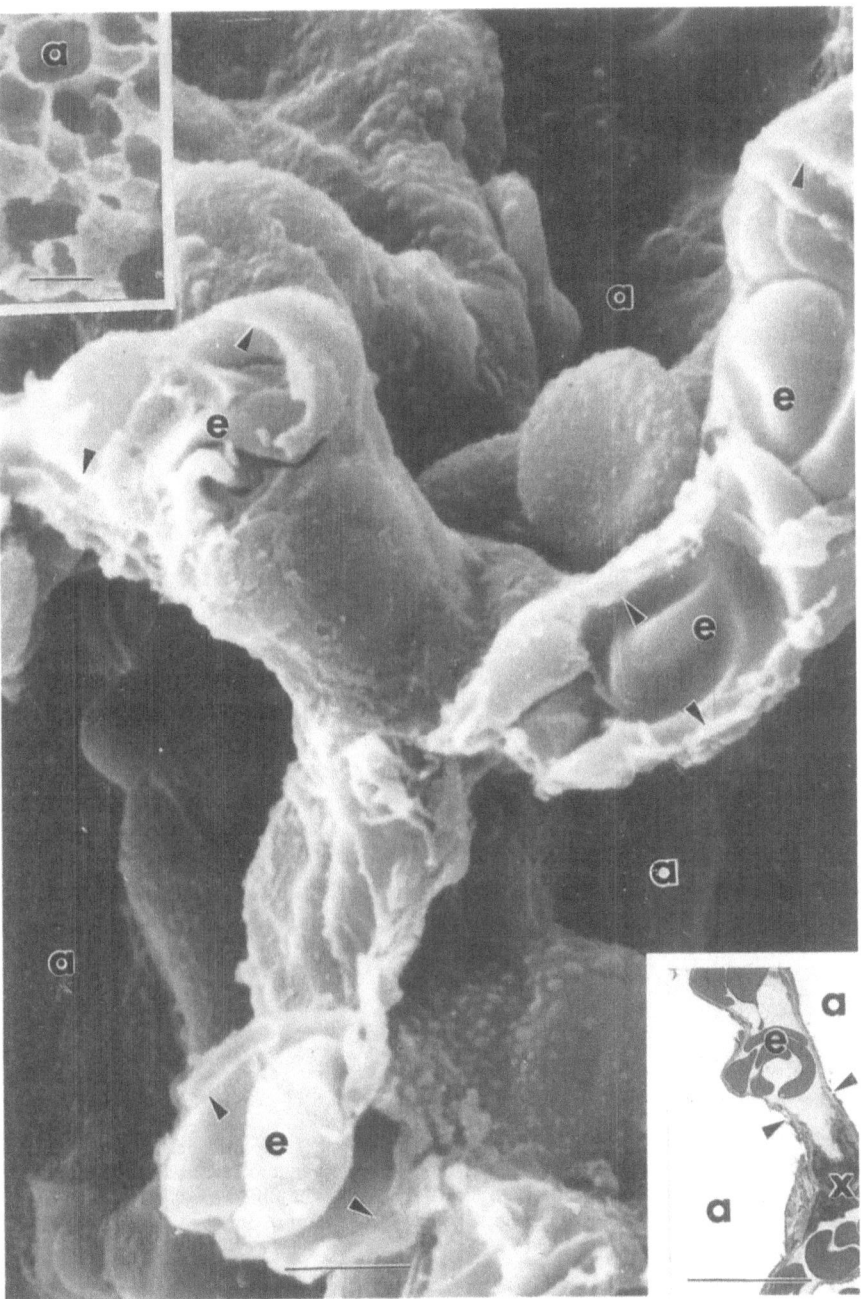

Fig. 12. Sheet blood flow pattern in a mammalian lung. *Top inset* The alveoli, *a*, separated in a honeycomb manner by interalveolar septa. *Main figure* Closeup of the interalveolar septum which is formed by two parallel epithelial cell layers, ➤, between which the blood capillaries are contained. *Bottom inset* Parallel epithelial cell layers, ➤; *e* erythrocytes; *a* alveolus; *x* endothelial cells. *Top inset bar* 21 μm; *main figure*, 3 μm; *bottom inset* 3 μm

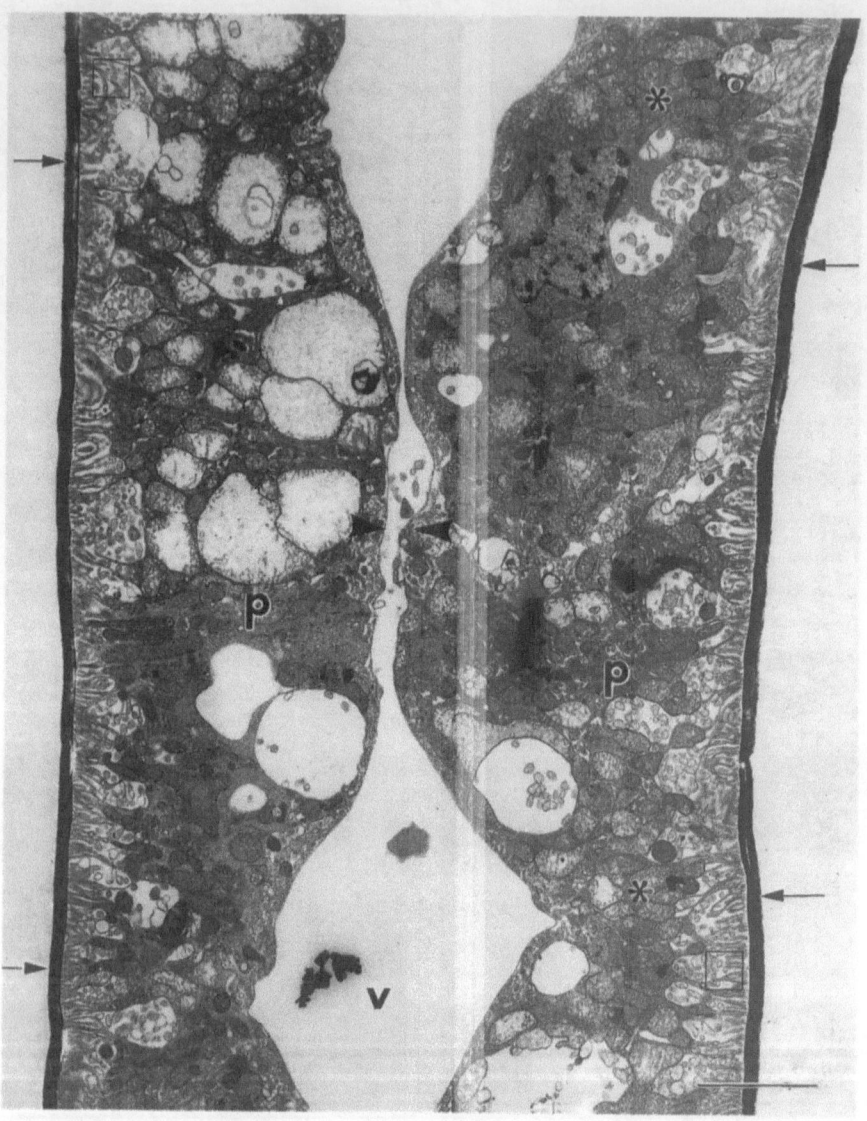

Fig. 13. Sheet blood flow pattern in a gill lamellar of a crab showing epithelial cells, ✳, which line the blood space, *v*. The high energetic demands for ionic exchange are evinced by the abundance of mitochondria in the epithelial cells, *p*, and the highly amplified basal infoldings, □. →, cuticular lining; ➤, points where the epithelial cells closely approximate to regulate the rate of blood flow. *Bar* 3.5 μm. (Maina 1990b)

(Fig. 14), mammalian (Fig. 12), and avian (Fig. 29) lungs. The constructional plan comprises thin parallel epithelial cell layers which are joined by connective tissue or pillar-like cell struts. This architectural configuration produces fine channels through which blood flows, spreading out into an extremely thin film over an

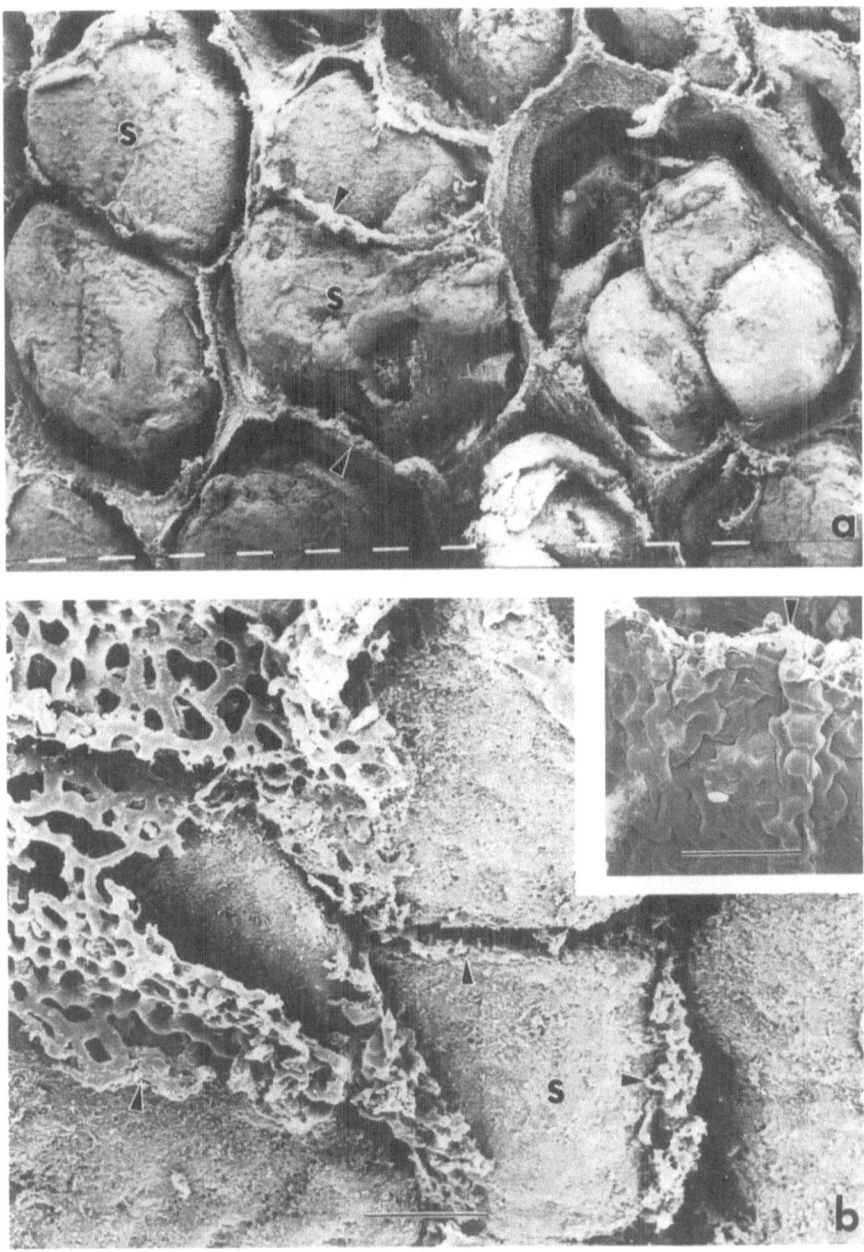

Fig. 14. **a** Semimacerated latex rubber cast of the gas exchange air spaces, s, of the lung of the snake, the black mamba (*Dendroapis polylepis*) showing the blood capillary bearing septae, ➤, **b** A completely macerated double latex cast of the gas exchange region of the snake lung showing the sheets of septal blood capillaries, ➤, which surround the air spaces, s. *Inset* Critical point dried material showing a septum, ➤, lining the air spaces. **a** *Bar* 100 μm; **b** 50 μm; *insert* 25 μm. (Maina 1989e)

extensive respiratory surface. This appears to be the only plan which adequately meets the structural requisites for efficient gas exchange. It provides optimal exposure of hemoglobin to the external milieu for maximal gas transfer. The external and internal gas exchange media are brought into as close proximity as possible, interfacing over an extensive surface area in a highly dynamic organ. In the human lung, about $213\,cm^3$ of the blood in the capillaries is spread over a respiratory surface area of $143\,m^2$ (Gehr et al. 1978; Weibel 1989), i.e., about $1.5\,cm^3$ of capillary blood per m^2, generating an extremely thin film (a sheet) of blood. While the mammalian fetal pulmonary circulation comprises less than 10% of the biventricular cardiac output, pulmonary vascular resistance is higher than the systemic and the PO_2 of blood is low. Soon before birth, the blood flow increases eight to ten times and a transition to low resistance circulation occurs (e.g., Dawes et al. 1953; Cassin et al. 1964; West 1974; Teitel et al. 1987). These changes appear to be influenced by the changes in the PO_2 in blood both in utero and after birth, the process being mediated by changes in the K^+ and Ca^{2+} channels in the smooth muscles of the pulmonary artery (e.g., Lewis et al. 1976; Sheldon et al. 1978; Accurso et al. 1986; Cornfield et al. 1994). Pulmonary vascular resistance is a product of pulmonary blood flow and the difference between the pulmonary artery pressure (average $2\,kPa$) and the left atrial pressure (about $0.7\,kPa$). Even at rest, all the capillaries of the lung are perfused with plasma within 2 min of blood leaving the heart (König et al. 1993). Failure in reduction of the pulmonary circulatory resistance after birth is a consequential clinical problem sometimes termed persistent pulmonary hypertension of the newborn and leads to various neonatal respiratory disturbances (Heymann and Hoffman 1984). In the fish gills, blood flow is slowed down as it percolates through the narrow vascular channels formed by the pillar cells (Fig. 15) improving O_2 uptake by the erythrocytes (e.g., Hughes and Wright 1970; Soivio and Hughes 1978; Farrell et al. 1980; Nilsson et al. 1995). While in transit, the erythrocytes are greatly compressed (Fig. 86) and exposed to the external respiratory medium on all sides across a thin tissue barrier. In the mammalian lung, it takes less than 1 s (West 1974; Lindstedt 1984; Swenson 1990) for the erythrocytes to pass through the alveolar blood capillaries, a period within which the erythrocytes are fully saturated with O_2 (Weibel 1984a). Because of the two-phase nature of blood, the erythrocytes and the plasma may take different paths through the capillary network (Okada et al. 1992; König et al. 1993) presumably depending on the resistances offered across different capillary segments (Okada et al. 1992) and the preponderance of leukocyte sequestration in the lumen of the blood capillaries (Perlo et al. 1975; Lien et al. 1987; Hogg et al. 1988; Yoder et al. 1990). In the skeletal muscle, 1 ml of mitochondria relates to about 14 km capillary length and $0.22\,cm^3$ of capillary blood (Conley et al. 1987). At term, the human placenta contains $45\,cm^3$ of capillary blood which is spread over an area of about $11\,m^2$ (Aherne and Dunnill 1966), generating a film of blood $0.41\,\mu m$ thick. In the book lungs of arthropods (Fig. 15), air rather than blood flows through the thin external conduits.

The similarity in the design and construction of the gas exchangers could be attributed to the plausibility that gas exchangers arose from a common ancient

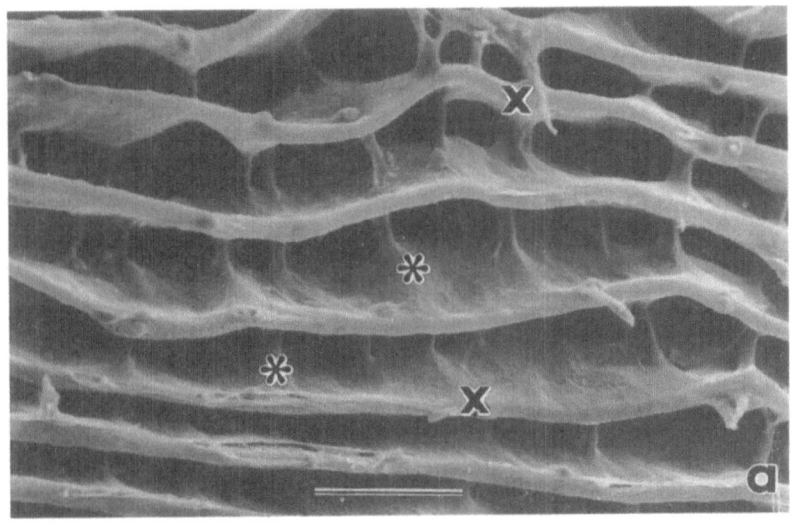

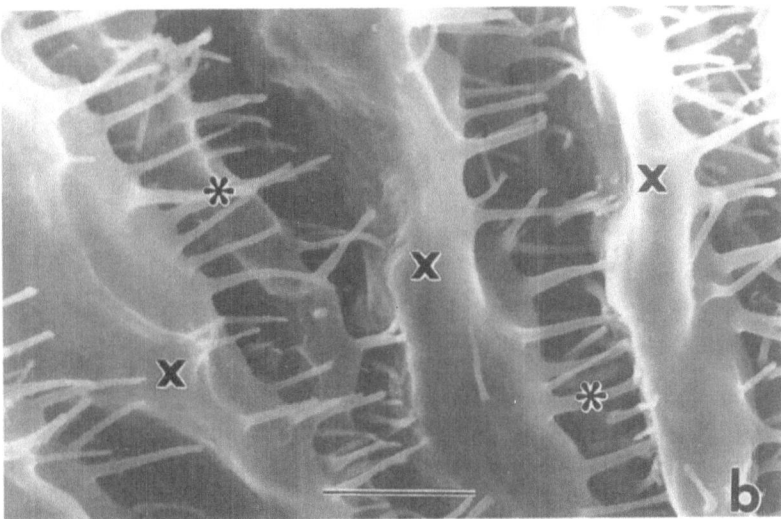

Fig. 15. a Book lungs of the desert scorpion, *Paruroctonus mesaensis*. The lamellae, *x*, are kept apart by vertical struts, ∗. **b** A higher view of the lamellae, *x*, and vertical struts, ∗. **a** *Bar* 40 µm; **b** *bar* 13 µm. (Farley 1990)

structure, whose basic plan has developmentally been highly conserved. Alternatively, as a process, gas exchange may enforced some basic, invariable structural attributes which all organs had to meet. Biological evolution does not occur by attrition but by parsimonious remodeling of common ancestral plans (Meyer 1988; Lauder and Liem 1989; Atchley and Hall 1991). In the ostariophysans (Order

Siluriformes), the accessory respiratory organs of the air-breathing fish developed directly from the branchial tissue in the immediate concavities above the gills (Hughes and Munshi 1968). In the African catfish, *Clarias mossambicus*, the suprabranchial chamber membrane is well vascularized (Fig. 66) and, where necessary, the labyrinthine organs developed as outgrowths from the gill arches (Maina and Maloiy 1986; Fig. 65c). Most bimodal-breathing fish have utilized default (already existing) organs like the stomach, intestines, and the anus. Only one siruriid, *Pangasius*, has evolved a gas bladder primarily for respiration (Browman and Kramer 1985). In principle, whether at molecular, cellular, or organismic level, when change must be made, existing structures are overhauled and improved. Rarely do totally new structures have to develop. For example, by evolving the appropriate enzymes, the evolution of aerobic respiration in the eukaryotes entailed addition of the citric acid (Krebs) cycle onto the original anaerobic (glycolytic) pathway of the anaerobic prokaryotes (Chapman and Schopf 1983). Homologous structures (similar constructional plans enforced by natural selection) illustrate the conservativeness of evolution. Investigating the basis of the different aerobic capacities of the athletic and nonathletic animals, Hoppeler et al. (1987) observed that in all species, the maximal average O_2 consumption of the mitochondria was the same (3.4 to 4.6 ml O_2 min^{-1} ml^{-1}) independent of the aerobic capacity of the species: the greater oxidative capacity of the athletic species was brought about by "building more mitochondria of similar kind, rather than by modification of the metabolic rate of the mitochondria". It is, however, recognized that at a certain level of mitochondria-muscle fiber ratio, further increases in the mitochondrial volume density may not only be futile but may compromise muscle function (e.g., Pennycuick and Rezende 1984; Weibel 1985a; Hochachka 1987; Hochachka et al. 1988; Suarez 1992). Once established, morphologic characteristics appear to be retained for indefinite periods of time until there are imperatives for change. For example, mammalian orders have retained a basically similar body form for 50 to 60 million years (e.g., Eldredge and Gould 1972; Carroll 1988). Compared with the total longevity of a species, however, changes take a much shorter period to be effected (Carroll 1988). Only a very small proportion of the genome appears to be directed towards morphological restructuring. The amino acid sequence of 12 varied proteins differs by only 1% between humans and chimpanzees (King and Wilson 1975).

Though argued to the contrary (e.g., Rose and Bown 1984; MacFadden 1985; Chaline and Laurine 1986), evolutionary changes are not progressive nor are they necessarily gradual improvements on earlier designs (e.g., Carroll 1988; Gould 1994; Kardong 1995). The apparent disparity of form between the various gas exchangers has resulted from the singular fact that animals at different phylogenetic levels of development have had to respond to common selective pressures (e.g., Murdock and Currey 1978), in such circumstances, individual solutions being found using different strategies and resources. To a certain extent, phylogenetic plasticity will allow correspondent structures to develop, but in most cases, alternative solutions are pursued. For example, long lungs (which are difficult to ventilate) have evolved in the thin cylindrical animals such as the snakes and the caecilians (Renous and Gasc 1989). The ventilation of such lungs is greatly hindered by locomotion. In lizards, breathing has been uncoupled from locomotion

(Carrier 1984, 1991). On the other hand, in an energy-saving strategy, bats have adopted a 1:1 ratio between wing beat and breathing (Thomas 1987). Clearly, what may constitute a constraint in one animal may be beneficial in another. Amidst these shifts, some structures which may appear to be of no evident biological value to organisms may evolve (e.g., Bock and von Wahlert 1965; Kimura 1983; Pierce and Crawford 1997).

In biological systems, structure encompasses the qualitative and quantitative characteristics of the constitutive components and their geometric features and arrangements. By altering the proportions, positions, and configurations, new polarities and states are created and different functional states are established. The components of complex structures constitute an integrated pattern of up- and downregulation of diverse functional capacities. For example, the arrangement of the mineral crystals of $CaCO_3$ in the eggshell determines the porosity of the shell and hence its diffusing capacity for O_2, a feature which, in turn, ensures proper development of the embryo. The permeability of the fish's swim bladder to gases depends on the orientation of guanine crystals in the bladder wall (Lapennas and Schmidt-Nielsen 1977). Swim bladders of fish which operate at depths greater than 1000 m have a greater concentration of guanine per unit area of the wall (Denton et al. 1970). The principles of homology and analogy are fundamental to understanding the correlation between structure and function as modified by natural selection and effected by the process of adaptation. Those structures which undergo irreversible deconstruction become vestigial and eventually disappear while some may be commissioned to perform roles different from those for which they were initially configured. Such adaptive traits were called exaptations by Baum and Larson (1991). For example, the surfactant evolved in the ancestral piscine lungs (e.g., Todd 1980) mainly to protect the epithelial surface (Liem 1987a). However, with the development of the more complex lungs in the tetrapods (where the buccal force pump was no longer adequate to ventilate such lungs), compartmentalization of the gas exchanger was necessary to enhance respiratory efficiency. With this modification, the surfactant (Sect. 6.9), by increasing the lateral stability of the phospholipid layer (Cochrane and Revak 1991), assumed the important role of reducing surface tension (e.g., Wilson 1981; Golde et al. 1994). In the anaconda, *Eunectes murinus*, the surfactant lining occurs in the alveolated and the succular parts of the lung (Phleger et al. 1978). While in the protochordates the pharygeal region serves as a filter-feeding apparatus (Sect. 6.10.1), in the chordates, with the addition of a respiratory role, the trophic one was phased out. The complex relocation, erosion, and eventual disintegration of some of the aortic arch blood vessels (which serviced the gills) on the formation of the lungs, is another example of such drastic transformations. Among vertebrates, the pulmonary arteries make their first entry in the Dipnoi (lungfishes) as branches of the sixth pair of aortic arches. Early in the development, the blood vessels supplied the swim bladder which is thought to have given rise to the lungs (Sect. 6.2). Only the necessary early structures (e.g., some of the blood vessels of the branchial arches) were retained as the primal gill arch blood vessels were reconfigured. The degeneration of one lung in, e.g., snakes (Ophidia) and the caecilians (Gymnophiona), is thought to have been one of the sacrifices they had to make to develop thin, limbless, cylindrical bodies which are important for slithering through confined spaces.

Unlike human-made machines, which are configured to carry out specified functions, biological structures are dynamic, multifunctional, composite entities designed to continuously absorb and respond to the fluidity of the external pressures of natural selection. In what de Beers (1951) called mosaic evolution, the different parts of an animal are variably affected by natural selection. Needs, to a greater extent than phylogenetic level of development, dictate the direction, nature, and magnitude of adaptive change. For example, the similarity of the lung-air sac system of birds (Sect. 6.7.5) and the tracheal-air sac system of insects (Sect. 6.6.1), animals separated by over 200 million years of evolution, indicates a morphological convergence for flight. Furthermore, compared with the respiratory system of birds and that of insects, the lung of the human being at the acme of evolutionary development is not as efficient. In the benign evolution of the *Homo sapiens*, natural selection appears to have targeted the nervous and musculoskeletal systems, leading to development of complex mental capacity, bipedal locomotion, opposable thumb, articulated sound (speech), etc. In an adult human being, the brain utilizes 25% of the overall resting O_2 consumption. While it would be anticipated that these developments would impart particular specializations, especially with respect to the mechanisms and control of breathing, the pulmonary system appears to have been disregarded during these transformations as long as it was adequate to support the ongoing changes. Except for the Hering-Breuer reflex, which appears to be much less well developed, the human respiratory characteristics are similar to those of all other mammals (Dejours 1990).

2.4 Scopes and Limitations in the Design and Refinement of the Gas Exchangers

Animals are aphoristically said to be structurally and functionally well constructed to meet the adversities of life (e.g., Olson and Miller 1958; Frazzetta 1975). Though easily conceivable, dedicated biological engineering is not easy to experimentally test, empirically prove, and convincingly demonstrate. This is largely due to our anthropocentric approach and misconception that evolutionary change is determined and driven by the same rules (to serve the same purposes) as human technological innovations (e.g., Basalla 1989). Whereas for human insight advancement means improvement and improvement means more sophisticated products, in nature, changes are generally highly resisted. If they occur, they are strictly survival-oriented and are configured specifically around the existing structures and prevailing conditions to engage known loads. Paleontological studies do not support the popular belief that through the evolutionary continuum, organismal design and function have undergone appreciable refinement and complexity (e.g., Rudwick 1964; Hickman 1987; McShea 1991): the ancient organisms were no less exquisitely designed than the modern ones. Indeed, some ancient structures of now extinct animals seem to have dealt with complex problems that are unsolved in the present animals. Stipulation for originality in technical inventions is purely a property of human ambition and competitiveness: evolution advances without purpose or direction, resulting in unpredictable changes. Adaptation entails cumulative selection of innovations that build on top of primeval ones. Mosaic evolution, i.e., selective degeneration combined with progressive specialization, e.g., in the acoustic adaptations of the fossorial rodents (e.g., Nevo et al. 1982; Heth et al. 1985) is an occurrence which shows nature's conservativeness and yet quest for optimization. It has been suggested that human technological progress may be proceeding 10 million times faster than natural evolution (Arthur 1997). Experiments on guppies, *Poecilia reticulata* (Reznick et al. 1997), however, demonstrated that evolution can occur very rapidly: in a 4-year observation, the rate of change of certain features was some 10 000 to 1 million times faster than the average rates estimated from fossil record. Thompson (1911) pointed out (but see counterviews by, e.g., Cody 1974 and Howell 1983) that "biology does not necessarily make progress toward perfection by mechanical analysis of changes that go on in living bodies". In nature, even where radical changes such as mutations occur, designs are altered through reconfiguring existing structures or enforcing new roles. Figuratively speaking, animals appear constantly to "reinvent the wheel" (though they have yet to evolve a real one!) as they look for the most relevant and least traumatic solutions to the demands prevailing in their ecological settings. Future developments are not anticipated. By combining the different assets they have gathered along the way, new "adaptive functional complexes" of high selective value are configured, conferring greater survival potential to an organism. S.A. Kauffman (cited in Ruthen 1993) asserts that "by selecting an appropriate strategy, organisms tune their coupling to their environment to whatever value fits them best". Comparative biology reveals the parameters in the "primitive" life forms which have been conserved during the evolvement of the complexity that characterizes the most "advanced" kinds (Fishman 1983), defining in broad terms the pathways followed and the strategies adopted in the quest for survival and self-perpetuation.

Establishment of optimum states calls for sound analysis of the alternative strategies, evaluation of the costs incurred, the benefits which accrue, and the difference between the level of operation with the theoretical maxima. In bats, for example, a typical mammalian lung has been structurally and functionally refined to supply the enormous amounts of O_2 needed for flight (Thomas 1987; Maina et al. 1991); (Sect. 6.7.4.1). Unequivocally, this shows that the lung-air sac system of birds and the tracheal-air sac system of insects, the gas exchangers which have evolved in the only other two volant taxa, are not prerequisite respiratory designs for flight. While the structural parameters are fixed, the functional ones are more flexible. Bats have shrewdly utilized combinations of these parameters and promoted the efficiency of a plainly inferior gas exchanger to rival and in a manner equal that of the distinctly superior bird lung (Maina 1998). After sojourn at high altitude (e.g., Bard et al. 1978; Heath et al. 1984; Durmowicz et al. 1993), factors such as ventilation, hematocrit, hemoglobin concentration, erythrocyte count, and blunted hypoxic pulmonary vasoconstrictor response change (some within a matter of hours or days) to avoid cellular hypoxia and to correct for adverse changes in the blood pH. The structural parameters, e.g., the thickness of the

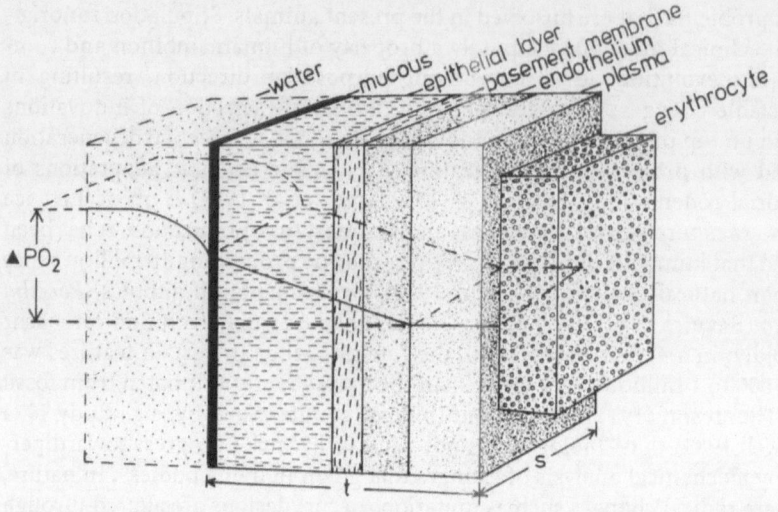

Fig. 16. Schematic drawing showing the resistance barriers across which O_2 diffuses in a water breather and the principal structural features, namely the barrier thickness, t, and the respiratory surface area, s, which influence the diffusion process. The partial pressure gradient of O_2 (ΔPO_2) decreases with the diffusion distance

blood-gas barrier and respiratory surface area, take months to change (e.g., Weibel 1984a). Whether in a water- or an air breather, O_2 procurement depends on parameters such as the diffusional distance, respiratory surface area, physical permeative properties of the respiratory barriers, and functional properties such as ventilation and perfusion states (Fig. 16).

For optimal energy production, in organisms, gas transfer is aligned with the metabolic needs. The correlation between an animal's environment and its respiratory needs is a very intimate one. Whereas O_2 need at rest is moderate, during exercise, e.g., flight, it increases tremendously (Weis-Fogh 1967; Tucker 1972; Thomas 1987). When exposed to a hypoxic environment, to support their normal metabolic activities, organisms still need to extract and transfer to the tissues the same quantity of O_2 as in a normoxic one. The respiratory system must be designed with adequate flexibility to accommodate the extreme demands which may occasionally be loaded on the organism (e.g., Maina 1998). Coordination of external factors (i.e., those involved in the procurement) and the internal ones (i.e., those involved in the uptake, transport, and distribution) ensures a satisfactory supply of O_2 to the tissues. In the event of hypoxia, it is the external factors rather than internal ones which become limiting. The high metabolic capacities, especially in the endothermic homeotherms, were achieved through evolution of more efficient means of O_2 procurement. This entailed development of features such as double circulation, greater blood pressure, higher tissue capillarity, higher mitochondrial volume density, and larger hemoglobin and myoglobin concentrations.

2.5 Optimal Designs in Biology and Gas Exchangers in Particular

2.5.1 Symmorphosis: the Debate

Form and function are inextricably interrelated. Indeed, it is axiomatic that animals should be constructed reasonably in order to be able to efficiently carry out the essential activities of life. The perception of a form-function correlation alludes to some advantages for the animals that possess the attribute (Gans 1988). Vogel and Wainright (1969) put it that "function without structure is a ghost; structure without function is a corpse" and Trivers (1985) bluntly stated that "organisms are designed to do something". The concept of optimal design of biological structures is based on the recognition that natural selection continuously regulates every aspect of structure and function. Though perceptively simple to conceive, the actual manner in which structure and function actuate each other is not that simple to read. This is particularly well known from the disappointing attempts to conceive form from phylogenetic reconstruction based on fossils (e.g., Cutler 1995). A correlation between structure and function can only be meaningfully made if both structure and function are simultaneously observed and quantified.

Based on their "firm belief that animals are built reasonably" and their confidence that "structural design is optimized for the role it plays", the rationale of structure-function interdependence for adept performance of the lung was formulated by Taylor and Weibel (1981) and Taylor et al. (1987b) and termed symmorphosis. Symmorphosis was defined as a state of structural design commensurate to functional needs resulting from regulated morphogenensis, whereby the formation of structural elements is regulated to satisfy but not exceed the requirements of the functional system. Direct morphological and physiological data on the mammalian lung and the skeletal muscle mitochondria, the primary O_2 sink, were adduced to show that the gas exchanger was reasonably well constructed to meet the metabolic needs. However, scaling the morphometric diffusing capacity (DLo_2m) and maximum O_2 consumption (VO_{2max}) with body mass indicated a paradox (Gehr et al. 1981; Weibel et al. 1983; Weibel 1989): VO_{2max} and DLo_2m scaled differently with body mass – VO_{2max} with a scaling factor (mass exponent b) of 0.8 and DLo_2m with that of unity. This suggested that large animals have a DLo_2m far in excess of their metabolic requirements and called in question the validity of what was initially conceived to be an encompassing concept (Taylor et al. 1987b; Weibel et al. 1991). Considering these observations, Taylor et al. (1987b) concluded that the principle of symmorphosis is only partially satisfied in the lung. This departure from the expected perhaps serves as a timely reminder of the intrinsic complexities of the biological systems. Such entities are multidimensionally regulated and when integrated, singular normally inconsequential events are greatly accentuated. Mathematical models with their intrinsic assumptions cannot possibly inclusively describe the arrangements and define the capacities of such complex matter. In complete departure from the Newtonian predictability in physical sciences, the cause and event interactions in complex systems are not necessarily linear and the output may not be anticipated. In fact, the predictive power in the physical sciences is only good for simple and single events operating under very narrow circumstances. Where multiple and possibly conflicting parameters interact and compromises and trade-offs permutate in space and time with unpredictable probability, except for within very broad limits, there may not be a single best solution to a problem. Maynard-Smith (1968) observed that "we rarely know enough about laws governing the components of biological systems to be able to write down the appropriate equation with any confidence". Wilkie (1977) pointed out that "biological systems are to some degree mixed regimes whose behavior is affected by physical quantities other than those assumed to be dominant". Compared with the physical structures, biological structures are complex and possess intrinsic hierarchical stratifications with certain resolutely restricting boundary conditions (Bartholomew 1982a; Brown 1994). Living entities are more difficult to conceive mentally while physical ones are generally simple enough to be adequately expressed in mathematical terms. Despite the limitations of applications in biology, mathematical reasoning is fundamental in defining the logical consequences of a situation at intellectual depths and expressing and understanding what is theoretically contemplated or observed.

Symmorphosis is not a totally new concept. Actually, the idea is as old as science itself: the great Greek philosopher Aristotle (3 B.C.) observed that 'nature does nothing to no purpose". The concept of symmorphosis corresponds with the theory of optimality (e.g., Rosen 1967; Maynard-Smith 1978; Alexander 1982b; 1996; Kramer 1987) and the principle of minimum work (e.g., Murry 1926; Mandelbrot 1983; Rossitti and Löfgren 1993a) which contend that, operating on finite resources, self-regulating systems strive to optimize their operations. Through cost-benefit trade-offs, some factors are subordinated while others are augmented as an optimal match between structural design and functional performance is established (e.g., Dullemeijer 1974; Wainright et al. 1976; Barel et al. 1989; Barel 1993). Remarking on the variability of the thermostability of proteins, an observation which may apply to all levels of biology, Wedler (1987) observed that "there is no single or universal mechanism (in nature) . . . it appears that nature has utilized every imaginable means, in different combinations, to achieve the same end". Evolution by natural selection is a continuous process of cultivating optimization: ways and means of enhancing fitness are aggressively pursued. Of all forms of life which have evolved during the about 4 billion years life has existed on Earth, about 99.99% are now extinct (e.g., Pough et al. 1989). Extant organisms represent a very narrow range of the designs which have developed during the long period of trial-and-error evolutionary experimentation and even less of the infinite theoretical possibilities. The history of life occurs in form of a complex, sporadic pattern of past evolutionary expansions and contractions as assemblages of organisms in a particular time frame respond to the prevailing conditions (Gans 1988): useful functions are intensified and improved while the less beneficial ones are relegated or even eliminated. For instance, the evolution of the complex, more efficient tetrapod lung entailed suppression and eventual elimination of the hydrostatic role of the ancestral piscine lung (Joss et al. 1991), creating a favorable situation for transition from aquatic to terrestrial life (Liem 1987a).

The structures which comprise the gas exchange pathway, namely the lung, heart, blood vessels, and mitochondria, have determinate functional capacities (Weibel 1984a, 1987). Pulmonary design is coadjusted with functional needs to supply O_2 at rates consonant with prevailing demands. Since animals have a definable capacity to bear natural loads (be they from within or without), overdesign and redundancy of biological systems is costly and untenable. An optimum structural design is one which requires the least metabolic (energetic) cost to work and sustain (e.g., Thompson 1959; Rosen 1967). Such a design may operate at an optimum level, though it may not necessarily be optimally adapted. While endeavoring to optimize their operations, biological systems appear never actually to attain an optimum condition (e.g., Gans 1983; Lindstedt and Jones 1988). Optimization is a process which appears to be continuously aspired for. There are, however, assertions that such states exist. For example, in their geometry, the cerebral arteries (Rossitti and Löfgren 1993b) and the medium pulmonary airways (Hammersley and Olson 1992) are thought to be optimized in respect to the ratio respectively between blood/air flow and blood vessel/airway radius minimizing resistance. Intuitively, a definitive optimal state should never be consummated. Considering the dynamical nature of environments and the plasticity of biological systems, optimum states cannot possibly be satisfied. As a matter of fact, it would be imprudent to achieve such a condition since an organism's capacity to respond to further changes would be undermined. Joyce (1997) asserts that "evolutionary search is a parellel process in which every individual in a population has an opportunity to give rise to novel variants with increased fitness": this is thought necessary in order to minimize the real possibility of the population "getting trapped in an evolutionary blind alley from which further improvements in the fitness are precluded". In constant pursuit of optimization, the best level of "workmanship" possible under certain constraints and opportunities is established, affording the best level of fitness. In this context, Gans (1983) and Lindstedt and Jones (1988) define optimization as "the best level of improvement possible under the evolutionary circumstances". Williston's law, which asserts that serially repeated structures are reduced to fewer sets of undifferentiated organs, illustrates an evolutionary process of optimization where through a rigorous process of trial and error superfluous features are pruned, leaving the bare essentials for life. The quest for optimization is a universal process which encompasses all self-organizing structures. Howell (1983) pointed out that "optimization does not belong exclusively to biologists, but to any discipline where the subject has some effect on fitness".

As a philosophical definition of the correlation between structure and function in the design of biological systems, the concept of symmorphosis is heuristically useful and as a working hypothesis implicitly valuable. It constitutes a useful tool for bridging the different disciplines of biology. The breakdown of the concept at some levels and in certain cases (e.g., Gehr et al. 1981; Weibel et al. 1983) is a clear manifestation of the vicissitudes of evolution. Paradigms of pulmonary structural-functional reciprocity occur in the nonmammalian gas exchangers but refinements to the extent envisaged in the dictum of symmorphosis cannot be proved owing to paucity of data. Birds, in particular, exhibit remarkable species-specific variations in the degrees of pulmonary structural refinements. The lungs of the nonflying species, e.g., the domestic fowl, *Gallus gallus* variant *domesticus* (Abdalla et al. 1982), emu *Dromaius novaehollandiae* (Maina and King 1989), and the penguin, *Spheniscus humboldti* (Maina and King 1987), have subordinate parameters compared with those of the more energetic ones, e.g., the passeriforms (Maina 1984) and the hummingbirds (Dubach 1981) (Sect. 6.7.5). Bats, the only volant mammals (Thewissen and Babcock 1992), have remarkably highly specialized lungs (Maina et al. 1982a; Maina et al. 1991) which enable them to provide the large amounts of O_2 needed for flight. Natural selection does not seek efficiency for its own sake but only in so far as it can improve fitness for survival and ensure self-perpetuation. Interestingly, in the construction of animal bodies, gravity has spared the very small. The structural integrity of a body which takes a jump from a height of 10 m is not threatened until the animal attains a size of a puppy or is larger (Went 1968). Interspecific deviations of the aspects which are envisaged to constitute safety factors, redundancies, or limitations exist: they may explain why different biological systems manifest different physiological threshold pressures.

2.5.2 The Operative Strategies for Optimization in the Gas Exchangers

The erythrocytes are packages which are said to provide an optimum environment for the function of the hemoglobin (Horvath and Borgia 1984). They present an excellent example of the "multiple tradeoffs" or "shared adaptive effort" as termed by Taylor et al. (1987b) of the morphohological, physiological, and biochemical processes involved in enhancing O_2 uptake and transfer (e.g., Edsall 1972). In vertebrates, the hemoglobin concentration of about 5 mM is close to the saturation mark (e.g., Riggs 1976). Acclimatization to hypoxia, e.g., at high altitude (Petchow et al. 1977; Wood and Lenfant 1979), and increase in aerobic capacity (Carpenter 1975; Balasch et al. 1976) are largely accompanied by an increase in the hemoglobin concentration so as to increase the O_2 carrying capacity of blood. In an exercising horse and steer, as O_2 consumption increases, circulating hemoglobin concentration rises, thereby enhancing the delivery of O_2 to the tissues through the circulatory system (Jones et al. 1989). While compared with the little auk, *Plautus alle*, the Arctic tern, *Stema paradisaea* has a lower hematocrit due to smaller red blood cells and hence a lower total repiratory surface area of the RBCs, the hemoglobin content per unit area of the RBCs in the two species is similar (Kostelecka-Mycha 1987). Upward regulation of the hematocrit (Hct) in particular serves a useful purpose only up to a certain point, when the returns start to diminish. The flow of blood is exponentially related to the blood viscosity, which in turn relates with the hematocrit. Change in the blood O_2 carrying capacity through increase in the Hct contracts extra cost of pumping work on the heart muscle. With a Hct of 65%, the blood viscosity of the elephant seals is three times that of the rabbit, which has a Hct of 35% (Hedrick and Duffield 1986). Birds with higher Hct such as the pigeon (Hct = 52%) adaptively have low plasma protein concentration in blood to maintain the viscosity of blood at a level similar to that of the birds with lower Hct, e.g., 32% in the domestic fowl (Viscor et al. 1984). There are diverse adaptations involved in optimizing gas exchange in blood. While the hemoglobin-hematocrit-blood viscosity intercourse appears to suggest an exhaustive process of refinement (e.g., Clarke and Nicol 1993), other evolved innovations continue to be discovered. For example, during hypoxia and/or hypercarbia, fish erythrocytes swell, resulting in intracytoplasmic dilution of the hemoglobin and nucleoside triphosphates, increasing the O_2 affinity (Lykkebone and Weber 1978). The rheology of the erythrocytes (Merrill 1969; Schmid-Schönbein 1975) and the hemodynamics of the blood flow in the blood capillaries can improve the delivery of O_2 to the tissues (Zander and Schmid-Schönbein 1973; Kon et al. 1983; Nilsson et al. 1995). This mainly occurs through diminution of the diffusion boundary layer of blood plasma around the erythrocytes and possible intracellular physical agitation of the hemoglobin molecules (Skalak and Branemark 1969; Secomb 1991; Maeda and Shiga 1994). The nucleated erythrocytes, e.g., those of birds, are more resistant to shear deformation than the nonnucleated mammalian ones (Gaehtgens et al. 1981; Nikinmaa and Huestis 1984; Nikinmaa 1990). The shear modulus of the membrane of the nucleated erythrocytes, which have a better-developed cytoskeletal system, is 5 to 15 times higher than that of the nonnucleated mammalian erythrocytes (Waugh and Evans 1976; Chien 1985). Decreased deformability causes the erythrocytes to be trapped in organs like the spleen, lung, and liver (Groom 1987; Simchon et al. 1987), reducing the microcirculatory blood flow in the tissues. Although experimentally more resistant to deformation than the nonnucleated erythrocytes, the nucleated erythrocytes are more deformed as they pass the blood capillaries (e.g., Chien et al. 1971; Akester 1974). Despite having a diameter 30% greater than the human erythrocytes, the fish erythrocytes are as deformable as the human ones (Hughes et al. 1982; Hughes and Kikuchi 1984). Deformation of the erythrocytes reduces the apparent viscosity of the blood (Chien 1970) and provokes convective mixing of blood (Bloch 1962). Contrary to anticipation, the size of the erythrocytes in mammals does not affect the lung diffusing capacity for O_2 (Betticher et al. 1991): small erythrocytes have a greater surface-to-volume ratio, thinner plasma boundary layer (Vandegriff and Olson 1984), and shorter intracellular diffusion distance (Yamaguchi et al. 1988), factors which would be expected to favor O_2 transfer. However, since a sphere is not deformable unless it is squeezed, the less spherical an erythrocyte, the more malleable it is without changing the surface area. Betticher et al. (1991), conceptualized that as the small erythrocytes are more spherical, the plasma around them is less well mixed than in larger cells, thus decreasing O_2 transfer into the cell. The elephant seals are thought to increase O_2 storage at the expense of aerobic scope which results from viscosity-limited perfusion efficiency (Hedrick and Duffield 1986). The normal hemoglobin concentration of 150 g/l of the mammalian blood is the value at which maximum amount of O_2 (210 cm^3), equivalent to the relative atmospheric concentration of the gas (Davenport 1974), is transported with least circulatory work (Schmidt-Nielsen 1984). Adaptive increase in the hemoglobin concentration at high altitude and in small animals such as shrews (Ulrich and Bartels 1963) and bats (Jürgens et al. 1981), values which may respectively be as high as 170 and 244 g of hemoglobin per l may set the operational limit to which hemoglobin and hematocrit levels may be usefully applied to improve respiratory efficiency. In the tracheolar system of insects (Sect. 6.6.1), the mean free path of O_2 molecules in air (i.e., the average distance a molecule travels in air before colliding with another) which is about $0.008\,\mu m$ (Pickard 1974) sets the limit of the smallest tracheolar diameter. Due to the remarkable variability and functional lability of the respiratory pigment system (Sect. 2.8), among the evolved sections of the integrated gas exchange system, the respiratory pigments appear to be the most recent addition and hence the least conserved parts. As opined by Jones (1972), the more ancient factors such as the morphological features of the gas exchanger and the vascular system, which are less flexible, would to a greater extent constitute limiting factors in respiration.

Metabolically, animals operate at two extremes, a steady-state resting condition and under maximal stress. Contrived reserves enable them to harmonically perform under the two different sets of conditions. To ensure structural and functional integrity, the body systems must be designed to withstand the maximal stresses they are likely to be subjected to. Digressing a little from the gas exchangers, in all evolved biological structures, the spider's dragline perhaps best illustrates the process of optimization in biology. A spider's dragline, a tool which determines survival by providing means for procurement of food and escape from predators (a lifeline in the true sense of the word!) is a multiphase material which consists of double filaments (Vollrath 1992). Individually the filaments can support the weight of the animal (if one is accidentally cut). A single line will break at a stress equivalent to that generated by about six times the spider's weight (Osaki 1996). Activities such as movement, jumping, and rapid descent and ascent (when greater stress is exerted on the dragline) must be accommodated in the design. The elastic-limit of a dragline gives the maximum safety (a safety coefficient being the ratio between the mechanical strength of a dragline and a spider's weight) for supporting a spider's weight. The mechanical properties of the spiders' draglines have been refined over their 400 million years of evolution to this level of sophistication. Turning to the respiratory organs, in the rainbow trout and the lingcod, only about two thirds of the secondary lamellae are perfused at rest (Booth 1978; Farrell et al. 1979; Nilsson et al. 1995) but the fish can increase O_2 consumption during exercise eight to ten times (Jones and Randall 1978). Hypoxia, e.g., in warm, bottom, and standing waters and during low tide (in intertidal animals), e.g., in the blue crab, *Callinectes sapidus* is avoided by increased ventilation and perfusion of the gills (Tuurala et al. 1984; DeFur and Pease 1988), lamellar recruitment (Booth 1979; Farrell et al. 1980), shortening of the thickness of the water-blood barrier (Farrell et al. 1980; Soivio and Tuurala 1981), and effecting changes in the blood O_2 binding characteristics (Jensen and Weber 1985). In extreme circumstances, especially where hypoxia is accompanied by high temperatures, the metabolic rate may drop to conserve O_2 and/or the fish may relocate to less poorly oxygenated waters (Jones 1952; Whitmore et al. 1960). Cephalopods, e.g., *Loligo* and *Octopus*, can increase their O_2 consumption by a factor of 2 to 3 from rest (O'Dor 1982; Wells et al. 1983). Squids which live in cold, deep seawaters have a large gill surface area and a thin water-blood barrier compared with those which live in shallow waters (Roper 1969; Madan and Wells 1996). The large gill surface area in the pelagic cephalopods may enable them to cope with the hypoxia prevalent in their habitat and could give them a competitive edge over fish (Madan and Wells 1996). This view was, however, disputed by Seibel and Childress (1996) on the basis of the fact that many fish sympatrically coexist with the squids and in some cases may even displace them. The cephalopod heart mainly works aerobically, relying largely on amino acids as substrates for oxidative metabolism (Hoeger and Mommsen 1985). By retracting into the shell, the bivalve mollusk, *Pholas dactylus*, can effect a complete shutdown of the posterior parts of the gills and, at maximal extension, the gills may be three times as long as the shell itself (Knight and Knight 1986). Bats and birds increase their O_2 uptake from rest to flight by a factor of 10 to 20 times (e.g., Thomas 1987; Butler 1991b) and insects by as much as 120 to 400 times (Weis-Fogh 1967). In gills, ion pumping and gas transfer can be regulated by modulating gill ventilation (Randall et al. 1972; Wood and Randall 1973). The mammalian lung maintains a remarkable excess of diffusing capacity up to a factor of 2 (Weibel 1984a). Even under hypoxic conditions, goats attain maximum O_2 consumption and only the smallest mammals use all of their diffusing capacity under such conditions (Taylor et al. 1987a).

Those animals which have relatively greater O_2 demands during exercise, e.g., birds and insects, are in most respects endowed with more efficient respiratory designs than the human alveolar lung. In a radical departure from the norm, in some respects, it could be argued that, compared with the other organ systems, the respiratory processes have been less sensitive to phylogenetic changes. Adequacy for specific metabolic needs in a given habitat appears to be the primary factor which has determined the refinement and construction of the gas exchangers. Fluid designs have made animals at completely different levels of phylogenetic development coexist in the same general habitats by varying their metabolic demands to meet individual requirements. From the perspective of respiration, animals have had few choices. There are only two naturally occurring respirable fluids – water and air – and three livable spaces, namely the aerosphere, the hydrosphere, and the lithosphere, i.e., the livable superficial part of the Earth's crust generally called soil.

2.5.3 Symmorphosis and Optimization: are they Logical Outcomes of Evolution?

The concepts of optimization and symmorphosis may have been accepted by biologists basically for their instinctive appeal to simple intuitive logic. Recently, the concepts have been ardently debated in depth (see Diamond 1992; Weibel et al. 1998). Conceived as perfect matching between structure and function, symmorphosis has been questioned from phylogenetic and developmental considerations (e.g., Gould and Lewontin 1979; Lewontin 1979; Garland and Huey 1987; Gans 1988). Optimization for particular conditions and circumstances, though logically desirable, curtails the potential and the range of adjustments which organisms can mobilize to counter external assaults. As a practical rule, every adaptive refinement for a particular circumstance that an organism or organ system undergoes leads to exclusion of capacity to take on others.

Moreover, what may be an optimal solution for a particular situation at a particular time may be a limiting factor in another. Moore (1990a) asserts that "organisms cannot be complete specialists and complete generalists at the same time". Natural selection refines each and every lineage towards effective utilization of a specific quota of the available resources. Faunal structuring, partitioning of dietary and habitat resources, and the different morphological specializations acquired for procuring them minimize competition, providing optimal survival conditions. Animals which face similar ecological and developmental constraints evolve similar features, as demonstrated by the phenomena of convergent evolution (e.g., Lauder 1981; Lauder and Liem 1989; Wagner 1989). In the process of perfecting for a particular lifestyle in a given environment, each incremental increase in fitness is accompanied by progressively narrower scope of structural and functional adaptive flexibility. This predisposes a species to collapse and extinction through what has been termed overadaptive meltdown (e.g., Minkoff 1983; Stuart 1991). Borrowing from the comment made by Wells (1962) on the cephalopods, remarking on the extreme marine predatory capacity of the elasmobranchs, Tota and Hamlett (1989) observed that "in doing so (adapting) they have become specialized to the point where their own structure and physiology will preclude further adaptive radiation". Gans (1985) contends that "we should not be surprised to find that no structure is perfect and that few structures are optimized to any particular role". Through intense genetic breeding, a process termed directed evolution by (Joyce 1992), the horse and the domestic fowl, *Gallus gallus* variant *domesticus*, have respectively been exceedingly manipulated for speed and weight gain. While better feed and feeding regimen may contribute, in 1960 it took 70 days for a table bird to reach a live weight of 1.8 kg and in 1985 it took only 40 days (Smith 1985). In the course of this enforced productivity, the functional integrity of some organ systems appears to have been severely compromised. Death due to aortic rupture (e.g., Carlson 1960) and vascular pathology (e.g., Julian et al. 1984) is a common problem in the turkey industry. A worldwide increase in occurrence of ascites has been reported in young broilers by Julian and Wilson (1986), Maxwell et al. (1986a,b) and Julian (1987). The syndrome was associated with right ventricular hypertrophy (Huchzermeyer and De Ruyk 1986; Julian and Wilson 1986). In an attempt to explain the pathogenesis of the condition, Huchzermeyer (1986) and Huchzermeyer et al. (1988) contended that hypoxia may cause pulmonary vasoconstriction with consequent pulmonary hypertension, resulting in a right ventricular hypertrophy. In the mammalian fetal lung, vasoconstriction due to hypoxia is thought to restrict blood flow to the developing lung in utero (Morin and Egan 1992). Environmental factors such as cold and altitude as well as nature of food have been associated with ascites in birds (Maxwell et al. 1986a,b; Julian 1987). Respiratory inadequacies in the domesticated birds may result from the poor pulmonary morphometric parameters which generally characterize the group (Abdalla and Maina 1981; Abdalla et al. 1982; Maina and King 1982a; Vidyadaran et al. 1987, 1988, 1990). The growth of the lung of a highly selected line of turkey did not match the increase in body mass (Timwood and Julian 1983; Timwood et al. 1987). The free-ranging village chickens are not susceptible to ascites (Pizarro et al. 1970). For similar reasons, i.e., intense selection for productivity in total disregard of the necessity for commensurate adaptation of the supporting organ systems, domestic fowls (particularly the males) are totally incapable of attaining VO_{2max} on treadmill exercise (Brackenbury and Avery 1980; Brackenbury et al. 1981; Brackenbury 1984). In the horse (an animal which has been fiercely genetically manipulated for speed performance), the pulmonary capillary blood pressure increases from 2.4 kPa to about 6 kPa during exercise (Sinha et al. 1996). Exercise-induced pulmonary hemorrhage has been reported to affect more than 40 to 80% of horses during high intensity exercise (Mason et al. 1983; Burrell 1985; West et al. 1991, 1993). Experimentally, the capillary transmural blood pressure which causes stress failure in the horse's pulmonary blood capillaries ranges between 10 and 13 kPa (Birks et al. 1997). Even at maximum exercise, a safety margin appears to exist in the lung's capacity to tolerate transmural mechanical stress at the alveolar level.

Based on the fact that many complex designs and patterns can be generated by a few simple natural algorithms, an almost infinite number of different designs of gas exchangers should perchance have evolved in animals. Certain constraints, however, must have enforced convergence and adoption of similar designs, even in animals of remarkably different ancestry. Employing different strategies and resources, such animals have found similar solutions to common challenges. Like most organs and organ systems, the gas exchangers are known to carry out multiple functions which include feeding, sexual display (inflation of the lung, e.g., in frogs), osmoregulation, and secretion and metabolism of certain pharmacological factors (Sects. 6.10.1 and 6.10.2). Interspecific differences in the roles which similar structures play and the many roles carried out by the same structure in an individual animal occur (e.g., Gans 1985, 1988; Bennett 1988). The conflict between the roles may keep some parameters from optimizing. Moss (1962) called the aggregate number of roles an individual structure plays a functional matrix. For performance of multiple functions, such organs have integrated (compromised) designs. Over time, the individual structural components have been refined to carry out a particular role best. Uncoupling the roles that certain structures play eliminates the constraints placed on their improvement and hence imparts opportunity for greater diversity and refinement in form and function (Liem and Wake 1985). To illustrate this point, the emergence of suctional breathing from the buccal force pump dissociated the feeding from the breathing functions, leading to diversification and greater specialization of the two processes in reptiles. Biological structures evolve to satisfy the immediate needs. They do not have to be optimal to be conserved by natural selection. The difference between the cost of operation of an organ during unstressed state and under severe stress can be considered to be its adaptive phenotypic plasticity. Unless the maximal loads a system can bear are well known, the available scope of operation can be read as an overdesign, overconstruction, or even redundancy. It is envisaged that overdesign results from a genetic programming which is based on an unpredictable environment (Gans 1988). Though normally operating at a lower scope (e.g., Bennett 1988), organisms appear to preadapt for the worst-case scenario (Gans 1979). The apparent paradox in the scaling of the morphometric pulmonary diffusing capacity (DLo_2m) of mammalian lung and the Vo_{2max} with body mass (Gehr et al. 1981) suggested that, compared with the small animals, large ones have overconstructed lungs, with DLo_2m being 10 to 20 times that which they actually need. A number of explanations have been offered to explain this discordance. Federspiel (1989) suggested that erythrocyte redistribution in the pulmonary capillaries during exercise may bring the physiological diffusing capacity closer to the morphometric one. Heusner (1983) asserted that redundancy in the gas exchangers increases with body size. It is plausible that, compared with the small ones, the large animals may possess a greater capacity to generate multifarious solutions to different functional needs. Alternatively, rather than being a redundancy, this feature may be indicative of a safety margin of operation in the lungs of the large mammals in order to support their higher mass-specific expenditure of energy in locomotion (e.g., Hill 1950). Dejours (1990) contended that time-dependent events such as the duration of exercise performed to determine VO_{2max} are longer in the larger than in the small animals. If this were so, the larger animals would need greater gas exchange potential than the small animals to support aerobic metabolism over the longer duration. At sea level, Wagner (1993) concluded that the parameters which are involved in O_2 uptake and delivery to the tissues (e.g., ventilation, hemoglobin concentration, cardiac output, lung, etc.) have been optimized (i.e., there is very little reserve in them) such that further adjustments do not affect VO_{2max}: under such circumstances, the most important parameter which determines the VO_{2max} is the cardiac output – not the lung. Pennycuick (1992) suggested that the surface area fractal dimension of the mammalian lung, which may be as high as 2.5, differs between the lungs of the small and large animals, a feature that may impart a better gas exchange capacity in the lungs of the large mammals. The extinction episodes that have befallen multicellular animal life indicate that in cases of sudden catastrophic events, the existing safety reserves are easily overwhelmed. Animals have no inbuilt contingencies for such rare and sudden events of great magnitude. Raup and Sepkoski (1984) argued that major extinction events recur with a periodicity of 26 million years. Rather than pay the cost of supporting superfluous structures which may not be utilized for many generations, it would seem that animals have gambled away their survival and invested on features such as diversity and numerical density in hope that these would see them through such occurrences. It is possible that during the long periods of relative stability, new species with better fitness characteristics would have arisen from ill-prepared ancestral ones. Gould (1994) envisages that the reason that mammals survived the dinosaur demise of the end-Cretaceous (though the two groups had coexisted for over 100 million years) is not, as has been widely argued, that the mammals had evolved special adaptations – as animals cannot possibly anticipate and prepare for future events. Their relatively small size (size of a rat or smaller) enabled mammals to fit into less hostile ecological niches which were out of reach to the more robust reptiles. Janis (1993) appropriately calls the succession of mammals victory by default. Like mammals, birds appear to have endured the pressure very well: at least 20 mammalian groups and 22 avian lineages predate the Cretaceous-Tertiary catastrophe (Coope and Penny 1997).

The thickness of the blood-gas barrier and the diameter of the erythrocytes are two factors in the gas exchangers which have been pushed close to optimization if this has not already occurred. While mammals span the enormous range of body mass from the 2.5-g shrew to the about 150-t whale, a factorial difference of

60 million, the average thickness of the blood-gas barrier of the lung of the shrew (Gehr et al. 1980) of 0.334 µm is comparable to that of 0.350 µm of the lung of the whale, *Balaena mysticetus* (Henk and Haldman 1990). Although no estimations of stress failure of the pulmonary capillary blood vessels are available on these two animals, the alveolar wall in the lung of the rabbit, *Oryctolagus cuniculus*, fails above a transmural blood pressure of 3.3 kPa (Tsukimoto et al. 1991; Costello et al. 1992), a value which is astounding since the harmonic mean thickness of the blood-gas barrier is only 0.50 µm (Weibel 1973). The pulmonary capillary pressure in the rabbit (Maarek and Grimbert 1994), dog (Okada et al. 1992), and human being (Hellems et al. 1949; Comroe 1974), animals of notably different body sizes, is about 1.1 kPa, giving a safety factor of 3. In the dog, Okada et al. (1992) observed that the resistance in a segment of an alveolar blood capillary remains stable even after large changes in the transmural pressure. In the human being, pulmonary capillary blood pressure elevation has been associated with rupture of the alveolar blood-capillary barrier (Severinghaus 1971) and presence of erythrocytes on the alveolar surface in cases of high altitude pulmonary edema (Schoene et al. 1988; Heath and Williams 1989). The pulmonary blood flow is pulsatile (e.g., Wiener et al. 1966; Milnor 1982; Maarek and Chang 1991). It is envisaged that dampening of the pressure wave occurs in the capillary system (Wiener et al. 1966). The pulsatility of the pulmonary microvascular pressure may influence filtration of fluid across the capillary wall and gas exchange at the alveolar level (Maarek and Chang 1991). In mammals, the size of the erythrocytes presents a good example of structural optimization. The erythrocytes of the shrew, the smallest extant mammal, have a diameter of 7.5 µm and compare with those of the humpback whale of 8.2 µm (Altman and Dittmer 1961). Since the smallest blood capillaries should set the limit for the erythrocyte diameter (though the cells fold greatly as they traverse these narrow conduits), the diameters of the blood capillaries in the smallest and largest mammal appear to be comparable. This suggests an optimal setting of the capillary diameter for satisfactory supply of O_2 to the tissues.

The criteria of categorizing a biological structure as inconsequential, superfluous, excessive, subservient, nonfunctional, or constituting a controlling, regulatory, limiting, constraining, or even being a triggering factor is oftentimes highly subjective. In some cases, by acting differently, opposing selective pressures greatly alter the physiological profiles of the functional components, whereby some processes may be suppressed, some may synergize, while others may become totally obliterated or remain functionally neutral. The operational definition of what is optimal is oftentimes biased and misleading. The interpretations may depend on circumstances and even on anthropocentric perceptions and personal preference of what may be deemed beneficial. For example, while intrapulmonary air and lighter bones may be considered to be useful in increasing the buoyancy of a flying animal, in an aquatic one, both parameters are a vulnerability rather than a benefit. Though the lung is considered to be specifically designed in view of efficient uptake of O_2 and discharge of CO_2 (e.g., Weibel 1983a, 1985b), the organ plays important defensive, pharmacological, and endocrine functions (Sect. 6.10.2). To yet unknown extents, such roles must be accommodated in the overall design of the organ. The composite nature of biological tissues and the multiplicative effects between the various integral components may explain why the sum of the individual functional processes often exceeds the aggregate value expressed by the whole organism (Hoagland and Dodson 1995). For example, while the full energy budget of an organism in normoxia is totally aerobic, even the well-oxygenated mammalian cell cultures invariably manifest a partly anaerobic state (e.g., Gnaiger 1991).

2.6 Fractal Geometry: a Novel Approach for Discerning Biological Form

Geometry, as advanced by Euclid (300 B.C.) and Pythagoras (6 B.C.), uses straight lines and smooth and regular curves to make flat shapes and figures such as squares, triangles, and circles to model structures. When applied to natural things, however, it idealizes form in inherently chaotic structures (e.g., Olsen and Degn 1985; West 1990). The classical geometry defines space in terms of discrete dimensions, e.g., a point has no (zero) dimensions, a line has one dimension, a plane (area) has two, and a solid (volume) has three dimensions. These integer dimensions are unsatisfactory in describing the highly complex natural forms and the dynamical physiological processes that do not have specific scales of length and time. To adequately define the topological characteristics of dynamical structures and processes, fractional power dimensions (fractals) are necessary. The development of biological states and events are regulated by nonlinear, iterative algorithms which program morphologies and physiologies that lack absolute spatial and temporal boundaries (e.g., West 1985, 1987; Nonnenmacher 1987; Voss 1988; Nelson et al. 1990; Bassingthwaighte et al. 1994). The designs and operations present scale-invariant properties and self-similarity (Horsfield and Woldenberg 1986; Barnsley et al. 1987; Bassingthwaighte 1988; Giaver and Keese 1989; Goldberger et al. 1990). Discovered by Mandelbrot (1977, 1983), fractal geometry provides a powerful tool for rationally investigating form and function (e.g., Barnsley et al. 1987; Goldberger and West 1987; Tsonis and Tsonis 1987; West and Goldberger 1987; Glenny et al. 1991). Fractal characteristics have been reported in viruses (e.g., Briggs 1992) and tissue cells (e.g., Nonnenmacher 1988; Smith et al. 1989; Losa et al. 1992). They are integral to the functions of organs like the heart (Goldberger et al. 1985; van Beek et al. 1989; Goldberger 1991) and the gastrointestinal system (Pennycuick 1992). Structures such as the bronchial tree of the lung (e.g., Mandelbrot 1983; West et al. 1986; Nelson et al. 1990; Weibel 1991, 1994; Bates 1993, but see a dissenting view by Phillips et al. (1994) and the pulmonary arterial tree (Krenz et al. 1992) present fractal attributes. Blood flow in the lung (Glenny and Robertson 1990; Caruthers and Harris 1994), in the middle cerebral artery (Rossitti and Stephenson 1994), and in the myocardium of humans (Bassingthwaighte et al. 1989) has fractal properties.

Fractal geometry provides means for realistically analyzing dynamic forms and processes (e.g., Nelson and Manchester 1988) and studying physiological properties (Goldberger and West 1987). A corrugated structure in three-dimensional space is physically in transition between a smooth surface and a volume. Depending on the degree of amplification, such structures should have a fractal dimension between 2 and 3. A fractal dimension defines to what extent the topological details of an object fit between the Euclidean dimensions. The surface fractal value of the mammalian lung is about 2.5 (Weibel 1991). Mandelbrot (1977, 1983) observed that both the fractal dimension, D, and the diameter component, Δ, of the bronchial tree were about 3. Such high values indicate that the surface area of the lung has been sufficiently highly folded and the peripheral airways sufficiently regularly branched as to nearly fill a three-dimensional space. The two-dimensional surface area of the human vascular system is so highly folded that it has an effective fractal dimension of 3, with the arteries alone having a value of 2.7 (Briggs 1992). It has been argued by Goldberger et al. (1990), Nonnenmacher (1989), Weibel (1991, 1994), and Rossitti and Stephensen (1994) that fractal designs should permit biological systems to operate over a wide range of perturbations without failure. Fluidity in the configuration imparts greater error tolerance (i.e., safety margin) to a biological structure and may be decisive in the trial-and-error process of evolution (West 1987). The inbuilt fractal algorithm reduces the probability of error during morphogenesis since it is dependent on a well-tested self-similar, repeating (iterative) process. Functionally, a fractal attribute eliminates the need for drastic constructional overhauls in a system operating within reasonable boundaries and exposed to moderate assaults. Pennycuick (1992) envisages that the use of fractal structures can allow a particular plan to be scaled over a wide range of sizes without consequential allometric change. Functional plasticity should make adaptation and the more or less trial-and-error evolutionary process by genetic programming much easier to effect, against stochastic settings (e.g., Doebeli et al. 1997).

Unique to most organ systems, and perhaps an indication of the importance of respiration for survival, the gas exchangers have been configured by essentially molding together three characteristically fractal entities, namely the pulmonary arterial system (e.g., Lefevre 1983; Glenny and Robertson 1991a; Krenz et al. 1992), pulmonary venous system, and the bronchial-alveolar system (Weibel 1986). Though it may appear passive, the lung is intrinsically a highly dynamic organ. It is subjected to continuous cardiovascular hemodynamic changes as well as biomechanical ventilatory rhythms. It acts as an interface between blood and air, media which are physically remarkably different. The lung is the only organ in the body which transmits the entire volume of blood in the systemic circulation. The multifunctionality of the lung was pointed out by Bakhle (1975), who declared that "the lung should now be considered not merely as an apparatus for gas exchange or mechanical filtration of blood but also that of providing an essential control of the blood levels of many biologically active substances". The fractal characteristics of the surface of the lung allow the large internal surface area to be homogeneously ventilated and perfused at low energy cost (Weibel 1983a, 1991, 1996) and the circulatory system to be highly distensible (Caro and Saffman 1965; Yen 1989a,b; Bshouty and Younes 1990) to contend with the fluctuating hemodynamic blood pressures (Maloney et al. 1970; Zhuang et al. 1983; Al-Tinawi et al. 1991). Although only about 9% of the total blood volume is contained in the heart (Dock et al. 1961; Milner 1980), Hainsworth (1986) observed that the "distensibility of the pulmonary circulation is of particular importance as it permits transient imbalance between the outputs of the left and right hearts". Pulmonary blood flow is pulsatile from the entrance of the pulmonary circulation to its outlet in the left atrium (Morkin et al. 1965; Wasserman et al. 1966; Milnor 1982; Maarek and Chang 1991). The fractal dimensions of the diameter element of the arterial and the venous tree in the human lung are 2.71 and 2.64, respectively, while equivalent values for the length element are 2.97 and 2.86, respectively (Huang et al. 1996). In the dog lung, the fractal dimension of the blood flow is 1.22 (Barman et al. 1966). Pennycuick (1992) suggested that the morphological complexity of the avian pulmonary system (Sect. 6.7.5) may be due to a high fractal surface (with a dimension of about 2.5), enabling birds to achieve exercise capacities such as sustained flights at altitude without the need to vary the general plan of the respiratory system. The morphological features of the avian respiratory system bear this suggestion out: except for differences in the sizes and locations of the air sacs, pneumatization of the long bones, and development of the parabronchial systems (Sect. 6.7.5), the configuration of the lung-air sac system in birds is uncommonly uniform in such a numerically large and ecologically diverse group. A small fractal dimension in the design of the characteristically mammalian bat lung (Sect. 6.7.4.1) may have called for multiple extrapulmonary compensatory changes (Maina 1998) in order to provide the large volume of O_2 needed for flight. This line of reasoning is supported by a number of observations which include: (1) bats have enormous lungs which occupy much of the celomic cavity (e.g., Maina and King 1984), and (2) the mass of the heart and parameters such as the hematocrit and hemoglobin concentration are some of the highest values among mammals (Jürgens et al. 1981). The high demands imposed by flight on a gas exchanger of moderate efficiency may explain why the heaviest bats, the flying foxes (pteropodids), weigh only about 1.5 kg, a value which is an order of magnitude smaller than the weight of the heaviest flying bird (about 15 kg). For gas exchange, birds operate on what was termed a broad-based low-keyed strategy where many moderately refined parameters are variably utilized in combination, affording a large functional reserve (Maina 1998). The highly efficient tracheal system of insects (Sect. 6.6.1) is thought to have fractal surfaces of high dimension (Pennycuick 1992). The fractal dimension of 2 suggested for the fish gills by Pennycuick (1992) may be due to the assumption that the secondary lamellae of fish gills are smooth. Lamellar microridges (Fig. 50b) are characteristic of secondary lamellae of most fish gills, even those of the most ancient fish such as the coelacanth, *Latimeria chalumnae* (Hughes 1995), and the sturgeon, *Acipenser transmontanus* (Burggren et al. 1979). Microridges are, however, poorly developed in the gills of *Trachurus mediterraneus* (Hughes and Mondofino 1983) (a carangid fish) and appear to be lacking in the secondary lamellae of the hill-stream fish such as *Danio dangila* (Ojha and Singh 1986). It would seem that generally, like other gas exchangers, the gill surface is a fractal construction. Microridges increase the respiratory surface area of the gills, provide anchorage for the mucous lining, generate turbulence, and reduce drag forces at the water-gill interface (e.g., Sperry and Wassersug 1976; Hughes and Mondolfino 1983; Hughes 1979, 1984). Recently, applying fractal geometry to the allometric scaling of animal size, West et al. (1997) concluded that the enormous size disparity in the evolved animal life has been possible due to the intrinsic fractal nature of tissues and organ systems.

Cope's rule (see Cope 1896, generally considered a pervasive evolutionary pattern, but see dissenting views by Fenchel 1993 and Jablonski 1997), asserts that phylogenetically, animal lineages tend to evolve larger body size for reasons that they achieve greater mating success, better defense capacity, predatory ability, and resistance to environmental extremes (Bonner 1988). In biology, time-dependent events, e.g., ventilatory cycle, incubation and gestation periods, and duration of exercise needed to reach maximal O_2 consumption are longer in large than in small animals (Dejours 1990). The mass specific cost of transport is lower in larger animals than in smaller ones (Boulière 1975; Taylor 1977).

2.7 From Diffusion, Perfusion, and Ventilation to Respiratory Pigments

2.7.1 Diffusion

The evolution of respiration unfolds critical moments in the past when at certain times momentous developments occurred. Some of the most profound changes took place during the transformation of the anaerobiotic prokaryotes to the aerobiotic eukaryotes (the inquisition of molecular O_2 into respiration), accretion of cells into the Metazoa, shift from water- to air breathing, transition from water to land, and change from ectothermic-heterothermy to endothermic-homeothermy. At each of these stages, the metabolic needs of organisms would have exceeded those that could be serviced by the default gas exchanger. Changes which resulted in greater diversity and specializations of the gas exchangers were thereby instituted. In the theory of chaos (e.g., Stewart 1990) such phenomenal moments when conditions occur to create dynamical, self-driving structures by amplifying and locking the system's feedback together are called bifurcation points (Briggs 1992).

A distinct respiratory system cannot be clearly delineated below the level of mollusks and arthropods. The most primitive gas exchanger in such simple metazoans is an unspecialized epidermis. Gas transfer occurs by diffusion across the general body surface and in some cases through inbuilt modifications (e.g., Mangum 1994). Diffusion is a natural phenomenon which dictates spontaneous flow of molecules from places of high to low concentrations free of energy, just as water flows downhill. This must have been the earliest mode of gas transfer which was initiated by the increase of molecular O_2 in the primeval biosphere, resulting in the evolution of aerobic processes. When it is the only means of supplying O_2 to the cells, diffusion determines the shape and size of organisms (e.g., Burggren and Roberts 1991). A small and flat body form enhances gas exchange by diffusion (e.g., Jell 1978; Runnegar 1982). When O_2 needs are high, at 1 atm, diffusion can only be adequate in organisms up to about 1 mm in diameter and in animal tissues 2 to 5 mm thick (e.g., Comroe 1966; Schmidt-Nielsen 1990). Larger animal forms must possess a low rate of metabolism if they are to rely wholly on diffusion as the only means of securing O_2 and voiding CO_2. The evolution of complex energetic animals obliged the development of diverse specialized systems to deliver O_2 from the external environment to the tissue cells (Comroe

1966). As a spherical body provides the lowest surface to volume ratio, to overcome diffusional limitations, animals have deviated from this design by evolving long, highly attenuated bodies. Factors such as the habitat occupied and the mode of life determine the definitive shape and size of animals (Gould 1966) as well as the type of a gas exchanger needed to effectively service the total cytoplasmic mass. The innermost part of the developing embryos in an egg mass of the sand snail, *Polinices sordidus*, which weighs as much as 210 g (Shepherd and Thomas 1989) and has a radius of as much as 40 mm, experiences extreme hypoxia ($PO_2 < 1$ kPa) though the PO_2 in the outer layer of water may be moderately high at $PO_2 > 10$ kPa (Booth 1995). Water inside a spawn of *Rana temporaria* was found to be only 3 to 16% saturated while the saturation about 50 cm away from the egg cluster was 136% (Savage 1935). By adopting appropriate morphological designs and subsisting in a well-oxygenated medium, an organism can attain greater body size while relying on diffusion only. Through an assemblage of irregular shapes which increase the surface area, the diffusional distances can be reduced. In the sponges, where there is no internal perfusive mechanism, the body has flagellated cells (choanocytes) which move water (by the beating of cilia) through numerous incurrent pores or ostia into ramified water channels of about 1 mm diameter. This brings the tissue cells close to water enhancing gas exchange and nutrient uptake (Fig. 17). As much as 90% of O_2 is extracted from the water passing through the pores of sponges during maximal activity (Hazelhoff 1939). The largest known sponge is the barrel-shaped loggerhead sponge (*Spheciospongia vesparium*) which is found in the West Indies and in the waters off the coast of Florida: it stands at a height of 1.5 cm and is 9 cm in diameter. Sponges have been recovered from depths up to 5.6 km. In the coelenterates, a group which has a remarkably low metabolic rate, a steady water flow by ciliary movements across the gastrovascular canals is maintained. Pelagia,

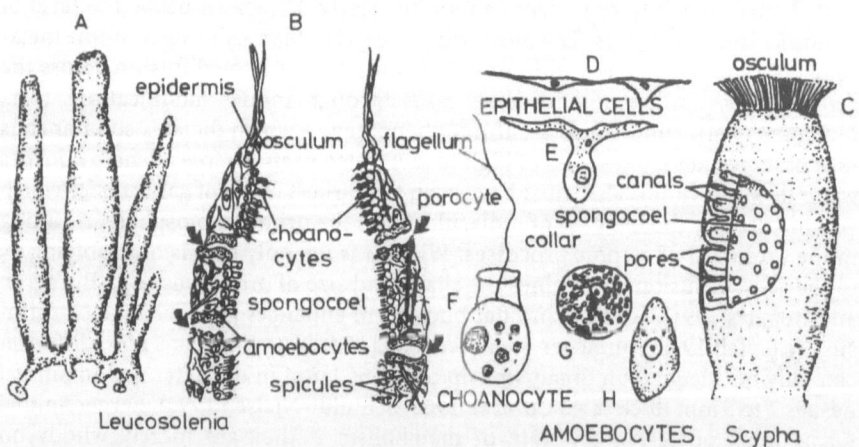

Fig. 17. A External morphology and sectional profiles B and C of an asconoid sponge, *Leucosolenia* showing the canals (↘, B) through which water passes into the body cavity. D, E, F, G, and H show the various types of cells which comprise the body wall. (Jessop 1995)

which can grow to a diameter of about 6 cm (Henze 1910), and the coelenterate *Cynea*, which can grow to a diameter of 2 cm, are said to rely sorely on diffusion for their O_2 needs (Krogh 1941). The tropical earthworms, animals which are as large as 1 kg in body mass, e.g., *Rhinodrilus fafner* and *Megascolides australis* have been recorded to attain lengths of 2.2 m (but have a diameter of about 24 mm), rely entirely on cutaneous diffusion for gas exchange (Stephenson 1930). In helminths, Fry and Jenkins (1984) estimated that to depend entirely on diffusion to maintain aerobic respiration, the critical thickness is 0.4 mm for the nematodes and 0.75 mm for the cestodes. The interstitial animals (animals which live in the water-filled spaces between the aquatic sediments) like the polychaete *Stygocapitella subterranea* and oligochaete *Marionina achaeta* do not have specialized respiratory surfaces. They acquire their O_2 needs entirely by diffusion through their great surface-to-volume ratio achieved from their thin, cylindrical body configurations. In an adult gastrotrich 500 μm long, Colacino and Kraus (1984) estimated that the O_2 transfer across the body surface was three times greater than its O_2 consumption when the PO_2 between the water and the mitochondria was only 0.13 kPa. Due to the much greater rate of O_2 diffusion in air than in water (Table 4) and the more constant level of O_2 in air, for the same concentration difference, the mass transfer attained over 1 μm in water can occur over a much longer distance of 1 cm in air. The PO_2 in the hemolymph of the gooseneck barnacle, *Pollicipes*, is elevated in air (e.g., Petersen et al. 1974). The air can support a spherical animal 100 times greater than water at the same rate of metabolism. However, such animals do not exist due to the parallel complications which would arise in air such as risk of desiccation and lack of mechanical support. In simple organisms, there exists a delicate compromise between the

Table 4. Some physicochemical properties of water and air[a]

Parameter	Unit	Water	Air	Water/air	Air/water
Density	g ml^{-1}	1.000	0.0012	833	–
Viscosity	Centipoises	1.00	0.02	50	–
O_2 Content	ml 100 ml^{-1}	0.66	20.95	–	32
CO_2 Content	ml 100 ml^{-1}	0.03	0.033	0.91	–
Thermal capacity	Cal ml^{-1}°C	1.00	0.0003	3333	–
Diffusion coefficients	cm^2 s^{-1}				
O_2		0.000025	0.198	–	8000
CO_2		0.00008	0.155	–	2000
Capacitance coefficients	nMol ml^{-1} mmHg^{-1}				
O_2		1.82	54.74	–	30
CO_2		51.89	54.73	–	1
Krogh's constants	nMol cm^{-1} s^{-1} mmHg^{-1}				
O_2		0.000046	10.84	–	20000
CO_2		0.00093	8.46	–	9000

[a] Measurements made at 20 °C and at 1 atmosphere pressure.

need to have an extensive and thin respiratory barrier or a thick and less water-permeable one. By living in nondesiccating (cryptozoic = humidic) habitats, risk of desiccation is minimized and a thin barrier can prevail. In most organisms, exceeding a diameter of about 1 cm, a circulatory system becomes a necessity. From hypothetical mathematical computations (e.g., Harvey 1928; Rashevsky 1960), in a normobaric environment and at moderate O_2 consumption, the maximum radius of a spherical cell, where anoxic state would not occur except at the center (i.e., the farthest point from the surface), was estimated to be 0.5 mm. Krogh (1941) calculated that for a homogenous spherical organism of a radius of 1 cm, at an O_2 consumption of $100 \, ml \, kg^{-1} h^{-1}$ (about half of the O_2 consumption of resting man) an external PO_2 of 25 atm (2533 kPa) would be required for O_2 to diffuse to all parts of the organism. He calculated that an organism cannot have a radius of more than 0.5 mm if it were to rely on diffusion alone, even assuming that it lives in water almost fully saturated with air at 1 atm of pressure. Where the external medium is separated by a 50-μm barrier from an internal circulating medium, a PO_2 of one quarter of an atmosphere would be required for satisfactory diffusion. Relying entirely on diffusion, a paramecium (volume $0.0006 \, cm^3$, O_2 uptake 1.3 ml per g per h and diameter 0.11 cm) would need a PO_2 of 0.73 of an atmosphere (Prosser and Brown 1962). Clearly, such high PO_2 rarely occurs in natural environments. Denney (1993) estimated that at the normal metabolic rate of a typical protozoan of about $0.1 \, mol \, m^{-3} s^{-1}$, a cell would have to be almost 7.5 cm in radius to experience hypoxia. The disparities in the estimated theoretical maxima of organisms indicate likely flaws in the idealized models used to calculate the largest possible sizes that microorganisms can attain. Except probably for eggs, there are no organisms which are absolutely spherical. At the organismal level, scaling with size is nonisometric (Schmidt-Nielsen 1984). The largest protozoans that have ever lived were the now extinct nummulites which had a diameter of about 2.4 cm and the largest extant protozoan is *Pelomyxa palustris*, which can attain a length of up to 1.5 cm. This size difference may be accounted for by the emerging possibility that the PO_2 in the biosphere at certain time(s) in the evolutionary past was higher than the present level (e.g., Graham et al. 1995). One such case occurred in the late Paleozoic (between the Carboniferous and the Permian), when for about 120 million years, O_2 level was 1.7 times greater than in the present atmosphere (Berner and Canfield 1989; Landis and Snee 1991; Graham et al. 1995; Fig. 9). Extraordinarily, the endothermic plants, e.g., *Philodendron selloum*, which produce heat at a rate surpassing that of the insect flight muscles and can maintain a temperature gradient of 30 °C with the ambient (Seymour 1997; Koch et al. 1983) rely entirely on diffusion for supply of the O_2 needed for generation of energy. In a single floret of the inflorescence of *P. selloum*, O_2 diffuses across a distance of about 1.2 mm through about 170 stomates (Koch et al. 1984): the average diffusional length from the surface to the individual cells, which is less than 0.75 mm, compares with that of the trachea of most small diffusion-dependent insects (Sects. 6.5 and 6.6.1). This reveals nature's amazing congruent solutions to similar needs!

Diffusion over an undifferentiated surface is the method of respiration in the simple organisms, e.g., Protozoa, Rotatoria, Planaria, Nematoda, eggs and young embryos, copepods and ostracodes, while the more complex ones (which utilize this process) include Spongia, Cerripedia, Coelenterata, arthropods like Tardigrada and Pauropoda, eggs, and early developing embryos (e.g., McMahon and Wilkens 1983). Larvae of many insects rely on diffusion across the integument (Fraenkel and Herford 1938) with the flux of gases essentially being regulated by O_2 consumption and CO_2 production within the organism. The tracheal system of the *Cossus* larvae delivers adequate O_2 entirely by diffusion (Krogh 1920a). The occurrence of air in the tracheoles of the young adult which emerges after the larval gas-filled tracheoles have been shed (there having been no previous contact with air) is probably due to passive diffusion of O_2 (Keilin 1924; Buck and Keister 1955). Surface-to-volume ratio decreases with body size since volume increases as the cube while surface area increases as the square of the radius. Assuming that the metabolic rate remains constant, O_2 transfer by diffusion should decrease with size. Since some organisms (including the amphibian eggs) develop to sizes greater than those theoretically predictable and the PO_2 inside the cell exceeds that which would be expected from diffusion alone, it was conceived that another process must promote gas transfer (e.g., Longmuir and Bourke 1960; McDougal and McCobe 1967). In the protozoa, protoplasmic streaming, a normal circulation-like process in living cells, enhances intracellular gas transfer (Seifriz 1943; Andrews 1955). Mechanical vibrations appear to intensity permeability of tissues to gases (Longmuir and Bourke 1960). Dynamic organs such as the heart, lung, and diaphragm, as well as activities such as change in muscle tone and physical interaction between the erythrocytes and the endothelial wall of the blood vessels may to an unknown extent influence gas transfer at the tissue level.

2.7.2 Convective Flows

2.7.2.1 Perfusion

The inadequacy of diffusion as a means of gas exchange necessitated development of auxiliary respiratory processes as organisms became larger, more complex, and their O_2 needs increased. A progressive development of a circulatory system occurred, promoting the efficiency of the respiratory processes (Figs. 18,19,20). Mature animals having an elementary circulatory system close to the surface of the body, e.g., earthworms and echinoderms, where the blood is moved by a heart and not cilia, can grow to a body mass of a few grams and a length of 30 to 40 cm without calling for development of special respiratory organs as long as they subsist in water or in a humidic environment. In such animals, a directional flow of the lymph may not exist and hence a circulatory system strictly does not exist. Churning of the fluid underlying the respiratory surface through contraction of body muscles should enhance the flux of the respiratory gases. In organisms such as Chaetopoda, Synapta, and Pantopoda, there is no regular circulation but the coelomic fluid is kept in motion by cilia (Lindroth 1939). The blood of the earthworms has a high O_2 affinity (P_{50} = 0.3 to 1.1 kPa), is highly sensitive to temperature (Laverack 1963) and has a high O_2 carrying capacity of 8 to 12 vol % (Haughton et al. 1958). The blood of the giant earthworm

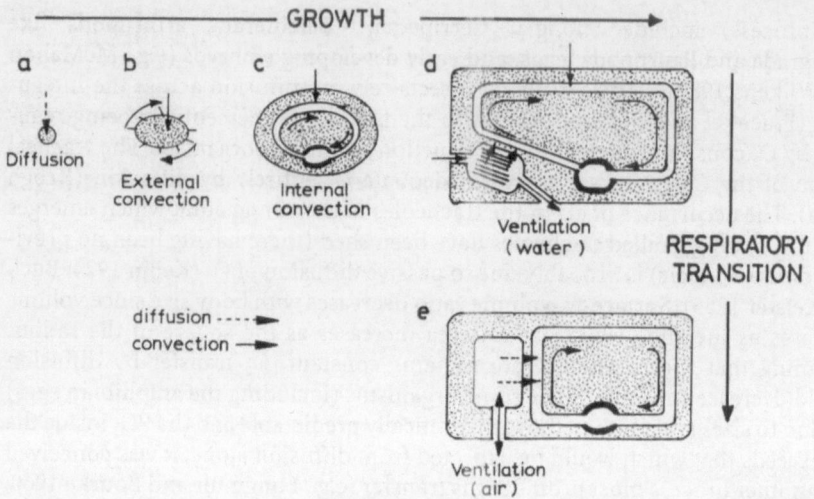

Fig. 18a–e. Development of the respiratory processes from unicellular to multicellular organisms. Diffusion (**a**) is the main process of gas exchange in simple animals with external (**b**) or internal convection (**c**). More complex animals combine perfusion and ventilation (**d** and **e**). While the gills are configured for continuous and unidirectional ventilation (**d**), the lungs and their derivatives are tidally ventilated (**e**). (Burggren and Pinder 1991)

(*Glossoscolex giganteus*), which can attain a body mass of 600 g, a length of 120 cm and a width of 2 to 3 cm, has a P_{50} of 0.9 kPa (at 20 °C and pH of 7.5) and a small Bohr shift (Johansen and Martin 1966). By regulating cutaneous perfusion (Burggren and Feder 1985) and surface area (Noble 1925), the hairy frog, *Astylosternus robustus*, can adjust gas exchange across the skin. In the higher vertebrates, the skin has been rendered virtually impermeable to O_2 and in the human being only 0.2% of the total O_2 need is acquired through it (Krogh 1941). The plethodontid salamanders (Collazo 1993; Ruben et al. 1993; Tilley and Bernado 1993; Wake and Marks 1993), which live in cold, well-oxygenated habitats and have adopted long and cylindrical body forms, rely entirely on the skin for gas exchange: this group, which is extremely successful (Pough et al. 1989), is presumed to have originated from torrential mountain streams (Beachy and Bruce 1992). In the freeze-tolerant frogs which can endure temperatures of between −3 and −7 °C, e.g., *Hyla versicolor*, *Rana sylvatica*, *Hyla cricifer*, and *Pesudacris triseriata macurata*, the heart does not beat in the frozen state (Lotshaw 1977; Storey and Storey 1988). In some species, intracellular glucose level increase and appear to serve both as a cryoprotectant and a metabolic fuel (Storey and Storey 1986). At natural wintering temperatures of 3 °C, freshwater turtles can remain submerged without O_2 for months (Carr 1952; Ultsch and Jackson 1982). Nemertines (ribbon worms) have only two main longitudinal blood vessels and in the species which have hemoglobin, reversal of blood flow is common (Hyman 1951). The bootlace worm, *Lineus longissimus*, which is found in the North Sea, can reach a length of 55 m. In some annelids, a closed circuit

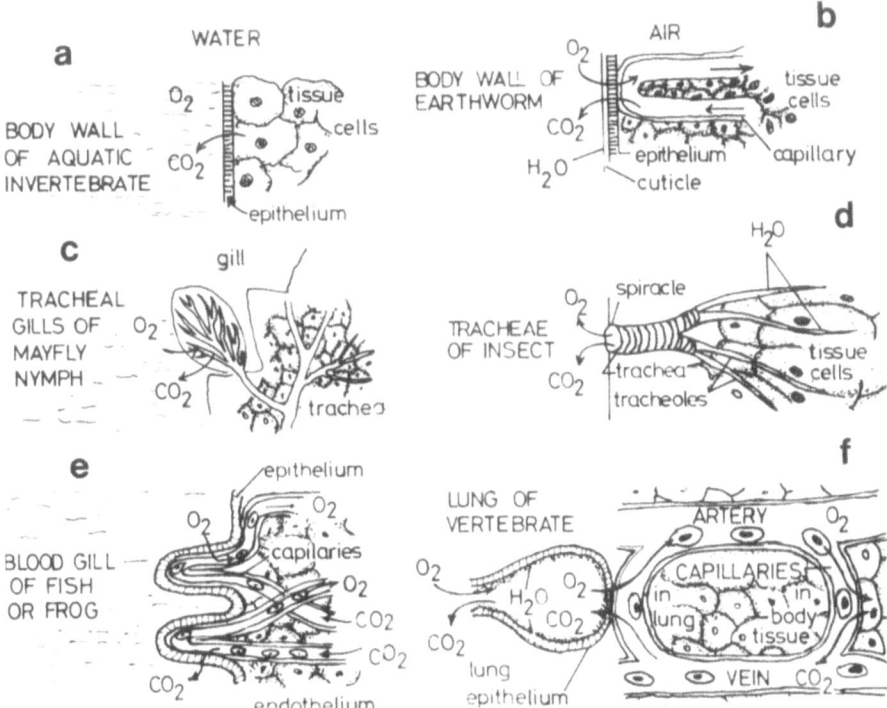

Fig. 19a–f. Mechanisms of gas exchange. The most basic gas exchange design is that which occurs across a cell membrane, e.g., in protozoa or across an unperfused skin (a). Perfusion of a respiratory site improves gas transfer by maintaining a concentration gradient (b). In the tracheal gills, e.g., of the mayfly, air rather than blood is contained in the gills (c). In the tracheal system of insects, air is delivered directly to the tissue cells by diffusion (d) and in the larger species by convection through abdominal pumping. In the conventional gills, e.g., of fish, the organs are well perfused with blood (e) and ventilated with water. The elaborate vertebrate gas exchangers combine tidal ventilation with perfusion (f). (Jessop 1995)

with well-developed blood vessels and pulsatile ancestral "hearts" exist in form of modifications along the blood vessels. In certain teleosts and elasmobranchs (sharks, skates, and stingrays), caudal hearts, which are located near the tail and powered by skeletal muscle, aid in venous return (e.g., Satchell 1992). Auxiliary hearts occur in the circulatory system of decapod crustaceans (Steinacker 1975) and locomotor movements generate large pressure differentials which promote the flow of the hemolymph (Belman 1975). The oscillations in the dorsal aortic blood pressure in the Atlantic hagfish, *Myxine glutinosa*, are associated with contractions of the gill musculature, a process which may be involved in propulsion of blood (Johansen 1960; Strathmann 1963). Such a process, however, does not seem to occur in the gills of the Pacific hagfish, *Eptatretus stoutii* (Chapman et al. 1963). At 10 °C, *M. glutinosa* has a heart rate of about 22 beats min^{-1}, a mean ventral aortic blood pressure of 1 kPa and a cardiac output of 9 ml per min per g

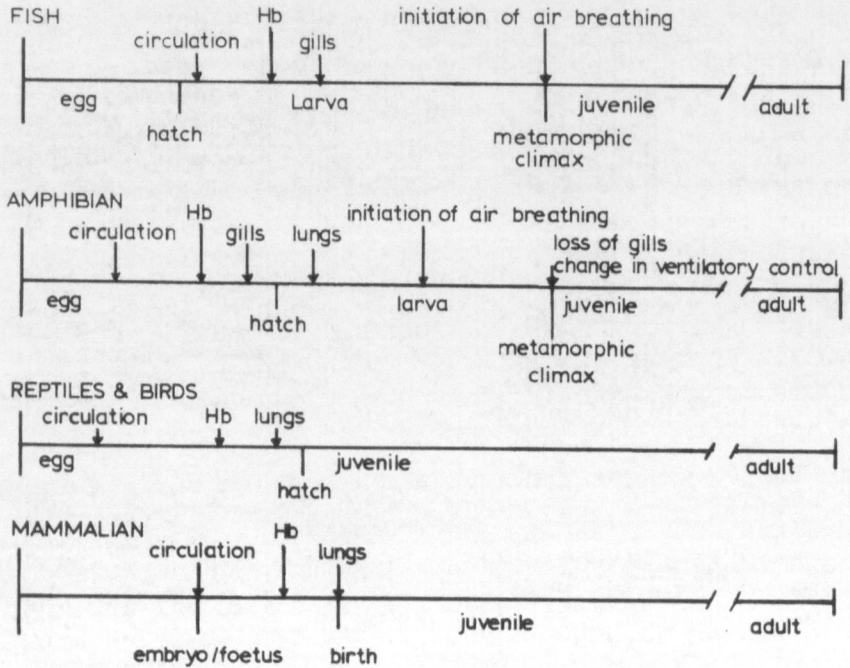

Fig. 20. Comparative schematic illustration of the stages at which cardiovascular and respiratory systems develop in relation to life histories of different vertebrate taxa. Some stages may fail to develop in some groups. For example, not all fish attain capacity for air breathing and not all amphibians develop lungs. Variably, the sequence in the increasing complexity and efficiency of the gas exchangers appears to progress from the simple membrane gas exchanger, a perfused one, a carrier-supported one, and finally to a ventilated one. Of note is the fact that hemoglobin (*Hb*) develops before respiration starts. (After Burggren and Pinder 1991, with the processes in birds and mammals added)

heart mass (Forster et al. 1988, 1991; Axelsson et al. 1990). While buried under the soft mud, *Myxine glutinosa* can remain without O_2 for at least 1 h (Strathmann 1963) and feed inside the body cavity of dead animals for a long period of time (Hardisty 1979): at 5 °C, the hagfish can survive in anoxic water for at least 20 h (Hansen and Sidell 1983). In most gastropods, to ensure forward blood flow, passive valves exist throughout the body (Jones 1983). In the open circulatory system, a large quantity of fluid is located in the intercellular space, providing mechanical support for locomotory activity and feeding movements (e.g., Jones 1983; Russell and Evans 1989). A fast circulatory return in a closed system with a smaller blood volume should be a more efficient design compared with a sluggish one with a large volume. Open circulations without hearts suffice where the respiratory demands are not high and where the diffusional distances are not great (Farrell 1991a). Heart rate in the terrestrial slug, *Deroceras reticulatum*, increases during feeding, a feature attributable to the need for substantial hemocoelic pressure required to protrude the odontophore (Duval 1983). Respiratory gas exchange across the body surface coupled with a simple circulatory system occurs in leeches, all oligochaetes, and some polychaete annelids. The organization is more complex in the more advanced animals (Fig. 19). In some fish embryos and larvae, circulation develops before the respiratory organs. Blood pressures as high as 2.5 kPa and fairly fast flow velocities have been recorded in some large earthworms (Johansen and Martin 1966). Ventricular pressure measurements in the black-lip abalone, *Haliotis ruber* (Russell and Evans 1989) ranges from 0.4 to 1.2 kPa and heart rate increases with water temperature, the maximum rate being reached at 22 °C. In teleosts and elasbobranchs, typical heart rates are 10 to 60 beats min^{-1}, cardiac outputs are 6 to 40 ml min^{-1}kg^{-1} body mass and the mean ventral aortic pressure ranges between 3 and 6 kPa (Farrell 1984, 1991b; Lai et al. 1990; Axelsson et al. 1992).

The gastropod (Jones 1983; Andrews and Taylor 1988) and cephalopod mollusks (Wells 1983) were the first taxons to acquire a distinct circulatory system. The heart is well organized internally for directional blood propulsion and has regular beats. In the pneumonate gastropod mollusks, the systolic pressure is as high as 4 kPa (Jones 1983; McMahon and Wilkens 1983). In some terrestrial crabs, the systolic peak blood pressure can be as high as 6.7 kPa (Cameron and Mecklenburg 1973), gradually dissipating to zero after the gills which contribute 40% of the peripheral resistance (Bourne and Redmond 1977). The closed circulation in the cephalopods constitutes the threshold towards circulatory adaptation for more efficient gas transfer. With certain exceptions, the most common response to hypoxia in gastropods and bivalves (DeFur and Mangum 1979; Russell and Evans 1989) is bradycardia. In fish, a similar response occurs (Randall and Shelton 1963; Farrell 1982a; Gehrke and Fielder 1988; Fritsche 1990): the heart rate may drop by as much as 50%. Interestingly, hypoxic bradycardia is weak or does not occur in fish such as the sea raven, *Hemitripterus americanus* (Saunders and Sutterlin 1971), winter flounder, *Pseudopleuronectes americanus* (Cech et al. 1977), and the rockpool fish, *Gobius cobitis* (Berschick et al. 1987). While these differences may be genuine interspecific adaptive responses to hypoxia, it cannot be ruled out that the experimental approaches and the depth of hypoxia at which the tests are being carried out may contribute. Compared with other teleosts, e.g., the eels and the goldfish, which can withstand hypoxia for hours at elevated temperatures (e.g., Walker and Johansen 1977; Waarde et al. 1983), the salmonids have a particularly limited capacity of coping with hypoxia (Doudoroff and Shumway 1970). In aquatic amphibians where cutaneous respiration is particularly important (Johansen and Burggren 1980), a marked bradycardia, accompanied by decreased blood pressure and cardiac output, occurs with submergence (Shelton and Jones 1965). Cardiac output may be reduced to 5% of the predive levels and heart rate may drop to as low as 4 to 8 beats min^{-1} in turtles (White and Ross 1966; Penney 1977; Herbert and Jackson 1985). The mass specific volume of blood is greater in animals with an open circulatory system compared with those with closed ones (Prosser 1961). In the former, the peripheral resistance and blood pressures are generally low (Jones 1983). In crustaceans, blood volume constitutes about 30% of the body weight (Prosser 1973), in gastropod molluscs 25 to 30%, in the bivalves as much as 60%, and in the dog only 8.3%. The relative blood volume decreases with body size in mammals (Gregersen and Rawson 1959). In most insects (Sect. 6.6.1.), the circulatory system serves no consequen-

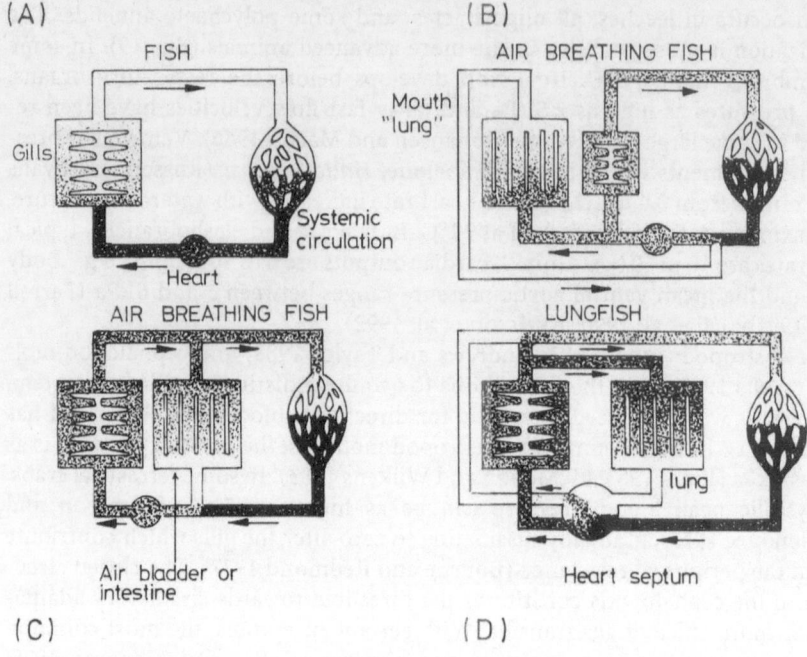

Fig. 21A–D. Arrangement of the vascular system relative to the accessory respiratory organs. In the single arch of a typical fish (**A**), the entire cardiac output is directed towards the gills. In various air-breathing fish, e.g., *Synbranchus* (**B**), *Hoplosternum* (**C**), and lungfish (**D**), to varying extents mixing occurs. The process is, however, minimal in the dipnoans (lungfish). (Johansen 1968)

tial role in gas exchange. In the lungless salamanders, respiration is entirely cutaneous. In the group, the left auricle is lacking. Invertebrates such as the crustaceans and mollusks have an open circulation where a capillary system between the arteries and veins is largely lacking and the blood returns to the heart more or less at random through a system of tissue spaces unbounded by endothelial cells (McMahon and Wilkens 1983; Burggren and McMahon 1988a).

The pinnacle of development of circulation, the double circulatory system took nearly 300 million years to configure from a single circulation. The lungfishes (Dipnoi), where a pulmonary vein and a partly divided heart are first encountered (Fig. 21), present a vital point in the evolution of the double circulation (Bugge 1960; Satchell 1976). All amphibians with a lung have a pulmonary vein and a complete or partial septum which separates the right and left atria. In air-breathing toads, at 22 °C, heart rate is 26 beats min^{-1} with an arterial pressure of 3 kPa and a systemic arch blood flow of 36 ml min^{-1} kg^{-1} body mass (Withers et al. 1988): the total cardiac output is about 30 ml per min per g heart mass (Driedzic and Gesser 1994). The crocodiles, a relatively remarkably advanced reptilian group (Densmore and Owen 1989; Norell 1989; Tarsitano et al. 1989), are the only ectothermic group which has virtually advanced to the stage of a four-chambered heart. In air-breathing turtles at 20 to 25 °C, the heart rate is 30 to 40 beats min^{-1}, cardiac output about $50\,\mathrm{ml\,min^{-1}\,g^{-1}}$ heart mass and the aortic pressure about $3\,\mathrm{kPa}$ (Driedzic and Gesser 1994). At $35\,°\mathrm{C}$, both the savanna monitors (*Varanus exanthematicus*) and the green iguana (*Iguana iguana*) have a resting heart rate of 40 beats $\mathrm{min^{-1}}$ with cardiac output of about $40\,\mathrm{ml\,min^{-1}}$ per g heart mass in *V. exanthematicus* and $70\,\mathrm{ml\,min^{-1}}$ per g heart mass in *I. iguana* (Gleeson et al. 1980). In a resting *V. exanthematicus*, the systemic blood pressure may approach $9\,\mathrm{kPa}$ (Burggren and Johansen 1982). It is only in the postembryonic endotherms, mammals and birds, where the heart is completely divided and the pulmonary and systemic circuits are anatomically distinct. The resting heart rate of a 500-g ectothermic vertebrate at 15 to $20\,°\mathrm{C}$ under normoxia is 20 to 60 beats $\mathrm{min^{-1}}$, the arterial blood pressure ranges from 3 to $5\,\mathrm{kPa}$ and the cardiac output is in the order of 10 to $50\,\mathrm{ml\,min^{-1}\,kg^{-1}}$ body mass (Driedzic and Gesser 1994): a mammal of similar size at $37\,°\mathrm{C}$ has a heart rate of 280 beats $\mathrm{min^{-1}}$, a left ventricular output of $125\,\mathrm{ml\,min^{-1}\,g^{-1}}$ and an aortic blood pressure of $13\,\mathrm{kPa}$. The pulmonary circuit, which handles the entire systemic venous blood, is a low resistance circulation where the pressures on average are $1.7\,\mathrm{kPa}$ (Rushner 1965; West 1974). In the human lung, the resistance to blood flow across the blood capillaries of the lung (diameter 10 to $14\,\mathrm{\mu m}$) which comprise a surface area of nearly $150\,\mathrm{m^2}$ (Gehr et al. 1978) is so low that 5 to $10\,\mathrm{l}$ of blood can flow through the lung each minute with a pressure of less than $1.3\,\mathrm{kPa}$ (Comroe 1974). In fish, gill vascular resistance is one half to one third that of the systemic circuit (Cameron et al. 1977), with the difference being much greater (about ten times) in mammals and birds (Langille and Jones 1975). In the amphibious ghost crab, *Ocypode saratan*, the gills are perfused both in submerged and air-breathing crabs. The lungs are preferentially perfused in air but not during submergence (Al-Wassia et al. 1989).

2.7.2.2 Ventilation

The need to reduce cutaneous water loss, especially with advent of terrestrial habitation, necessitated development of an impermeable skin. This rendered much of the body surface nonrespiratory. Specialized respiratory sites, where soft and well-vascularized parts could be exposed to the ambient respiratory medium, formed. For such areas to be effective, convective movements of air or water through mechanical effort were necessary in order to create and sustain a satisfactory partial pressure to maintain O_2 influx. Ventilation entails mass renewal of the environmental medium in the immediate proximity of a respiratory surface. Except for the freshwater limpets, *Ancylus fluvialis* and *Acroloxus lacustris* (Berg 1951), invertebrates which subsist in running water (i.e., are passively ventilated) show a higher metabolic rate than those from stagnant water (Fox et al. 1935; Walshe 1948). The convective systems which have evolved differ remarkably and reflect heavily on the restrictions imposed by the physical characteristics of the respiratory media on the design of the exchangers (Fig. 18). In simple aquatic animals, e.g., the bivalve mollusks and amphioxus (e.g., Baskin and Detmers 1976), and ascidians, the movement of water across the gills is effected by cilia. The much more sophisticated branchial pumps move the water across the gills in the advanced fish. Tubiculous polychaetes rely on ciliary currents (e.g., *Nephtys*), peristalsis (e.g., *Arenicola*) or undulating movements of the body (e.g., *Chaetopterus*). In mollusks, where the gills are largely used both for feeding and respiration, water is moved across the gills by the beating of cilia which are located on the gills (Borradaille et al. 1963). The echinoderms respire through movable tubes (podia) which extend from basal dilatations (ampullae), structures which protrude through openings in the calcaneous outer covering of the body (Hyman 1955; Steen 1965). The hemolymph, which contains no respiratory pigments, is moved through ciliary action into the microcirculatory units. In the sea urchin, *Strongylocentrotus droebachiensis*, the podial respiratory surface area becomes limiting to O_2 consumption only at higher temperatures (Steen 1965): at 19 °C, O_2 consumption ($2 \, ml \, h^{-1}$, 70 g body mass) is directly proportional to the available respiratory area, while at 6 °C, only 20% of the available surface is utilized to transfer O_2. In *S. droebachiensis*, a 70-g specimen has some 100 podia which are about 20 mm long and 0.4 mm in diameter. The overall respiratory surface area is about $250 \, cm^2$ and the thickness of the diffusional pathway is about 15 μm. The soft-bodied cucumbers (Holothuroidea) use an internal respiratory tree-like organ which they rhythmically ventilate through muscular contractions. *Holothuria tubulosa* renews the water of the respiratory tree every 1 to 4 min, with the expelled water having an O_2 content of about 50 to 80% of that of the surrounding water (Hazelhoff 1939). The marine annelid, *Chaetopterus variopedatus*, a burrow-dwelling filter feeder, has a high mass-specific ventilatory rate of the burrows of about $110 \, ml \, g^{-1} h^{-1}$ at 15 °C, a low O_2 extraction coefficient (the ratio of the amount of O_2 taken up to that available in the inspired medium) of about 30% and an O_2 consumption of $11 \, \mu mol \, g^{-1} h^{-1}$ in an actively ventilating organism of 4 g wet weight (Dales 1969). Fish embryos develop a rhythmic contraction of the tail muscles before the respiratory movements begin and show motor response to hypoxia (Polimanti 1912). To a slight extent, beating of cilia moves water currents into the molluskan mantle cavity. Some burrowing annelids, e.g., the marine echiuran worm, *Urechis caupo* (Wells 1949; Mangum 1985), generate a water current over their bodies and in the tubes by waving their bodies in water through peristaltic contractions of their muscular body wall. The mud shrimp, *Callianassa truncata*, a species which inhabits sand sediments in the Mediterranean Sea makes burrows which may be as much as 8 m deep (Ziebis et al. 1996): at a depth of 48 cm, the shrimp can maintain burrow O_2 concentration at 3 to 12% of air saturation by generating a water current of 10 m per second. Rhythmic movements of the external gills in the urodele, *Necturus*, renew the water on the surface of the gills. Increases in the PO_2 of the water in the burrows were recorded after a short period of irrigation (mean duration 21 s) produced by body undulations of the snake blenny, *Lumpenus lampretaeformis* (Atkinson et al. 1987): a flow rate of $40 \, ml \, min^{-1}$ in a burrow of a diameter of 2 cm was produced. Depending on the PO_2 in the water, the red band fish, *Cepola rubescens*, irrigates its burrow by body movements (Pullin et al. 1980) creating a water flow of $10 \, ml \, min^{-1}$ in a burrow of a diameter of 5 cm. In areas where the water flow is fast, the design of the burrows and the locations of the opening(s) relative to the direction of water flow may create pressure gradients which may passively suck water into the burrow (Vogel 1977). Passive ventilation is thought to occur in the burrows of the tile fish, *Lopholatilus chamaeleonticeps* (Grimes et al. 1986), and the mud-shrimp, *Callianassa truncata* (Ziebis et al. 1996). At a PO_2 of about 5 kPa, the lug worm, *Arenicola marina* stops ventilating the burrows (Toulmond 1991). Below the critical PO_2, it is clearly no longer cost-effective to expend energy ventilating the burrows. In the shelled *Nautilus* (a paleontological relic of more than 2000 extinct genera of nautiloids and ammonoids) where mantle movement is not possible, the ventilatory stream is generated by movements of fused collar and funnel folds, the "wings", which create small pressure gradients of the order of 0.1 kPa (Wells and Wells 1985). The ventilatory frequency, which is 35 times $\min^{-1}$ at 16 °C, increases with temperature and the stroke volume ranges from 5 to 22 ml for a 395-g animal. The volume of the mantle cavity in a fixed specimen of *Nautilus* (470 g) is 75 ml (Packard 1972). In *Octopus vulgaris*, the number of respiratory movements per minute decrease with increasing body mass (Polimanti 1913). Through a process called apneic oxygenation (Malan 1982; Szewczak and Jackson 1992), O_2 moves from the atmosphere down the trachea to the lung by diffusion or by bulk convection. If the respiratory quotient is greater than 1, in some hibernating animals with a low metabolic rate and long apneic periods, a significant amount of the resting O_2 needs can be met by diffusion down the respiratory tract through an open glottis during nonrespiratory periods. Ventilation by oscillatory movements of special appendages occurs in some polychaetes (e.g., *Chaetopterus*), amphipods (e.g., *Gammarus*), isopods (*Idotea*), and crustaceans, e.g., *Cancer pagurus* (Bradford and Taylor 1982). Such currents may deliver food, e.g., in *Chirocephalus*, *Artemia*, and *Daphnia*. In animals which live in torrential, air-saturated waters (rheophilic species) where the PO_2 of the water next to the gas exchanger is equal or almost equal to the atmospheric one, e.g., the hill-stream fish like *Danio dangila* which subsists in hyperoxic water with a concentration of O_2 of 9 mg $O_2 l^{-1}$ (Ojha and Singh 1986), the gills and skin are passively ventilated with water in an energy-saving process. In the oegopsid cranchid squids, which are known to store ammonia in the enlarged coelomes to regulate buoyancy (Denton et al. 1958), the mantle contractions do not participate in respiration. The flow of water over the gills is effected by movements of the coelom (Clarke 1962). The urodele salamanders (Plethodontidae), which lack lungs or gills, rely entirely on cutaneous respiration. Movements which stir the external respiratory medium and subcutaneous perfusion act as the only aids to the diffusive conductance across the skin (e.g., Gatz et al. 1974; Piiper et al. 1976).

The convection requirements are high in water breathers compared with air breathers. The requirements correlate inversely with the concentration of the molecular CO_2 of the medium. Due to the relatively low concentration of CO_2 in water, aquatic animals exhibit ventilatory rates 10 to 30 times those of the air breathers (White 1978). Increased ventilation in response to declining environmental O_2 has been observed in marine and freshwater bivalves (Zinkler 1966; McMahon 1988), crustaceans (Hughes et al. 1969a; Taylor 1982), polychaetes and oligochaetes (Mangum 1963), and holothurians (Newell and Courtney 1965). In cephalopods, hypoxia results in reduction of respiratory movements (Fredericq 1878) while activity (Ghiretti 1966) and hypercapnia (Winterstein 1925) elevate it. In some crustaceans, e.g., the crayfish, *Astacus leptodactylus* (Angersbach and Dekker 1978), and the crabs, e.g., *Cancer productus* (McMahon and Wilkens 1977)

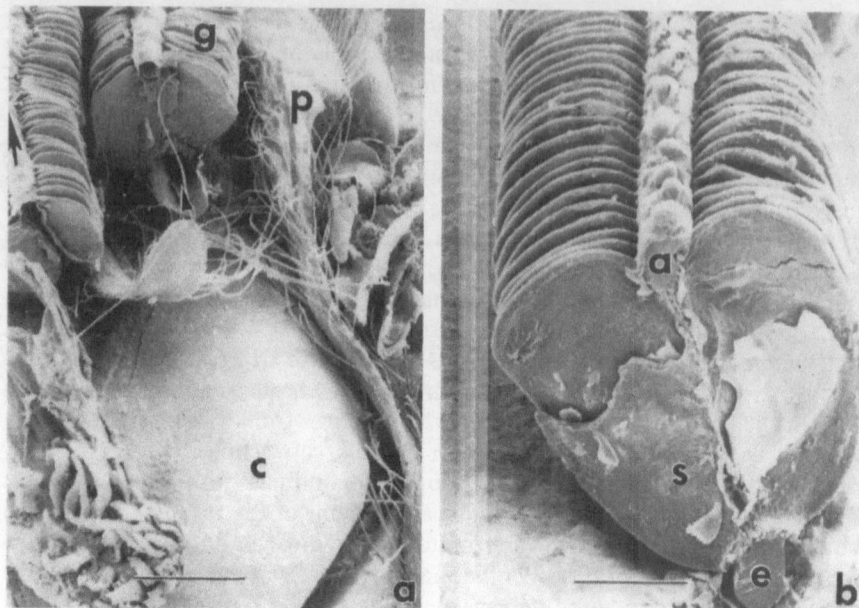

Fig. 22. (a) Gills of the freshwater crab, *Potamon niloticus* (*g*), showing a scaphognathite, *p*, the afferent gill artery, →, and the efferent artery, ➤; *c* skeletal mass. (b) A closeup of a gill arch showing gill lamellae, *s*, and afferent, *a*, and efferent, *e*, blood vessels. (a) *Bar* 500 μm; (b) 200 μm. (Maina 1990b)

and *Cancer pagurus* (Bradford and Taylor 1982), special mouth part appendages (the scaphognathites – flattened exopodities of the maxillae) ventilate the gills which are covered by lateral extensions of the carapace lined by a membrane, the branchiostegite (Lockwood 1968; Burggren et al. 1974; Fig. 22). The flow of the water through the branchial chamber is adequately effective to maintain a PO$_2$ of 15 to 20 kPa at the respiratory surface of the gills in normoxia (Butler et al. 1978; Wheatly and Taylor 1981). The PO$_2$ in the arterialized haemolymph is as high as 10 to 13 kPa (Butler et al. 1978; McMahon and Wilkens 1983). The direction of air flow can be reversed in some species of crabs, e.g., *Carcinus maenas* and *C. guanhumi* (Taylor and Butler 1978; Burggren et al. 1985a). It has been established that in some species of crabs (e.g., Yonge 1947; Hughes et al. 1969a), and the larvae of air-breathing fish (Liem 1981) the interaction between water flow and that of blood (in the gill lamellae) is countercurrent. In the species which burrow into the substratum, the direction of the ventilatory current is reversed when the animal is buried in the sediment (Arudpragasm and Naylor 1964a,b; Dyer and Uglow 1978). Except in *Holthuisana transversa* (Greenaway and Taylor 1976; Taylor and Whitley 1979), where tidal ventilation is achieved by movements of the membranous thoracic wall, the scaphognathites are effective in ventilation of the gills and the lungs of the land crabs while in air (Taylor and Butler 1978; Burggren et al. 1985a; Al-Wassia et al. 1989) especially during hypercapnia. Depending on ambient temperature (Taylor and Wheatly 1979), the shore crab,

100

Carcinus maenas, occasionally partially emerges from water to bubble air through the branchial chamber, aerating the water it holds in it (Taylor and Butler 1973; Taylor et al. 1973; Taylor and Whitley 1979). The movements of the scaphognathites of *Coenobita clypeatus* in air create pressure wave forms corresponding to those generated by the scaphognathites of water breathers (McMahon and Burggren 1979). The ventilatory mechanism in *H. transversa* (Taylor and Greenaway 1979) is unique among land crabs. Its efficiency (of which the operational definition is mechanical work output per metabolic work input) approximates to that of vertebrate respiration of about 10%. In *Ocypode saratans*, the scaphognathites beat at a rate of 53 times min^{-1} in a submerged crab, 218 times min^{-1} when active in air, 43 times min^{-1} when inactive in air, and 235 times min^{-1} when exposed to hypercapnia (Al-Wassia et al. 1989). The intermittent beating of the scaphognathites is neurogenically synchronized with that of the heart (Young 1978; Young and Coyer 1979). By changing the dimensions of the gill lamellar blood vessels, scaphognathite movements which generate pressures of -0.53 to $0.93\,kPa$ may play a significant role in the perfusion of the gills and the lung (Blatchford 1971). In the crab, *Carcinus maenas* (Taylor et al. 1973; Wheatly and Taylor 1979) and the crayfish, *Orconectes rusticus* (McMahon and Wilkes 1983), air is bubbled through water held in the branchial cavity by the scaphognathites beating in the reverse direction. In the amphibious ghost crab, *Ocypode saratan*, heart rate varies with ventilation (Al-Wassia et al. 1989): the lungs are not perfused while in water, but in air the lungs are perfused at four times the rate of the gills. The routine breathing frequencies in the juvenile and adult fish range from 30 to 70 times per minute (Roberts 1975) and ventilatory flows range from 100 to $300\,ml\,kg^{-1}\,min^{-1}$ (Wood et al. 1970; Johansen 1982). Functional coupling between respiration and locomotion occurs in some tetrapods. This may be a means of improving gas exchange efficiency or a saving on respiratory work. When a lizard runs, the left and right lungs are alternately compressed, pumping the air between the two lungs (Carrier 1987a,b). However, no significant movements of air from outside occur. In bats, during flight, synchronization between wing beat, respiratory rate, and heart rate has been reported (e.g., Thomas 1987) but in birds, this occurs in only a few species (Torre-Bueno 1985). In the horse and many mammals, breathing is closely coupled with locomotion (Bramble and Carrier 1983).

Gas exchangers have evolved in form of invaginations or evaginations from the body surface (Figs. 4,5). The former are generally categorized as gills and are largely used in aquatic respiration while the later are termed lungs and are used for aerial respiration. Gills are unidirectionally ventilated while tidal ventilation occurs in the lungs (Fig. 6,18). Compared with other vertebrates, fish exhibit diverse ventilatory mechanisms. These range from active and passive (ram) ventilation, continuous or intermittent unidirectional flow of water across the gills, to tidal process in the air-breathing organs of the bimodal breathers. In a four-phase serial pressure suctional buccal force pump, the gills are ventilated with a constant flow of water which is taken into the mouth and forced out through the opercular flap (Liem 1985; Brainerd 1994). Pressure differences of about $0.4\,kPa$ fill the mouth and those of 0.7 to $1.3\,kPa$ move the water across the gills (Ogden 1945; Hughes and Shelton 1958). Fish which swim strongly and much of the time such as the mackerel, the tuna, and some sharks move with their mouths open. Such fish have lost the capacity for mechanical gill ventilation and rely on ram effect (passive ventilation of water powered by swimming) for ventilating their gills. Because of the atrophy of the brachiomeric muscles, active ventilatory rate is very low in the group. Such fish cannot maintain an optimal level of oxygenation of the blood if they are held in restricted enclosures where forward movement is prohibited. The bimodal breathing fish use a slightly altered buccal force pump (analogous to that utilized on the gills) to ventilate their accessory respiratory organs. Air is forced into the organ(s) and exhalation is thought to be a passive process, especially in those air-breathing fish which have a fixed air space (e.g., a suprabranchial chamber) or through pulmonary elastic tissue recoil (DeLaney and Fishman 1977; Farrell and Randall 1978; Liem 1980). In some air-breathing fish, e.g., *Anabas testudineus* and probably in *Clarias mossambicus* (Maina and Maloiy 1986), where the inhalant and exhalant apertures are respectively contained in the pharyngeal and opercular cavities, the buccal and opercular pumps effect a unidirectional air flow across the labyrinthine organs which are found in the suprabranchial chamber (Peters 1978; Liem 1980). The cephalopods and mollusks have evolved a throughflow ventilatory mechanism which falls between the bidirectional one of the air breathers and the unidirectional one of fish: the inspired and expired streams flow through different openings over the gills, with some mixing probably occurring in the mantle cavity (e.g., Packard 1972; Gosline et al. 1983; Bone et al. 1994). With an O_2 extraction factor of 33 to 72%, e.g., in the scallops (van Dam 1954), and a possibility that the arterial PO_2 may in some cases be higher than that in the effluent water in the octopus (Johansen and Lenfant 1966), the throughflow ventilatory mechanism is exceptionally efficient. In *Nautilus*, there are two pairs of gills instead of one and the funnel is the main contractile structure (Ghiretti 1966). In the holothurians, e.g., sea cucumber and the cephalopods, the cloaca pumps water tidally across the ramified diverticula of the hind gut which forms the respiratory tree. In the sea cucumber, *Holothuria forksali*, 60% of the total O_2 needs is taken from O_2-saturated water across the cloaca, the remainder passing over the skin (Newell and Courtney 1965): O_2 consumption decreases with arrest of cloacal respiration. About ten successive cloacal contractions, each of which transfers 1 ml of water into the animal, are followed by body contractions which inject it out. Below an ambient PO_2 of 12 kPa, the organisms respond by respiratory inhibition and relocation to better oxygenated sites (Lutz 1930). Tidal breathing of water across the hind gut also occurs in annelids, e.g., *Urechis caupo* (Pritchard and White 1981; Menon and Arp 1992a), and in insects, e.g., dragon-fly nymph *Aeshna* (Fig. 23), via the branchial openings in lamprey eels (Johansen 1971) and through the mouth in some amphibians and reptiles, e.g., the soft-shelled turtle, *Amyda*. Such rather simple organs, of which the primary functions are olfactory and/or gustatory, are collectively called water-lungs (Sect. 4.6). Contrary to what one would expect from morphological evidence, a countercurrent exchange may occur in the lamprey during slow inspiratory phase (Johansen 1971). At normoxia, hind gut ventilation with seawater in *U. caupo* is about 0.7 ml per g body mass and at a PO_2 of 4 kPa rises to about 2 ml of seawater g^{-1} body mass (Julian and Arp 1992; Menon and Arp 1992a). The mucosa of the hind gut contains collagenous and

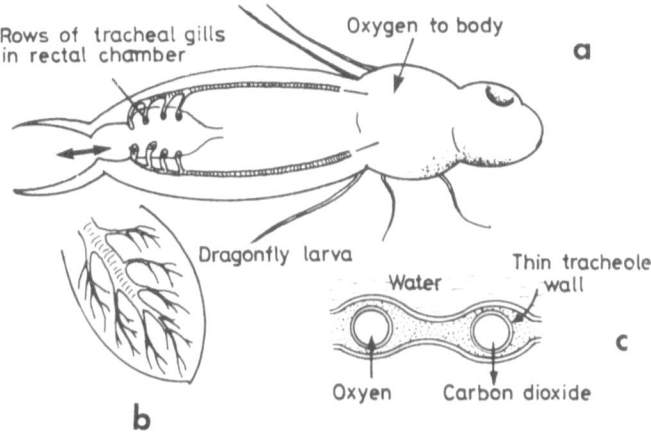

Rows of tracheal gills in rectal chamber

Oxygen to body

a

Dragonfly larva

Water

Thin tracheole wall

c

Oxyen   Carbon dioxide

b

Fig. 23. a Rectal tracheal gills of the dragonfly. **b** Enlargement of a gas exchange site and **c** the gas exchange mechanism between water and air in the trachea. The rectum is tidally ventilated, ⟷. (Hughes 1982)

Table 5. Distribution of the ventilatory mechanisms in vertebrates and the respiratory media in which they occur. (Randall et al. 1981)

Ventilatory process	Respiratory milieu	Vertebrate taxon	Environment
Buccal and opercular pumps	Water	Fish (gills)	Aquatic
with unidirectional flow of the medium	Air	Some air-breathing fish (anabantids)	Aquatic
Modified buccal force pump with tidal flow of medium	Air	Most air-breathing fish	Aquatic
		Dipnoi	
		Amphibians	
Aspiration pump with tidal flow of medium	Air	*Arapaima*	Aquatic
		Reptiles	Majority
		Birds	terrestrial,
		Mammals	but some aquatic forms

elastic fibers which may allow greater stretching during filling under hypoxic conditions (Menon and Arp 1992a). A similar rectal-gill mechanism exists in the echiuran worm, *Arhynchite pugettensis* where extensive cloacal diverticula occur (Manwell 1960). In the diving turtle, *Stemothaerus minor*, 30% of the O_2 need is met by rhythmic gular movements which maintain movement of water in and out of the bucco-pharyngeal cavity (Belkin 1968).

In aquatic animals, buccal pumping is the ancestral mode of ventilation of the gills (e.g., Brainerd et al. 1993; Brainerd 1994). It may initially have been used to ventilate the accessory respiratory organs on attainment of utilization of atmospheric O_2 (Randall et al. 1981). Aspirational (suctional) breathing characterizes air breathers while buccal pumping occurs in both water and air breathers (Table 5). Buccal and aspirational breathing coexist in the obligate air breather *Arapiama gigas*, where the gas bladder is ventilated by aspiration and the gills by a characteristic teleostean buccal force pump (Farrell and Randall 1978; Randall et al. 1978a). During buccal pumping in the frog *Rana pipiens*, work is done by the floor of the mouth (against the elastic forces of the lung) at an average efficiency of about 8%. The O_2 cost of breathing constitutes about 5% of the total energy budget (West and Jones 1975). In *Rana*, the cost of respiration is about two times that in the human being but falls within the same range of fish.

2.8 Blood and the Respiratory Pigments

The increase in the complexity and efficiency of the gas exchangers has been a gradual process which has developed in response to specific needs that have called for greater and more efficient means of O_2 uptake and delivery to the tissues to support higher aerobic capacities. As in practically all transformations in biology, improvements were made on former simpler designs (e.g., Schaeffer 1965a,b; Riggs 1976). With the evolution of a circulatory system followed by a ventilatory mechanism, O_2 was initially carried in solution in the body fluids. In the ensuing period, demands must have exceeded supply, rendering the process inadequate. Respiratory pigments have evolved widely in the Animal Kingdom (e.g., Antonini 1967). They reversibly bind, store, and transport O_2 (and to an extent CO_2), increasing the O_2 carrying capacity of blood (Bauer 1974; Jensen 1991). The pigments are found dissolved in the plasma or in a two-phase system where the carrier is packaged in corpuscles, especially in those animals which require efficient internal fluid transport of O_2 to the body tissues (Lamy et al. 1985; Burggren et al. 1991). The primitive type of blood or hemolymph which contains no respiratory pigments has essentially the same respiratory capacity as salt water and can carry only about 0.2 ml O_2 per 100 ml water while vertebrate blood can carry 5 to 45 ml O_2 per 100 ml blood. In fish, the physically dissolved O_2 usually constitutes less than 5% of the total O_2 carried in blood (Boutilier et al. 1984). The development of hemoglobin increased the O_2 carrying capacity of blood by about 100 times. In the crustaceans, the presence of hemocyanin increases the O_2 carrying capacity of the hemolymph above the dissolved levels by a factor of 2 to 4 (Taylor 1982; Mangum 1980, 1983a,b; Shiga 1994). The arterial PO_2 (PaO_2) in resting crustaceans is reported to be generally low (1 to 3 kPa at 13 to 15 °C) independent of the blood pigment concentration (Forgue et al. 1992a,b). This baseline value corresponds with that of the mussel *Anodonta cygnea* which lacks blood pigment (Massabuau et al. 1991). Compared with the high values of PaO_2 (range from 2 to 13 kPa) that have been reported by Shelton et al. (1986) and McMahon and Wilkens (1983) in the water breathers, hemocyanin gives crustaceans a large functional reserve for uptake and transfer of O_2. Theoretically, in the human being, if O_2 was carried in physical solution in blood instead of by the hemoglobin, the circulatory rate would have to be 30 times more to meet the metabolic demands. Interestingly, the O_2 content of blood is equal to that of air, i.e., 20 ml O_2 per 100 ml (Davenport 1974; West 1974). This suggests a possible optimization of chemical binding and transfer of O_2 by the blood.

The development and refinement of the respiratory carriers added a significant factor to the gas exchange capabilities in animals, increasing the adaptability of the respiratory system to different environments. This entailed impressive molecular creativity directed at meeting the special needs of organisms (Perutz 1970, 1990a; Manning et al. 1990). For example, in the dimeric hemoglobin of the blood clam, *Sapharca inaequivalvis*, the hemes and the heme-linked helices E and F of adjacent subunits are in contact instead of facing outwards as is the case in the vertebrate hemoglobins (Royer et al. 1985). Certain invertebrate species have hemoglobins in which each polypeptide relates with multiple heme binding domains. Polymeric globin is thought to have occurred through fusion of multiple monomeric globin transcriptional units at the gene level in organisms ancestral to the invertebrate lineage some 200 to 500 million years ago (Manning et al. 1990). The tetrameric hemoglobin of the fat inn-keeper worm, *Urechis caupo*, shows neither cooperativity of O_2 binding nor a Bohr effect, while the hemoglobin of the brine shrimp, *Artemia*, the blood clam, *Sapharca*, and the earthworm, *Lumbricus*, displays both properties (see Perutz 1990a,b). Every vertebrate hemoglobin is an oligometric protein comprising four polypeptide subunits (protomers), each of which possesses an O_2 binding protoheme (Bauer 1974). While the fish hemoglobins are $\alpha_2\beta_2$ tetramers with a tertiary and quaternary structure similar to that of other vertebrate hemoglobins, among the teleosts, the amino acid sequences differ greatly between different groups (Kleinschmidt and Sgouros 1987). Many of the differences are functionally neutral (Perutz 1983).

Physiologically, the most important properties of hemoglobin are the cooperativity of O_2 binding and the effect of H^+, CO_2, and organic phosphate components on the affinity of hemoglobin for O_2. Cooperativity arises from the change of hemoglobin from low – to high-affinity forms with the binding of O_2 (e.g., Hewitt et al. 1972; Ten Eyck 1972; Perutz 1979). The limit to O_2 storage capacity in blood is set by the product of the respiratory pigment concentration and the blood volume (e.g., Davenport 1974). In general, diving animals have a larger blood volume than the nondivers (e.g., Bond and Gilbert 1958; Butler 1991a). They have a greater O_2 storage capacity of blood (Ridgway and Johnston 1966; Lenfant et al. 1969; Hedrick and Duffield 1986). During submersion, the concentration of the hemoglobin increases by a factor of 60 to 70% in the blood of the Weddell seal (Qvist et al. 1986). In the winter months, when the muskrat has to dive under water to look for food, its hemoglobin concentration in blood is at the high of 20 g per 100 ml of blood, while in summer, when the animal stops diving, the value drops to 14 g per 100 ml (Aleksiuk and Frohlinger 1971). In the aquatic pneumonate gastropod, *Planorbis corneus*, and several species of *Daphnia*, a low PO_2 initiates almost instantaneous synthesis and increase in the concentration of the hemoglobin (Fox 1955), a feature which has great survival value. Hemoglobin plays an important role in O_2 transport especially in hypoxic habitats (e.g., Johnson 1942; Cosgrove and Schwartz 1965). The lugworm, *Arenicola marina*, which has adapted to the anoxic intertidal sediment mainly because of its efficient anaerobic metabolic pathways, has a high O_2 affinity hemoglobin

(Toulmond 1985) and two kinds of body wall myoglobins of very high O_2 affinity (P_{50}, 0.1 and 0.2 kPa, at 20 °C) (Weber and Pauptit 1972). Presence of a hemoglobin with very high O_2 affinity (P_{50} about 0.13 kPa) was reported in specialized cells of gastrotrichs by Colacino and Kraus (1984). The alvinellids, polychaete annelids which live in the deep-sea hypoxic hydrothermal vents, have complex morphometrically well-adapted gills (Jouin and Toulmond 1989; Toulmond 1991). *Alvinella pompejana* has a closed vascular system (Toulmond 1991) and a high molecular mass extracellular hemoglobin (Terwilliger and Terwilliger 1984) with a high O_2 affinity (P_{50}: 0.02 to 0.3 kPa – measured at 20 °C, 1 atm pressure and pH 7.6 to 6.6; Toulmond 1991). Complex molecular mechanisms which included the effect of pH, CO_2, organic phosphates, and temperature developed to regulate O_2 uptake and transport by the hemoglobin (Nikinmaa 1990). Of the total proteins in the vertebrate erythrocytes, hemoglobin constitutes 95% (Antonini 1967).

The evolution of carrier pigments in blood (Fig. 24) constituted a significant improvement on the respiratory function and is probably one of the most recent innovations in the complex gas exchange adaptive strategies. The presence of blood pigments enhanced the rate of diffusion of O_2 even at low PO_2s (e.g., Hemmingsen 1963; Moll 1966). In the mammalian skeletal muscle (Wittenberg and Wittenberg 1989) and the heart muscle (Braulin et al. 1986), myoglobin facilitates diffusion of O_2 and in the fish cardiac muscle, it enables the hearts of some fish to extract O_2 at lower levels of ambient PO_2 than would otherwise be possible (Driedzic and Gesser 1994). Polar icefish (e.g., Douglas et al. 1985) and interestingly even nonpolar fish such as lumpfish (*Cyclopterus lumpus*), monkfish (*Lophius piscatorius*), and ocean pout (*Macrozoarces americanus*) have insignificant myoglobin content in their tissues (Driedzic and Stewart 1982; Sidell et al. 1987). On the other hand, some species of fish such as tuna (*Thunnus thynnus*), mackerel (*S. scombrus*), and the carp (*Cyprinus carpio*) have remarkably high levels of myoglobin in their hearts, respectively 580, 332, and 488 nmol per g wet heart mass (Giovane et al. 1980; Sidell et al. 1987). High myocardial myoblobin characterizes fish that have high swimming capacities and those tolerant to hypoxia (Giovane et al. 1980; Driedzic 1988; Driedzic and Gesser 1994). Presence of myoglobin enables the heart muscle to function at a lower level of extracellular PO_2 than would otherwise be possible (Braulin et al. 1986; Taylor et al. 1986).

Bohr and Root effects of the hemoglobin (decrease in O_2 affinity of blood) boost diffusion by increasing the blood-to-tissue O_2 gradient at the sites of CO_2 production. Blood hemoglobin concentration is increased in hypoxic mammals (Lenfant 1973). In fishes, both hypoxia and exercise highly elevate the blood hemoglobin level (Kiceniuk and Jones 1977; Weber and Jensen 1988). A high hemoglobin concentration in blood achieves a high blood O_2 capacitance coefficient, decreasing the pumping requirement of the heart (Jensen 1991). In the diving animals, myoglobin enhances the amount of O_2 which can be stored in muscle tissue. The concentrations increase with the increasing diving ability (Butler 1991a). In the physoclistic swim bladders, secretion of lactic acid from glucose metabolism at the gas-gland of the rete mirabile (Pelster and Scheid 1991, 1993) forces O_2 out of the hemoglobin (and N_2 out of solution) compressing them into the bladder (e.g., Pelster and Scheid 1992a,b). Through still unknown mechanisms, the Root effect (pH-dependent release of O_2 from the hemoglobin) is

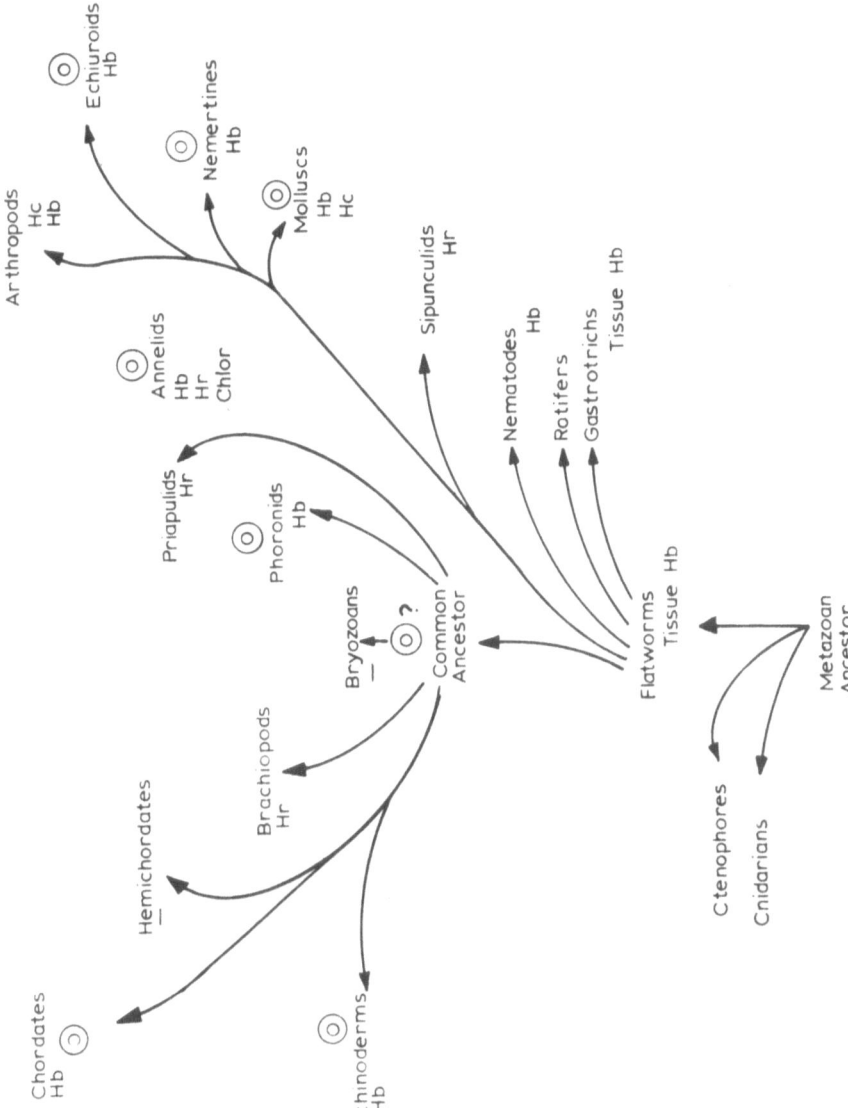

Fig. 24. Phylogenetic distribution of the blood respiratory pigments. The pigments found in a group are listed next to the particular names. A *circle logo* is appended to the animal groups with erythrocytes. *Hb* Hemoglobin; *Chlor* chlorocruorin; *Hr* hemerythrin; *Hc* hemocyanin. *Question mark* indicates a debatable point as to whether the common ancestor had erythrocytes. The hemichordates and bryozoans, –, have erythrocytes and the carrier (*Hb*) is intracellular. (Cameron 1989)

107

Table 6. Distribution of oxygen-carrying pigments in different groups of animals. (Cameron 1989)

Pigment	Environment	Occurrence	Comments
Hemoglobin	Intracellular (RBCs)	Nemertines, annelids, mollusks (bivalves), Phoronida, Echiurida, Echinoderms, Hemichordates, Chordates	Small: monomers to octamers
	Extracellular	Annelids, mollusks, arthropods	Highly variable
Chlorocruorin	Extracellular	Annelids	Formyl substitution on protoporphyrin
Hemerythrin	Intracellular (RBCs)	Brachiopods, annelids, sipunculids, priapulids	Nonporphyrin iron
Hemocyanin	Extracellular	Mollusks, arthropods	Probably of different origin in the two taxa

controlled by proton-induced allosteric conformational changes in fish hemoglobins (Howlett 1966). Hemoglobin may have evolved from the ubiquitous cytochrome molecule, a complex multicatalyst (Bernhardt 1995), with which it shares a common porphyrin nucleus and which has been implicated with facilitated diffusion of O_2 in some tissues (e.g., Scholander 1960; Burns and Gurtner 1973; Longmuir 1976). The incorporation of hemoglobin in the convective transport of blood in the vertebrates (Fig. 20) fully exploited the O_2 delivery role of the molecule. The large interspecific differences in the blood O_2 capacity, hemoglobin affinity, the extent of Bohr shift and erythrocyte morphometry and morphology (e.g., Riggs 1979; Dejours 1988), features which do not strictly correlate with phylogeny, environment, and mode of respiration (Fig. 24, Table 6) are strongly indicative of an ongoing intricate optimizing process. The carriers as well as other molecular factors are probably still being refined and integrated into the cardiovascular and respiratory systems for O_2 uptake and transfer. For example, as a means of ascertaining efficient O_2 delivery to the tissues by regulating peripheral blood pressure, while the erythrocytes are being oxygenated in the lung, the hemoglobin is S-nitrosylated to form S-nitrosohemoglobin (Jia et al. 1996; Perutz 1996). Subsequently, nitric oxide (NO), a recently recognized potent endothelium-derived relaxing factor (e.g., Palmer et al. 1987; Koshland 1992; Nathan 1992) is released during arterial-venous transit, causing vasodilation. The heme-bound NO [Hb(FeII)NO] is detectable in the venous blood when animals are subjected to oxidative stress (Kagan et al. 1996). Nitric oxide has, moreover, been reported to increase glucose transport in skeletal muscle (Balon and Nadler 1997). By increasing blood flow, nitric oxide ameliorates the effect of carbon dioxide on the brain (Meilin et al. 1996). Since the reaction between hemoglobin and O_2 is exothermic, increase in temperature results in a corresponding decrease in O_2 affinity, an adaptive feature especially in the homeotherms where high temperatures prevail in exercising tissues (e.g., skeletal muscle). A decrease in O_2 affinity promotes O_2 unloading. The hemoglobin in the tuna, *Thunnus*, a fish with exceptionally well-developed myogenic endothermia (e.g., Carey and Teal 1966) and high body temperature differentials, is insensitive to heat (Johansen and Lenfant 1972). Chum salmon have multiple hemoglobins, one group with a low Bohr effect and low temperature sensitivity and the other with normal proton and temperature sensitivity (Hashimoto et al. 1960). These forms are adaptively mobilized depending on the ambient temperature fluctuations. In facultative air breathers such as *Hypostomus* species and *Pterygoplichthys*, O_2 binding properties depend on the mode of breathing (Weber et al. 1979). While breathing air, the blood O_2 affinity increases and the Bohr effect decreases mainly owing to a drop in the erythrocytes organic phosphate concentration. Reptiles, a group which typically experience sudden body temperature changes especially as they bask in and withdraw from the sun, have a somewhat temperature-insensitive hemoglobin (Sullivan and Riggs 1967; Wood and Moberly 1970).

The effect of body size on the magnitude of Bohr effect and hemoglobin-O_2 affinity has been highly debated. Conflicting results ranging from a direct relationship (e.g., Schmidt-Nielsen and Larimer 1958), an inverse one (e.g., Clausen and Ersland 1968), and a weak or no correlation (e.g., Hilpert et al. 1963; Lahiri 1975) have been reported. The divergence may be characteristic of a nonoptimized state. The amphibians, a group which is highly instructive in the study of evolution of respiratory processes, show dramatic ontogenetic transformations in hemoglobin function (e.g., Gahlenbeck and Bartels 1970; Wood 1971). Changes in O_2 availability accompany the metamorphosis of a water-breathing tadpole to the adult air breather (Wood 1971; Broyles 1981). A notable decrease in O_2 affinity, increase in O_2 carrying capacity of blood, and metabolic rate occur with changes from larval to adult stages. With some exceptions, e.g., in the tuna and the lugworm, *Arenicola*, the concentration of the pigments is higher in the air breathers than in water breathers (Toulmond 1975; Dejours 1988). The O_2 affinity of the hemoglobin of the aquatic breathers which live in well-oxygenated water, e.g., the mackerel and salmon, are similar to those of air-breathing forms in the same habitat. This indicates that availability of O_2 rather than the nature of the respiratory medium influences O_2 affinity. Air breathers generally have a larger blood O_2 capacity, lower hemoglobin-O_2 affinity, and a larger Bohr shift than water breathers, at least in closely related species (Johansen and Lenfant 1966; Johansen et al. 1978; Fig. 63). Based on a wide cross section of species from the two taxa, a discordant view has, however, been expressed by Powers et al. (1979). Environmental PCO_2 and pH appear to determine the blood O_2 capacity of particular fish. At the physiological blood pH of 7.8 to 8.4, the Bohr effect is totally nonexistent in the blood of *Tilapia grahami*, which lives in the alkaline (pH 9.5 to 10.5) Lake Magadi of Kenya (e.g., Lykkeboe and Johansen 1975). In comparison with terrestrial mammals, the hemoglobin in diving animals has a low O_2 affinity, a high O_2 carrying capacity, and a large Bohr effect (Andersen 1966; Lenfant et al. 1970a; Wood and Johansen 1974). In the tench, *Tinca tinca*, hypoxia-hypercapnia causes an increase in hematocrit (due to an increase in the erythrocyte volume) associated with reduced intracellular concentration of the hemoglobin (Jensen and Weber 1985).

Increased capillarization of the tissues, a process which minimized the transcapillary-tissue diffusional distance and development of tissue-based high affinity O_2-storing noncirculating hemoglobin factor, myoglobin (found in some tissues) (e.g., Manwell 1963; Kreuzer 1970), were important innovations which enhanced delivery of O_2 to the cells. In the chiton, *Chryptochiton stelleri*, myoglobin has a lower P_{50} (0.4 kPa) than circulating hemocyanin, 2.7 kPa (Manwell 1958). In the buccal muscle of the mollusk, *Aplysia deplians*, the hemoglobin content is 6 mg% (Rossi-Fanelli and Antonini 1957) and in the human muscle tissue, myoglobin comprises about 2.5%. In his bucket-bridge model, Scholander (1960) envisaged that O_2 passes from one hemoglobin molecule to another. In this way, the O_2 flux can be increased eight times. It is, however, interesting that intracellular hemoglobin does not exist in the tissue barriers of the lung and fish gills. The role of hemoglobin in O_2-facilitated transfer has been questioned by Hemmingsen (1965) and remains an unsettled issue. The early proposition that the O_2 affinity of bird blood is lower than that of mammals (e.g., Jones 1972; Prosser 1973; but see dissenting views, e.g., Scheid and Kawashiro 1975 and Baumann and Baumann 1977) has been attributed to methodological error due to the time lapse between the collection of blood and analysis (Lutz et al. 1973, 1974; Holle et al. 1977). This may arise from the high metabolic rate of the nucleated avian erythrocytes. Though the erythrocytes in 90% of the vertebrate species are nucleated (e.g., Nikinmaa 1990), annucleation is thought to be the more evolved feature. Nucleated erythrocytes generate energy aerobically and their transmembrane pathways are more diverse than those of the nonnucleated mammalian cells. Nonnucleated erythrocytes have been described in some fish (Hansen and Wingstrand 1960). Compared with annucleate erythrocytes, nucleated erythrocytes have lower deformability, greater orientation instability, and show a greater propensity to interact with each other during steady capillary blood flow (Gaehtgens 1990). In these respects, nonnucleated erythrocytes which encounter less resistance due to low viscosity provide a distinct advantage in O_2 transfer.

Paradoxically, in some organisms, the contribution of the respiratory pigments in gas exchange is highly questionable. Some choronomid larvae and insects lack an O_2 carrier. Pigments appear to be of significance for life only during circumstances of high O_2 demand. Goldfish behave normally in water equilibrated with 80% carbon monoxide and 20% O_2 at temperatures below 20 °C (Anthony 1961). In the abalone, *Haliotes corrugata*, the concentration of the hemocyanin in different specimens was reported to differ by a factor of 900 times (Pilson 1965). In water equilibrated with air containing enough carbon monoxide to make hemoglobin totally inefficient as an O_2 carrier, the eel, the carp, and the pike will live for hours (Nicloux 1923). Extended exposure to reduced PO_2 results in manufacture of respiratory pigments in some animals but not in others (Fox 1955). In *Artemia*, hemoglobin synthesis is stimulated by high external salinity (Gilchrist 1954). The respiratory properties of the copper-containing hemocyanin are highly labile (Mangum 1980a) with the blood of the cephalopods being very sensitive to pH changes (i.e., having a marked Bohr shift), temperature, ionic, and osmotic composition of blood (Houlihan et al. 1982). By increasing hemocyanin-O_2 affinity, *Octopus vulgaris* can cope with hypoxia even if 30 to 50% of the gill surface area is surgically removed (Wells and Wells 1984). In *Sepia officinalis*, the P_{50} is 0.4 kPa at a pH of 7.97 but is 9.3 kPa at a pH of 7.24 (Wolvekamp et al. (1942). Although a notable degree of evolutionary refinement of the hemocyanin-O_2 transport has occurred in the cephalopods (Mangum 1980b, 1990), O_2 transport in blood is the major limiting factor for power output in the taxon (Brix et al. 1989; O'Dor and Webber 1991). The O_2 carrying capacity of the blood in *Loligo pealei* is only one-half that of the hemoglobin-based bloods of vertebrates (Mangum 1990). The O_2 affinity of the hemocyanins in the crabs, *Cancer magister* (Terwilliger and Brown 1993) and *Cancer productus* (Wache et al. 1988), and the lobster, *Momarus americanus* (Olson et al. 1988, 1990b), changes with the stage of development. Hemolymph inorganic ions particularly divalent cations change with development (Brown and Terwilliger 1992) and influence the O_2 affinity and cooperativity of decapod hemocyanins (van Holde and Miller 1982).

The evolutionary biology of the respiratory pigments is an unsettled subject (e.g., Wells 1990; Nikinmaa 1990; Mangum 1992; Weber 1992). Their inter-taxonomic distribution and functional diversity are intriguing (Jensen 1991). The best-known metalloprotein O_2 carriers are hemoglobin, hemocyanins, chlorocruorins, and hemerythrins (Fig. 24, Table 6). With some exceptions, chlorocruorins are found in some polychaetes, hemocyanins predominate in the mollusks but not bivalves (Ghiretti 1966; Mangum 1980b) and crustaceans (McMahon 1985; Lallier and Truchot 1989) while hemerythrins are restricted in distribution, occurring in unrelated taxa such as in the polchaete, *Magelona*, most sipunculids, some brachiopods, and some priapulids. Some mollusks have hemo-globin while some have both hemoglobin and hemocyanin. Polychaetes have either chlorocruorin or hemoglobin while yet others, e.g., *Serpula*, have both types of pigments (Jones 1972). Among the gastropods, in contrast to a large number of Prosobranchia and Pulmonata, Ophisthobranchia do not seem to have a respiratory pigment (Ghiretti 1966). Lack or presence of a particular pigment does not appear to correlate with either the type of respiratory medium a species utilizes or its phylogenetic level of development. It is conjectured that the evolu-tion of the respiratory pigments is polyphyletic (Mangum 1985). Depending on species, hemoglobin is found in blood corpuscles or dissolved in plasma, chlorocruorin and hemocyanins are found only in solution, and hemerythrin is found in cells (Ghiretti 1966; Mangum 1980a; Cameron 1989). The occurrence of pigment carriers in cells, an ubiquitous feature in the higher vertebrates, has been interpreted to be an evolutionary innovation enabling more efficient O_2 uptake and transfer to support the higher aerobic states. The intracellular location of the carriers is thought to maintain an optimal colloidal osmotic pressure and viscos-ity of plasma for an equivalent amount of protein (e.g., Snyder 1977), provide a more efficient intracellular control of allosteric modulators of the pigment's O_2 affinity (Gillen and Riggs 1973), and avoid loss of the small hemoglobin molecule through renal filtration. Mangum (1985) asserted that the erythrocyte evolved several millions of years before it became associated with the roles used to explain its derivation. Some animal groups have even lost the erythrocyte in preference for dissolved extracellular pigments (Cameron 1989). As argued elsewhere in this account, from their remarkable plasticity (a feature characteristic of least conserved parameters), the carrier pigments and especially the erythrocytes appear to be the terminal parts of the respiratory evolutionary chain. In the trial-and-error process of refining O_2 uptake and transfer, the erythrocytes may have been suddenly mobilized to contribute to the respiratory process. This apparently was not without certain costs. Schmidt-Nielsen and Taylor (1968; but see contrary conclusions by Snyder 1973 and Gaehtgens 1990) observed that at equal O_2 carrying capacity, the viscosity of the red cell suspension is higher than that of the hemoglobin solutions. The diffusing capacity of the rabbit lung for O_2 is higher when the lung is perfused with a hemoglobin solution compared with when perfused with a washed human red cell suspension (Geiser and Betticher 1989). This is attributable to an extraerythrocytic diffusion resistance which arises from an unstirred plasma boundary layer (Coin and Olson 1979; Huxley and Kutchai 1983; Vandegriff and Olsen 1984; Yamaguchi et al. 1985). The resistance depends on the hydrodynamic conditions of the erythrocyte flow (Rice 1980) and the physical resistance offered by the lipid cell membrane and the hemoglobin (Fischkoff and Vanderkooi 1975; Kon et al. 1980). The stagnant plasma boundary layer of blood is as thick as $4.2\,\mu m$ (Weingarden et al. 1982). In a trade-off process, in animals with particulate blood, i.e., where the hemoglobin is contained in cells, gas exchange efficiency may have been sacrificed for proper osmoregulation of the blood. Optimal O_2 uptake and transport in blood is established through compromises between factors such as hematocrit, hemoglobin concentration, body temperature, metabolic state, and erythrocyte morphology (Guard and Murrish 1975; Weathers 1976; Shepherd and Riedel 1982; Hedrick et al. 1986). The carriers which are found in cells have low molecular weights whereas the extracellular ones have high ones. The O_2 transporting function of respiratory pigments is dependent on interactions with organic and inorganic cofactors (e.g., Mangum and Lykkeboe 1979). The O_2 affinity of the vertebrate hemoglobins is decreased by the erythrocytes' organic phosphates (e.g., Weber and Jensen 1988). In the crustacean hemocyanin, L-lactate (Truchot 1980; Bridges and Morris 1986) and urate (Morris and Bridges 1986) increase O_2 affinity. In a self-regulating process, L-acetate and urate, products of anaerobiosis in decapod crustaceans (e.g., Bouchet and Truchot 1985; Czietrich et al. 1987), sustain O_2 transport in conditions of hypoxia or intense activity. Through resisting Bohr shift with pH change (Truchot 1987), the arterial blood remains highly oxygenated (Mangum 1983; Lallier and Truchot 1989): over some range of pH, especially in the marine gastropods, a similar phenomenon, i.e., a negative or reversed Bohr effect, occurs. During hypoxia, the Bohr effect improves O_2 loading and transport since hypoxia induces acidosis and hemocyanin-O_2 affinity increases with decrease in pH (e.g., Brix 1982).

When integrated with the circulatory system, the presence of blood pigments facilitates greater uptake, transport, and delivery of O_2 between the gas exchanger and the tissue cells, supporting high metabolic rates. A complex chain of respiratory adaptations was observed by Green et al. (1973) in the volcanic crater Lake Borambi Mbo in the Cameroon (West Africa), where one of the endemic species of cichlid fish, *Konia dikume*, which feeds on the larvae of *Chaoborus*, migrates to the anoxic layer to feed. A high mean concentration of the hemoglobin of 16 g per 100 ml was far above the range of the values of 5.4 to 8.7 g per 100 ml in the other ten sympatric cichlids. Fish from high latitudes generally have low concentrations of hemoglobin compared with tropical ones (e.g., Everson and Ralph 1968). Many fish possess multiple hemoglobins which in some cases (e.g., in the carp) are functionally similar (Tan et al. 1972; Weber and Lykkeboe 1978). The more energetic species, however, show functional differences in their O_2 binding capacities, allowing O_2 transport under different circumstances (Powers 1972; Brunori 1975; Weber et al. 1976). The hearts of some species of fish, e.g., the tuna, *Thunnus thynnus*, mackerel, *S. scombrus*, and carp, *Cyprinus carpio*, have high levels of myoglobin, respectively of about 580, 332, and 488 nmol per g tissue wet weight (Sidell et al. 1987). The maximum rates of O_2 consumption in active, free-swimming skipjack tuna (2.5 mg O_2 $g^{-1}h^{-1}$) is more than two times the values reported for other teleosts (Stevens and Carey 1981). Rainbow trout can survive and even swim up to 70% of their maximal capacity with the coronary vasculature ligated (Farrell 1993). The microaerophilic parasitic nematode, *Ascaris*, has a muscle myoglobin with high O_2 affinity and the perienteric fluid has an abundant hemoglobin that binds O_2 25 000 times more tightly than its mammalian homologue (Sherman et al. 1992).

2.8.1. Hemoglobinless Fish

Although hemoglobin is widely distributed in vertebrates, a few animals exist without it (Ruud 1954). The well-known group is that of the Antarctic icefish of the family Chaenichthyidae, e.g., *Chaenocephalus aceratus* and *Pseudochaenichthys georgianus* (Rudd 1965; Steen and Berg 1966; Jakubowski et al. 1969) which lack it. Furthermore, the fish generally lack myoglobin, except small amounts in the ventricles of the heart (e.g., Feller and Gerdy 1987; Sidell et al. 1997). The Perciform (suborder: Notothenioidei) which includes the icefish evolved around the Antarctica within the last 25 million years (Eastman 1993), a time during which the continent became separated on the opening of the Drake Passage and formation of circumpolar currents which produced rapid cooling of the Southern Ocean. In the icefish, O_2 is carried essentially physically dissolved in plasma: the O_2 capacity of blood in the icefish (0.7 vol%) is no higher than that of seawater (Ruud 1954) and is substantially lower (onetenth) that of about 8% by volume of the fish having hemoglobin (Holeton 1970). Icefish have evolved in well-oxygenated waters whose temperature fluctuates between +0.3 and −1.87 °C (Eastman 1991). To survive in such habitats, the fish have evolved manifold adaptive features. Special blood serum glycoproteins which lower their freezing temperature below that of the near or below zero sea temperature protect them from imminent death (DeVries 1971; Ahlgren et al. 1988). In *C. aceratus*, a relatively high arterial PO_2 of 16 kPa exists (Holeton 1970): since the water has a high concentration of O_2, less of it is passed over the gills in order to secure the required amount of O_2. This and the absence of erythrocytes, a feature which lowers the blood viscosity, reduces the cardiac work of pumping blood. The O_2 consumption of *C. aceratus* compares with that of other Antarctic fish (Ralph and Everson 1968; Holeton 1970).

Under similar conditions and circumstances, in absolute terms, the Antarctic icefish display higher rates of O_2 consumption than temperate ones by about two fold (Holeton 1974; Somero 1991). Delivery of O_2 is effected through immense cardiac outputs (per unit body mass) facilitated by particularly large hearts (0.3% body mass), and large blood volumes even though heart rate (14 times min^{-1}) is particularly low (Holeton 1970; Feller and Gerdy 1987; Harrison et al. 1991; Tota et al. 1991). In *C. aceratus*, at 1 °C, heart rate is about 16 beats min^{-1}, the mean ventral aortic pressure is about 2 kPa and the cardiac output is 20 to 40 ml min^{-1} per g heart mass (Holeton 1970; Hemmingsen et al. 1972). The large cardiac output may be a compensation for the relatively small gill surface area in the species (Hughes 1972a), a favorable feature which results in low resistance to branchial water flow. In order to enhance the flow of O_2 from the water to the tissues, the hemoglobinless Antarctic icefish have highly aerobic muscles (Johnston et al. 1983; Johnston and Harrison 1985; Harrison et al. 1991), remarkably profuse muscle blood capillary supply (Fitch et al. 1984), and the muscle mitochondrial volume density compares with that of the flight muscles of insects (e.g., Elder 1975; Ready 1983; Londraville and Sidell 1990). The activities of the mitochondrial enzymes that are essential for aerobic fatty acid catabolism are significantly greater in the Antarctic icefish compared with other ectotherms (Sidell et al. 1987). Although the percentage utilization of O_2 in water is relatively low, the efficiency of oxygenation of blood is very high in the icefish gills (Hughes 1972a) and the group survives very well in absence of hemoglobin (Andriashev 1962). The icefish, however, are not able to tolerate hypoxia as well as the red-blooded fish (Holeton 1970). *C. aceratus* succumbs when the PO_2 falls to below 6.7 kPa while the sympatric red-blooded species, e.g., *Notothenia neglecta*, and *N. gibberiforms* can extract O_2 down to a PO_2 of 2 kPa.

2.9 Energetic Cost and Efficiency of Respiration

In biology, O_2 is an essential resource which must be procured from outside at a cost. In bimodal breathers, the expense of procurement of O_2 from the surface can be gauged from the duration of stay at the surface, the distance traveled, and the intervals between surfacing. The proportion of O_2 acquired from aerial respiration decreases as the cost of traveling to the surface increases (e.g., Kramer 1988; Shannon and Kramer 1988). Bimodal breathers can regulate the cost of procuring O_2 to match the level of metabolism and the availability of O_2 from the environment (e.g., Anderson 1978). The respiratory patterns are determined by factors such as the efficiency of the water-breathing organ, the ambient temperature, and economical utilization of O_2 during submergence. Insects exhibit what has been termed cyclic CO_2 release (e.g., Buck 1962; Kanwisher 1966), a process which entails irregular discharge of CO_2. The frequency of spiracular opening corresponds with the energetic demands, with the bursts being more frequent during activity. Although no ventilatory movements accompany such events, as gas transfer takes place essentially by diffusion across the spiracles (at least in the small species), some degree of energy saving must be gained when the intervals of spiracular opening are widely spaced. At an ambient temperature of below $10\,^{\circ}C$, when O_2 diffusion across the closed or fluttering spiracles is adequate to support the low rate of metabolism, in *Cecropia*, bursts stop altogether (Kanwisher 1966).

Respiratory efficiency is a measure of the performance of a gas exchanger and should express the ratio of gas procurement and transfer of O_2 against that of the energy expended in the process. The amount of O_2 consumed during respiration is required in two main areas: (1) to physically move the respiratory medium over the respiratory site(s) and (2) to power the contractions of the muscles which drive the process. Energy is needed for both ventilation and perfusion of the gas exchanger(s), complex processes of which the absolute costs remain largely unknown (e.g., Scheid 1987). In fish, the cost of breathing water (per O_2 uptake) has been reported to range from a mere 0.5 to 10% during routine ventilation to 70% of the overall O_2 consumption during exercise (e.g., Schumann and Piiper 1966; Alexander 1967; Edwards 1971; Jones and Schwarzfeld 1974; Jones and Randall 1978; Steffensen and Lomholt 1983; Milsom 1989; Rantin et al. 1992). The very high values in some species may be due to technical and experimental problems such as strenuous experimental conditions entailing physical restraint and effect of anethesia (e.g., Cameron and Cech 1970). Compared with those animals which breath air, e.g., in the healthy human being where the value is 1 to 2% (Dejours 1975), the high cost of breathing water is attributable to the greater density (about 800 times that of air), viscosity of water (50 times air) (Table 4) and the low solubility of O_2 in the medium: more work is needed for pumping water. Under normal circumstances, fish faced with hypoxia move from it (Dandy 1970) or reduce their O_2 consumption to conform with the availability (Hughes 1981). Positional changes of the gill filaments during water breathing and strict coupling of the buccal and opercular pumps, factors which improve the efficiency of the hydrodynamic flow of water, reduce ventilatory work (Ballintijn 1972). Faced with hypoxia, as an energy-saving strategy, most fish (e.g., Smith and Jones 1982; Fernandes and Rantin 1989; Rantin et al. 1992) increase gill ventilation by augmenting tidal volume and not ventilatory rate. A mechanical ventilator working at three times per minute was used to ventilate the human lung with hyperoxygenated saline (Kylstra and Schoenfisch 1972). It is probably due to the enormous cost of reversing water flow that tidal ventilation in aquatic respiration is a rare process in the evolved animals. It occurs naturally only in a few primitive organisms. In experimental liquid breathing, water has to be mechanically moved through the air passages (e.g., Kylstra et al. 1966; Sect. 6.11). The mean efficiency of the buccal force pump as it inflates the anuran lung is 8%, the O_2 consumption per $100\,g$ of the respiratory muscle is $0.89\,mlO_2$ per min, and the O_2 cost of breathing at rest is about 5% of the total resting O_2 consumption (West and Jones 1975). The piston pump ventilatory mechanism of the lugworm, *Arenicola marina*, which occurs by means of peristaltic movements of its dorsal body wall musculature (Wells 1966) takes about 40% of the total O_2 uptake in animals ventilating normoxic or slightly hypoxic seawater (Toulmond 1975; Toulmond and Tchernigovtzeff 1984). At a hypoxia below $5.3\,kPa$, the worm is unable to pump water at an adequately high rate to overcome the reduced water O_2 content (Toulmond et al. 1984) and can withstand several days of experimental anoxia (Schöttler et al. 1984). During routine activity, in normoxic water, the gill convection requirement is 200 to 400 ml H_2O per ml O_2 (Johansen 1982). The cephalopod mollusk, *Nautilus*, which lives at depths of 100 to 300 m and regularly encounters areas of low O_2 concentrations, overcomes such episodic occurrences of hypoxia by remarkable suppression of aerobic metabolism to as low as 4 to 8% of that at the normoxic level (Boutilier et al. 1996), a value comparable to that of well-known facultative anaerobes (Hochachka and Guppy 1987). A hypometabolic-hypoxic animal conserves energy by decreasing the level of activity and by extending ventilatory and circulatory pauses. Water breathers have to irrigate their gills with a 40 times larger volume of water than the air breathers (with air) to extract an equivalent amount of O_2 (Block 1991a). Since water contains substantially less dissolved O_2 than air, to extract the same quantity of O_2, the ventilation-perfusion ratio in aquatic breathers is ten times or more that for air breathers. In mammals, the ratio is about 1. Deviations from unity result in dramatic impairment of O_2 diffusion (e.g., Escourrou et al. 1993). Ventilatory requirements for fish are four to eight times higher than those of terrestrial ectotherms with similar metabolic demands (Milsom 1989). In active species, ventilatory requirements can increase 10 to 15 times above resting while O_2 consumption rarely increases by a factor greater than 5 (Brett 1972). In juvenile to adult fish (body mass <100 g), ventilatory flow rates range from 100 to 300 ml per kg per minute (Johansen 1982). Indirect estimation of the cost of water breathing in fish as a factor of the drop in O_2 consumption when the fish changes from active to passive (ram) ventilation indicates that in the trout, O_2 consumption falls by more than 10% and in the sharksucker about 5% (Steffensen and Lomholt 1983). During emergence into air, the crayfish, *Austropotamobius pallipes*, ventilates its branchial chambers at very low rates (5 ml per kg per min; Taylor and Wheatly 1980).

To minimize the energy expended on ventilatory work, fish have adopted different strategies of optimizing gas exchange to fit particular circumstances as well as to relate the respiratory process to activities such as osmoregulation. In the flounder, *Platichthys fleus*, about 7 ml per kg per h of O_2 and in the rainbow trout, *Oncorhynchus mykiss*, $6 \, ml \, kg^{-1} h^{-1}$ of O_2 are utilized for osmoregulation, values which constitute 10 to 15% of their standard metabolism (Kirschner 1993). In absolutely resting fish, ventilation can be intermittent (Perry and McDonald 1993). Benthic fish such as the bullhead catfish, *Ictalurus nebulosus*, exhibit apneic periods which may last for as long as 1 min, whereas others show periods of very shallow breathing alternating with strong ones (Roberts and Rowell 1988). Ventilatory pauses may be as long as 30 min in some crabs (Taylor 1984). In the intermittent ventilatory pattern of the reptiles as well as during hibernation and sleep in endotherms, i.e., where the ventilatory periods alternate with breath-holding ones, the nonventilatory period may constitute an energy-saving strategy for minimizing ventilatory cost (e.g., Glass and Wood 1983; Milsom 1991; Wood and Glass 1991; West et al. 1992) and for reducing convective respiratory water loss (Innes et al. 1986). Pulmonary blood flow increases during the breathing phase (e.g., Shelton and Burggren 1976; Burggren 1977; Burggren et al. 1977).

The nonventilatory period in the crocodile, *Alligator sinensis*, may last from a few seconds to 20 to 30 min (Zhao-Xian et al. 1991): the pattern of "discontinuous" or "intermittent" ventilation has been reported in the air breathing brachyurans, e.g., *Cancer paguras* (Burnett and Bridges 1981), *Pseudothelphusa garhami garhami* (Innes et al. 1986), *Cardisoma* (Wood and Randall (1981) and *Holthuisana* (Greenaway et al. 1983), terrestrial insects (Schneiderman 1960), and air-exposed bivalves (Jokumsen and Fyhn 1982). In such animals, to optimize gas transfer during the ventilatory period, heart rate, cardiac output, and pulmonary perfusion are synchronized with ventilation (e.g., Shelton and Burggren 1976; White 1978; Zhao-Xiao et al. 1991). In a quiescent, unstressed state, the crab, *Cancer magister*, utilizes unilateral ventilation whereby only one scaphognathite beats (McDonald et al. 1977). During such a state, variable levels of oxygenation of the postbranchial blood are attained (McMahon 1985). In the active state, the combined ventilatory volume is 288 ml per kg per min and the PO_2 in the hemolymph is 10.8 kPa while in a quiescent one the values are 50% less and the O_2 consumption is reduced by 30% (McDonald et al. 1977). The displacement of the limb girdles into the rib cage and fusion of the skeletal parts into the protective armor prohibits turtles from utilizing abdominal or thoracic movements to ventilate the lungs. In the group, respiration entails rotation of the limb girdles and hence expiration and inspiration are active processes (Gans 1976). By estimating the O_2 consumption during artificial ventilation and normal ventilation, at 22 °C, the energetic cost of moving air in the lung of the turtle *Pseudemys floridana* was found to be 0.0047 ml O_2 ml^{-1} of gas ventilated (Kinney and White 1977).

In general, fish respond to hypoxia by increasing gill ventilation, reducing heart rate, and increasing the stroke volume. The cardiac output is held constant or is even elevated (Randall 1970). The two main costs of gill ventilation are energy loss due to the resistance offered by the gills to water flow and the loss of energy from cyclic acceleration and deceleration of water as it is pumped through the branchial chambers. Gill resistance to water flow occurs both during branchial and ram ventilation while loss of kinetic energy is abolished during ram ventilation. Frequently, ram ventilation occurs in the large pelagic predatory fishes, e.g., sharks, tuna, striped bass, and mackered which swim constantly (e.g., Roberts and Rowell 1988; Burggren and Bemis 1992). In very active fish, e.g., the mackered, the bucco-pharyngeal movements are not sufficient to provide the required volume of water for adequate O_2 to be extracted. The fish has to be continuously in motion to enhance ventilation. In the juvenile paddlefish, *Polyodon spathula*, as swimming speed increases, buccal ventilation becomes intermittent and continuous ram ventilation occurs above a speed of 0.6 to 0.8 body lengths s^{-1} (Burggren and Bemis 1992). In the torrential hillstreams, a fish need only physically place itself along a moving water current for the water to flow over the gills. In most fish, active buccal ventilation changes to ram ventilation at swimming speeds of 20 to 60 cm s^{-1}, with the actual transitional velocity depending on factors such as environmental PO_2 and temperature (e.g., Roberts and Rowell 1988). A transition speed between the two ventilatory processes is well marked in fishes such as bluefish (*Pomatomus saltatrix*), striped bass (*Morone saxatilis*), and rainbow trout (*Oncorhynchus mykiss*) (Freadman 1981; Steffensen and Lomholt 1983). During locomotory ram ventilation, the work of respiration is essentially transferred from the buccal and opercular muscles to the swimming muscles of the trunk (e.g., Burggren and Bemis 1992). In the bluefish and striped bass, the transition to ram ventilation is accompanied by as much as 50% increase in swimming speed without an increase in metabolic rate (Freadman 1981). Although constant swimming is energetically expensive, the energy conserved appears to justify adoption of ram ventilation. *Polyodon spathula* and fish such as anchovies and menhaden combine raw ventilation with filter feeding (James and Probyn 1989; Burggren and Bemis 1992), a process which results in energy saving on the cost of gill ventilation over the combined one which would be required for buccal pumping and foraging. Exercising fish increase their O_2 consumption five times and the ventilation of the gills about 15 times, raising the O_2 consumption of respiratory muscles to 15% of the total O_2 consumption (Hughes and Shelton 1958). The response to hypercapnia entails an increase in gill ventilation (Jansen and Randall 1975) but hyperoxia can alleviate or even abolish the ventilatory response (Randall and Jones 1973). In contrast to air breathers, which utilize a medium with high O_2 concentration, fish have lower heart rates than respiratory rates, a higher ventilatory requirement being necessary in a medium of low O_2 content. The ventilation-perfusion ratios in fish are greater than unity. In aquatic breathers, the ventilatory response is mainly directed to O_2 availability rather than to elimination of CO_2 (Randall and Cameron 1973). The bimodal breathers perceive and react to both hypoxia and hypercapnia in water and air to similar levels of sensitivity (Johansen 1970), with the degree of perfusion and ventilation depending on the actual role of a particular structure in CO_2 elimination or O_2 uptake. Some air-breathing fish, e.g., *Trichogaster*, however, seem to be incapable of distinguishing O_2 levels in the inspired air and water (Burggren 1979). In the continuous air breathers, where the lung is ventilated by a single medium (air) and the intrapulmonary CO_2 and O_2 levels are somewhat fixed, compared with the bimodal breathers, respiratory control is a rather simple and direct process. Peripheral as well as pulmonary afferent inputs are integrated in the brain to make the necessary cardiovascular and pulmonary adjustments in response to O_2 needs and CO_2 levels in the body. Fetal respiratory movements (e.g., Cooke and Berger 1990) are well known and their significance in lung developments has been suggested (e.g., Alcorn et al. 1980; Liggins et al. 1981). Fetal breathing is said to be energetically expensive, constituting as much as 30% of the overall fetal O_2 consumption (Rurak and Gruber 1983). This may explain the reduction in "respiratory" frequency towards term (Berger et al. 1986), the long periods of fetal apnea in utero, and the total apnea during labor when O_2 is conserved for utilization by the contracting uterine muscles of the mother.

In assessing and comparing the efficiencies of biological systems, it is important to estimate the extreme performances since biological systems have different inbuilt safety margins (e.g., Alexander 1981, 1996; Currey 1967; 1984; Wainright et al. 1976; Vogel 1988). In the case of gas exchangers, at the maximum O_2 consumption, all the reserve is exhausted as the functional and structural parameters are fully committed in procuring and delivering molecular O_2 to the tissue cells (e.g., Gehr et al. 1981; Weibel and Taylor 1986). The cost and efficiency of respiration will depend on the rate and nature of propulsion of the external respiratory medium to the respiratory site as well as the geometrical pattern of presentation of the internal and external gas exchange media. Using different means, the gas exchangers have been variably optimized. For example, the rectification of the air flow in the bird lung (whereby the highly efficacious unidirectional and continuous flow of air in the paleopulmonic parabronchi is generated) is passively achieved – there being no evident anatomical valves (Jones et al. 1981). The geometry of the secondary bronchi (especially the mediodorsal ones) relative to the primary bronchus and a recently described constriction of the primary bronchus (just before the origin of the first mediodorsal secondary bronchus) called segmentum accelerans (Banzett et al. 1991; Wang et al. 1992) are associated with the throughflow of air in the bird lung. Inspiratory aerodynamic valving has been shown to be dependent on factors such as gas density and the convective inertial forces generated by the air in motion (Banzett et al. 1987, 1991; Butler et al. 1988; Kuethe 1988; Wang et al. 1988).

It is now well recognized that in biological systems, geometric configuration and spatial disposition of the constitutive components can in some cases constitute limiting factors in function (e.g., West et al. 1986). Due to the great diversity in design of the gas exchangers, as well as the complexity of the respiratory process itself, the factors required to estimate the cost of respiratory work are presently uncertain and imprecise. Assumptions have to be made even on some fundamental aspects. Parameters such as O_2 extraction from the environment and the arterio-venous difference in O_2 content are used as approximate measures of

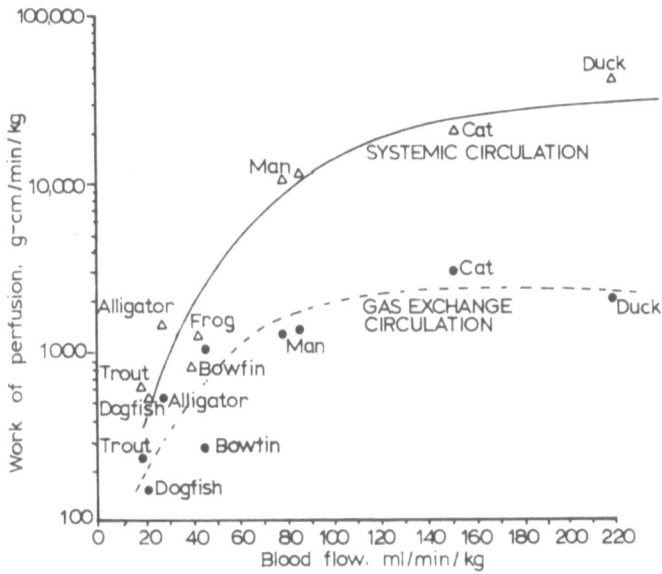

Fig. 25. Work of perfusing the systemic and gas exchange circulations in some vertebrate animals. The low work entailed in perfusing gas exchangers compared with the systemic circuit constitutes a substantial energy saving. (Johansen 1972)

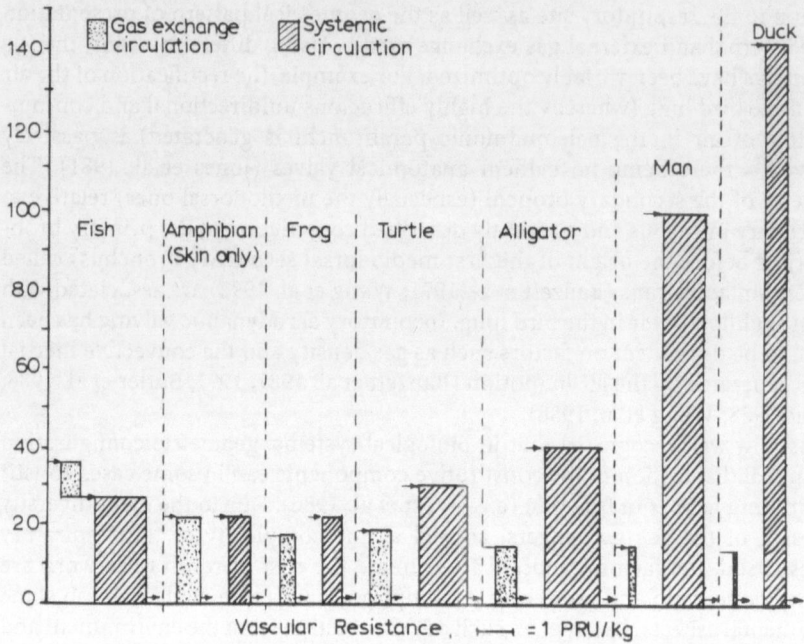

Fig. 26. Perfusion pressures and vascular resistance (VR) of the gas exchange and systemic circulations in some vertebrates. The *bar widths* show VR. Compared with water breathers, VR is very low in the circulatory system of the gas exchangers. This constitutes immense energy saving on respiratory work in the air breathers. *Arrows* show the scope of increased vascular resistance; *PRU* physiological resistance unit – a relative measure of resistance in the different animal groups. (Johansen 1972)

respiratory efficiency. In general, the greater specific gravity, viscosity, and lower O_2 content of water compared with air make the cost of aquatic respiration per unit volume of O_2 extracted greater than the aerial one (Johansen 1972). To optimize the gas transfer process, the vascular resistance of the gas exchangers is particularly low compared with the systemic one, especially in the air breathers (Figs. 25,26). In water, the ventilatory rate of aquatic crabs is three times that in air and the ventilatory cost is 30% that of O_2 consumption (Herreid and Full 1988). In the eel, a drop in the concentration of O_2 from saturation to $4\,\text{mll}^{-1}$ results in a 40% increase in O_2 uptake ($=30\%$ of the total metabolism) largely due to a five fold increase in ventilatory rate resulting from greater activity of the branchial muscles (Jones 1972). The energetic cost of aquatic ventilation in the crab, *Cancer*, may be as high as 76% (McMahon and Burggren 1979). The cost of breathing in the human being has been determined by increase in O_2 consumption due to isocapnic voluntary hyperventilation (e.g., Cournand et al. 1954) or by increased hyperventilation consequent to additional instrumental dead spaces, e.g., Milic-Emili (1991). At rest, the cost ranges from 1 to 2% of the total O_2 consumption but increases in hyperbaria because of increase in gas density and during exercise due to the work needed to overcome inertia and gas turbulence.

Deep sea divers need to substitute the lighter helium for nitrogen at depths in excess of 50 m in order to reduce respiratory work (Lanphier 1969). In fish, the respiratory cost, which ranges between 10 and 25%, is appreciably higher than in the human being (e.g., Hughes 1965; Dejours et al. 1970). The ratio of ventilatory requirement per unit O_2 consumption in the human being is ten times lower than in a goldfish (Dejours et al. 1970). In a hypocapnic turtle, *Pseudemys floridana*, the cost of breathing ranges from 10% (at 37 °C) to 40% (at 10 °C) of total O_2 consumption (Kinney and White 1977). The inverse relationship between the ventilatory cost and the body temperature (White 1978) indicates that respiration becomes more efficient with increasing metabolic capacity.

Air breathing has the salient advantage of energy saving on respiratory work and gives flexibility in the adjustment of respiratory rate without undue increase in energetic demands. The energy thus saved can be utilized for growth and development and to secure newer ecological opportunities. In air breathers, the cost of breathing is a fixed fraction of the total O_2 consumption and is in general about 5% or slightly less (Tenney 1979). The respiratory cost in the diving marine mammals, e.g., the cetaceans where the lungs are capable of emptying to 10 to 15% of total lung capacity (compared with only 20 to 40% in man) in as short a time as 0.3 s (1.5 to 2.0 s in man) have not been estimated (Olsen et al. 1969; Kooyman and Sinnett 1979; Kooyman and Cornell 1981). In fish, myocardial power output is a useful indicator of the O_2 cost of the cardiac pumping (Farrell 1993). Myocardial O_2 consumption is about $0.3 \, \text{ml s}^{-1}$ per mW^{-1} of cardiac pumping in various species of fish (Davie and Farrell 1991). The cost of cardiac pumping at rest constitutes 0.6 to 4.6% of resting O_2 consumption but in the hemoglobin-free Antarctic fish, *Chaenocephalus aceratus*, the cost may be as high as 23% (Farrell and Jones 1992).

2.9.1 The Requisites for Efficient Gas Exchange

The process of gas exchange is governed by various structural elements and functional events. These include: (1) the partial pressure gradient prevailing across the tissue barrier separating the blood and/or body fluids from the external milieu (Fig. 27), (2) the thickness of the blood-gas barrier, (3) the surface area available for gas transfer (Fig. 16), (4) the material properties (i.e., the intrinsic biophysical attributes) of the barrier tissue, properties which determine the permeability of the respiratory gases (Roughton and Forster 1957), (5) capillary transit times across the gas exchanger, (6) O_2-hemoglobin binding kinetics, (7) the physical properties of the external respiratory medium, e.g., air, water or artificial atmosphere or liquid, (8) an efficient neuroregulatory mechanism for coordination of ventilation and perfusion of the gas exchanger to establish optimal gas fluxes, and (9) dynamics and geometric pattern of presentation of the internal and external respiratory media. The limitations inherent in a particular respiratory medium and the nature of transfer of O_2 can be best assessed on the basis of the differences in the PO_2 and PCO_2 across a gas exchange surface. Oxygen and CO_2 molecules are convectively transported at equal rates whereas diffusion

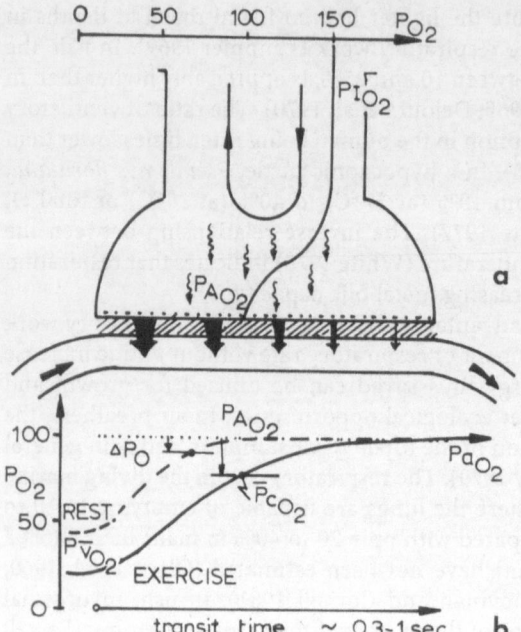

Fig. 27. a Schematic diagram showing the diffusion of O_2 in an alveolus across the blood-gas barrier, ↓, and the oxygenation profile of pulmonary capillary blood, ↗, in transit through the mammalian lung. The PO_2 in the alveolus, PAO_2, decreases as O_2 diffuses into the capillary blood, **b**. The pulmonary capillary transit time is longer during rest than during exercise when the capillary length is almost fully utilized. In normal cases, capillary transit times are adequate for complete arterialization of blood both during exercise and rest. PIO_2, partial pressure of O_2 in the inspired air; PaO_2, partial pressure of O_2 in the arterial blood; PcO_2, average capillary partial pressure of O_2; PvO_2, partial pressure of O_2 in the venous blood; ΔPO, partial pressure of O_2. (Weibel 1984a; reprinted by permission of Harvard University Press; copyright 1984 by the President and Fellows of Harvard College)

transport favors O_2 over the larger CO_2 molecule both in gaseous and aqueous media (e.g., Rahn et al. 1971). Since the ratio of the diffusion rate of CO_2 to O_2 is 0.78, for the same CO_2 tension, the O_2 gradient at a respiratory surface will always be higher (Wangensteen et al. 1971). This difference is particularly important in some organisms such as insects at various stages of development and avian, reptilian, and insect eggs, which depend wholly or partly on the diffusion of O_2 and CO_2 across the gas exchange surface. The extant amphibian eggs are restricted to a diameter of 9 mm (Carroll 1970) whereas in the much large eggs of reptiles and birds, where diffusion would not suffice in transfer of O_2, adaptations such as development of a well-vascularized chorioallantoic membrane which promotes uptake and transfer have evolved (e.g., Luckett 1976; Sect. 6.13). For historical interest now, at the beginning of this century, it was believed by physiologists as eminent as Christian Bohr and J.S. Haldane (see Haldane 1992), apparently from erroneous measurements which indicated a higher PO_2 in the pulmonary capillary arterial blood than in the alveolar air, that gas exchange in the vertebrate lung occurred by an active process. It was envisaged that the lung was able to "absorb" O_2 from the air spaces and "secrete" it into the blood. This was thought to occur particularly during exercise when the O_2 demand was high and during adaptation to hypoxia when there was deficiency of the same. The irreconcilable morphological observations of the day did not help to resolve the matter. Until the application of electron microscopy, debate ranged as to whether an epithelial lining covered the lung. Among others, Albert Policard (e.g., Policard 1929) championed the "naked alveolar capillaries" concept asserting that the respiratory surface of the lung was "like the flesh of an open wound" and that the inhaled air came into direct contact with blood. More accurate estimations of the PO_2 and PCO_2 in the alveolar and pulmonary capillary arterial blood (Krogh 1910) and better instrumentation and tissue processing when complete alveolar epithelium was discovered (Low 1953) dispelled these notions. In different animals, the alveolar PO_2 is consistently higher than the mean capillary blood PO_2 by an appreciable margin (e.g., Lindstedt 1984; Karas et al. 1987b; Constantinopol et al. 1989). In the lung of the Arctic fox, *Alopex lagopus* (Longworth et al. 1989), the pressure head (the pressure gradient = the driving force) for diffusion of O_2 (ΔPO_2) across the blood-gas barrier was estimated to be about 5 kPa, a value greater than that in the lungs of other highly aerobic animals, e.g., the pony (3.4 kPa) and the dog (4.6 kPa), the goat (1.9 kPa), and the calf (2.4 kPa) (Weibel and Taylor 1986; Karas et al. 1987b). In the horse, ΔPO_2 is 2.6 kPa and in the steer 2.1 kPa at VO_{2max} (Contantinopol et al. 1989). In muscles, during severe hypoxia, a condition when the ΔPO_2 between the blood capillaries and the mitochondria is drastically reduced, a diffusion limitation for O_2 occurs in animals with a normal hemoglobin O_2 affinity (Stein and Ellsworth 1993): left-shift of the oxyhemoglobin dissociation curve provides the means for the tissue to preserve its level of O_2 extraction within the blood capillary network. It has been shown that VO_{2max} can be increased by raised O_2 supply to the body through increased PO_2 in the inspired air (e.g., Welch 1987; Knight et al. 1992; Gonzalez et al. 1993). Hemoglobin concentration affects VO_{2max}, with changes as little as 1 to $2 g dl^{-1}$ being significant (Woodson 1984). Infusion of autologous frozen erythrocytes, a process which resulted in elevation of hemoglobin concentration from 15 to $16.3 g dl^{-1}$, produced a 5% increase of VO_{2max} 24 h later and 4% 7 days later in the human being (Buick et al. 1984). In a similar procedure, Williams et al. (1981) reported an increased running capacity on a treadmill while Robertson et al. (1982) observed a 13% increased of VO_{2max} and 16% of endurance time in human subjects injected with 375 ml of frozen autologous erythrocytes on 2 successive days when hemoglobin concentration rose from 13.8 to $17.6 g dl^{-1}$ and the hematocrit from 43 to 55%. Carrier-mediated transfer of O_2 in the lung (Burns et al. 1975) and the placenta (Burns and Gurtner 1973) are known to occur, but play an insignificant role in the overall gas exchange process. Diffusion and convection are the only known processes involved in the transport of O_2 into the respiratory organs and within organisms. It is only in rare, specialized organs such as the swim bladder (e.g., Fänge 1966; Sect. 6.2) and the choroid rete of the eye of teleosts

(Wittenberg and Wittenberg 1962), where O_2 is known to be secreted against a concentration gradient. The respiratory roles of such organs, if any, is secondary. Since the solubility of CO_2 in water and tissues is much greater than that of O_2, its elimination and transportation in solution presents no problems to an aquatic organism and the need for its active secretion is even less necessary. The transfer of gases from the ambient milieu to the tissue cells takes place essentially by convection (the mass transfer of O_2 by the moving air in which O_2 is contained) and by diffusion, the molecular flux of O_2 through air or liquid in a manner essentially related to the continuous random Brownian motion. High rates of O_2 uptake and CO_2 elimination are germane to the high metabolic rate of the endothermic terrestrial vertebrate endotherms. These processes are only rivaled in rate and magnitude by the transfer of water vapor between an animal and its immediate environment. The efficiency and plasticity of the gas exchange processes is underscored by the fact that during exercise or exertion, the levels of flux of respiratory gases may increase more than ten times above the resting level.

Gas exchange is constrained by temporal limitations consequent to finite rates of convection, diffusion, and reaction kinetics of CO_2 and O_2 with the carrier pigments. Inert gases, i.e., those gases which dissolve in blood without undergoing any chemical reaction, equilibrate much faster than do CO_2 and O_2, which are chemically bound to the hemoglobin and other carriers. Such gases require a longer transit time at the respiratory site for full equilibration to occur. Respectively, O_2 and CO_2 require 430 and 210 ms to reach 99% equilibration while inert gases of equivalent molecular weights require only 15 to 20 ms (Wagner 1977). Capillary transit times depend on factors such as tissue, species, metabolic rate, temperature, blood flow rates, pressure, viscosity, and capillary architecture, i.e., density, number, length, and geometric configuration (Karas et al. 1987b; Fig. 27). At rest, capillary transit times (in s) scale disproportionately to body mass (kg) to a power of 0.20 in the mammalian lungs (Lindstedt 1984). In mammals, capillary transit times are shorter in small animals than in the larger ones (Swenson 1990; Table 7). Oxygen binding to hemoglobin may limit maximum O_2 consumption in the smallest mammals (Lindstedt 1984). Birds have capillary transit times always less than 1 s. Under experimental conditions, transit times as short as 0.3 s have been reported (Henry and Fedde 1970). Adversely short transit times are prevented by capillary recruitment (e.g., Malvin 1988) and opening of arterio-venous anastomoses. Since on average the transit times are longer than the saturation times and the pulmonary capillary distance necessary for exposure and full saturation of blood is more than adequate (Karas et al. 1987b), it is conceivable that the inert and the respiratory gases are fully equilibrated as the blood traverses the pulmonary capillaries, at least under resting normoxic state. The pulmonary capillary transit time of about 0.5 to 3.0 of a second in the fish gills (Perry and McDonald 1993) is the same order of magnitude as in mammals, i.e., 0.75 of a second (Roughton 1945; Constantinopol et al. 1989) but in birds (e.g., chicken) the average transit time appears to be much shorter (e.g., Burton and Smith 1967; Henry and Fedde 1970), a value as short as 0.31 of a second having been estimated by Henry and Fedde (1970) in the chicken. This may explain the lower arterial O_2 tensions in birds compared with mammals (Jones and Johansen 1972). Relative to

Table 7. Capillary transit times in various vertebrate species and organs. (After Swenson 1990)

Species	Organ	Condition	Average transit time (s)
Human	Lung	Rest	1.8
		Exercise	0.4
		Rest-hypoxia	0.3
Cow	Lung	Rest	4.6
		Exercise	0.9
Dog	Lung	Rest	1.8
		Exercise	0.3
	Muscle	Rest	0.9
		Exercise	0.13
	Heart	Exercise	0.6
Fox	Lung	Exercise	0.12
Mouse	Lung	Rest	0.6
		Exercise	0.12
Frog	Lung	Rest	1.3
	Skin	Rest	1.0
Fish	Gill	Rest	1.0
Bird	Lung	Rest	0.8
		Exercise	0.5

mammals, birds have relatively larger hearts (e.g., Hartman 1954, 1955; Tucker 1968) and faster heart rates and hence greater cardiac outputs and stroke volumes off (Lasiewski and Calder 1971; Grubb 1983). A flying budgerigar, *Mellopsitacus undulatus*, has a cardiac output of about $41 kg^{-1} min^{-1}$, a value which is seven times greater than that of a mammal of equivalent body mass (Tucker 1968) and a heart rate as fast as 1020 beats min^{-1} has been reported in the giant humming-bird, *Patagona gigas* (Lasiewski et al. 1967). The cardiac blood pressure in birds is very high. Systolic pressures as high as 53 kPa have been determined in the domestic turkey, *Melleagris gallopavo* (Speckman and Ringer 1963). In contrast, among the living vertebrate ectotherms (i.e., fish, amphibians, and reptiles), systolic blood pressure rarely exceeds 7 kPa (Lindsay et al. 1971). Poikilothermic vertebrates store lipids primarily as tricylglycerols (Sheridan 1994). In birds, occasional deaths commonly attributed to fright are largely due to hemorrhage consequent to arterial rupture resulting from sudden excessive hypertension arising from such encounters (Walkinshaw 1945; Carlson 1960; Hamlin and Kondrich 1969). In *Nautilus*, during the hypometabolic state, i.e., after the PO_2 in the arterial blood has fallen to below 0.7 kPa, a reversal of the normal arterial-venous blood PO_2 gradient occurs with the venous PO_2 exceeding that of the arterial blood (Boutilier et al. 1996). This is thought to be brought about by an extremely slow intermittent blood flow during severe hypoxia when the large vena cava which is located on the roof of the mantle cavity has time to equilibrate with the ambient PO_2. Most interestingly, during such episodes, the vena cava serves as a gas exchanger, transferring respiratory gases between the blood and the air in the mantle cavity!

Morphometric studies have shown that the erythrocytes offer the greatest resistance to O_2 diffusion in the gas exchangers (e.g., Hallam et al. 1989; Maina et al. 1989a). Due to lack of carbonic anhydrase in the capillary endothelial cells of the gills (e.g., Perry and Laurent 1990; Lessard et al. 1995), unlike in the mammalian lung, in fish, the erythrocytes appear to be the only site for HCO_3^- ion dehydration (Perry and McDonald 1993). The relatively long transit time of the blood in the gills, the high capacitance of water for CO_2, and the efficacious countercurrent system of the gills may account for the effectiveness of the gills in CO_2 elimination. The factors which may inhibit end capillary PO_2 equilibration include: (1) an overly high cardiac output, e.g., during exercise and disease conditions factors which cause loss of the functional capillary bed, resulting in shorter transit times (Wagner et al. 1986), (2) thicker tissue diffusional barriers, e.g., in pathological conditions such as edema (Staub et al. 1967; Staub 1974), (3) severe hypoxia (reduction in O_2 driving gradient), e.g., during high altitude exercise (Wagner et al. 1986), (4) large erythrocytes which are associated with greater unstirred layers, and (5) pathological conditions which are accompanied by reduction of the concentration of the hemoglobin or those characterized by reduced erythrocytes count, a factor which increases the thickness of the unstirred layer resulting in reduction of O_2 binding velocity (e.g., Nguyen-Phu et al. 1986). The erythrocytes of the bullfrog, which have a volume of $700 \mu m^3$, take up O_2 at a rate five times slower than the smaller ones of $20 \mu m^3$ of the goat (Holland and Forster 1966). Diffusional resistance during severe exercise may inhibit end capillary O_2 equilibration as the capillary transit times fall to as low as $200 \, ms$ (Groebe and Thews 1987). Owing to the shape of the haemoglobin-O_2 dissociation curve which causes large O_2 loading during the early part of intracapillary transit (see. e.g., Comroe 1974; Shiga 1994), even when the end capillary complete equilibration is not attained, the amount of O_2 taken up is adequate for the respiratory needs. The differences in the equilibration time courses in the inert gases and in the biochemically active gases such as CO_2 and O_2 can be attributed to the ratio of the solubility of the gas in the membrane barrier to that in blood. In inert gases, the ratio is unity while the value is several orders of magnitude less for CO_2 and O_2 (Piiper and Scheid 1980). The greater solubility of O_2 and CO_2 in blood relative to the membrane is due to the reversible chemical binding and reaction with hemoglobin and other proteins in the erythrocytes and plasma (Shiga 1994). The rate at which the partial pressure of a particular gas builds up is exclusively dependent on its solubility in blood. The more soluble a gas, the slower will be its rate of partial pressure change for a given quantity of gas transferred. The rate of equilibration of CO_2 in any solution is strongly dependent on the buffering capacity. The rate-limiting steps in O_2 transfer are ascribed both to diffusional resistances and chemical reactions in an inhomogeneous medium. Unlike in CO_2 exchange and in the Bohr-Haldane effects, these fast processes do not require enzymes or membrane transport carriers. Even during exercise, in mammals, O_2 transfer and equilibration with the capillary blood are usually complete by the end of the transit time (e.g., Karas et al. 1987b; Longworth et al. 1989; Fig. 27).

Ventilatory and perfusive processes are particularly important in gas transfer during episodes of extreme exertion, hypoxia, and respiratory acidosis. In the course of its passage from the external environment to the mitochondria, O_2 travels across several anatomical spaces which offer different degrees of resistance (Figs. 2,3). According to Fick's Law, the diffusing capacity of a gas through a barrier depends on surface area, the concentration gradient, the thickness of a barrier (the diffusional distance), and the diffusion (permeation) coefficient across the barrier (Figs. 16,28,29). Among air breathers, the thickness of the blood-gas barrier of the gas exchangers decreases from the amphibians, reptiles, mammals, to birds (e.g., Maina 1994). The water-blood barrier in the aquatic gas exchangers is in general thicker than the blood-gas barrier of the lung (e.g., Hughes and Morgan 1973). The diffusion rate in the mammalian lung is very efficient. Determinations by Wagner and West (1972) and Hill et al. (1973) showed that in the human alveolar-capillary barrier whose normal thickness is 0.62 μm (Gehr et al. 1978), the thickness would have to be increased four to ten times before it becomes a limiting factor for end capillary PO_2 equilibration. From theoretical considerations (e.g., Scheid 1978) and experimental work by Powell and Wagner (1982a), in the avian lung the diffusion of O_2 in the gas phase of the air capillaries which radiate from the parabronchial lumen (Figs. 30,31) is not thought to constitute a limiting factor at rest (Powell 1982; Powell and Scheid 1989). Owing to the long diffusional distance (20 to 50 μm), cutaneous respiration is accompanied by limitations of diffusion (e.g., Malvin 1988) but movements of air over the skin may enhance the diffusional process. Increased perfusion may be utilized to enhance O_2 uptake particularly during hypoxic episodes (Malvin and Hlastala 1986). Movements by skin-breathing aquatic organisms may reduce

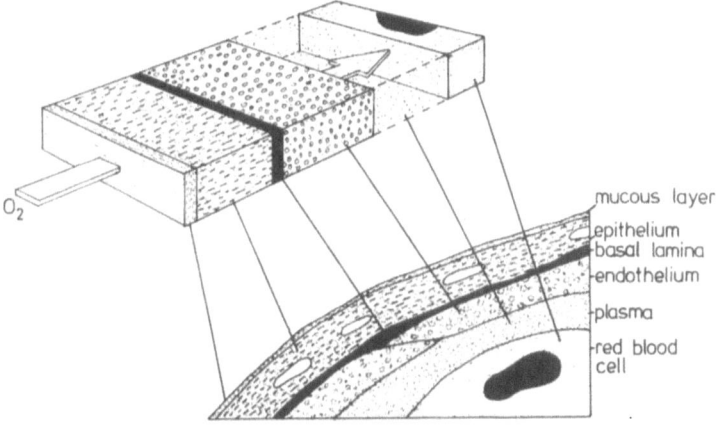

Fig. 28. Schematic view of the oxygen diffusional pathway in an aquatic breather. The diffusing capacity of the gas exchanger depends on the thickness and surface area of the respiratory barrier. In the water breathers, the unstirred layer of water over the gills may constitute a significant part of the resistance to O_2 diffusion and may need to be considered in modeling the diffusion capacity of the gills

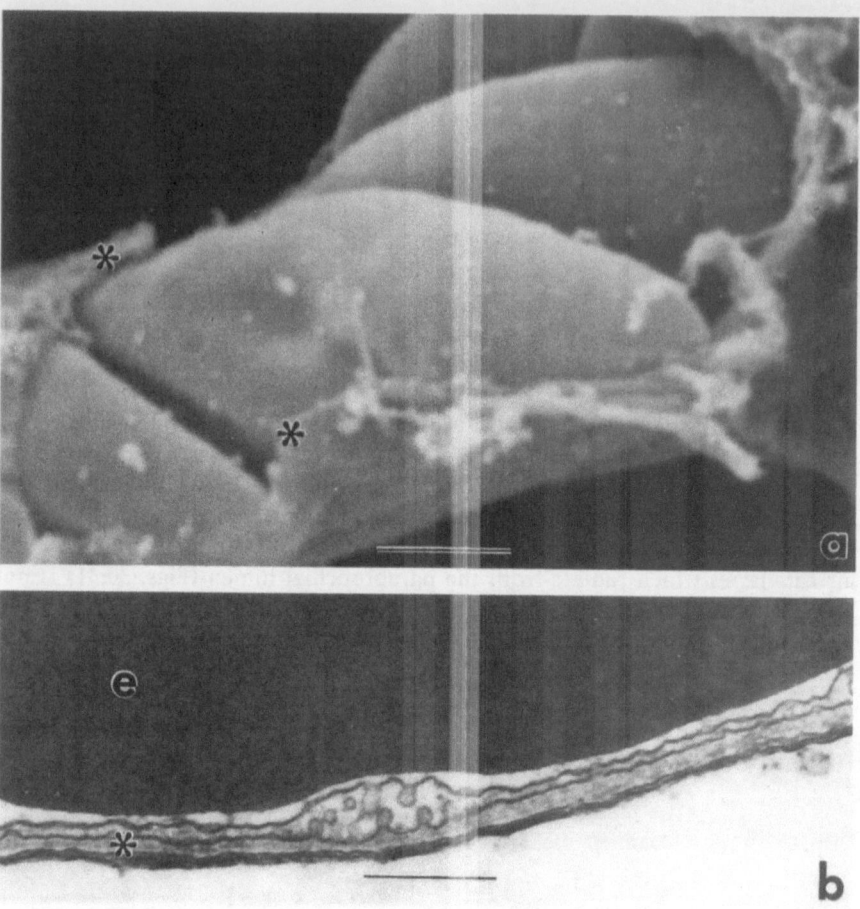

Fig. 29. a Blood capillary in the lung of the domestic fowl, *Gallus domesticus*, showing a file of erythrocytes in a blood capillary. *, blood-gas barrier. b The extremely thin blood-gas barrier, *, shows sporadic attenuations. e erythrocytes. a *Bar* 0.6 μm; b 0.6 μm. (Maina 1993)

diffusional limitations in the layer of water next to the skin (Feder and Burggren 1985a,b). In fish, factors such as low body temperature, thick water-blood barrier, and presence of an unstirred water layer over the secondary lamellae reduce O_2 transfer. They may cause significant diffusion limitations and curtail end-capillary equilibration particularly during resting normoxic conditions (Randall and Daxboeck 1984). At a simulated altitude of about 12 km, the resting bar-headed goose, *Anser indicus*, tolerated inspired PO_2 of about 3 kPa and no difference was observed between the PO_2 in inspired air and that in the arterial blood (Black and Tenney 1980): the arterial PCO_2 was as low as 1.1 kPa. Due to the large arteriovenous shunts which cause notable arterial O_2 desaturation in the amphibian and reptilian lungs (Glass and Wood 1983; Wood and Glass 1991), diffusional limitations are masked.

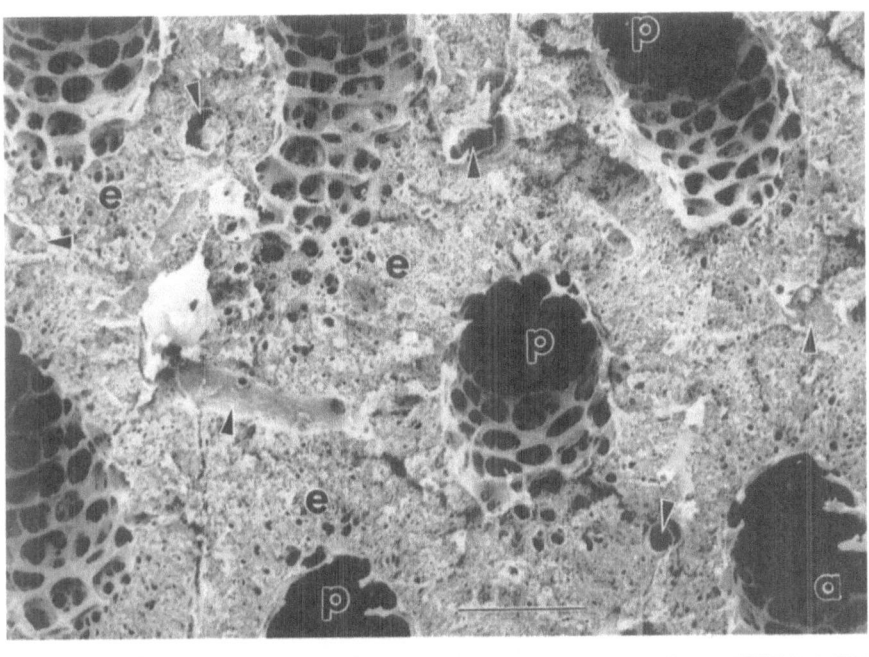

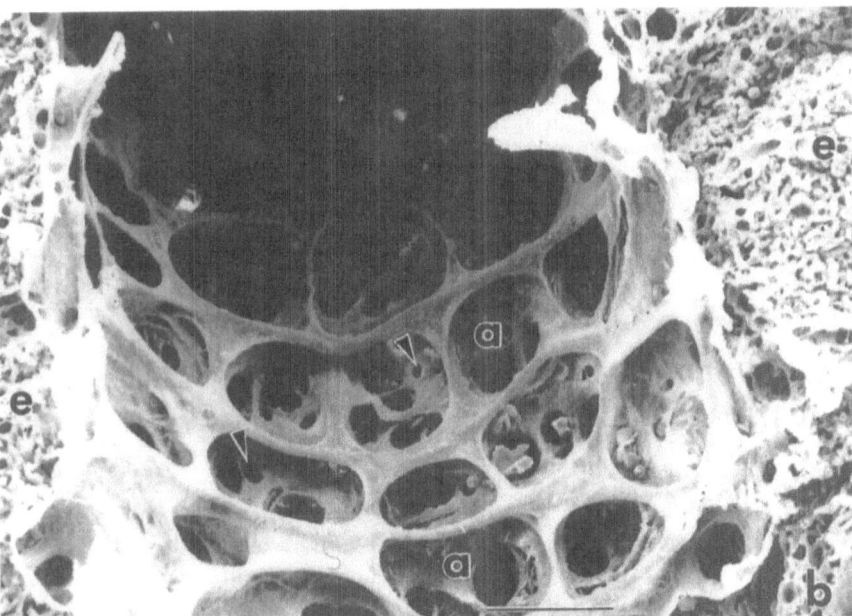

Fig. 30. **a** Lung of the domestic fowl, *Gallus domesticus* showing parabronchi, *p*, surrounded by gas exchange tissue, *e*, and blood vessels, ➤, located in the interparabronchial septa. **b** Closeup of a parabronchial lumen showing atria, *a*, and infundibulae, ➤, which channel air to the exchange tissue (*e*). **a** *Bar* 70 µm; **b** 20 µm

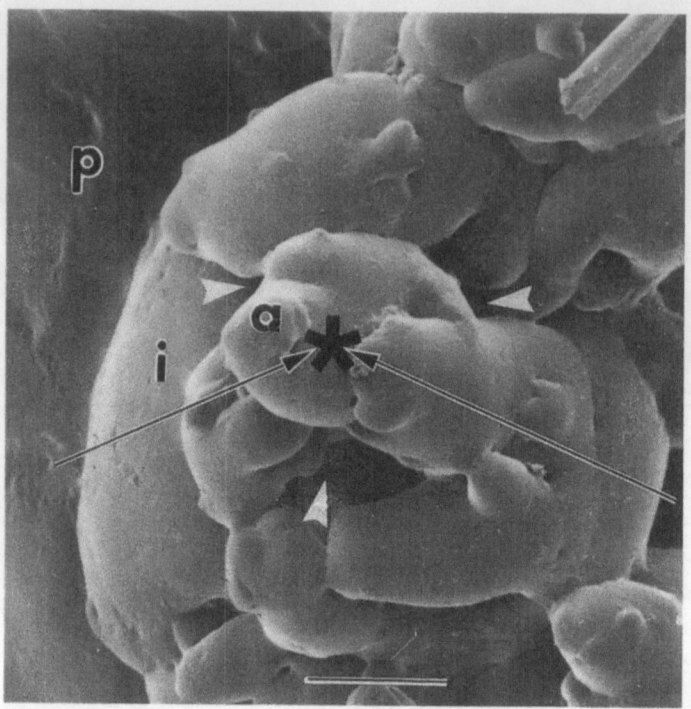

Fig. 31. Latex cast of the airways of a bird lung, *Gallus domesticus*. This closeup shows an atrium arising from a parabronchus, *p*, giving rise to an infundibulum, *i*, terminating in an air capillary, *a*. The perpendicular to the arrows, →, converging onto the star, *, illustrate the distance O$_2$ has to diffuse from the parabronchus to the air capillary. ➤, spaces occupied by blood capillaries. *Bar* 40 μm

2.9.2 Efficient vs. Inefficient – Primitive vs. Advanced Gas Exchangers: the Contention

From cladistic principles, the morphological, physiological, and biological character states of organisms are considered either to be primitive or derived. Commonly, these interpretations are subjective. For example, some organisms (e.g., bacteria) which are notably speciose and highly adapted to particular habitats where they thrive and even outcompete supposedly better-adapted forms are deemed primitive. Strictly, in a particular animal lineage, the primeval (plesiomorphic) features should be taken to be those which characterize the ancestral forms. The derived (apomorphic) features typify the more recent animal groups. A character which is unique to an animal (an autamorphic mark) indicates a unique adaptive state. The primitive features are assumed to be inefficient while the advanced (= derived) ones are taken to be adaptive (= more efficient). The fact that the so-called primitive features have been saved for millennia amidst intense selective pressures emphasizes either their neutral selective value or the exceptional advantage(s) they confer for survival in a particular habitat. Gould (1994) opines that "our impression that life evolves towards greater complexity is probably only a bias inspired by parochial focus on ourselves and consequent over attention to complexifying creatures while we ignore just as many lineages adapting equally well by becoming simpler in form". Along the same argument, Thompson (1959, p. 41) points out that "however important and advantageous the subdivision of the tissues into cells may be from the constructional, or from the dynamic, point of view, the phenomenon has less fundamental importance than was once, and is often still, assigned to it".

Primitive and derived features exquisitely allow different kinds of animals to subsist in the same habitat. The different resource procuring capacities enable the congeners to fill species-specific spatial and/or temporal ecological niches without undue competition. Special environments are characterized by low biodiversity though the numerical densities of the occupying species may be very high. Pough et al. (1989) assert that "the primitive character state is not inferior or necessarily less adaptive, it is simply older". These sentiments are echoed by Kardong (1995), who observes that "the notion of better does not apply to biological changes: primitive and advanced species represent different ways of surviving not better ways of surviving". From these sentiments, the terms primitive and derived should be used to define only the temporal appearance, i.e., the chronological advent of certain attributes in the phylogentic history of a particular animal lineage and not the polarity of a change. The primitive features are those that occur in the ancestral members of the group while the derived ones are found in the modern forms and differ in notable morphological and physiological ways from the primordial condition. Gas exchangers are designed to service given cytoplasmic masses operating within certain metabolic boundaries. Overdesign is costly and wasteful. In the context of functional capacities, strictly, primitive gas exchangers do not exist.

The simplest design and apparently the most efficient evolved gas exchanger has occurred in form of the insectan tracheal system (Sect. 6.6.1; Figs. 74,75). Few animals have successfully emulated it. Gas exchangers present certain properties which are fundamental in explaining the remarkable diversity of design. These are: (1) the transfer of gases from the external milieu to the body and vice versa occurs by the physical process of diffusion, (2) there is no ubiquitous structural plan: one or more pathways may be utilized simultaneously or in phase, (3) there is only a weak correlation between the constructional complexity, efficiency, and the phylogenetic level of development, and (4) the diffusion of gases across the tissue barriers occurs in solution. To enhance the diffusing capacity, efficient gas exchangers must have an extensive respiratory surface area, a thin blood-gas barrier, and must be well perfused with blood and ventilated with the ambient respiratory medium. Organs as diverse as the anus, stomach, mouth, and skin are used as gas exchangers. The advanced animals, e.g., mammals, do not necessarily have the most efficient gas exchangers. Invagination of the gas exchangers not only afforded a greater respiratory surface area but also provided better control and regulation of the physical characteristics of the inhaled air which must be cleansed and particularly moistened to curtail water loss from the alveolar surface.

2.10 Modeling: Utility in Study of Integrative Construction of the Gas Exchangers

Gas exchangers are complex, dynamic, high fidelity systems. Respiration entails intricate interplay between physiological features such as convection, diffusion, and chemical reactions (e.g., Piiper et al. 1971; Wagner 1977) and anatomical factors such as the architecture and geometry of the airways and the respiratory barriers. Morphological observations and quantification of the components of gas exchangers makes it possible to construct realistic models and also advance understanding of the evolutionary processes that have precipitated the diversity in form and function in the process of adaptation to environments and adoption of different lifestyles. Through selective pressure acting on the genotype, the gas exchangers have been honed to meet species-specific, design-oriented needs. The final constructional plans present rational, multifaceted, integrated engineering (Beament 1960; Dullemeijer 1974; Gutman 1977). An intractable number of variables (of which the individual functional capacities may not be sufficiently known) contribute to the ultimate fabrications of biological systems. Biology has long been mathematically emancipated and somewhat changed from how it was once perhaps legitimately described, as a descriptive (inexact) science (compared with physical science). In the gas exchangers, the pulmonary diffusing capacity, defined as the flow rate of O_2 from air to blood under partial pressure gradient between alveolar and mean O_2 tension in the alveolar capillary blood (Bohr 1909; Roughton and Forster 1957; Forster 1964; Weibel 1970/71) is the outcome of the integration of the functional and structural parameters. The functional anatomies of composite structures (e.g., the gas exchangers) can be adequately described through appropriate modeling, i.e., by integrating the relevant parameters central to an organ system, e.g., the respiratory (Singh et al. 1980; Federspiel and Popel 1986; Bozinovic 1993) and the cardiovascular systems (Zagzoule and Marc-Vergnes 1986; Wang et al. 1989; Melchior et al. 1992). Gutman and Bonik (1981) defines a model as "an abstraction of a real situation which describes only the essential aspects of the situation". Implicit if not explicit assumptions underlying various parameters and properties are often made. A mathematical model aims to "dismember" a structure into the formidable number of individual components, isolates those components that are relevant in answering the question at hand, and then reassembles it. The model must endeavor to be accurately descriptive of the complex reality of nature, must be simple and easy to conceptualize, and must be theoretically and practically testable. By applying appropriate physical and physiological constants, a static structural model can mathematically be converted to an acceptably dynamic one. The net worth of a model can be judged from its predictive potential. In modeling highly dynamic systems, e.g., the wing beat kinematics of an insect in flight (Weis-Fogh 1973; Ellington 1981, 1984; Wilkin and Williams 1993) and atmospheric turbulence (e.g., Hollinger et al. 1994), a defined state (e.g., a steady- or a quasisteady state) has to be assumed. Complex systems are intrinsically chaotic and are hence unpredictable (Gleick 1987). Experimental manipulation of a functional model gives an insight into the cause-and-effect relations between the individual and the covariant factors that drive a system. Outcomes can be evaluated by altering one or more factors or conditions while holding others constant. Constraining, potentiating, and super-fluous components can be identified and the relative performances determined. Based on structural-functional integrative studies, encompassing observations can be made on the gas exchangers. Depending perhaps on the approach adopted and the emphasis laid on the different parameters which are considered, in some cases the conclusions have turned out different. Assessing the factors which determine O_2 flow from the lung to the muscle tissue during exercise, Weibel et al. (1987a) and Karas et al. (1987a) concluded that "cardiac output is the most important variable controlling flow of O_2 through all steps of the respiratory system". However, evaluating the factors which limit the delivery of O_2 to the skeletal muscles in an exercising human being, Saltin (1985) resolved that "pulmonary diffusing capacity is the ultimate limiting factor" during extreme conditions. In a theoretical model, Wagner (1993) concluded that at sea level, cardiac output has the greatest effect on VO_{2max} while at altitude, muscle diffusing capacity is the most important parameter. Explaining the 2.5-fold higher maximum O_2 consumption in the athletic species (pony and dog) over that of the less athletic ones (calf and goat), Karas et al. (1987b) attributed one half of it to the higher PO_2 at the blood-gas barrier (1.7-fold greater in the athletic species) and the other half to the diffusing capacity which was 1.5-fold larger. It is most instructive to be able to accurately predict, infer the consequences of, or explain how one or a group of respiratory variables affects the others. Physiologic and morphometric estimations of diffusing capacities of different gas exchangers, e.g., fish gill (Scheid and Piiper 1976; Piiper et al. 1977, 1986; Hughes et al. 1986b), mammalian lung (Gehr et al. 1978; Weibel et al. 1983), and bird lung (Maina et al. 1989a; Vidyadaran et al. 1988; Burger et al. 1979) are now available. In all cases, the morphometric value exceeds the physiological one. This has been interpreted to indicate a safety margin of operation. In estimating the morphometric diffusing capacity, an ideal state is assumed, i.e., that O_2 flux occurs across the entire water-air /blood barrier under a ventilation-perfusion ratio of unity. This state is only approached at maximum O_2 consumption (e.g., Powell and Wagner 1982a,b). Gans (1985) holds that "animals cannot afford one-function/one-structure designs "and" each activity tends to involve multiple aspects of the phenotype and each aspect of the phenoytpe may be involved in multiple activities" (Gans 1988). Empirical prototypes in form of biological models provide useful insights (even though in theoretical way) into the underlying control mechanisms, the multidimensional utilization of the phylogenetic characteristics, and the effect altered parameters have on the entire system (Cameron 1989; Anker and Dullemeijer 1996). Simplification, abstraction, and generalization of biological processes and concepts have contributed greatly to the advancement of scientific thought (Homberger 1988; Vogel 1988).

Gas exchangers, like most organs, present a multilevel organizational character. The various parts of a structure are arranged in discrete functional units which are, in turn, intricately connected to ensure integrity in the overall performance. For optimal transfer of gases at the exchange site, the convective (ventilatory and perfusive) pumps must be coherently linked so that the gas delivered across the respiratory barrier is immediately removed to maintain a concentration gradient. Overperfusion relative to ventilation or vice versa incurs unnecessary metabolic cost. The actual quantity of respiratory gases transferred across the barrier in such episodes is lower than that prevailing during the optimal and less expensive conditions. The ideal ventilation-perfusion ratio in mammals is about $1:1$ (West 1977a) and in the chelonian reptiles it is $2:1-5:1$ (Burggren et al. 1977). In aquatic fish, the equivalent value is $9:1-35:1$ (Randall 1970). By refining and multiplicatively integrating the structural and physiological parameters such as lung volume, cardiac output, hemoglobin concentration, muscle mass, and mitochondrial number, land vertebrates such as the pronghorn antelope, *Antilocapra americana* (Lindstedt et al. 1991), and the cheetah, *Acionyx jubatus* (Hildebrand 1959, 1961), can attain and sustain speeds of 75 to 100 km per h. Lindstedt et al. (1991) estimated that the pronghorn antelope can take up O_2 at a late of 3.2 to $5.1 \, ml \, O_2 \, kg^{-1} s^{-1}$, a value 3.3 times greater than that predicted for a typical mammal ($1.5 \, ml \, O_2 \, kg^{-1} s^{-1}$) of a similar body mass of 32 kg. On the other hand, the naked mole rat, *Heterocephalus glaber*, a highly inbred (Reeve et al. 1990), eusocial (Burda 1993), heterothermic, low metabolism animal (McNab 1966) which lives in a stable environment (Cossins 1991) has neotenic lungs (Maina et al. 1992). The pulmonary morphometric diffusing capacity of the lung of the mole rat, *Spalax ehrenbergi* was found to be 40% greater than that of an equivalently sized surface-dwelling white rat (Widmer et al. 1997).

2.10.1 Evaluation of the Functional Efficiency of the Gas Exchangers

Analysis of the performance of a gas exchanger entails physiological measurements of O_2 tensions prevailing in the gas exchange media or that between the arterial and venous blood. This is done by taking cognizance of the inherent construction of the exchanger and the fluid flow dynamics within and without the gas exchanger. Respiratory efficiency can also be gauged from the quantity of O_2 transferred against the energy expended during the process (e.g., Piiper and Scheid 1992; Shelton 1992). By varying or holding certain factors constant while integrating or suppressing others, the contributions of the various variables to the overall function can be evaluated and the limiting or potentiating factors identified. The efficiencies of different or same gas exchanger(s) under different operating conditions can be compared. Though gross mathematical abstractions of a complex biological system which is regulated by an infinite number of variables, all of which are difficult to individually assess, manipulatable theoretical analogs are highly instructive in comparative experimental and analytical studies: they generate "artificial" didactic situations which afford a perception of a complex outlay. Physiological phenomena are cumulative effects of interactions between numerous minuscule, individually noneventful, nonlinear, but coupled local unit processes which interact at various levels with a potentiating or even a depreciative effect. Though with intrinsic limitations, functional and morphological models are extremely valuable in the understanding of the complex relations in multidimensional systems (e.g., Riggs 1963; Penry and Jumars 1987; Usry et al. 1991; Horn and Messer 1992). From the results of functional and structural analyses, a piecemeal synthesis of an organ system can be made. The efficiency of

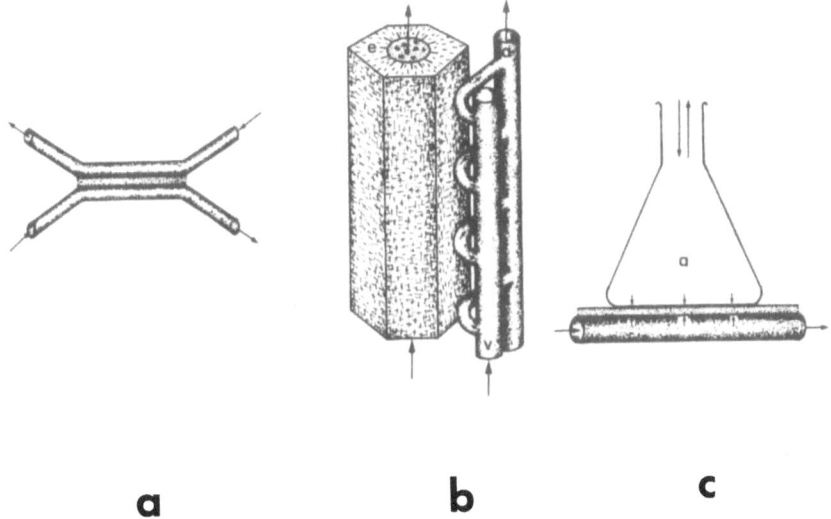

a   **b**   **c**

Fig. 32a–c. Schematic drawing showing the spatial relationship between air and blood flow, →, in the fish gills (**a**) and bird (**b**) and mammalian (**c**) lungs. In fish, air and blood flowing in a countercurrent manner. In birds, the flow between the parabronchial blood (*small arrows*) and the air flow in the parabronchial lumen is crosscurrent. In the mammalian lung, the flow is described as uniform pool. *e* Parabronchial gas exchange tissue and air flow in the lumen, ↑; *v* venous blood; *a* (in **b**) arterial blood; c *a* alveolar air; ↓, diffusion of O_2 across the blood-gas barrier: →, direction of alveolar capillary blood flow; ⇌, tidal ventilation of the mammalian lung. To a large extent, the efficiency of the gas exchangers depends on the presentations between the gas exchange media

a gas exchangers is largely determined by the nature of presentation of the respiratory media (Fig. 32) which is dependent on the intrinsic refinements and the geometric arrangements of the structural components (Figs. 29,33,34). Capillary geometry and tortuosity has a profound effect on the efficiency of O_2 transfer in muscles (e.g., Ellis et al. 1983; Groom et al. 1984a,b; Mathieu-Costello 1990; Mathieu-Costello et al. 1992). Biologists can appropriate the mathematical and conceptual contrivances of engineers to discern how animals function and why they have acquired certain morphologies.

2.10.2 Modeling the Gas Exchangers

In its most simplistic form, the respiratory system can be conceptualized as a set of compartments interposed between the body tissue cells and the environment. Across the cascade, which comprises the gas exchanger (ventilatory convection), water/air-blood barrier (diffusion), blood (circulatory convection), systemic capillary bend, interstitial fluid, and cells (diffusion), O_2 and CO_2 flux at intensities essentially determined by the metabolic states and environmental circumstances. It is mainly in the chordates where adequate structural and functional data have

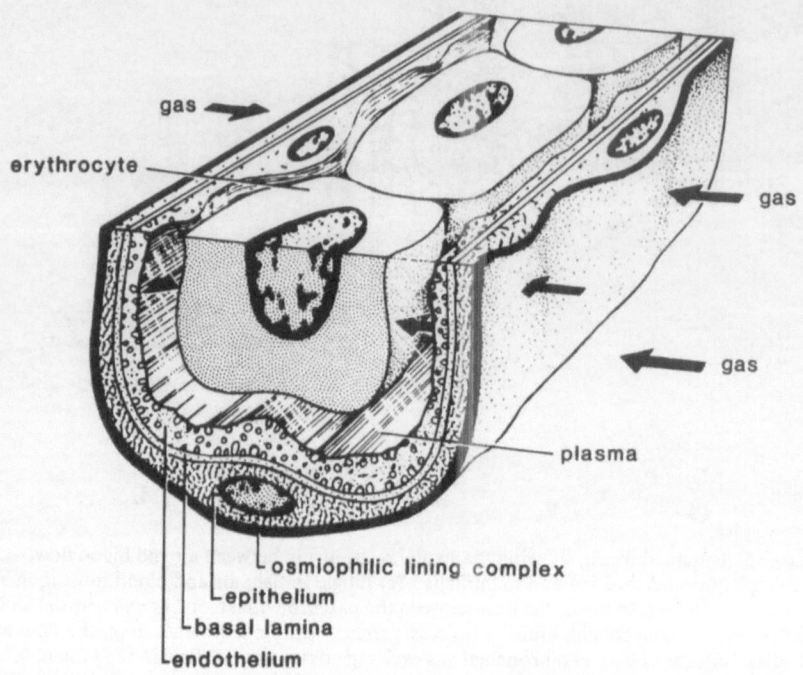

gas →

erythrocyte

gas

gas

plasma

osmiophilic lining complex
epithelium
basal lamina
endothelium

Fig. 33. Schematic view of a pulmonary blood capillary showing the in-series components of the air-hemoglobin pathway, namely the blood-gas (tissue) barrier (which comprises a surface lining, an epithelial cell, a basement membrane, and an endothelial cell), plasma layer, and the erythrocytes. O_2 uptake by the erythrocytes constitutes the greatest point of resistance to flow. (Maina 1994)

been gathered to warrant satisfactory modeling of the gas exchangers. The general model relating diffusion of gases across the blood-gas barriers and the water-blood barriers to ventilation and perfusion (e.g., Piiper and Scheid 1975) has been extensively used in different gas exchangers (e.g., White and Bickler 1987). Morphometric pulmonary modeling after Weibel (1970/71) has been carried out in many gas exchangers, e.g., fish gills (Hughes 1972b), reptilian (e.g., Perry 1983), avian (e.g., Maina 1989a), and mammalian lungs (e.g., Gehr et al. 1981).

When the gas exchange media, i.e., the external (the ventilatory) and the internal (the perfusive) media run parallel to each other and in the same direction, the design is termed concurrent. If the media run in opposite directions (e.g., water and blood in the fish gills), it is called countercurrent and when the media run at right angles to each other, e.g., the blood capillaries of the parabronchial exchange tissue relative to the flow of air in the parabronchial lumen of the bird lung), the arrangement is designated crosscurrent. When the external medium is held constant against a gas exchanger (e.g., skin) or is ventilated with a medium of which the gaseous partial pressures are fairly uniform (e.g., mammalian lung where the alveolar gas exerts a steady-state concentration)

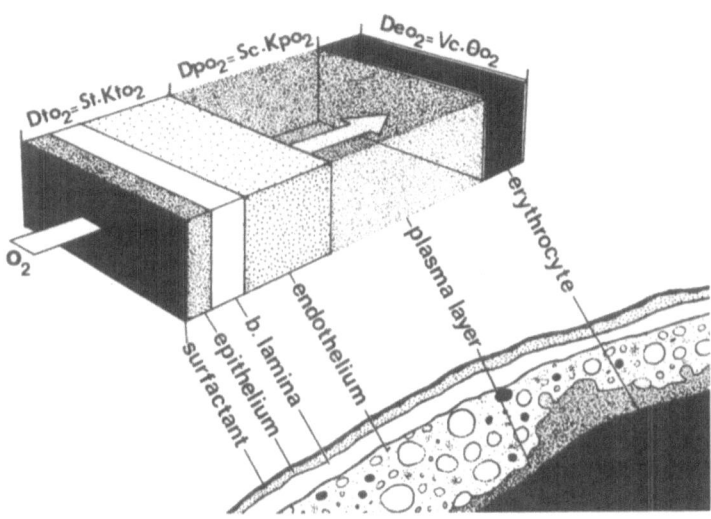

Fig. 34. Model for estimating pulmonary diffusing capacity or conductance of a lung for O_2 (DLo_2). The principal barriers which influence Do_2 are the tissue barrier [epithelium + basal lamina (interstitium in some cases) + endothelium], the plasma layer, and the erythrocytes. DLo_2 is directly proportional to the surface area and the permeation constant and inversely proportional to thickness. Respectively: Dto_2, Dpo_2 and Deo_2 diffusing capacities of the tissue barrier, plasma layer, and erythrocytes; St and Sp surface areas of the tissue barrier and plasma layer; Vc volume of the pulmonary capillary blood; Kto_2 and Kpo_2 Krogh's oxygen permeation coefficients of the tissue barrier and plasma layer; o_2 – O_2 uptake coefficient by the whole blood. The products of St and Kto_2 and Sc and Kpo_2 must be divided (not shown on the figure) by the harmonic mean thickness (τht) of the respective barriers. (Maina 1993 courtesy Pergamon Press). For a detailed account on the model as applied to the bird lung, see Maina et al. (1989a)

(Milhorn et al. 1965), respectively, the configurations are termed infinite pool and uniform pool (White 1978; Figs. 30,32,35,36). In the small insects, O_2 is conducted directly to the tissues by diffusion through trachea (Maina 1989b; Figs. 36,37) and by convection through the air sacs in the larger ones (Fig. 38). Delivery of O_2 directly to the tissue cells is a more efficient means of gas transfer (Sect. 6.6.1) compared with the conventional one where the molecule is convectively transported through a circulatory system (Levi 1967; Bromhall 1987).

By fitting or approximating a gas exchanger to any of the above main models, the efficiency and the limitations (be they developmental, structural, or functional) inherent in a particular gas exchanger under different circumstances can be gauged (Hills 1972; Piiper and Scheid 1972, 1975; Sikand et al. 1976). The relative efficiencies between different gas exchangers can be assessed. In a concurrent design, the gas tension in the internal medium leaving the exchanger cannot possibly exceed that of the external medium in normal and about equal fluid flows. However, under special conditions, e.g., when large quantities of water flow very fast over a small volume of slowly moving film of blood, a high degree of oxygenation with low utilization can be realized (Hughes 1963) but after

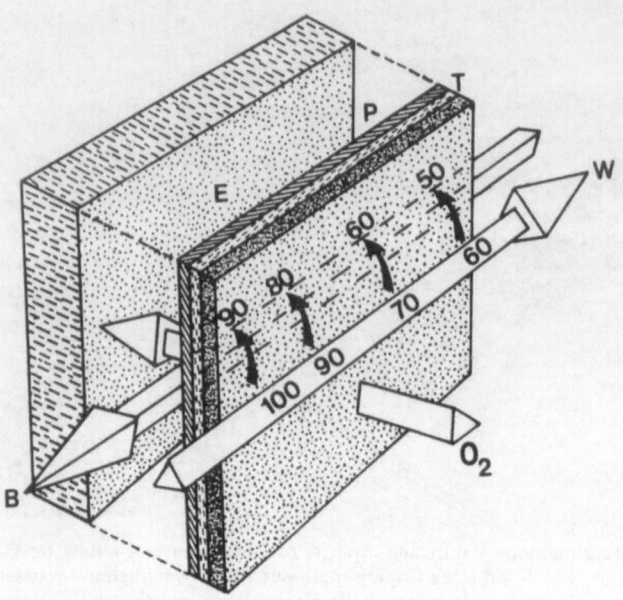

Fig. 35. Countercurrent system. The exchange media which contain O_2 at different concentrations flow in opposite directions. A diffusional gradient prevails over the duration and length of contact. A countercurrent arrangement occurs in the fish gills where O_2 extraction has been reported to be as high as 96%. *Numbers* Percentage concentrations of O_2 in the respiratory media; ↗ O_2 flows down a concentration gradient; *T* tissue barrier; *W* water; *E* erythrocytes; *P* plasma; *B* blood. (Maina 1994)

great energy expenditure on ventilatory work. In such cases, the gain in real terms is very small. In a countercurrent system, since the efferent blood has been exposed to the inhalant water, the PO_2 between the two media is high. Unlike in the mammalian lung, where theoretically ventilation-perfusion homogeneity may allow alveolar PO_2 to equilibrate with the arterial PO_2, in the fish gills arterial blood and inspired water never reach equilibrium (Randall 1970). This may be attributed to factors such as the thick water-blood barrier (e.g., Hughes and Morgan 1973), the large ventilation-perfusion inequalities (Booth 1978), and added diffusion limitations resulting from the boundary layer of water close to the surface of the secondary lamellae (Hills 1972). The ventilated uniform-pool design in most vertebrate lungs is more efficient than the nonventilated one (e.g., the skin and the buccal cavity). However, behavioral changes such as exposure to air currents and movement may initiate passive ventilation enhancing gas exchange. In the human lung, due to the tidal nature of air renewal, only about 12% of the intrapulmonary air volume is exchanged at the alveolar level per inspiratory cycle. Due to the low efficiency of the concurrent system, the design has only rarely been invoked in evolutionary biology. In contrast, the countercurrent system is widely encountered in biology (Scholander 1958) especially in form of heat exchangers, e.g., cephalic and ophthalmic rete (Jackson and Schmidt-Nielsen 1964; Kilgore et al. 1979; Baker 1982; Midtgärd 1983, 1984; Kamau et al. 1984;

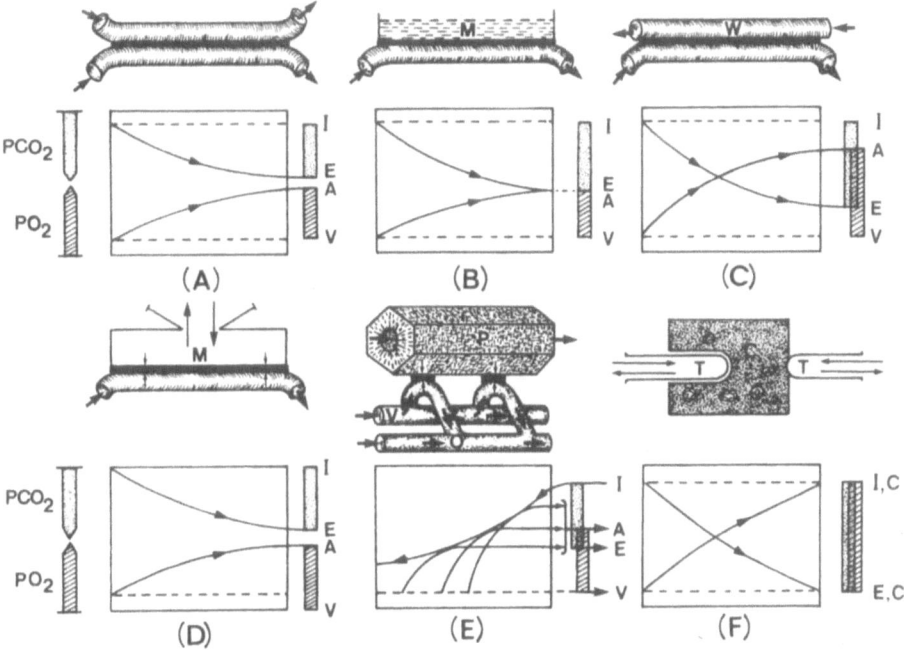

Fig. 36A–F. Schematic illustrations showing the geometric interactions between the respiratory media in different gas exchangers. The respiratory efficiencies are shown by the PO_2 and PCO_2 profiles in the inspired air, I, expired air, E, venous blood, V, and arterial blood, A. **A** Cocurrent system: gas exchange media flow in the same direction and O_2 flux is very low; **B** The skin: the gas exchanger is perfused but not ventilated. **C** Countercurrent arrangement: the PO_2 in the arterial blood exceeds that in the end expired air. **D** Uniform-pool: the influx of O_2 depends on ventilation and perfusion inequalities. **E** Crosscurrent system in the bird lung: through the multicapillary serial arterializational arrangement, the PO_2 in the arterial blood may exceed that in the expired air. **F** The insect tracheal system: O_2 is delivered directly to the tissue cells (by the tracheoles, T) with the PO_2 at the cell level being only slightly lower than that in the ambient. The PO_2 in the arterial blood may exceed that in the expired respiratory medium only in models **C**, **E**, and **F**. m Respiratory medium which depending on the type of animal could be air or water: p parabronchus; c cell; v blood (venous); w water. In the schematic drawings, the *single arrows* show the directions of flow of respiratory media and the diffusion of O_2; $\rightleftarrows$ *arrows* show the gas exchangers which are tidally ventilated. (Maina 1994)

Pinshaw et al. 1985; Block 1987), salt-concentrating systems such as kidneys and salt glands of birds (Kokko and Tisher 1976; Schmidt-Nielsen 1990), and in gas secretion organs, e.g., the choroid rete (Wittenberg and Wittenberg 1962) and the rete mirabile of the swim (air) bladder of fish (Fig. 39; e.g., Kuhn et al. 1963; Kobayashi et al. 1989a,b; Pelster and Scheid 1992a). Chemical (e.g., Coulson and Richardson 1965) and mechanical engineers (e.g., Carslaw and Jaeger 1959) exploit countercurrent systems to enhance concentration gradients in order to improve mass transfer. The functional efficiency of the countercurrent system depends on factors such as a thin and extensive surface area for interaction of the gas exchange media, optimum flow rates and physical states and characteristics.

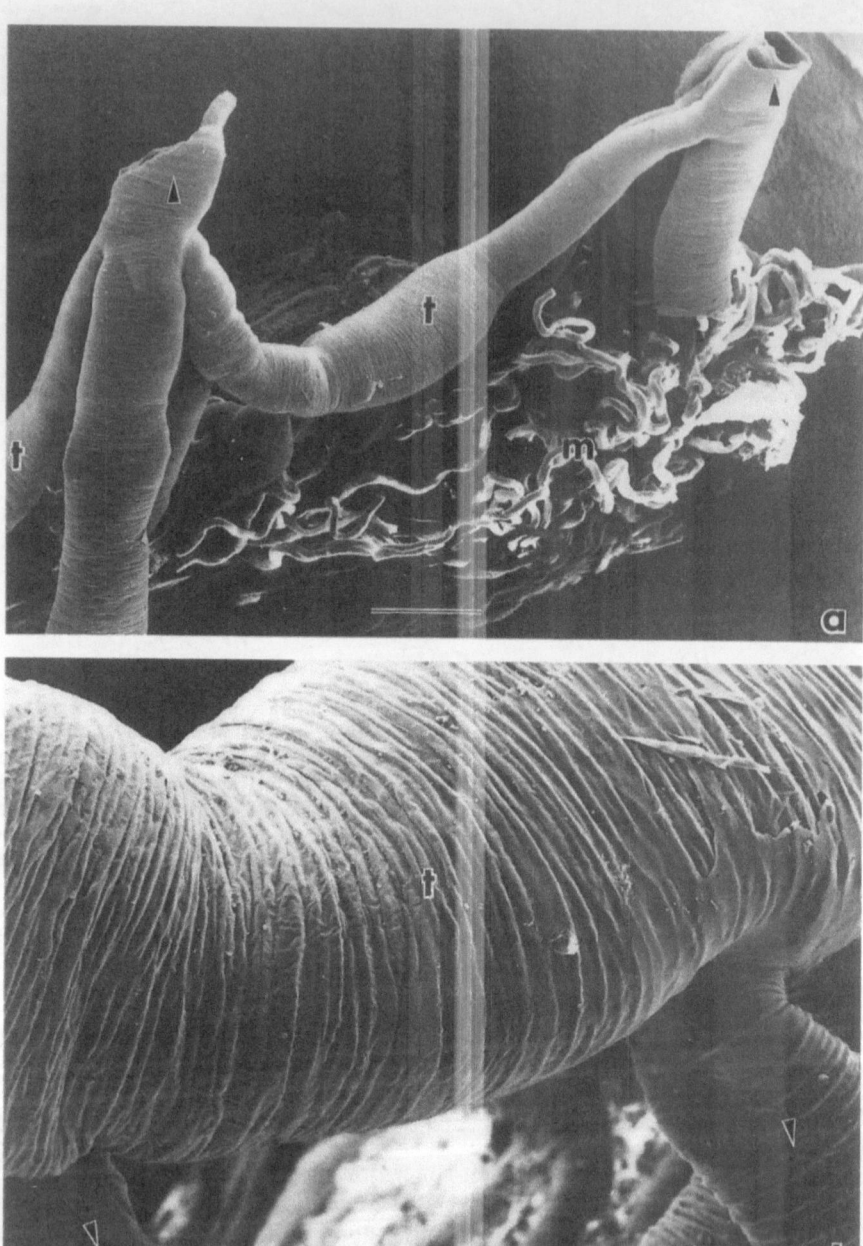

Fig. 37. a Tracheal system of a grasshopper, *Chrotogonus senegalensis*. The primary trachea which start near the spiracles, ➤, are connected through anastomotic chains, *t. m*, Malpighian tubules. b Closeup of a primary trachea, *t*, showing the spiral taenidia which maintain patency. ➤, secondary trachea. (Maina 1989b.) a *Bar* 700 μm. b 80 μm

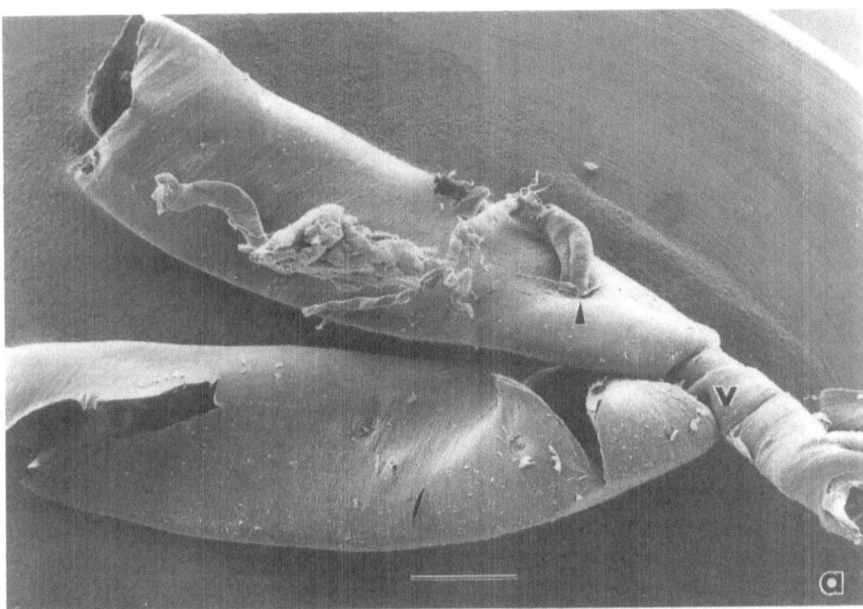

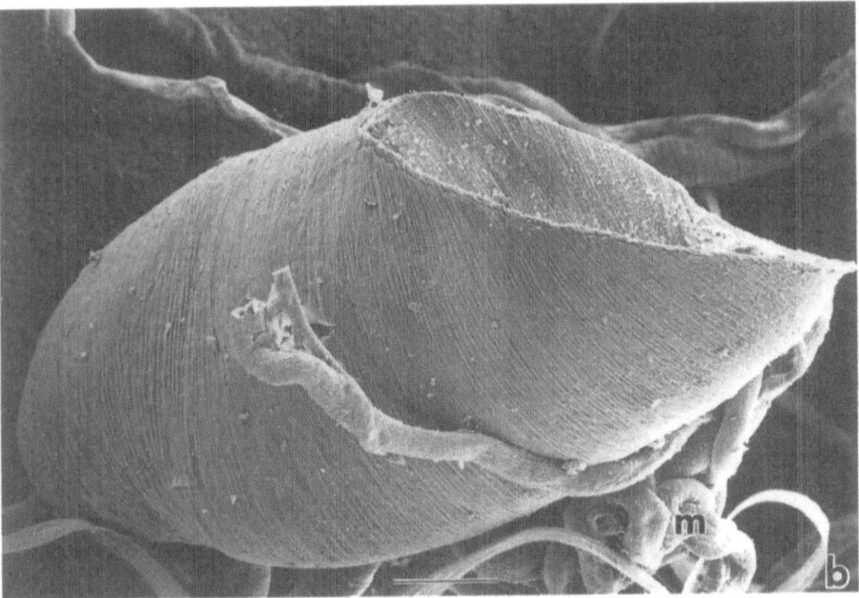

Fig. 38a,b. Air sacs of the grasshopper, *Chrotogonus senegalensis*. The air sacs occur in pairs (a), singly (b) or in clusters. They are characteristic of the large and the energetic insects which use abdominal pumping to enhance movement of air along the tracheal system. ➤, efferent trachea; *v*, possible location of a valvular apparatus which maintains a unidirectional flow of air in the tracheal system; *m*, Malpighian tubules. Like the trachea, the air sacs are made up of helical taenia, indicating that they are simply dilatations of the trachea. (Maina 1989b) **a** *Bar* 700 μm; **b** 220 μm

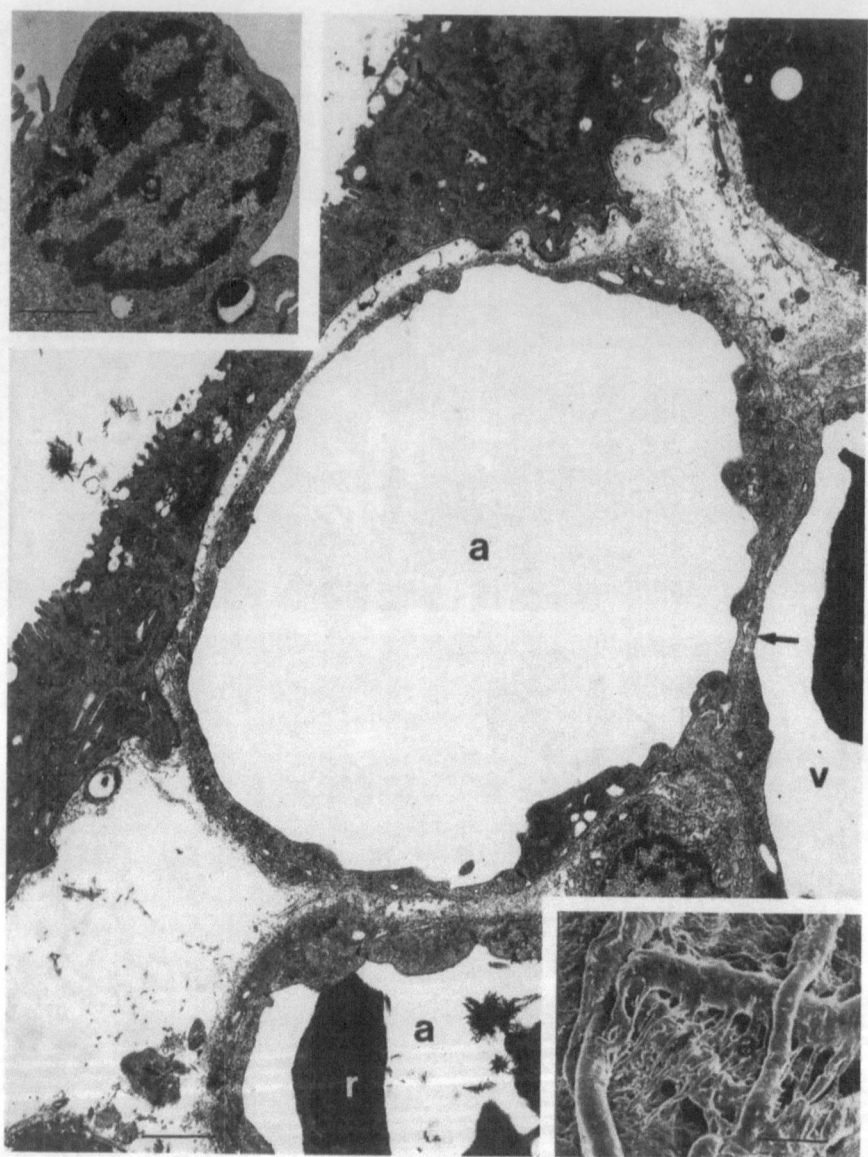

Fig. 39. The gas gland cell, *g* (*top inset*) and the rete mirabile of the swim bladder of a tilapiine fish, *Oreochromis alcalicus grahami* (*main figure* and *bottom inset*). The *bottom inset* is a latex cast SEM preparation of the arterial vasculature at the rete. The venules (*v* in the *main figure* and *arrow* in the *bottom inset*) and arterioles, *a* (*main figure*) run parallel to each other. Secretion of the acidic metabolites produced by the gas gland cells into the arterial blood and subsequent transfer into the venous blood causes release of blood gases into the bladder. ← (*main figure*) shows an extremely thin area of separation between a venule and an arteriole, sites which may promote the transfer of the acidic metabolites released by the gas gland cells. *r* Red blood cell. *Bars top inset*, 0.25 µm; *main figure* 0.8 µm; *bottom inset*, 16 µm. (Maina et al. 1996a)

In biology, however, countercurrent heat exchangers cannot simultaneously serve as efficient gas exchangers. To enhance heat transfer by bulk flow, the blood vessels of the heat exchanger have a diameter in excess of 1 mm. Gas exchange by diffusion can only efficiently occur at the capillary level where the walls are adequately thin. In fact, if O_2 were to diffuse equally well in the heat exchangers, very little, if any, O_2 would reach the peripheral tissue. Such arrangements (countercurrent heat exchangers) would have to be very sparingly used (if at all) in biology. The low O_2 tension found in the kidney tissue (e.g., Landes et al. (1964), an organ supplied with a large volume of blood, may be due to diffusion of O_2 between the blood capillaries supplying the renal papilla and those leaving it, i.e., the descending and ascending versa recta. In birds, interestingly, while the ophthalmic rete is considered a countercurrent exchanger (e.g., Kilgore et al. 1979; Midgärd 1983), it has been shown to supplement O_2 supply to the brain during high altitude flight (Bernstein 1989, 1990; Bernstein et al. 1984): the arterial blood passing through the rete en route to the brain exchanges O_2 with the venous blood, which has a higher PO_2 after draining the highly vascular eye, nasal cavity, and upper respiratory passages. Pigeons have been reported to increase cerebral blood flow during hypoxia (Pavlov et al. 1987). In mammals, direct estimations of the PO_2 in the arterial vessels of the carotid rete and systemic arterial blood indicate that a gas exchange process does not occur in the organ (Duling et al. 1979). The great efficiency of the countercurrent system of the fish gills (Fig. 36) may have been necessary for survival in a medium usually deficient in O_2 and where the levels may fluctuate dramatically spatially and temporarily. In birds, the interaction of the pulmonary capillary blood in the parabronchial gas exchange tissue (Maina et al. 1982b) occurs across a thin blood-gas barrier and an extensive surface area (Maina 1984, 1987a; Fig. 40). A continuous unidirectional air flow enhances the efficiency of the gas exchange process (Figs. 32,36). Through the multicapillary additive serial arterialization system which is built in the crosscurrent geometric arrangement between the parabronchial air flow (the air in the parabronchial lumen) and the pulmonary venous blood (Figs. 41,42), in some conditions (e.g., hypoxia and exercise), O_2 level in the arterial blood may exceed that in the end expired air (Scheid and Piiper 1972, 1989; Fig. 36). The avian lung-air sac system is more efficient in arterializing blood than all the other evolved air breathing organs of the vertebrates (Powell 1990). The arterial PCO_2 in birds may be as low as 0.93 kPa (equivalent value in mammals = 5.6 kPa) compared with that in the expired air of 5.3 kPa (Scheid and Piiper 1972; Scheid 1990). The O_2 extraction ratio in fish and cephalopods ranges between 50 and 80%, in polychaetes 30 and 70%, in crustacea 43 and 76%, sponges 6 and 40%, lamellibranchs 3 and 10%, and ascidians 4 and 7% (Benedict 1938; Hazelhoff 1939; Randall et al. 1967; Hanson and Johansen 1970). A remarkably high value of 92% has been reported in the triggerfish, *Balistes capriscus*, by Hughes (1967). In fish inhabiting well aerated water, the arterial PO_2 is generally high and usually exceeds the values needed to totally saturate the blood (Stevens and Randall 1967). Expectedly, the ventilation-perfusion ratios in fish are very high and range from 9 in the shark, *Scyliorhinus* (Baumgarten-Schumann and Piiper 1968), to as high as 70 in the trout (Randall et al. 1967): the markedly high value in the trout is due to a high ventilatory rate occurring with low O_2 extraction from water.

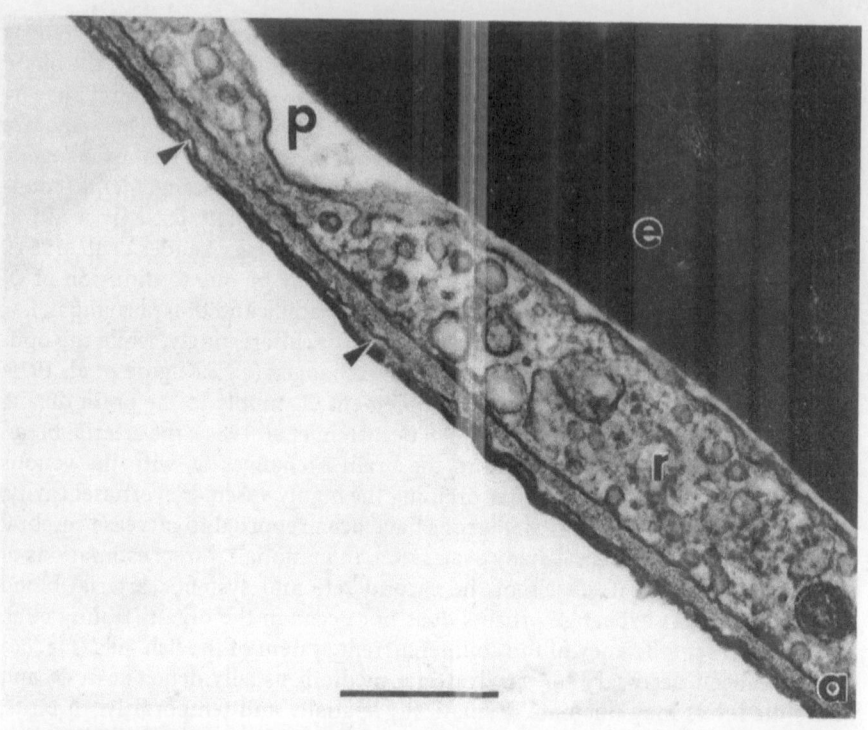

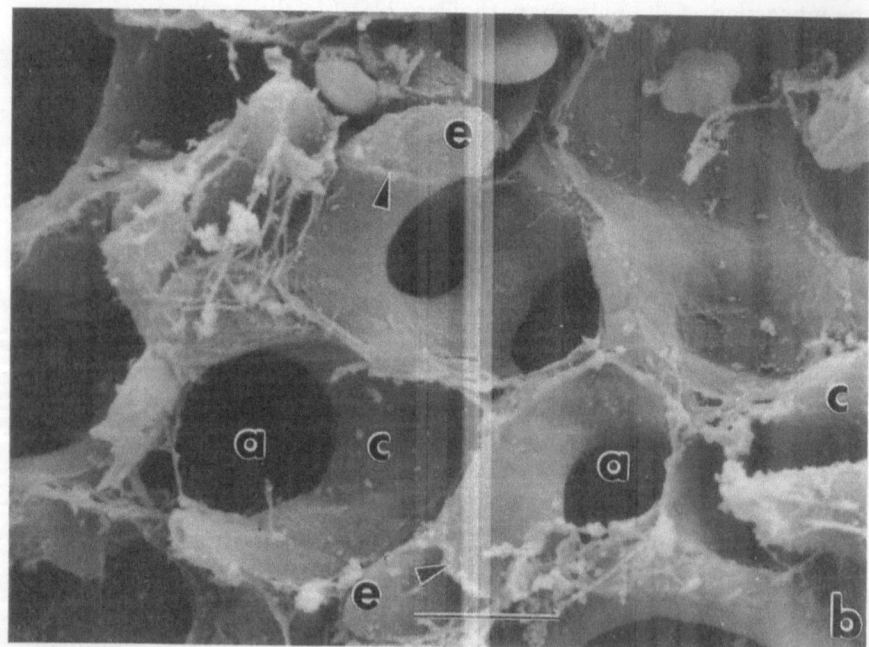

Fig. 40a,b

144

Oxygen extraction decreases with increasing ventilation-perfusion ratio in the dogfish, *Squalus suckleyi* (Hanson and Johansen 1970), and the water boundary layer provides about 80% of the overall resistance to O_2 exchange (Hills and Hughes 1970). The low O_2 extraction values in the gills with a trophic role (Table 8) may be due to the very high rate of flow of water necessary to procure food. In the human lung, where the concentration of O_2 in the inhaled air is 20.9% and in the expired air 16.4%, the O_2 extraction factor is about 22% while in birds (where the concentration of O_2 in expired air 14.5%), the value is 31% (Sturkie

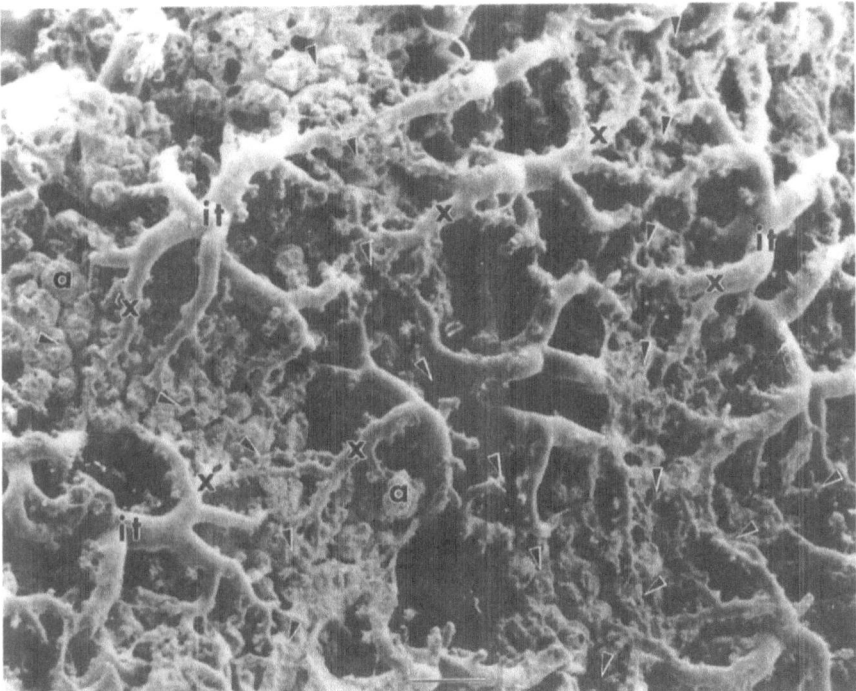

Fig. 41. Double latex cast preparation, i.e., simultaneous injection of cast material into the vasculature and into the airways of the lung of the domestic fowl, *Gallus domesticus* showing the crosscurrent relationship between the blood flow in the intraparabronchial arteries, *x*, and the airflow in the parabronchi of which the locations and the directions are shown by *arrow-heads*. Through this arrangement, gas exchange is effected along the parabronchial lengths, the concentration of O_2 in the pulmonary vein being an aggregate effect of that effected at many sites. The efficiency of the bird lung is largely attributable to this design. *it* Interparabronchial artery; *a* atria. *Bar* 100 µm. (Maina 1988a)

Fig. 40. **a** Blood-gas barrier of the lung of a bird, the house sparrow (*Passer domesticus*) showing the extremely thin epithelial cells, ➤, and a much thicker endothelial one, *r*; *p* plasma layer; *e* erythrocyte. **b** View of the gas exchange region of the lung of the domestic fowl, *Gallus domesticus* showing the interdigitation of the blood, *c*, and air capillaries, *a*; *e*, erythrocytes; ➤, blood-gas barrier. **a** *Bar* 10.3 µm; **d** 2 µm. (**a** Maina and King 1982; **b** Maina 1982)

145

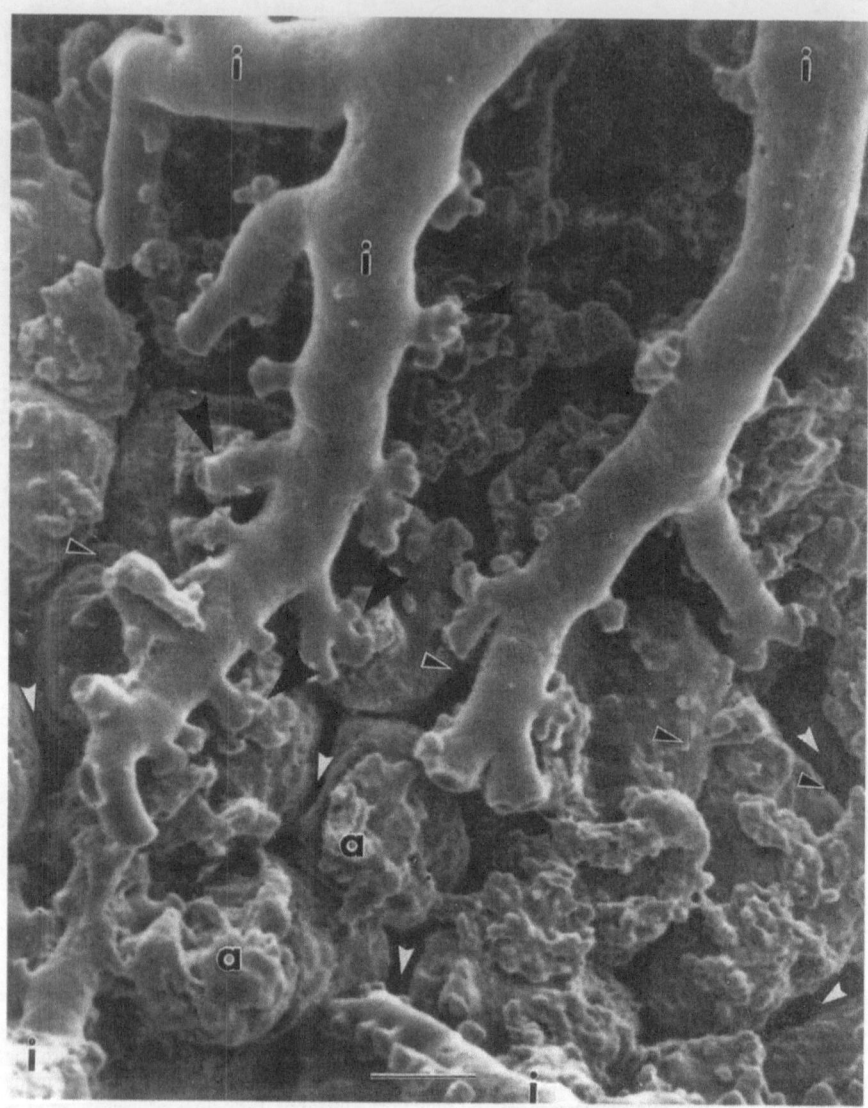

Fig. 42. Closeup view of a double latex cast preparation, i.e., injection of cast material into the vasculature and into the airways of the lung of the domestic fowl, *Gallus domesticus*, showing the crosscurrent relationship between the blood flow in the intraparabronchial arteries, *i*, and the air flow in a parabronchus (direction shown by *black arrowheads outlined in white*) which are studded by atria which terminate in air capillaries, *a*. The relationship between the intraparabronchial arteries with the parabronchi is crosscurrent, i.e., is perpendicular, while that at the gas exchange site (*large black arrowheads,* ➤) is countercurrent. The efficiency of the bird lung is largely attributable to the multicapillary serial arterialization system integral to the crosscurrent arrangement. *white arrowheads,* interatrial septae. *Bar* 20 μm. (Maina 1988a)

Table 8. Coefficient of oxygen extraction (FIO_2) from water in filter-feeding and non-filter-feeding animals. (Hazelhoff 1939)

Filter-feeding animals		Non-filter-feeding animals	
	FIO_2		FIO_2
Sponges	0.19 (0.06–0.57)[a]	Annelids	0.41 (0.30–0.66)
Lamellibranchs	0.07 (0.05–0.09)	Crustaceans	0.49 (0.29–0.76)
Ascidians	0.06 (0.04–0.07)	Gastropods	0.66 (0.38–0.70)
		Echinoderms	0.53 (0.49–0.55)
		Fish	0.62 (0.46–0.82)
Mean	0.11		0.55

[a] Values in parentheses show the range of the values.

1954). These values are on the lower side of the maximal ones of the water breathers given above. Although a countercurrent flow was established in the gills of the shore crab, *Carcinus maenas*, the O_2 extraction factor of 7 to 23% is relatively very low (Hughes et al. 1969a) compared with the values reported by Hazelhoff (1939) of 60 to 90% in *Caloppa granulata*. This is probably due to factors such as presence of a large branchial dead space, low permeability of the lamella cuticular lining, and inadequate lamellar perfusion in *Carcinus*. The importance of the countercurrent arrangement in the efficiency of gill function is evident from the fact that if the direction of flow of either of the respiratory media is reversed (i.e., to establish a concurrent system), the O_2 extraction ratio falls to below 10% (Hughes 1963). In the crosscurrent system, however, a similar procedure has no appreciable effect on the O_2 extraction ratio (Scheid and Piiper 1972). As expected from the geometric configuration of the parabrochial vasculature and blood flow to the exchange tissue relative to the air flow (Figs. 41,42), only the sequence of capillary arterialization is changed. The degree of arterialization remains the same.

Chapter 3

Gas Exchange Media, Respiratory States, and Environments

"The interaction between organisms and their environment is an old but very important problem to biologists. Organisms respond to environmental change in different ways according to the time during which the environmental change persists and according to the magnitude of the stress". Prosser (1958)

3.1 Water and Air as Respiratory Media: General Considerations

Regarding the part of the biosphere they occupy, animal life is classified into aquatic, terrestrial, and aerial groups. Among vertebrates, fish are predominantly aquatic, amphibians are transitional, and reptiles, birds, and mammals are fundamentally terrestrial. Overlaps in occupation of various ecosystems occur. Among mammals, the cetaceans have reinvaded water while some amphibians live in highly desiccating deserts (McClanahan et al. 1994). The insects, the now extinct pterosaurs, the birds, and the bats, chronologically in that order, are the only groups which have evolved powered flight. The assortment of animals such as the flying squirrels, lemurs, snakes, lizards, and flying fish which can momentarily remain air-borne are essentially gliders. They use a part of their body to delay the fall and did not have to grapple with the aerodynamic and aerobic challenges which beset the active flyers.

The biophysical properties of water and air have profoundly influenced life patterns and body forms of animals (e.g., Alexander 1990; Giorgio 1990; Strathmann 1990). These features are fundamental to understanding the evolution of form, function, and the divergence between terrestrial and aquatic animals (e.g., Bliss 1979; Graham 1990; Table 9). In particular, the differences between water and air have so greatly influenced the respiratory processes that the mechanisms for obtaining O_2 and for eliminating CO_2 which are efficient in water often fail in air. Water has been abundantly and intricately incorporated in the composition of the living tissues and cells. It is required for life by practically all living things. This ubiquitous molecule makes up as much as 90% of the total protoplasmic mass. It is unequivocally the sine qua non of life. It has been worshipped (e.g., River Ganges in India), wars have been fought over it, and civilizations have risen from it or collapsed after losing or mismanaging it (e.g., Leopold and Davies 1968). While some very simple organisms can live without O_2, none can grow without it. The fluctuations in the levels of O_2 and CO_2 in water as well as in air have greatly determined the biomass and the species composition and distribution of animal communities (Davies 1975; Graham 1990).

Among animals, remarkable differences in the tolerance, response, and susceptibility to low O_2 and high concentrations of CO_2 exist. The primary factors which govern respiratory adaptation in both air and water include: (1) the

Table 9. Main differences between water and air and their physiological consequences. (Dejours 1988)

Parameter	Water	Air	Air/Water	Consequences
O_2 and CO_2 diffusivity	+	++++	≈8000	O_2 and CO_2 tensions; acid-base balance
O_2 capacitance	+	++	30	
CO_2 capacitance	++	++	1	
NH_3 capacitance	++++	++	1/700	N end products
Viscosity	++	+	1/60	Work of breathing
Density	+++	+	1/800	Circulation, skeleton, locomotion
Kinematic viscosity	+	++	13	Buoyancy, gravity
Water availability	+++	Very variable	–	Water turnover
Ionic environment	Very variable	–	–	Osmoregulation, ionoregulation
Sound velocity	++	+	1/4	Audition
Sound absorption	++	+	1/4	
Light refractive index	1.33	1	0.75	Vision
Light absorption	++	+	1/12	
Dielectric constant	+++	+	80	Electroreception
Solubility of molecules	Variable	–	–	Distance
Volatility of molecules	–	+		Chemoreception
Diffusivity of molecules	+	++++		
Heat capacity	++++	+	1/3500	Heat dissipation
Heat conductivity	++	+	1/24	Body temperature
Heat evaporation	≈2450 kJl^{-1}			

molecular characteristics of the respiratory medium, (2) the solubility of the respiratory gases in a medium, and (3) the mode of transfer of the respiratory gases in a medium. Since other than subsistence in the different fluid media (water and air) there have been no physical barriers in the evolution of the possible gas exchangers, as envisaged by Perry (1989), water gills, air gills, water lungs, and air lungs should have evolved to the same extent. However, owing to the relatively low solubility of O_2 in water (0.031 mlO_2 per ml water), high viscosity of water, and low vapor pressure in air, water lungs and air gills have only rarely evolved. The difference in the physicochemical attributes of water and air prohibited direct conversion of the gills to lungs: a transitory gas exchanger which had to function equally well in both media was necessary. Air gills and water lungs occur (but rarely) in simple organisms, e.g., in the respiratory pleopods of the terrestrial isopods (e.g., Marsh and Branch 1979; Hoese 1983), in arachnids (Kaestner 1929), and in cases of retrogression from air- to water breathing, e.g., in the aquatic pneumonate gastropods. The gills nurtured the development of the air-breathing organs. In-series vascular connection followed. There was subsequently a gradual deemphasis of aquatic respiration, leading to subsidence of the gills. The gills were directly replaced by the air-breathing organs only where the skin served as a bridging organ.

3.2 Physical Characteristics of Water and Air

For utilization as respiratory media, only fluids are configured to offer the necessary convective transport of the respiratory gases. Water, a liquid over the biological range of temperature and pressure, and air, a gas under similar conditions, are the only two naturally occurring respirable fluids. The structural and functional consequences of the interactions between the respiratory media with the gas exchangers have depended on the magnitudes of the changes in the levels of the respiratory gases and the time scale over which the interfacing has occurred (Table 9). Water is an exacting environment to survive in. In saturated water, at 20 °C, 1 ml of O_2 is contained in 200 g of water while 1 ml of O_2 is present in 5 ml of air (weight, 7 g). The rate of diffusion of O_2 in water of $3.3 \times 10^{-5} cm^2 s^{-1}$ (Grote 1967) is lower by a factor of 10^5 compared with that in air of $1.98 \times 10^{-1} cm^2 s^{-1}$ (Reid and Sherwood 1966) while the capacitance coefficient, i.e., increment of concentration per increment in partial pressure of O_2, in water is only $1.82 nmol min^{-1}$ per 0.133 kPa compared with the much higher one in air of $54.74 nmol min^{-1}$ per 0.133 kPa (Dejours 1988; Table 4).

As a respiratory medium, the general properties of air are more obliging than those of water. Convection requirements are high in water breathers compared with the air breathers. An octopus ventilates 17 l of water for each mmol of O_2 consumed (Dejours et al. 1970). In air, all other factors held constant, diffusion facilitates supply of greater quantities of O_2. Less energy is expended in the convective transfer of the gas to the respiratory site. Aquatic organisms have evolved within the constraints of an O_2-deficient environment. Zaccone et al. (1995) contend that the lungs phylogenetically evolved as adaptations to the hypoxic conditions in the aquatic medium in which fish lived. Due to the high solubility of CO_2 in water, the molar concentration of the free gas is about equal to that in air while the concentration of O_2 in water is only about 5% of that in air. Diffusion rather than concentration reduces the rate of transfer of CO_2 in water. Both decreased diffusion and low concentration slow down the transfer of O_2 in water. Many accounts dealing with adaptations of organisms to dissolved gas levels in water address themselves to O_2 availability rather than the concentration of CO_2. Owing to the high $CO_2 : O_2$ solubility ratio, if originally normoxic water was to be rendered anoxic from aerobic metabolism alone, the PCO_2 would only increase by about 0.9 kPa (Rahn 1966). Owing to the high PO_2 in air, diffusion across the blood-gas barrier is highly efficient. The arterial partial pressures of O_2 reasonably approach those prevailing in the respiratory medium. While the O_2 extraction factor in the water breathers is high, e.g., 90% in sponges, 60 to 90% in the crab (*Caloppa granulata*), 33 to 70% in the octopus, 85% in the eel (van Dam 1938, 1954), and on average 85% in fishes (Steen and Kruysse 1964), in the air breathers, the O_2 extraction factor very rarely exceeds 25%.

3.3 The Distribution of Water and Air on Earth

There is ample evidence from geological records that water was available on Earth in abundant quantities as early as 3.8 to 3.5 billion years ago (Schopf and Walter 1983). The hydrosphere, which presently covers about 72% of the Earth, was much more extensive during the early times (e.g., Handerson 1913), reaching the greatest extent in the early Paleozoic (Bray 1985). The Earth has a geometrical surface area of about $5 \times 10^8 \text{km}^2$ and the seas and oceans cover $3.6 \pm 10^8 \text{km}^2$ of this area and contain $1.3 \times 10^9 \text{km}^3$ of water. Practically all the Earth's water is contained in the oceans, with the water in the lakes and rivers constituting only 0.01% of that which is available to life. Freshwater constitutes about 2.5% of the total volume of the Earth's water, with nearly 75% of it locked in a frozen state in the polar ice caps and in glaciers while about 0.5% of it is held in aquifers as underground water (Gross 1990; Shiklomanov 1993). Rivers and lakes are complex, biologically highly productive habitats with short histories on geological time scale. For example, Death Valley in California, the hottest and driest place in the United States, was covered by a lake some 60 m deep only about 20 000 years ago (Leopold and Davies 1968). About one third of all bony fish have evolved and live in freshwater. While it is possible to accurately determine the limits of the aquasphere, the dimensions of the aerosphere are more difficult to define. The air gradually rarefies with altitude, approaching the total vacuum of the vast outer space at an altitude of between about 100 to 1000 km (e.g., Denney 1993). From the highest altitude at which aerial animals, especially birds, operate, which with some exceptions is about 10 km, the biologically utilizable fraction of the atmosphere is $5 \times 10^9 \text{km}^3$ in volume. This is nearly four times the volume of the oceans. The physical characteristics of the ocean and seawater are fairly constant while those of the freshwater lakes and rivers are more variable (e.g., Clarke 1991). Mass movement of air by convection and diffusion equalizes the gas tensions within and between habitats. In the open seas, circadian variations in the respiratory gas tensions and concentrations are common in the superficial layers of the water (Riley and Skirrow 1975) but are much less than in ponds or rock pools.

By definition, a normoxic medium is one in which the PO_2 at sea level is about 20 to 21.3 kPa. For any PO_2, air contains more O_2 than water. The difference between O_2 production through photosynthesis and utilization in respiration determines the net levels and changes in the concentration of O_2 while CO_2 uptake by plants (during photosynthesis) and production during respiration by both plants and animals regulate its level (Fig. 10). Ensuing from the high diffusivity of CO_2 in air, increases in the PCO_2 to a level that would cause respiratory stress in the free atmosphere rarely occur except in certain microhabitats, e.g., in the waterlogged soils where local hyperbaria may occur. In free air, hyperoxia does not arise. However, in ponds or seawater pools, it may develop as a localized phenomenon. Peak levels as high as 80 kPa have been reported (Truchot and Jouve-Duhamel 1980; Heisler 1982a). In our presently stable aerial environment, hyperoxia is only studied as a means of understanding the respiratory control mechanisms and management of some clinical problems.

Although O_2 has been produced by photosynthetic organisms for the last 3.5 billion years (Chapman and Schopf 1983; Schopf et al. 1983; Fig. 8), anaerobic habitats have existed continuously during the entire Earth's history. The marine detrital sediments constitute the most extensive global continuum of an anaerobic niche. Oxygen penetrates the sediments to only a few millimeter (Revsbech et al. 1980a). The widespread distribution of black shales is a firm indicator of the anoxic conditions in the seas at particular depths in the early periods of the Earth's geomorphosis (Berry and Wilde 1978). Much of the CO_2 in the natural waters is derived from carbonates in solution and very little from the atmosphere. Acidity increases the PCO_2 in water while alkalinity, even in form of carbonates, lowers it. Even in carbonate-free water, the PCO_2 cannot be very much due to the high solubility of the gas in water, a factor which reduces its occurrence in a free molecular state to almost zero. It is presumably due to this singular fact that CO_2 does not constitute a regular respiratory stimulant in gill and other water breathers. In some habitats, putrefactive processes which entail bacterial anaerobic breakdown of organic matter may produce enormous quantities of CO_2, which may result in adverse tensions of it. In water which is free of or contains scarce aquatic plant life, dissolved O_2 is the most important respiratory factor. At a critical level, an organism may be unable to procure adequate amounts of O_2 for aerobic metabolism. In such cases, it has to evoke certain behavioral, physiological, morphological, and biochemical measures (essentially in that order) (e.g., Bartholomew 1988; Carroll 1988; Gans 1988). These may include relocation to more favorable habitats, increase in O_2 uptake from the inimical environment through physiological adjustments, and reduction in O_2 need by entering an ametabolic state. In long-standing cases of hypoxia, the animal is driven to evolve a capacity to procure O_2 directly from the atmosphere. In some organisms, extreme hypoxia results in reversible cessation of respiration especially in those animals like the lugworm, *Arenicola marina* (Toulmond and Tchernigovtzeff 1984), and the prawn, *Palaemon elegans* (Morris and Taylor 1985), which experience large diurnal fluctuations in O_2 levels in the rockpools. Hypoxia increases ventilation in all animals which have been studied and variably, hyperoxia results in hypoventilation inducing hypercapnia (Dejours 1988).

3.4 Water: a Respirable Medium and an Integral Molecule for Life

It is believed (e.g., Jervis 1995) that the quantity of water presently found on Earth already existed when the planet was formed some 5 billion years ago, but in the vapor form. With the gradual cooling of the planet to below 100 °C, liquid water was formed. In biological systems, the fundamental life processes like ionic and gas fluxes take place in aqueous solution. The suitability and fitness of the water molecule as the habitat in which life evolved on Earth was well affirmed by Handerson (1913). Over millions of years water has shaped and continues to shape the Earth. As a chemical compound, water (H_2O) is unique and paradoxical in many ways. It is a remarkably stable odorless, colorless and tasteless liquid with powerful solvent properties. Until 1783 (some 200 years ago) when Henry

Cavendish synthesized the water molecule by igniting hydrogen and oxygen, it was believed that water was an indestructible element rather than a chemical compound. The basis of the unique chemical and physical properties of water (e.g., powerful solvency power, high capillarity, expansion of liquid water between 0 and 4 °C, high thermal capacity) is the covalent bonding between O_2 and H_2 producing a dipolar molecule. Water boils at about 162 °C higher than its analog hydrogen sulfide. Its density at 4 °C is $1\,g\,cm^{-3}$. At 1 atm pressure, water melts at 0 °C and boils at 100 °C (the thermometric fixed points). It is a good ionizing solvent, a property connected with its high dielectric constant and its ability to donate and share electrons. Water acts as a catalyst, reactant as well as a solvent. For example, in some reactions dry ammonia will not react with hydrogen chloride and neither will dry carbon monoxide and dry O_2. Under geologic conditions of time, heat, and pressure, water has an important role in the conversion of plant and animal matter into organic fuels. At high temperatures, water carries out condensation, cleavage, and hydrolysis reactions (Siskin and Katritzky 1991). Ethers and esters, compounds which are not susceptible to heat alone, suffer facile cleavage and hydrolysis respectively in water at 250 and 350 °C. As the temperature of the water rises from 25 to 300 °C, some of its physical characteristics change dramatically: the density decreases from 0.997 to 0.731 g per ml, the dielectric constant decreases from 78.85 to 19.66, and the solubility parameter from 23.4 to $14.5\,cal\,ml^{-1}$. Between 250 and 350 °C, the water solvent properties approach those of polar organic solvents at room temperature.

Among the nine planets in our solar system, the Earth is endowed with large quantities of water naturally occurring in the three fundamental states of matter, i.e., liquid, solid, and vapor. It is debatable how liquid water was maintained on early Earth and Mars (e.g., Carr 1996; Sagan and Chyba 1997) despite the solar luminosity being 25 to 30% lower than at present (e.g., Newman and Rood 1977). It is popularly believed (e.g., Owens et al. 1979) that high levels of CO_2 produced by action of the carbonate-silicate cycle provided a greenhouse effect adequate to warm the early Earth. Though now cold and dry and having a thin atmosphere from constant loss of water and CO_2 into space, from study of the surface topography and the geometry of outflow channels, Mars appears to have experienced episodes of massive flooding (Baker et al. 1991). As recently as 300 million years ago, the conditions on Mars appear to have been amenable to life (Kargel and Strom 1996), at least as we know it. Compared with air, water is a more dynamic habitat. Aquatic animals may face wide extremes in O_2 availability due to factors such as ice cover, plant respiration, animals burrowing into the substratum for food or protection, and high environmental CO_2 and hence low pH, usually during nighttime plant metabolism.

Some of the unique physical features of water which are fundamental to animal and plant physiology include: (1) a high specific heat (0.9988 calories per g per °C) which affords a stable temperature, (2) relatively weak intermolecular forces which enable heat to be efficiently transferred by convection (thermal conductivity, $5.14\,cal\,h^{-1}\,cm^{-1}\,K^{-1}$ at 20 °C) within the medium, (3) high surface tension which at 72.8 dynes cm^{-1} is one of the highest among liquids, (4) high wetting property and capillarity, features which are crucial in protoplasmic-cell organelle interfacing, (5) neutral pH which promotes fast reaction kinetics in most

154

biochemical processes, (6) great solvency power and high dielectric constant (80.1 in pure form) which enable it to accommodate different molecular and ionic factors, (7) maximum density (at 4 °C) which makes it possible for water to sink to the bottom while still in a liquid form. The top layer, which cools to below 0 °C, expands and floats to the top, providing a surface cover of ice which prevents the entire water mass from freezing and killing most of the organisms (which take refuge at the bottom), and (8) in vapor form, water has high diffusivity and is thus important for thermoregulatory processes. The presence of dissolved salts, e.g., in seawater, lowers the temperature at which the maximum density is attained. In brackish water (salinity 18 ppt), the maximum density occurs at 0 °C and in the ocean water (35 g of salt per kg of liquid), it occurs at -3.5 °C. Within ordinary conditions of life, water is incompressible. The intermolecular distance is about ten times less than the average value for most gases. Owing to the much weaker cohesive forces (van der Waal's forces), CO_2 and O_2 are gaseous at the physiological range of temperature while in water, where the forces are much stronger, the gas/liquid transition (when the cohesive forces are overcome by the kinetic energy of the molecules) takes place at the relatively high temperature of 100 °C (373 K) at 1 atm (101.3 kPa) pressure. It is only at extreme pressures, e.g., at great depths of the seas and oceans, that compression of water becomes significant in biology (Somero 1992). At a depth of 10 km, the PO_2 is 0.8 atm (Enns et al. 1965). Gases such as CH_4, H_2S, N_2, NH_3, SO_2, H_2, and CO_2 are found dissolved in some aquatic habitats. Many flatfishes burrow into the bottom mud, a process likely to reduce the flow of water across the gills impeding excretion of CO_2 and O_2 uptake (Lennard and Huddart 1989). A concentration of H_2S of 12 mg per l was reported in a lake in Cyrenaica by Smith (1952). Despite its abundance in the atmosphere, N_2, which constitutes about 78.08% by volume at normal pressures, plays no known respiratory role. At high pressures, it may be lethal by inducing narcosis.

3.4.1 Oxygen and CO_2 Content in Water: Effect on Respiration

In natural circumstances, O_2 and CO_2 are the only gases of biological interest in water. The concentrations of these gases have dramatically fluctuated during the past geological epochs (Graham et al. 1995; Fig. 9). While CO_2 tension decreased from a peak values of about 100 times of the Precambrian period to the present low level, O_2 has exhibited remarkable changes (e.g., Kasting et al. 1979). Mainly owing to the complex interrelationship between the total concentration of CO_2 and its partial pressure in water, particularly if water contains CO_2 fixing (buffering) factors such as carbonates, in many freshwaters and to a slight extent in marine water, it is difficult to accurately predict the amount of CO_2 in a given kind of water. The levels of CO_2 in waters (except in distilled water, which is devoid of any buffer system and hence CO_2 exists in a dissolved form) differ remarkably as opposed to its rather stable level in the atmosphere. Large quantities of CO_2 may exist at very low tensions due to formation of bicarbonates. Changes in CO_2 levels greatly affect its excretion and hence pH regulation in water breathers. In the ocean water, which has a high environmental buffering capacity, the PCO_2 does not vary much from the value of 0.03 kPa which is the equilibrium point with the atmosphere. When exposed to water, the atmospheric gases are taken up at the surface by diffusion and convection until an equilibrium (PO_2 = 20 to 21.3 kPa) is reached: the enrichment is effected mainly by the currents which are produced by the winds as well as by the variations in the temperatures which influence the specific gravity of the water. Over a distance of a few mm, diffusion is a very slow process. Krogh (1941) estimated that it would take as long as 42 years for an O_2 molecule to reach a depth of 250 m of water. In oceanic inlets, notable vertical stratifications in O_2 levels attributable to changes in temperature and salinity, horizontal variation associated with pockets of photosynthesis, and respiration and decay of organic materials occur (e.g., Platt and Irwin 1972): while the surface water is close to saturation, the bottom one is virtually anoxic, with the O_2 levels generally being below $1 \, ml \, l^{-1}$ at depths exceeding 100 m. In seas and oceans, as temperature and salinity increase, O_2, CO_2, and NH_3 content decrease (Gameson and Robertson 1955). In most cases, anoxia is accompanied by increased concentrations of CO_2 and H_2S (Powell et al. 1979). With increased temperature, O_2 content drops due to reduced solubility while the PO_2 drops only moderately owing to compensatory increase in the kinetic molecular diffusivity. The effect of temperature on the solubilities of CO_2 and NH_3 is rather complex owing to the chemical reactions with the water molecule. It is not as easily predictable as is the case for the relatively inert O_2.

Dependent on factors such as temperature, barometric pressure, photosynthetic activity of the plant matter, respiratory processes of the microorganisms, circulatory and mixing processes, and concentration of dissolved solids, at sea level, a liter of distilled and surface sea water contains only 6.34 and 5.11 cm^3 of O_2 respectively. These values are equivalent to the O_2 content of the rarefied air at an altitude of over 20 km. At 15 °C, 1 l of pure water dissolves 1 l of CO_2 at 1 atm pressure. The great solubility of CO_2 in water compared with air suggests that the greatest challenge towards the evolution of air breathing was not that of acquisition of O_2 but rather that of elimination of CO_2 into the air (Sect. 1.18 and 5.4). The respiratory epithelia of the air breathing organs, e.g., gills in fish (Randall et al. 1981), skin and lungs of amphibians (Toews et al. 1978), and lungs of most higher vertebrates (e.g., Bidani and Crandall 1988) contain appreciable quantities of carbonic anhydrase which accelerates the dehydration of HCO_3^- ions in blood, enhancing CO_2 clearance across the gas exchanger (Burnett et al. 1981; Gros 1991). In torrential rivers and streams, as well as the surface waters of lakes, seas, and oceans, the PO_2 and PCO_2 are at or near equilibrium with the atmosphere due to turbulence and agitation, which enhances the solubility of air (Piiper et al. 1962). However, in a variety of closed or stagnant collections of water such as the Black Sea, (Sverdrup et al. 1949), the inner continental shelves like the Gulf of Mexico (Leming and Stuntz 1984), bottom or even surface waters of tropical marshes (Carter and Beadle 1931; Jones 1961), small ponds and intertidal pools (Sverdrup et al. 1949; Truchot and Jouve-Duhamel 1980), interstitial water held between sediments (Brafield 1964; Revsbech et al. 1980b; Reimers et al. 1986), and the hydrothermal vents of the midocean ridges (Johnston et al. 1986), the water is hypoxic and even virtually anoxic. In such waters, the depletion of O_2 usually correlates with elevated CO_2 concentration and low pH status especially at night

Table 10. Oxygen and carbon dioxide capacitances in distilled water, in seawater, and air at various temperatures. (Dejours 1981)

T	Distilled water			Seawater[a]		Air
	βWO_2	βWCO_2	$\beta WO_2/\beta WCO_2$	βWO_2	βWCO_2	βg
°C	$\mu mol\,l^{-1}\,mmHg$			$\mu mol\,l^{-1}\,mmHg$		
40	1.35	31.33	0.0431	1.10	29.40	51.23
37	1.40	33.52	0.0418	1.17	30.40	51.73
35	1.43	34.99	0.0409	1.21	31.30	52.06
30	1.53	39.30	0.0389	1.32	34.47	52.92
20	1.66	44.86	0.0370	1.43	39.01	53.81
15	2.01	60.23	0.0334	1.67	51.58	55.68
10	2.23	70.57	0.0316	1.83	60.00	56.66
5	2.52	84.17	0.0299	2.03	70.79	57.68
0	2.87	101.25	0.0283	2.28	85.00	58.74

[a] Data on seawater based on chlorinity of 19 ppt; in air, the capacitance of O_2 and CO_2 are identical.

when photosynthesis stops and respiration continues (Heisler et al. 1982; Truchot and Jouve-Duhamel 1980). Water PCO_2 values as high as 8 kPa (Heisler et al. 1982) and PO_2 values beyond 67 kPa (Dejours et al. 1977) have been recorded in natural waters. By suppressing gas ventilation and hence CO_2 elimination, hyperoxia may elevate PCO_2 in blood and precipitate acidosis. Respiratory gases are less soluble in salt solutions (Table 10). Whereas pure water at 0 °C contains 10.29 mlO_2 per l, at a salinity of 10 ppt, it contains 9.13 mlO_2 per l and at 20 ppt, 7.97 mlO_2 per l. At 30 °C, the quantities of dissolved O_2 are 5.57 mlO_2 per l (pure water), 5.01 mlO_2 per l (10 ppt salinity) and 4.46 mlO_2 per l (20 ppt salinity). Contingent on location, size of the water mass, and physical mixing, convective movements occur up to a particular depth, the thermocline. Below the critical level, the water is stagnant and is held at a relatively low temperature. In oceans, convective currents equilibrate the respiratory gas tensions to an appreciable depth but at 100 to 1000 m, O_2-poor areas exist (Harvey 1957). In physically isolated water masses such as the Black Sea and the Gulf of Panama, where conditions for circulation are restricted, hypoxia may exist at all levels. Due to the relatively slower rate of diffusion and convective movement of the respiratory gases, compared with air, aquatic environments are subject to greater spatial and temporal variations in O_2 and CO_2 levels resulting from biological activities of plant and animal life (Garey and Rahn 1970). In isolated water basins and fjords, convective mixing may be prohibited by flow and thermal differentials and/or salinity-induced layering of water leading to hypoxic and even anoxic conditions (e.g., Seliger et al. 1985). In some coastal estuarine areas, due to its lower density, the riverine water floats on top of the heavier seawater (as a result of its less salinity and higher temperature) causing differences in O_2 and CO_2 tensions between surface and deep waters (e.g., Gnaiger 1983; Officer et al. 1984). The diversity of benthic species decreases with depth and varies seasonally with the available O_2 (Hoss 1973).

Fluctuations in the levels O_2 and CO_2 are common in water. The frequency and amplitude are determined by factors such as depth and rate of water circulation, transparency and presence of dissolved and particulate substances, temperature, density of flora and fauna, eutrophication of organic matter, and presence of ice or plant cover (e.g., Carter 1955). The penetration of light through water is an important factor which determines productivity and hence availability of food to aquatic life. It depends on factors such as surface cover, turbidity from presence of suspended mineral particles, organic matter and microorganisms, and the wavelength of light. The blue-green light travels to a depth of 50 m and the red light is least penetrative (Harvey 1957). Solar light is entirely extinguished at a depth of 1000 m in the clearest of the oceans (Pough et al. 1989). In the turbid water of Lake Victoria, long wavelengths penetrate into the water layers more than the shorter ones (van Oijen et al. 1981; de Beer 1989): during the day, at a depth of 2.5 m, the light is about 5% of the intensity at the surface. During the day, owing to photosynthetic activity of algae and other simple plants, O_2 production may exceed the respiratory demands with the concentration increasing to very high levels, while during hours of darkness the water may be rendered totally anoxic from the ensuing respiration (Ultsch 1973; Kramer et al. 1978; Fig. 43). Compared with freshwaters, O_2 saturation levels are generally more stable in marine waters where, due to turbulent mixing with air, surface water is often at equilibrium with the atmospheric gases. In the sea, O_2 levels in the photic zone occasionally rise to 130% saturation or above owing to photosynthetic activities

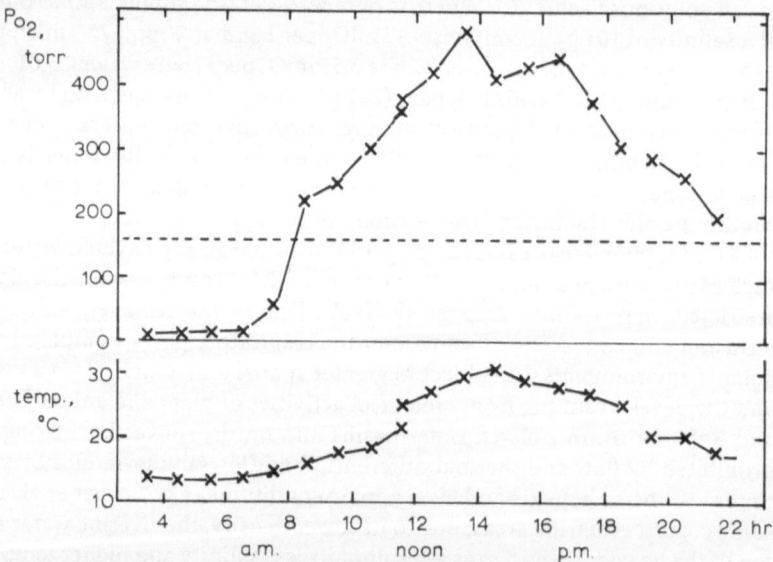

Fig. 43. Diurnal fluctuations of the partial pressure of O_2 and that of temperature in a pool of water containing aquatic plants. The water is saturated with O_2 at about midday and is virtually anoxic after midnight. *Dashed line* indicates the PO_2 in air. (After Dejours 1988)

(Fairbridge 1966). At the bottom, most water masses are asphyxic. In the brackish Japanese Lake Nakanoumi, at the surface, owing to algal photosynthetic activity, the water is 100% saturated with O_2 while at the bottom (depth 6.5 m), the concentration is almost zero due to the biodegradation of organic matter (Kimoto and Fujinaga 1990). In the marine intertidal rockpools, the PO_2 ranges from 0.27 kPa at night to more than 66.7 kPa during the day while the PCO_2 and the pH respectively range from 0.36 to 1.3×10^{-4} kPa and 7.3 to 10.2 (e.g., Truchot and Jouve-Duhamel 1980). Resulting from photosynthetic activity, over a 24-h period, in rockpools, the temperature reaches a high of 24°C and a minimum of 14°C. During the day, the concentration of O_2 is in excess of 20 mg per l (more than 300% saturation) dropping to less than 1 mg per l (about 3% saturation) during the night (Daniel and Boyden 1975): the concentration of CO_2 is reduced from 100 mg per l to about half of the value while the pH drops from a night time value of 7.5 to a daytime maximum of 9.5. In warm sunny days, the shallow ditches of the Dutch polder lands present a diurnal PO_2 peak level of 66.7 kPa dropping to 2.7 kPa at night (Jones 1961). Supersaturation with O_2 as high as 364.5% in the upper portion of a Wisconsin lake was attributed to the photosynthesis of the algal growth (Welch 1952). In Lake Waubesa (Wisconsin, USA), a concentration of O_2 as high as 30 mg per l which resulted in sudden fish mortalities was reported by Woodbury (1942). Similar observations were made by Wiebe (1933). Seasonal low levels of O_2 in freshwater masses can lead to massive deaths of fish (winter kills) which result from decay of organic materials especially where atmospheric O_2 recharge is prohibited by surface ice cover (Gnaiger and Forstner 1983). In such cases, aquatic hypoxia is exacerbated by reduced level of photosynthesis due to the decrease of the solar (shortwave) radiation available to the plant life. Atlantic lobster kills are sometimes attributed to low dissolved O_2 (Young 1973). Where light can get through, however, the bubbles which are released by underwater plants and accumulate under the ice may contain as much as 45% O_2 by concentration. The air may be utilized by aquatic organisms (Krogh 1941). In most air-breathing vertebrates, hypoxia is the main drive for respiration, but at sea level CO_2 and H^+ are the basic biochemical regulators. Such animals maintain CO_2 in a steady state leading to an average arterial PCO_2 of about 5.3 kPa. In the aquatic animals, PCO_2 levels are very low. The importance of CO_2 in the regulation of breathing increases with dependence on air breathing. Ventilatory activity in fish which breath water is driven by O_2 levels and is virtually insensitive to CO_2. Diving (e.g., Andersen 1966) and fossorial (e.g., Augee et al. 1970/71) air-breathing vertebrates (birds and mammals) have blunted sensitivity to CO_2.

3.4.2 Density and Viscosity of Water

Water is about 1000 times denser than air. The molecular diameter of water is about 2 Å while the intermolecular distance is about 3.1 Å. The average intermolecular distance in gases is about 33 Å while the molecular diameters of most gases ranges from 2 to 5 Å, i.e., about one tenth of their intermolecular spacing. Water, a medium with a high specific gravity and viscosity, provided the necessary support which promoted the development and evolution of the delicate inverte-brate life to amazing sizes: a specimen of the Atlantic giant squid (*Architeuthis* sp.) weighing 2 t has been captured. However, the metabolic cost of convective transfer of water in the gas exchangers as well as the restrictive physical move-ment in it sets a limit to the exercise and metabolic capacities of aquatic organ-isms. The fastest fish, the tuna, can only briefly attain and sustain a maximal speed of about $20\,\mathrm{m\,s^{-1}}$ (for 10 to 20 s) while a bird, e.g., the swift will attain and sustain a speed of 40 m per s (Gray 1968). The cost of aquatic respiration is increased by the fact that although the PO_2 may be similar in water and air, water contains 30 times less O_2. To extract an equivalent quantity of O_2, a water breather has to expend much more energy than does an air breather (Sect. 2.9).

3.4.3 Thermal Capacity and Conductivity of Water

The heat transfer properties, caloric capacity, and conductivity of water is about 3 orders of magnitude greater than that of air (Tables 4,9). With exception of liquid ammonia, water has the highest specific heat ($4200\,\mathrm{J\,kg^{-1}}$) of any substance in liquid form at room temperature. The thermal conductivity and capacity of water are respectively 24 and 3000 times greater than that of air. These features present constraints for survival and respiration in water. In aquatic breathers, the metabolic heat carried by the blood to the gills is soon lost to the environment (Carey 1973; Carey and Lawson 1973). The temperature of the body tissues is within 1 °C of the ambient water temperature (Carey et al. 1971; Reynolds et al. 1976). The limitations precipitated by the water with respect to heat conservation and body temperature regulation obliged terrestrial location for evolution of endothermic-homeothermy to develop. For similar reasons, except for the endothermic fish, the extant members of the vertebrate classes Agnatha, Chondrichthyes, and Osteichthyes are obligatively poikilothermic (Hazel 1993).

For the period over which the multicellular organisms have been on Earth (Schopf et al. 1983; Fig. 8), the average temperature of the tropical surface waters has varied by only about 5 °C while that of air has changed by perhaps 15 °C (e.g., Cloud 1988). Surface temperatures may change dramatically within short inter-vals and distances. Annual temperature fluctuations of as much as 60 to 70 °C on land are occasionally recorded. The annual latitudinal fluctuations in tempera-tures throughout all temperate and subtropical seas range between 0 to 28 °C and at no place in the open sea is the annual range of temperature more than 10 °C (Nicol 1960). In the North Atlantic Ocean, the surface temperature varies by only 8 °C while at the equator, the mean yearly temperature variation is only 0.5 °C (Sverdrup et al. 1949): deeper waters exhibit less variations in temperature. The mean annual temperature of the ocean at the equator is about 27 °C, at 30°S latitude it is 20 °C, and at 30°N it is about 17.5 °C (Wüst et al. 1954). The narrow range of temperatures found in the ocean is attributable to constant convective circulation (F.G.W. Smith 1957) and the high specific heat of water (Dorsey 1940). Freshwater lakes exhibit greater stratification and fluctuation of temperature than oceans, the gradient depending on factors such as drainage, mass of water, latitudinal location, surrounding terrain, and depth (Beadle 1974; Hutchison 1975). The temperature characteristics of a lake particularly in the warmer parts of the year are more complicated than those in the seas and oceans (Beadle 1974): the temperature from the surface to a depth of about 6 m is fairly stable at about 20 °C, from 6 to 10 m it drops suddenly to about 5 °C (the thermocline) and stabilizes below that depth. Very small masses of water may show a range of temperature of 0 to 42 °C (Young and Zimmerman 1956). In the salt water pools in southeast England, Mardsen (1976) noted that in summer, diurnal temperature dropped from 24 to 14 °C over a 24-h period and in winter the temperature ranged from a minimum of −0.5 to 5 °C. Many aquatic animals are able to regulate their body temperatures within a very narrow range by ascending or descending in a water column or moving to shallower or deeper waters (e.g., Feder et al. 1982). As the O_2 content of water varies inversely with the water temperature, a compromise has to be established between the needs for O_2 and the preferred body temperature.

3.4.4 Derelict Waters: Respiratory Stress from Hypercapnia and Hypoxia

Derelict waters reproduce the inimical environmental conditions in which air breathing evolved. The adaptations by which animals are able to survive in hypoxic and hypercapnic waters are hence of relevant scientific interest. They demonstrate the factors which enforced and the strategies which animals adopted for transition from water- to air breathing. Current geoclimatic data suggest that the environment during the Devonian period, a time during which most of the bony fish first evolved lungs (Pough et al. 1989), was in all likelihood similar to a tropical swamp. It was characteristically anoxic, reducing, and CO_2 occurred at high concentrations (e.g., Valentine and Moores 1976). During the continental formative years, the drainage systems of most tropical swamps particularly the Central ones were not well formed (Bishop and Trendall 1967). Especially in the major land mass located at the equator, which is often called the Old Red Continent (e.g., Livermore et al. 1985), intense putrefaction of the plant matter resulted from the prevalent high temperatures. Pressure and the need to evolve potential for air breathing and even relocation to land was very intense for survival in such pernicious and noxious habitats. While the quantity and the PCO_2 in air is generally low, in some waters, a fairly large amount of CO_2 occurs at low tension but in a few cases it may be found in such a high tension as to suppress respiration (Table 11). Such conditions are common in the warm waters of the tropical regions of the world (Beadle 1974; Munshi and Hughes 1992). Apart from the ephemeral ones which exist only in wet seasons, there are few inland water masses which have not been radically changed in form and drainage pattern during the past million years. Such habitats are associated with abundant and distinct animal and plant communities. The presence of amphibious vertebrates in a swamp corresponds with a radical transformation of one ecosystem to another, a process which may be enhanced by natural and human factors such as drought, siltation, excessive infestation with aquatic flora, drainage, and land reclamation.

Table 11. Environmental features of some representative swamps of the tropical world. (Beadle 1932; Carter 1935; Rzoska 1974; Dehadrai and Tripathi 1976)

Swamp	Area	T	pH	cO_2	CO_2	Alkalinity	Conductivity
Nile swamps							
Swamp rivers	Bahr el Gabel	25	7.5	0.66–6.2	3–18	–	112–550
	Bahr el Ghazal	25	7.4	1.3–8.0	–	–	40–370
	Bahr el Sobat	25	7.3	3.4–6.2	–	15–50	110–280
	Bahr el Zeraf	25	7.8	2.2–5.9	–	92–116	245–370
Standing waters	Shambe Lagoon	25	7.5	2.4–6.0	–	32	–
	Lagoon RP 12	25	7.8	6.1–102	–	21–31	200–250
	Lake Ambandi	25	6.8	8.0	–	25	40–55
	Khar Perboi	25	–	1.8–7.2	–	–	–
	Khar Atar	25	8.3	6.9–12	–	29	–
Guiana swamps	Grass Swamp A	28	4.5	0.22–1.2	8–9	0	–
	Other Swamps	28	4.5	0.65	14	0	–
	Pool B	28	4.1	0.6–1.5	5.0	0	–
East African swamps	Lake Naivasha (Papyrus swamp)	25	7.8	2.5	–	–	–
	Kazinga Channel (papyrus swamp)	27	5.9	0	–	7.5	–
	Kitoma (papyrus swamp)	–	6.7	0	–	72	
Indian swamps	Polluted swamp	26	8.8	6.5–8.2	16–20	148 ppm	–
	Semi-senescent swamp	26	8.3	0.6–3.2	20	143 ppm	–
	Senescent swamp	31	8.0	1.5–2.7	12	94–162	–

Symbols and units: T – temperature (°C); cO_2 – dissolved $O_2 \, mg l^{-1}$; CO_2 – free CO_2, $mg l^{-1}$; alkalinity – total alkalinity, $10^{-4} N$; conductivity, $\mu \, mho \, cm^{-1}$ (20°C).

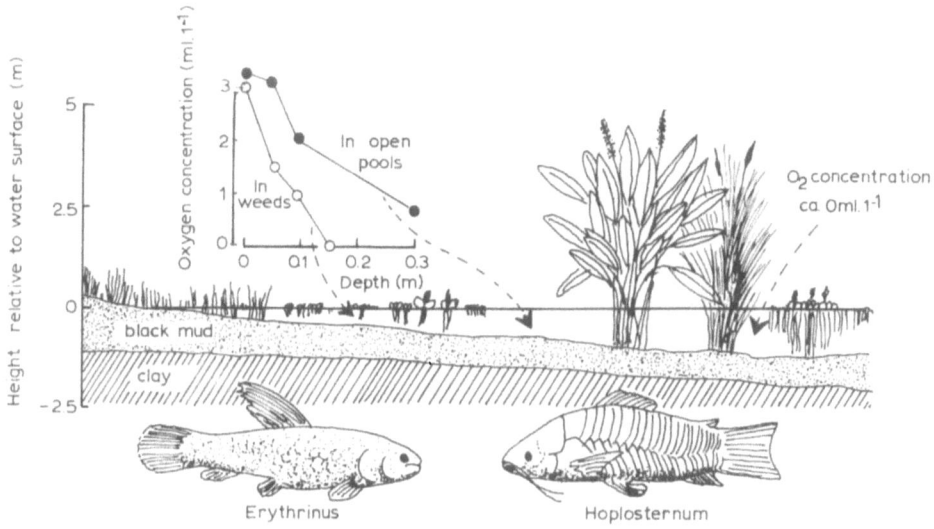

Fig. 44. Cross-section of the edge of a Paraguayan swamp showing the effect of aquatic plant growth on the O_2 levels in water. The O_2 tensions both in the open pools and the weeds decrease with depth but even close to the surface, the levels are very low. The commonest fish are *Hoplosternum litorale* and *Erythrinus unitaeniatus*. Both are air breathers. (Carter and Beadle 1931)

The tropical derelict masses of water which are shallow and stagnant or slow moving present adverse respiratory conditions. In standing waters, mixing of surface and deeper waters is inhibited by thermal stratification. The bottom water may be virtually anoxic while the top layer is saturated with O_2. In a Danish lake, Nielsen and Gargas (1984) observed that O_2 saturation was as low as 5% in the near-bottom water layers. In the tropical Paraguayan swamps, the Chaco, which are dominated by emergent macrophytes, *Thalia* and *Typha*, both of which grow to a height of 5 m and floating vegetation such as *Pistia*, *Azolla*, and *Aichhornia*, the surface O_2 levels are seldom above 50% saturation and the bottom ones are perpetually anoxic (Carter and Beadle 1931; Fig. 44). In some swamps, the dense plant canopy reduces the photosynthetic production of O_2 by algal activity and prohibits water stirring, factors which compounded by the intense utilization of O_2 in the putrefactive processes make the water anoxic within a fraction of a centimeter of the surface (Carter 1955; Beadle 1957). In the vegetation-covered Floridan swamps, the PO_2 is less than 0.67 kPa, $PCO_2 = 8.3$ kPa and the pH is 5.6 (Heisler et al. 1982). The waterlogged mat and the bottom peat are not only totally devoid of O_2 but are also highly reducing. Redox potentials (Eh) of -100 mV have been recorded within 30 cm of the water surface (Beadle 1957) and in most such cases, no measurable O_2 was detected within 2 cm of the surface. Due to the fact that the decomposing organic matter in the Lake Victoria basin swamps (largely covered by papyrus overgrowth) is mainly carbohydrate, the gaseous end product is mainly CH_4 (60%) with CO_2 consitituting only 30% and the remaining 10%

163

being made up of H_2, carbon monoxide, and ethylene (Visser 1963; Dehadrai and Tripathi 1976). In the extensive Sudd swamps of the Upper Nile (Rzoska 1974), a high concentration of H_2S has been reported (Talling 1957), a feature associated with the much greater reducing conditions where the O_2 levels deep inside the vegetation are only about 10% saturation (Rzoska 1974). The peculiar aspects of the swamps (scarcity of O_2, highly reducing environment and high CO_2 levels; Table 11) determine the variety, nature, and the biomass and productivity in such habitats (Carter and Beadle 1931). To complicate matters, toxic gases such as H_2S occur in high concentrations (Somero et al. 1989). In an ecosystem in which the conditions are highly adverse to life, adaptation for air breathing is intense. Such a change confers great selective advantage (Carter and Beadle 1931; Dehadrai and Tripathi 1976). In the Ugandan swamps, aquatic insects (Hemiptera, Coleoptera, mosquito larvae), pulmonate gastropods (e.g., *Biomphalaria sudanica*), oligocha-ete worms (e.g., *Alma emini*), and fishes (e.g., *Protopterus aethiopicus, Polypterus bichir, Clarias lazera, Ctenopoma muriei, Gymnarchus niloticus*), all of which are air breathers coexist (Beadle 1974; David et al. 1974). The swamps are character-ized by a rapid growth of macrophyte cover and subsequent decomposition of the luxuriant organic matter leading to intense putrefaction and anaerobic decompo-sition of the overgrowth. These processes generally result in a hypoxic and hypercarbic habitat (Nassar and Munshi 1971; Dehadrai and Tripathi 1976; Ultsch 1976), except during episodes of peak photosynthesis.

The tropical swamps shrink during the hot spells or become muddy to the extent that the gills become unsuitable for gas exchange due to clogging by masses of floating and suspended detritus. Such waters are slightly acidic (with a pH of 6 to 6.5) mainly from the high levels of dissolved CO_2. The Guiana swamps have very low pH ranging from 4.3 to 4.4 (Carter 1935) while those in India are alkaline (pH 8 to 10; Dehadrai and Tripathi 1976; Munshi and Hughes 1992). Dissolved and bound CO_2 in the papyrus Ugandan swamps was found to be 148 mg per l at the end of a rainless season (Milburn and Beadle 1960). Up to 30 ppm free CO_2 was recorded in some Indonesian swamps by van Vass and Vaas (1960). During summer, in the water hyacinth-infested Floridan and Louisianian swamps, Lynch et al. (1947) measured a concentration of dissolved CO_2 as high as 80 ppm (about 213.3 kPa). A zero concentration of O_2 at a depth of about 30 cm, 0.6 ppm (about 1.2 kPa) at a depth of 5 cm and dissolved CO_2 level as high as 8 kPa were deter-mined in a hyacinth infested pond in Gainesville, Florida (Ultsch 1976). In the tropical swamps, the CO_2 levels can rise to a level which would adversely affect the oxygenation of the hemoglobin and disrupt acid-base balance. In such desolate aggregations of water, in circumstances where the dissolved O_2 is low, that of CO_2 is usually high (Lynch et al. 1947). High concentration of O_2 alleviate the effects of high CO_2 levels: a 230-g salamander, *S. lacertina*, can tolerate a PCO_2 of 28 kPa for as long as 70 h when held in water in which the PO_2 ranges from 17.9 to 37.3 kPa (Ultsch 1976). Owing to their low metabolic rate (Whitford and Hutchison 1967) and capacity to tolerate high concentration of CO_2 (Ultsch 1976), the sirenids are well adapted to subsist in the hyacinth-infested waters where they constitute the dominant vertebrate fauna (Ultsch 1973). In the vegetation-covered lakes of the southeastern United States inhabited by two species of salamanders and the Congo eel, Heisler et al. (1982) reported a PO_2 of less than 0.67 kPa, a PCO_2 of

8.3 kPa, and a pH of 5.6. In a hyacinth (*Eichhomia crassipes*)-infested swamp, zero concentration of O_2 and a high one of CO_2 of 80 ppm were recorded (Lynch et al. 1947; Ultsch 1973). The temperate tidal pools examined by Truchot and Jouve-Duhamel (1980) had a temperature which ranged from 12 to 24 °C and the O_2 content amounted from virtual anoxia to hyperoxia within a 24-h cycle.

High concentration of CO_2 in water leads to certain physiological stesses on the aquatic animal life which call for specific adaptations for subsistence in such a habitats (e.g., Dubale 1959). These include: (1) low O_2 affinity of the blood (the Bohr effect), (2) decrease in O_2 capacity of blood (Root effect), (3) acid-base imbalance due to the decrease in pH of water consequent to an increase in PCO_2, and (4) reduced capacity to discharge CO_2 into the surrounding water owing to a reduction of the PCO_2 between the blood and the surrounding water. In *Anabas testudineus*, pH influences the pathway adopted for respiration (Hughes and Singh 1970s). Both aquatic and aerial respiration occur in neutral waters (pH 8 to 6.85) while aerial respiration predominates and aquatic respiration is depressed between pH 6.5 and 6.25. Gill ventilation completely stops (and is replaced by aerial respiration) in very highly acidic water (pH < 6.25). Similar behavior was observed in the yellow (*Erythrinus erythrinus*), an air-breathing fish, where gill ventilation virtually stopped and was replaced by aerial respiration in hypercarbic waters with a concentration of CO_2 above 39 ml per l (Willmer 1934). Such a response may be a safeguard against excessive transfer of CO_2 from water into blood across the gills, avoiding possible respiratory acidemia (Singh 1976). By switching to air breathing, the fish can tolerate high concentrations of CO_2 without adverse effects. Air-breathing fish, most of which subsist in hypoxic waters, are confronted with a real problem of losing O_2 through the gills into the surrounding hypoxic water (e.g., Smith and Gannon 1978). This is, however, minimized or avoided by vascular reflexes which lead to shunting of blood from the gills with a momentary elimination of branchial ventilation making the accessory respiratory organ(s) the only pathway(s) for transfer of O_2 to the body. Air-breathing fish, e.g., *Anabas* (Munshi 1968), *Channa* (Wu and Chang 1947; Hakim et al. 1978; Wu 1993), *Amphipnous* (Munshi and Singh 1968). and *Monopterus* (Liem 1961), have very poorly developed gills, perhaps to curtail O_2 loss through the gills during hypoxia. In the lungfish, *Protopterus*, the arteries of the embryonic 3rd and 4th branchial arches are devoid of gill filaments and form shunt vessels which correspond with the carotid and systematic arches of the Amniota (e.g., Wood and Lenfant 1976).

3.5 Terrestrial Habitation and Utilization of Atmospheric O_2

The degree of specialization and the survival strategies of most animals are determined by the efficiency with which they can in the first instance procure O_2 and secondly eliminate CO_2. In open aerial environments, production and utilization of CO_2 and O_2 are relatively small compared within the enormous volumes and high capacitances of the gases in the atmosphere (Dejours 1988). The constant level of O_2 and CO_2 in the atmosphere is of biological importance as major deviations from the tolerable range have harmful effects on unacclimatized animal life. With respect to composition and availability of respiratory gases, air is a much simpler medium to handle than water. Biological, physical, and chemical processes greatly influence the O_2 content of water. In stagnant waters and especially those covered by plant matter, there is reciprocity of environmental O_2 and CO_2 levels. As a result of organic respiratory processes, hypoxia will invariably be associated with some degree of hypercapnia. Except for fossorial habitats, terrestrial habitats are not normally liable to hypoxia or hypercapnia due to the great diffusivity of the gases in air.

3.6 Hydrogen Sulfide Habitats: Tolerance and Utilization

The energy-rich hydrogen sulfide (H_2S) is common in the hypoxic marine sediments and around hot springs at concentrations of 1 to 300 ppm (e.g., Berner 1963). Hydrogen sulfide is highly toxic even at very low molar concentrations. A concentration greater than 1 ppm is lethal to most organisms (Oseid and Smith 1974; Smith et al. 1976). The gas inhibits several heavy metal containing enzymes especially by binding to the heme of mitochondrial cytochrome c oxidase much as cyanide does. This prevents O_2 transport by the hemoglobin, arresting aerobic respiration (e.g., Somero et al. 1989). About 90% of cytochrome c oxidase is inhibited by a 5-molar sulfide solution (e.g., Julian and Arp 1992) and a sulfide concentration of 17 mol initiates maximal O_2 consumption (Eaton and Arp 1993). At moderately low concentrations (3.63 mol per l), specimens of *Rivulus marmoratus* were observed to leap from H_2S-contaminated water (Abel et al. 1987). Some animals, however, have acquired a capacity to tolerate H_2S and even utilize it for metabolic processes producing water and various sulfates (e.g., Felbeck et al. 1981). Others, e.g., in the Phylum Gnathostomulida (Farris 1976) and the turbellarian families Solenofilomorphidae (Crezee 1976) and Retronectidae (Sterrer and Rieger 1974), have adapted so well that they are virtually confined to the extreme anoxic H_2S-rich habitats (Fenchel and Riedl 1970). They survive by adopting different strategies which include: (1) physical exclusion of sulfide from the body, (2) possession of sulfide-insensitive cytochrome c oxidase, (3) direct detoxification whereby sulfide (H_2S, HS^- or S^{-2}) is chemically converted to less toxic products such as sulfate (SO_4^{-2}), sulfite (SO_3^{-2}), and thiosulfate ($S_2O_3^{-3}$) at the superficial tissue layers of the body or in certain specialized organs (Vetter et al. 1987; Menon and Arp 1992b), (4) strategic coexistence with chemoautotrophic endosymbiotic bacteria which break down H_2S, utilizing the energy thus acquired for production of ATP (e.g., Felbeck 1983; Fisher and Hand 1985; Powell and Somero 1985; Belkin et al. 1986; Firsher 1990), and (5) dependence on anaerobic metabolism. The marine echiuran worm, *Urechis caupo*, subsists in U-shaped burrows in intertidal mud flats where the concentration of H_2S may be as high as 66 μM and the concentration of O_2 as low as 3.3 kPa (Eaton and Arp 1993). Hydrogen sulfide passes through the body wall and the hind gut (Julian and Arp 1992), where hematin, which is contained in coelomic fluids and in the coelomocytes (Arp 1991), catalyzes its oxidation to nontoxic sulfur compounds as O_2 continues to be utilized for aerobic respiration (Powell and Arp 1989; Eaton and Arp 1993). A correlation between hematin concentration in the coelomic fluid and sulfide oxidizing activity was reported in *U. caupo* by Powell and Arp (1989). *U. caupo* has a very high tolerance for H_2S (Somero et al. 1989; Julian et al. 1991). The worm remains aerobic at concentrations above those which suppress the process in most aerobic organisms. In the intertidal lugworm, *Arenicola marina*, a specialized heme compound called brown pigment is thought to catalyze sulfide oxidation (Patel and Spencer 1963). The hemoglobins of the lugworm and the deep-sea hydrothermal vent-living polychaete, *Alvinella pompejana* (Desbruyères and Laubier 1986), are highly resistant to oxidation (Toulmond et al. 1988).

Extreme physiological conditions have been described in the submarine geothermal springs which have been discovered in the Galapagos Rift and East Pacific Rise fracture zones (e.g., Childress et al. 1989; Nisbet 1988). The water emerging from the fissures is as hot as 200 °C, is anoxic, and contains a high concentration of H_2S (Desbruyères et al. 1982; Johnston et al. 1988). The hot plume of water interfaces with the remarkably different sea-water at a temperature of about 2 °C, with a PO_2 6.7 kPa, pH about 7.5, and is totally devoid of H_2S (Johnston et al. 1986). This creates the greatest temperature gradient in any known environment occupied by animal life. Simple photosynthetic life is thought to have evolved in the submarine hydrothermal vents where chemotrophic organisms acquired the capacity to detect light (e.g., Russell et al. 1994; Nisbet et al. 1995). The interface between the hydrothermal plume and the bottom seawater constitutes a highly dynamic and complex ecosystem with fluctuating PO_2 (Johnston et al. 1986). A large number of different organisms such as sulfur oxidizing bacteria and animal species such as giant pogonophoran tube worms, crabs, shrimps, giant crams, fishes, and mussels flourish in complete darkness some 3 km below the suface (Grassle 1985; Fustec et al. 1987). Some of these species were unknown to science more than 20 years ago! The animals which stray from the vent risk starvation while those which get too close die of poisoning or heat. The hydrothermal plume crab, *Bythograea thermydron*, which has no bacterial symbionts, can tolerate high concentration of H_2S through having a particularly efficient sulfide oxidation capacity in the hepatocytes (Vetter et al. 1987) while *Riftia pachyptila* and *Solemya velum* possesses hemoglobins which hold and transport H_2S preventing toxicity (e.g., Doeller et al. 1988). In the deep sea hydrothermal vent cram, *Calyptogena pacifica*, which has symbiotic bacteria in the gills, O_2 consumption increases on exposure to concentrations of H_2S as high as 130 to 160 µM (Childress and Mickel 1982). The endosymbiotic bacteria in the gutless cram, *Solemya reidi*, which live in H_2S-rich sewage outflows, are able to maintain aerobic metabolism in the tissues through oxidation of H_2S to thiosulfate in presence of a concentration of H_2S of 100 µM (Anderson et al. 1990) and a much higher one of 500 µM in *Solemya velum* (Chen et al. 1987). The granules in the cytoplasm of the gill cells in *Solemya reidi* (Powell and Somero 1985; Powell and Arp 1989) and the osmiophillic electron dense organelles in the epithelial cells of the hind gut of *U. caupo* (Menon and Arp 1992b) are associated with H_2S detoxification. Similar granules have been observed in the epithelial cells of the "lung" groove (Fig. 45) of the oligochaete swamp worm, *Alma emini* (Maina et al. 1998)

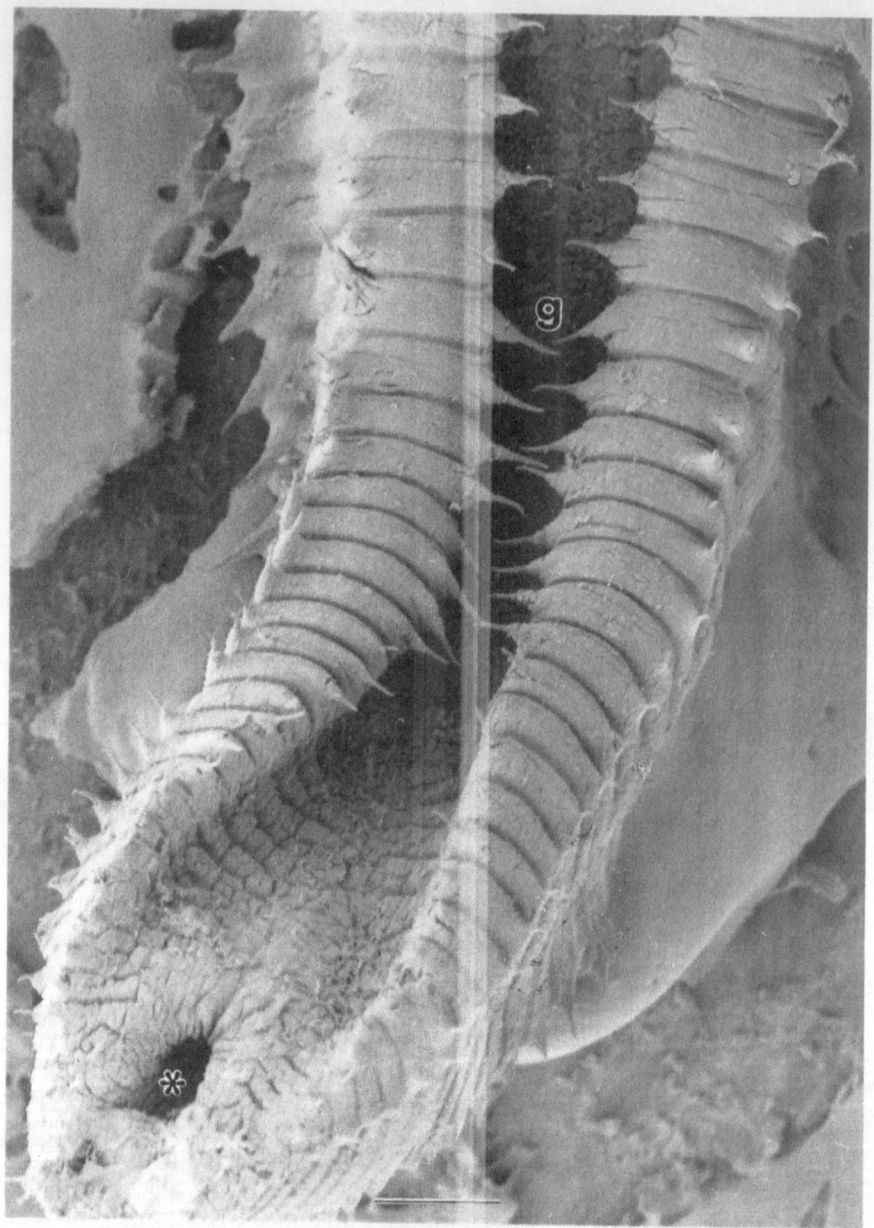

Fig. 45. Dorsal view of the temporary respiratory groove, *g*, of the oligochaete worm, *Alma emini*, which lives in putrefying plant matter of the East and Central African tropical swamps. The worm occasionally surfaces, forms a lung, and exchanges gases with the atmosphere. * cloaca. *Bar* 0.95mm. From Maina et al. (1998).

which lives in waterlogged soils where intense putrefaction of plant matter with possible release of H_2S occurs.

3.7 The Porosphere and Fossorial Respiration

In addition to the aerosphere and the hydrosphere, soil, the thin skin of the lithosphere, offers an important natural habitat to many animals. About 150 mammalian species temporarily or permanently live underground (e.g., Nevo 1970). The respiratory properties of the soil are very fluid. Its characteristics fluctuate between those of water (when the soil is wet or waterlogged) to those of air when it is dry. In three-dimensional space, the porosphere can be conceptualized as an intricate maze of fine air conduits between the soil particles. The air spaces open to the free atmosphere. Gas diffusivity in the soil may be up to one third that of free air (Ar 1987) but moist compact soil may offer a severe hindrance to gas diffusion (Currie 1984). Depending on the global temperature changes, soils offer an important source or sink of CO_2 (Susan et al. 1996). The principal factors which determine O_2 and CO_2 content of the porosphere are: (1) the intensity of respiration of the organic matter in the soil (Vannier 1983), (2) moisture content of the soil (e.g., Wilson and Kilgore 1978), and (3) the chemistry of the soil (Verdier 1975).

In general, the underground microenvironments are characterized by a high temperature, reduced light intensity, and in most cases perpetual darkness, low PO_2, high PCO_2, and frequently high humidity (e.g., Kennerly 1964; McNab 1966; Arieli 1979). These variables are translations of the surface (free atmospheric) factors except for CO_2, which is intrinsically produced by the soil fauna. The subsurface microclimates are more stable compared with the more variable and cyclic characteristics above. However, subterranean features such as O_2 and CO_2 levels and, in some cases, moisture content show greater range and oftentimes faster shifts than the corresponding surface ones. The stimulus of light as a physiological phenomenon, which to varying extents influences many of the surface-dwelling animals, is virtually eliminated under the soil. Since the gas requirements of an animal living in an underground burrow must in the first instance be derived from the free atmosphere by diffusion through the soil, many underground conditions, the physical features of the soil especially its porosity, are important in the distribution, abundance, and respiratory activity of fossorial animals (Kennerly 1964).

While many animals spend only part of their time underground in open burrows, only a few of them permanently live underground in closed burrows. In some habitats, subterranean (fossorial) animals constitute an important part of the total fauna. Burrows provide fossorial animals with protection against environmental extremes and predators and enable them to gain access to subterranean parts of plants particularly geophytes (roots, tubers, bulbs, and corms) and soil invertebrates. However, fossoriality imposes certain constraints on the animals which face huge energetic costs in subterranean excavation during foraging (Vleck 1979). Depending on soil hardness and burrow diameter, the cost of

169

burrowing may be 360 to 3400 times greater than that of traveling the same distance on the surface (Vleck 1981). Pocket gophers can dig well over 200 m in 48 h (Hill 1944). The burrows of the naked mole rat (*Heterocephalus glaber*) may be as long as 3 km (Brett 1986). Wet soils constitute a limitation for thermoregulation in fossorial animals (Wierenga et al. 1969). As a consequence of their having had to adapt to a similar mode of life, remarkable convergence of form and function (e.g., Jarvis and Bennett 1990) has occurred in a remarkably phylogenetically diverse group.

The rate of diffusion of a gas or vapor through the soil is determined by the porosity, type of gas, and the prevailing concentration gradient between the air in the burrow and the free atmosphere (Penman (1940a,b). High concentration of CO_2 correlates with the soil moisture content (Kennerly 1964). When the content is high, water rather than air occupies the interstitial spaces between the soil particles thereby impeding the diffusion of gases. Respectively, moisture contents as high as 13 and 18% have been reported in fields occupied by the pocket gopher, *Geomys bursarius* (Kennerly 1964) and *Thomomys bottae* (Miller 1948). After heavy rain, 30 cm below the surface, Ege (1916) observed that the PCO_2 increased from 1.5 to 6.1 kPa and PO_2 dropped from 20.4 to 8.5 kPa. When the soil is waterlogged, gas diffusion may become a limiting factor, the soil becoming virtually anoxic (Currie 1962). Some animals living underground (e.g., the earthworms) may succumb to hypoxia or react to it by surfacing. The distribution of the pocket gopher, *Geomys pinetis*, correlated with that of the soils with a high water-holding capacity (McNab 1966). Acidic soils (podzols) are unable to chemically fix CO_2 while calcareous ones (e.g., clay soil) are able to buffer some CO_2. Due to the greater solubility and capacitance coefficient of CO_2 in water, O_2 rather than CO_2 diffusion should be the limiting factor in moist soils. Kennerly (1964), however, observed that in soils with different moisture contents, O_2 and CO_2 diffused through the soil at different rates, with O_2 reaching equilibrium much faster than CO_2. While CO_2 is organically produced within the soil, there is no subterranean source of O_2, which must be derived by diffusion from the surface. With the PO_2 falling to between 16 kPa in *Caretta* and 12 kPa in *Chelonia* and PCO_2, respectively, at 3.3 and 4 kPa in the nests, sea turtles hatch into a subterranean hypoxic and hypercapnic environment (Maloney et al. 1990): within 3 to 5 days, they must reach the surface or die of suffocation.

3.7.1 Gaseous Composition in Burrows

In addition to soil chemistry and porosity, the other factors which determine the composition of the air in the burrows include the size, length, location, and geometry of the burrow (Wilson and Kilgore 1978), and the numerical density of the burrow congeners. The burrows of the pocket gophers tend to be deeper in summer when the porosity and hence the diffusion capacity of the soil for O_2 is higher (Kennerly 1964). In most soils, the sum of the concentrations of O_2 and CO_2 remains almost constant and for soils with air porosities greater than 10% by volume, the rate of exchange of O_2 and CO_2 with the surface comes to equilibrium up to depths of 30 cm within a period of 1 h (Collins-George 1959). In the burrows of the pocket gopher, *Geomys bursarius*, burrows up to 30 cm in depth permit diffusion of O_2 from the surface at rates adequate for the respiratory needs of the gopher. Thermal equilibrium occurs up to a depth of 60 cm (Kennerly 1964). In a forest, 30 cm below the surface, Ege (1916) reported a CO_2 concentration of 0.2% (0.2 kPa) and one of O_2 of 20.6% (20.3 kPa). Aeration of the burrow must be most critical after rains when diffusion of O_2 and CO_2 between the burrow and the ground surface is greatly hindered. Under such circumstances, a net O_2 deficit in the burrow might occur. In the pocket gophers, *Thomomys* (Miller 1948) and *G. bursarius* (Kennerly 1964), burrows are opened in the early morning and late evening or night, a time when the cooler layer of air next to the warm ground enhances the diffusion of the air into the extensive burrow systems which range from 60 to 200 m in length. With an average diameter of 7.5 cm, a total burrow roof exposure area of about 60 to 200 m^2 is attained. This was considered by Kennerly (1964) to be more than adequate for the transfer of the respiratory gases while providing access to sufficient vegetation cover with subterranean plant parts, insects, and worms for food.

Concentrations of CO_2 as high 10% and O_2 levels as low as 10% have been reported in burrows (Boggs et al. 1984). The burrows of the pocket gopher (*Thomomys bottae*) had a concentration of O_2 as low as 6% and CO_2 as high as 3.8% (McNab 1966; Darden 1970; Chapman and Bennett 1975). Concentrations of O_2 as high as 20% and as low as 12.1% have been reported in the burrows of birds, with corresponding values for CO_2 ranging from 1.2 to 9% (e.g., Wickler and Marsh 1981). In ant hills, in summer, Portier and Duval (1929) found a PCO_2 of up to 1.9 kPa while Ege (1916) observed that while the PO_2 in the field was 20.4 kPa, in an ant hill it was 19.3 kPa and in a decaying beech trunk the value was 17.7 kPa. Concentrations of O_2 of 10 to 15% have been reported in some mammalian burrows by Kennerly (1964). In the burrows of five fossorial rodent species, the concentration O_2 ranged from 15 to 20% and CO_2 from 0.5 to 2.0% (McNab 1966). Concentrations of CO_2 as high as 2.3% and O_2 levels as low as 6% were estimated in the burrows of the pocket gopher, *Geomys bursarius* (Kennerly 1964), and in dens of hibernating mammals, concentrations of CO_2 as high as 13.5% and O_2 as low as 4% were reported by Williams and Rausch (1973). Soil flora (especially algae and bacteria) is thought to be the main source of CO_2 in the burrows (Kennerly 1964) though the intensity of activity, especially during mound building, must greatly elevate it. The gas concentrations in the burrows are remarkably different from those which can be tolerated by nonburrowing mammals. Extreme hypoxia as well as hypercapnia affect cardiac function in most mammals (e.g., Tucker et al. 1976), induce artificial hypothermia and in some cases torpor (Hyden and Lindberg 1970), have a general depressing effect on growth (Xu and Mortola 1989), and lower the ventilatory rate (Arieli and Ar 1979). The hypoxia that the mole rat, *Spalax ehrenbergi*, can withstand is comparable to an altitude of more than 9 km (Arieli et al. 1977). The respiratory physiology for subterranean subsistence is outlined by, e.g., Arieli and Ar (1979), Boggs et al. (1984), and Webb and Milsom (1994). Insensitivity of the pulmonary vasculature to hypoxia (Walker et al. 1982, 1984) may reduce resistance and ensure adequate pulmonary perfusion. Compared with those of the adult mammals, the lungs of the eusocial naked mole rat, *Heterocephalus glaber* show a remarkable degree of pedomorphy/ neoteny (Maina et al. 1992). Perhaps what appears like a primitive (ancestral = plesiomorphic) state constitutes what has been called an evolutionary stable strategy by, e.g., Maynard-Smith (1996) and McNamara and Houston (1996) where the phenotype has been specially selected for the particular habitat the mole rat lives in. A highly refined resource use and niche occupation may reduce or even eliminate sympatric competition, decreasing fitness cost (e.g., Martin 1996).

3.7.2 Burrowing Aquatic Annelids, Crustaceans, and Fish

For reasons basically similar to those of the fossorial mammals, i.e., escape from predators and occupation of a more stable micromileu, a number of annelids (Mangum 1964; Myers 1972; Toulmond 1991), crustaceans (Little 1983; Atkinson and Taylor 1988), and fish (Pelster et al. 1988a,b; Taylor and Atkinson 1991) have adopted a subaquatic burrowing life-style. Among crustaceans, burrow construction is commonest in members of the infraorders Astacidea, Thalassinidea, and Brachyura. Fish such as the tile fish (Branchiostegidae) (Able et al. 1982) and the garden eels (Anguilliformes) construct mucus-lined tubular burrows (Taylor and Atkinson 1991). The marine polychaete, *Chaetopterus variopedatus*, lives in elaborate U-shaped tubes which are said to be impermeable to water and gases (Fauvel 1927; Dales 1969; Brown and McGee-Russel 1971) while the lugworm, *Arenicola marina*, an intertidal polychaete, lives in an L-shaped mucus-lined burrow system (Wells 1966). The mud shrimp, Mediterranean shrimp, *Callianassa truncata*, builds extensive burrows in the sediments to a depth of 48 cm (Ziebis et al. 1996): the shrimp can maintain burrow O_2 concentration at 3 to 12% of air saturation by generating water currents of a flow rate of $10\,\mathrm{m\,s^{-1}}$.

Like the surface-based fossorial mammals, the subaquatic burrowing animals face unique respiratory challenges that call for special adaptive strategies to overcome hypoxia and hypercapnia. High concentrations of H_2S are a common occurrence (Bridges 1987; Atkinson and Taylor 1988; Toulmond 1991). Burrowing fish such as *Periophthalmus cantonensis* may encounter extreme hypoxia (PO_2, 0.7 to 3 kPa) at a depth of 5 to 15 cm (Gordon et al. 1978) and may face total anoxia at a depth of about 1 m (Gordon et al. 1985). In the burrows of the snake blenny, *Lumpenus lampretaeformis*, the PCO_2 of the water in the burrow is above 0.2 kPa compared with that at the surface of the mud (Atkinson et al. 1987). To alleviate adverse respiratory stress, the crabs use their pleopods (e.g., Farley and Case 1968; Hill 1981) and to a small extent their scaphognathites (Taylor and Atkinson 1991) to intermittently ventilate their burrows and their gas exchangers with better oxygenated water. In most species, e.g., *Callianassa californiensis*, the beating of the pleopods increases with the level of hypoxia (e.g., Felder 1979).

Certain physiological adaptations have been reported in the aquatic burrowing crustaceans and fish. Greater O_2 affinity and high Bohr values, favorable parameters for O_2 uptake in hypoxia, have been reported (e.g., Innes 1985; Brigdes 1986; Pelster et al. 1988a,b). These features, however, do not appear to be specific to the burrowing mode of life but rather to the habitat they occupy (Taylor and Atkinson 1991). The greatest adaptive strategy that the burrowing crabs and fish possess is their overt capacity to tolerate hypoxia (e.g., Hagerman and Uglow 1985; Swain et al. 1987) and even withstand anoxia (Hill 1981; Mukai and Koike 1984). The red band fish, *Cepola rubescens*, and *Lumpenus lampretaeformis* can maintain their normal aerobic metabolism constant down to a PO_2 of 7 to 9 kPa (Pullin et al. 1980; Pelster et al. 1988a,b). Hypoxia causes increased opercular ventilation in the burrowing decapods and fish (Pelster 1985; Bridges 1987).

3.8 Living at High Altitude: Coping with Hypoxia and Hypobaria

With some exceptions, the majority of living organisms have evolved at or close to sea level. They have hence adapted to an O_2-enriched environment with an ambient atmospheric pressure of 1 atmosphere (101.3 kPa; Table 12). Including other factors such as cold, rarefied atmosphere, and harmful cosmic radiation, hypoxia is the most significant factor which limits survival at high altitude. Since the demands for O_2, whether at sea level or at altitude, are essentially the same, all other factors held constant, movement to a hypoxic environment imposes great metabolic demands on an animal and calls for definite physiological adaptations for procurement of the necessary amounts of O_2. Acute exposure to severe hypobaric-hypoxia equivalent to that at the top of Mt. Everest (altitude 8848 m, barometric pressure 33.7 kPa), where only 30% of the initial maximum aerobic capacity (maximum O_2 consumption) at sea level is lost, would be fatal to most unacclimatized human beings without an auxiliary source of O_2. With adequate adjustments, however, fit human subjects performing at the limits of their aerobic capacity have reached some of the highest peaks (Dejours 1982; West 1983; Houston et al. 1987). Animals living at sea level benefit from a considerable head

Table 12. Oxygen concentration in inspired air in animals of various body temperatures at different barometric pressures. (After Dejours 1981)

T (°C)	Alt. (m)	BP (mmHg[a])	K1	K2	FIO_2	CIO_2 ml STPD/ml BTPS	CIO_2 mmol/lBPTS
41	150	747	0.787	35.13	0.2095	0.165	7.37
15	150	747	0.915	40.83	0.2095	0.192	8.55
37	0(SL)	760	0.826	36.84	0.2095	0.173	7.72
37	3400	500	0.525	23.40	0.2095	0.110	4.91
37	7200	300	0.293	0.293	0.2095	0.061	2.74

T, body temperature; BP, barometric pressure; K1, reduction factor from values BTPS to volumes STPD; K2, reduction factor from 1 li of volume BTPS to the dry quality of substance in mMol; FIO_2, fractional concentration of O_2 in normal dry air; CIO_2, concentration of O_2 in inspired air; BTPS, body temperature, pressure, saturated with water; STPD, standard temperature, pressure, dry air (°C, 1 atmosphere pressure).
[a] To convert to kPa multiply by 0.133.

pressure of O_2 in the atmospheric air ($PO_2 = 21.2\,kPa$). At the respiratory site, the pressure drives O_2 in adequate amounts past the tissue barriers into the blood and finally to the mitochondria where the prevailing PO_2 is between 0.1 to 0.3 kPa (Figs. 2,3). Diffusion of O_2 across the blood-gas barrier is the limiting factor for survival at high altitude (West and Wagner 1980; Piiper and Scheid 1981).

About 30% of the Earth's surface lies above 1 km altitude. There are many areas with an elevation in excess of 2.5 km (Webber 1979). Natural acclimation to moderate hypoxia is thus a common feature to many animals. Above an altitude of 3 km, most unacclimatized subjects will show overt signs of high altitude respiratory distress such as shortness of breath and increased pulse rate (e.g., Heath and Williams 1981). Physiological disturbances such as reduced aerobic capacity (Squires and Buskirk 1982) or night eversion (McFarland and Evans 1939) may start at the much lower altitudes of 1.2 to 1.5 km. At 11 km, an auxiliary source of O_2 will be needed while at about 19 km, without compensatory hyperventilation, little if any O_2 reaches the alveoli even when pure O_2 is breathed. This is because at that altitude the total barometric pressure of 11.6 kPa equals the sum of the partial pressure of the water vapor of 6.3 kPa (at 37 °C) and that of CO_2 of 5.3 kPa. Above this altitude, when the external vapor pressure falls below that of the body fluids at body temperature, in homeotherms, evaporation which may be accompanied by life-threatening formation of vapor bubbles (boiling) in the tissues, blood vessels, and body cavities occurs (e.g., Armstrong 1952). Just as the rarefaction of the atmosphere with altitude has a significant effect on the aerodynamics of flight, it has considerable influence on the tidal movement of air in the respiratory tract and the forces generated in the mechanical ventilation of the lung (e.g., Luft 1965). Since the resistance to turbulent flow scales with the square of the flow velocity, the effect of rarefaction of air with altitude is more predominant at high flow rates, e.g., during hyperpnea, which is one of the initial responses to high altitude hypoxia. The lungs and the chest, which function as pneumatic pumps, operate at a disadvantage at altitude while the driving force of the respiratory muscles is somewhat fixed (Johnson 1964). The overall respiratory work at altitude is, however, less than that at sea level mainly owing to the reduction of turbulent flow in favor of laminar one (Ulvedal et al. 1963). This may account for the fact that at 6.1 km, in acclimatized men, spontaneous pulmonary ventilation may exceed 200 l per minute during strenuous exercise (Pugh 1962). Though the fractional concentration of O_2 in the dry atmosphere stays constant at 20.93%, at least until an altitude up to 110 km, due to the compressible nature of gases, at sea level, the PO_2 is greater. With ascent to high elevations, the molar concentration of O_2 in air decreases in proportion to the decrease of PO_2. Long-term and immediate human physiological adjustments at high altitude are now well known. The subject has recently been reviewed by West (1991). High altitude hypoxia does not appear to restrict the distribution of reptiles and amphibians (Hock 1964). Eleven species of frogs live between a height of 3.7 and 5.2 km. The highest living known reptile, *Leiolopisma ladacense*, is found at an altitude of 5.5 km. *Telmatobius culeus*, a frog found in Lake Titicaca (altitude 3.8 km) is adapted to low aquatic PO_2 by combining behavioral, physiological, and morphological adaptations to the cool 10 °C O_2-saturated (13 kPa) water (Parker 1940; Monge and Monge 1968; Hutchison et al. 1976). Its skin is well vascularized and folded (to increase the surface area) and the blood has small, numerous erythrocytes (the highest count among amphibians), high hemoglobin concentration, and high hematocrit. These factors give rise to a high O_2 capacity. When swimming, the frog increases the convective movement of water over the skin through what has been described as bobbing behavior by Hutchison et al. (1976), a violent locomotory agitation of the water. Since the inhaled air in transit through the respiratory passages is warmed up to the body temperature and is maximally humidified, the PO_2 in the moisture saturated air is about 20% lower than that of the ambient dry air. The possible respiratory advantages of a lowered partial pressure of the water vapor in the ectotherms, where due to the lower body temperature the "alveolar" PO_2 should be higher compared with that in the endotherms, has not been fully investigated. The water vapor partial pressure is 6.2 kPa at the body temperature of 37 °C in mammals and is even higher in birds (7.3 to 8.3 kPa, Tb = 40 to 42 °C).

3.8.1 Tolerance of Arterial Hypocapnia in Birds

Migrating birds have been tracked at altitudes of over 6 km (e.g., Richardson 1976). The highest authenticated record of a flying bird is that of a Ruppell's griffon vulture, *Gyps rueppellii*, which was sucked into the engine of a jet craft at an altitude of 11.3 km over Abidjan (Côte d'Ivoire, West Africa; Laybourne 1974). At that altitude, the barometric pressure is about 24 kPa (i.e., 20% of that at sea level), the PO_2 in the expired air is less than 5.3 kPa (closer to 2.7 kPa if hyperventilation could bring the PCO_2 to about 0.67 kPa) and the ambient temperature is about −60 °F (Torre-Bueno 1985). The capacity of birds to survive, let alone exercise, under such circumstances is unmatched among animals. Flapping flight is energetically a very expensive mode of exercise (e.g., Tucker 1972). An actively flying animal consumes O_2 at two times the rate of a ground-dwelling one at maximum exercise (e.g., Thomas 1987). Hummingbirds hover with a muscle power output of nearly 100 to 120 W kg^{-1} at a 9 to 11% mechanical energy (Wells 1993a). In the housefly, a sure sign of approaching death is that of inability to fly (Sohal and Weindruch 1996). In a rarefied atmosphere, as occurs at high altitude, for a particular speed, the cost of flight is much greater than at sea level (Torre-Bueno 1985).

A fundamental respiratory physiological difference between birds and mammals is that birds can withstand greater hypocapnia where the arterial PCO_2 may drop to below 0.8 kPa (as the pH rises to 7.96) during panting (e.g., Faraci and Fedde 1986). Part of the respiratory efficiency and tolerance to hypoxia at altitude in birds must indirectly be related to this particular attribute which enables birds to hyperventilate and thus acquire sufficient amounts of O_2 without respiratory complications. A house sparrow at a simulated altitude of 6.1 km and ambient temperature of 5 °C has a respiratory frequency 38% above the sea level value and the ventilation of the parabronchi increases by 75% (Bernstein 1990). Birds can withstand the high level of arterial hypoxemia because brain blood flow is not affected by arterial hypocapnia (e.g., Faraci 1990). In fact, in species such as pigeons, the bar-headed geese, *Anser indicus*, and the duck, *Anas platyhynchos*, blood flow to the brain is not affected or actually starts to increase when the arterial blood PO_2 drops to below 10 kPa: the flow can increase sixfold (Grubb et al. 1978; Faraci and Fedde 1986; Pavlov et al. 1987). In the human being, lowering the arterial PCO_2 to 1.3 kPa results in hypocapnic cerebral vasoconstriction, causing a reduction of the flow of blood to the brain by about 50% (Wollman et al. 1968). In the bar-headed goose, *Anser indicus*, cerebral blood flow is not affected by reduction of the arterial PCO_2 to 0.9 kPa (Faraci and Fedde 1986). The remarkable efficiency of the bird lung particularly during hypoxia is brought about by a multiplicity of cardiovascular and pulmonary factors: the most significant advantage is imparted by the crosscurrent arrangement between the parabronchial air and blood flows (Fedde et al. 1985; Maina 1994, 1996, 1998). At altitude flight, birds can hyperventilate without the risk of experiencing respiratory alkalosis and cerebral vasoconstriction from the resultant hypocapnia. Using measurements and estimates of man at the altitude of Mt. Everest (Dejours 1982; West 1983), Scheid (1985) calculated that if the human alveolar lung were replaced by the avian parabronchial one, for the same arterial blood gases, the person would ascend 780 m higher in altitude.

3.8.2 Flying over Mt. Everest: the Bar-Headed Goose, *Anser indicus*

Perhaps the most astounding high altitude flight behavior in birds is that exhibited by the bar-headed goose (*Anser indicus*) during its annual trans-Himalayan migration from the wintering grounds of the Indian subcontinent to the breeding grounds around the large lakes in the south-central regions of Asia which are at elevations of about 5.5 km (Swan 1970). The birds take off from virtually sea level and cross the Himalayan mountains almost directly, reaching an altitude of about 10 km over the summits of Mt. Averest and Annapurna 1 (Swan 1961, 1970; Black et al. 1978; Black and Tenney 1980). The barometric pressure at these altitudes is about 31 kPa and the PO_2 in dry air is 6.5 kPa (West et al. 1983). Assuming that during these excursions the geese maintain a constant body temperature of 41 °C and the inhaled air is warmed to that of the body and is fully saturated with moisture, the PO_2 in the humid inhaled air which arrives at the gas exchange surface of the lung would barely exceed 4.9 kPa. Experimentally, the bar-headed goose withstands hypoxia at a simulated altitude of 11 km (Black and Tenney 1980) and up to an altitude of 6.1 km, it maintains normal O_2 consumption without need to hyperventilate. On ascending to 11 km, where the concentration of O_2 is only 1.4 mmol l^{-1}, the bird takes in adequate O_2 to necessitate only a minimal increase in ventilation. At 39 °C and pH 7.4, the blood of the bar-headed goose has a much higher O_2 affinity (P_{50} = 3.9 kPa) than that of the greylag goose (*Anser anser*) (P_{50} = 5.3 kPa), a close relative which subsists at lower altitudes (Petschow et al. 1977; Black et al. 1978).

Extreme high altitude-adapted animals show a left shift of the O_2-hemoglobin curve (Perutz 1990b; Weber et al. 1993). In *A. indicus* (Weber et al. 1993) and the Andean goose, *Cleophaga melanoptera* (Hiebl et al. 1987), where unusually high blood O_2 affinity is crucial for survival, the amino acid residues α-119 and β-55, which form an $\alpha_1\beta_1$ contact in human hemoglobin, are respectively altered in these two birds. The loss of contact appears to increase O_2 affinity (Weber et al. 1993). In *A. indicus*, proline is replaced by shorter amino acid alanine (Weber et al. 1993) and in *Cleophaga*, leucine is replaced by short-chained serine, resulting in the loss of a single intramolecular contact (Hiebl et al. 1987). These observations suggest that adaptive changes in protein function evolved by substitution of one or a number of amino acids at specific sites, but a large proportion of such mutations are selectively neutral (Perutz 1983). The similarities in the strategies adopted by the bar-headed and the Andean geese, i.e., two single-point amino acid mutations that alter intramolecular contact enhancing hemoglobin-O_2 affinity in two geographically separated species (Weber et al. 1993) is a classic case of convergent evolution at the molecular level. In the bar-headed goose, cardiopulmonary parameters indicate that muscle blood supply and O_2 loading from the blood capillaries rather than ventilation or pulmonary gas transfer are the limiting steps in the supply of O_2 to the contracting flight muscles under hypoxia (Fedde et al. 1989). This conforms with the observation made by Black and Tenney (1980) that the PO_2 in the arterial blood at a simulated altitude of 11.6 km is only 0.13 kPa less than that in the exhaled air. Weinstein et al. (1985) observed that "the evolution of hypoxia tolerance in birds may have developed secondary to that of the effective gas exchange and transport features needed for flight".

3.9 Gravity: Effects on Respiratory Form and Function

Land animals evolved from aquatic forms which subsisted in virtual weightlessness. Within historical times, humans have experienced the almost constant gravity of about 9.81 N kg^{-1} on the Earth's surface. Weightlessness or zero gravity (0-G) presents a completely new dimension in biology. Compared with the other three forces that govern matter in the Universe, i.e., electromagnetism and the weak and strong nuclear forces, gravity is unique in many ways: it only pulls but does not push and is a relatively very weak force that has an infinite range. Thompson (1959, p 32) observed that "gravity not only controls the actions but also influences the forms of all save the least of organisms" and humorously cites "sagging wrinkles", "hanging breasts" and "many other signs of old age" as part of gravity's slow, remorseless creations. Due to its assumed invariability since the formation of the Earth (but see different views from geological, e.g., Carey 1976 and paleontological, e.g., Holden 1993, studies that argue that the parameter has varied in the past) and lack of easy means and ways of manipulating it for purposes of testing, gravity has largely been considered to be of little consequence in biology. Experimentally, increase in gravitational field has been shown to increase the metabolic rate (Smith 1976, 1978; Economos 1979; Pace and Smith 1981). Like the other features which influence aspects such as body size, shape, and locomotion, and hence prescribe the metabolic scope of an animal, gravity must have greatly influenced the definitive designs, the allometric scaling, and the structural and functional parameters in all evolved life forms. The lung and chest, which on the ground have to be elastic to be able to carry out ventilatory activity, are easily deformed by their own weight just as the gills collapse out of water. Hypogravity greatly modifies the pleural pressure, shape, and regional distribution of air and blood in the respiratory system (e.g., Engel 1991). Both the structural and functional features of the pulmonary system such as alveolar and blood capillary size and volume (Glazier et al. 1967; Hogg and Nepszy 1969; Gehr and Erni 1980), ventilatory distribution (Michels and West 1978), and blood perfusion (West 1977a), factors which cause deformation of the parenchyma (e.g., West and Mathews 1978), are variably affected by gravity. The lungs of a variety of mammals and reptiles show remarkable vertical stratification of blood flow with the "highest" regions of the lung receiving substantially less blood than the dependent ones (e.g., West 1977b; Seymour et al. 1981). In the horse and the dog, there is a fourfold difference in the volume of alveoli from the apex to the base of the lung (Glazier et al. 1967; Hogg and Nepszy 1969; Gehr and Erni 1980). Recent studies, e.g., those by Glenny and Robertson (1991b) and Hlastala et al. (1996), however, indicate that the effect of gravity on the pattern of pulmonary perfusion is less important than was earlier thought. In the dog, topographical differences in pleural pressure exist (e.g., Proctor et al. 1968) and the intrapleural pressure is more negative on the dorsal than on the ventral aspect. The dependent regions of the lung have a small resting volume but a large increase in the inspiratory volume (Milic-Emili et al. 1966). The spatial characteristics of the pulmonary arterial tree play an important part in the perfusion of the lung (Glenny 1992). In the land crab, *Holthuisana transversa*, stratification of the hemolymph in the branchiostegal circulation occurs, with the ventral parts containing more hemolymph than the dorsal ones (Taylor and Greenaway 1984). Though the concentration of the vascular units in the apical and basal regions of the lung is the same (McGrath and Thomson 1959), hydrostatic differences lead to a decrease of blood flow from the gravity dependent parts of the lung to the apical parts (in an upright human lung), there being a nine times difference between the two regions of the lung (West et al. 1964). Ventilation decreases from the lower to the upper lobes of the lung though the decrease is not as much as that of the blood (e.g., West and Dollery 1965). The ventilation-perfusion ratio increases five times from the "bottom" to the "top" of the upright human lung, leading to regional differences in gas exchange efficiency (West and Jones 1965; Wilson and Beck 1992). Gravity-independent ventilatory inhomogeneities are as large as gravity-dependent ones (Verbanck et al. 1996). Such differences are to be expected in an organ that separates blood from air, fluids of very different densities, over a thin barrier.

Practically all the regional ventilation-perfusion inequalities which are observed at 1-G state are significantly reduced at 0-G (e.g., Michels and West 1978) or in a lung filled with and immersed in a fluid of which the specific gravity is the same as that of blood (West et al. 1965). At 0-G, increased redistribution of pulmonary perfusion to the upper parts of the lung occurs (e.g., Michels et al. 1979), increasing the diffusing capacity of the lung (Prisk et al. 1993; Vaïda et al. 1997). Acceleration (i.e., increased G), a process which is accompanied by compressive stresses, results in increased regional differences in lung expansion in the dog (Glazier et al. 1967) and in the human being (Crosfill and Widdicombe 1961).

Table 13. Effects of weightlessness during parabolic flight and of +2Gz on central venous pressure and heart rate in humans. (After Norsk et al. 1987)

Parameter	Ground			Inflight		
	Supine (+1 Gx)	Upright (+1 Gz)	Supine (1 Gz)	Upright sitting (+1 Gz)	0G	Upright
CVP, mmHg	7.7	1.9	5.0	2.6	68	2.8
HR, beats min^{-1}	–	–	65	70	79	80

CVP, central venous pressure; HR, heart rate; +Gz, headward acceleration with a head to foot direction of resultant inertial force; +Gx, forward acceleration with a chest to back direction of resultant inertial force.

Redistribution of the blood and extracellular fluid due to removal of the hydrostatic pressure gradient coupled with normal tissue elasticity and muscle tone in the lower limbs is an important response to acute weightlessness which induces a significant increase in the central venous pressure (CVP) in humans compared with that in the supine and upright sitting positions at 1-G (e.g., Pendergast et al. 1987; Engel 1991; Table 13). In the human being, the venous hydrostatic indifference point is located below the heart (Blomqvist and Stone 1983). The increase in the CVP leads to congestion of the intrathoracic circulation, including that of the respiratory system due to the negative intrathoracic pressure as the blood is removed from peripheral circulation and concentrated in the central systemic circulation. At 0-G, the CVP is 0.9 kPa compared with a value of 0.3 kPa on the ground (1-G, sitting position) and 0.35 kPa in flight (1-G, upright sitting position) (Norsk et al. 1987). The blood volume shift decreases the functional residual capacity while leading to an increase in the thoraco-abdominal volume (e.g., Kimball et al. 1976): the changes in blood volume cause reciprocal changes in lung volume and the chest wall. Increased intrathoracic pressure may lead to congestion and subsequent engorgement of the pulmonary vasculature, a feature which, all factors being normal or nearly so, would lead to increased pulmonary diffusing capacity of O_2 due to the greater capillary surface area. However, elevated pulmonary blood volume results in increased microvascular pressure which is likely to cause interstitial edema of the blood-gas barrier. This causes a reduction in the diffusing capacity of the lung especially during exercise, when pulmonary arterial and venous pressures rise in correspondence with the increase in pulmonary blood flow. Interstitial edema resulting from increased pulmonary capillary blood pressure due to redistribution of blood to the central circulation in microgravity leads to rapid shallow breathing as a result of stimulation of the J-receptors (Engel 1991) and may be life-threatening. Excesssive transmural pressure (TMP) may lead to stress failure of the blood-gas barrier, causing disruption of capillary endothelial cells, alveolar epithelial cells, or both (Tsukimoto et al. 1991; West et al. 1991; Costello et al. 1992). The blood-gas barrier is designed to withstand high intramural stress by having a thin layer of the strong collagen-IV in the extracellular matrix (Costello et al. 1992). Stress failure of the pulmonary capillaries leads to some pathophysiological conditions such as edema, exercise-induced pulmonary hemorrhage, and emphysema in horses (e.g., West et al. 1991). Above a TMP of 3.3 kPa, stress failure in the alveolar wall in the rabbit lung enhances exudation onto the alveolar surface (Costello et al. 1992). In the horse, the capillary transmural blood pressure which causes stress failure in the pulmonary capillaries ranges between 10 and 13 kPa (Birks et al. 1997).

Chapter 4

Water Breathing: the Inaugural Respiratory Process

> *"Of all the substances that are necessary to life as we know it on Earth, water is by far the most important, the most familiar, and the most wonderful; yet most people know very little about it." Thomson (1961)*

4.1 The Design of the Gills

By way of the hydrologic cycle, water on Earth is believed to have remained unchanged in amount and character for about 3000 million of years (Leopold and Davis 1968). From the current concepts of paleobiology, it is popularly considered that life started in water (e.g., Thompson 1980; Selden and Edwards 1989). Currently, as many as 21 000 species of fishes (e.g., Nelson 1976; Gilbert 1993), the largest extant vertebrate taxon, live in it. More than half of the living vertebrates have arisen from evolutionary lineages which still inhabit water (Pough et al. 1989). For the first 150 to 200 million years of life on Earth, owing to the harmful effects of the UV light, life was consigned to water. The earliest complete fish fossils, members of the long extinct group named ostracoderms, date to at least 425 million years ago (Repetski 1978). As an ecosystem, water presents greater microhabitat diversity than air and land. Some extreme aquatic habitats include the hot geothermal springs at the floor of the deep oceans 3 km from the surface (e.g., Childress et al. 1989) and the volcanic, hot, alkaline lakes, e.g., Lake Magadi of the Kenyan Rift Valley where the osmolarity of the water is 600 mOsm l^{-1}, pH 9.6–10.5, O_2 level 2.2 mg l^{-1} and temperature about 43 °C (Reite et al. 1974; Johansen et al. 1975). During the millions of years that they have lived in water, fish have adapted very well. Presently, they have a cosmopolitan distribution, occupying diverse ecological niches.

Compared with air breathing, water breathing is the more ancient mode of respiration (e.g., Kämpfe 1980). Pharyngeal gills are a characteristic feature of the phylum Chordata (e.g., Gutman and Bonik 1981). Except for detailed morphology, for such a heterogenous taxon, the basic structure of the gills in the agnathan and gnathostomatous fishes is strikingly similar. The gills combine simplicity of design with functional complexity. Interestingly, some species such as *Hypopomus*, a tropical swamp fish from Paraguay, and *Gillichthys mirabilis* take in air bubbles and hold them over their conventional gills (Carter and Beadle 1930, 1931; Todd and Ebeling 1966; Gans 1971). This may have been the first but futile attempt to extract O_2 directly from air before respiratory concavities and pouches which subsequently evolved into lungs developed. The open aquatic environment contains respiratory gases in equilibrium with the aerosphere. Evaporative water loss as a factor in respiration does not arise. While the gills

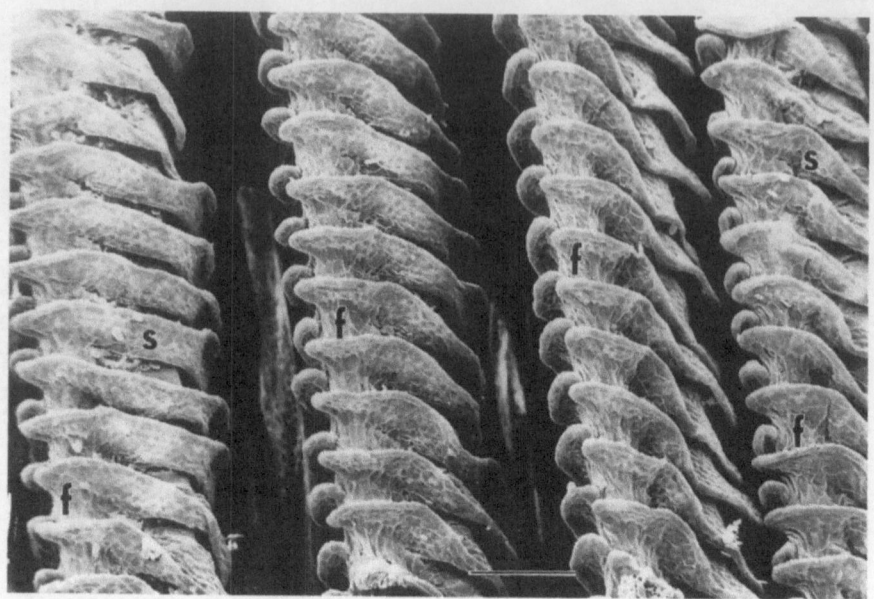

Fig. 46. Gill filaments (f) of a tilapiine fish, *Oreochromis alcalicus grahami*, showing the bilaterally located secondary lamellae (s). The secondary lamellae are an efficient site of gas exchange where the blood and water flow in a countercurrent manner. *Bar* 80 µm

are highly efficient in O_2 uptake in water, except in a few specially adapted air-breathing fish, the lamellae, the numerous closely packed leaf-like plates which are arranged around the gill filaments (Fig. 46), after drying out in air, become impermeable to gases. Furthermore, out of water, the lamellae cohere due to surface tension and collapse under the force of gravity, in the manner of shore weeds at low tide. This reduces the respiratory surface area, creates large diffusional dead air spaces, and increases branchial vascular resistance, ultimately drastically lowering the gas exchange capacity of the gills (Burnett and McMahon 1987; Graham 1990). The animal becomes anoxic, hypercapnic, and eventually succumbs to asphyxia though exposed to a medium rich in O_2. Moreover, during exposure to air, CO_2 levels in blood increase, precipitating a hypercapnic and subsequently an acidotic state due to the inefficiency of the gills in discharging CO_2 into the air. The transfer of O_2 in the gills and lungs of the coconut crab, *Birgus latro*, decreases 30-fold if the chitin is allowed to dry up (Harms 1932). The decline of O_2 consumption of the intertidal limpet, *Patella granularis*, when held in air is due to a functional impairment of the external pallial gills which dry up after evaporative water loss (Marshall and McQuaid 1992). When out of water, the gas exchange capacity of the gills of aquatic crustaceans is severely reduced by a factor of 3 in *Callinectes sapidus* (e.g., O'Mahoney and Full 1984) and up to five times in *Cancer productus* (DeFur and McMahon 1978). After adaptations such as greater rigidity of the gills through increased sclerotization and wider spacing of the secondary lamellae, features which minimize adhesion and col-

182

lapse, *Carcinus maenas* can use its gills in air (Taylor and Butler 1978) and may tolerate air exposure for several days (Truchot 1975). In the terrestrial crabs, e.g., *Geograpsus grayi, G. crinipes, Cardisoma hirtipes*, and *Gecarcoidea natalis*, and in fish like mudskippers, e.g., *Periophthalmus*, which live in shallow mudflats (Gordon et al. 1969, 1985; Mutsaddi and Bal 1969; Tytler and Vaughan 1983; Clayton and Vaughan 1986), the gills are diversely adapted through wider gill filament spacing, gill lamellae stiffening by thicker chitinous deposition and presence of nodules which physically keep the lamellae apart (e.g., Low et al. 1990; Farrelly and Greenaway 1992) enabling the gills to be used in air (Tamura and Moriyama 1976; Tamura et al. 1976). Terrestrial vertebrates which can permanently use gills in air have not evolved and probably never will since, in addition to the gas exchange role, the gills serve other important functions such as ionic regulation and ammonia excretion (e.g., Wendelaar-Bonga and Meis 1981; Laurent and Perry 1991) which can only be effectively carried out in water. Over 90% of the NH_3 excreted in aquatic fish takes place across the gills (Smith 1929). The development of accessory respiratory organs was utilized to circumvent this evolutionary dead end. Some fish which are well adapted to terrestriality, e.g., the mudskipper, *Periophthalmus* (Tytler and Vaughan 1983), can convert ammonia to the less toxic urea (Gregory 1977) and thus extend their survival out of water. Interestingly, insectan wings are thought to have evolved from ancestral gills which were used by aquatic insects for ventilation and swimming (Marden and Kramer 1994).

4.2 Adaptive Diversity and Heterogeneity of Gill Form

Developmentally, gills are gas permeable evaginated outgrowths from the body of aquatic animals (Figs. 4,5). They constitute an interface between two compartments filled with aqueous solutions, an external medium (water) and internal extracellular fluid (hemolymph or blood). Gills occur in different sizes, forms, and locations. They range in complexity from the simple tegmental evaginations like the tube feet and pupullae of some classes of echinoderms which provide only a minor supplement to gas transfer (much of which takes place across the skin), the external gills of the annelid polychaetes, some molluskan nudibranchs, larvae of teleosts (e.g., *Gymnarchus*), and the tadpoles to the more elaborate multifunctional internal ones contained in the branchial chambers, e.g., in crustaceans and teleosts or accommodated in a mantle cavity, e.g., in mollusks (e.g., Laurent 1982; Hughes 1984). The fragile and yet elegant structure of the gills (Ojha and Singh 1986; Olson 1996) has fascinated biologists for a long time. They constitute a paradigm of an efficient external (evaginated) aquatic gas exchanger and provide a fundamental model for the study of the manifold respiratory processes and mechanisms adopted by the higher vertebrates. Gills carry out diverse seemingly unrelated functions which include respiration, osmoregulation, acid-based balance, ammonia excretion, regulation of circulating hormones like catecholamines and angiotensin, and detoxification of plasma-borne harmful substances (e.g., Neckvasil and Olson 1986), locomotion (e.g., Septibranchs), and feeding. The definitive design of the gills must be a compromise between all the functional requirements (McDonald et al. 1991). The trophic (filter-feeding) role of the gills is exhibited by the simple chordates in the subphyla Tunicata and Cephalochordata, e.g., *Amphioxus* and *Petromyzon* (e.g., Baskin and Detmers 1976; Youlson and Freeman 1976) and actinopterygian fish, *Polypterus* (Hughes 1980), bivalve mollusks, in Ascidia, in ammocoete larva of lampreys (Youlson and Freeman 1976), and in embryos and larvae of amphibians (Billet and Courtenay 1973). Divergent views as to whether the gills initially evolved as trophic or respiratory organs persist (e.g., Willmer 1970; Gutman and Bonik 1981; Hickman 1984). In the lamellibranch mollusks, the gills have shifted from the primary respiratory function to filter feeding (Hazelhoff 1930; Jorgensen 1952). In those animals where the gills serve a dual function of respiration and filter feeding, ventilation is generated by cilia. Movement of the water in contact with the surface of the gills is an important part of the gas exchanging process and may occur in simple animals in form of ciliary action, e.g., on the branchial crown of sabellids, on the parapodial gills of polychaetes, e.g., *Nephtys*, and in the ctenidia of aquatic mollusks and ascidians. While enhancing procurement of food, the fast flow of water necessary for food uptake reduces the O_2 extraction capacity due to the short transbranchial transit time of the water (Table 8). Larval caecilians use gills for respiration in the egg but are lost soon after hatching (Welsch 1981). Some invertebrate organisms such as the terebellid worms (e.g., Weber 1978) present some localized areas of the body where the thickness is drastically reduced, gas exchange occurring to the same degree of efficiency as in the gills proper.

Vertebrate gills are categorized into external gills, i.e., those which dangle freely into the water, e.g., in the larval lungfishes and many amphibians and internal ones, i.e., those which are covered by various forms of cutaneous modifications, e.g., the opercular flap of the teleosts or are contained deep in the mesenchymal tissue mass, e.g., in sharks and the cyclostomes (e.g., Emery and Szczepanski 1986). Except for the cuticular gills of some insect larvae (Wigglesworth 1950), the external gills are highly susceptible to physical and chemical damage. Furthermore, they physically restrict locomotion in water. Typically, teleost fish have four pairs of gill arches. Gills have to be selectively permeable and well tuned to control and regulate transepithelial flux of ions and water between the internal and external milleu to curtail excessive loss or overload (Kirschner 1982). A conflict in the design of the gills is brought about by the need to optimize gas exchange by fully exposing extremely delicate organs to the elements. Any protective cover in form of a corneum cuticle or mucus would drastically limit the permeability and hence the transfer, especially of gases across the gills. In many animals with gills, however, the organs are housed in enclosures such as a mantle cavity, a shell (e.g., mollusks), outgrowths of the carapace (e.g. arthropods) and crustaceans, or are contained within a branchial cavity enclosed by the integument, e.g., in teleost fishes. While providing protection to the internal gills, this modification impedes water flow over the gills. This limitation was circumvented by evolution of an energetically expensive buccal force pump for moving a viscous O_2-deficient medium, through narrow spaces. In the larval amphibians, external and/or internal gills together with the general body surface play a major role in respiratory gas exchange. Except in two groups of animals, holothurians and cephalopods, aquatic respiratory ventilation is largely circulatory rather than tidal (Figs. 6,18). However, in sea cucumbers, the cloaca pumps water tidally whereas in the octopus and squid, the mantle serves as both a locomotory device and a respiratory one by pumping water into and from the mantle cavity. Tidal breathing of water across the anus occurs among annelids (e.g., *Urechis caupo*) and insects (e.g., dragon fly nymph, *Aeshna*; Fig. 23), via the branchial pores in lamprey eels and via the mouth in some amphibians and reptiles, e.g., the soft-shelled turtle, *Amyda* (McCutcheon 1954). The energetic cost of oscillatory ventilation in water explains why tidal ventilation rarely evolved in the active water breathers where throughflow designs predominate. Contraction of the gills, ciliary movements and mechanical ventilation (e.g., in the decapods and fish), placing the body in a moving current of water or active body movement maintain a PO_2 gradient across the respiratory surface. Fish which move constantly like the mackerel and the tunas ventilate their gills passively simply by keeping their mouths open. They have lost the power to mechanically pump water across the gills and hence must swim continuously. To avert physical damage during the fast forward movements, the gill filaments of such fish are fused to each other (Muir and Kendall 1968). By hitching a ride on a shark, the gills of *Remora remora* are passively ventilated but at rest, the fish utilizes active respiratory movements (Muir and Buckley 1967). In the siren, *Necturus maculosus*, the muscles associated with the gills are well developed and play an efficient respiratory role. Waving the external gills (in water) generates convective movement of the water. The process is stimulated both by increased water temperature and reduced O_2 content (Guimond and Hutchison 1972). The development of the external gills of larval amphibia depends on O_2 tensions in water (e.g., Foxon 1964). When placed in hypoxic water, the gills of *Rana temporaria* hypertrophy while in well-oxygenated water they atrophy. The external gills of the larvae of the salamanders which live in well-oxygenated waters are less well-developed, i.e., the respiratory surface area is small and the water-blood barrier is thicker than in those which live in O_2-poor waters (Dratisch 1925; Bond 1960). In fish living in water with a PO_2 of 10.7 kPa, the gills are much larger than in those at 100 kPa (Dratisch 1925). In *Xyelacyba myersi*, a fish which lives in the deep water of the Pacific Ocean (at depths of 1400 m), the gills are poorly developed (Hughes and lwai 1978). Unlike in some fish, amphibian gills are not specialized for exchange of gases in air. They never constitute the exclusive respiratory organs at any one stage of development. The contribution of the gills to the overall respiratory process in the amphibians differs from within and between species, depending on the habitat occupied and the level of development (e.g., Shield and Bently 1973a, b; Burggren et al. 1983; Malvin 1989). In the early stages of development, e.g., in the neotenic urodele larvae, the skin, lungs, and buccopharyngeal cavity function as adjunct respiratory organs. When exposed to low O_2 tensions ($2 \text{ ml} \text{l}^{-1}$), ventilation in *Nereis virens* may become continuous and at very low values (0.6 ml per l) stop altogether (Lindroth 1938a). In the octopus, *Octopus vulgaris*, the O_2 extraction factor may be as high as 80% (Winterstein 1925), with the ventilatory rate of the gills increasing ten times above normal during hypoxia.

4.3 The Functional Innovations of the Gills for Aquatic Respiration

In all evolved gas exchangers, the structural adaptations which favor O_2 transfer are a thin barrier between the respiratory media and extensive surface area. In many aquatic animals, these features engender new problems of maintaining ionic and osmotic steady states in the body fluids. Opercular breathing movements in fish have been reported to occur before hatching. They appear to assist in the hatching process itself (S. Smith 1957). Except for the external gills of the neotenic urodeles, the amphibian gills (Fig. 47) can be considered to be transient or disposable gas exchange organs. Vascular shunt (anastomotic) vessels between the afferent (dorsal segment of the aortic arch) and efferent (ventral segment of

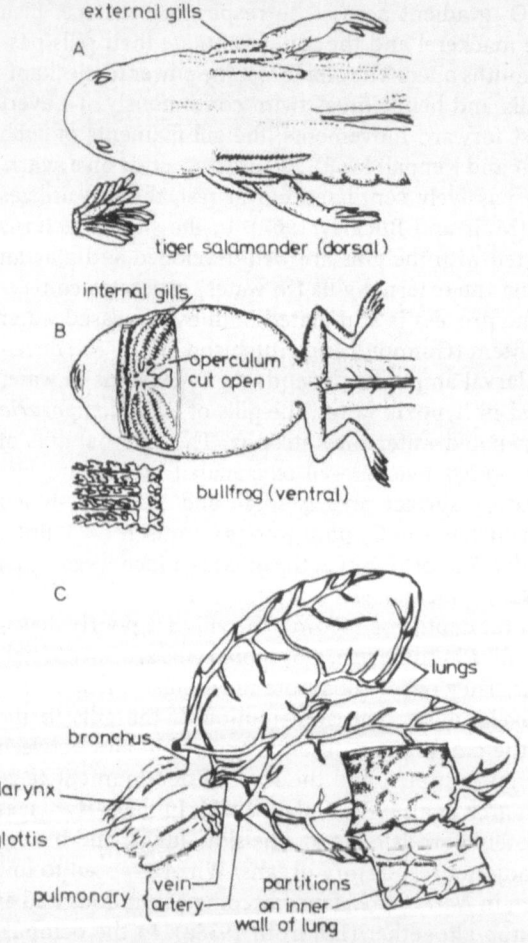

Fig. 47A–C. Respiratory organs of amphibians at different stages of development. **A** External gills of the larvae enlarged in the inset. **B** Internal gills of a bullfrog tadpole. **C** Lungs of an adult frog showing the large peripherally located air cells. (Jessop 1995)

the aortic arch) gill arteries which bypass the external gills occur in both *Rana temporaria* and *Bufo bufo* (De Saint-Aubain and Wingstrand 1981) and in the gills of the urodele amphibians, e.g., *Amblystoma tigranum* (Baker 1949). The shunts are presumed to sustain circulation after metamorphosis when the external gills atrophy on changeover to pulmonary respiration.

Although aquatic animals procure O_2 through the gills, the majority of them take up a substantial fraction across the skin and lose a significant amount of CO_2 through it. In normoxic diving, the soft-shelled turtles, *Trionyx spiniferus asperus*, and musk turtles, *Stemotherus odoratus*, can remain submerged under water for 100 days and maintain normal acid-base status (Belkin 1968). Although the skin of fish has been considered to be impermeable to gases, measurements made on the eel, trout, and tench indicate that substantial uptake of O_2 occurs transcutaneously (Kirsch and Nonnotte 1977). Much of the O_2 is, however, used by the skin itself. The partial pressure of CO_2 in the blood of the lungfish, *Protopterus* declines after vasodilation of the cutaneous vasculature owing to enhanced loss of CO_2 (DeLaney et al. 1974). Nonpulmonary CO_2 elimination is substantial in many aquatic reptiles but only a small fraction of the total O_2 need is acquired in that way. Of the total CO_2 output in the aquatic turtle, *Trionyx mucita* (Jackson et al. 1976), 65% and as much as 94% in the sea snake, *Pelamis platurus* (Graham 1974), is lost cutaneously. In the aquatic snake, *Acrochordus javanicus*, 8% of the total O_2 uptake and as much as 33% of CO_2 excretion occurs through the skin (Standaert and Johansen 1974). Invertebrates such as the oligochaetes, platyhelminthes, mollusks such as Scaphoda, some Aplacophora, and some Gastropoda which have no distinct respiratory organs rely entirely on the skin for O_2 uptake. Such animals have a large surface-to-volume ratio which allows enough O_2 to be taken up by diffusion. Blood capillaries arranged in form of loops are present in the outer (circular) muscle layers in the gill-less maldanids (Pilgrim 1966), terebellid worms (e.g., Weber 1978), and on the parapodia of the annelid sandworm, *Nereis succinea* (Mangum 1982a). In such areas, the diffusional distance is thin enough to facilitate satisfactory transcutaneous movement of the respiratory gases. In *N. succinea*, the diffusional distance from the medium to the blood ranges from only 1.5 to 2.3 µm (Mangum 1982a). The naked gastropods of the genera *Lymnea* and *Helicostoma* can stay under water for a long period acquiring O_2 through the skin (Cheatum 1934). In the limpet, *Siphonaria zelandica*, 25% of gaseous exchange occurs across the side of the foot (Innes et al. 1984). Some salamanders such as *Desmognathus quadramaculatus* (Gatz et al. 1974) rely entirely on the skin for gas exchange. Oxygenation of the tissues which lie beyond the limiting diffusional distance must be effected by the coelomic fluid though in most species the fluid has a very low O_2 carrying capacity. In *Arenicola cristata*, at the microenvironmental PO_2, the PO_2 in the coelomic fluid is less than 6.7 kPa (Mangum 1982a). The metabolic rates of the deeply located and poorly vascularized tissues is relatively low and is finely tuned to O_2 availability. Respiration in *Arenicola* is intermittent with pauses of about 20 min. During the activity periods which last for about 10 min (at 20 °C), about 90 ml of water is passed across the body with an O_2 extraction factor of 50% (van Dam 1935, 1938). In amphibians, those species with smooth, poorly vascularized internal surface have a thinner epidermis and a much denser capillary network on the skin than those with well-developed lungs (Foxon 1964; Czopek 1965). The relative length of the cutaneous capillaries expressed as percentage of the total length of respiratory surface-associated capillaries is about 50 to 80% (Czopek 1965).

4.4 The Simple Gills

4.4.1 Morphological Characteristics

The general features of both the simple and the complex gills have been described by Kennedy (1979). Not all gills or gill modifications play a significant role in gas transfer (e.g., Fox 1921; Thorpe 1930). Certain hair-like or plate-like external outgrowths from the body which have been called gills play no determinate respiratory role. Such forms are found externally in echinoderms, mollusks (e.g., Nudibranchia), arthropods (e.g., Brachiopoda) (Kikuchi 1992), fishes (e.g., adult lungfish, *Lepidosiren*), blood gills of *Chironomous*, and in some amphibians (e.g., adult salamanders). In the larval stages, such gills may function as appendages and/or filter feeders with respiration being a secondary process. The freshwater brachiopod, *Branchinella kugenumaensis*, has ten pairs of gills (metepipodite segments of thoracic legs) and the neck organ (located on the cephalothorax) is an important respiratory organ (Kikuchi 1992). The anal gills of Diptera larvae are largely salt-absorbing organs (Koch 1938). Among the Cirripedia, the extensions attached to the cirri in the four genera of Lepadidae, though described as gills by Darwin (1851), appear to play no role in respiration. This was, however, refuted by Kaestner (1970) and Burnett (1972): a large number of species in the group flourish without them. On the other hand, some structures generally considered to be feeding organs such as the tentacles of the sea anemones may play a notable role as gas exchangers (Sassaman and Mangum 1973; Mangum 1994). The so-called ventral gills or blood gills of the aquatic Chironomid larvae take up a smaller quantity of O_2 from the water than the skin (Harnisch 1937). The annelids are among the simplest animals which have an organ morphologically recognizable as a gill. In the ancestral members of this taxon, the gill is a simple, smooth tubular evagination of the body wall (e.g., Nakao 1974; Mangum 1976a) where, e.g., in *Glycera* and in *Nereis succinea* the water outside and in the coelomic fluid is kept in motion by cilia, a closed circulatory system lacking. In others such as the oligochaete, *Alma nilotica*, the comparably simple gills (Gresson 1927; Khalaf El Duweini 1957) are perfused through a closed circulatory system (Stephenson 1930). The parapodia in many polychaetes, e.g., *Nereis*, *Arenicola*, and *Dasybranchus*, are well vascularized and serve as gills (Nicoll 1954). Relatively simple gills which are located inside the branchial skeleton rather than outside (Randall 1972), as is the case in the gnathostomatous fishes, are found in the lower animal forms, e.g., the Cyclostomata, i.e., the hagfish (Jensen 1966; Bardack 1991) and the lampreys (Youlson and Freeman 1976; Lewis and Potter 1982; Mallatt and Paulsen 1986). The larval forms of the viviparous aquatic caecilian, *Typhlonectes*, exchanges gases by apposition of the fetal gills to the highly vascularized wall of the oviduct (Wake 1977): the gills are absorbed before birth.

4.4.2 Ventilation and Functional Capacities

The advantage of having respiratory organs on locomotory parts of an aquatic animal, e.g., in some crabs (Fig. 48) is great (Maitland 1986). Mechanical displacements of the appendages confer a ventilatory advantage. This may be of particular importance especially in standing hypoxic water. The development of gill modifications on the locomotor parapodia in polychaetes (e.g., *Arenicola* and *Dasybranchus*) guarantees efficient ventilation. The primary role of respiratory pumping in the marine snail, *Aplysia californica*, has been presumed to be that of enhancing O_2 uptake into the hemolymph through the gill epithelium (Kanz and Quast 1992). A negative correlation between respiratory pumping and O_2 consumption under hypoxic conditions has, however, been reported by Levy et al. (1989) in related species, i.e., *A. depilans* and *A. fasciata*. Gills without any specific ventilatory adjuncts are encountered in echinoderms where cilia produce fluid movement, e.g., in mollusks (Nudibranchia), arthropods (Brachiopoda and insect larvae), fishes (e.g., adult lungfish *Lepidosiren* – in the male during breeding and in larvae of teleosts and elasmobranchs), and amphibians (e.g., mud puppy, *Necturus*, and a number of other salamander adults and in numerous larval and tadpole stages of frogs). In *Necturus maculosus*, when the animal is stationary,

Fig. 48. Sand-bubbler crab, *Scopimera inflata*, has pairs of membranous disks (gas windows) in each leg meral segment, ★, which are utilized for gas exchange during low tide. The gas windows are mechanically ventilated during locomotion, enhancing the gas exchange process. (Maitland 1986; reprinted by permission from *Nature*, vol. 319, pp. 493–495; copyright 1986; Macmillan Magazines Ltd.)

muscles at the base move the gills especially when the animal is at rest (Guimond and Hutchison 1972) but many other urodeles, though possessing similar gills, are not known to execute such maneuvers. The gills hang out in water, where they are passively ventilated by the physical movement of the animal in water or when the gills are placed in a water current. Agitation of the immediate (boundary) water layer over the gills or the skin renews the water, thereupon increasing the PO_2 (Feder and Burggren 1985a, b; Feder and Pinder 1988). Consequently, diffusion of O_2 across the tissue barrier is enhanced. Some species, e.g., *Dasybranchus*, exhibit sporadic gill contractions presumably effected by movements of the coelomic fluid (Mangum et al. 1975). Ventilatory activities can be effected by ciliary movements (e.g., in nudibranchs), by physical movement (e.g., in Phyllopoda), muscular contractions of the branchial outgrowths (e.g., in Amphibia and Ephemend larvae) and in others, e.g., crustaceans, by the beating of special appendages (e.g., pleopods) which maintain a water current across the gills. The high shore littoral crustacean, *Ligia*, obtains about 50% of its O_2 needs through the pleopods and the rest through the ventral surface of the abdomen (Edney and Spencer 1955). Terrestrial woodlice breath through the pleopods, with *Porcellio scaber* obtaining as much as 70% of their O_2 needs through them. The book gills of the chelicerate arthropod, *Limulus polyphemus*, are borne on five pairs of modified appendages which are located on the ventral surface of the opisthosoma (Mangum 1982a). Each modified appendage consists of about 150 rounded lamellae which in a 1-kg specimen are 3 to 4 cm wide. The thickness of the water-blood barrier is on average 5.6 µm thick. When *Limulus* is in hypoxic water, the frequency of the ventilatory movements of the opisthosomal appendages increases. On exposure to extremely hypoxic or hypercarbic water, the totally aquatic and solely cutaneous breathing amphibians, *Cryptobranchus alleganiensis* and *Telmatobius culeus*, rock or sway their bodies back and forth in water (Hutchison et al. 1976; Boutilier and Toews 1981), passively ventilating their skin, In *Necturus*, at 25 °C, through agitation of the gills in water, 60% of the O_2 needs can be transferred across the gills (Guimond and Hutchison 1976) while at the same temperature, in the inactive gills of *Siren lacertina*, the gills meet less than 5% of the O_2 needs (Guimond and Hutchison 1973a, 1976). *Necturus* is found in diverse habitats which range from clear waters of lakes, streams, weed-choked canals, and drainage ditches (Bishop 1943). Adaptive features such as physical movement of gills, capacity to undergo long periods of anaerobiosis, and lowering metabolic rate some 30 to 40% below normal are some of the adaptive features which enable animals to withstand hypoxia.

4.4.3 Gas Exchange Pathways and Mechanisms

Amphibians possess gills during their larval stages of development. Caudata and caecilians (Gymnophiona = Apoda), have external gills, while in Silentia they are internal. In the neotenic forms, e.g., *Necturus* and *Ambystoma*, the gills persist throughout life. Many aquatic salamanders utilize a trimodal gas exchange strategy in which pulmonary, branchial, and cutaneous surfaces variably contribute to gas exchange. In the *Siren*, of the total amount of O_2 taken up by the animal (at 25 °C), the gills account for only 2.5% while they eliminate 12% of the total CO_2 (Guimond and Hutchison 1972). A 42% reduction in the total gas exchange capacity occurs on ligation of the gills of *Necturus* at 22 °C (Shield and Bentley 1973a,b). The external gills contribute as much as 54% of the total O_2 need and void 61% of the total CO_2 at 5 °C, with the O_2 consumption through the gills increasing to 60% at 25 °C (Guimond and Hutchison 1972). *Necturus* quickly lowers the aquatic PO_2 from 20 to 5.3 kPa (Guimond and Hutchison 1972). An arterial PO_2 of about 4.7 kPa was reported by Lenfant and Johansen (1967) in the blood of *Necturus*, a value indicative of a high efficiency of the gills in O_2 extraction at very low ambient PO_2. Fanelli and Goldstein (1964), however, observed that the gills of *Necturus* are of no consequence in respiration. This may, nevertheless, only be true in an inactive animal at low ambient temperatures in O_2 saturated water. In cool (5 to 15 °C) aerated water (PO_2 = 17 to 20 kPa), in undisturbed animals, the gills do not move and are retracted and held to the side of the head. However, as temperature increases, the animal becomes restless and the gills oscillate (Guimond and Hutchison 1976). When removed from water, at 20 °C, the gills of *Necturus* collapse (Lenfant and Johansen 1967). This is followed by a dramatic rise in the arterial PCO_2 and a marked reduction in arterial PO_2 in spite of evident pulmonary ventilatory and gulping movements. The gills of *Siren* are less involved in respiration than those of *Necturus* and are notwithstanding more efficient in CO_2 elimination than O_2 uptake (Shield and Bentley 1973a,b; Bentley and Shield 1973). Features such as a thick branchial epithelium (e.g., Cope 1885), abundant arterial venous shunts which bypass the gill filaments (e.g., Darnell 1949), small size, structural simplicity, and immobility of the gills (Guimond and Hutchison 1976) explain the respiratory inefficiency of the gills of the *Siren*. The branchial beating frequency in *Necturus* depends on factors such as metabolic rate and environmental hypoxia or hypercapnia. The rate increases from about 10 times per min to more than 50 times per min (with temperature increasing from 10 to 25 °C (Guimond and Hutchison 1972, 1973a). In insect and amphibian larvae, extended hypoxia induces gill growth and cutaneous vascularization (Bond 1960). The role of the external gills of the larval anuran amphibians in gas exchange is not known (Boutilier 1990) but the internal ones of the lungless tadpoles of *Rana catesbeiana* and *R. berlandieri* account for as much as 40% of the total O_2 and CO_2 exchange at 20 °C (e.g., West and Burggren 1982). In some anuran species, ventilatory rates as high as 90 times per min at 20 °C have been reported by Burggren and West (1982) and in some, ventilatory frequency appears to be synchronized with the heart rate (Wasserzug et al. 1981).

In the different amphibian species, the levels of development of the external gills correspond with the availability of O_2 in the environment in which they subsist and the functional needs (Noble 1931). During the development of the lungs, the unidirectional ventilation of the internal gills with water through bucco-pharyngeal muscular activity decreases by as much as 50% as the skin assumes a prominent respiratory role (Burggren and West 1982). *Siren*, which is endowed with capacity for efficient cutaneous and pulmonary respiration, has moderately developed gills while *Necturus*, which relies on pulmonary exchange, has more elaborate gills to supplement cutaneous gas exchange (Guimond and Hutchison 1976). Gills of *Necturus* kept in cool well-oxygenated water atrophy compared with those kept in warmer nonaerated water (Guimond and Hutchison 1976). After being forcefully held under water for over 2 weeks, the gills of *Necturus* enlarge. The lugworm, *Arenicola cristata* which lives in burrows, a fairly anoxic habitat, has 11 pairs of gills which are located in the midregion of the body. The main trunks of the gills branch four times, the terminal branches which account for much of the respiratory surface being 0.6 mm long and 0.07 to 0.10 mm in diameter while the diffusional distance is 2 to 4 μm thick (Mangum 1982a). Features such as high hemoglobin-O_2 affinity and great cooperativity of O_2 binding (Mangum 1976b) may enable such species to subsist in a hypoxic environment. In the polychaete families (e.g., Wells et al. 1980) and in the filamentous gills of other taxa such as crustaceans (McLaughlin 1983), the gills are organized in the same general pattern as in *Arenicola*. The gills of the members of the family Sabellidae, which inhabit vertical tubes either within soft sediment or, as in the case of *Eudistylia vancouver*, attached to firm substratum, are considered to be more advanced (in some respects) than those of the lugworms (Mangum 1982a). The gill, a pinnate structure known as a branchial crown, is not metamerically arranged but is confined to the anterior end where the respiratory gases can be easily exchanged with water: the gill serves as a filter-feeding organ while the organism is still confined to the tube. In *E. vancouveri*, the gill consists of two principal trunks or branchioles which divide at the first level to give rise to numerous long ciliated pinnae giving rise to a double row of filaments called pinnules at the second level of branching (Vogel 1980). A single branchial crown consists of about 54 000 pinnules which provide a surface area accounting for more than 70% of the surface area of the body. This translates into a mass-specific surface area of 30 cm^{2}l^{-1} (wet wt.). Individual branchioles are associated with a single blood vessel which divides in the same bifurcating plan, terminating blindly at the pinnules. Owing mainly to the absence of the afferent and efferent blood vessels, the blood flow pattern in the branchioles is believed to be tidal in nature (Fox 1938; Ewer 1941). In the cephalochordates and tunicates, the blood flow in the gills is irregular. In the tunicates, the flow reverses in direction (Remane et al. 1980). In both groups, hemoglobin is lacking but the tunicates have a vanadium-based pigment which may carry O_2. The respiratory differences between the sabellids and the lugworms include: (1) the ventilatory rate in the sabellids is three times greater than in arenicolids (Dales 1961; Mangum 1976b), (2) the O_2 extraction factor from water is only 6 to 10% in sabellids and 30 to 60% in arenicolids (Dales 1961; Mangum 1976b), and (3) the affinity of the sabellid respiratory pigment (chlorocruorin) is generally low, ranging from 5.9 kPa in *Myxicola* (Wells and Dales 1975) to 6.9 kPa in *Spirographis* (Antonini et al. 1962) at a pH of 7.4 and a temperature of 15 °C. Of all respiratory organs which have been studied, in *E. vancouveri*, blood may pass in and out of the gill a number of times before reaching the ventral blood vessel which distributes it to the body tissues (Ewer 1941). The sebellid gills present a unique model of a microcirculatory configuration in the animal kingdom in which multiple cycling of the blood occurs (within a gas exchanger) to maximize O_2 uptake by the blood.

In modern tadpole larvae of Anura, internal gills (Fig. 47) from rows of branched lamellae supported by gill bars which are separated by four gill slits (e.g., Uchiyama et al. 1990). With the onset of pulmonary respiration, the gill slits close up except in some Urodela which have readapted to aquatic life where some slits remain open throughout life. External gills are rare in adult fish but play an important respiratory role in amphibian larval forms of the tadpoles and neotenic larvae (e.g., in *Amblyostoma* – Axolotl), in *Discoglossus*, and in water-breathing adults (e.g., *Necturus maculatus*). In the larval forms of elasmobranchs and some larvae of Chondrostei and Teleosti, external gills form as long filaments floating in the albuminous fluid within the egg case. True external gills occur in the larval forms of fishes, e.g., Polypteridae and Dipnoi, and in amphibians in form of threads or fine feathers (Dunel-Erb and Laurent 1980b). In the less advanced invertebrate gill-breathing life forms, the circulatory system is not very well connected to the branchial respiration: the gills are located on the venous side of the heart which receives arterial blood. In most Gastropoda, Cephalopoda, advanced Crustacea, and fishes, the venous blood is collected and passed to the gills and then distributed to the rest of the body. The amphibian external gills differ remarkably from those of fish in that macroscopically they form arborescent organs and are not arranged in a refined hierarchical order.

In many lower forms of animals, the regulation of branchial respiration is absent or very poorly developed. In *Ligia oceanica*, the beating of the pleopods, which produce water currents across the gills, does not change with O_2 tension in water except at very high or low concentrations. In *Gammarus locusta*, the pleopods stop beating at a concentration of O_2 above 5.6 ml per l while a drop causes rapid beating (Fox 1921). In the aquatic insect larvae, *Cryptochaetum iceryae* (Diptera, Agromyzidae) and *Icerya purchasi* (Coccidae, Monophlebini), Thorpe (1932) observed that CO_2 efflux occurred at specific areas of the body. In the fish larvae, before the gills develop, the body surface serves as the only gas exchange pathway. To maintain a high PO_2 in the layer of water next to the skin, the larvae cause convective movement of the surrounding water either by positioning themselves in moving water currents or by executing physical movements (Liem 1981). The external gills of newly hatched larvae of *Protopterus* have cilia which move water across the gills before the ventilatory muscles develop (Whiting and Bone 1980). During the postlarval development, in *Protopterus ampibius*, the fraction of the total O_2 which is acquired through the external gills and the skin declines with the development of the lung (Johansen et al. 1976). In the bimodally breathing teleost *Monopterus*, the large muscular fins generate a posteriorly directed water current which interacts with a well-vascularized region below the epithelial surface of the fins and the yolk, creating a highly efficient countercurrent gas exchange system (Liem 1981). This adaptive property may enable the fish to subsist in hypoxic water. The tuft-like structure on the pelvic fins of the male *Lepidosiren paradoxa* which is highly vascularized has been considered to be important in the parental rearing of the eggs and larval forms in burrows constructed for that purpose (Kerr 1898; Agar 1908): the gill-like organ has been assumed to be involved in O_2 uptake from water. This prospect may be supported by the fact that during the husbandry period, the fish does not surface as frequently as it normally does to exchange gases in air. It is thought that the pelvic fin respiratory elaboration may be utilized for transferring O_2 from the blood of the male fish to the immediate vicinity of the eggs and the developing larvae in the frequently hypoxic tropical swamps in which the fish lives (Kerr 1898; Cunningham and Reid 1932). On hatching, the young larvae of *Lepidosiren* respire through the external gills which start to atrophy after 45 days of life (Carter and Beadle 1930; Krogh 1941). At about the same time, the pelvic gills of the male lungfish begin to regress as the larvae surface to breath air (Agar 1908).

4.5 The Complex Gills

4.5.1 Structure and Architectural Plans

The structure and arrangement of the internal gills varies remarkably. The most complex ones are found in the bony fish (class: Pisces) which typically have four pairs of gill arches. In fish, a single ventricle receives venous blood that is subsequently pumped to the gills for the uptake of O_2. The gills of the elasmobranchs lack a skeletal support. In adult cyclostomes, the design of the gills fundamentally differs from that of fish: the main skeletal mass lies external to the gill epithelium (Jarvik 1968; Youlson and Freeman 1976; Lewis 1980; Lewis and Potter 1982). In the hagfish, coronary arteries are lacking: the heart, which is thin-walled, receives nourishment from the venous blood it pumps. Gills are generally poorly developed in air-breathing teleosts (e.g., Munshi 1976), mudskippers (e.g., Low et al. 1988, 1990), and the lungfishes, Dipnoi (Laurent 1982). The bimodal breathing fish face the singular problem of losing O_2 (acquired during air breathing) to the surrounding hypoxic water as the blood traverses the gills. Shunting of the blood away from the gills (Fig. 21) and reduction in the gill respiratory surface area avert such losses. In the most terrestrial of these fish, the gills are retained mainly for elimination of CO_2 which is more readily discharged into the water. The gills of the lungfishes are remarkably different from those of other classes of fish (Laurent 1982, 1996). They do not form regular arrays of primary and secondary lamellae but look like the external arborescent gills of a tadpole rather than the gills of Teleosti and Chondrichthyes (Laurent et al. 1978). Moreover, pillar cells (Fig. 49) are lacking in the dipnoan gills (Laurent et al. 1978; Laurent 1982). The gills of the hemoglobinless Antarctic ice fish, *Chaenocephalus aceratus* and *Chamsocecephalus esox* (Steen and Berg 1966), and *Chaenichthys rugosus* (Jakubowski et al. 1969), much as they are structurally similar to those of other fish which have hemoglobin, have fewer secondary lamellae and the skin is very well vascularized. Fast-swimming fish, e.g., the tuna, show certain modifications of the gills which include presence of calcified flattened gill rays which offer the filaments better physical support (Iwai and Nakamura 1964) and fusion of the gill filaments to the lamellae, preventing lamellar deformation during high speed cruising especially during passive or ram ventilation (Muir and Kendall 1968). The bowfin, *Amia calva*, a freshwater fish which differs remarkably in habit and habitat from the marine fish, has independently arrived at lamelar fusion (Bevelander 1934). It has been speculated that the attribute supports gas exchange

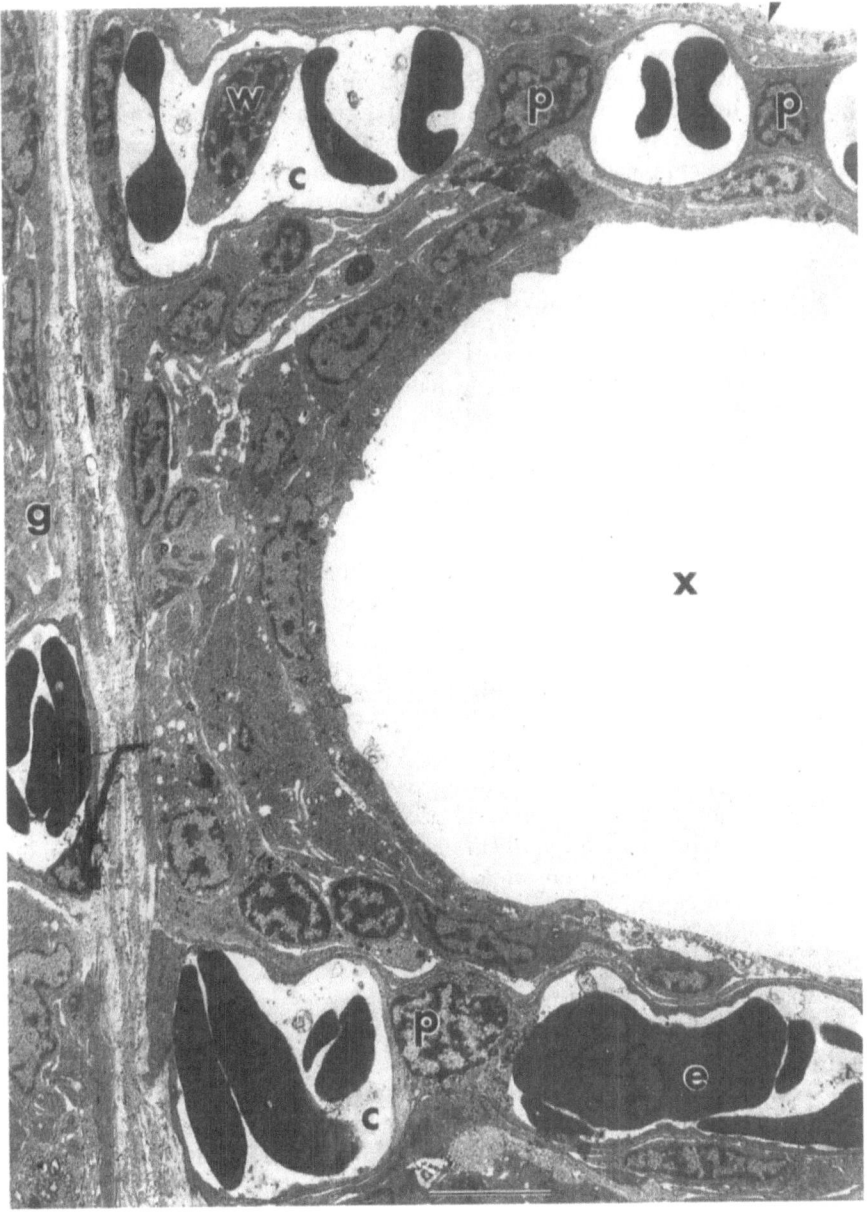

Fig. 49. Gills of a tilapiine fish, *Oreochromis alcalicus grahami* showing a gill filament, *g*, and secondary lamellae separated by an interlamellar space, *x*; *p* pillar cells; *c* vascular channels; *e* erythrocytes; *w* white blood cell; ➤ intercellular junction of epithelial cells. Note the extremely thin water-blood barrier over the vascular channels. *Bar* 17 μm. (Maina 1990)

in the O_2-deficient turbid water and prevents gill collapse during air breathing. Cutaneous gas exchange contributes about 40% of the total O_2 uptake in these fish (Hemmingsen and Douglas 1970). The high efficiency of the skin in O_2 transfer compared with that of the gills may be a means of economizing on respiratory work through conserving the activities of the bucco-pharyngeal pump. In the cold polar water where O_2 levels are normally high, to increase gas exchange, a fish only needs to physically move to sufficiently "ventilate" the skin. The resting O_2 consumption of the ice fish is one half to one third that of hemoglobin-carrying fish of similar size found in the same habitat (Hemmingsen and Douglas 1970). Further compensations for the low O_2 capacity of the ice fish include a high blood volume (7.5% of the body weight) compared with 2 to 3% in normal fish and a reduced viscosity of blood due to absence of erythrocytes, factors which enhance O_2 transport by blood. In fish, during growth, the number of gill filaments increases more rapidly than that of the secondary lamellae (Hughes 1982).

The filament is the functional unit of a branchial arch system (e.g., Olson 1996). Gas exchange, however, occurs across the secondary lamellae which are thin semicircular flaps which are bilaterally set on the filament perpendicular to its longitudinal axis (Figs. 11,46). An elaborate epithelium (the primary epithelium) covers the gill filament while a less complex one, the secondary epithelium, lies over the secondary lamellae (Laurent and Dunel-Erb 1980; Laurent 1984; Maina 1990a, 1991). Three types of cells, the pavement cells, the chloride (= mitochondria-rich = ionocytes) cells, and the mucous cells are most prevalent on both types of epithelia. The different functions of the gills, i.e., osmoregulation (e.g., Leatherland et al. 1974; Eddy et al. 1981), acid-base balance (e.g., Heisler 1984), elimination of products of nitrogen metabolism (Goldstein 1982; Randall et al. 1989), and respiration appear to occur at specific sites of this highly well-organized and differentiated epithelium. The first two processes take place in the composite primary epithelium while the last two occur in the thin, less elaborate secondary epithelium. Like the vertebrate lung (Sect. 6.10.2), the gills perform certain metabolic functions whereby they modify the plasma hormones in the arterial blood before they pass to the systemic circuit (Neckvasil and Olson 1986; Olson et al. 1986a). The pavement cells are simple squamous broad cells which are characterized by surface microridges (e.g., Olson and Fromm 1973; Kendall and Dale 1979; Hughes 1979; Hughes and Umezawa 1983; Hossler et al. 1986): the patterns of the microridges differ between species. Lamellar epithelia of pelagic fish such as bluefish, *Pomatomus saltatrix*, Atlantic mackerel, *Scomber scombrus*, and Atlantic bonito, *Sarda sarda* (Olson 1996) and the hill stream fish, *Danio dangila* (Ojha and Singh 1986), are virtually devoid of microridges. The microridges decrease in size and frequency from the gill arch, the gill filament to the secondary lamellae (Dunel-Erb and Laurent 1980a; Karlsson 1983), and have been associated with diverse roles which include trapping and holding mucus, providing structural integrity to the gill epithelium, and increasing the surface area at the water-epithelial surface interface (Sperry and Wassersug 1976). Olson (1996) contemplated that microridges generate an unstirred boundary layer of water over the gill epithelium hindering gas transfer. The presence of a mucous cover, which streamlines the gill surface by evening out the crypts between the microridges, should reduce the resistance of the flow of water across the gills (Daniel 1981; Hughes and Mondolfino 1983). The presence of microridges on the nonrespiratory surfaces of the accessory respiratory organs (AROs) in fish such as the climbing perch, *Anabas testudineus* (Munshi and Hughes 1991; Wu 1993), and the snake-head fish, *Channa striata* (Hughes and Munshi 1986), supports the suggestion that developmentally, the AROs arise from in situ modifications of the gills. Microridges have been said to enable the pavement cells to greatly distend without engendering mechanical disruption (Knutton et al. 1976). This may be crucial for the integrity of cells which are exposed to a medium whose ionic composition and osmotic pressures may change over short distances. The size and shape of microridges are thought to be influenced by electrolytes, salinity, hormones, and hydrodynamic flow of the water over the gills (Hughes 1979; Schwerdtfeger 1979; Wendelaar-Bonga and Meiss 1981). The mucus covering of the gills has been associated with numerous functions which include protection from mechanical damage and invasion by pathogens, absorption and expropriation of toxic heavy metal ions (e.g., Varansi et al. 1975; Varansi and Markey 1978), and regulation of O_2 (Ultsch and Gros 1979), and electrolyte (Handy 1989) transfer across the epithelial lining of the gills. From morphological characteristics and location within the epithelium, factors which may depend on the stage of cell maturity, two types of chloride cells have been described in fish gills (e.g., Laurent 1984; Maina 1991). An accessory chloride cell has been described in the gills of the saltwater fish (Dunel-Erb and Laurent 1980a; Laurent et al. 1995). The epithelial cells of the gills, especially the chloride cells, are highly sensitive to changes in ambient conditions. Movement of euryhaline fish from freshwater and seawater generates reversible changes in chloride cell morphology, location, and numerical density (e.g., Hossler 1980; Laurent and Hebibi 1989; Laurent and Perry 1991). Hyperoxic-effected hypercapnic acidosis causes an increase in the apical surface area of the chloride cells (Laurent and Perry 1991) while ambient hypercapnia increases the chloride cell number in the catfish, *Ictalurus punctatus* (Cameron and Iwama 1987). In what was construed to be an adaptive process, injections with cortisol for 10 days increased the number of chloride cells by a factor of 3 (Laurent and Perry 1990). The organization of the gill microcirculatory pathways in fish gills is far from settled (e.g., Steen and Kruysse 1964; Boland and Olson 1979; Dunel-Erb and Laurent 1980b; Olson 1996). The lamellae are divided into vascular channels which are lined by polygonal endothelial pillar cells (Figs. 11, 49). In the gills of some land crabs like *Carcinus maenas* (e.g., Farrelly and Greenaway 1992) and some teleosts, e.g., *Oreochromis alcalicus grahami* (Fig. 11), the pillar cells which contain the contractile microfilamental actomyosin elements and collagen (Bettex-Galland and Hughes 1973; Youlson and Freeman 1976) may play an important role in regulating translamellar blood flow. No shunts that bypass the secondary lamellae en route from the ventral to the dorsal aorta occur in the common water breathing teleosts (Olson 1996). In the gills of the European eel, *Anguilla anguilla*, the principal sites of gill vascular resistance are at the level of the afferent lamellar arterioles and the secondary lamellae (Bennett 1988).

The epithelial cell layer, basement membranes, interstitial space and endothelial cell (Fig. 49) constitute the water-blood barrier. Though typically thick in most fish (Hughes and Morgan 1973), the barrier may be as thin as $0.2\,\mu m$ in some regions of the secondary lamellae of some species (Maina 1990a, 1991; Laurent et al. 1995). The morphometric features of the gills, especially the respiratory surface area, correlate with the metabolic demands of fish as well as the environment in which they live (e.g., Hughes and Morgan 1973; Maina et al. 1996a). The pillar cells are arranged as struts which span the width of the space between the two parallel epithelial sheets. They maintain the structural integrity of the secondary lamellae and the blood spaces by preventing overdistension under undue intramural blood pressures which may be as high as 12 kPa (Bettex-Galland and Hughes 1973; Hughes 1976). Fish can regulate the surface area of their gills as well as the exposure of the chloride cells to water (Fig. 50). In that way, gas exchange and ionic regulation can be optimized to suit the prevailing needs and circumstances (e.g., Randall 1982; Butler and Metcalfe 1983). The position of the gill filaments can be varied by contraction of smooth muscles which are enervated by adrenergic nerves (e.g., Nilsson 1985). In a single gill filament, the blood flow across the lamellae at the tip is less than in those at the base (Hughes 1980) and can be regulated by certain pharmocological agents, e.g., serotonin, adrenaline, and noradrenaline (Östlund and Fänge 1962; Nilsson 1986; Fritsche et al. 1993; Sundin et al. 1995). In an individual secondary lamella, by contraction or relaxation of the pillar cells, the blood/hemolyphatic flow may be shifted to and from the larger marginal channels (Fig. 11). In the gills of the mudskipper, *Boleophthalmus boddarti*, a species which lives on the surface of the mudflats of the Arabian Gulf where it makes U-shaped burrows which may be as deep as 1 m (Clayton and Vaughan 1986), the water-blood barrier is thinner around the marginal channels (Hughes and Al-Kadhomiy 1996). Blood remains in contact with water for about 0.5 s, a duration considered to be adequate for complete oxygenation of the erythrocytes (Hughes et al. 1981).

4.6 The Water Lungs

Gas transfer from water by ventilated lung-like (invaginated) structures is utilized by only a small group of animals. It presents a very simple mode of respiration. In the relatively more advanced organisms, it illustrates a retrogressive use of an earlier air breathing organ. In the marine snail, *Aplysia califormica*, respiratory pumping (e.g., Kandel 1979) or interneuron II response as it was called by, e.g., Eberly et al. (1981), is an activity which entails synchronized contraction of the gills, parapodia, and siphon. The seawater is taken in and injected from the mantle cavity. A more complex respiratory pattern which is depressed by long-term (24 h) exposure to hypoxia (about $2\,mlO_2$ per 1 of water) occurs in *A. califomica* (Kanz and Quast 1990). Hypercapnia (Croll 1985), hypoxia, or anoxia (Levy et al. 1989) increase the respiratory frequency in *A. califormica*. Aquatic animals like the sea cucumber (holothurians) have water lungs which occur in form of branched thin-walled tree-like perivisceral tubes that stretch from the

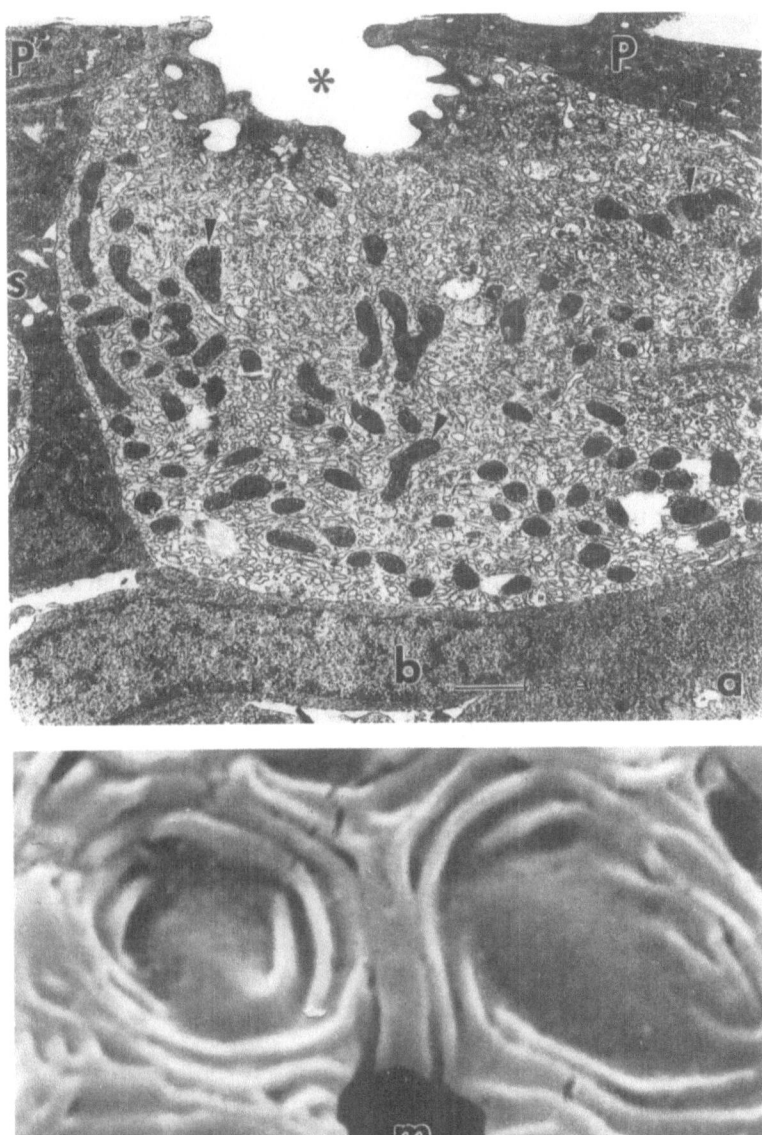

Fig. 50. **a** Chloride cell on the gills of a tilapiine fish, *Oreochromis alcalicus grahami*. The cells are involved in ionic exchange between the blood and water. ➤ mitochondria; *s* supporting cells; *p* pavement cells; ✳ pore; *b* basement cell. **b** A surface view showing closed chloride cells (*c*) and an open mucus cell (*m*). **a** *Bar* 1 μm; **b** 6 μm. (**a** Maina 1990a)

cloaca, running through the whole length of the body. Expulsion of the seawater is produced by irregular contractions and relaxations of the body wall muscles and inspiration by the pumping action of the cloaca (Newell and Courtney 1965): gas exchange takes place between the inspired sea water and the hemoglobin containing coelomic fluid, a process enhanced by the stirring action of the peristaltic waves. In *Holothuria forksali*, cloacal O_2 uptake contributes about 60% of the total need. Gas exchange is reduced by 50% when cloacal ventilation is arrested. Although circulation of the body fluids in the Holothuria has been described (e.g., Kawamoto 1927), it appears to occur to a very small extent. The ventilatory activity of the lung is a more significant factor in effecting movement and mixing of body fluids. In a number of annelids, respiration through the walls of the gut has been suspected though the actual mechanism is not known (Krogh 1941). In a large number of Oligochaetes and polychaetes, antiperistaltic and ciliary movements in the anus are considered to be respiratory (Stephenson 1930). With the exception of *Owenia*, however, Lindroth (1938b) refuted that any significant per rectal movement of water occurs in the polychaetes to warrant the gastrointestinal system being considered to be of any functional consequence in respiration. Stephenson (1930) described a water-swallowing process and expulsion through the anus in the large worm, *Aphrodite aculeata*. In the small thread-like freshwater tubiform worms, e.g., *Tubifex* and *Limnodrilus*, antiperistatic movements of the hind gut occur (Alsterberg 1922). The worm, *Limnodrilus*, builds tubes in the very soft organic mud (which is virtually free of O_2) and stretches its hind parts further into the surrounding water and makes strong undulating movements. When the O_2 tension drops to below 0.08 ml per l, the worms become immobile. In well-aerated water, *Tubifex tubifex* completely retracts into its burrow and, in a hypoxic condition, the worm waves its tail freely in the water (Alsterberg 1922; Palmer 1968). High O_2 concentration is said to be toxic to *T. tubifex* (Fox and Taylor 1954; Walker 1970). The worm has been shown to be highly tolerant to H_2S toxicity (Degan and Kristensen 1981). The tracheal gills on the abdomen of nymphal Plecoptera and Ephemeroptera and larval Trichoptera consist of a panoply of tracheae which are ventilated by rhythmic movements in water while in some larval Odonata (e.g., *Aeschna*), similar structures are found in the hind gut where they are ventilated by contractions of the muscles of the body wall (Fig. 23).

The freshwater pulmonate gastropods of the Order Basommatophora after evolving air breathing have readapted remarkably well to aquatic life (Macchin 1974). This is demonstrated by the fact that the group lacks ctenidia and the mantle cavity has been totally converted to a water lung (Hunter 1953). In *Planorbis corneus*, the mantle is regularly extended into the water to serve as an accessory gill while in *Lymnaea pereger*, air is taken in as the animal dives. This presents a gas-gill function. In specimens living at considerable depths, the mantle cavity functions as a water lung (Ghiretti and Ghiretti 1975). In most pulmonate gastropods, a substantial quantity of O_2 is normally absorbed through the skin. The lung can remain closed for a long period of time and some of the aquatic species, e.g., *Limnea* and *Helicosoma* can remain indefinitely submerged at considerable depths at low temperatures (Cheatum 1934). If the O_2 drops below a critical level, the animal surfaces to breath. When exposed to a hypoxic environment or as the tide recedes, the bivalve *Pholas dactylus* collects water in the aperture on the inhalant siphon from which it draws 47% of its O_2 needs (Knight 1984; Knight and Knight 1986). The lungs of aquatic pulmonate gastropods, *Lymnaea pereger* and *Physa fontinalis*, which live away from water, may be filled with water throughout life (Hunter 1953). *Arhynchite pugettensis* has extensive cloacal diverticula (Manwell 1960) which may serve a similar role. The gephyrean worm, *Urechis caupo*, lives in U-shaped burrows in the intertidal flats along the Pacific coast of America. The levels of H_2S may be as high as 25 to 30 µl and the water becomes hypoxic during low tide, with O_2 levels dropping to 46% of the air saturation. Through muscular contraction of the body wall, the worm ventilates its hind gut (which serves as the primary site for gas exchange) with seawater (e.g., Julian and Arp 1992): the maximum inflation of the hind gut is 2 ml of water per g body weight. At normoxia, the mean hind gut ventilatory volume is 0.70 ml of water per gram body mass and increases to $1.4 \, \text{ml g}^{-1}$ at 4 kPa, the critical PO_2 at which O_2 uptake ceases (e.g., Eaton and Arp 1990). At maximal inflation, the mucoserosal thickness at the respiratory hind gut decreases from its resting size of 283 to 21 µm (Menon and Arp 1992a), a change which reduces the diffusional resistance to O_2, enhancing gas transfer but adversely increasing the permeability of H_2S into the tissues (Julian and Arp 1992). A similar rectal gill mechanism exists in the echiuran worm, *Arhynchite pugettensis*, which has an extensive cloacal diverticulum (Manwell 1960). The skin of the echiuroid worm is remarkably thick to reduce O_2 loss to the ambient hypoxic water. To compensate for this, the hind gut is thin-walled and extends along the whole length of the body (e.g., Redfield and Florkin 1931). In a 60-g animal, the total ventilation of the burrow is 29 ml per min when the animal is feeding, and when it is not, it is 13 ml per min (Hall 1931): one half of the water is taken into the hind gut where one third of O_2 is utilized. Over and above causing mixing of water in the hind gut and hence ensuring more efficient delivery of O_2 into the tissues, the peristaltic waves also lead to movement of the coelomic fluid (which contains a large number of corpuscles which have hemoglobin), further promoting O_2 transfer to the tissue cells. The soft-shelled river turtles of the family Trionychidae, namely *Amyda mutica* and *Aspidonotus spirifer*, ventilate their mouths and pharynx with water at a rate of 16 times min^{-1} when submerged and absorb O_2 through villus-like highly vascularized processes which cover the mucous membrane of the pharynx (Simons and Sussana 1886). The well-vascularized mouth of the cuchia eel, *Amphinous cuchia*, is utilized for gas exchange in both air and water (Singh et al. 1984). Through rhythmic ventilation of the cloacal bursae of the Amazon turtle, *Podocnemys*, 90% of the O_2 needs are met (Steen 1971). Due to the very small quantities of water taken up by the peristaltic processes, doubts have been expressed on the effectiveness of the rectal and intestinal gas exchange process particularly in hypoxic water (Krogh 1941). In such organisms, the skin is utilized for respiration to a greater extent.

A number of air-breathing aquatic snails successfully subsist in the tropical African swamps. The best-known one is the large amphibious *Pila ovata* (Ampullariidae), of which the mantle cavity is divided into water- and air-breathing chambers, and *Biomphalaria sudanica* (Planorbidae), an entirely aerial pulmonate gastropod. The European snails, *Planorbis* (*Biomphalaria*) *corneus*

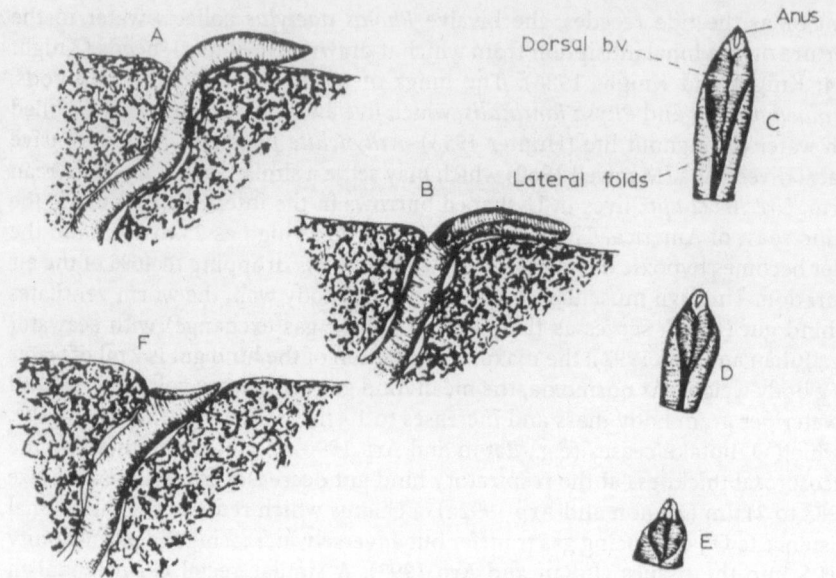

Fig. 51A–F. Respiratory mechanisms and structure of the "lung" of the swampworm, *Alma emini*. **A** Surfacing from the soil. **B** Formation of a respiratory groove. **C, D, E** Closing up the respiratory groove. **F** Retracting into the mud. (Beadle 1957)

and *P. ovata* (Jones 1964), are, however, only marginally adapted for survival in a hypoxic environment by having a slightly higher blood O_2 affinity. Compared with the well-established fish, the tenuous hold of the snails and the insects on the anoxic tropical swamps is a reflection of their much recent reinvasion of this habitat, conceivably after earlier perfection for life on dry land. The swamp worm, *Alma emini*, presents a particularly fascinating adaptation for respiration in the tropical African swamps. Its ecological success in the habitat is reflected in its numerical abundance in the floating mats of papyrus swamps (Stephenson 1930; Beadle 1974). Like most African species of the genus *Alma*, the worm can lead a successful amphibious existence (Beadle 1957; Wasawo and Visser 1959). It can extract O_2 from both air and water: in water, the respiratory groove functions as a water lung. *Alma* subsists in a habitat which is both anoxic and highly reducing. The worm can survive for at least 2 days in total absence of O_2 (Beadle 1957). The dorsal surface which is highly vascularized is spread out on emerging from the soil (Fig. 51) to form a temporary tubular "lung" (Figs. 45,52) through which gas exchange occurs (Maina et al. 1998). Air bubbles are thought to be trapped in the lung and drawn down into the soil for use during the subterranean sojourn. Mangum et al. (1975), however, noted that much of the air captured by the lung at the surface is lost during descent but some bubbles may be held by the hydrophobic cuticle, leaving a volume of about 0.2 µl in an average specimen. The rate of O_2 uptake by the lung is 50 to 60% of that which occurs across the total

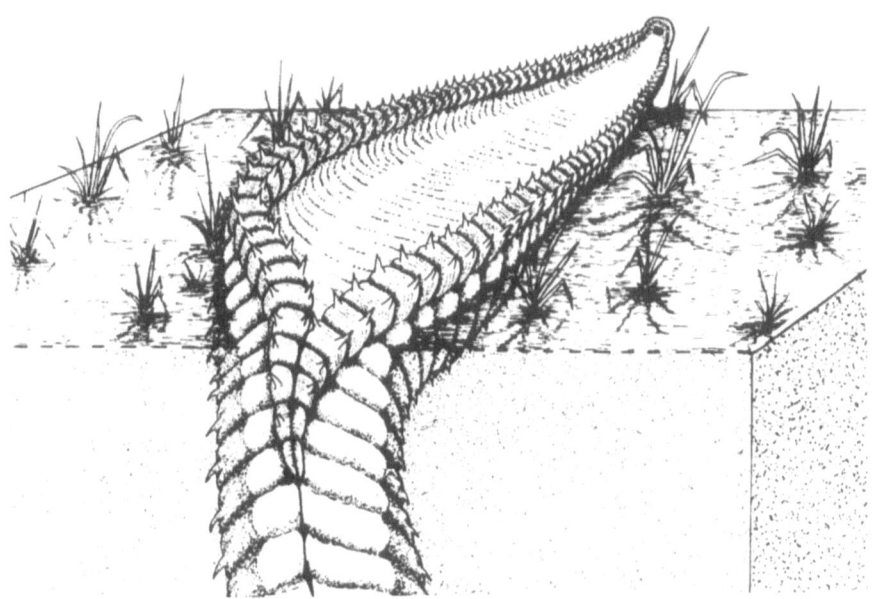

Fig. 52. Schematic view of the respiratory groove of the swampworm, *Alma emini*, which subsists in waterlogged hypoxic soil with abundant putrefying plant matter. The capacity to form a temporary lung for exchange of gases with air enables the worm to live in a hypoxic and hypercarbic habitat

body surface while in water or in air though the lung constitutes only 1.5% of the surface area of the body (Mangum et al. 1975): a 0.75-g worm was reported to have a respiratory surface of $11.4 \, mm^2$. Greater values (lung volume $67 \, cm^3$) and respiratory surface area $245 \, mm^2$) were estimated by Maina et al. (1998). The blood of *Alma* contains an extracellular high molecular weight hemoglobin with a remarkably high O_2 affinity (Mangum et al. 1975). The hemoglobin is fully oxygenated at a PO_2 of 0.3 kPa in absence of CO_2 and at less than 1.3 kPa at very high PCO_2 (27 kPa) (Beadle 1957). The high mortality rate when the worms are prohibited from gaining access to air (Beadle 1957) indicates that O_2 uptake occurs exclusively through the "lung". Metabolism in *Alma* may be mainly anaerobic with aerobic gas exchange serving only to neutralize the toxic end products of glycolysis. The biochemistry of aerobic metabolism in *Alma* is essentially similar to that of other multicellular animals (Beadle 1957; Coles 1970). Coles (1970) estimated that O_2 consumption in air is only $10.8 \, \mu l \, g^{-1} h^{-1}$ at 23 °C but much higher values were reported in both water ($123 \, \mu l \, g^{-1} h^{-1}$) and air ($230 \, \mu l \, g^{-1} h^{-1}$) by Mangum et al. (1975). These observations correspond with those made by Laverack (1963) on related aquatic tubificid worms. In the juvenile stages, the suprabranchial chamber membranes of the climbing perch, *Anabas testudineus*, are used for aquatic respiration, i.e., as water lungs (Munshi and Hughes 1986), a process which persists in adult anabantoids (e.g., Peters 1978).

4.7 The Placenta: an Ephemeral Liquid to Liquid Gas Exchanger

Though ubiquitous among the metatherian and eutherian mammals, in lineages where viviparity is ancient and appears to have evolved only once from a common inceptive ancestor and was from then conserved (e.g., Guillette and Hotton 1986; Packard et al. 1989), the placenta has developed in practically all vertebrate groups except in the agnathan and avian species. In some vertebrate classes, viviparity has evolved repeatedly (Hamlett 1986, 1989; Wourms and Callard 1992). For example, among reptiles, in the squamates, e.g., the lizards *Sphenomorphorus quoyii* (Grigg and Harlow 1981) and *Niveoscincus metallicus* (Stewart and Thompson 1994), and the snakes, e.g., the adder, *Vipera berus* (Bellairs et al. 1995), and the garter snake, *Thamnophis sirtalis* (Hoffman 1970), the process has evolved many times (Weekes 1935; Blackburn 1993). In the sphenodontids, crocodilians, and turtles, however, it does not appear to have ever evolved (e.g., Blackburn 1982; Shine 1985). The lizard, *Sceloporus aeneus*, which lives at high altitudes in Mexico, exhibits a bimodal mode of reproduction, i.e., both viviparity and oviparity can occur (Guillette 1982; Guillette and Jones 1985). Among amphibians, a few anurans, 15% of the urodeles, and 50% of the caecilians (e.g., *Typhlonectes compressicauda*; Garlick et al. 1979) exhibit viviparity (Wake 1989, 1993). Fish like the teleost, *Zoarces viviparous* (Weber and Hartvig 1984; Hartvig and Weber 1984), sharks, e.g., *Scyliohinus settaris* (Wourms et al. 1988; Hamlett 1989; Wourms 1993), and insects (e.g., the tse-tse fly) have functional placentae. Viviparity affords protection of the embryo from adverse environmental conditions and predation (Blackburn 1982; Shine 1983; Shine and Guillette 1988). Extended internal fetal nutrition supports development to a more advanced state (e.g., Lillegraven et al. 1987; Shine 1989; Guillette 1993), improving chances of survival. Oviparity is thought to be ancestral to viviparity (Hamlett 1989). It is envisaged that reduction of the thickness of the eggshell, e.g., in the lizard, *Sphenomorphus fragilis* (Greer and Parker 1979) where the thickness of the shell is only 10 μm, the egg membrane in sharks (Hamlett 1987, 1989) and reduction of the number of eggshell glands, and increase of oviductal vascularity, e.g., in the shark, *Squalus acanthias* (Jollie and Jollie 1967) may be the initial morphological preparations that preceded the evolution of viviparity (Guillette 1989, 1991). Decrease in the eggshell thickness brings the embryonic and maternal circulations closer, predisposing implantation, egg retention, and formation of the placenta, an organ that serves both nutritive and respiratory roles. The transition from oviparity to viviparity is a gradual process which entails both morphological and endocrine changes (Guillette 1991; Hamlett 1989).

Like the gills, the placenta is a liquid-to-liquid gas exchanger. Unique to practically all evolved gas exchangers, the placenta is a secondary gas exchanger, in that it relies entirely on another organ (gill or lung) for gas transfer. While the placenta performs other important roles which include synthesis of hormones, transfer of ions and metabolites (e.g., Faber 1993), and protection of the fetus from adverse ambient pertubations (Laburn et al. 1994), its primary function is unequivocally that of gas exchange (Mayhew 1992). As in other multifunctional organs, the ultimate design of the placenta must accommodate all the constitutive roles. Faber et al. (1992) observed that, despite the purpose of the placenta in all species that have evolved it being essentially the same, "there is no other mamma-lian organ whose structure and functions are so species-diverse". The respiratory challenges faced by the mammalian fetus are to an extent similar to and in some ways different from those of the bird embryo (Sect. 6.13): both operate within a limited and fixed space. Compared with the eggshell, however, the placenta is a dynamic organ which grows and changes with gestation to meet the increasing fetal demands for O_2. The placental function is determined by the growth and development of the terminal villi, the sites where materno-fetal exchange of respiratory gases and metabolites occurs (e.g., Teasdale 1980; Fox 1986; Mayhew et al. 1986; Stoz et al. 1988; Jackson et al. 1992). The human placenta has been conceived as a spongy medium whose porosity is set by the spatial interdigitation and configuration of the maternal and fetal villous systems (Schmid-Schönbein 1988), giving a villous surface area of about $11 m^2$ contained in an average volume of about $500 cm^3$ (Aherne and Dunnill 1966). The geometry of the maternal and fetal vasculature determines the perfusive (hemodynamic) characteristics and the diffusive capacities of the placenta to the respiratory gases (e.g., Lee and Mayhew 1995). With a few falling in between, dependent on factors such as molecular weight, electrical charge, vascular geometry, and concentration gradient, the transfer of substances across the placenta is either entirely flow-limited or diffusion-limited (Faber et al. 1992). During normal human placental develop-ment, villous maturation is characterized by increased tissue and blood volume. The cross-sectional surface area of the terminal villi increases gradually stabiliz-ing at $2000 \mu m^2$ between the 28th and 36th week of gestation (Ruckhäberle et al. 1977; Teasdale 1980; Stoz et al. 1988; Karsdorp et al. 1996). As in all gas exchang-ers, a short diffusional distance and an extensive surface area are the structural features for efficient gas exchange (Figs. 53,54). In the placenta, the distance between the two blood streams, i.e., the materno-fetal placental barrier, ranges between 2 to $6 \mu m$ in man, 6 to $8 \mu m$ in the cat, and 1.5 to $3 \mu m$ in rodents and

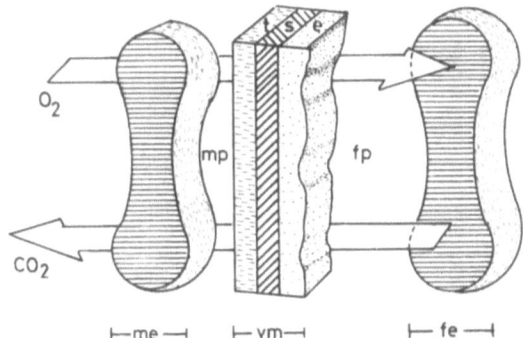

Fig. 53. Schematic drawing of the maternal-fetal gas exchange pathway. The maternal placental circulation is separated from the fetal one by a tissue barrier, *vm*, composed mainly of tropho-blast, *t*, syncytiotrophoblast, *s*, and endothelial cells, *e*. The placental barrier is highly attenuated to enhance gas exchange; *mp* maternal plasma; *fp* fetal plasma; *me* maternal erythrocytes; *fe* fetal erythrocytes

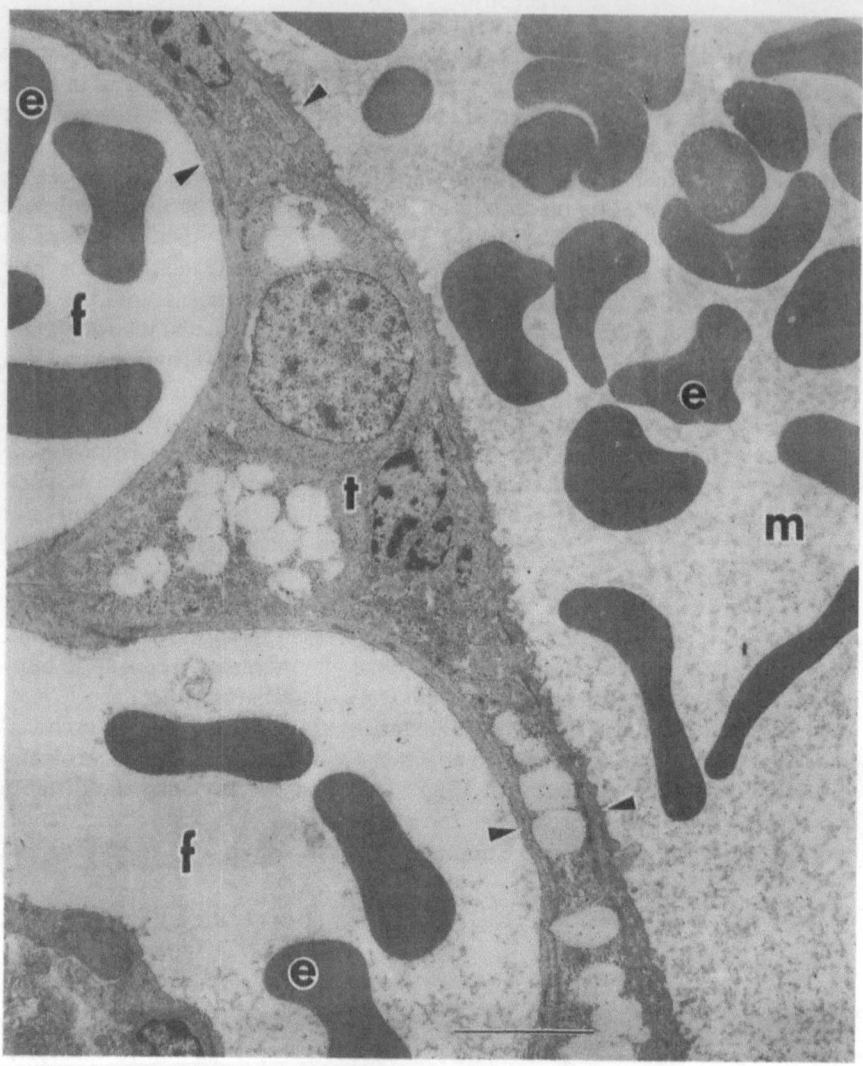

Fig. 54. Maternal-fetal barrier, ➤, of the placenta of the spotted hyena, *Crocuta crocuta*, showing the fetal capillaries, *f*, and the maternal blood spaces, *m*; *e* erythrocytes; *t* trophoblast. The
white spaces in the barrier are lipid aggregations which have been removed after tissue processing. *Bar* 3 μm. (Courtesy Prof. D.O. Okello, Department of Veterinary Anatomy, University of
Nairobi, Kenya)

Leporidae (Bartels 1970). The transplacental O_2 transfer is effected by diffusion
driven by prevalent partial pressure gradient between the maternal and fetal
blood streams. A human placenta with a surface area of 15 m^2 and a thickness of
5 μm will transfer 0.113 ml O_2 s^{-1} mbar^{-1} (Bartels 1970). The microsomal membrane carrier, cytochrome P$_{450}$ which reversibly binds with O_2 has been implicated

in promoting O_2 transfer across the placenta (e.g., Burns and Gurtner 1973). The importance of P_{450} in the flux of O_2 across the materno-fetal placental barrier, however, appears to be very small. No increase in the concentration of P_{450} occurs during hypoxia (Gilbert et al. 1979). In dog and sheep lungs (Burns et al. 1975, 1976), in the liver (Rosen and Stier 1973), and in body tissues in general (Longmuir and Sun 1970), P_{450} has also been incriminated with O_2 transfer. There is presently no hard evidence to support occurrence of active transport of O_2 in any evolved gas exchanger (Dawes 1965). Present experimental evidence strongly indicates that the process occurs entirely by simple physical diffusion.

The placental diffusional pathway is complicated, even in the much simpler hemochorial human placenta. The maternal-fetal placental barrier is neither uniform in thickness nor is it homogenous in composition (e.g., Jackson et al. 1985). Oversimplified physiological models fail to take into account factors such as tissue inhomogeneity (e.g., Laga et al. 1974), maternal-fetal placental perfusion inhomogeneities (Bøe 1954), and placental O_2 consumption and vascular shunts (Barcroft and Barron 1946). About 40% of the human placenta consists of maternal blood (hematocrit, 36%) which is three to five times greater in volume than that of the fetal blood (Ht, 50%) (Mayhew et al. 1984): the mean harmonic mean diffusional distances of the maternal blood plasma, the villous membrane, and the fetal plasma are respectively 0.92, 4.08, and 1.88 µm, giving an overall placental thickness of 6.8 µm. Aherne and Dunnill (1966) reported a thinner maternal-fetal barrier of 3.5 µm. The villous membrane exhibits remarkable sporadic attenuation (Aherne and Dunnill 1966; Jackson et al. 1985), a property similar to that presented by the blood-gas barrier of the mammalian (Gehr et al. 1981) and avian lungs (Maina and King 1982a; Figs. 29b,49a), where it is said to enhance O_2 transfer by generating an overall thin boundary without compromising the mechanical integrity of the gas exchanger. In the human placenta, the maternal-fetal barrier, which constitutes as much as 58 cm³ of the trophoblast (Aherne and Dunnill 1966), may constitute a significant sink for O_2.

The fundamental structural features of any gas exchanger include an extensive surface area and a thin barrier between the respiratory media. These features are achieved in the placenta through different processes which include: (1) plasmodial activity which results in nuclear aggregation of the trophoblastic cell masses at syncytial knots leading to attenuation in some regions, especially those overlying the blood capillaries – such sites from the extremely thin vasculosyncytial membranes which promote gas and nutrient transfer (Bender 1974; Jones and Fox 1977; Heijden 1981), (2) through a mechanistic process which causes distension and margination of fetal capillaries and their relocation to the overlying trophoblast (Amaladoss and Burton 1985; Jackson et al. 1988a,b; Mayhew and Wadrop 1994), and (3) enlargement of the surface area through increase in the number of microvilli (Firth and Farr 1977; Heijden 1981). Attenuation of the trophoblast occurs in the guineapig placenta (Bacon et al. 1984) and in the human cultured one after long-term maternal exposure to hypoxia (Burton et al. 1989). Some parts of the placenta may be concerned with gas exchange while others are involved in processes which require greater tissue density such as hormonal synthesis and nutrient transfer (e.g., Bartels and Metcalfe 1965) as well as mechanical support. Interestingly, a similar engineering process appears to occur in the bird lung where the cell bodies of the pneumocytes are largely confined to the penultimate gas exchange sites, the atria and infundibulae: the extremely thin blood-gas barrier is virtually lined by the attenuated cytoplasmic extensions (Maina and King 1982a). In the mammalian lung, thin and thick sides of the interalveolar septae occur (Fig. 86). The thin parts are utilized for gas transfer while the thick ones render mechanical support and provide pathways for lymphatic drainage (Fishman 1972).

It is widely assumed that the structural complexity of a placenta, i.e, the number of tissue layers comprising it, determines the thickness of the materno-fetal placental barrier and hence the diffusing capacity of the organ for O_2. The less elaborate hemochorial placentae are hence thought to be more efficient than the more elaborate epitheliochorial ones. There are, however, no reliable data to indicate that the simple placentae are in a way more efficient in gas exchange and in promoting fetal growth and development (e.g., Dempsey 1960). Furthermore, it is known that in certain epitheliochorial placentae, the overall barrier thickness is smaller than in some hemochorial ones, which specific areas of the barrier being only about 1 μm thick (Ludwig 1965). The maternal-fetal placental barrier thickness varies greatly in different species and even in the same species during different stages of gestation. Moreover, it has been shown that placental forms may be mixed, i.e., epitheliochorial and hemochorial parts may coexist in the same placenta (e.g., Starck 1959). The diffusing capacity of the placenta depends not only on the path length of the barrier and permeability but also on factors such as the mean diffusion gradient and the placental surface area, features which can be altered according to needs. The transplacental O_2 gradient in the llama (about 2 kPa) is remarkably low, a state perhaps compensated for by the extensive placental gas exchange surface (Barron et al. 1964). The mean diffusional gradient of O_2 in the hemochorial placenta of the rabbit at 27 to 30 days gestation (term about 31 days) ranges from 1.2 to 1.9 kPa and in the syndesmochorial placenta of sheep between 126 and 137 days of gestation (term about 147 days) ranges between 4.3 and 6.4 kPa (Barron and Meschia 1954). From estimates of O_2 consumption of the pregnant human uterus and PO_2 in the maternal and fetal blood streams, Metcalfe et al. (1967) calculated that the physiological diffusing capacity of the human placenta ranges from 0.014 to 0.018 ml $O_2 s^{-1} mbar^{-1}$. Calculations based on the diffusing capacity of CO_2 in pregnant women (Forster 1973) gave higher value of 0.025 ml $O_2 s^{-1} mbar^{-1}$. On average, data indicate that the physiological diffusing capacity of the human placenta lies between 0.013 and 0.038 ml $O_2 s^{-1} mbar^{-1}$ and is appreciably lower than the morphometric one, which ranges from 0.055 to 0.072 ml $O_2 s^{-1} mbar^{-1}$ (Mayhew et al. 1984). The placental membrane of the physiologist (= villous membrane + serial blood plasmas) accounts for 86 to 94% of the total placental resistance with only minor contributions made by the O_2-hemoglobin interactions with the erythrocytes (Mayhew et al. 1984, 1986). In the human lung (Gehr et al. 1978), the physiological and the morphometric diffusing capacities of the lung differ by a factor of 2. This is taken to constitute a functional reserve (Weibel 1984a). The value of the physiological diffusing capacity approaches the morphometric one at the maximum O_2 consumption (Vo_{2max}). The difference between the morphometric and the physiological diffusing capacity of the placenta may functionally be accounted for by vascular shunts, placental O_2 consumption, and regional inequalities of perfusion (Metcalfe et al. 1967; Mayhew et al. 1984, 1990), as well as the lack of uniformity of the thickness of the villous membrane. The latter feature may lead to local inhomogeneities of diffusion resistances across the sporadically attenuated barrier (Mayhew et al. 1984; Jackson et al. 1985). In the vertebrate lung, anatomical and functional shunts and regional inhomogeneities in gaseous diffusion and vascular perfusion contribute to the discordance between the physiological and morphometric diffusing capacities (e.g., Crapo and Crapo 1983). Aherne and Dunnill (1966) envisaged that the mass of the placenta and hence the development of the chorionic villous area (transfer area) correlate directly with the total fetal metabolism.

While the lung is subjected to sudden increases in functional demands from rest to maximal O_2 consumption during exercise, the needs of a fetus tend to be fairly stable and gradually increase with gestation. In the human being, during the 3rd month of fetal life, the O_2 capacity is 12 ml O_2 per 100 ml blood, increasing to 20 to 22 ml O_2 per 100 ml blood during the next 5 months: the maximum O_2 capacity is reached at 6 to 8 weeks (e.g., Betke 1958). From the evident progressive degenerative changes such as infarcts, fibrinoid deposition, thickening of the trophoblastic basement membrane, partial obliteration of decidual arteries, and endothelial cell proliferation and calcification (Mayhew et al. 1984), the morphometric diffusing capacity of the placenta would be expected to remarkably increase towards term. This deterioration is, however, counteracted by a gradual decrease in the thickness of the villous membrane which occurs with gestation when the remarkably attenuated vasculosyncytial regions become more pronounced (Fox 1964a; Jackson et al. 1988b). The arithmetic mean thickness of the trophoblast in the human placenta decreases from about 11 μm at 12 weeks to 5 μm at 38 weeks (Jackson et al. 1988b), a change brought about by displacement of the trophoblastic cell masses, leading to better exposure of the fetal capillaries to maternal blood (Teasdale 1978; Jackson et al. 1988a,b).

4.7.1 The Functional Reserves of the Placentae

The placenta is widely taken to be a progressively aging organ of which the functional capacity declines gradually towards term (Winick et al. 1967). This process is oddly accompanied by increased fetal demands which occur with growth and development. Towards term, fetal requirements come close to totally eroding the functional reserves of the placenta, in some cases leading to intrauterine growth retardation (Hellman et al. 1970; Garrow and Hawes 1971). Winick et al. (1967) found that no further increase in placental DNA content occurred once the fetus reached a body mass of 2.4 kg or the placenta reached a weight of 300 g. According to Rolschau (1978; but see a contrary observation by Sands and Dobbing 1985), growth of the human placenta levels up at 35 weeks of gestation. At altitudes above 3 km above sea level, due to hypobaric hypoxia (Mayhew 1991), human fetal growth is retarded (e.g., Haas 1976). Increased fetal erythropoiesis expressed as high hematocrit, hemoglobin concentration, and proportion of hemoglobin-F suggest that the fetus experiences hypoxia during the last stages of pregnancy (Ballew and Haas 1986). Maternal adjustments, among others hyperventilation and elevated hematocrit, during high altitude pregnancy (Moore et al. 1982), appear to fall short of providing the necessary driving pressure of O_2 across the placenta. Mayhew (1991) observed that with altitude, while adaptive diffusive changes occur on the maternal side of the placenta, fetal conductances especially of the erythrocytes and plasma do not adjust to the same degree, leading to low fetal birth weights at altitude (Mayhew et al. 1990). The difference between the maternal and fetal adjustments for high altitude hypoxia is greatest at birth weights greater than 3 to 3.3 kg, which is about the average birth weight at altitude (Mayhew 1991).

The O_2 transferred by the placenta and made available to the fetus must initially be procured by the maternal lung. The fetus is essentially a temporary addition to the maternal gas transfer cascade (Figs. 2,3). On this account, it is plausible that at some critical point, the fetus would be indirectly affected by extreme states and circumstances which may occasion inadequacy in maternal pulmonary gas flux and cardiovascular transfer. These may result from exposure to hypoxia, i.e., reduction of the driving pressure of gases across the lung and the placenta or may be due to pathological conditions at either or both sites. An increase in the fetal O_2 saturation is generated on maternal exposure to high PO_2 (e.g., Dawes and Mott 1962; Cassin et al. 1964; Assali et al. 1968). Overdesign and/ or plasticity to respond to circumstances when the functional capacity of the maternal gas exchanger may become insufficient is necessary (e.g., Becker 1963; Bender 1974). Such instances include maternal exposure to acute life-threatening hypoxia and pathological conditions such as edema, atelectasis, and pulmonary infarction (e.g., Staub et al. 1967; Staub 1974; Heijden 1981), changes which constrain gas exchange. Perfectly normal reproduction can occur when the maternal blood PO_2 is subnormal, e.g., in women with congenital heart disease (Bartels 1970): in most cases, the consequences largely affect the mother more than the fetus (Burwell and Metcalfe 1958). The Korean Ama (sea women) who voluntarily dive to harvest food at the bottom of the ocean (at depths as much as 30 m) work up to the last day of pregnancy and after giving birth nurse the babies between shifts (Hong and Rahn 1967). Maternal diabetes mellitus does not appear to affect placental development (Teasdale and Jean-Jacques 1986). Oxygen tension in the umbilical vessels was found to be similar in sheep fetuses at sea level and those of ewes living at high altitude (3.5 to 4.0 km) (Metcalfe et al. 1962), indicating that the mother offers efficient protection against changes in blood PO_2. During the birth process, the fetus is protected from hyperthermia: in sheep, during parturition, the maternal body temperature rose at $0.70\,°C\,h^{-1}$ in the final stages of labor but the fetal one rose at a significantly lower rate of $0.45\,°C\,h^{-1}$ (Laburn et al. 1994). In normally developing human pregnancies, a drop in the peripheral vascular resistance in the placenta occurs after a gestational age of 16 weeks. This results in positive end diastolic flow velocity waveform in the umbilical artery (van Zalen et al. 1994). In pregnancies complicated by hypertension and/or intrauterine growth retardation, however, placental flow resistance is elevated. This may lead to absent or even reversed (negative) end diastolic flow velocities in the umbilical arteries during the second and third trimester of pregnancy. Such a condition may result in preterm delivery, neonatal death, and lower birth and placental weights (e.g., Aherne and Dunnill 1966; Trudinger et al. 1991; Pattinson et al. 1993; Karsdorp et al. 1994, 1996).

Like the avian and mammalian lungs, the placentae possess a substantial functional reserve. Large fetal lambs (a few days to term) can maintain an O_2 consumption of 5 to $6 \, \text{ml kg}^{-1} \text{min}^{-1}$ (values within the normal range) even when the umbilical venous and arterial saturations are respectively reduced to 54 and 35% at an umbilical flow rate of $180 \, \text{ml kg}^{-1} \text{min}^{-1}$ (Dawes et al. 1953). Perhaps it is as a part of an inbuilt safety margin of operation that the morphometric diffusing capcity of the human lung ($2.47 \, \text{ml} \, O_2 \text{s}^{-1} \text{mbar}^{-1}$; Gehr et al. 1978) is about 33 times greater than that of the placenta of $0.075 \, \text{ml} \, O_2 \text{s}^{-1} \text{mbar}^{-1}$ (Mayhew et al. 1984). Structurally, the placenta is a highly adaptable organ with a considerable functional reserve capacity (e.g., Karsdorp et al. 1996). Perfusion of the fresh human placenta with varying concentrations of O_2 causes obvious thinning of the trophoblast from 0.44 to $3.3 \, \mu\text{m}$ in a matter of 6h and when the O_2 levels are brought back to normal, the dimensions are reversed (Tominage and Page 1966). The proliferation of the cytotrophoblast appears to be sensitive to the prevailing O_2 levels (Fox 1964b; Kaufmann 1972): cytotrophoblast decreases when oxygenation is good and increases in conditions associated with intrauterine hypoxia (Fox 1964b, 1970; Kaufmann et al. 1977). Chronic maternal exposure to hypoxia in guinea pigs leads to thinning of the trophoblast (Bacon et al. 1984). The rate of flow of the placental blood at term, which is 500 to $600 \, \text{ml min}^{-1}$, is in excess of that of 300 to 400 ml required to supply 16 ml of O_2 to the human fetus per minute (Gahlenbeck et al. 1968). A maternal-fetal PO_2 of 0.3 to 0.4 kPa is adequate to supply the required amount of O_2 to the fetal tissues (Bartels 1970) but a much higher gradient of 2 to 4 kPa has been determined in the larger blood vessels (Metcalfe et al. 1967). Infections, underlying pathological conditions, and exposure to severe conditions such as extreme hypoxia appear to accelerate placental development (Jackson et al. 1995; Lee and Mayhew 1995; Karsdorp et al. 1996). Terminal villi of placentae, of which the pregnancy is accompanied by absent or reversed end diastolic blood flow in the umbilical artery, show a more homogeneous pattern of small villi (Karsdorp et al. 1996). Hitschold et al. (1992) suggested that the accelerated development of the terminal villi may be a compensatory mechanism but the potential advantages gained from it may be curtailed by the concomitant reduction in the blood flow rate in the umbilical artery (Erskine and Ritchie 1985; Karsdorp et al. 1996). The harmonic mean thickness of the placentae of women living at high altitude (average thickness $4.5 \, \mu\text{m}$) was 8 to 19% thinner than that of those living in the lowlands (average thickness $5.2 \, \mu\text{m}$) and the morphometric diffusing capacity of O_2 in the former was higher than in the later (Mayhew et al. 1984; Table 14). Elevated fetal hematocrit, high hemoglobin concentration, low O_2 affinity, and high O_2 carrying capacity of blood (e.g., Ballew and Haas 1986; Tables 15,16) are vital physiological adjustments in highland pregnancy and increase the diffusive conductance of the placenta in a hypobaric and hypoxic circumstance. The most critical point of the development of the human placenta when there is substantial increase in volume,

Table 14. Diffusive conductances in low and high altitude placentae. (After Mayhew et al. 1990)

Variable	Low altitude	High altitude
Dme	1050	1700
Dmp	620	1190
Dtr	28.6	28.7
Dst	49.8	70.8
Dfp	726	666
Dfc	290	237
Overall Dp	15.7	17.5
Specific Dp	4.7	5.79

Units: $ml\,O_2\,min^{-1}\,kPa^{-1}$; specific value, $ml\,O_2\,min^{-1}\,kPa^{-1}\,kg^{-1}$.
Symbols: me, maternal erythrocytes; mp, maternal plasma; tr, trophoblast; st, stroma; fp, foetal plasma; fe, foetal erythrocytes; Dp, diffusing capacity of the placenta.

Table 15. Oxygen affinity (P_{50}) and oxygen carrying capacity of maternal and fetal blood of various species (After Novy and Parer 1969)

Species	P_{50} at pH 7.40 (mmHg)		O_2 capacity $ml\,O_2\,100\,ml^{-1}$	
	Maternal[a]	Fetal[a]	Maternal	Fetal
Man	26	22	15	22
Rhesus monkey	32	19	15	18
Rabbit	31	27	15	14
Sheep	34	17	15	17
Goat	30	19	13	12
Pig	33	22	13	13
Elephant	24	21	20	17
Camel	20	17	15	17
Llama	21	18	14	19
Cat	36	36	12	16

[a] To convert to kPa multiply by 0.133.

surface area, length of villi, and overall thinning of the trophoblast lies between 17 to 21 and 22 to 26 weeks of gestation (Jackson et al. 1992; Simpson et al. 1992; Mayhew and Simpson 1994).

The performance of the placenta falls between that of the lung, which on average operates well below its maximal capacity but can respond to sudden demands placed on it, and the eggshell, in which the structural parameters are firmly incorporated. There are fundamental limitations intrinsic to the placenta as a gas exchanger: the O_2 consumption of the human hemochorial placenta (2 to $10\,ml\,O_2\,kg^{-1}$ wet mass min^{-1}) at term indicates that placental tissue utilizes as much O_2 as the fetus itself (e.g., Nyberg and West 1957). The placental O_2 con-

Table 16. Maternal and fetal oxygen capacities of a number of spacies. (After Bartels 1970)

Species	O_2 Capacity ml O_2 100 ml blood^{-1}		Half saturation pressure (mmHg)	
	Maternal	Fetal	Maternal[a]	Fetal[a]
Man	15	22	26	22
M. mulatta	15	18	–	–
Rabbit	15	14	31	27
Sheep	15	17	34	17
Goat	13	12	30	19
Guineapig	16	16	30	19
Elephant	20	17	23	17
Camel	15	17	21	16
Llama	14	19	21	18
Cow	15	12	21	22
Chicken	14	12	49	34
Seal	32	28	29	21

[a] To convert to kPa multiply by 0.133.

sumption in the more complex placentae like the epitheliochorial ones which have a greater tissue density would be expected to have an even greater O_2 consumption. In sheep, the placenta and fetal membranes consume as much as one third of the fetal O_2 uptake (e.g., Longo et al. 1973). This suggests that in cases of anoxia, the fetus may be critically deprived of its normal O_2 needs. In the cow, the physiological diffusing capacity between the 5th and 9th month of gestation is 0.0016 ml O_2 s^{-1} mbar^{-1} kg^{-1} (Gahlenbeck et al. 1968) but the placenta increases in mass three to ten times during the same period. Experimental Swiss mice at a simulated altitude of 4.3 to 6.1 km show normal mating behavior, fertility, and reproduction, but early embryo resorption occurs at a greater frequency than in the control (sea level) group (Baird and Cook 1962). The resorption takes place at an embryo size of about 7 mm crown rump length, presumably due to deprivation of O_2 (by the placenta) at a critical stage of development. In an attempt to overcome these limitations, the epitheliochorial placentae of sheep and goat and the hemochorial ones of rabbit and guinea pig have developed the efficient countercurrent arrangement between fetal and maternal blood flows (e.g., Kaufmann and Davidoff 1977; Faber et al. 1992). In the rabbit and the guinea pig, the PO_2 in the umbilical vein exceeds that in the uterine vein. The placentae of the rabbit and the guinea pig are categorized among the efficient group of exchangers with the hemochorial placentae of the rhesus monkey and human being including the epitheliochorial placentae of goat and sheep being placed in the inefficient category (Faber et al. 1992; Table 17). From studies of compensatory mechanisms of the injured guinea pig placenta, Heijden (1981) observed that the organ has a very small functional reserve capacity and its potential to respond to reduced capacity is very poor. The countercurrent system of the placenta in sheep and goat is notably inefficient as evinced by the lower physiological diffusing capacity of the organ in the two species (Metcalfe et al. 1967). Having an efficient placenta has its

Table 17. Placental type, vascular geometry, foetal oxygen consumption (Vo_2), oxygen pressure gradient between maternal and fetal blood (PO_2) and diffusion capacity (Dpo_2) of placentas of several species. (After Bartels 1970)

Species	Placental type	Vascular geometry	Vo_2	PO_2 (mmHg)[a]	Dpo_2[b]
Human	Hemochorial	Multivillous	7.4	23	0.32
M. mulatta	Hemochorial	Multivillous	10.0	33	0.30
Cow	Epitheliochorial	Countercurrent	9	38	0.24
Sheep	Epitheliochorial	Countercurrent	9	40	0.22
Goat	Epitheliochorial	Countercurrent	12	40	0.30
Rabbit	Hemochorial	Countercurrent	7.7	10	0.70
Guineapig	Hemochorial	Countercurrent	7.0	10	0.70

[a] To convert to kPa multiply by 0.133.
[b] Dpo_2 calculated using the carbon monoxide method of Longo et al. (1967).
Units: Vo_2, $ml\,O_2\,min^{-1}\,kg\,fetus^{-1}$; PO_2, mmHg; Dpo_2, $ml\,O_2\,min^{-1}\,mmHg^{-1}\,kg^{-1}$ fetus.

Table 18. Oxygen partial pressure in maternal arterial and uterine vein blood as well as in venous and arterial fetal blood for a number of species. (After Bartels 1970)

Species	PO_2[a]				Placenta type	Flow pattern
	Maternal		Fetal			
	Art.	Ven. Ut.	Ven. Umb.	Art. Umb.		
Cow	70	34	21	12	Epitheliochorial	Countercurrent
Sheep	86	52	29	18	Epitheliochorial	Countercurrent
Goat	84	46	33	14	Epitheliochorial	Countercurrent
Man	100	33	29	17	Hemochorial	Multivillous
M. mulatta	88	30	16	12	Hemochorial	Multivillous
Rabbit	80	25	46	17	Hemochorial	Countercurrent
Guinea pig	92	20	30	8	Hemochorial	Countercurrent

[a] Units: mmHg – to convert to kPa multiply by 0.133.
Art., arterial; Ven., venous; Ut., uterine; Umb., umbilical.

price and risk. The rate of flow-limited transfer diminishes much more steeply in the countercurrent exchanger than in the concurrent one when the flow rate of one of the blood streams decreases (Faber et al. 1992). After a flow reduction of 50%, gas exchange in the countercurrent system falls by as much as one half but in the concurrent system this drops by only one third. In this respect, the rabbit and guinea pig placentae possess a lower safety margin or operation, as has been noted by workers who have observed the remarkable fragility of the rabbit and guinea pig placentae, compared with the more resilient ones of the sheep. Despite the intrinsic structural differences, the mature hemochorial placentae show notable similarity in diffusional permeabilities regardless of whether they belong to the hemomonochorial (guinea pig and human), hemodichorial (rabbit), or hemotrichorial (rat) groups (e.g., Metcalfe et al. 1967; Štulc 1989; Tables 17,18).

Dawes (1965) observes that "there is a rugged quality about the way these machines (placentae) are put together which appears to give a wider safety margin than is usually supposed". Placental morphology very poorly reflects the phylogenetic and systematic affinities, especially in mammals (Mossiman 1987; Luckett 1993). The morphological disparity of the placenta defies simple logic. It is a showpiece of the remarkable intrinsic plasticity of biological entities for solving different challenges. Different animal groups have achieved viviparity through manifold strategies and with it design-specific placentae. In a transient organ on which enormous resources are invested to ensure proper growth and development of the fetus, perhaps need to evolve a common architectural plan, a process which would be limiting to some animals, has not arisen. As for the evolution of the blood pigment carriers (Sect. 2.8), those aspects of biology which show great diversity of form and function may be symptomatic of conditions and processes which are far from attaining optimal designs and states. In such cases, encompassing solutions are untenable due to immutable phylogenetic constraints.

Chapter 5

Bimodal Breathing: Compromise Respiration

"For animal lines moving from water to land, the changes in physical and chemical characteristics of the environment are immense. These changes affect all possible life processes, from respiration and excretion to methods of movement, the functioning of the sense organs and reproductive mechanisms. The transition must therefore have been made very gradually, and some of the transition stages are repeated by present day groups." Little (1990)

5.1 The Water-Air Interface: an Abstract Respiratory Rubicon

The division of the Animal Kingdom into aquatic and terrestrial life is ancient: it is still relevant to contemporary life. This distinction is ascribed to the different structural and functional attributes which have been imposed by the physical characteristics of the two different fluid media. While looking at this broad picture, it is too often forgotten than a rich assortment of animals regularly commutes between water and air. While some largely subsist in water and extract their O_2 needs from the air above, some start their development in water and end up on land at maturity. The mutual assemblage and the distribution of the animals that share this complex lifestyle demonstrates the highly pragmatic strategies which animals have adopted to overcome the ceaseless selective pressures that have beset them (Fig. 55). While hypoxic conditions are a rare occurrence in terrestrial environments except at high altitude and in deep, compact wet soil, many marine and freshwater habitats are characteristically hypoxic and/or hyperbaric (Carter and Beadle 1931; Carter 1935, 1955; Hora 1935; Saxena 1963; Rhoads and Morse 1971). Depending on nature and rate of organic putrefaction of plant and animal matter, such habitats may contain noxious gases such as H_2S, a potent inhibitor of cytochrome c oxidase systems, and ammonia (e.g., Theede et al. 1969).

One of the pivotal developments which promoted the adaptive radiation in the Animal Kingdom was that of realization of capacity to procure molecular O_2 directly from air in the Paleozoic fish (e.g., Romer 1967; Johansen 1968; Packard 1974; Dejours 1994; Smatresk 1994). Once having attained air breathing, the aquatic animals seized the opportunity for terrestrial habitation. In evolutionary terms, such milestone events have been called evolutionary novelties (e.g., Miller 1949; Mayr 1960; Riedl 1978), key innovations (Lauder and Liem 1989), or broad adaptations (e.g., Schaeffer 1965a). Baum and Larson (1991) defined such monumental events as "derived traits instrumental in acquiring entry into a novel adaptive zone featuring novel selective regimes". Those innovative animals that underwent the transition to air breathing and life on land are by inference thought to have formed the stem reptiles and are all now certainly extinct. The ones caught at the water-air interface include the extant air-breathing amphibi-

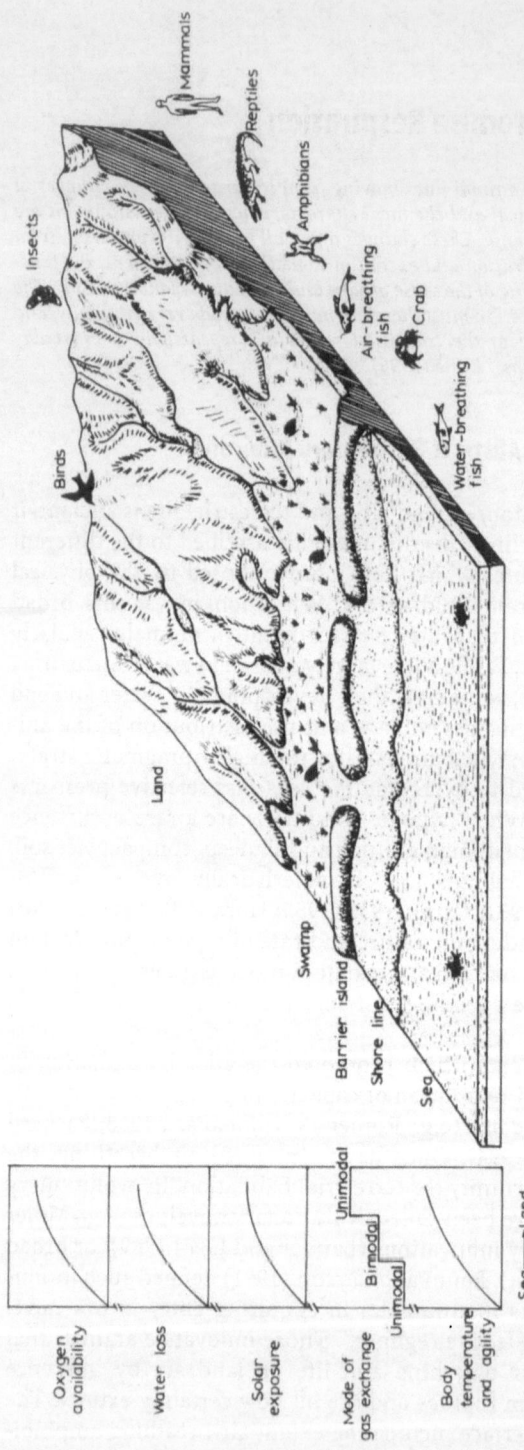

Fig. 55. Schematic view showing the evolution of air breathing and transition from sea to land. Also shown are some of the benefits gained and challenges faced. Representative taxa at various levels of respiratory development are given. Air breathing evolved in hypoxic stagnant waters which were claimed by plants to form swamps. The amphibians still rely on water for many of their physiological processes such as respiration and reproduction while reptiles, mammals, birds, and insects, after developing a waterproof cover, rely entirely on internalized gas exchangers for O_2 procurement. Hypoxia was the main driving pressure which prompted sojourn on land

218

ous forms and the aquatic air breathers (Rahn and Howell 1976). The contemporary transitional (= amphibious = intermediate = bimodal = dual) breathers, animals which are able to exchange gases in both air and water (using the same or multiple organs), however, do not constitute the direct progenitors, i.e., the bridging animals, between the gill breathers and animals with a modern lung (Rahn and Howell 1976). The early stem forms of animals with a derived (modern) lung would have been the Devonian amphibians (Romer 1972). They had a simple lung capable of taking up ample amounts of O_2 and eliminating a substantial measure of metabolic CO_2, as the gills gradually regressed. The surviving bimodal breathers constitute a contemporary paradigm of the processes and pathways through which air breathing and terrestriality evolved: they occupy a central position in the ecology and physiology of the evolved animal life. There is no physiological or paleontological evidence to indicate that direct passage from water to land has ever occurred (Dejours 1994). The contemporary transitional vertebrate breathers, in particular the lungfishes, comprise a provocative group that occupies a pivotal point in the evolution of the terrestrial tetrapods (Joss et al. 1991; Meyer and Dolven 1992). The adaptive diversities and the extents to which animals went to attain air breathing suggest that the selective pressures that launched the change were extremely severe and pervasive (Carter 1957, 1967). The great flexibility required to adapt to the transitional habitats may explain the dearth of intermediate breathers in many taxa. On exposure to a new habitat, adaptive measures are instituted in the order of behavioral, physiological, and structural shifts, transformations, and modifications. The intensity and the frequency of the changes in the environment and the capacity to adapt to the variations may determine the difference between survival and extinction.

The subject of the evolution of air breathing is of particular interest to both physiologists and morphologists. It presents a model for understanding some fundamental changes in the structure and function of one of the central organ systems in an animal's body. Air breathing was a monumental event in the sequence of different preadaptations for terrestrial habitation, a process which culminated in the emancipation of animals from water (Das 1940; Dejours 1989; Moore 1990b). The archaic fish, e.g., lungfishes (Dipnoi) and the bichirs and three quarters of the modern amphibious fish that inhabit the tropical and subtropical regions (Table 19) breath air (Burggren et al. 1985a; Munshi and Hughes 1992). This suggests that the factors which inspired air breathing may have been most severe in such regions. Fish in the Amazonian basin, the largest freshwater equatorial basin, have been widely studied (e.g., Carter and Beadle 1931; Junqueira et al. 1967; Johansen 1968; Kramer et al. 1978; Stevens and Holeton 1978a; Cala 1987): the majority of them have evolved air breathing. The induction of air breathing in the aquatic breathers is widely ascribed to intrinsic changes in the aquatic environment, especially hypoxia and, by extension, ambient temperature (e.g., Townsend and Earnest 1940; Davis et al. 1963; Tulkki 1965; Moshiri et al. 1979; Kutty and Saunders 1973). High metabolic demands for O_2 and intense putrefactive processes precipitated a hypoxic crisis which was exacerbated by hypercapnia (Johansen 1968; Table 11). This combination of events appears to have constituted a decisive force which prompted search for an alternative source

Table 19. Systematic position, habitat, and types of accessory respiratory organs (AROs) of air-breathing teleosts. (After Dehadrai and Tripathi 1976)

Species	AROs[a]	Family	Habitat
Electrophorus electricus	BPA	Electrophoridae	Rivers and swamps – S. America
Opicephalus (=Channa) punctatus, marulius, striatus, gachua	PL	Ophicephalidae	Tropical ponds and rivers – Asia and Africa
Amphipnous cuchia	PL	Ophicephalidae	Tropical ponds and rivers – Asia & Africa
Hypopomus brevicostris	OC	Sternarchidae	Swamps – S. America
Symbranchus marmoratus	OP	Symbranchidae	Swamps – S. America
Monopterus javanensis	OP	Symbranchidae	Freshwaters – Southern Asia
Pseudapocryptes lanceolatus	OP	Gobiidae	Pools and swamps – S. Asia
Heteropneustes (=Saccobranchus) fossilis	OL	Saccobranchidae	Ponds and swamps – Sri Lanka, India, Burma, Laos, Thailand, and Vietnam
Clarias (C. lazera, C. magur and C. mossambicus)	OL	Clariidae	Ponds and swamps – Africa, S. and W. Asia
Macropodus cupanus	OL	Anabantidae	Tropical ponds – Asia
Colisa fasciata	OL	Anabantidae	Freshwater – S. Asia
Betta	OL	Anabantidae	Freshwater – S. Asia
Osphronemus gorami	OL	Anabantidae	Freshwater – S. Asia
Anabas testudineus	OL	Anabantidae	S. Asia, IndoMalaysian Archipelago, Tropical and S. Africa
Ancistrus (A. anisitsi, A. chagresi)	SM	–	Swamps – S. America
Plecostomus plecostomus	SM	Loricariidae	Swamps – S. America
Misgurnus fossilis	I	Cobitidae	Rivers and pools – Europe and Asia
Lepidocephalichthys guntea	I	Cobitidae	Freshwaters – Asia
Doras	I	Doradidae	Rivers and swamps – S. America
Hoplosternum litorale	I	–	Swamps – S. America
Arapaima gigas	SB	Arapaimidae	Swamps and R. Amazon – S. America
Gymnarchus	SB	Gymnarchidae	Swamps and rivers – S. Africa
Erythrinus unitaeniatus	SB	Characinidae	Swamps – S. America
Umbra	SB	Umbridae	Stagnant waters – Europe and N. America
Notopterus (N. notopterus, N. chitala)	SB	Notopteridae	Freshwater – Asia
Phractolaemus ansorgii	SB	Phractolaemidae	Tropical – W. Africa
Anguilla (A. anguilla, A. bengalensis, A. japonicus)	SK	Anguillidae	Rivers – Europe, Asia, Africa, and N. America

[a] BPA, buccopharyngeal apparatus; PL, pharyngeal lung; OC, opercular chamber; OL, opercular lung; SC, stomach; I, intestine; SB, swimbladder; SK, skin.

of molecular O_2 (e.g., Das 1927, 1940; Saxena 1960; Johansen 1968; Kramer et al. 1978; Kramer 1980; Randall et al. 1981; Davenport 1985). Dramatic spatial and temporal variations in aquatic hypoxia in local inland water masses, especially those formed after flooding, have been a common feature of the Earth's surface since the Cambrian period (e.g., Barrell 1916; Fish 1956; Street and Grove 1976; Jenkyns 1980; Bray 1985; Little 1990). At various geological times, increases in environmental temperatures resulted in low solubility of O_2 especially in the tropical freshwater ponds, prompting the quest for an alternative source of molecular O_2 (e.g., Graham 1949; Grigg 1969; Graham et al. 1978a,b). Drying up of the shallow and extensive continental shelves not only aggravated the respiratory conditions but caused overcrowding and competition for finite resources. Physicochemical changes such as increase in salinity and turbidity must have acted as additional stimuli for abandoning water for land (Sayer and Davenport 1991). Evidently, severe respiratory episodes have occurred frequently in the past. In the Silurian-Devonian periods (e.g., Inger 1957; Berkner and Marshall 1965; Thompson 1971), O_2 levels are thought to have dropped to about 10% compared with the present-day levels (Fig. 9): arising from microbial as well as animal respiratory processes, a reciprocality of environmental O_2 and CO_2 levels occured. Invariably, environmental hypoxia is associated with some degree of hypercapnia especially in standing, plant-infested waters. Hypoxia, especially when accompanied by hypercapnia, constitutes a powerful force which induces air breathing. In the tench, *Tinca tinca*, hypoxia-hypercapnia reduces routine O_2 consumption and causes a swelling of the erythrocytes (Soivio and Nikinmaa 1981; Jensen and Weber 1985). Though morphologically similar (Godoy 1975), the respiratory physiology of the two ecologically distinct erythrinid fish, *Hoplias malabaricus* and *H. lacerdae*, which occur in the South American tropical and subtropical shallow waters and streams, is remarkably different (Rantin et al. 1992, 1993): compared with *H. lacerdae* which lives in well-aerated streams, *H. malabaricus*, which inhabits stagnant O_2-deficient environments, is characterized by low metabolism, higher O_2 extraction, tolerance to low O_2 tensions, low gill ventilation (Rantin and Johansen 1984; Rantin et al. 1992), and large respiratory surface area (Fernandes et al. 1984; Fernandes and Rantin 1985) and high aerobic capacity (Driedzic et al. 1978; Hochachka et al. 1978). The critical O_2 tensions for *H. malabaricus* is 2.7 kPa and for *H. lacerdae* 4.7 kPa (Rantin et al. 1992). Around the Gulf of Mexico, during the summer months when the eastwards winds push the surface waters offshore and make the hypoxic deeper water flow inshore (May 1973), the marine crab, *Callinectes sapidus*, emerges into the air (Loesch 1960). An El Niño effect, a major oceanographic change related to shift in global weather patterns that originated from the eastern Pacific during the winter of 1982–1983 and the spring and summer of 1983, caused massive extensive upwelling of apparently hypoxic warm tropical equatorial water along the shores of South and North America (Cane 1983; Philander 1983) resulting in massive mortality of the Oregon's coho (*Oncorhynchus kisutch*) and chinook (*O. tshawytscha*) salmon (Johnson 1988): El Niños have been associated with larval dispersal of and southward displacement of tropical species (DeVries et al. 1997). When exposed to hypoxic water (4 to 5.3 kPa), the freshwater crayfishes, *Austropotamobius pallipes* (Taylor and Wheatly 1980) and *Orconectes rusticus* (McMahon and Wilkes 1983), surface to ventilate their (branchial chambers) "lungs". The air-breathing fish *Erpetoichthys calabaricus* senses and deliberately avoids hypoxic areas of the water in which it lives (Beitinger and Pettit 1984). No examples of air breathing have evolved in animals which subsist in well-aerated waters such as fast-flowing streams where the O_2 levels are perpetually at or near saturation. Krogh (1941) considered the transition to air breathing to have been stimulated by "emergency respiration" resulting from withdrawal of water (from which animals extracted O_2) from the intertidal animals. From the morphological similarity and close topographical relationships between the gills and the accessory respiratory organs (both are located around or open in the pharynx), the evolution of air breathing appears to have been a carefully crafted slow process. It did not involve an ovehaul of the gills but a gradual phasing out. In the mud-eel *Amphipnous cuchia*, for example, the air sac which develops from the gill arches 2 to 5 and the ectoderm cells (derived from the integument bordering the pharyngeal openings) receives venous blood through the afferent branchial vessels of the second and third gill arches and the vascular papillae which develop in the buccal cavity and in the air sacs exchange gases with air (Singh et al. 1984).

The exigency to procure O_2 directly from air has evolved many times in response to different environmental pressures (e.g., Randall et al. 1981; Graham 1994). In *Polpterus bichir*, the gas bladder functions as an accessory respiratory organ only when the O_2 content of the water is low (Budgett 1900). A few air-breathing fish, however, show anomalous air-breathing behavior. About 40 marine species in 6 families regularly breath air in a habitat which is not characterized by hypoxia or hypercarbia (Graham 1976). In some species of aquatic bimodal breathers, e.g., young tarpons, *Megalops atlantica*, air breathing is socially regulated (Böhlke and Chapline 1968; Kramer and Graham 1976). Interestingly, not all the present-day bimodal air breathers are phylogenetically ancient forms which have survived. This indicates that the quest for air breathing is an ongoing covert process which is being aggressively pursued by some of the contemporary aquatic animals especially those which experience hypoxia or sudden fluctuations of O_2 in their habitats. Oxygen is both an important factor for aerobic metabolism and a necessary resource in growth and development. Bader (1937) demonstrated that normal development of the accessory respiratory organs of *Macropodus* (Belontiidae) was hindered if the fish was refused access to air. It has been interestingly postulated that by reducing the risk of desiccation through respiratory water loss, atmospheric hyperoxia, as occurred during the mid-Devonian and Carboniferous periods (Fig. 9), may have enhanced terrestrial habitation: breathing hyperoxic air reduces respiratory frequency and hence lowers respiratory water loss (Withers 1992). The evolution of air breathing, however, did not have a direct causal relationship with terrestrial colonization (e.g., Carter and Beadle 1931). Occupation of land happened to offer one of the many solutions to the prevalent respiratory stress in water. This is evinced by, among others, the air-breathing fish which even after acquiring a significant preadaptation for terrestriality (i.e., capacity to breathe air) permanently live in water. In water, a highly dynamic habitat, animals are behaviorally and physiologically adapted to cope with sudden changes in O_2 availability, temperature, and salinity (Horn and Gibson 1989): at high tide they take O_2 from the water and during low tide from air. The dragon-fly larva surfaces to breath when the PO_2 in water falls below 7.3 kPa (Wallengren 1914). Similar behavior occurs in the fish, *Leuciscus erythrophthalamus*, when the PO_2 falls to below 2 kPa (Winterstein 1908). Behavioral, functional, structural, biochemical, and molecular changes, plausibly in that order, were utilized to accommodate aquatic hypoxia as animals switched from water to air breathing (Hiebl et al. 1987; Weber et al. 1993) and then gravitated towards land. The gills, the archetype aquatic gas exchangers honed for respiration in water proved deficient in air, a physically remarkably different medium (Tables 4,9). The gills were gradually phased out as extraction of O_2 from water was deemphasized and air breathing consolidated. In terrestrial crabs (Sect. 5.6.2), a great deal of this change entailed expansion of the branchial cavity, modification of the branchial epithelium, and mechanical ventilation of the same. In the pneumonate gastropods (Sect. 5.6.1), the ctenidia in the mantle cavity was replaced by a lung.

5.2 Strategies and Adaptive Convergence for Air Breathing

The intermediate breathers use the accessory respiratory organs or lungs to procure about two-thirds of their O_2 needs and eliminate only about one quarter of the metabolic CO_2. The nonpulmonary gas exchanger(s) void roughly three fourths of the metabolic CO_2 irrespective of whether the animal lives in water or air (Rahn and Howell 1976). Some of the primary attributes of an accessory respiratory organ are that: (1) it must possess regular or irregular means for renewal of air, (2) the gas voided from the organ must contain less O_2 and more CO_2 compared with the inspired (atmospheric) air, and (3) the epithelial surface must (in most cases) present conspicuous morphological modifications such as good vascularization and/or surface amplifications. An inverse correlation occurs between the level of commitment of an accessory respiratory organ to air breathing and the degree of regression of the alternative respiratory sites such as the skin, buccal cavity, and gills. The capacity of air breathing in many animal groups and the remarkable uniformity of the morphological, physiological, and biochemical features inaugurated in so phylogenetically different taxa is a model case of convergent evolution. Those features which are common to a wide cross-section of animals contributed the foundations to the comprehensive gas exchange process and those traits unique to a particular group present a solution to a specific problem. By identifying the pressures to which the animals were subjected, by analogy, reconstruction of the events which initiated and sustained the momentous change can be made. There is now ample evidence indicating that the need to breath air evolved essentially because the O_2 levels in water were unstable and largely low (e.g., Carter and Beadle 1931; Schmalhausen 1968; Randall et al. 1981). Due to the remarkable differences in the physiochemical properties water and air (Sect. 3.2), the shift drastically affected the form and function of practically all the biological processes (e.g., Edney 1960; Young 1972; Mangum 1982b; Greenaway et al. 1983; Bridges 1988; Val et al. 1990; Morris 1991; Morris and Bridges 1994). These included locomotion, respiration, reproduction, excretion, and sensory perception of external stimuli. The impact was, however, greatly minimized since need for direct switch from water (a fairly stable environment) to air (land), a highly fluid one, rarely arose. The aquatic and terrestrial habitats intergraded extensively especially during the formative years of the continental land masses when massive uplifting, subsidence, and displacement of the plates caused dramatic shifts in the sea level and flooding of land (Takeuchi et al. 1970; Ben-Avraham 1981; Fig. 55). The intermediate zone had adequate water vapor pressure stability to sustain the development of the accessory respiratory organs without the risk of desiccation in predominantly aquatic animals. In such humidic habitats, the animals adapted to hypoxia tolerance in water and insti-tuted measures for air breathing and terrestrial existence. Among the erythrinids (Order: Cypriniformes) which inhabit shallow O_2-poor freshwaters of the tropical and subtropical regions of South America (Dickson and Graham 1986), *Hoplias malabaricus*, which lives in stagnant hypoxic water, is more tolerant to hypoxia (threshold PO_2 for onset of bradycardia = 2.6 kPa) than *H. lacerdae*, which inhab-its well-oxygenated rivers (threshold PO_2 for onset of bradycardia = 4.7 kPa; Rantin et al. 1993). Factors such as the larger respiratory surface area (320 mm^2 per g; Fernandes et al. 1984) and high O_2 affinity of blood (Wood and Lenfant 1979) in *Hoplias malabaricus* compared with *H. lacerdae* may explain the differences in hypoxia tolerance in the two species. Some erythrinid fish, e.g., *Hoplererythrinus unitaeniatus* and *Erythrinus erythrinus* (Randall et al. 1981), have adopted air breathing. Surface skimming for air and/or well-oxygenated top water layer is a common strategy of overcoming hypoxia in the tropical freshwa-ters (Kramer and McClure 1982). Gulping air at the surface elevates O_2 transport during aquatic hypoxia in the goldfish, *Carassius auratus* (Burggren 1982a). Of the 20 000 or so species of fish, only a relatively small number has evolved the capacity for air breathing (e.g., Dehadrai and Tripathi 1976; Sayer and Davenport 1991; Graham 1994; Table 19). An even much smaller number has adopted terrestrialness. Clearly, air breathing and subsequently shift to land were at-tempted at the extremes of circumstances, e.g., at the threshold of failure of normal O_2 transfer and/or when there were particular benefits to be derived from the shift. The least drastic and most economical solution to air breathing in the Devonian fish would have been to utilize the gills, the highly engineered aquatic gas exchangers, for procuring O_2. The probable sequence of events utilized for adaptation to hypoxia should have entailed: (1) physical avoidance of it by relo-cating to more hospitable areas (Whitmore et al. 1960; Cook and Boyd 1965; Costa 1967; Gamble 1971), and (2) skimming the top 1 to 2 mm surface of water which is richer in O_2 (Burggren 1982a; Kramer and McClure 1982; Barton and Elkins 1988; Horn and Gibson 1989), gulping in air, e.g., in the gobies (Gee 1976; Graham 1976; Gee and Gee 1995) and holding it in vascularized buccal and/or pharyngeal cavities (Coutant 1987; Kramer 1987). In the interim, long-term physiological tolerance to hypoxia would have become established. In the bivalve mollusk, *Pholas dactylus*, air gaping occurs in the laboratory after a drop in PO_2 in the surrounding water and in nature during low tide (Knight 1984). The Amazonian freshwater ray (*Paratrygon* sp.) surfaces when the PO_2 drops to below 2 kPa and utilizes the O_2-rich surface water (Steen 1971). Increased CO_2 in water depresses branchial respiration and stimulates air breathing. Like the fossorial mammals and birds (Sect. 3.7), fish which live in derelict waters are less sensitive to CO_2 than those from well-aerated ones (Hughes 1963). It has been argued that it was not lack of molecular O_2 per se but factors such as availability of new food sources, reproductive needs, escape from predators (Little 1990; Sayer and Davenport 1991), and the enormous energetic advantages derived from the switch (Bennett 1978; Fig. 56) which led to the development of air breathing, especially in the shore tidal areas. The gastropod mollusk, particularly those of the family Ampullariidae, provide excellent examples of the extents to which animals went to attain and maintain an air-breathing capacity. Some, e.g., the amphibious prosobranch, *Pomacea depressa*, which lives in the Everglades swamps in Florida, a rather extensive stagnant mass of warm water where intense organic putrefaction of the massive vegetational growth occurs, have evolved a divided mantle cavity with a gill in one half, the other half acting as a lung (McClary 1964; Little 1990). Such animals are able to breathe air and water simultaneously or switch from one medium to another depending on needs and circumstances.

Water characteristically constitutes an environment where O_2 is scarce (Table 4). In some habitats, survival is compounded by the presence of a high concentration of CO_2 and other gases such as H_2S and NH_3. Owing to the effect of CO_2 on the central control, O_2 consumption falls drastically with rising PCO_2 particularly when the PO_2 is low (Saxena 1962; Tenney 1979). When exposed to hypoxic water, *Gillichthys* gulps air as its oxidative metabolism decreases (Todd 1971). In *Pseudapocryptes* (Das 1934), hypoxia induces air breathing and in *Tomicodon* (Eger 1971), stagnation of water elicits air breathing. When held in hypoxic water, *Clarias* and *Heteropneustes* show metabolic rates which are 60% below normal (Hughes and Singh 1970b; Singh and Hughes 1973). Depending on the ambient temperature and level of activity, the bowfin, *Amia calva*, an ancestral halecomorph North American actinopterygean fish which is an active swimmer and subsists in an environment which is cold and covered with ice in winter and warm in summer, relies on the gills and a vascularized gas bladder for gas exchange (Johansen et al. 1970a; Liem 1987a; Hedrick and Jones 1993): at temperatures between 10 and 30 °C, O_2 consumption is shared equally between the gills and the air bladder, at above 30 °C the air bladder accounts for more than two thirds of the O_2 consumption while at 10 °C and below, the gills meet all the O_2 needs. Breath holding can last from 5 to 55 min depending on temperature, amount of light, and the O_2 concentration in the water (Liem 1987a). Ambient temperature (Burggren et al. 1983) and PO_2 (Burggren and Mwalukoma 1983; Burggren and West 1982) influences the pathway utilized for procuring O_2 in the larval *Rana berlandieri* and *R. catesbeiana*.

The development of the amphibians from eggs through tadpoles to air-breathing adult forms is accompanied by dramatic changes in the respiratory strategies (Fig. 47) and offers a highly instructive model in understanding the change from water to air breathing and transition from water to land. The amphibian eggs acquire O_2 entirely by diffusion across the surface, the tadpoles initially have external gills which are followed by internal ones, and later in life the adults develop fully functional lungs. Metamorphosis starts in water, a medium from which O_2 is less available, with the animal proceeding to the much better oxygenated aerial habitat. The access to a more O_2-rich medium is accompanied by a

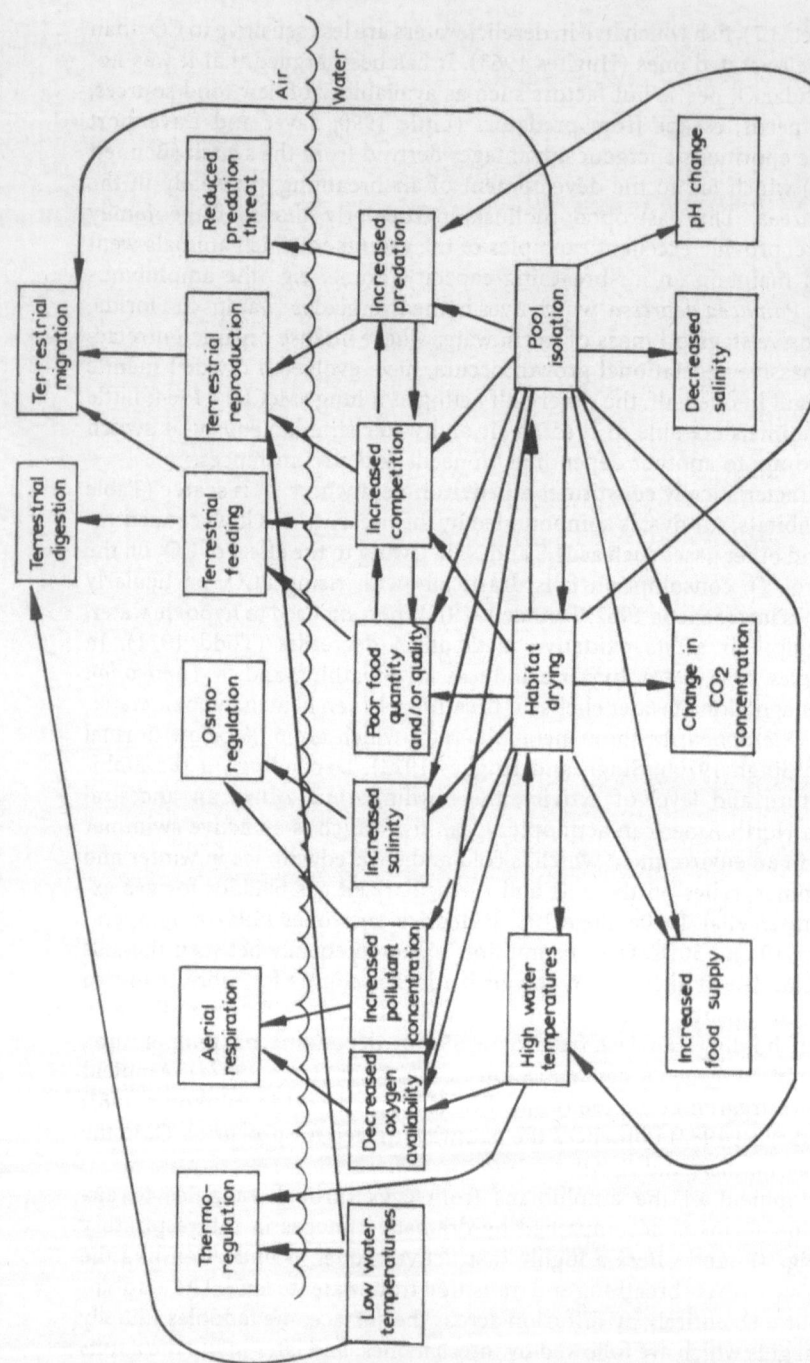

Fig. 56. Principal biotic and abiotic factors which elicited emergence behavior from water by amphibious fish in closed systems (e.g., freshwater and intertidal pools). While the prompting factors differed from habitat to habitat, the pervasive features which prompted air breathing and shift to land included decreasing O_2 and food availability in isolated, shrinking aquatic habitats. In this complex and highly dynamic process, other factors included increasing temperatures, CO_2, and pH. While definite rewards were reaped such as access to abundant O_2 and newer resources, certain costs such as adaptations for reproduction independent of water, more efficient capacity for osmoregulation, and excretion of products of nitrogen metabolism, e.g., urea and uric acid, were called for. (Sayer and Davenport 1991)

decrease in the O_2 affinity on metamorphosis from a tadpole to an adult amphibian (Broyles 1981; Burggren and Wood 1981). The P_{50} of the blood of the tadpole of a bullfrog, *Rana catesbeiana*, is 0.7 kPa (McCutcheon 1936), a PO_2 at which in the adult frog the blood is only 5% saturated. After giving rise to the successful amniotes, the amphibians have literally lingered on with one foot in water and the other on land. Though the first animals to conquer terra firma, owing to their reliance on water for crucial physiological processes such as reproduction, excretion of waste products, osmoregulation, and gas exchange, in general, the amphibians, a group recalcitrant to change, are a defeated group. They constitute an inconspicuous taxon among the extant tetrapods. The contemporary amphibians can be looked on as a relic of an evolutionary stage between air-breathing fish on one hand and reptiles, the first animals with a true lung, on the other.

5.3 Risks, Costs, and Benefits in the Change to Air Breathing

In changing residence from water to land, animals faced fundamentally different environmental and physiological challenges (Fig. 7). They had to procure O_2 from air and void CO_2 into the same, avoid desiccation, store or excrete different toxic products of nitrogen metabolism, somewhat regulate body temperature, and avoid or confront completely new predators. Whereas aquatic habitats offer rather stable and predictable "climatic" features, terrestrial ecosystems show greater spatial and temporal diversity. The terrestrial macrophytes, particularly gymnosperms and angiosperms, generate remarkably different microclimates over short distances (e.g., Geiger 1965). In order to derive the best of two worlds, the aquatic (water-residing) air breathers have chosen to physically remain in water and periodically surface to extract O_2 from the air. Such animals lack specializations like means for aerial vision, terrestrial locomotion, adaptations to curtail cutaneous water loss, and specific means for elimination of products of nitrogen metabolism in form of urea and uric acid, as occurs in amphibious fish (e.g., Gordon et al. 1978; Table 9). Although most aquatic animals will cope with hypoxia by appropriate microhabitat selection, this strategy is not very effective in dealing with long-term (i.e., diurnal and seasonal) fluctuations in O_2 levels. In such cases, permanent solutions are obligatory. Aquatic animals, especially those which live in highly dynamic microhabitats such as tidal pools, thermoclines, or shallow coastal waters display different mechanisms for coping with the extremes of hypoxia. Though indirect, the effect of temperature on respiration is far-reaching. Elevated temperature leads to reduced solubility of O_2 in water and increased metabolic rate. In a hypoxic condition, raised temperature makes life highly precarious, and access to the atmosphere, where O_2 is available in large quantities, becomes a necessity (Serfaty and Gueutal 1943). In the freshwater fish, temperature preference increases with exposure to hypoxia (Bryan et al. 1984; Schurmann et al. 1991). By selecting a lower temperature, the animal takes advantage of reduced metabolism and increased blood O_2 affinity (Schurmann and Steffensen 1992). Similar strategies have been adopted by the reptiles (Hicks and Wood 1985) and salamanders and crayfish (Dupré and Wood 1988). Owing to the more stable PO_2 in air, habitat relocation in a heterothermal environment with changing O_2 saturations is particularly important in aquatic ectotherms, especially in coastal and standing waters. When the circulation of water is limited or respiratory demands of aquatic organisms are high, anaerobiosis may result. In the bowfin, *Amia calva*, air-breathing frequency increases with aquatic hypoxia (Johansen et al. 1970a). This also occurs in most other air-breathing fish (e.g., Shelton et al. 1986). Due to factors such as different capacities to tolerate hypoxia and the variability in the levels of hypoxia in different habitats, the physiological traits acquired during evolution of air breathing appear to have followed different and independent pathways which were dictated entirely by need. The designs of the contemporary gas exchangers and the existing respiratory adaptations cannot be accurately used to discern the systematic affinities between taxa.

The transition to air breathing and, subsequently, terrestrialness called for profound changes in the respiratory strategies. Animals aggressively mobilized resources and changed their habits and habitats with the specific goal of acquiring O_2 from "above". Switching from one respiratory medium to another and from one gas exchange pathway to another enabled the transitional breathers to utilize the most convenient and efficient method(s) for extracting O_2 from the alternative respiratory media. This provided the necessary flexibility to optimally meet the changing conditions. Plainly, the evolution of air breathing was not ipso facto the attainment of terrestriality. Procurement of molecular O_2 directly from air, a medium eloquent of the gas, was simply the immediate, most convenient, and permanent solution for surmounting the critical problem of hypoxia prevailing in the aquatic biotope. Residence on terra firma, a venture first attempted by the now extinct rhipidistian crossopterygians some 350 million years ago (in the Upper Devon) (Pough et al. 1989), was a costly, risky process which was approached parsimoniously using different strategies (Figs. 57,58). It had to await decisive preadaptions which included: (1) redesigning of the existing gas exchangers, (2) development of an impermeable surface cover, (3) solution of problems of acid-base and osmotic balance, and (4) development of more appropriate ways and means of eliminating products of nitrogen metabolism (e.g., Little 1990). Furthermore, animals had to cope with factors such as thermal instability, exposure to new predators, and locomotory problems resulting from the greater gravitational effect on the body. Air breathing and subsequently relocation to land arose when and only if there was absolute need for it or where tangible advantages and rewards, e.g., acquisition of more livable space and greater ecological opportunities, were to be reaped to offset the enormous risk and cost. Due to the abundance of O_2 in the free air, when expressed in terms of ventilatory requirement per unit of O_2 consumption, an air breather expends much less energy to extract an equivalent volume of O_2 compared with an aquatic one (Dejours et al. 1970; Sect. 2.9). When subjected to intra- or interspecific competition, amphibious fish, e.g., the climbing perch, *Anabas testudineus* and *Monopterus albus*, will embark on overland excursions (Liem 1987b). The highly aerial behavior of the pearl blenny, *Entomacrodus nigricans*, is thought to have arisen as a result of competition amongst the intertidal fish (Graham et al. 1978a). Air breathing and subsequent transfer to land opened new ecological opportunities which resulted in remarkable adaptive radiation. Among the marine teleost fish, the family Gobiidae, which has extensively evolved air breathing (Lewis 1970), contains the largest number of the present-day species in this taxon. Generally, the obligate air-breathing fish, which, due to greater access to O_2, are apparently more agile and can hence acquire more nutriments, are larger (25 to 30 cm) than the facultative ones (at 7 to 15 cm) (Munshi and Srivastava 1988).

The extant bimodal breathers provide modern analogs of the transitional animal forms in evolution of air breathing. In the lungfish, contact with air is a more powerful stimulus than a tactile or painful stimulus (Johansen and Lenfant 1968). The intertidal clingfish, *Sicyases sanguineus*, increases the number of exposures to air as the dissolved concentration of O_2 decreases (Ebeling et al. 1970). As the PO_2 in water drops, *Mnierpes macrocephalus* spends 92% of the time out of water, returning briefly at regular intervals (Graham 1970). The shanny, *Blennius pholis*, totally avoids water at low partial pressures of O_2 (Davenport and Woolmington 1981). The catfish, *Eremophilus mutisii*, uses its accessory respiratory organ (the stomach) in both normoxic and hypoxic water (Cala 1987) by periodically dashing to the surface to swallow air. In fish which surface more often in normoxic than hypoxic water, air breathing may be utilized for buoyancy control rather than gas exchange (e.g., Gee 1976; Gee and Graham 1978). Since the water breathers have evolved definite mechanisms for regulating hydrogen ions in face of high

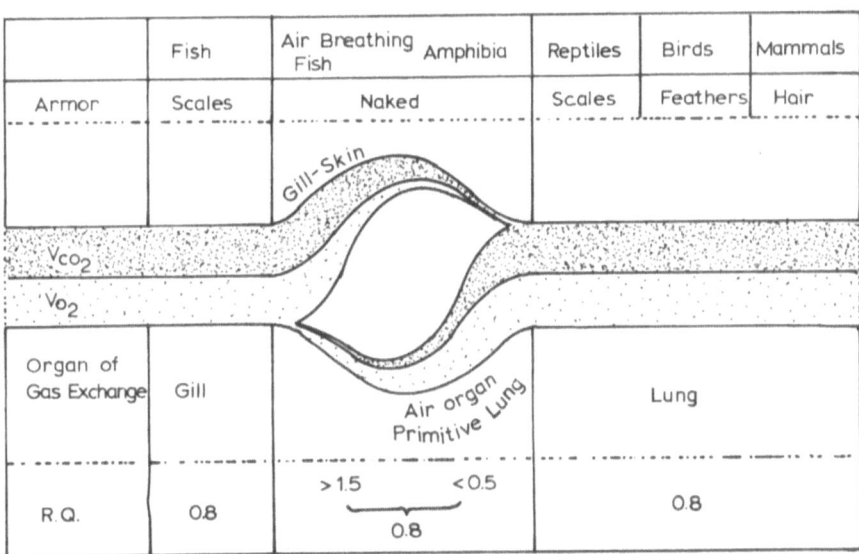

Fig. 57. Change from gill system in fish to the modern lung in the higher vertebrates. The transitional animal forms went through stages of bimodal gas exchange with a large gas exchange ratio initially in the gill-skin system and a low ratio initially in the primitive lung (air organ). During this stage, the skin was naked and acted as a "bridging" respiratory organ. Such animals were highly susceptible to dehydration on land. The development of surface covers like scales, feathers, and hair had to await the development of the modern lung. In essence, the switch over from water to air breathing was made very slowly and cautiously. *R.Q.* Respiratory quotient. (Rahn and Howell 1976)

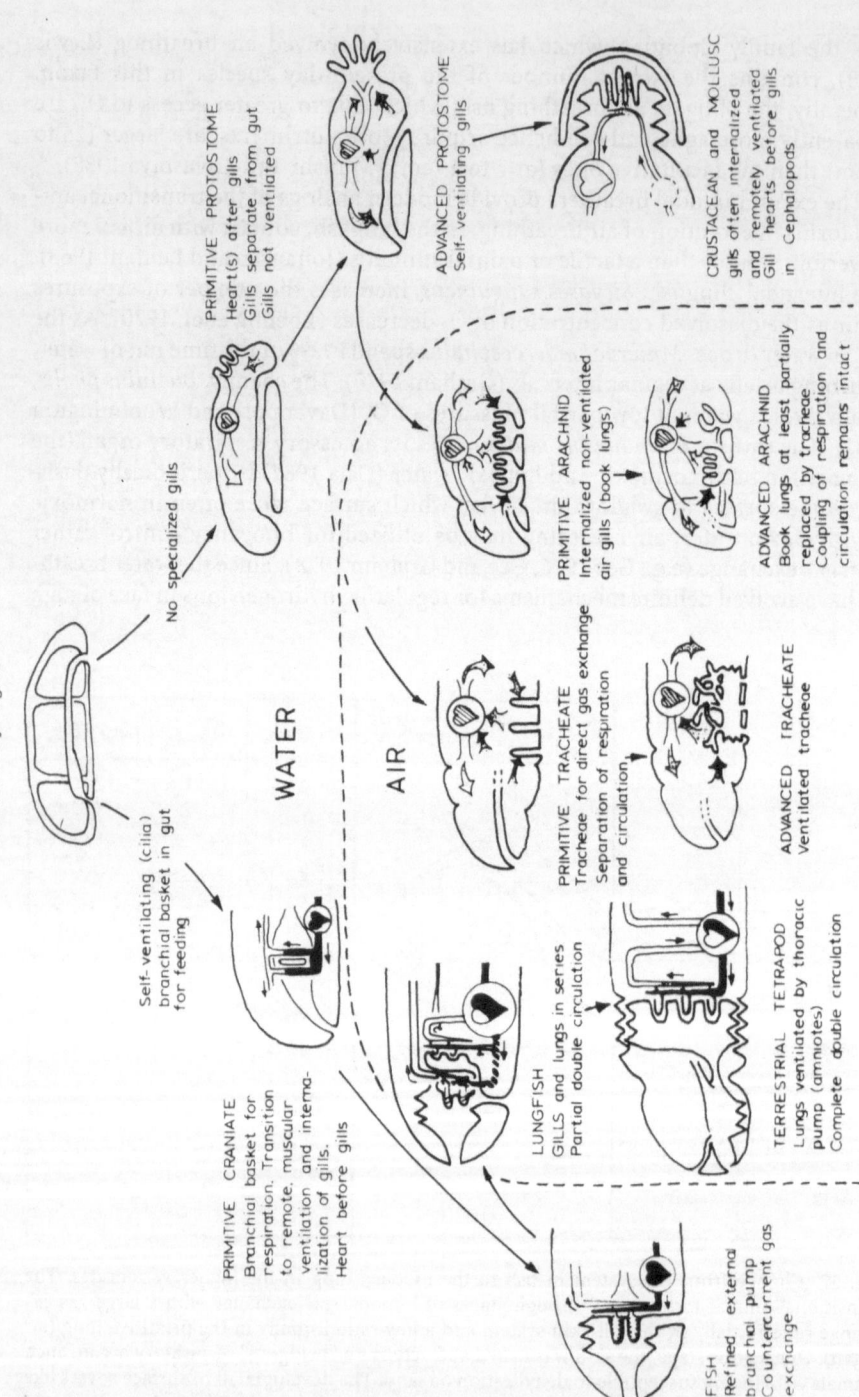

Fig. 58. Suggested pathways for evolution of air breathing, the increasing organizational complexities of organisms, and the corresponding sophistications of the gas exchangers. Endothermy, as evolved in birds and mammals, could only be supported by ventilated and perfused aerial gas exchangers. (Perry 1989)

PRIMITIVE COELOMATE
No respiratory organs
No circulatory organs

WATER

AIR

Self-ventilating (cilia)
branchial basket in gut
for feeding

No specialized gills

PRIMITIVE PROTOSTOME
Heart(s) after gills
Gills separate from gut
Gills not ventilated

ADVANCED PROTOSTOME
Self-ventilating gills

CRUSTACEAN and MOLLUSC
gills often internalized
and remotely ventilated
Gill hearts before gills
in Cephalopods

PRIMITIVE ARACHNID
Internalized non-ventilated
air gills (book lungs)

ADVANCED ARACHNID
Book lungs at least partially
replaced by tracheae
Coupling of respiration and
circulation remains intact

PRIMITIVE TRACHEATE
Tracheae for direct gas exchange
Separation of respiration
and circulation

ADVANCED TRACHEATE
ventilated tracheae

PRIMITIVE CRANIATE
Branchial basket for
respiration. Transition
to remote, muscular
ventilation and internali-
zation of gills.
Heart before gills

FISH
Refined external
branchial pump
Countercurrent gas
exchange

LUNGFISH
GILLS and lungs in series
Partial double circulation

TERRESTRIAL TETRAPOD
Lungs ventilated by thoracic
pump (amniotes)
Complete double circulation

230

ambient concentration of CO_2, deficiency of O_2 rather than elevation of CO_2 appears to have been the most important stressor in the evolution of air breathing. Changing pH levels and/or raising CO_2 levels do not elicit emergence from water in *Boleophthalmus pholis* (Davenport and Woolmington 1981). Low ambient PO_2 does not cause *Monopterus albus* and *Anabas testudineus* (Liem 1987b) or the mangrove forest fish, *Rivulus marmoratus* (Abel et al. 1987), to immerse their bodies into water. In the bimodally breathing fish, the accessory respiratory organs are located away from contact with water. In this way, such organs can be utilized simultaneously with the gills. The bimodal breathers avoid desiccation and CO_2 accumulation by subsisting in water while exploiting air as a source of O_2: they hence enjoy the benefits which accrue from the two habitats. While movement to land may be induced by a definite need to feed or may simply be behavioral and purposeless (e.g., Gordon et al. 1978; Sacca and Burggren 1982), abiotic factors such as high CO_2 and NH_3 levels and biotic ones such as predation and interspecific aggression may, to varying extents, prompt migrations to land (Fig. 8). Compared with the many complex integrated processes which were essential for terrestrial habitation (Little 1990), though an important initial step, the attainment of air breathing was a rather simple and direct affair.

5.4 CO_2 Elimination: Impediment to Evolution of Air Breathing and Terrestriality

Oxygenation is the primary purpose of respiration and in case of conflict, CO_2 clearance is unequivocally subordinate (e.g., Dejours 1988). Adaptations favoring O_2 uptake are automatically adopted in preference to those aiding CO_2 elimination. In this sense, O_2 plays a more central role in the process of gas exchange. In the bimodal breathers, while the accessory respiratory organs supply O_2 to the body, they are unable to eliminate CO_2 as efficiently (e.g., Randall et al. 1978a). Indeed, this particular limitation appears to have been a fundamental obstacle in the transition from water- to air breathing as the modern lung evolved (e.g., Gans 1970, 1971; Withers 1992; Olson 1994; Fig. 57). Even after some bimodal breathers could secure as much as 96% of their O_2 needs from the air, the skin and/or the gills continued to be the major organ(s) for CO_2 elimination (Tables 20,21). This restricted such species to water. In the totally land-based solely air-breathing crustaceans, appropriation of the enzyme carbonic anhydrase into the membrane fraction of the branchiostegites may have been one of the vital molecular events which enabled pulmonary CO_2 elimination into the air (Morris and Greenaway 1990; Henry 1994). During activity, in the highly terrestrial coconut crab, *Birgus latro*, CO_2 accumulates in the hemolymph if the gills are surgically removed (Smatresk 1979). Since CO_2 is not a highly toxic molecule, at least not as much as O_2 (Sect. 1.18), and concentrations can moderately rise without causing irreparable physical damage, regulation of the concentration of CO_2 in the body is utilized strictly to adjust body pH and not CO_2 levels. Unlike the charged molecules such as HCO_3^- and H^+ ions, which are transferred through ionic exchange for Cl^- and Na^+ ions, respectively, the uncharged molecular CO_2 easily moves

Table 20. Partitioning of gas exchange between the lung and the skin system in terrestrial amphibia. (After Rahn and Howell 1976)

Taxon	Species	T (°C)	Lung (%)		Skin (%)	
			VCO_2	VO_2	VCO_2	VO_2
Urodela	Ambystoma maculatum	25	29	69	71	31
	Taricha granulosa	25	32	68	68	32
Anura	Rana clamitans	25	24	74	76	26
	Rana pipiens	25	43	68	57	32
	Hyla gratiosa	25	19	68	81	32
	Hyla versicolor	25	16	67	84	33
	Bufo americanus	25	17	66	83	34
	Bufo boreas	25	31	69	69	31
	Bufo cognatus	25	30	44	70	56
	Bufo marinus	25	24	50	76	50
	Bufo terrestris	25	25	78	75	22
Mean			26	65	74	35

Table 21. Gas exchange partitioning between water-breathing organs (gill, G and skin, S) and air-breathing organs of some air-breathing fish. (After Singh 1976)

T (°C)	Species	VO_2 (%)			VCO_2 (%)		
		From water	Organ	From air	Into water	Organ	Into air
20	Amia calva	63	G	37	75	G	25
25	Heteropneustes fossilis	59	G,S	41	94	G,S	6
25	Anabas testudineus	46	G	54	91	G	9
25	Clarias batrachus	42	G,S	58	94	G,S	6
22	Lepisosteus osseus	27	G	73	92	G	8
23	Monopterus albus	25	G,S	75	–	–	–
26	Electrophorus electricus	22	S	78	81	S	19
25	Cobitis fossilis	20	–	80	34	–	66
20	Protopterus aethiopicus	11	G	89	70	G	30
24	Protopterus aethiopicus	10	G	90	68	G	32
20	Lepidosiren (Juvenile)	64	G,S	36	76	G,S	24
20	Lepidosiren (Adult)	4	G,S	96	41	G,S	59

across the cell membranes. During the transition from water to land, the skin, though often considered evolutionarily a dead end or a failed experiment, functionally acted as an important respiratory/acid-base bridging organ during the development of air breathing and subsequently terrestrial habitation (Fig. 57).

With the gradual involution of the gills, ionic exchange and CO_2 clearance were translocated to the skin (Rahn 1967; Randall et al. 1981). In amphibians, blood flow to the mid-dorsal skin is 1.8 times that to the ventral thoracic skin (Moalli et al. 1980). This indicates that the former site may be more important in CO_2 excretion than the latter (Talbot and Feder 1992). In the contemporary bimodal breathers, the gill-skin system removes about 76% and the lung 24% of CO_2 while O_2 uptake varies with the species, the gas exchanger utilized, and habitat occupied (Emilio et al. 1970; Rahn and Howell 1976). In aquatic amphibian species, *Siren lacertina* and *Amphiuma means*, the lung takes up 65% of the O_2 needs but the gills and/or skin eliminate nearly 75% of the CO_2 (Guimond and Hutchison 1973a, 1976). The utilization of the skin for gas exchange in many animals in different habitats attests to its great importance of having served as a bridging organ during the development of the aerial gas exchanges. To compensate for its inherent limitations as a gas exchanger, cutaneous respiration is much less energetically expensive (Feder and Burggren 1985a,b). Interestingly, though the marine air-breathing fish do not have very well-developed accessory respiratory organs, they appear to have developed the capacity to release CO_2 into the air so efficiently that most of them will maintain the same level of metabolism when exposed to air (Graham 1976).

Relatively, the absolute values of PCO_2 and HCO_3^- in blood and tissues are lower in the water breathers than in the air breathers. Weighted against the air breathers, generally, challenges for acid-base balance are less severe in water breathers (e.g., Driedzic and Gesser 1994). In the plasma of fishes, PCO_2 ranges between 0.1 to 0.5 kPa and HCO_3^- from <5 to 15 mM (Heisler 1984). Fish gills contain a high concentration of carbonic anhydrase which catalyzes breakdown of plasma HCO_3^- ions to CO_2 (e.g., Haswell and Randall 1978) with a small amount being voided across the kidneys (e.g., Wood and Cadwell 1978). About 90% of the total CO_2 in *Anabas*, *Clarias*, and *Heteropneustes* is voided through the gills (Hughes and Singh 1970a,b, 1971): the ratio of aquatic to aerial CO_2 removal is 10:1. Characteristically, the drop in pH induced by hypercapnia is corrected by elevation of HCO_3^-. Exposure of *Conger conger* and *Scyliorhinus stellaris* to an environmental PCO_2 of 1 kPa lowers plasma pH by about 0.4 units as plasma PCO_2 rises from about 0.25 to about 1.3 kPa (Heisler 1982b): over a period of 5 to 10 h, the pH was partially corrected by an elevation of HCO_3^- from <10 to about 20 mM. In exercising *S. stellaris*, H^+ released from skeletal muscle caused the arterial pH to drop from 7.8 to 7.2, the PCO_2 to rise from about 0.25 to 0.67 kPa, and the HCO_3^- to drop from 7 to 3 mM (Holeton and Heisler 1983). In the facultative air-breathing fish, *Symbranchus marmoratus*, change from water to air breathing alters PCO_2 from about 0.75 to 3.5 kPa with an accompanying drop in pH of about 0.6 units and a fourfold increase in intracellular HCO_3^- (Heisler 1982b). Vertebrate air breathers excrete much of their CO_2 in molecular form across the lung although a little of it is removed as HCO_3^- ions through the kidneys (Boutilier et al. 1979a,b). In fish, the processes of CO_2 elimination in pH regulation are different from those of mammals: HCO_3^- ions regulation, rather than molecular CO_2, is the more important factor in the process (e.g., Cameron 1978). Unlike in mammals, in fish, ventilation affects the CO_2 concentration of blood and hence pH only in the extremes of circumstances, e.g., during hyperoxia when ventilation is suppressed (Randall and Jones 1973) or during extreme hypoxia (Dejours 1973). Fish have evolved in a demanding habitat naturally deficient in O_2. Ventilation has been configured solely for delivery of O_2 to the gills, rendering CO_2 clearance somewhat of secondary importance.

The transition to air breathing and residence on land presented a challenge towards acid-base balance. It necessitated an increase in blood PCO_2 and a corresponding increase in plasma HCO_3^- ions for maintenance of normal pH (e.g., Rahn 1966; Hughes 1966; Howell et al. 1970; Lenfant and Johansen 1972; Tables 2,22). Plasma PCO_2 tends to increase as an animal shifts from water- to air breathing (Driedzic and Gesser 1994) though there is no clear correlation between the mode of breathing and the total CO_2 (Toews and Boutilier 1986). The high capacitance coefficient of CO_2 in water compared with O_2 curtails the ability of aquatic animals to regulate internal acid-base status by ventilatory adjustments of blood PCO_2 (e.g., Cameron 1979). In aquatic breathers, pH control is largely effected by exchange of acidic or basic equivalents between the extracellular fluid and the environment. This leads to changes in plasma HCO_3^- ion levels. In transitional breathers, ventilation does not correspond with CO_2 levels in blood as in the air breathers. The significance of ventilation in pH regulation is of little importance as the gills and the skin are the main pathways for CO_2 excretion. Fish appear to possess a mechanism for enzymatically curtailing CO_2 and H^+ ion loss across the gill epithelium (e.g., Haswell and Randall 1978) with the respiratory rate being mainly responsive to the PO_2 in the arterial blood (e.g., Randall et al. 1981). On the other hand, in the air breathers, CO_2 has to be voided before it accumulates to the level where the lowered pH interferes with the O_2 binding properties of the hemoglobin. Elimination of CO_2 across the vertebrate lung and hence the significance of ventilation in pH regulation was enforced onto the lung with the decrease in the CO_2 diffusing capacity of the gills. This resulted in high levels of CO_2 in blood (Randall et al. 1981). The increase in the permeability of the erythrocytes to HCO_3^- ions comprised a further means of enhancement of CO_2 clearance through the lungs.

Table 22. Some respiratory variables in some water and air breathers. (After Dejours 1988, from where individual sources of data should be consulted)

Taxon	T (°C)	PIO_2 (mmHg)[a]	PEO_2 (mmHg)[a]	PbO_2 (mmHg)[a]	$PbCO_2$ (mmHg)[a]	pHb	(HCO_3^-)[b] (meq l^{-1})
Water breathers							
Lugworm							
Arenicola marina	19	160	35	–	1.7	7.3	1.6
Octopus dofleini	11	127	94	78	3.1	7.1	–
Crab							
Carcinus maenas	10	156	–	–	2.3	7.8	7.2
Cancer magister	8	141	97	75	1.7	7.9	–
Dogfish							
Scyliorhinus stellaris	16	149	56	49	2.0	7.8	–

Table 22. *Continued*

Taxon	T (°C)	PIO_2 (mmHg)[a]	PEO_2 (mmHg)[a]	PbO_2 (mmHg)[a]	$PbCO_2$ (mmHg)[a]	pHb	(HCO_3^-)b (meql^{-1})
Transitional breathers							
Frog							
Rana catesbeiana							
Tadpole	23	150	–	–	4.3	7.8	6.8
Adult	23	150	–	–	17.2	7.8	29
Salamanders							
Cryptobranchus alleganiensis	25	154	–	27	6.2	7.8	–
Desmognathus fuscus	19	150	–	40	7.9	7.4	9.2
Lungfish							
Neoceratodus forsteri	18	131	–	40	4.7	7.6	5.3
Protopterus aethiopicus	25	130	–	27	26	–	–
Gar-fish							
Lepisosteus osseus	25	155	–	–	13	7.4	10.2
Electric eel							
Electrophorus electricus	27	150	–	12	27	7.6	30
Amphipnous cuchia	25	155	–	–	12	7.5	–
Air breathers							
Snail							
Otala lactea	15	154	144	144	8.6	8.1	–
Crab							
Gecarcinus lateralis	21	155	–	–	8.9	7.5	11.4
Uca pugilator	20	155	–	–	10.6	7.7	21.2
Birgus latro	29	150	–	27	9	7.7	12.7
Turtle							
Pseudemys scripta	25	155	–	–	27.4	7.6	33.0
Pekin duck	41	144	10.8	91	33	7.5	23
Toad (Adult)							
Bufo marinus	25	155	–	80	11.1	7.8	21.4
Human being	22	149	119	93	20.4	7.7	–

Symbols: PIO_2 and PEO_2 partial pressure of O_2 in the inspired and expired air/water; PbO_2 and $PbCO_2$, partial pressures of O_2 and CO_2 in the arterial blood; pHb, acidity of blood; (HCO_3^-) bicarbonate concentration in blood.

[a] To convert ot kPa multiply by 0.133.

In order to disengage the primary functions of the gills (respiration, ionic exchange, and pH regulation), a transitional respiratory stage was necessary (Fig. 57). The accessory respiratory organ was charged with O_2 uptake and the gills or any other respiratory surface, e.g., the skin and the buccal cavity with CO_2 removal and hence pH regulation and ionic homeostasis. A relocation of CO_2 excretion to the accessory respiratory organs had to await development of adequately efficient gas exchangers and ventilatory mechanisms. Osmoregulation and excretion of nitrogenous waste products was consigned to the kidneys (Schwartz 1976), the urinary bladder (Sachs 1977), and to a lesser extent to the skin (Ehrenfeld and Garcia-Romeu 1977). In the higher vertebrates, the lungs play an important part in acid-base regulation by tuning PCO_2 levels in blood with ventilation and, to a smaller extent, through elimination of plasma HCO_3^- ions by the kidneys (Davenport 1974; Burg and Green 1977). Such animals can dissociate themselves from water for ion and pH regulation. The highly terrestrial Trinidad mountain crab, *Pseudothelphusa garhami*, which has evolved very efficient lungs (Innes and Taylor 1986a,b), has very a low PCO_2 in the hemolymph. The passive buffering capacity of blood with respect to both HCO_3^- and non- HCO_3^- ions is five to six times smaller in water breathers than in mammals and birds, with the buffer value of the skeletal muscle tissue being only 50 to 70% that of mammals (e.g., Heisler 1984). The arterial PCO_2 in the rainbow trout (*Oncorhynchus mykiss*) is about 0.3 kPa, in the facultative air-breathing jeju (*Hoplerythrinus unitaeniatus*) it is 1.6 kPa, and in the obligate air breather piracucu, *Arapaima gigas*, it is 3.7 kPa (Randall et al. 1978a,b). The absolute values of PCO_2 and HCO_3^- ions in blood and tissues are lower in water breathers than in air breathers (Tables 2,22). The PCO_2 characteristically ranges between 0.1 and 0.5 kPa and HCO_3^- ions from 5 to 15 mM in the plasma of fishes (Heisler 1984): a drop in pH induced by hypercapnia is counteracted by elevation of HCO_3^- ions. Exposure of the eel, *Conger conger*, to an environmental PCO_2 of 1 kPa lowered plasma pH by about 0.4 units as plasma PCO_2 rose from 0.25 to 1.3 kPa (Heisler 1982b): over a duration of 5 to 10 h, the concentration of HCO_3^- ions increased from under 10 to about 20 mM, partly adjusting the pH. In the facultative air breather, *Synbranchus marmoratus*, movement from water to air changes plasma PCO_2 from about 0.75 to 3.5 kPa: pH falls by about 0.6 units and the intracellular concentration of HCO_3^- ions rises by a factor of 4 (Heisler 1982b).

The changes in the blood PCO_2 which occurred in animals moving from water to air are well demonstrated by the contemporary amphibians which start their life (larval stage) in water where they are obligate water breathers and metamorphose to become perfect air breathers. In *Rana catesbeiana*, at the tadpole stage, where the gills and the skin are the main gas exchangers, the blood PCO_2 is about 0.3 kPa while during and after metamorphosis, when the gills disappear and the lung assumes a prominent respiratory role, the blood PCO_2 increases notably (Erasmus et al. 1970/1971; Just et al. 1973; Burggren and Wood 1981; Tables 2,22). In the garfish, *Lepisosteus osseus*, the blood PCO_2 fluctuates with the ambient concentration of O_2 and the use of the gills or lungs for respiration (Rahn et al. 1971). During the summer months when the concentration of O_2 decreases due to the elevated temperature and its demands are high owing to increased metabolism, the blood PCO_2 increases. In winter (when the dissolved concentration of O_2

is high and metabolic rate lower) the gills are largely utilized and the blood PCO_2 drops. At such times, the fish rarely comes to the surface to breath and the diving durations are longer (e.g., Kruhoffer et al. 1987). In the bimodal air breathers, long respiratory intervals are possible because the greatest porportion of CO_2 is eliminated through the skin. The lungfish, e.g., *Protopterus amphibius*, and the bowfin, *Amia calva* (Dence 1933), regularly estivate during unpropitious circumstances, with the former staying in this state for as long as 5 years. When in water, the fish loses CO_2 and NH_3 through the gills and on land (during estivation), CO_2 accumulates internally and NH_3 is converted to urea and stored. This strategy allows the fish to survive during a time when water breathing is impossible.

5.5 Control and Coordination of the Bimodal Gas Exchange Process

In both air- and water breathers, ventilation is generally a constant and rhythmic process. Endogenous respiratory rhythm in mammals might emanate from specialized pacemaker cells located in the respiratory neural system (e.g., Onimaru and Homma 1987; Smatresk 1990; Smith et al. 1991). Response to hypoxia in fish and mammals is mediated by peripheral chemoreceptors (Sinclair 1987; Burleson and Milsom 1990; Smatresk 1994). In the air breathers, the central ventilatory control is more responsive to PCO_2 and/or pH while in the water breathers there does not appear to be a direct correlation between PCO_2 and ventilation (e.g., Batterton and Cameron 1978; Hedrick et al. 1991). Only after successfully switching the major CO_2 clearance role to the lung could a bimodal breather dispense with the need to use the skin for respiration and develop an impermeable surface cover to avoid dehydration on land. Such a change, however, had to await development of a more efficient lung and respiratory ventilatory mechanism to clear CO_2. Aspiratory breathing (a costal suction pump effected by action of the ribs and the intercostal muscles) first evolved in reptiles (Hughes 1963). The modern lung (first encountered in reptiles) has adopted the dual role of O_2 and CO_2 exchange and acid-base regulation. The cost and risk in air breathing was that of increased pulmonary water loss, a process accentuated by the fact that the air breathers relied exclusively on respiratory ventilation for O_2 acquisition and CO_2 clearance. Furthermore, except in reptiles, respiration in mammals and birds is regular and continuous. Exposure of the eel, *Anguilla*, to air causes the blood CO_2 level to rise and the arterial PO_2 and the O_2 consumption to decrease (Berg and Steen 1965). In *Synbranchus*, aerial respiration inhibits CO_2 elimination leading to increase in the blood PCO_2 (Johansen 1966). The inability to lose CO_2 into the air through either the gills or simple lungs is one of the main factors which constrained emergence from water (Hughes 1966; Thompson 1969). This explains why some modern air breathers with lungs and/or special respiratory organs inhabit water for the singular purpose of ease of CO_2 elimination. The air-breathing crabs which cannot eliminate CO_2 across the hard carapace have retained the gills (Innes and Taylor 1986a,b). This, while enabling them to maintain a low arterial PCO_2, confines them to habitats where water is easily accessible. In the robber crab, *Birgus latro*, much of the CO_2 is eliminated across the gills during rest, but during activity about one half of it is voided via the lungs (Greenaway et al. 1988).

On adopting air breathing, the bimodal breathers were not only faced with the problem of detecting respiratory stimuli from both internal and external milieus but also that of controlling physiologically distinct modes of respiration (e.g., West and Burggren 1983; Shelton and Croghan 1988; Smatresk 1988; Boutilier 1990). The respiratory control changed from a centrally modulated process to a diffusely coordinated one: the central control on the branchial ventilation was attenuated and the aquatic and air breathing integrated. The complexity of the plan depended on the relative importance of each mode of breathing to the overall respiratory needs and on factors such as the degree of adaptation to aquatic or terrestrial subsistence, the level of activity, and external factors such as hypoxia, temperature, and humidity. Fast, multiple respiratory responses are essential for animals subsisting in highly dynamic habitats characterized by precipitate fluctuations in O_2 and CO_2 levels. In the unimodal air breathers, e.g., birds and mammals, the control of ventilation is mediated centrally with the peripheral inputs simply modulating centrally produced impulses. However, in the intermittent air breathers, the ventilatory effort and pulmonary blood flow appear to be mainly regulated and synchronized through chemical and/or mechanical information initiated by the prevailing needs for O_2 (e.g., Johansen et al. 1997; Milsom 1990; West et al. 1992). Mechanoreceptors, which detect the gradual decrease in lung inflation (as O_2 is used up to some baseline value) have been described in the lungs of the bowfin (*Amia calva*) (Milsom and Jones 1985) and in the lungfish (*Protopterus aethiopicus*) (DeLaney et al. 1983). In *Amia*, aquatic hypercapnia at a concentration of CO_2 of up to 3% increases branchial ventilation (Johansen 1970). The absence of ventilatory or cardiovascular responses after intracranial perfusions of hypercapnic and low pH solutions shows that a central chemoreceptor involvement of O_2 or CO_2/pH in ventilatory regulation is lacking in *A. calva* (Hedrick et al. 1991). In terrestrial vertebrates, central chemoreceptive loci sensitive to CO_2 levels and/or pH changes are known to include the ventral region of the medulla oblongata (e.g., Hitzig and Jackson 1978). Internal O_2 chemoreceptors which stimulate air breathing have been described in the "lung" of the gar-fish (Smatresk et al. 1986) and in *P. aethiopicus* (Lahiri et al. 1970). The characteristic response of bimodal breathers to hypoxia is a notable increase in the ventilation of the accessory respiratory organs and a reduction in gill ventilation (Willmer 1934; Johansen et al. 1970b): in the gar-fish, a ten-times increase in ventilatory volume occurs consequent to increases in the ventilatory frequency and tidal volume (Smatresk et al. 1986). In *Protopterus*, respiratory responses appear to be highly sensitive to arterial gas tensions or pH (Lenfant and Johansen 1968; Johansen and Lenfant 1968). In *Neoceratodus forsteri*, the Australian lungfish, branchial ventilation increases remarkably as the PO_2 in the "inhaled" water falls below 10 kPa while aerial respiration is induced at a water PO_2 level of 11.3 kPa (Johansen et al. 1967): the chemoreceptors responsible for these respiratory responses are apparently external (branchial) as injection of urea into the lung does not elicit any compensatory response. In adult *Protopterus aethiopicus*, water PO_2 does not affect branchial or aerial breathing (Johansen and Lenfant

238

1968) but reduced lung PO_2 and systemic PO_2 stimulate the frequency of air breathing (Johansen and Lenfant 1968; Lahiri et al. 1970). Sectioning the nerves to the branchial arches reduces the responsiveness of *P. aethiopicus* to hypoxia (Lahiri et al. 1970), suggesting that chemoreceptive area(s) may exist in the branchial arches of the species. Such site(s) have, however, not been morphologically identified (Fishman et al. 1989). Using chemical probes such as sodium cyanide for O_2-sensitive elements (Lahiri et al. 1970) or nicotine (Johansen 1970), the existence of intravascular chemoreceptors in *Protopterus* has been demonstrated. Juvenile *Protopterus* rely more on aquatic respiration and respond to low water PO_2 (8% O_2 by volume) with a 50% increase in branchial and 300% increase in aerial breathing.

In the air-breathing fish, the ambient and internal CO_2 tensions are important regulatory factors of aquatic and aerial respiration. There are indications that water facing chemoreceptors occur in the gills of the air-breathing fish, e.g., *P. aethiopicus* (Johansen and Lenfant 1968) and the gar-fish, *Lepisosteus* (Smatresk et al. 1986; Smatresk and Azizi 1987). In *Protopterus*, the CO_2 receptors are located in the branchial region (or arterial side of the branchial circulation) (Johansen and Lenfant 1968), a feature which shows that increased gill PCO_2 inhibits branchial breathing before the arterial PCO_2 increases (Jesse et al. 1967). Externally situated O_2 receptors are important in monitoring the ambient PO_2 while the internal ones monitor the O_2 levels in blood. Switching to the most economical mode of gas exchange, i.e., from aquatic to aerial modes and vice versa, optimizes O_2 consumption to suit different metabolic needs and circumstances. Air-breathing fish are particularly sensitive to CO_2. In *Neoceratodus*, increased gill concentration of CO_2 reflexively inhibits branchial respiration while stimulating air breathing (Johansen 1966). A biphasic response to CO_2 similar to that in the terrestrial vertebrates was observed in *Protopterus* by Johansen and Lenfant (1968) and Jesse et al. (1967): at physiological levels of water concentration of CO_2 of 0.5% by volume or lower, both gill and lung ventilation were stimulated and at higher levels of CO_2 (1 to 5%), gill ventilation declined steadily. In both air and water breathers, the main response to hyperoxia is reduction of ventilation in correspondence to increased PO_2 in blood. The ventilatory inhibition in air breathers leads to accumulation of CO_2, resulting in respiratory acidosis which stimulates ventilation to correct the acid-base status. In many surviving air-breathing fish, maximal heart rates occur soon after the ventilation of the aerial gas exchanger (Singh and Hughes 1973). This optimizes transfer and utilization of the inhaled air. Due to the apparent ventilatory insensitivity of the water breathers to PCO_2, a persistent hypercapnic acidosis develops (Heisler 1982b). Mechanical inflation of the accessory respiratory organ in many air-breathing fish and amphibians often causes depression of the air-breathing activity (Lenfant et al. 1970b; Pack et al. 1984, 1992). In *Amphiuma*, inflation of the lung with pure nitrogen delays the onset of breathing, suggesting that mechanical distension may be an important factor in regulating O_2 ventilatory patterns (Toews 1971). Following an interbreath period of 4 to 5 min, in *Protopterus* (Lenfant and Johansen 1968), soon after inhalation, cardiac output is increased four times over the apneic level and almost three quarters of the injected blood is directed to the lungs (Johansen et al. 1968a). In the frogs, *Xenopus laevis* (Shelton 1970) and

Rana pipiens (Jones and Shelton 1972), the relative flow of blood to the gas exchangers depends on the breathing pattern. As in lungfishes and other air-breathing fish, apneustic breathing in amphibians determines the blood flow through the different circuits. During sustained breathing, the pulmocutaneous blood flow exceeds the systemic one, a condition reversed during apnea (Shelton 1976).

In *Bufo marinus* (e.g., West and Burggren 1984) and the lungfish, *P. aethiopicus* (DeLaney et al. 1983), mechanoreceptor elements are an integral factor in regulating and effecting ventilation-perfusion equality through regulated vasodilatation and constriction of the pulmonary vasculature. During air breathing, in lung-fishes, vasomotor activity maintains a lower pulmonary than systemic vascular resistance, enhancing lung perfusion. In the most aquatic of all lungfishes, the Australian *Neoceratodus forsteri*, branchial vascular resistance is highest in the taxon and compares with that in the elasmobranchs and teleosts (Johansen et al. 1968a). Branchial vascular resistance in *Protopterus* as well as in *Lepidosiren* is very low due to the presence of low resistance vascular shunts which allow much of the cardiac output to bypass the branchial arches which have virtually lost all the gill filaments (Johansen and Reite 1968; Laurent et al. 1978). The same regulatory mechanism prevails in the clawed toad, *Xenopus laevis*, where pulmocutaneous blood flow increases two to four times over the prebreath level compared with the systemic flow which almost remains constant (Shelton 1970). In the snakehead fish, *Channa argus*, aerial hypoxia (a rather rare occurrence) results in a five-times increase in "lung" ventilation as the PO_2 decreases from 20.7 to 5.3 kPa, but hypercapnia produces little change in lung ventilation (Glass et al. 1986). In intermittently breathing animals, e.g., *Xenopus* (Emilio and Shelton 1974), the diving turtles, *Pelomedusa subflava* (Glass and Wood 1983), *Chrysemys* (=*Pseudemsy*) *scripta* and *Testudo graeca* (Burggren and Shelton 1979), *Chelonia mydas* (West et al. 1992), and the garter snake, *Thamnophis* (Burggren 1977), O_2 exchange is greatest soon after inspiration when the blood flow is highest. In some air-breathing fish, e.g., the gourami, *Trichogaster* (Burggren 1979), aquatic hypercapnia results in increase in ventilation but in the spotted gar-fish no such a significant response seems to occur (Smatresk and Cameron 1982a,b). While in air, *Trichogaster* can efficiently eliminate CO_2 across the accessory respiratory organ (the labyrinthine organ) because the epithelial lining which is derived from gill tissue contains a high concentration of carbonic anhydrase (Randall et al. 1981).

5.5.1 Ventilatory Modalities of the Gas Exchangers in the Bimodal Breathers

During transition to air breathing, the general pattern of breathing changed from regular transfer of water across the gills to the episodic breathing pattern which typifies a large number of air-breathing fish, amphibians, and reptiles. A determinate breathing pattern (of air) was thereafter reverted to (under resting conditions) in mammals and birds (Table 5). In the Dipnoi (McMahon 1969; Burggren and Johansen 1986) and the amphibians (Burggren and Doyle 1986), the air-breathing mechanism was derived from the aquatic mechanism with inspiration occurring by a buccal force pump and expiration by elastic recoil of the lung. In the bimodally breathing teleosts, air ventilation is less well understood (Peters 1978; Kramer 1978; Liem 1980; Hellin and Chardon 1981; Ishmatsu and Itazawa 1981. In *Channa*, air breathing is complex and differs remarkably from water breathing (Liem 1984). In what was described as a cough mechanism, convergently, phylogenetically different air-breathing teleosts, the Channidae (Liem 1984), Anabantoidae (Peters 1978), and Clariidae (Hellin and Chardon 1981) have adopted a primeval aquatic process used to clear debris from the gills to ventilate the air-breathing organ(s). Since coughing is a water-dependent mechanism, fish utilizing the process only to ventilate their accessory respiratory organs are restricted in the period during which they can stay out of water (Liem 1984). The pressures which induced selection for interrupted ventilation and hence a lower respiratory frequency in bimodally breathing fish include: (1) need to reduce the number of vertical migrations to breath at the air-water interface, providing a substantial energy saving (e.g., Kramer and McClure 1981; Kramer 1983), and (2) avoidance of risks of predation during travels to the surface. When breathing air, *Gillichthys* becomes positively buoyant and has to hold onto substrate with its sucker-like pelvic fin (Todd and Ebeling 1966) while *Pseudapocryptes* is reported to float on the surface of the water with the opercular chambers inflated (Das 1934). During a dive, turtles, e.g., *Chrysemys* can breath-hold for as long as 1 to 2h even when they have access to air: active breathing takes only 2 to 15% of the total activity when undisturbed (Belkin 1964; Burggren 1975). Intermittent breathing occurs in the bimodal breathers due to the large O_2 stores in their body and the fact that during the breath-holds, gas exchange can be effected through secondary respiratory site(s) (e.g., Farrell and Randall 1978). Animals which do not have these features, e.g., fish and vertebrates like birds and mammals, though with large O_2 stores, have a high O_2 consumption and must continuously ventilate their gas exchangers. Those animals whose O_2 needs are low, e.g., amphibians, reptiles, and air-breathing fish, can: (1) acquire supplemental O_2 through multiple pathways, (2) reduce their O_2 consumption by adopting anaerobic metabolism, or (3) reduce their overall energy needs, e.g., by estivating or displaying rhythmic breathing. Since they contain O_2-rich air, the accessory respiratory organs of the bimodal breathers increase body O_2 stores by an order of magnitude. This lowers the ratio of O_2 consumption to O_2 store in the body. Episodic respiration in bimodal breathers rarely influences CO_2 and pH levels in the body as CO_2 is eliminated through pathways such as the skin and the gills. However, the intermittent air breathers have higher body CO_2 stores than the continuously breathing aquatic animals. Most bimodal breathers increase the ventilation of the accessory respiratory organ when the arterial or mixed venous blood PO_2 falls. This occurs, e.g., after exposure to hypoxia, when the ambient temperature increases beyond a critical point or when the activity level is raised. In the larva of the dragonfly (*Aeschna*), when the O_2 content is reduced to 2.5 ml per liter ($PO_2 = 7.3$ kPa), the organisms move to the surface of the water. Such a reflex is not initiated in the fish, *Leuciscus erythrophthalmus*, until the concentration of O_2 drops to a low of 0.6 ml l^{-1} (Winterstein 1908). In *Protopterus*, exposure to 8% O_2 increases lung ventilation three times but gill ventilation is raised by a factor of only one half. This suggests that different receptors are involved in the control and coupling of respiration in the air breather (Jesse et al. 1967). The buccal force pump used to ventilate the gills of the water breathers has been carried over to the air-breathing fish and the amphibians to ventilate the accessory respiratory organs. The pump in *Rana catesbeiana* is particularly efficient since remarkably high arterial PO_2 levels (values as high as 12.7 kPa) are attained (Lenfant and Johansen 1967). In fish, it is only in the Amazonian osteoglossid, *Arapaima gigas* (Farrell and Randall 1978), and the lungfish, *Protopterus amphibius*, during estivation (Lomholt et al. 1975) that aspirational breathing has been alleged. Suctional breathing may have evolved in water but is clearly a design characteristic of air breathers (e.g., Brainerd 1994). The respiratory frequency in air-breathing fish can vary from 1 breath min^{-1} to less than 1 breath h^{-1} (Johansen et al. 1971a; Smatresk and Cameron 1982a,b). It has been suggested that the buccal force pump in modern amphibians may have evolved secondarily after loss of ribs (Foxon 1964).

5.5.2 The Circulatory Patterns in the Gas Exchangers of the Bimodal Breathers

The transition from water to air breathing entailed drastic reorganization of the cardiovascular system. An elaborate mechanism of regulating the perfusion of the aerial gas exchanger(s), especially where they were arranged in parallel with the systemic circuit, was formed (e.g., Olson 1994). The occurrence of anatomical gill shunts is unique to the bimodal breathers (Laurent 1985). They have not been reported in the exclusively water-breathing fish (Laurent 1984). A number of attendant physiological problems arose from the superimposition of the accessory respiratory organ(s) on the branchial and systemic vascular circuits. These changes are most evident and better understood in the vertebrate transitional breathers. In the single arc circulatory pattern of the ancestral water-breathing fish, the entire cardiac output is directed to and perfuses the gills. The oxygenated blood passes to the dorsal aorta for distribution to the rest of the body. Attainment of air breathing necessitated integration of a separate circulatory network onto the original single circuit (Fig. 21). After the relocation of the ionic and gas exchange from the gills to the kidneys and the lungs, respectively, the aquatic vertebrates (which by now had developed into the terrestrial tetrapods) reverted to the original plan where all the cardiac output was driven through the only gas exchanger, the lung. The parallel arrangement in the bimodal breathers was converted into two circuits, pulmonary and systemic circuits.

The attendant gas exchange complications which arose with the evolution of bimodal breathing included: (1) undersaturation of arterial blood resulting from mixing of deoxygenated systemic blood with the efferent (oxygenated) blood from the accessory respiratory organ, (2) an apparent drop in respiratory efficiency consequent to perfusion of a gas exchanger with blood which had already passed through the gills, (3) shunting of the efferent oxygenated blood from the accessory respiratory organ(s) resulting in admixture with the systemic venous blood, (4) risk of transbranchial efflux of O_2 from the oxygenated blood if the gills were ventilated with hypoxic ambient water, and (5) undue tissue blood pressure arising from the addition of extravascular resistance pathways. These limitations were overcome or minimized by: (1) a drastic modification of the vascular layout, (2) axial blood flow separation, and (3) development of vascular shunts. In the lungfishes (Dipnoi), which have gills and lungs, the blood pumped from the ventricle is separated into two streams. One stream perfuses the anterior gill arches which are devoid of gill filaments (Parker 1892) and the other passes through the posterior gills and then onto the pulmonary artery (Johansen and Lenfant 1967; Burggren and Johansen 1987; Fig. 21). The latter pathway is accentuated when the ambient PO_2 is low (Johansen et al. 1968a). Laurent et al. (1978) described a vascular shunt in the posterior gill arches (IV and VI) of *P. aethiopicus* which bypassed the secondary lamellae. The arrangement corresponds with that of the external gills of the amphibians (e.g., Fige 1936) and *Lepidosiren paradoxa* (Robertson 1913). As the blood returns to the heart, mixing of the pulmonary arterial blood with the systemic deoxygenated blood is avoided by the presence of a spiral valve in the conus arteriosus. By establishing and maintaining laminar flow, the valve effectively separates the two blood streams in the heart. The oxygenated blood passes through the gill arches to the body while the deoxygenated blood passes to the lungs (Johansen and Hol 1968; Satchell 1976) for oxygenation. In the lungfishes, the degree of intracardiac separation of the arterial and venous blood depends on the degree of reliance on air for O_2, with the process being least effective in the obligate water breather, *Neoceratodus*, and best developed in the obligate air breathers, *Lepidosiren* and *Protopterus* (Johansen and Lenfant 1967; Johansen and Hanson 1968; Lenfant and Johansen 1968). The partial separation between the pulmonary and systemic circuits similarly occurs in amphibians although no distinct intracardiac anatomical modifications are evident (DeJong 1962; Johansen and Hanson 1968). Unlike in fish, in amphibians, the cutaneous circulation is arranged in parallel with the systemic and the pulmonary circuits and can be greatly varied without affecting the other two pathways (Johansen 1979). In the air-breathing teleosts such as *Channa* (Ishmatsu et al. 1979; Munshi et al. 1994; Olson et al. 1994), *Anabas testudineus* (Munshi et al. 1986a; Olson et al. 1986b), *Monopterus cuchia* (Munshi et al. 1989, 1990), *Heteropneustes fossilis* (Munshi et al. 1986b; Olson et al. 1990a; Hughes et al. 1992), and the lungfishes (Fishman et al. 1989), cardiovascular remodeling occurs as a part of a gross reorganization of the vascular connections between the gills, the accessory respiratory organs, and the systemic circuit. To a less conspicuous but significant extent, modifications in the form of vascular shunts (preferential perfusion channels), changes in microcirculation, and vascular endothelium (Hughes and Munshi 1979) develop.

Hemodynamic considerations and physiological functions are the main factors which initiated, regulated, and determined the changes in the circulatory patterns in bimodal breathers (Olson et al. 1994). In *Channa*, the vascular organization allows partial separation of blood flow to the gills, accessory respiratory organs, and systemic tissues (Munshi et al. 1994) resulting in basically two functional circuits: (1) heart → gills (1st and 2nd arches) → the accessory respiratory organ(s) → heart, and (2) heart → gills (3rd and 4th arches) → systemic tissue → heart. The accessory respiratory organs are essentially in series with the branchial vasculature and in parallel with the systemic circuit. The total cardiac output must hence first pass through the gills before perfusing the accessory respiratory organ or systemic tissues and both the deoxygenated systemic venous blood and oxygenated blood from the accessory respiratory organs returns to the heart. The first pathway (at the gill level) performs functions like osmoregulation, acid-base balance, CO_2 clearance, and metabolic N_2 excretion: Oxygen is subsequently taken up at the level of the accessory respiratory organ. The second pathway serves the role of shunting blood past the gills and delivering oxygenated blood directly to the systemic tissues: this curtails O_2 loss to a hypoxic aquatic medium. In the holostean, *Amia calva*, to minimize O_2 efflux from the gill blood, the blood flow in the gills is altered so that much of it is shunted to the nonrespiratory parts of the gills (Johansen et al. 1970a). In having elaborate gill filaments and secondary lamellae with an adequate number of chloride cells (ionocytes), the 1st and 2nd pairs of gill arches of the bimodal breathers retain features characteristic of those of the water-breathing teleosts and the well-developed gills in air-breathing species. The 3rd and 4th arches are less elaborate (e.g., Munshi et al. 1990; Olson et al. 1990a; Olson 1991). In most species, e.g., in the cuchia eel *Amphipnous cuchia* (Lomholt and Johansen 1976), and the electric eel, *Electrophorus electricus* (Johansen et al. 1968b), the 3rd and 4th gill arches are used as a bypass of the vestigial gills to avoid loss of O_2. Vascular and intracardiac modifications close to the heart and gills reduce the mixing of the arterial and venous streams of blood. The deoxygenated systemic venous blood and the oxygenated blood from the accessory respiratory organ returns to the heart, facilitating delivery of the deoxygenated fraction to the accessory respiratory organs. The systemic tissues are perfused with the oxygenated fraction (Ishmatsu et al. 1979; Ishmatsu and Itazawa 1983; Munshi et al. 1994).

The intracardiac physical separation of the venous and arterial streams of blood was unequivocally demonstrated by Ishmatus and Itazawa (1983): in *Channa argus*, it was found that the PO_2 in the anterior ventral aortic blood is lower than that in the posterior ventral aortic blood. The overall efficiency of the gas exchange process is to a great extent dependent on the effectiveness of minimization of the mixing of these two blood streams in the heart. In air-breathing fish, the perfusion of the accessory respiratory organ is well matched to the O_2 available for aerial gas exchange. In the cuchia eel, *Amphipnous cuchia* (Lomholt and Johansen 1976), and electric eel, *Electrophorus electricus* (Johansen et al. 1968b), when breathing air, 75% of the cardiac output, respectively 80 and $70 \, ml \, kg^{-1} min^{-1}$, is directed to the accessory respiratory organ: during apnea, the perfusion of the organ drops to 20%. This marks an incipient stage in the evolution of synchronization of ventilation and perfusion and functional separation of pulmonary and systemic perfusion. The process is better refined in the lungfishes, where a partial separation has been attained (Johansen at al. 1968a; Fig. 21). Total separation is reached in the higher vertebrates, birds and mammals. The low pressure pulmonary circuit (Figs. 25,26) which appears to have evolved very early, e.g., in the Dipnoi (Johansen et al. 1968a) enables extreme refinements of the structural parameters of the gas exchangers accommodating the development of a very thin blood-gas barrier in a highly dynamic organ.

5.6 Taxa with Notable Propensity for Bimodal Breathing

5.6.1 Mollusks

The Phylum Mollusca, with about 100000 species, is one of the largest and most successful in the Animal Kingdom (e.g., Jones 1983; Seed 1983). It is exceeded in specific diversity and numerical density only by Arthropoda and perhaps by Protozoa and Nematoda. Ecologically, the mollusks have penetrated a wider range of habitats than any other animal group (Ghiretti 1966). The gastropods, which constitute about 80% of all living mollusks, are the only group in the taxon which has colonized land. After the arthropods, the slugs are probably the most successful land invertebrates (Jones 1983). The success of the pulmonate gastropods is largely attributable to the development of certain physiological and behavioral adaptations which have enabled them to cope with problems of water balance, temperature, osmotic and ionic regulation, and gas exchange (Riddle 1983). Mollusks, which have a respiratory pigment with low O_2 unloading tension, can subsist in hypoxic environments and to a certain point may show metabolic independence to external O_2 levels (Borden 1931; Chaetum 1934).

Air breathing in pulmonate gastropods probably arose in O_2-deficient waters such as estuaries, swamps, and muddy rivers (Seed 1983). In the prosobranches. e.g., *Littorina rudis*, there is a tendency towards formation of a lung on exposure to air. In the intertidal pulmonate limpets constituting the Siphonariidae, on return to the sea, secondary gills may appear within the former lung (Yonge 1952). The related *Trimusculus (Gadinia)* continues to breath air while residing in water (Yonge 1958). In the pulmonate gastropods, the mantle cavity has been converted into a highly vascularized lung (Maina 1989c; Figs. 59,60) where a large volume of "blood" is brought into close proximity to a respiratory medium for gas exchange in air or secondarily in water. Pulmonate gastropods are structurally the most conservative gastropods and their evolutionary advances have been committed essentially to face the challenges of terrestrial colonization. In a notable deviation from the normal pattern, some pulmonates, e.g., the siphonarid limpets, possess secondary gills which are housed in a dorsal mantle cavity (e.g., McMahon 1983; DeVilliers and Hodgson 1987). The gills of the pulmonate, *Siphonaria capensis*, bear ciliary tufts (about 400 cilia per mm^2). The surface is covered by a single layer of epithelial cells with interspersed mucus-secreting cells (DeVilliers and Hodgson 1987). From the ciliated gills and other anatomical features, *Siphonaria* is believed to be one of the least developed pulmonates (Marcus and Marcus 1960; Hyman 1967). While Marcus and Marcus (1960) contend that the secondary gills have always been a feature of siphonariids, it has been suggested by, e.g., Yonge (1952) and Morton (1979) that the gills were initially lost as the taxon evolved terrestriality but redeveloped when it reverted to water. In the freshwater pulmonate gastropods, *Lymnaea stagnalis* and *Planorbis corneus*, which have secondarily taken to water after the extirpation of the gill-like ctenidia, the mantle cavity was converted into a well-vascularized water lung which is rhythmically filled with water (Precht 1939; Hunter 1953). As the ambient PO_2 drops below a critical level, the snail comes to the surface to breath air

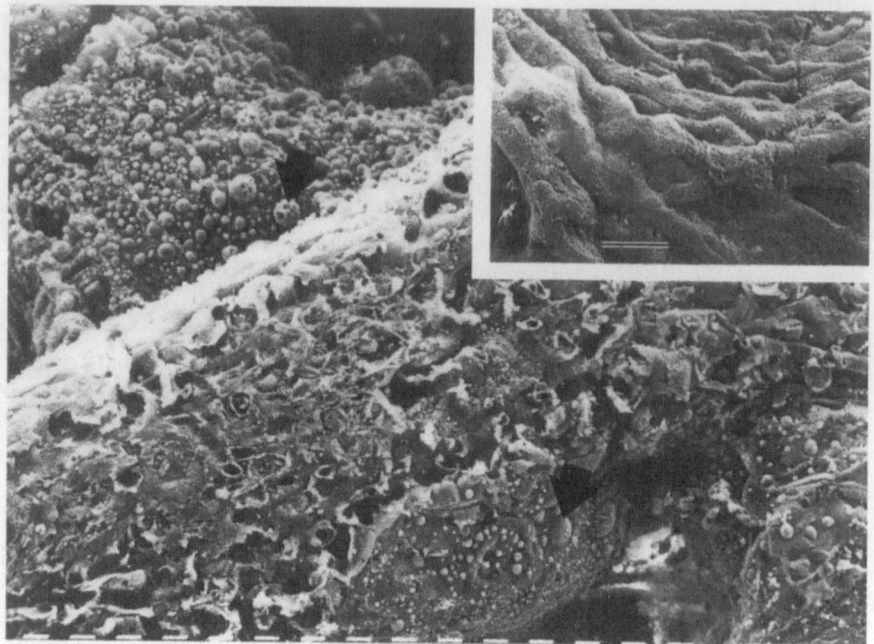

Fig. 59. Surface of the lung of a pneumonate gastropod, *Trichotoxon copleyi*, showing the vascular network of blood capillaries (*inset*). The *main figure* shows one of the blood capillaries and clusters of mucus cell, ▸, which lie on the epithelial surface. *Bar* 10 μm; *inset* 90 μm. (Maina 1989c)

(Cheatum 1934). When access to air is prohibited, *Lymnaea* responds by elevating the concentration of hemocyanin in blood to enhance extraction of O_2 from the lung (Jones 1972). The opening of the pneumostome is regulated to optimize gas exchange while limiting water loss. Low ambient concentration of O_2 stimulates the opening of the pneumostome, which may remain permanently open at concentrations below 10% (Ghiretti 1966). The opening of the pneumostome in the terrestrial gastropods *Limax*, *Arion*, and *Helix* is influenced by the level of CO_2. It remains permanently open at a concentration of CO_2 of 3 to 5% (Dahr 1927; Prosser 1961). In *Helix pomatia*, at 20% O_2, the pneumostome remains closed and only opens when O_2 drops to 10% or less (Ghiretti 1966). Other factors which regulate pneumostomal size include temperature and humidity (Ysseling 1930; Wit 1932; Mass 1939). In the inactive slug, *Limax maximus*, the lung contributes 20% of the O_2 consumption, the value rising to 50% during activity (Prior et al. 1983). In all mollusks, even where specific respiratory organs exist, gas exchange through the body surface occurs to varying extents. Transcutaneous gas exchange may be of critical importance under certain circumstances. In the land snail, *Otala lactea*, a significant amount of CO_2 is eliminated across the skin (Barnhart 1986a). In the aquatic air-breathing snails, *P. corneus* and *L. stagnalis*, where CO_2 is soon lost into the water through the skin, CO_2 does not act as a respiratory

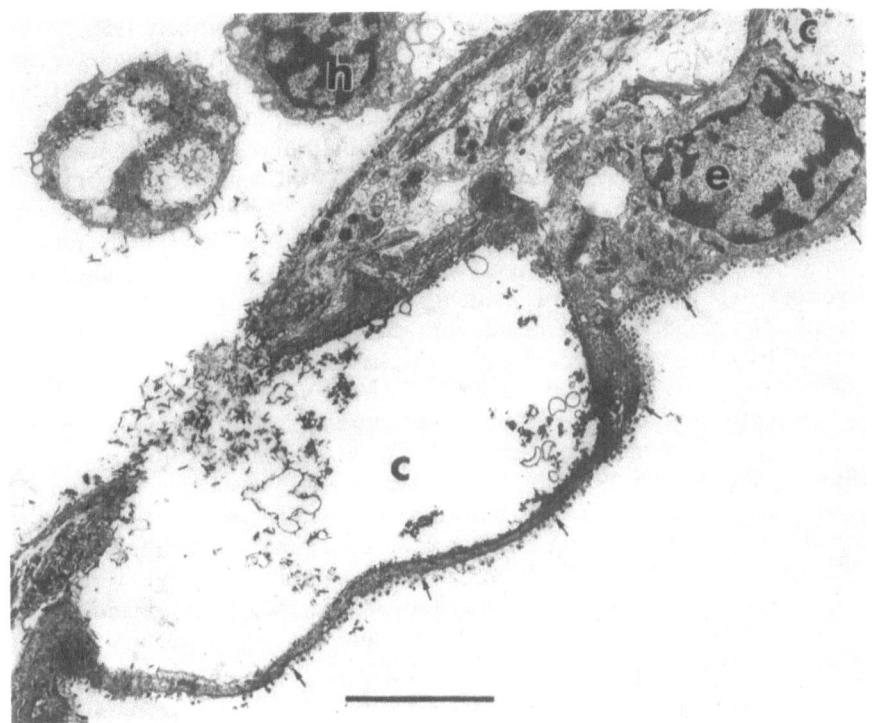

Fig. 60. Blood capillaries, *c*, on the surface of the lung of a pneumonate gastropod *Trichotoxon copleyi*; *e* epithelial cell; *h* hemocyte; → microvilli on the epithelial cell surface. *Bar* 2 μm. (Maina 1989c)

stimulus. Factors such as PO_2 and loss of buoyancy regulate respiration (Jones 1961). In the Athoracophoridae, the lung is small. Many fine tubules lead from it to the roof of a blood sinus in a manner resembling the trachea of insects (Runham and Hunter 1970). Accordingly, the taxon has been termed the Tracheopulmonata.

The respiratory surface area of the lung in pulmonate gastropods ranges from 7 to $13.5 \, cm^2 g^{-1}$ (wet body mass) (Yonge 1947). The dimensions compare and exceed those of some vertebrate air breathers, e.g., the lizard, *Tupinambis nigropunctus* $3.2 \, cm^2$ per g (Perry 1983) and the domestic fowl, *Gallus domesticus*, $10 \, cm^2 g^{-1}$ (Abdalla et al. 1982). Perhaps due to the large respiratory surface area of the lung and the thin blood-gas barrier (Maina 1989a), diffusion is adequate for gas exchange even at the very low PO_2 of 0.3 kPa characteristic of the habitats in which the slugs live (Ghiretti 1966). Depending on temperature, slugs can increase their O_2 consumption by as much as 400 to 500% (Mackay and Gelperin 1972; Prior et al. 1983). In *O. cygnea*, in the process of repaying the O_2 debt after activity, maximum O_2 consumption is reached after 1 h (Hers 1943). In active snails, PO_2 and PCO_2 in the lung of the land snail, *Otala lactea*, differed by less than 2 kPa from air, the arterial PCO_2 was similar to that in the gas in the lung but the arterial PO_2 was 8 kPa lower than that in the lung (Barnhart 1986b). This bespeaks a strong diffusional limitation of O_2 transfer across the blood-gas barrier. Only about 10% of the hemocyanin-O_2 carrying capacity is utilized at 5 °C but rises above 70% at 25 °C in the active land snail *O. lactea* (Barnhart 1986c). In *H. pomatia*, O_2 affinity is 1.5 to 2 kPa at 20 °C and pH 7.6 to 8.2 (e.g., Konings et al. 1969), in *Arion ater* it is 2.2 kPa at 20 °C, pH 7.94 (Wells and Weber 1982) and in *O. lactea* 5.3 kPa at 25 °C, pH 7.9 (Barnhart 1986b). In *Trichotoxon copleyi* (Maina 1989c) and *H. pomatia* (Pohunkova 1967), the lung has developed in the form of a modified mantle cavity. The roof, which is highly vascularized, is lined by a respiratory epithelium which is made up of squamous cells bearing stubby microvilli. The blood vessels protrude into the air space (Figs. 59,60). Scattered among the squamous cells are goblet mucus cells (Fig. 59). Contrary to the observations made by Pohunkova (1967) on *H. pomatia*, the air-blood pathway lacks a continuous endothelial lining (Maina 1989c). In the pneumonate gastropods in general, the thin blood-gas barrier ranges in thickness from 6 to 10 μm (Runham and Hunter 1970). In *Trichotoxon* (Maina 1989c), some parts of the barrier are as thin as 0.2 μm. The diffusion potential for O_2 in the lung of *Agriolimax agrestis* is six times greater than the animal's total O_2 requirement (Runham and Hunter 1970). As in the lung-breathing exothermic vertebrates (e.g., Jackson 1978), in the arterial blood of *O. lactea*, PCO_2 increases and pH and PO_2 decrease with temperature, presumably due to a hypoventilatory response (Barnhart 1986b).

5.6.2 Crustaceans

The crustaceans have exploited habitats ranging from totally terrestrial ones to the deep sea (Bliss 1979; Hartnoll 1988; Henry 1994). The decapod crustaceans present remarkable respiratory diversity with evident progressive sophistication occurring from aquatic through to air breathing and terrestrial species (e.g., Bliss 1979; Cameron 1981; McMahon and Wilkens 1983; Innes and Taylor 1986a,b; Burggren and McMahon 1988b; Henry 1994). Land crabs are members of the Anomura and Brachyura which can to varying extents survive on land. The majority belong to the families Coenobitidae, Gecarcinidae, Grapsidae, Potamoidea, and Ocypodidae (Hartnoll 1988). Among the terrestrial species, respiratory efficiencies correlate with the O_2 demands which are imposed by factors such as habitat and level of activity (Johnson and Rees 1988). Most of the non-malacostracan crustaceans have no specialized respiratory organs: gas exchange occurs across the integumentary surfaces (Wolvekamp and Waterman 1960; McLaughlin 1983). In brachiopods and phyllopods, it occurs through appendages and their modifications such as the gills which come in remarkably different forms (Eriksen and Brown 1980a,b; McLaughlin 1983). The internal surfaces of the carapace are well adapted for gas exchange in the Brachyura (Cameron 1981; McLaughlin 1983).

Crustaceans developed in the sea and have evolved air breathing and invaded land severally (e.g., Bliss and Mantel 1968; Bliss 1979; Cameron 1981). The notably terrestrial species come from the groups Amphipoda, Isopoda, and Decapoda (Bliss and Mantel 1968; Powers and Bliss 1983). Land crabs are of particular interest in comparative respiratory physiology as they display relevant adaptations associated with transition from water- to air breathing and terrestrial colonization (Innes and Taylor 1986a,b; Taylor and Innes 1988). Air-breathing crabs have complex lungs which are contained in the branchial chambers formed from branchiostegal and sometimes thoracic walls (e.g., Farrelly and Greenaway 1987, 1992; Maitland 1987; Fig. 61). In crustaceans, terrestrialness and air-breathing efficiency correlate with the elaboration of the lung (e.g., Diaz and Rodriguez 1977). Though restricted to burrows and demonstrating nocturnal activity, isopods inhabit some of the most xerix habitats such as the Sahara desert (Cloudsley-Thompson 1977). The notably terrestrial species of crabs such as *Birgus latro*, *Holthuisana* (*=Austrothelphusa*) *transversa*, *Pseudothelphusa garhami*, and *Ocypode saratan* have evolved an elaborate well-perfused, highly amplified epibranchial chamber lining which constitutes for all practical purposes a lung (e.g., Farrelly and Greenaway 1987, 1992; Maitland 1987; Al-Wassia et al. 1989; Fig. 62). Such species can stay away from water for a long time. The anomuran coconut crab, *Birgus latro*, can survive in air with its gills removed (Harms 1932; Smatresk 1979) and, like *Ocypode*, will drown if forcefully submerged in water. The respiratory physiology of *Birgus* differs from that of aquatic crabs in certain ways (Cameron and Mecklenburg 1973). As in the vertebrate lung breathers, the acid-base balance of its hemolymph is effected by respiratory exchange rather than by ionic transfer mechanisms. The blood pressure in most terrestrial crabs is two to four times greater (up to 6.6 kPa) than aquatic crabs and ventilation of the branchial cavity is efficient and continuous, a mechanism which, though the O_2 extraction factor of the lung is only 2 to 8%, promotes gas exchange rate greatly (McMahon and Burggren 1988; Fig. 63). The PCO_2 of the venous blood of *Birgus latro*, which is as low as 1.2 kPa (Burggren and McMahon 1981), may be explained by high ventilatory rates (Dejours and Truchot 1988). During severe hypoxia or hypercapnia, continuous scaphognathite beating (at a

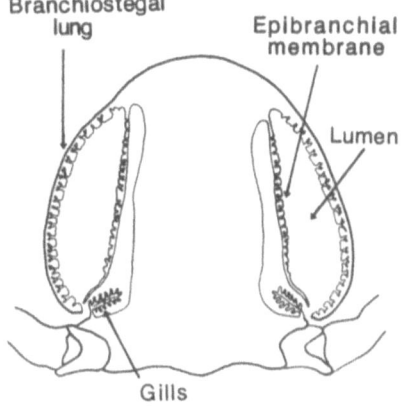

Branchiostegal lung

Epibranchial membrane

Lumen

Fig. 61. Transverse section of the body of a soldier crab, *Mictyris longicarpus*, through the branchial chambers which are divided into an inner gill space and an outer space which has been converted into a lung. (Farrelly and Greenaway 1987)

Gills

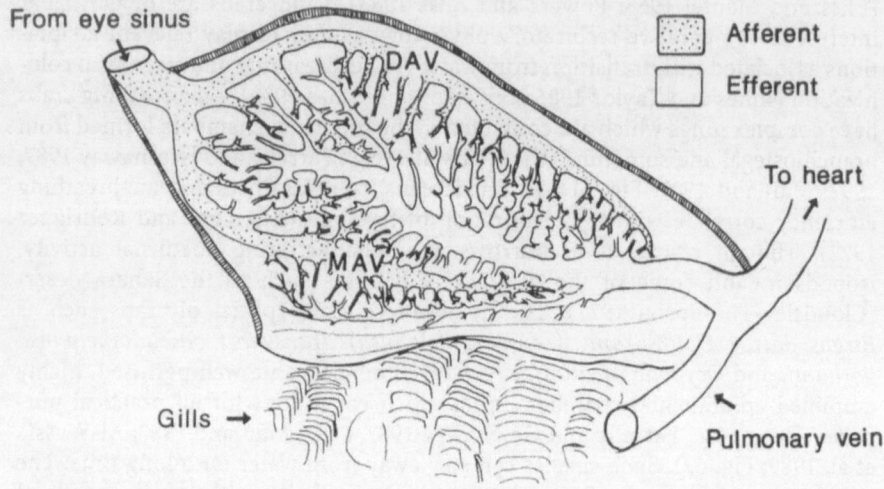

From eye sinus

DAV

MAV

Gills →

Afferent

Efferent

To heart

Pulmonary vein

Fig. 62. Schematic diagram of the vascularization of the epibranchial membrane of the lungs of the soldier crab, *Mictyris longicarpus*. The lung is very elaborate, providing a very highly efficient gas exchange capacity and hence ability to subsist over long periods on land. *DAV* Dorsal arterial vessel; *MAV* middle arterial vessel. (Farrelly and Greenaway 1987)

rate of 50 to 300 beats min^{-1}) produces a ventilatory volume of 100 $cm^3 min^{-1}$ in a 1- to 2-kg specimen of *Birgus iatro*. In such circumstances, O_2 extraction factor may increase to 20% (Maitland 1990a). The gills provide 29 µl O_2 $g^{-1} h^{-1}$ while the balance of about 60 to 110 µl O_2 $g^{-1} h^{-1}$ in a routinely active land crab (at 25 °C) is transferred across the lungs.

As in the air-breathing fish (Hughes and Morgan 1973; Pelzenberger and Pohla 1992), in crabs, the evolution of alternative air-breathing sites is associated with reduction in the gill surface area (Table 23). While the active aquatic crabs have a respiratory surface area of about 900 to 1400 $mm^2 g^{-1}$ (Gray 1954, 1957; Veerannan 1974, Hawkins and Jones 1982; Johnson and Rees 1988; Henry et al. 1990), that in the intertidal ones ranges from 500 to 900 $mm^2 g^{-1}$ (Hawkins and Jones 1982; Rabalais and Cameron 1985; Santos et al. 1987) and in the fully terrestrial species the values range from 12 to 500 $mm^2 g^{-1}$ (Cameron 1981; Farrelly and Greenaway 1992). The branchial chambers of the bimodal breathing crabs are larger than those of the aquatic ones (Henry 1994). Internal modifications in form of intense foldings and invaginations (e.g., Innes and Taylor 1986a; Farrelly and Greenaway 1987; Maitland 1987; Fig. 62) provide a large respiratory surface area (Greenaway and Taylor 1976; Diaz and Rodriquez 1977). Combined with a remarkably thin blood-gas barrier of 0.2 to 0.4 µm (Farrelly and Greenaway 1992; Table 23), a high diffusing capacity of the lung for O_2 is generated in the terrestrial crabs. Air channels in *P. garhami* run from the lung to the air sacs beneath the carapace and are actively ventilated in a throughflow manner (Diaz and Rodriguez 1977; El Haj et al. 1986) giving a high arterial PO_2 (16 to 18.7 kPa)

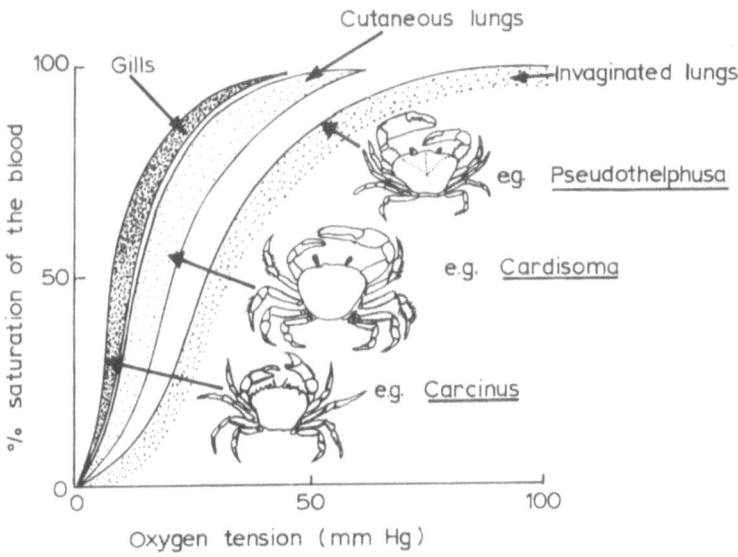

Fig. 63. Correlation between O_2 affinity of blood of air-breathing crabs and mode of gas exchange. In an aquatic breather, e.g., *Carcinus*, O_2 uptake through the gills is aided by the high affinity of the hemocyanin to O_2 (i.e., values of P_{50} – the PO_2 at which 50% of the Hc is saturated – are low). In terrestrial crabs with well-vascularized gill chambers, e.g., *Cardisoma* and species such as *Pseudothelphusa* which can efficiently ventilate their lungs, a high arterial O_2 tension can be attained. In such species, P_{50} is much greater. (Innes and Taylor 1986a)

and a low PCO_2: gas exchange does not seem to be diffusion-limited (Innes and Taylor 1986a,b). The arterial PO_2 in the soldier crab, *Mictyris longicarpus*, is 12.8 kPa (Farrelly and Greenaway 1987) and in the shore crab, *Carcinus maenas*, is 14.2 kPa (Lallier and Truchot 1989). Many semaphore and grapsid intertidal and supratidal crabs carry some water on the lower parts of the branchial chambers (e.g., Burnett and McMahon 1987; Maitland 1990b). On average, crabs carry 0.35 to 0.5 ml of water per 7-g live body mass. The pumping action of the scaphognathites (aided by setae) circulates 0.2 ml of it (per 3-g body mass) around the body across the branchial chambers while on excursion on land (e.g., Hawkins and Jones 1982; Felgenhauer and Abele 1983). The water keeps the gills moist (avoiding desiccation and enabling aquatic respiration in air), functions as a sink for the excreted CO_2 (Burnett and McMahon 1987), and moistens the food during eating (Fielder 1970; Maitland 1990b). In some species, air is bubbled through the water in the branchial cavity to oxygenate it (McMahon and Burggren 1988). In the fully terrestrial land crabs, e.g., *Birgus latro*, CO_2 is excreted through the gills (Greenaway et al. 1988).

Although the factors which have limited the terrestrial radiation of the land crabs are not well known (e.g., Hartnoll 1988), compared with the most successful arthropod groups, e.g., Arachnida, Chilopoda, Diplopoda, Insecta, and terrestrial vertebrates like birds and mammals, as a group, the crustaceans display relatively weak adaptation for air breathing and terrestrialness. They manifest an extremely

Table 23. Gas diffusional distances and respiratory parameters in some terrestrial and aquatic crabs

Crab	Diffusion distance		VO_2 (μmol g^{-1}h^{-1})	PaO_2 (mmol l^{-1})	PVO_2 (mmol l^{-1})
	Gills (μm)	Lung (μm)			
Terrestrial					
Holothuisana transversa	5–8	0.2–0.3	0.80	–	–
Birgus latro	–	0.5–1.2	1.41	27	13
Pseudothelphusa garhami	5	0.4–1.0	1.51	124–140	–
Mictyris longicarpus	3.5–7.0	1.2–3.6	–	79–95	20–27
Ocypode ceratophthalmus	2.9–3.5	0.25–0.3	–	–	–
Cardisoma hirtipes	5–12	2.5–4.5	2.29	–	–
Aquatic					
Procambarus clarkii	3.15–8.7	–	–	–	–
Astacus pallipes	1.4–5.7	–	–	–	–

Respiratory parameters from McMahon and Burggren (1988) and thicknesses from Burggren and McMahon (1988b).

tenuous hold on land. In contemporary biology, the crustaceans are a living model for studying the prerequisites for air breathing, terrestrial colonization, and occupation of new habitats. It is not surprising that in the taxon, potential respiratory organs/sites are still being discovered and debated (e.g., Mangum 1994).

5.6.3 Fish

The common expression "like a fish out of water" conveys the general inability of most fish and other aquatic life forms to survive and effectively exchange gases in air. Phylogenetically, the lungs are thought to have evolved as adaptations in hypoxic or anoxic conditions in the aquatic medium in which the ancestors of the modern fish lived. A small number of the contemporary descendants of the ancient fish, especially those which continued to subsist in derelict hypoxic warm tropical freshwaters, have retained air breathing (Burggren et al. 1985b). In the entire ichthyoid fauna, only the Holostei and the tropical freshwater teleosts can efficiently breathe air (e.g., Munshi and Hughes 1992). The process of air breathing has become so perfected in some fish that they succumb on extended forced submersion. Fish, the most successful vertebrate taxon with 20 000 or so species, have lived in water for over 350 million years (Gilbert 1993). Only a handful, the majority of which are teleosts, have evolved air breathing (Bertin 1958; Fänge 1976). As many as 60 (Sayer and Davenport 1991) and, according to Graham (1994), as many as 370 species of fish are known to breath air to varying extents (Table 19). More than any other animal group, fish present the broadest perspective of the evolution of air breathing (Graham 1994) and, to an extent, terrestriality.

The extant teleosts are far from being ancestral to the terrestrial air-breathing vertebrates as they only arose from the holosteans during the late Triassic period (e.g., Jarman 1970). They are thus far removed from the mainstream progenitors of the land vertebrates. The air-breathing teleostean fishes present remarkable morphological, physiological, biochemical, and behavioral respiratory diversity. However, considering their very different systematic backgrounds, the similarity in the preadaptations and the strategies adopted to achieve air breathing are astonishing and overwhelm the differences. In general, larval and juvenile bimodal breathing fish, e.g., *Clarias*, *Colisa*, and *Anabas*, derive much of their O_2 needs from water utilizing the gill/skin system. Their dependence on water for O_2 decreases with age (Hughes et al. 1973; Prasad 1988). These fishes acquire the air-breathing capacity at about 8 to 20 days after hatching (Bruton 1979; Singh and Mishra 1980; Singh et al. 1982). In *Anabas*, the labyrinthine organs start developing at the 5th day of hatching but air breathing does not occur until the 13th to 14th day (Hughes et al. 1986a). Obligate air breathers, e.g., adult piracucu, *A. gigas*, die if refused access to the atmospheric air even when kept in well-oxygenated water (Stevens and Holeton 1978b). The gar-fish, *Lepisosteus*, is a facultative air breather at low temperatures but an obligate one at higher ones (Rahn et al. 1971). At high PO_2, in water, *Piabucina* derives 10% of its O_2 need from air but at a lower water PO_2 of 4.7 kPa, it acquires 70% from air (Graham et al. 1977, 1987). In the snake-headed fish, *Channa punctatus*, and the climbing perch, *Anabas testudineus*, the accessory respiratory organs develop during larval and juvenile stages. They leave water only when the development is complete (Singh et al. 1982; Hughes et al. 1986). Many air-breathing fish possess reduced scales and have a well-vascularized skin through which substantial gas exchange occurs (Lenfant and Johansen 1972; Romer 1972). At a temperature of 7 °C, buccal and cutaneous breathing in the common eel, *Anguilla vulgaris*, supports the metabolic needs (Faber and Rahn 1970). The buccal cavity of the eel is both profusely diverticulated and highly vascularized. It is ventilated every 2 to 5 min with air (Johansen et al. 1968b). At 26 °C, the O_2 uptake rate of 30 ml per kg per h, which constitutes 80% of the overall needs, is procured from the air across the buccal cavity and the rest across the skin. The highly atrophied gills contribute an insignificant amount of O_2. The gills are, however, responsible for elimination of as much as 94% of CO_2 (Table 21). The accessory respiratory organs in fish are diverse and are variably ventilated. They include the skin (e.g., *Amphipnous cuchia*), bucco-pharyngeal membrane (e.g., *Electrophorus electricus*), suprabranchial chamber membranes, and labyrinthine organs (e.g., *Clarias mossambicus*), gastrointestinal system (e.g., *Ancistrus anisitsi* and *Plecostomus plecostomus*), and the air bladders (e.g., *Arapima gigas* and *Amphipnous cuchia*). In *Plecostomus* and *Ancistrus* (both tropical Siluroidae), the stomach is a respiratory organ into which air is swallowed and regurgitated (Carter and Beadle 1931; Carter 1935). In the pond loach, *Cobitis* (=*Misgurnus*), the middle and distal parts of the gastrointestinal system (GIT) are respiratory (McMahon and Burggren 1987) and the residual air is passed out through the anus (Krogh 1941; Jeuken 1957). The blood-gas barrier in the respiratory sites of the GIT is remarkably thin (Jasinski 1973). Respiration using the gastrointestinal system may have been a forerunner to the development of the lung by evagination of the foregut. Morphometrically, the air sac of *Heteropneustes fossilis* is superior to that of *Aphipnous cuchia*, similar to that of *Lepidosiren*, but less than that of *Protopterus* (Hughes et al. 1992): the respiratory surface area of a 40-g specimen was reported to be 24 cm^2, the harmonic mean thickness of the blood-gas barrier 0.342 μm, and the morphometric diffusing capacity 3.2×10^{-5} ml O_2 s^{-1} mbar^{-1}. In those air-breathing fish which use isolatable diverticula such as the stomach, intestine, and gas bladder, the gills can simultaneously be ventilated to supplement O_2 uptake and carry out processes such CO_2 elimination, ionic transfer, and clearance of nitrogenous waste products. This cannot occur in those fish, e.g., the electric eel (*Electrophorus electricus*) and the knifefish (*Hypopomus*) which use the buccal cavity as an accessory respiratory organ, a time when aquatic respiration is impeded and hence bimodal respiration momentarily eliminated (e.g., Graham et al. 1987). Reckoning from the spatial and temporal diversity of the evolved accessory respiratory organs, air breathing is an independently evolved attribute. Different lineages endeavored to look for solution to common selective pressures, especially hypoxia. Bimodal breathing is estimated to have evolved at least 67 separate times (Graham 1994). Regression of the accessory respiratory organs has occurred in many fish which have radiated into habitats where need for bimodal breathing is not intense. This has occurred in fish which live in stable, well-oxygenated waters, e.g., in the swift mountain streams, expansive freshwater lakes, or at great depths (Greenwood 1961; Graham 1994). Together with the ancestral crossopterygians which gave rise to the primitive amphibians, the dipnoans are the closest extant relatives of the modern tetrapods (e.g., Rosen et al. 1981; Duellman and Traub 1986; Panchen and Smithson 1988; Pough et al. 1989; Joss et al. 1991; Meyer and Dolven 1992). For this reason, the group continues to arouse substantial interest in biology. Respiratory modifications for air breathing were extensively found in most or all of the late Paleozoic fishes (Romer 1972) when the O_2 content of the water, e.g, in the Silurian period, was only 1.9 kPa. The earliest functional lungs are presently found in only the three genera (and six living Southern Hemisphere freshwater species) of lungfishes (Dipnoi), namely *Lepidosiren, Protopterus*, and *Neoceratodus*, in the chondrosteans (*Polypterus* and *Calamoichthys*), and the holosteans (*Amia* and *Lepisosteus*) (e.g., Burggren et al. 1985b; Sacca and Burggren 1982). Developmentally, these fish have been extremely conservative. They have changed very little since they attained the capacity for breathing air and hence serve as living examples for understanding the adaptive strategies and the backgrounds against which air breathing evolved. The South American *Lepidosiren* and the African *Protopterus*, genera which evolved in separate continents after the splitting of the Gondwana Land in the early Mezozoic but continued to live in similar habitats (respectively in the poorly oxygenated waters of the Amazonian basin and the derelict inland freshwater masses of the continental Africa) have paired, symmetrical internally sudivided lungs (Fig. 64; Klika and Lelek 1967; Hughes 1973; Hughes and Weibel 1976; Maina and Maloiy 1985; Kimura et al. 1987; Maina 1987a) and vestigial gills (Laurent et al. 1978; Laurent 1996). Both are obligate air breathers (Table 25) and have a better CO_2 buffering capacity of blood than the Western Australian *Neoceratodus forsteri*, a facultative air breather. *Protopterus aethiopicus* acquires 89% of its O_2 needs and eliminates 40% of CO_2 through the lungs (Johansen and Lenfant 1968; Lenfant and Johansen 1968; Table 24). *Neoceratodus* has a single, unpaired, thin-walled but internally subdivided lung (Grigg 1965) and possesses fully functional gills (Johansen and Lenfant 1967). The diameter of the pulmonary blood capillaries of *Neoceratodus* is 20 µm, a value which corresponds with the large (40 $\times$ 30 µm) oval erythrocytes (Gannon et al. 1983). In *Protopterus*, the posterior branchial arches carry a notable number of lamellae while the anterior ones, which provide shunt pathways for systemic circulation, are vestigial (Laurent 1996). To support CO_2 clearance, the functional branchial arches are continuously perfused (Lenfant and Johansen 1968; Laurent et al. 1978) even though pulmonary ventilation is intermittent. Encased in a cocoon of hard soil as the water dries up, *Protopterus amphibius* has been reported to survive in a semianimated (estivated) state for months to years as it episodically breathes air through a "snorkel" which connects the cocoon with the surface (DeLaney et al. 1974). The harmonic mean thickness of the blood-gas barrier in *Neoceratodus* ranges from 1.5 (Hughes 1973) to 2.5 µm (Gannon et al 1983), that of *Lepidosiren* is 0.85 µm (Hughes and Weibel 1976), *Protopterus* 0.37 (Maina and Maloiy 1985) and 0.85 µm (Hughes and Weibel 1976), and in *Polypterus* 1.22 µm (Zaccone et al. 1995).

To varying extents, the air-breathing fish rely on air or water for their O_2 needs (Table 25). The more terrestrial species are obligate air breathers. Among the actinopterygian fishes (Lauder and Liem 1983), only the polypterids have devel-

Table 24. Gas exchange in the fish lungs. (After Wood and Lenfant 1976)

Species/T (°C)	Total VO$_2$ (ml min^{-1}kg^{-1})	Lung VO$_2$ (% of total)	Lung VCO$_2$ (% of total)	Tidal vol. (% of lung vol.)
Neoceratodus				
(18–20)	0.25	0	0	40–60
Protopterus				
(20)	0.3	89	40	60–80
Lepidosiren				
(18–20)	0.37	96	60	≦100
(25–28)	–	–	–	–
Lepisosteus				
(22)	0.65	73	40	40
(10)	0.30	0	–	–
Amia				
(10)	0.30	0	0	–
(20)	1.5	35	25	≦100
(30)	2.0	75	40	–

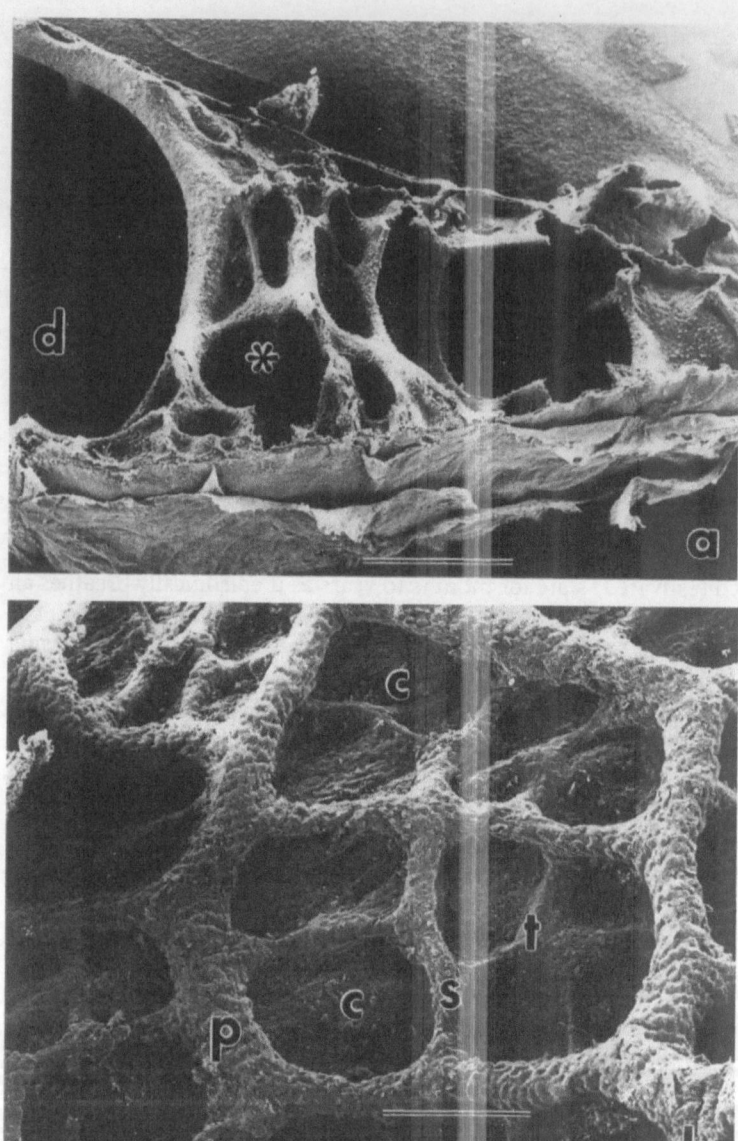

Fig. 64. a Cross section of the lung of the African lungfish, *Protopterus aethiopicus*, showing the eccentrically located air duct, *d*, and the gas exchange cells, *∗*. **b** Closeup of the gas exchange air cells, *c*, of the lung of *Protopterus* which are formed by hierarchically arranged septa, i.e., the primary septa, *p*, secondary septa, *s*, and tertiary septa, *t*. The septa support the blood capillaries and contain the contractile and supporting tissue elements like collagen, elastic tissue, and smooth muscles. **a** *Bar* 0.67 mm; **b** 0.28 mm. (Maina 1987a)

Table 25. Oxygen uptake in some air-breathing fishes. The proportion of Vo_2 from water and air and the quotient between aquatic and aerial Vo_2 are also given. Numbers in parentheses are % proportions. (After Singh 1976)

T (°C)	Species	O_2 uptake ($ml\,kg^{-1}h^{-1}$)			Aquatic/aerial
		Aquatic	Aerial	Total	
25	Anabas testudineus	52.62 (46.40)	60.80 (53.60)	113.42	0.865
25	Clarias batrachus	38.85 (41.60)	54.54 (58.40)	93.39	0.712
25	Heteropneustes (=Saccobranchus) fossilis	50.10 (59.25)	34.45 (40.75)	84.55	1.454
26	Electrophorus electricus	6.92 (23.07)	23.08 (76.93)	30.00	0.300
20	Amia calva	66.00 (68.75)	30.00 (31.25)	96.00	2.20
30	A. calva	30.00 (26.32)	84.00 (73.68)	114.00	0.357
22	Lepisosteus osseus	14.40 (24.66)	39.00 (75.34)	58.40	0.369
25	Cobitis fossilis	14.40 (19.51)	59.40 (80.49)	73.80	0.242
20	Protopterus aethiopicus	1.26 (11.05)	10.14 (88.95)	11.40	0.123
24	P. aethiopicus	5.80 (9.28)	56.70 (90.72)	62.50	0.102
18	Lepidosiren (Juvenile 150 g)	54.00 (63.83)	30.60 (36.17)	84.60	1.764
20	Lepidosiren (adult)	1.4 (4.36)	30.70 (95.64)	32.10	0.045

oped conspicuous lungs. The lungs are, however, relatively more primitive than those of the Dipnoi. The lungs of *Polypterus* are slender and lack discernible internal compartmentation (Klika and Lelek 1967; Hughes and Pohunkova 1980) with the right lung being better developed than the left. *Hoplostemum thoracatum* from the Paraguayan swamps, which travels overland across the grass from one pool to another (as they dry), swallows air continuously: O_2 is absorbed from the posterior part of the intestine which is profusely supplied with blood (Carter and Beadle 1931; Huebner and Chee 1978). The Indian catfish, *Clarias batrachus*, which subsists in shallow derelict waters at night emerges from the water to feed on earthworms (Dehadrai and Tripathi 1976). During such a time, it uses the suprabranchial chamber membranes and the labyrinthine organs for respiration. The organs are similar to those of *Clarias mossambicus*, an African catfish (Figs. 65,66; Maina and Maloiy 1986). Other Indian species such as the climbing perch, *Anabas testudineus*, and the Cuchia eel, *Amphipnous cuchia*, are highly terrestrial, spending much of their time out of water (Munshi and Hughes 1992). *Anabas* (weighing 29 to 51 g) acquires about 53.6% of its O_2 need from air (Hughes and Singh 1970b). Due to the lower metabolic rate and the relatively greater

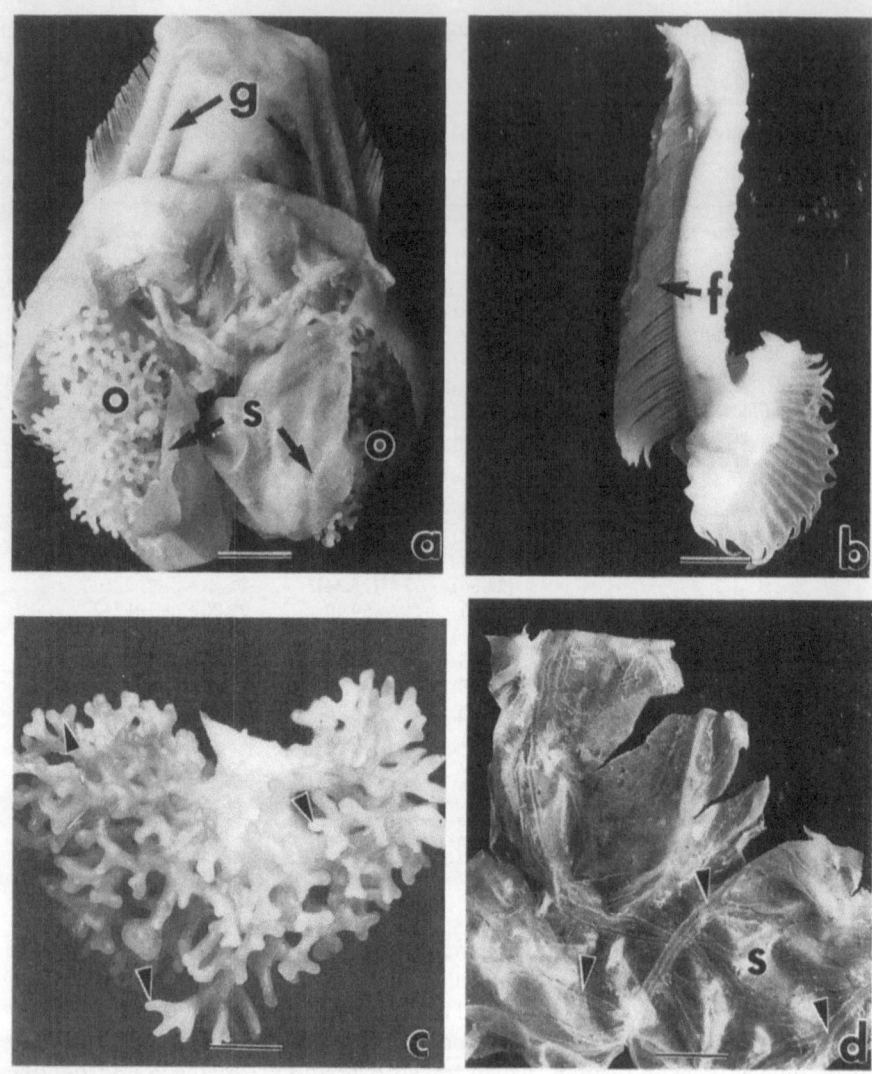

Fig. 65. a Respiratory organs of the catfish, *Clarias mossambicus*, showing the gills, *g*, labyrinthine organ (*o*) and suprabranchial chamber membrane, *s*. **b** A gill arch showing gill filaments, *f*. **c** Tree-like labyrinthine organ which terminates in rounded knobs, ➤. **d** Suprabranchial chamber membrane which has many blood vessels, ➤. **a** *Bar* 0.2 mm; **b** 0.1 mm; **c** 0.15 mm; **d** 0.15 mm. (Maina and Maloiy 1986)

amount of dissolved O_2, at lower temperatures (17 to 20 °C), specimens of *Anabas* weighing up to 30 g survive well for a few months without air breathing but at higher temperatures (30 to 31 °C) even smaller fish (10 to 15 g) succumb within 24 h (Dube 1972). *Amphipnous* has a particularly high O_2 capacity exceeding 20 vol% and high hemoglobin-O_2 affinity (Lomholt and Johansen 1976): after 30 min

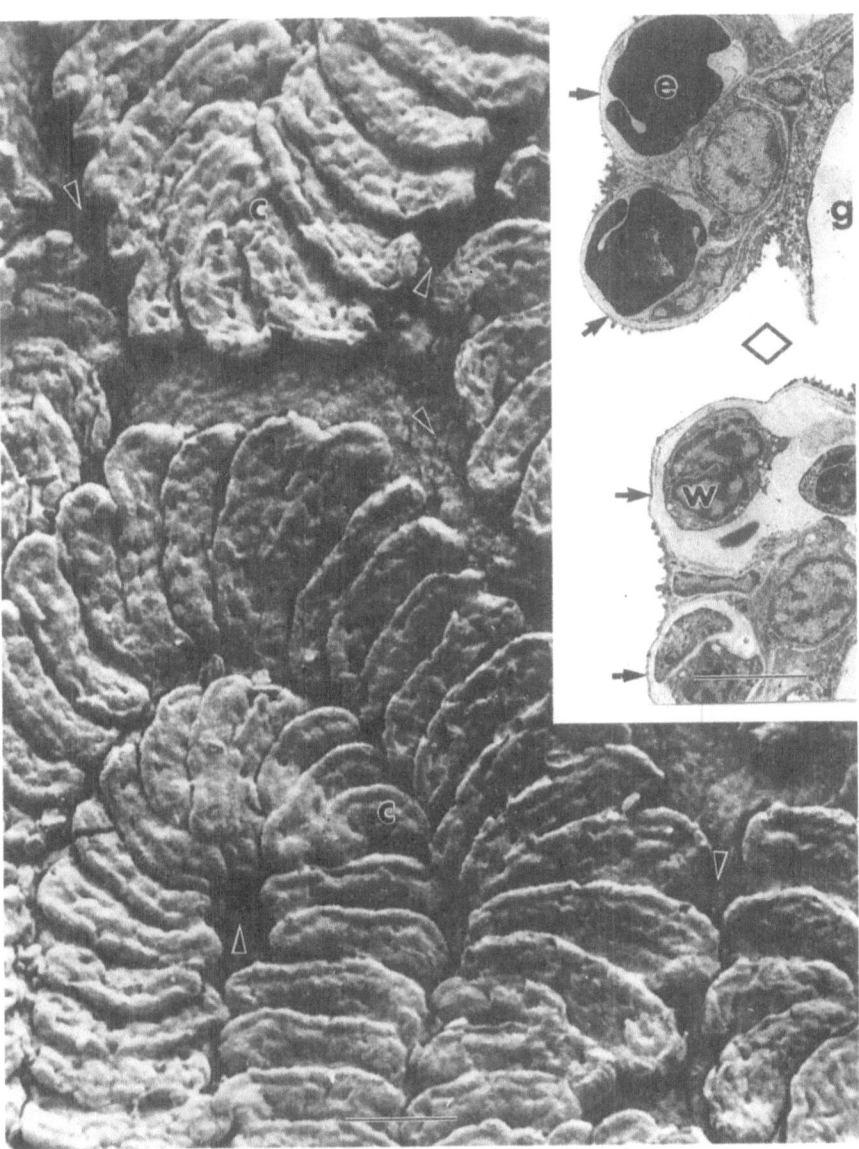

Fig. 66. Views of the surface of the labyrinthine organ of the African catfish, *Clarias mossambicus*, showing the intense vascularization in form of surface transverse capillaries (*c*) which are separated by bare tracts, ➤, and ◊ (*inset*), areas that contain mucus cells, *g* (*inset*). ➜ (*inset*), transverse capillaries, *e*, erythrocytes contained in the transverse capillaries; *w*, white blood cell. *Bar* 13 µm; *inset* 10 µm

of apnea, only a modest drop in arterial saturation (90 to 60%) occurs in the fish. The secondary reduction of the suprabranchial chambers and the increase in the gill respiratory surface area in some species of *Dinotopterus* (Clariidae) in the deep Lake Malawi and that in *Sandelia capensis*, an anabantoid fish in the South

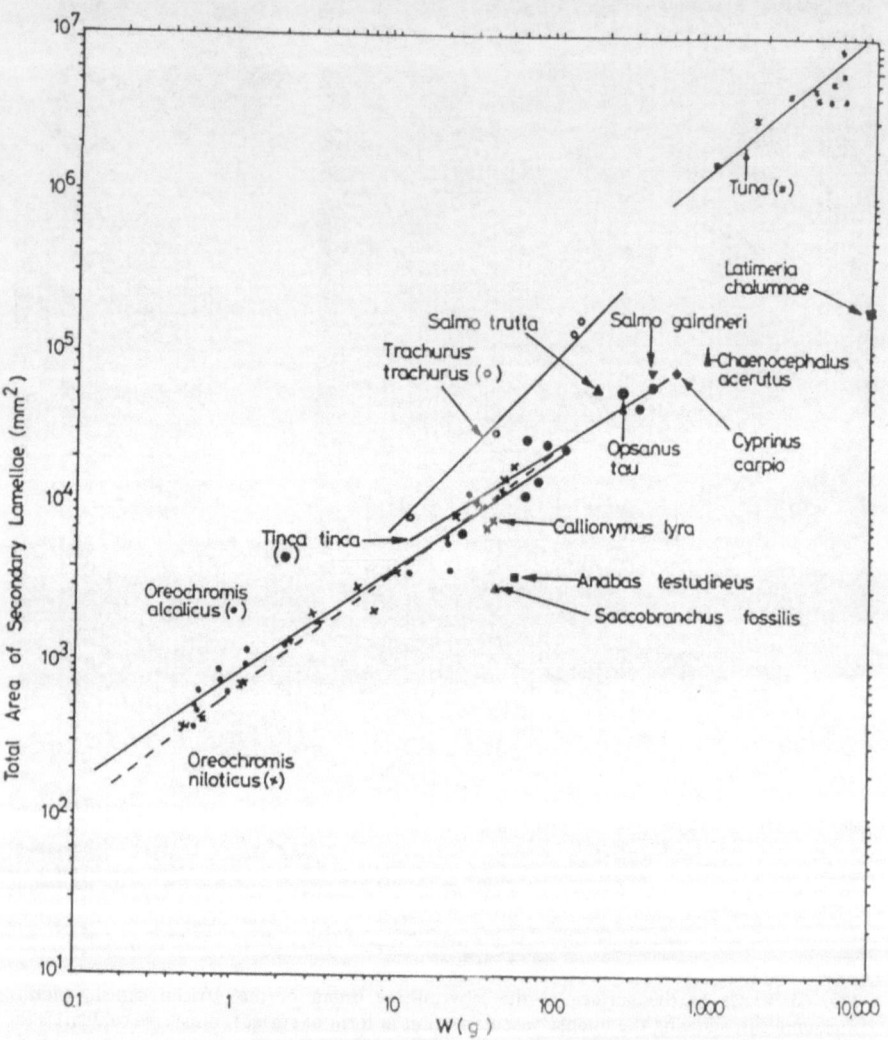

Fig. 67. Plot between total respiratory surface area and body mass in different species of fish. Energetics, habitat, and mode of respiration determine the development of the gas exchangers. The tuna, a highly energetic fish, has remarkably extensive gill surface area while in the air-breathing fish, gill development is generally inhibited. In the air-breathing fish, e.g., *Anabas testudineus* and *Saccobranchus fossilis*, the accessory respiratory organs have a high diffusing capacity of O_2. This compensates for the reduction in the development of the gills. Data on *Oreochromis* from Maina et al. (1996a) and those of the other species from Hughes and Morgan (1973) and other publications

African Cape region (Barnard 1943), may be viewed as a reversion to totally aquatic respiration in fish formerly adapted to aerial breathing. A greater spacing of secondary lamellae, an adaptation which prevents gill filament adherence and collapse in air, characterizes most air-breathing fish (e.g., Todd and Ebeling 1966; Hughes and Morgan 1973). In marine air-breathing fish, this situation occurs in *Sicyases* (Ebeling et al. 1970) and in the mudskipper, *Periophthalmus vulganris* (Singh and Munshi 1968). In *Periophthalmus schlosseri* (Schöttle 1932), the first gill arch is replaced by a well-vascularized epithelium and the fish has only three pairs of functional gill arches. In *Gillichthys* (Todd and Ebeling 1966), the gills are short and the total respiratory surface area is less than in those species which do not breathe air. The more amphibious fish, e.g., *Mnierpes* (Graham 1973), have stronger gills (with a smaller surface area) which are less susceptible to collapsing out of water than in aquatic ones, e.g., *Blennius pholis* (Milton 1971). A smaller gill surface area reduces the rate of loss of O_2 through the gills into the hypoxic external medium. Shunting of blood within the gill system reduces the energetic cost of branchial perfusion (Satchell 1976; Randall et al. 1981; Smatresk and Cameron 1982a,b). The bimodal breathers in general have smaller respiratory surface area than the entirely water-breathing types (Figs. 67,68, Table 26). The extremely thin blood-gas barriers of the accessory respiratory organs provide a greater diffusing capacity than the gills (Table 26; Maina and Maloiy 1986). The

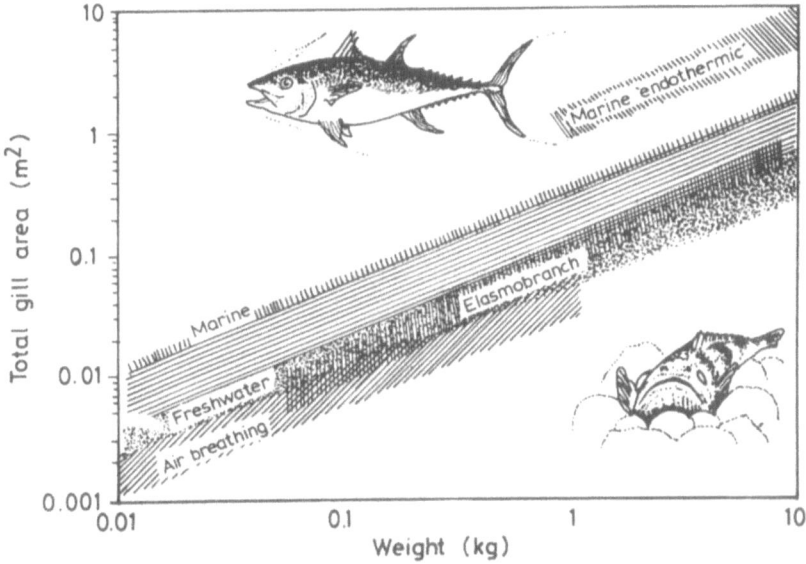

Fig. 68. Correlation between total gill surface area and body mass in fish living in different waters and exhibiting different modes of respiration. Marine endothermic fish which include the tuna have relatively high values while air breathers, especially the obligate ones, have low values. (Pelzenberger and Pohla 1992)

Table 26. Morphometric pulmonary diffusing capacity of the tissue barrier (Dt) and measurements for respiratory surfaces of the gills and air-breathing organs of some air-breathing fish. Data on some water-breathing fish and invertebrates are included for comparison

Species/organ	Body wt. (g)	Thickness of tissue barrier (μ)	Surface area ($mm^2 g^{-1}$)	Dt ($ml\,O_2\,min^{-1}\,mmHg\,kg^{-1}$)
Air-breathing fish				
Anabas testudineus	100			
All gills		10.00	47.2	0.0071
Labyrinthine organ		0.21	32.0	0.2286
Heteropneustes fossilis	100			
All gills		3.58	57.7	0.0242
Air sac		1.60	30.7	0.0288
Skin		98.00	200.0	0.0031
Clarias mossambicus	458			
All gills		1.97	19.30	0.0213
Suprabranchial chamber		0.313	7.79	0.050
Labyrinthine organ		0.287	4.65	0.070
Channa punctatus	–			
All gills		2.033	71.88	0.0530
Suprabranchial chamber		0.780	39.17	0.0753
Labyrinthine organ		–	–	–
Amphipnous cuchia	–			
Air sac		0.435	4.84	0.0165
Skin		0.44	227.5	–
Channa striatus				
Suprabranchial chamber	–	1.359	–	–
Clarias batrachus	–			
Labyrinthine organs		0.55	–	–
Water-breathing fish				
Tinca tinca	141	2.47	228.0	0.1493
Salmo gairdneri	35	4.30	260.0	0.1180
Tuna	100	0.50	2000.0	6.000
Opsanus tau	100	5.00	210.0	0.0630
Latimeria chalumnae	10 000	5.00	18.9	0.0057
Invertebrates				
Carcinus maena	100	5.00	744.0	0.0540
Nautilus macromphalus	135	10.00	930.0	0.1395

Sources of data: *Anabas*, Hughes et al. (1973); *Heteropneustes*, Hughes et al. (1974); *Clarias mossambicus*, Maina and Maloiy (1986); *Carcinus*, Scammell and Hughes (1981); *Nautilus*, Perseneer (1935); rest Hughes and Morgan (1973).

fact that the gills have been widely retained in virtually all transitional breathing fish shows their importance for CO_2 clearance in water. In amphibians, the change to terrestriality and air breathing is accompanied by greater dependence on the lungs for gas exchange (Lenfant and Johansen 1967).

Chapter 6

Air Breathing: the Elite Respiration

> *"The external gas exchangers that have evolved in higher organisms – fish gills, bird lungs, and alveolar lungs of amphibia, reptiles, and mammals have some basic features in common, irrespective of the different principles that determine their functioning in detail."* Weibel (1984a)

6.1 Is the Surface of the Lung Dry, Moist, or Wet? Do Real Air Breathers Exist?

Water forms an important structural and functional constituent of the intercellular and intracelular lung tissue (e.g., Bastacky et al. 1987). Furthermore, a hydrated layer lines the air spaces of the lung (e.g., Fishman et al. 1957; Cantin et al. 1987; Chinard 1992). In the larger air spaces, the aqueous layer is comprised mostly of mucus, a glycoprotein-containing phase which is about 98% water (e.g., Sturgess 1979). The mucus forms an important source of moisture which humidifies the inhaled air, traps solid particles, and protects the ciliated epithelium. At the alveolar level, the hydrated layer occurs in form of an aqueous subphase in which proteins, carbohydrates, ions, and surfactant are dissolved. In the vertebrate lungs, where detailed investigations have been carried out, gas exchange occurs across an extracellular alveolar fluid film which lines the surface. The lining has been lucidly demonstrated by Finley et al. (1968), Weibel and Gil (1968), Kikkawa (1970), Bastacky et al. (1987, 1993), and Hook et al. (1987). In the airways, the thickness of the surface liquid lining is 20 to 150 µm (Widdicombe 1997) while on the alveolar surface, the thickness ranges from 0.1 to 0.24 µm (Weibel and Gil 1968; Bastacky et al. 1993, 1995; Stephens et al. 1996). In the human lung, it has been estimated physiologically (e.g., Rennard et al. 1986) that the epithelial lining fluid (ELF) makes up 20 to 40 ml while through morphometric techniques (e.g., Untersee et al. 1971; Gorin and Steward 1979), the ELF was estimated to range from 15 to 70 ml. The alveolar fluid layer contributes significantly to the gas exchange function of the lung. During strenuous exercise, accumulation of extracellular fluid on the surface of the lung is thought to cause a transient decrease in the membrane-diffusing capacity of the lung (Manier et al. 1991). In a delicate process which includes regulation of hydrostatic and colloidal osmotic forces across the capillary wall, the pulmonary surface is kept moist but not flooded (Levine et al. 1965; Fishman 1972). In the mammalian lung, 1 to 4% of the blood fluids lost to the surrounding tissue is efficiently carried away by the lymphatic vessels (Comroe 1974). When introduced into the alveoli, water quickly passes into the pulmonary blood capillaries (Effos and Mason 1983; Jones et al. 1983; Effros et al. 1992; Grimme et al. 1997). This is because the pulmonary blood capillary (microvascular hydrostatic) pressure of about 1.1 to 1.2 kPa (which is inclined to filter blood into the alveoli) is always well below the colloidal (osmotic) pressure of the plasma proteins of about 3.3 to 4 kPa which draws fluid from the alveoli into the blood (Comroe 1974). Normally, the lung would be expected to be dry and to absorb but not filter fluid in its circulation.

Water is not a foreign factor in the lungs of the air-breathing vertebrates. Due to their extensive surface area, during prenatal development, the lungs play an important role in the production of the amniotic fluid (Setnikar et al. 1959). During fetal life, the lung is filled with liquid that flows into the developing air spaces in response to Cl^- secretion across the epithelium of the respiratory tract (Olver and Strang 1974). In sheep, the rate of production may be as high as about $2 ml kg^{-1} h^{-1}$ (Normand et al. 1971) and at birth the total pulmonary fluid is estimated to be as much as $30 ml kg^{-1}$ (Normand et al. 1971). Experimental or congenital obstruction of the trachea or the bronchi leads to intrapulmonary accumulation of fluid and overdistension of the lung (Potter and Bohlender 1941). In some animals, e.g., the turtles, which have relatively high arterial pulmonary blood pressures and low osmotic pressure of the plasma proteins (White and Ross 1965; Shelton Burggren 1976), water may constitute a significant portion of the O_2 diffusional pathway. In *Pseudemys scripta*, at a blood flow rate of 12 to $14 ml kg^{-1} min^{-1}$, during ventilation, 20 to 40% of the fluid may be left behind in the nonvascular part of the lung tissue (Burggren 1982b): the net loss of plasma into the tissues in the ventilated reptile lung is 10 to 20 times greater than in the mammalian lung. Well-developed intercellular junctions in the air-blood barrier were demonstrated in the lung of the turtle, *Pseudemys* by Bartels and Welsch (1983) by freeze fracture electron microscopy: they may account for the remarkable "leakiness" of the pulmonary capillaries of the turtle's lung. It is interesting to note that in insects, the terminal tracheoles contain fluid which is osmotically absorbed into the surrounding cells or released in the air tubules depending on the metabolic state of the tissues (Wigglesworth 1953). In mammals, about 2 to 3 days before birth (Kitterman et al. 1979; Dickson et al. 1986) or during labor (Bland et al. 1982; Brown et al. 1983), pulmonary filtration stops and absorption of the intrapulmonary fluid starts. Both processes are influenced by the level of circulating epinephrine or isoproterenol (e.g., Walters and Olver 1978), adrenaline (Brown et al. 1983), or Na^+ flux out of the lung lumen (Cotton et al. 1983; Bland 1990; Chapman et al. 1994). In the newborn calf, the mean pulmonary arterial pressure drops from 8.7 to 4 kPa in the first 6 h of birth (Reeves and Leathers 1964). The intrapulmonary fluid is physically expelled through: (1) the upper respiratory airways due to pressure exerted on the fetal thorax during transit through the pelvic canal (Borell and Fernstrom 1962; Karlberg et al. 1962); (2) absorption into the lymphatics (Humpreys et al. 1967; Strang 1967; Gonzalez-Crussi and Boston 1972); and (3) transfer by the pulmonary capillaries (Egan et al. 1975; Hutchison et al. 1985). The fluid remaining in the lung after delivery is cleared from the lung within a couple of hours (Humpreys et al. 1967; Fletcher et al. 1970; Adams et al. 1971; Bland et al. 1980).

The general perception of the surfaces of the internalized gas exchangers of the terrestrial air breathers as dry organs needs to be reconsidered. It is more realistic to consider them as moistened but not flooded. Oxygen does not diffuse efficiently across dry tissue barriers. In the toad (Dupre et al. 1991), dehydration reduces cutaneous gas diffusion capacity. Owing to a dependence on water for efficient respiration across the lungs, it could be argued that air breathing has not yet strictly evolved. However, since terrestrial animals seek and take up O_2 from air and discharge CO_2 into the same, according to the general definition, such animals are recognized as air breathers. This also includes those insects which have secondarily invaded water and extract O_2 from air bubbles regularly ferried from the surface (Sect. 6.12). It is only if an air breather was to "aspirate" water into its lungs, a rather rare retrogressive move which has, nevertheless, developed in some simple animals, e.g., aquatic pulmonate gastropods (Sect. 5.6.1) after the ctenidia (gills) were lost and the mantle cavity converted into a "water lung", that an animal is categorized a water breather.

About 1 week before hatching, in the loggerhead turtle, *Caretta caretta* (Perry et al. 1989a), and during stage one of development of *Salamandra salamandra* (Goniakowska-Witalinska 1982) the lungs are filled with water. The intrapulmonary fluid is removed soon after hatching (Gatzy 1975) and in sheep soon after birth (Strang 1977; Bland 1990). It has been postulated that the flooding of the lungs with fluid provides biomechanical support necessary for proper development of the air passages and spaces (Alcorn et al. 1977; Maloney 1984). It is envisaged that after the contraction of the smooth muscles, the pressure in the fluid-filled lung increases, leading to an outward formation of the air cells (e.g., Marcus 1937). In deviation from the moist surface requirement, some animals such as the minute aerial arthropods are known to exchange respiratory gases through a dry cuticle (Krogh 1941). The same process probably occurs in the book lungs of some spiders (Zoond 1931; Herreid et al. 1981; Strazny and Perry 1984). The low O_2 demands in such animals may account for the adequacy of O_2 transfer across a dry surface.

6.2 Lung and Swim Bladder – Which Developed Earlier and for What Purpose?

Swim bladders and lungs are said to be homologous (Fänge 1983). Lungs are envisaged to have evolved in the Paleozoic freshwater vertebrates as a means for adaptation to hypoxic stress (e.g., Romer 1972; Graham et al. 1978a,b). All bony fish (Osteichthyes) have a swim (air) bladder at least at one stage of their development (e.g., Marshall 1960). In most contemporary fish, the swim bladders are hydrostatic with no discernible respiratory role while some teleosts breath air by a lung or a swim bladder (Fänge 1983; Alexander 1993). The bladder is historically interesting for the following reasons: (1) in some fish, the organ serves as an auxiliary respiratory organ and hence its biology may shed some light on the evolution of the respiratory organs and processes, (2) its role as a hydrodynamic organ helps to explain some of the adaptive processes which were essential for subsistence in water (a medium of relatively higher viscosity) and the necessary subsequent postural and locomotory biomechanical adjustments for living on land, and (3) the capacity of the organ to passively secrete and concentrate gases to several hundred atmospheres of pressure far in excess of those in blood and in the external environment (e.g., Scholander 1954; Marshall 1960; Fänge 1976; Kobayashi et al. 1989a,b) is a most intriguing and unique biophysical phenomenon. Recent works and reviews on this aspect include those of Gerth and Hemmingsen (1982), Fänge (1983), Kobayashi et al. (1989a,b), and Pelster and Scheid (1992a,b). It is now generally thought that a multiplicative salting-out effect in the countercurrent system of the rete mirabile (e.g., Pelster et al. 1988c); (Fig. 39) explains the passive secretion of gases into the swim bladder. The lactate acid which is produced by glucose metabolism in the gas-gland epithelium (Pelster and Scheid 1991, 1993; Figs. 39,70) is secreted into the blood (D'Aoust 1970; Kobayashi et al. 1989a,b; Pelster et al. 1989), where it acidifies the blood, increasing the blood PO_2 by Bohr and Root effects, PCO_2 by conversion of HCO_3^- into CO_2, and the inert gases by reducing their physical solubility (= salting-out effect) (Pelster et al. 1988c). Back diffusion of CO_2 in the rete mirabile increases the blood pH, enhancing the secretion of O_2 into the swim bladder (Kobayashi et al. 1990). Oxygen constitutes the greater proportion up to 95% of the gas mixture secreted into the swim bladder (Wittenberg 1965) with N_2, CO_2, and rare gases like argon constituting lesser proportions (Tait 1956). The PO_2 in the swim bladders of fish which live in great depths may exceed 150 atm (Scholander and van Dam 1953). In such cases, the PO_2 in the swim bladder may be as much as 10 000 times that in the surrounding water (Kanwisher and Ebeling 1957). Considering that all the gases must in the first instance be derived from water, to generate such a pressure differential calls for a remarkably efficient concentrating process. In the European eel, *Anguilla anguilla*, under hypoxic conditions, gas deposition and blood flow through the rete mirabile were significantly reduced (Pelster and Scheid 1992c). Fish which live at depth may required a substantial amount of work to secrete air into a swim bladder against partial pressure gradients (Alexander 1982). In fish such as the swordfish, *Xiphius*, which have been recorded to surface from a depth of 100 m in less than 5 min (Carey and Robinson 1981) and the lantern fishes which are recorded to commute between 300 and 50 m at rates in order of 160 km h^{-1} (Barham 1966), the process must be remarkably efficient.

The chronology of the evolvement of the lung and the swim bladder and whether the two organs are homologous is an ongoing debate (e.g., Romer 1972; Fänge 1976; Liem 1987a). While the early evolution in potentially O_2-deficient waters may be a pointer to an essential initial respiratory role (e.g., Liem 1991), on the contrary, subsistence in the pelagic marine well-oxygenated waters emphasizes a predominant hydrostatic function (Liem 1987a). It is conceivable that swim bladders may have been necessary to offset the increase in weight with the development of the bony skeleton in fish (Alexander 1993). However, the hydrodynamic and respiratory roles of the swim bladder are by no means mutually exclusive and neither can the possibility that the two organs evolved independently of each other be discounted based on the presently available data (e.g., Liem 1989; Hedrick and Jones 1993; Alexander 1993). It is plausible that an initial simple saccular air-containing organ may have served both roles, as evident in some of the Central American catfishes (e.g., Gee 1976, 1981) and the subsequent functional and structural differences may have arisen to meet specific environmental demands. In *Hoploerythrinus unitaeniatus*, a small tropical fish, during the periods between the respiratory cycles, the concentration of O_2 in the swim bladder is reduced to 8%, indicating a very high degree of extraction of molecular

O_2 from the organ (Carter and Beadle 1931). The absorption rate of O_2 from the swim bladder of the mudminnow, *Umbra limi*, is very high (Safford-Black 1944). Some teleosts in which the swim bladder functions as a lung include *Arapaima* (Osteoglossidae), *Gymnarchus* (Mormyridae), *Erythrinus* (Characinidae), *Umbra* (Esocidae), *Notopterus* (Notopteridae), and *Lepisosteus* (Potter 1927; Dehadrai 1962; Fänge 1976). *Gillichthys* consumes O_2 from its physoclistous (closed) bladder to support metabolism during the short transitional period (average 17 min) from aquatic to aerial respiration (Todd and Ebeling 1966). Through a buccal force pump mechanism, physostomatous fish (i.e., those with open bladders) which live close to the surface of the water inflate their bladders with air gulped from the surface of the water (Fänge 1976). Some physoclistic fish are known to use a pneumatic duct during their larval stage for the first filling of their bladders (e.g., Jacobs 1938). In *Erythrinus unitaeniatus*, the middle region of the physostomatous bladder is well vascularized and is used for gas exchange (Carter and Beadle 1931). The tarpon, *Megalops*, regularly ventilates its physostomatous swim bladder even in normoxic water (Böhlke and Chaplin 1968). The tarpon's swim bladder has been described as being lung-like since it has alveolar-like tissue (Shlaifer and Breder 1940): the swim bladder facilitates aerial respiration, a process which augments the tarpon's aquatic respiration and enhances the fish's metabolic scope of activity (Johansen 1970). In the holostean air-breathing fish, *Amia calva*, and the teleost, *Arapaima gigas*, gas exchange is effected through a physostomatous swim bladder which is supplied with blood from the dorsal aorta, coeliac, or mesenteric arteries. While not directly involved in gas exchange, the physoclistous swim bladder of *Dormitator* in hypoxic water generates positive buoyancy, which enables the fish to expose its vascularized frontal skin to air (Todd 1973). The swim bladder of the teleost, *Oreochromis alcalicus grahami*, a small cichlid fish which lives in the alkaline lagoons of the volcanic Lake Magadi of Kenya (Wood et al. 1994; Maina et al. 1996b), has a well-vascularized swim bladder (Figs. 69,70); (Maina et al. 1996b). The bladder is probably used for gas exchange during the night when the water is virtually anoxic (Narahara et al. 1996): on exertion, e.g., after being chased with a net, the fish are observed to skim the surface of the water, probably taking air into the bladder or ventilating the gills with the better oxygenated top layer of water. In the physostomatous bladders, the neuromuscular mechanisms involved in the intake and expulsion of air are not well known. Inspiration may be effected by the buccal force pump while expiration may be brought about by the activity of the smooth muscles of the pneumatic duct of the bladder. In *Hoploerythrinus*, the pneumatic duct is notably wide and muscular (Lüling 1964) and in the eel, *Anguilla anguilla*, the duct is adapted for gas exchange by diffusion and has a distinct circulatory system (Fänge 1953). In most physostomatous fish, e.g., in the salmonids (Fahlén 1971), cyprinids (Plattner 1941), the electric eel (Evans 1929), and notopterids (Müller 1950), the entrance of the pneumatic duct to the esophagus is guarded by smooth and striated muscle sphincters. Dependent on species, gas is released through the pneumostome by deflation. This occurs by a reflex action, the so-called gas spitting reflex, which is initiated by a reduction of the external pressure and/or by nervous excitation (Fänge 1976). The elasticity of the bladder wall due to contraction of the well-developed smooth muscles (Maina et al. 1996b) and contraction

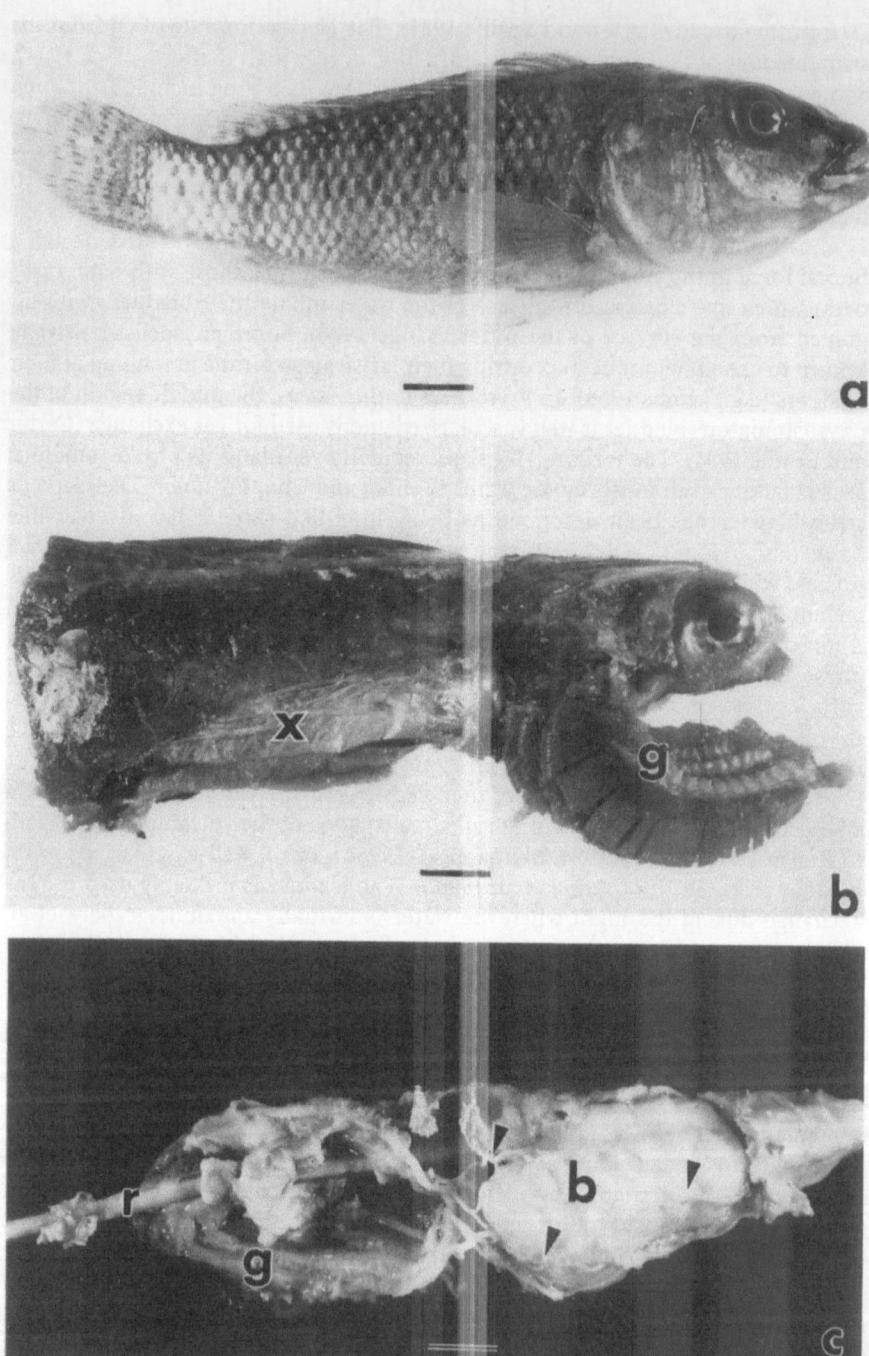

Fig. 69a–c

of body wall muscles (Evans and Damant 1928) may help in the release of gases. In an aquatic animal, the use of a gas exchanger for hydrostatic control is inefficient since when O_2 is removed and little if any CO_2 is secreted back, the animal becomes less buoyant and sinks (Gee 1976; Gee and Graham 1978). The animal then has to surface to refill the lung. Oxygen consumption (i.e., removal of O_2 from the lung) has to be synchronized both with hydrostatic requirements and respiratory frequency. During hypoxic episodes, the cephalopod mollusk, *Nautilus* (often referred to as a living fossil due to its multichambered shell which resembles that of extinct animal forms from the early Paleozoic) draws O_2 from the buoyancy chambers of the shell (Boutilier et al. 1996): in a 459-g (fresh weight) *Nautilus* at 18 °C, the shell volume is $100\,cm^3$, the O_2 store in the shell is $6.9\,cm^3$ and a partial pressure gradient of O_2 of 7.3 kPa exists. In a hypometabolic state, it is estimated that the O_2 stores in the shell should support metabolic rate for as long as 6 h. In the European eel, *Anguilla anguilla*, 83% of O_2 removed from the blood was secreted into the swim bladder and only 17% was used for metabolic purposes (Pelster and Scheid 1992c).

The swim bladder of the teleosts may have evolved independently from the gas chambers of *Polypterus* and the lungfishes (Dipnoi), a respiratory organ having been more urgent for gas exchange in the early hypoxic aquatic medium. On the other hand, it has been surmised that the swim bladder may have evolved entirely for buoyancy control and in some cases, e.g., in the Dipnoi, secondarily acquired a respiratory role. The lungs of the Dipnoi and the ancestral actinopterygian fish such as *Amia, Polypterus*, and *Lepisosteus* are considered to be homologous with the swim bladders of fish (Packard 1974). In teleosts, Dipnoi and Polypteridae, the swim bladder and the lungs originate as outgrowths from the wall of the foregut, the primodial air-sac/lung becoming evident very early in development (Packard 1974). Both organs are enervated by branches of the vagus and sympathetic nerves. However, this is as far as the similarity between the two organs goes. The swim bladder: (1) arises from the dorsal or lateral walls of the foregut (e.g., Goodrich 1930) and the lungs from the floor (ventral aspect) of the pharynx, (2) the lungs receive left and right pulmonary arteries which originate symmetrically from the last pair of epibranchial arteries but the swim bladder is supplied with arterial blood from the aorta (Goodrich 1930), (3) the lungs remain connected to the pharynx while physoclistic swim bladders totally lose connection with it, and (4) the lungs are generally lined by a surface active factor, a complex mixture of phospholipids, neutral lipids, and proteins. Due to their ventral location and the inherent problems of balance which would accompany such an organ, the lungs may have all along been respiratory while the dorsally located swim bladder could combine hydrostatic and respiratory roles. In teleosts, the swim bladder is mainly

Fig. 69. a *Oreochromis alcalicus grahami*, a small tilapiine cichlid fish which lives in the highly alkaline Lake Magadi of Kenya where there are dramatic diurnal fluctuations of levels of O_2 in water. b The swim bladder, *x*, and the gills, *g*, are very well developed. c Latex rubber cast of the gills, *g*, and the swim bladder, *b*, showing the air way, marked by a cannula, *r*, and vascular connections, ➤, between the two organs. a Bar 7 mm; b 5 mm; c 3 mm. (Maina et al. 1996b)

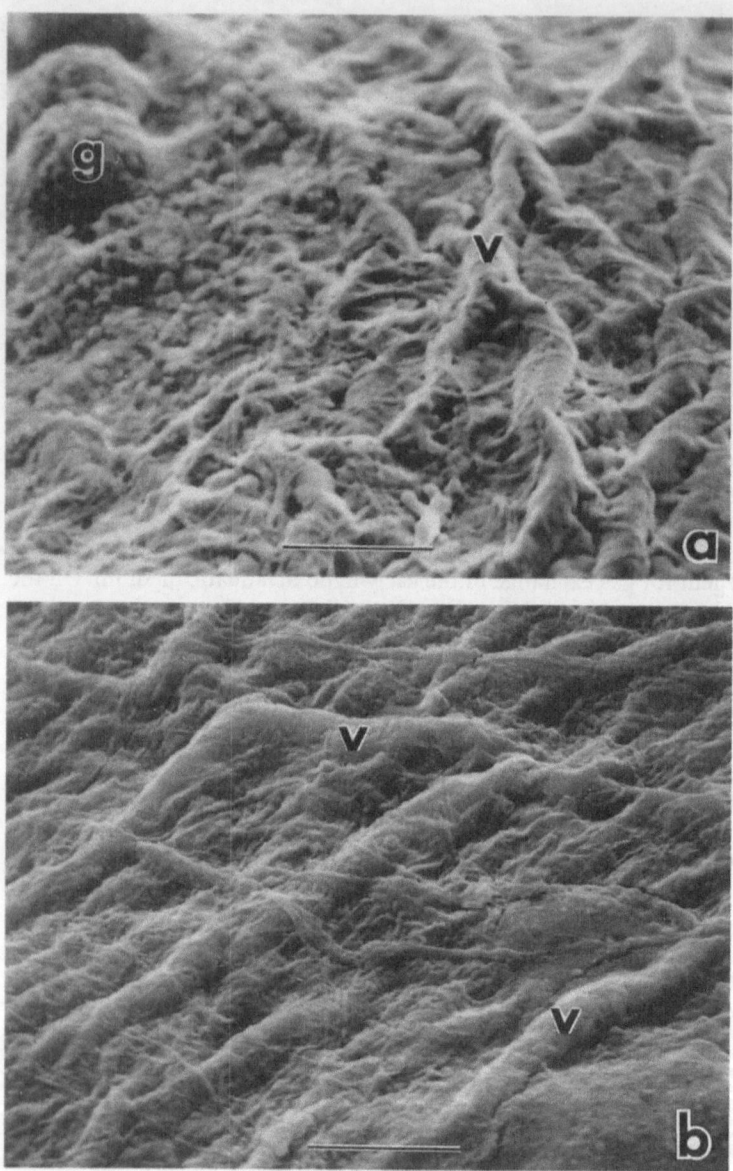

Fig. 70a,b. Views of the surface of the swim bladder of *Oreochromis alcalicus grahami*, showing the profuse vascularization, *v*; *g* gas-gland cells. It is probable that the physostomatous swim bladder in this species may be utilized for gas exchange during extreme hypoxia **a** Bar 20 μm; **b** 17 μm. (Maina et al. 1996b)

a hydrostatic organ and is thus structurally much less complex than the Dipnoan (Fig. 64) and Polypteridae lungs which are essentially respiratory (Klika and Lelek 1967). The nonseptate saccular lungs of the urodele amphibians, the salamanders, e.g., *Cryptobranchus* and *Necturus*, appear to be largely hydrostatic (Noble 1931; Guimond and Hutchison 1973b). However, the well-developed finely subdivided lungs of some aquatic frogs, e.g., *Xenopus laevis* and *Pipa pipa* (Czopek 1962a; Goniakowska-Witalinska 1995), which inhabit hypoxic waters and the salamanders, *Amphiuma tridactylum* (Stark-Vancs et al. 1984) and *Salamandra salamandra* (Goniakowska-Witalinska 1978; Meban 1979), which have an extensive surface area, are predominantly respiratory.

6.3 Evolution of Air Breathing and Terrestriality: the Limitations

In virtually all terrestrial habitats, the atmospheric air is never 100% saturated with moisture. On this account, terrestrial habitation is accompanied by an enduring conflict between the need to procure O_2 and the necessity to preserve water. Extended emergence from water was impossible until the two processes could be harmonized. Conservation of water entailed development of an impermeable surface cover and invagination of the respiratory organs (e.g., Hadley 1980; Loveridge 1980; Quinlan and Hadley 1993). Terrestrial air breathers are in constant danger of desiccation especially through respiratory water loss. No other internal organ is in more intimate contact with the external environment as the lung. Everyday, about 12 000 l of air are filtered by the human lung and 6000 l of blood perfuse it (Burri 1985). The interaction occurs over a surface area of nearly 150 m² across a blood-gas barrier 0.6 μm thick (Gehr et al. 1978). In most respects, air breathing is a less complicated process than water breathing. Whereas in air breathing only O_2 and CO_2 are exchanged, in water, superimposed on the gas exchange process is ionic exchange and elimination of end products of nitrogen metabolism (e.g., Goldstein 1982; Zadunaisky 1984). Even if all other factors were held constant, gill respiration should energetically be more demanding than air breathing owing to the fact that except for the animals which are iso-osmotic to seawater, energy has to be expended to selectively regulate ionic flux. A large number of mitochondria rich cells (= ionocytes = chloride cells), occur, e.g., in fish (Fig. 50; Maina 1990a, 1991) and the crab gills (Maina 1990b; Figs. 13,71). Aquatic animals are exposed to a medium which has the same heat capacity as blood compared with the air breathers where the heat capacity of air is 3000 times lower than that of blood (Table 9). Since for equivalent volumes water contains about one thirtieth of the volume of O_2 in air, to extract an equal amount of O_2, aquatic breathers expose their blood to a heat sink which has a heat transfer capacity 9×10^4 times greater than air (Steen 1971). This may explain why only some degree of endothermy has evolved in a few aquatic animals and in most cases only in specific tissues. In fish, endothermy occurs in a few fish, e.g., the swimming muscles of the tuna and some sharks (e.g., Carey and Teal 1966, 1969; Block 1991b). Endothermic-homeothermy, an evolutionary step which delinks an animal's physiological processes from the environmental thermogenic

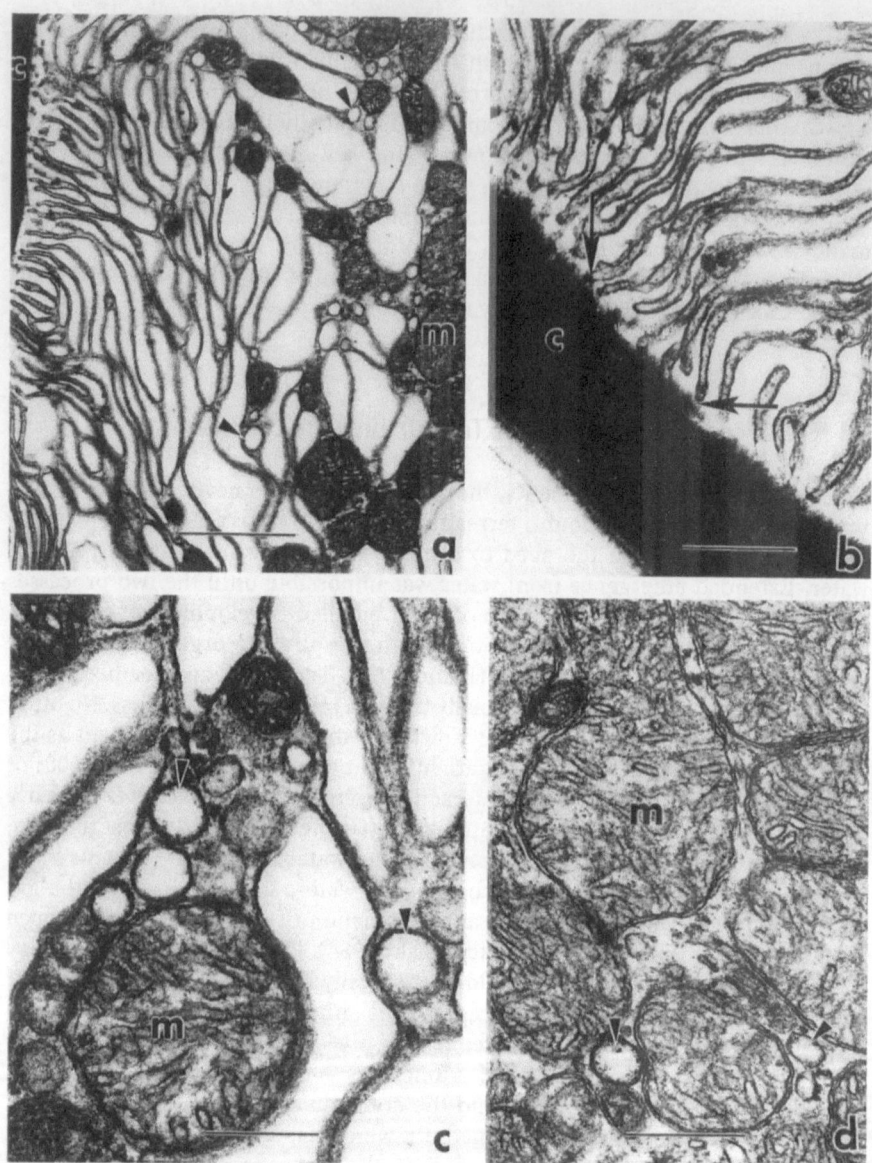

Fig. 71a–d. Gills of the freshwater crab, *Potamon niloticus*, showing: a the basal aspect of the epithelial cells with numerous membrane infoldings which attach onto the surface cuticle (*c*); mitochondria, *m*, and a micropinocytotic vesicles, ➤. b A closeup of the basal aspect of an epithelial cell showing the attachment of the basal infoldings, ➜, to the cuticle, *c*. c A view of the mitochondria, *m*, in close proximity to the micropinocytotic vesicles, ➤. d The numerous mitochondria, *m*, with profuse cristae in close proximity to micropinocytotic vesicles, ➤. The general structural features of the gills of the crab illustrate a compromise design between a thin water-blood (hemolymph) barrier and osmoregulatory exchange. **a** *Bar* 0.75 μm; **b** 0.5 μm; **c** 0.3 μm; **d** 0.3 μm. (Maina 1990b)

fluctuations, has been exclusive to the elite air-breathing animals, namely mammals and birds (Romer 1967; Crompton et al. 1978; Bennett and Ruben 1979; Hochachka 1979). At the extreme small body sizes, e.g., in the about 2-g Cuban bee hummingbird, the Etruscan shrew, and the Thai bumblebee bat, endothermic homeothermy is extremely expensive to maintain. Hummingbirds hover more than 100 times a day as they feed on nectar (Krebs and Harvey 1986), each bout lasts about a minute, and 20% of the daylight hours are spent feeding (Diamond et al. 1986). At night, some but not all hummingbirds undergo torpor to conserve energy (Lasiewski 1963b; Bucher and Chappell 1989). Though energetically expensive (Else and Hubert 1981, 1985), regulation of body temperature at a higher level imparts certain biological advantages. The biochemical reactions are faster, the rate of information processing is rapid (through quicker nervous coordination and response to environmental changes), and because diffusion is a physical process, gas transfer across the respiratory organs is faster. It is envisaged (e.g., Randall et al. 1981) that endothermy in terrestrial vertebrates evolved in temperate regions where the low ambient temperature may have called for dissociation of the body temperature from the ambient. In whichever way it evolved, the origin of endothermy had to await development of more efficient respiratory organs and processes to provide the large amounts of O_2 necessary to support a high metabolic rate (Wood and Lenfant 1979; Duncker 1991). Amazingly, for somewhat different purposes, some plants, e.g., aroid plants such as *Philodendron selloum* and *Arum maculatum* (Nagy et al. 1972; Seymour 1997) and the lotus plant (Seymour and Schultze-Motel 1996) have evolved endothermy. Heat production in the two aroid plants rivals the high levels generated by the flight muscles of bees (Seymour 1997)! In *P. selloum*, a 125-g spadix produces about 9 W of heat to maintain a temperature of 40 °C against an ambient one of 10 °C (Koch et al. 1983, 1984; Seymour 1991), surpassing heat output in some thermoregulating animals (Seymour 1997).

In abandoning liquid breathing and adjusting to air breathing, the chances of survival of animals on land depended on the level of structural development and functional adequacy of the gas exchangers as well as that of the necessary homeostatic changes, especially those relating to acid-base balance (e.g., Robin et al. 1969; Rahn 1974; Reeves 1977). Among the different animal lines, air breathing has evolved independently many times, following different pathways (e.g., Truchot 1990; Graham 1994). The possession of lung-like structures by non-teleostean groups of ray-finned fish is paleontologically well documented (e.g., Romer 1972; Jarvik 1980). Primeval lungs which developed from the floor of the junction of the esophagus and the pharynx are recognized in the ancestral armored fossil placoderm fish, *Bothriolepis* (Denison 1941), and among some of the oldest fishes, e.g., the bichirs, *Polypterus* (Lauder and Liem 1983; Burggren et al. 1985a). The elaborate dipnoan lungs (Fig. 64) are acclaimed to be the possible precursors of those of the tetrapods (e.g., Inger 1957; Romer 1967; Løvtrup 1977; Gardner 1980). The earliest definite lungs in the chordates occurred in the ancestral fishes. Lungs are present in all extant holosteans, dipnoans, and predominate in the chondrosteans. The evolution of the lungs enabled animals to fully emerge from water and enjoy the physical advantages of occupying an O_2-potent environment and an almost limitless habitat. So perfected has the process of air breathing become that most of such animals soon succumb on submersion in water irrespective of the concentration of O_2. Inundation of the pulmonary airway tree prohibits gas exchange in the lung, producing obstructive asphyxia. Physiologically, this is characterized by hypoxia, hypercapnia, and acidosis. Depending on the nature of the aspirated liquid, ionic imbalance also occurs (Moritz 1944) leading to complications such as hydremia, hemoconcentration, hemolysis of the erythrocytes, ventricular fibrillation, cardiac arrest, and death.

The colonization of land by the tetrapods at about the Early Devonian and that of the higher plants in the Upper Silurian or a little earlier (e.g., Romer 1967; Gray 1985a,b) formed important steps in the evolution of the major ecosystems on Earth. The problems and challenges which accompanied air breathing and transition to land included: (1) postural problems as a result of increased effect of gravity on the body as animals lost the buoyancy provided by the water; (2) potentially large ecological variability of water vapor pressure and hence risk of desiccation; (3) thermal instability due to the low thermal capacity of air; and (4) problems of reproduction and fertilization (in water the gametes could simply be released into the water). Evidence of a causal relationship between hypoxia and evolution of air breathing is overwhelming (Sects. 5.1 and 6.3). Air breathing in general and terrestrial habitation in particular entailed a conflict between O_2 uptake, water conservation, and acid-base balance. To effect the change, certain trade-offs and compromises were necessary. It is probably in an attempt to maximize on the advantages while giving up very little that a substantial number of animals after having evolved the capacity for aerial respiration, a major preadaptive step towards land habitation, opted to subsist in water. Indeed, it is only in arthropods and vertebrates where extensive evolutionary adaptation to air breathing has occurred. The main problem which faced the pioneering colonizers of the aerial biotope was elimination of CO_2. The challenges in acid-base balance are apparently less in water than in air due to the high solubility of CO_2 in water. The absolute values of PCO_2 and HCO_3^- ions in blood and tissues are lower in water breathers than in air breathers (Tables 2,22). In reptiles, the first exclusively lung breathers (except for the aquatic species which significantly use the skin), the PCO_2 and the concentration of HCO_3^- ions vary remarkably depending on ecological adaptations, morphological design, and lifestyle (Jackson 1986). In Squamata and Crocodilia, the mean concentration of HCO_3^- ions is $15\,mMl^{-1}$ at a PCO_2 of $2\,kPa$ and in Chelonia, the concentration of HCO_3^- ions is $39\,mMl^{-1}$ and the PCO_2 $4.5\,kPa$ (Howell 1969). The PCO_2 ranges between 0.1 and $0.5\,kPa$ and the concentration of HCO_3^- ions from 5 to $15\,mMl^{-1}$ in the plasma of fishes (Heisler 1984). Owing to the greater amount of O_2 in air, to maintain a PO_2 of $13.3\,kPa$ in the alveolar air (at a respiratory quotient of 1), an air breather need only move $17\,ml$ air min^{-1} ml^{-1} O_2 compared with an aquatic animal which must move $480\,ml$ water min^{-1} ml^{-1} O_2 to maintain an equivalent PO_2 in the gill water (Rahn 1967; Howell 1969, 1970): the alveolar PCO_2 in the air breather would be $6.7\,kPa$ while in a water breather the PCO_2 in the gill effluent would be only $0.24\,kPa$. The ventilatory rate of an aquatic animal at $20\,°C$ is 28 times (at that temperature the solubility of CO_2 is about 28 times greater than the solubility of O_2 in water!) that of an air breather: the PCO_2 of the blood of a fish is one-twenty eighth that of an air breather. Thus, animals which endeavored and succeeded in attaining air breathing could drastically reduce ventilatory rate but would be faced with a profound increase in arterial PCO_2 which in terms of acid-base balance results in a respiratory acidosis. The alternative strategies which were used to surmount these obstacles included: (1) renal mechanisms which resulted in increased blood concentration of HCO_3^- ions thus maintaining a constant OH^-/H^+ ratio and/or, (2) evolution of an alternative pathway other than the lung, e.g., the skin and the buccal cavity, for CO_2 clearance during the traumatic switchover of respiration from the gills to the modern lung (Fig. 57). Conceivably, the transition was interfaced by a relatively simple precursor of the modern lung which must have been, from perspectives of design and mechanical ventilation, relatively ineffi-cient. The forfeiture of the skin, after the integument was covered with scales or armor to reduce water loss, in reptiles (Mertens 1960) was the initial step in the transfer of gill to lung respiration (Hughes 1963). During this time, the lung and respiratory mechanics developed to a level where the air-breathing organ could eliminate the entire load of CO_2 produced in the body. The role of acid-base balance was wholly shifted to the modern lung and, to a smaller extent, the kidney. At 20 °C, only about 3% of the CO_2 output is cutaneous in the tortoise, *Testudo dendriculata* (Jackson et al. 1976) and in the desert lizard, the chuck-awalla (*Sauromalus obesus*) about 4% of CO_2 output and less than 2% of O_2 uptake occurs through the skin at 25 °C (Crawford and Schultetus 1970). In the land-dwelling box turtle, *Terrapene ornata*, a large proportion of CO_2 is voided through the skin at the low hibernating temperatures (Glass et al. 1976). Gans (1970) observed that cutaneous respiration is an acquired mode of respiration unique to the modern amphibians without any relationship to the Devonian ancestral lineage of the extant tetrapods. Romer (1972) and Colbert and Morales (1991) have espoused the theory that the Carboniferous amphibians were com-pletely heavily armored. If this is correct, the highly gas- and water-permeable skin characteristic of the extant amphibian species is an adaptive secondary condition. Clearly, in amphibious animals the skin was not as efficient as the gills in CO_2 elimination, as demonstrated by the fact that, despite the role the integu-mentary system played in gas exchange, it was still necessary for the level of the plasma HCO_3^- ions to be elevated in the air breathers. The increase in the arterial PCO_2 and the concentration of the HCO_3^- ions occurred as animals became more dependent on air for their O_2 needs, starting from the water-breathing fish through the various forms of bimodal breathers to the air breathers. Despite the differences in the PCO_2 and the concentration of the HCO_3^- ions in the arterial blood of the water- and air breathers (e.g., in the carp and the bullfrog), the pH (about 7.9) of these animals at the same temperature (20 °C) is essentially the same (Howell et al. 1970; Table 2). The pH of the water-breathing tadpole at 20 °C is the same as that of the air-breathing frog at the same temperature despite the fact that the arterial PCO_2 and the concentration of the HCO_3^- ions of the bullfrog are about five times greater than that of the tadpole (Erasmus et al. (1970/71). Conceivably, since the bimodal breathing occurred in O_2-deficient waters, the O_2 affinity of blood of such animal forms should be lower, i.e., P_{50} should be higher, the more aerial-breathing an animal is. The P_{50} of the blood of the air-breathing fish ranges from 2 to 2.7 kPa at 20 °C, values which are higher than those of water breathers in similar habitats. In the amphibians, the most aquatic species have a P_{50} about one half that of the completely aquatic aerial species (Lenfant and Johansen 1967). The differences are accounted by the fact that air is a respiratory medium much richer in O_2.

6.4 Aerial Gas Exchangers: Structural and Functional Diversity

The air-breathing organs are thought to have evolved as special adaptations to the prevalent hypoxic conditions, presumably those which prevailed in the warm tropical waters of the late Devonian (e.g., Zaccone et al. 1995). Confinement of gas exchange to specific site(s) of the body where the process could be well regulated, water loss restricted, the inhaled air "cleaned" and physically modified, and better protection from toxicants and trauma afforded were prerequisites for efficient aerial respiration and successful terrestrial habitation. If the human lungs were like the external gills of fish, i.e., were evaginated and exposed to air, even in a moderately desiccating environment, the water loss would be about 500 l day^{-1}. This value is about 1000 times greater than the normal loss (McCutcheon 1964). In arthropods, removing the cuticle (by scratching) increases cuticular water loss much more than it affects gas exchange (Ito 1953; Richards 1957). The reduction of the respiratory rate in the air breathers (owing to accessibility to a greater O_2 concentration) constituted an important step in the reduction of respiratory water loss. Most animals die if water loss exceeds 20 to 50% of their body mass (Adolph 1943). The majority of terrestrial vertebrates cannot tolerate a body temperature of 45 °C or more but, when provided with adequate water for evaporative cooling, they will withstand even higher temperatures (Calder and King 1974). The development of internalized gas exchangers (Figs. 4,5), however, was accompanied by certain functional limitations. While the gills, e.g., in crustaceans, mollusks, agnathans, and fish by virtue of their external location can be ventilated unidirectionally and continuously, the "dead-ended" lungs of the vertebrate air breathers can only be ventilated periodically and bidirectionally (i.e., tidally) through narrow opening(s) (Figs. 6,18). This constituted a major trade-off in the design of the aerial gas exchangers. While in the gills the configuration was compatible with the highly efficacious countercurrent system, as a consequence of their invagination, only the relatively inefficient uniform pool arrangement was tenable in the aerial gas exchangers (Figs. 6,35,36; Sects. 6.2 and 6.4). The internalized gas exchangers fail to maximally exploit the high ambient PO_2: the inspired air is greatly diluted by the residual air, reducing the head pressure from 21 kPa to about 13 kPa, a loss of about one third of the potential partial pressure gradient. The arterial PO_2 in *Amia calva, Synbranchus,* and *Neoceratodus* (an obligate water-breathing lungfish) are below 1.3 kPa in water but are higher when held in air (Lenfant et al. 1966; Johansen et al. 1970a). In a resting human being, where the dead space is about 140 cm^3, about 28% of the 500 cm^3 of the inhaled air does not reach the gas exchange region. Compensatory advantages such as low cost of ventilation in air, greater O_2 loading and transport due to presence of hemoglobin, and better ventilation-perfusion controls help overcome the apparent limitations intrinsic to the design. On the plus side, tidal ventilation makes it possible for a much more stable and well-controlled local respiratory condition to be established. The alveolar PO_2 is lower and the PCO_2 higher than in the ambient air, the elevated alveolar PCO_2 in the vertebrate lung being used in the HCO_3^- ion buffer system for pH regulation. Such microenvironments are impossible to create in those gas exchangers which are unidirectionally ventilated or directly exposed to the external respiratory medium. Mechanical ventilation is necessary for mass renewal of the air in contact with the respiratory surface(s) to maintain the partial pressure gradient essential for O_2 influx. The most elaborate of the respiratory processes are restricted to the large and phylogenetically more advanced animals. A highly developed nervous system and proper nervous integration were necessary to operate these complex organs and functions. Shrewdly, by isolating the lung from the air sacs, birds have developed a unidirectional continuous air flow in the parabronchial system of the lung within an invaginated, tidally ventilated organ! The tracheal system of insects has attained a remarkable state of efficiency through synchronization of spiracular opening with the ventilation of the air sacs (Weis-Fogh 1967). Intermittent breathing in reptiles (e.g., Glass and Wood 1983; Wood and Glass 1991), amphibians (De Jong and Gans 1969), and lungfishes (Lenfant and Johansen 1968) is possible because of their low O_2 needs and the great O_2 stores in their lungs. Episodic respiratory pattern constitutes an energy-saving strategy compared with the continuous mode of breathing in birds and mammals (Milsom and Jones 1979). Although incidental breathing is a common feature of the ectothermic vertebrates (Milsom 1988; Shelton and Croghan 1988), it occurs in hibernating nondiving mammals such as hedgehog, dormouse, and the little brown bat (e.g., D.W. Thomas et al. 1990) and in the ground squirrel, *Spermophilus lateralis*, during deep hypothermia (e.g., Garland and Milsom 1994). When in such a state, large fluctuations in lung and blood PO_2 and PCO_2 occur (e.g., Musacchia and Volkert 1971). Functionally, gas exchangers in terrestrial air breathers fall into three general categories. In the diffusive type which, e.g., occur in most small insects, pneumonate gastropods, and some terrestrial crabs, no ventilatory movements occur: diffusion is adequate to supply O_2 and remove CO_2. In the mixed type, as found in the large insects and terrestrial crustaceans, mechanical pumping aids convective movement of air at least in the principal respiratory pathways. Mass transport of air to the pulmonary surface is necessary to produce efficient transfer of respiratory gases in the convective (ventilatory) type gas exchangers. These are found in the air-breathing fish, amphibians, reptiles, mammals, and birds.

6.5 The Diffusive Type Gas Exchangers

As a means of delivering O_2 to a respiratory site, diffusion is only efficient over short distances. The process can hence only effectively service the extremely small, low-metabolism animals. Diffusion lungs occur in most pulmonate gastropods (Ghiretti 1966). A diffusive tracheal system has evolved in the terrestrial arthropods – in the uniramians (e.g., insects, centipedes, and millipedes), in chelicerates (e.g., scorpions and spiders), in isopods particularly the Porcellionidae (e.g., sow bugs) (Paul et al. 1987; Fincke and Paul 1989), and even in plants (Walsby 1972). The simplest lungs are found in snails and slugs where the mantle cavity has been converted into well-vascularized internal respiratory spaces (Fig. 59). Through muscular contractions when the pneumostome is closed, the intrapulmonary pressure increases to about 0.2 to 0.3 kPa. The periodic compression of the air inside the lung is thought to enhance the diffusion of O_2 into the blood (Ghiretti 1966). According to Dahr (1924, 1927), a very small PO_2 of 0.3 kPa is adequate to provide the animal with sufficient O_2. In *Arion* and *Helix*, at a PO_2 of 2 kPa, the pneumostome opens 15 to 30 times in 30 min, remaining open for 7 min (Dahr 1924). The opening of the pneumostome in the gastropods is determined by factors such as levels of CO_2, O_2, humidity, and temperature (Mass 1939). The arachnids are among the earliest animals to occupy land and acquire the capacity to breathe air (Paul 1992). Their respiratory organs, the book lungs, have literally been frozen in time. In having changed very little since the Devonian, they are highly instructive in the understanding of the design of the gas exchangers in the original air breathers. The book lungs of the spiders and the scorpions (Figs. 15,72) consist of stacks of parallel blood-filled lamellae or plates contained in invagination(s) of the abdominal wall (Pohunkova 1969; Moore 1976; Vyas and Laliwala 1976; Herreid et al. 1981; Strazny and Perry 1987). The spiracular muscles, which respond to CO_2 (or low pH) and PO_2 (Fincke and Paul 1989) regulate the size of the opening into the atrium, expelling CO_2, letting in O_2, and regulating water loss and pH (Angersbach 1978; Fincke and Paul 1989). Gas

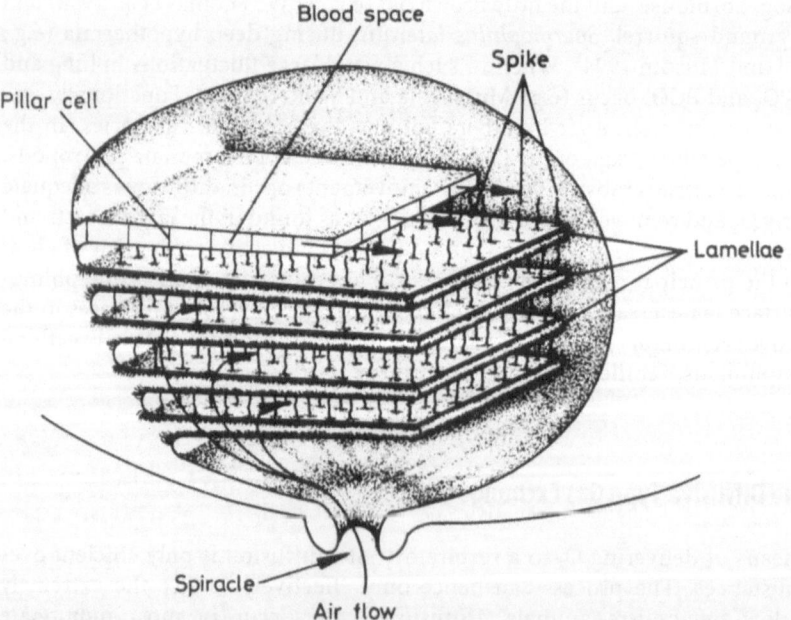

Fig. 72. Schematic view of the diffusive, sheet flow book lungs of a spider showing stacks of lamellae through which hemolymph flows, →, and spikes which keep the lamellae apart

exchange is effected with the hemolymph, which is pumped across the hollow chitinous lamellae (Anderson and Prestwich 1982; Fincke and Paul 1989; Farley 1990; Fig. 72): in *Eurypelma*, when the spiracles are open, arterial PO_2 rapidly rises from 3.7 to 9.9 kPa and hemocyanin is fully saturated (Angersbach 1978). The absence of ventilatory activity and control of the spiracle entrance area in the diffusion lungs minimizes water loss. It is compatible with low aerobic metabolic rate reflected in the low O_2 uptake especially in arachnids (Anderson 1970). Peristaltic movements in the book lung lamellae of some spiders have been reported (Moore 1976). The actual significance of these movements in respiration is, however, not well known (Paul et al. 1987). Anaerobic metabolism plays an important part in the activities of arachnids (Prestwich 1983). Their locomotory muscles lack mitochondria (Linzen and Gallowitz 1975). Only the heart, CNS, Malpighian tubules, and midgut glands work aerobically (Paul et al. 1987). The largest known spider is the bird-eating spider, *Theraphosa blondi*, of northern South America, which attains a body mass of about 55 g (Dresco-Derouet 1974).

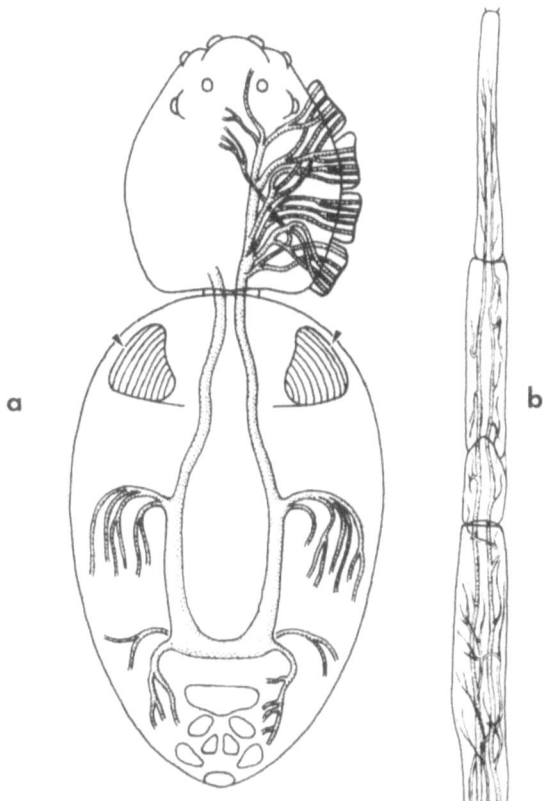

a   b

Fig. 73a,b. The well-developed tracheal system in the opisthosoma and prosoma (a) and first legs (b) of the web monitoring spider, *Uloborus glomosus*; ▶ book lungs. The trachea in the highly active legs are well developed. (Opell 1987)

Variably, spiders possess two kinds of respiratory organs, the book lungs and the trachea. The so-called pulmotracheates possess both organs (Fig. 73). The more metabolically active spiders and scorpions have book lungs and a tracheal system, while the ancestral ones (e.g., the orthognaths) have book lungs only (Paul et al. 1987; Strazny and Perry 1987). In some families of spiders, e.g., Caponiidae and Symphytognathidae, only the trachea exist (Paul 1992) but unlike in insects, the trachea do not directly contact the tissue cells. Hemocyanin is used to carry O_2 over the remaining distance (Foelix 1982). Some pulmonate snails of the family Athoracophoridae (e.g., tropical snails, Janellidae) and some other unrelated animals, e.g., the sow bug (an isopod) *Porcellio*, the house centipede (chilopods), and *Scutigera* (Krogh 1941) present a fairly complicated respiratory system which combines the features of gastropod lungs with the tracheal system in insects: the pneumostome/spiracles open into a common vestibulum from which an array of fine diverticula radiate into a blood sinus providing a large respiratory surface area. A pneumotracheal organization provides a more efficient gas exchange capacity compared with simple smooth respiratory invaginations. In analogy, the design is closer to the mammalian tracheobronchial lung than to the tracheal system of insects. The largest diffusive lungs are probably those of the African pulmonate snails, *Achatina* and *Bulimus*, which attain a volume of up to 500 ml (Krogh 1941). In the freshwater pulmonates, *Planorbis corneus* and *Lymnea stagnalis*, despite lack of ventilatory activity, the PO_2 rises to 18 kPa before the closure of the pneumostome (a value higher than the alveolar level of the mammalian lung of about 13 kPa. The PO_2 may drop to as low as 2.7 kPa before the pneumostome reopens (Precht 1939; Jones 1961). There is no respiratory system in the vertebrates which is adapted to diffusive respiration alone. In states such as hibernation (e.g., in the lungfish) and conditions such as exposure to high humidity (e.g., in the bullfrog), the skin may meet most, if not all, the O_2 requirements. Dogs have been experimentally kept alive for an hour on diffusive respiration alone. This, however, occurs only under conditions of minimal O_2 demands such as in states of anesthesia and while the animal is breathing pure O_2 (Lambertsen 1961).

6.6 The Mixed Type Gas Exchangers

6.6.1 The Insectan Tracheal System

Among the air-breathing animals, the insectan tracheal respiration is unique. In many ways, it is astonishing both for its intuitive architectural simplicity and functional efficiency. In a degree of refinement almost past belief, the insects have disengaged the circulatory and respiratory systems, totally relegating the former from any meaningful role in gas exchange. This is a radical deviation from the prevalent plan in the vertebrate and the invertebrate (nontracheate) air and water breathers where a circulatory system is intercalated between the gas exchanger and the body tissues (Figs. 1,2,3). In insects, O_2 is delivered by the trachea directly to the body tissues (Figs. 75,76). The trachea, the portals of entry of air to the body

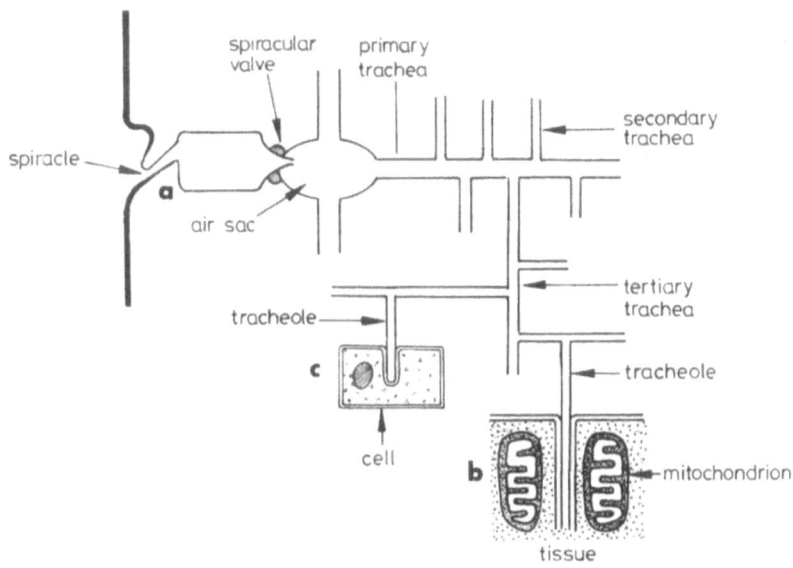

Fig. 74a–c. Schematic diagram of the air flow pathway in the tracheal system of insects. The vulvular spiracles (a) open to the outside while the tracheoles terminate deep in the animal's body (b) in some cases indenting some tissue cells (c)

(Figs. 37,74), form as ectodermal invaginations. With minimal drop in PO_2 along the way, virtually every individual cell in the body is served by a tracheole(s), structures which are analogous to the vertebrate blood capillaries (Fig. 77). The PO_2 between the tracheoles and the metabolizing tissue cells in insects is about 5.3 kPa (Weis-Fogh 1964a, 1967) compared with that of less than 0.3 kPa in the mammalian tissues. In adult *Aphelocheirus*, between the spiracles and the tracheoles, the PO_2 drops by only 0.3 kPa (Thorpe and Crisp 1941). Generally, the concentration of O_2 drops by only 1% from the spiracle to the tissues (Buck 1962). The tracheal system can supply ten times more O_2 g^{-1} tissue than the blood capillary system (Steen 1971). With the spiracular valve serving as a carburetor, in mechanical terms, the trachea operate simultaneously both as a compressor and an exhaust pipe, presenting the epitome in the design of the gas exchangers. Contrary to the cases in the branching tubular structures like the bronchial system of the mammalian lung (e.g., Horsfield 1981), where between the principal and the terminal bronchi the flow velocity decreases by a factor of 700 due to increase in the cross-sectional area (Horsfield and Thurlbeck 1981; Horsfield and Woldenberg 1986), the insect trachea are the only system of air conduits known where the cumulative cross-sectional area and hence the flow velocity remain constant with length (Krogh 1941). In the endothemic insects (Heinrich 1992), thermoregulation of the thorax during flight enables them to attain some of the highest known muscle power outputs in the Animal Kingdom (Harrison et al. 1996). The aerobic rates of the flight muscles are some of the highest reported for any tissue (Kammer and Heinrich 1978), values which approach those of pure microbial cultures (e.g., Hughes and Wimpenny 1969). Insect flight muscles do

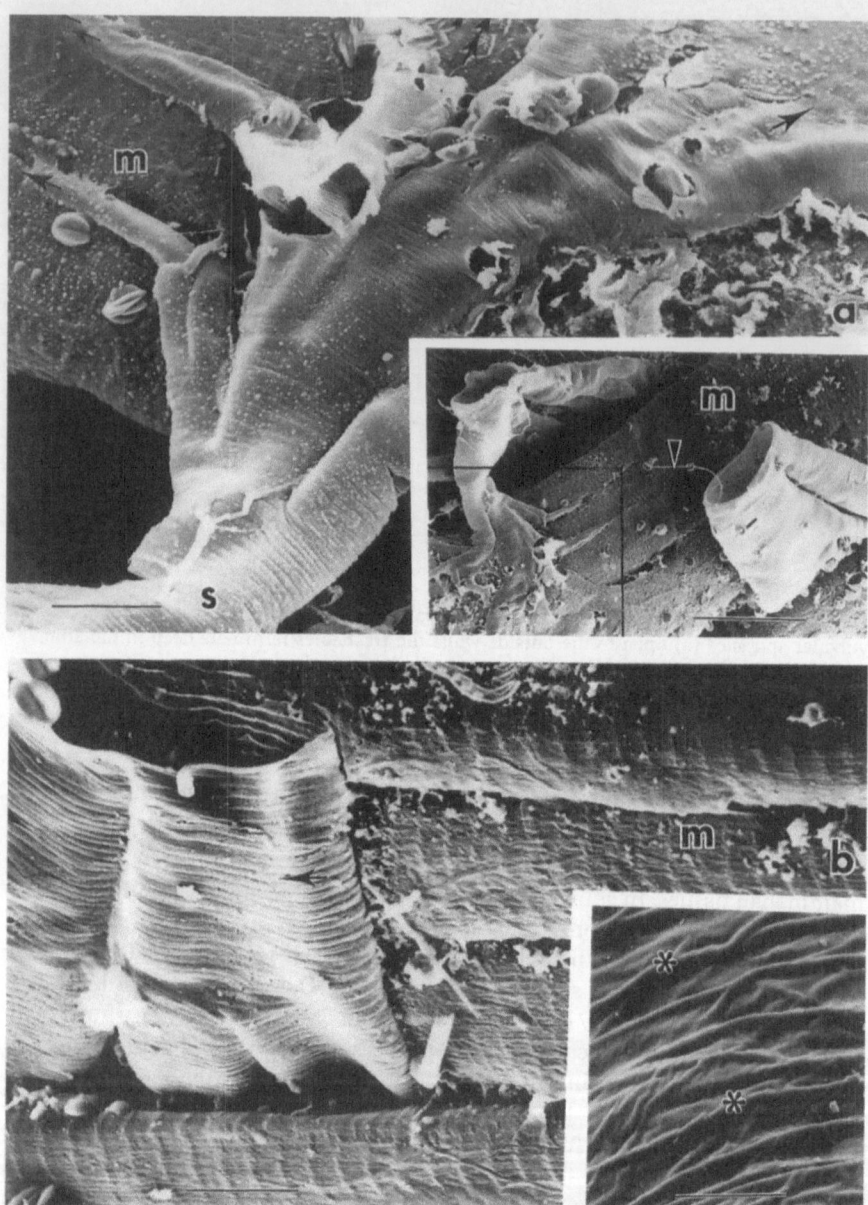

Fig. 75a,b. Air supply to the flight muscles of a grasshopper, *Chrotogonus senegalensis* showing: a secondary trachea, *s*, approaching the flight muscle, *m*, and tertiary trachea, ➜, indenting the muscle. The trachea are supported by the spiral taenidia (➤ *inset*) which keep them open. The main figure is an enlargement of the enclosed region in the inset. b Closeup of a tertiary trachea with distinct taenidia, ➜, about to indent the flight muscle, *m*. *Inset* *, taenidia. **a** *Bar* 20 μm; *inset* 80 μm; **b** 45 μm; *inset* 30 μm. (Maina 1989b)

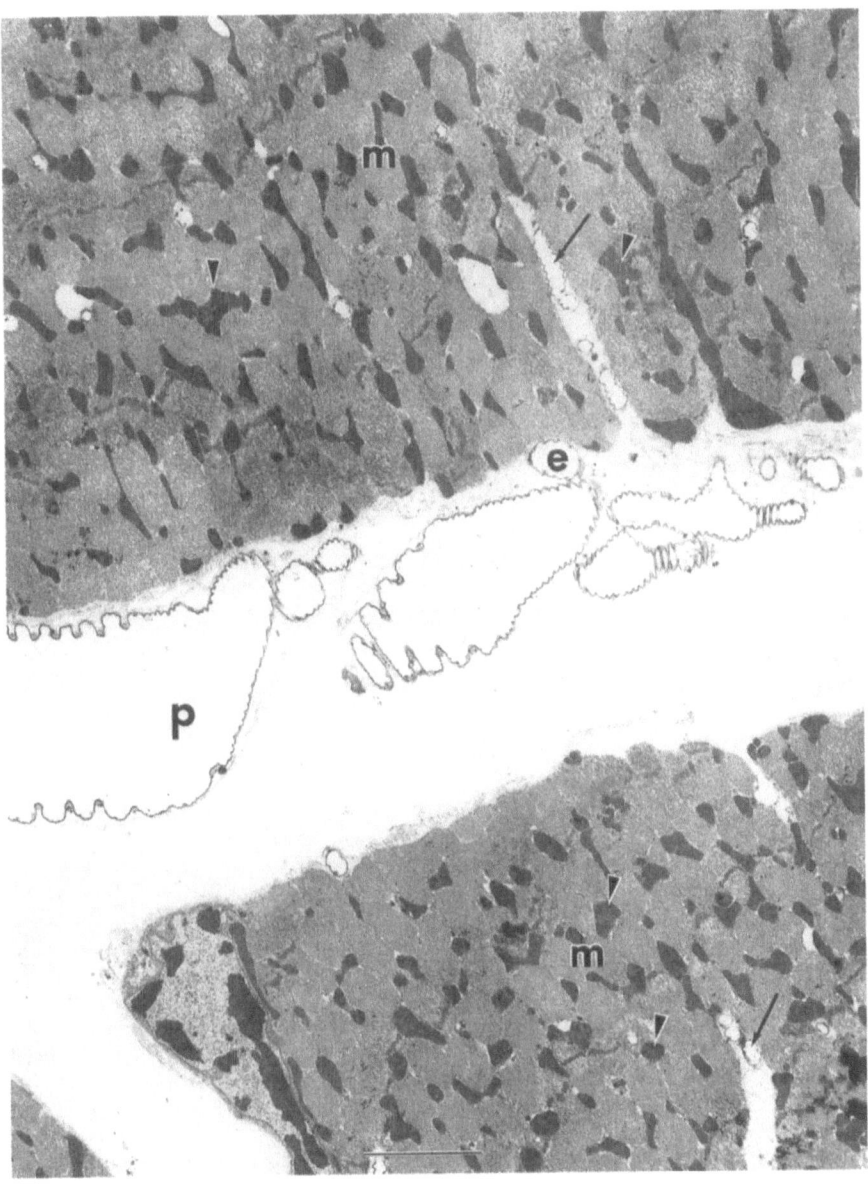

Fig. 76. Flight muscles, *m*, of the desert locust, *Locusta migratoria migratoria* showing secondary trachea, *p*, on the surface of the muscle and tertiary trachea, *e*, indenting the muscle. The terminal tracheal, →, lie in very close proximity to the muscle fibers and the mitochondria, ➤. *Bar* 3 μm

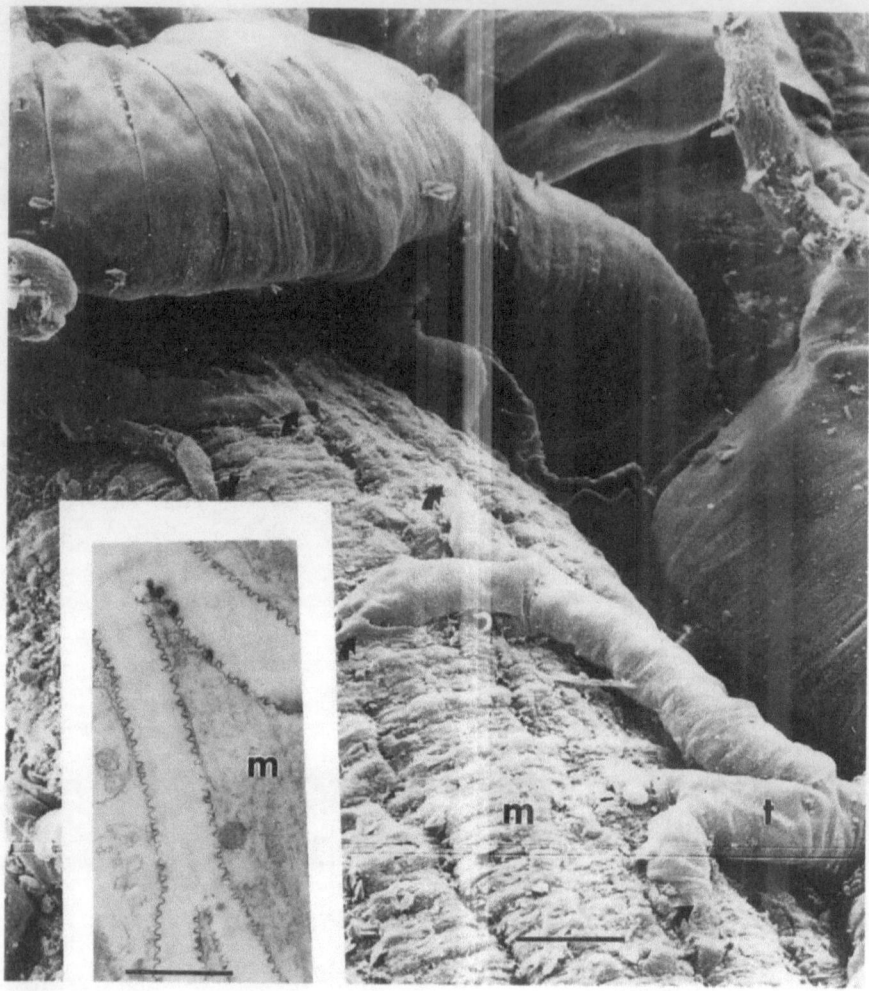

Fig. 77. Secondary trachea, *t*, giving rise to tertiary trachea, ↘, indenting the abdominal muscles of the grasshopper, *Chrotogonus senegalensis. Inset* View of a tertiary trachea in the flight muscle, *m. Bar* 185 μm; *inset* 0.6 μm. (Maina 1989b)

not contain myoglobin (Elder 1975). In some small insects the flight muscles may attain a wing contraction frequency of up to 1000 Hz (Sotavalta 1947), a process amazingly powered entirely by aerobic metabolic pathways. Such high metabolic rates are possible because of the direct delivery of O_2 to the tissues and the remarkably high volume density of the mitochondria (Weis-Fogh 1964a; Elder 1975; Ready 1983). In the process of attaining an optimal respiratory design, by exposing their body tissues to the ambient environment through a myriad of air conduits, the tracheates, and in particular insects, have had to pay a great price. They are particularly vulnerable to invasion by pathogens (e.g., Engelhard et al. 1994) and aerosol-based toxicants which pass the spiracles. It is common knowl-

edge that some of the insecticides which are commercially advertised as to knock
– down – insects – dead do exactly that! While vertebrate muscles can continue
contracting by anaerobic metabolism for a period of time after O_2 supply stops,
since insect muscles lack lactic dehydrogenase, to function, they have to have a
continuous supply of O_2 (Pringle 1983).

Although best studied in insects, tracheal respiration has evolved in animal
groups such as Onychophora (Peripatus), Solifugae, Phalangidae, some Acarina,
Myriapoda, and Chilopoda. The bodies of the tracheates are pervaded by fine air-
filled tubes, the trachea, which are simple noncollapsible hollow airways strength-
ened by endocuticular spiral or annular chitinous thickenings, the taenidia (Figs.
37,78). In some of the most metabolically active tissues, the finest branches of the
trachea (the tracheoles) as they approach the tissue cells may be as narrow as
0.2 μm in diameter and may indent the cells in the manner of a finger poked into
a balloon (Steen 1971; Fig. 74). In the case of the flight muscles, in some insects,
the very narrow unventilated tracheoles, which are about 1 mm long, must supply
O_2 at a tremendous rate of 6.5 mol per m^3 of tissue per s (Weis-Fogh 1964a). In the
highly metabolically active tissue such as the flight muscles, the terminal tracheoles are never more than 0.2 to 0.5 μm from a mitochondrion (Fig. 76) and in
some tissues they may be as close as 0.005 μm (Wigglesworth and Lee 1982; Maina
1989b). The mitochondria cluster around the terminal tracheoles forming what
has been termed mitochondrial continuum (Edwards et al. 1958). In the flight

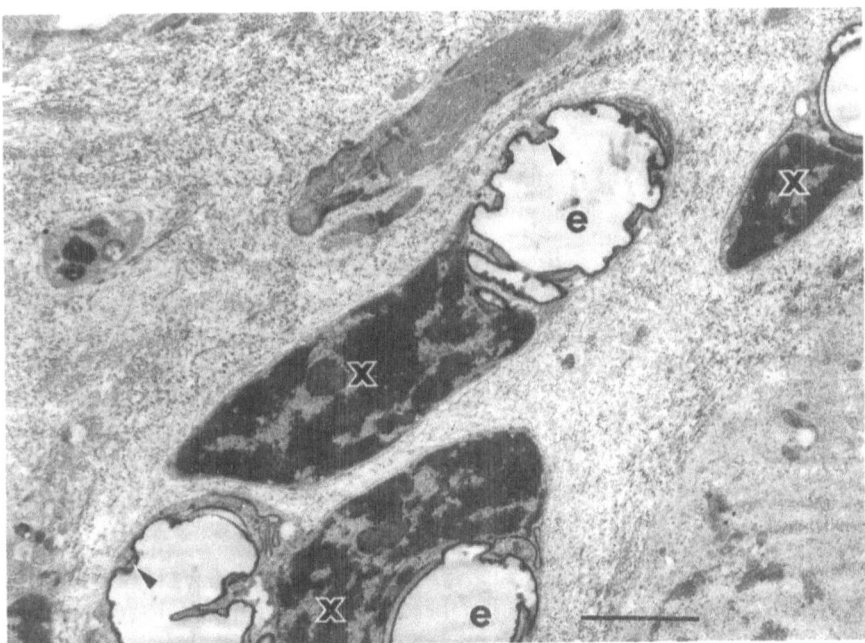

Fig. 78. Trachea, *e*, in the abdominal muscles of the desert locust, *Locusta migratoria
migratoria*, surrounded by tracheoblasts, *x*, cells that are thought to be involved in laying down
trachea. ➤, taenidia. *Bar* 0.3 μm

muscle, the tracheoles may invest single muscle fibrils (Krogh 1941). The tracheoles terminate blindly (Richards and Korda 1950) though possible anastomoses have been reported (Buck 1948). Estimations made on the tracheal system of the giant lepidopteran, *Cossus cossus*, larva (mass 3.4 g, length 60 mm) gave a total cross-sectional surface area of all trachea supplying the tissues of 6.7 mm^2 with an average length of 6 mm: O_2 diffuses at a rate of 0.3 mm^3 s^{-1} at a pressure head of 1.5 kPa, a value which is more than adequate even during muscular exertion (Krogh 1920b). Depending on factors such as age and stage of development, the tracheal system constitutes 5 to 50% of the volume of an insect. In the silkworm, *Bombyx mori*, the tracheoles are 1.5 m long (Buck 1962) and the volume which for a 5.7-g worm is 49 μl g^{-1} makes about 5% of the body volume (Bridges et al. 1980). In the adult cockchafer, *Melolotha*, the trachea constitute a volume of 585 μl g^{-1} (Demoll 1927). The tracheal volume of a 5-g *Cecropia* pupae is about 250 mm^3 (Kanwisher 1966).

The terminal tracheoles are filled with fluid, the degree of filling depending on the level of activity (e.g., Wigglesworth and Lee 1982). The endotracheal fluid is removed osmotically by the increased concentration of the end products of metabolism in the interstitial cell spaces during times of increased metabolic demands and exposure to hypoxia. The air/fluid interface is brought closer to the tissue cells as the interstitial fluid is drawn upstream of the peripheral tracheoles and into the cytoplasm of the surrounding cells (Wigglesworth 1953, 1965). The process shortens the diffusional pathway of the respiratory gases. When the acidic metabolites are eliminated and better aerobic conditions preponderate, the action is reversed. Such a sequence of events is not unique to insects as it is utilized by many other animals to open up capillary beds in specific body tissues during conditions of high O_2 demand. Evidently, the diffusion-based insectan tracheal system evolved entirely for aerial respiration. Because the diffusion of O_2 in water is 10^3 lower than in air, the tracheoles would need to be as small as 12 μm to adequately supply O_2 to tissues respiring at the same rate as the insect flight muscle (Denney 1993). Theoretically, aquatic insects relying on delivery of O_2 entirely by diffusion would need to greatly increase tracheolar density, lower their metabolic rate, and drastically reduce the distance between the body tissue cells and the surrounding water. This would call for extremely thin, minuscule, intensely tracheated bodies. Such requirements may not be compatible with the essential insectan morphology. Those insects which have reverted to living in water have retained air-filled trachea (the gas gills) which essentially act as internal plastrons (Sects. 6.6.1 and 6.12): the O_2 taken up from the surrounding water is transferred along a gas phase within the tracheal system in much the same way as in a surface-dwelling insect.

The insectan tracheal system provides a unique and perhaps the most cost-effective design for supplying body tissue cells with O_2. However, the limitation of diffusion and the large mechanical ventilatory forces which have to be generated to move air at extremely high rates in the countless fine conduits have consigned the insects to small body sizes. This may explain why such an efficient scheme was never adopted in the larger animals. The heaviest living insects are the Goliath beetles (family: Scarbaeidae) of the Equatorial Africa: *Goliathus goliathus* weighs between 70 and 100 g. The hairly-winged beetles of the family Ptiliidae (=

Trichopterygidae) and the battledore-wing fairly flies (parasitic wasps) of the family Myrmaridae which measure only 0.2 mm in length are the smallest insects: the smallest insects are smaller than some of the largest protozoa! The average tracheolar length for optimal diffusion appears to be 5 to 10 mm and the minimum diameter 0.2 μm (Krogh 1920a,b; Weis-Fogh 1964a). The tracheates which utilize diffusion as the main mode of gas exchange include the Onychophora (Peripatus), the tracheate Arachnoidea, Myriapoda, and Chilopoda, almost all terrestrial insect larvae, all pupae, and most of the small imagines. This is made possible by the relatively fast diffusion of O_2 in air compared with that in water. The largest ever known insect was the dragonfly-like *Meganeura* of the Carboniferous, which reached a length of 60 cm and was 3 cm in width (Krogh 1941). It is well known that the levels of atmospheric O_2 have greatly fluctuated over geological times (Sect. 1.11). A greater PO_2 in air as occurred in the Carboniferous period (Graham et al. 1995; Fig. 9) may have allowed the development of the giant insects. Presently, the largest extant insects are the tropical beetles which reach a length of 15 cm. Flying insects range in body mass from 1 μg to 20 g (Norberg 1990). The stick insects best demonstrate the compromises between size and shape which have occurred in insects, features which have been modulated for optimal tracheolar ventilation and diffusion. Some of the longest insects in the world are the tropical stick insect, *Phamacia serratipes*, and the Central and South American dragonfly, *Megaloprepes caeruleata*, which have very narrow bodies which are respectively 33 and 12 cm long. While a housefly, which weighs about 15 to 20 mg, does not need to ventilate the tracheal system, a bee, which is more energetic and weighs about 100 mg, does so regularly. In insects such as locusts, dragonflies, and cockroaches, at rest, well-synchronized abdominal and, to a smaller extent, thoracic ventilation occurs (Brocher 1931). Although during rest no ventilatory movements take place in the cockroaches, *Peripaneta* and *Blatella*, during flight when O_2 consumption increases 10 to 100 times, wing beats compress the thorax ventilating the trachea and the air sacs (Brocher 1920; Portier 1933). Ventilation in flight may be aided by direct inflow of ambient air at the ventral surface and the slightly reduced pressure over the abdominal spiracles due to a Bernoulli-Venturi effect. During steady flight in the desert locust, about $320 l kg^{-1} h^{-1}$ of air with an average tidal volume of $167 cm^3$ and frequencies of 30 to 60 times min^{-1} is ventilated into the tracheal system by abdominal and thoracic pumping. The intratracheal pressure increases to 0.9 to 3.3 kPa at the peak of an abdominal contraction (Miller 1960; Weis-Fogh 1967). The giant beetle, *Petrognatha gigas*, has a ventilatory rate of about $2000 l kg^{-1} h^{-1}$ (Miller 1966).

Through synchronized action of the spiracles, the trachea are ventilated continuously and unidirectionally, particularly among the Orthoptera (e.g., Fraenkel 1932; Weis-Fogh 1964a, 1967). In the honeybee, the flow is unidirectional during flight (Bailey 1954) and in *Sphodromantis*, 95% of the inhaled air passes unidirectionally while only 5% passes tidally (Miller 1974). Among the cockroaches, *Periplaneta* and *Blatella*, tidal ventilation only occurs during stress while in other roaches, *Byrsotria*, *Blaberus*, and *Nyctobra*, anteroposterior ventilation occurs during rest (Buck 1962). The unidirectional and continuous ventilation, as occurs in the parabronchial bird lung (e.g., Scheid 1979), minimizes or abolishes dead space air, ascertaining that the gas exchange site is supplied with air at the highest possible PO_2. Abdominal pumping is inadequate in supplying O_2 to the long muscles of the legs in some large insects. In the grasshopper, the concentration of O_2 in the tibial tracheae is fairly high (16%) in the resting state but drops to 5% during physical exertion (Krogh 1913). Special spiracles have developed on the legs of the harvestmen (Opiliones) apparently to overcome the diffusive and convective limitations (Hansen 1893). The smallest tracheoles in insects which measure about 0.2 μm are close to the mean free path (MFP) of O_2 molecules in air (the average distance a molecule travels in air before colliding with another) which is about 0.008 μm (Pickard 1974). Below the MFP, the effective diffusion coefficient is reduced, lowering the rate at which O_2 is delivered to the tissues. The minimum tracheolar diameter in insects appears to have been set by the MFP of molecules in air. In the most energetic species, this parameter seems to have been optimized in the most highly metabolically active tissues such as the flight muscles.

The development of the tracheal system appears to be partly determined by certain intrinsic factors in the target tissues (Locke 1958a,b,c) while the actual distribution to specific organs is determined by the local aerobic conditions (Edwards et al. 1958). In much the same way as occurs between the capillarization of the tissues and metabolic activity in vertebrates, in insects, tracheolar density is dependent on the metabolic activities and PO_2 levels in particular organs and parts of the body (Edwards et al. 1958; Wigglesworth 1965; Steen 1971). The trachea are particularly well developed in the legs, which are actively used for web-monitoring activity in the spiders of the family Uloboridae (Opell 1987). In larval meal-worms, *Tenebrio molitor*, hypoxia influences tracheal growth. Development at an ambient PO_2 of about 10 kPa leads to wider trachea (London 1989). In the wing muscle of the locust, between 10^{-1} and 10^{-3} (volume of the trachea per volume of muscle) is taken up by the tracheal system (Weis-Fogh 1967). The entire respiratory system in insects may form as much as 50% of the entire body volume (Steen 1971). In the small and relatively inactive insects and arachnids, the tracheal system may be simple but in larger and more energetic species (e.g., wasps and bees), it may be complex with the system comprising an intricate maze of longitudinal and transverse branches (Fig. 37) connected to air sacs (Fig. 38). The air sacs increase the tidal volume by as much as 70% of the total air capacity affordable by the trachea alone (Bursell 1970) and reduce the longitudinal diffusion gradient for O_2 along the gas exchange pathway. They are well developed in Diptera and Hymenoptera but are absent in the subclass Apterygota. In cicada, *Fidicina monnifera*, the air sacs together with the tracheal system constitute 45% of the body volume (Bartholomew and Barnhart 1984).

6.7 The Convective Type Gas Exchangers

6.7.1 Ventilatory Mechanisms and Organization of the Gas Exchangers

The ventilatory lungs have evolved only in the vertebrates. They were a major factor which provided the means for realization of large and complex body sizes and forms as well as high metabolic lifestyles. Different ventilatory processes have evolved in the air-breathing vertebrates. The bucco-pharyngeal pump, where air is literally swallowed, occurs in the amphibians and the dipnoans (Brainerd et al. 1993). The mechanism appears to have arisen as a compromise between respiration and feeding. The energy which operates the force pump arises from contraction of the muscles of the mouth (De Jong and Gans 1969; West and Jones 1975; Liem 1987a). The amount of air which can be transferred during a single breath is a function of the pressure differential between the lung and the buccal cavity and the potential change in the volume of the buccal cavity (Liem 1987a). In the aquatic habitat, during feeding, pressure changes far exceed those recorded during air ventilation (Lauder 1980; Bemis and Lauder 1986). The buccal force pump operates below its maximum potential during air breathing and is hence an overdesign for the role it plays in the medium. The tidal volume acquired by the buccal force pump is limited. The buccal force pump is ineffective for filling long narrow lungs like those of snakes (Gans 1971; Guimond and Hutchison 1976) and is a rather inefficient method which cannot support high levels of metabolism. The ventilatory inefficiency of the buccal force pump in the terrestrial settings is thought to have constituted a major obstacle in the evolutionary progression of the amphibians (Gans 1970; Liem 1987a). Suctional (aspirational) breathing occurs in most reptiles and all mammals and birds (Bainerd 1994), animals which operate at a higher level of metabolism and have long complex lungs which cannot be effectively serviced by a buccal force pump. The evolution of suctional breathing dissociated the feeding apparatus from the breathing one. It comprised a milestone in the development of more complex, efficiently ventilated gas exchangers which were necessary on transition to land and the resultant higher metabolic requirements. While suctional breathing occurs in various reptiles, positive-pressure inflation through the buccal force pump preponderates in the taxon (Gans and Hughes 1967; Gans 1971): breathing is exclusively effected by movement of the rib cage and complex changes in the volume of the visceral cavity, as a diaphragm is lacking (e.g., Perry and Duncker 1980; Gaunt and Gans 1969; Gans and Clark 1976, 1978). Modem amphibians have vestigial ribs. It is, however conjectured that the larger now extinct forms had functional ribs which they utilized to draw air into their lungs. In amphibians and reptiles, postural changes and presumably contraction of the pulmonary smooth muscles (Figs. 79,80; Stark-Vancs et al. 1984; Maina 1989b) and elastic tissue (Fig. 80) as well as hydrostatic forces (when the animal is immersed in water) may aid in expiration or air. The smooth muscles which line the central cavity of the lung of the tegu lizard, *Tupinambis nigropunctus* (Hlastala et al. 1985), as well as snake lungs (Maina 1989e; Fig. 80) have been associated with changing the shape of the lung and assisting in the convective mixing of the intrapulmonary air. In birds, the ventilation of the lung is unique among vertebrates. The virtually rigid lungs are ventilated by a synchronized activity of the air sacs which are interspersed in the abdominal cavity, pneumatizing some of the adjacent bones. The pressure changes in the air sacs are generated by contraction of muscles which attach to the thoracic wall adjusting the thoracoabdominal space. A partition between the thorax and the abdomen is lacking: the liver and not the lung (as is the case in mammals) surrounds the heart. The oblique and the horizontal septa (see King

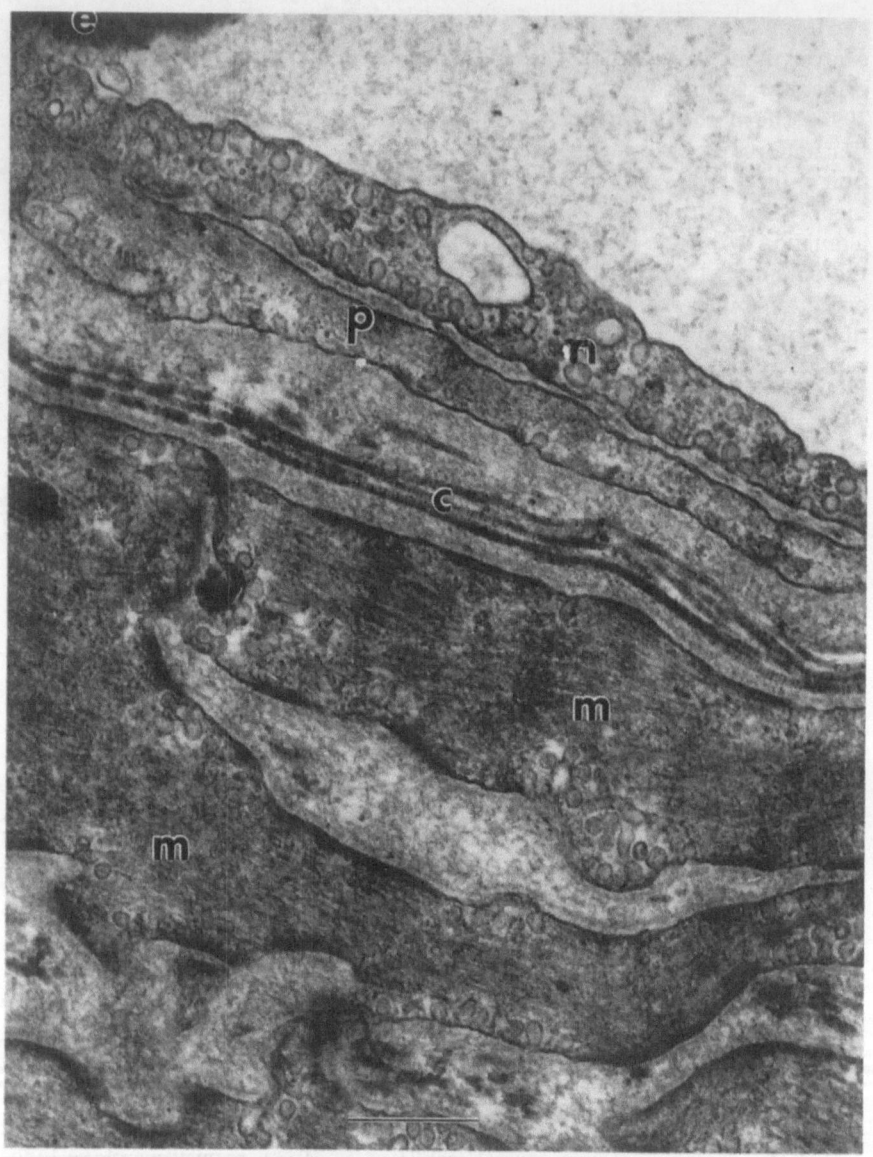

Fig. 79. Smooth muscles, *m*, and collagen, *c*, in the lung of the tree frog, *Chiromantis petersi*. *p* pericyte; *n* capillary endothelial cell; *e* erythrocyte. Contraction of the smooth muscle may assist in the expiration. *Bar* 1 μm. (Maina 1989d)

and McLelland 1975) are thought to be homologous to the mammalian diaphragm but fall far too short of anatomically dividing the coelomic cavity. The mammalian lung can be divided in two zones: the nonparenchyma (the conducting zone) of branching air conduits (the bronchi) and the parenchyma (the respiratory zone), made up of millions of alveoli. Together with the intercostal

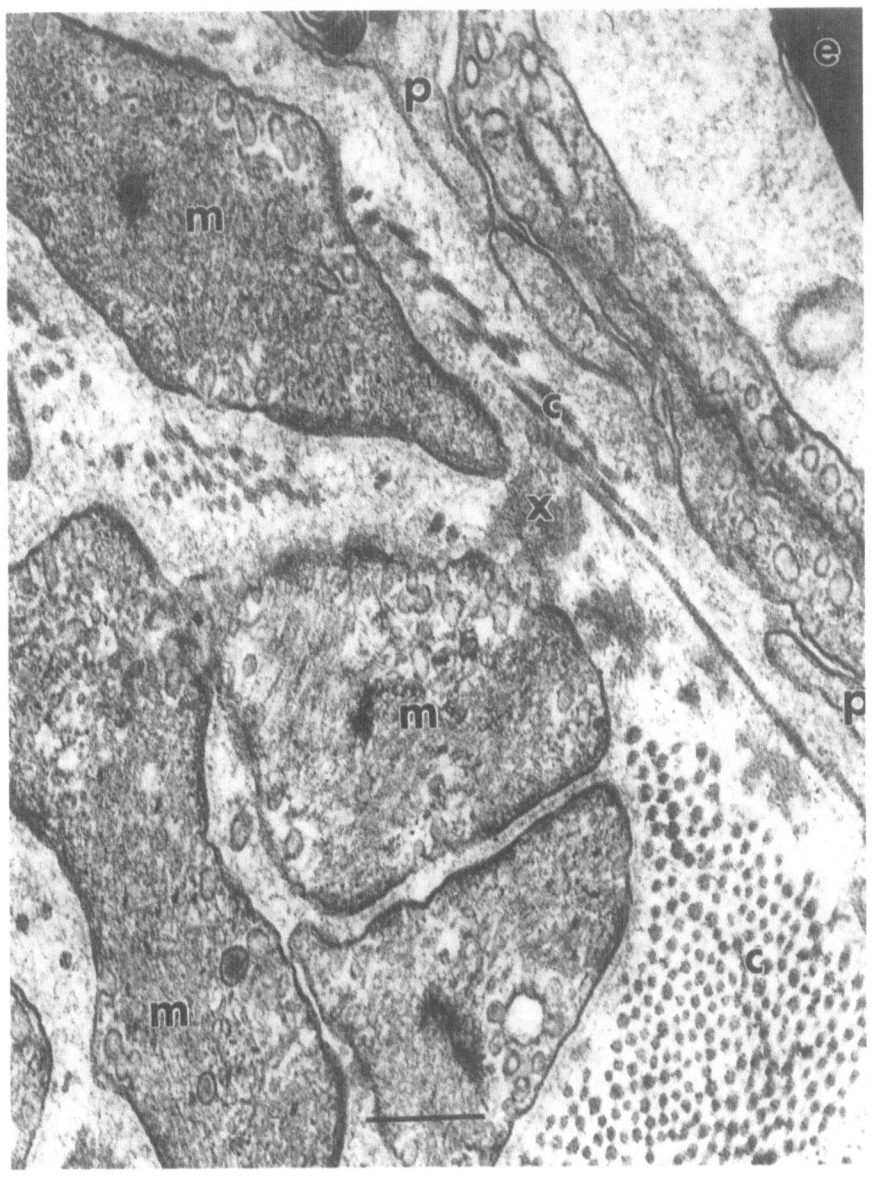

Fig. 80. Smooth muscles, *m*, in the lung of the black mamba, *Dendroapis polylepis. x* Elastic tissue; *c* collagen; *p* pericyte; *e* erythrocyte in a blood capillary. The smooth muscles may assist in intrapulmonary movement of air. *Bar* 1 μm. (Maina 1989e)

muscles, the diaphragm is an important respiratory muscle. By changing the volume of the thoracic cavity, air is moved into and out of the lung. The complete separation of the coelomic cavity into the thoracic and the abdominal cavities by the diaphragm was a key evolutionary development for efficient respiration in mammals. The evolved ventilatory mechanisms show varied levels of refinement which correspond with the phylogenetic levels of progression. The lungs in amphibians are filled with air under pressure from the bucco-pharyngeal cavity. In some reptiles (e.g., *Chamaeleo*), this process has been conserved and is used only in emergency cases to inflate the lungs. The various forms of suctional breathing which have developed in both terrestrial and aquatic animals allow greater flexibility in the tidal volume and have facilitated the development of the respiratory organs independent of the ventilatory mechanisms. Gas exchange has been totally disengaged from feeding. In general, those vertebrates which are adapted to respiration in xeric environments have the more specialized lungs (Fig. 81) and are capable of the highest sustained rate of gas exchange. This supports a high level of metabolic activity, enabling a greater degree of organizational and ecological progress. Such animals are found in only a few taxa which include insects,

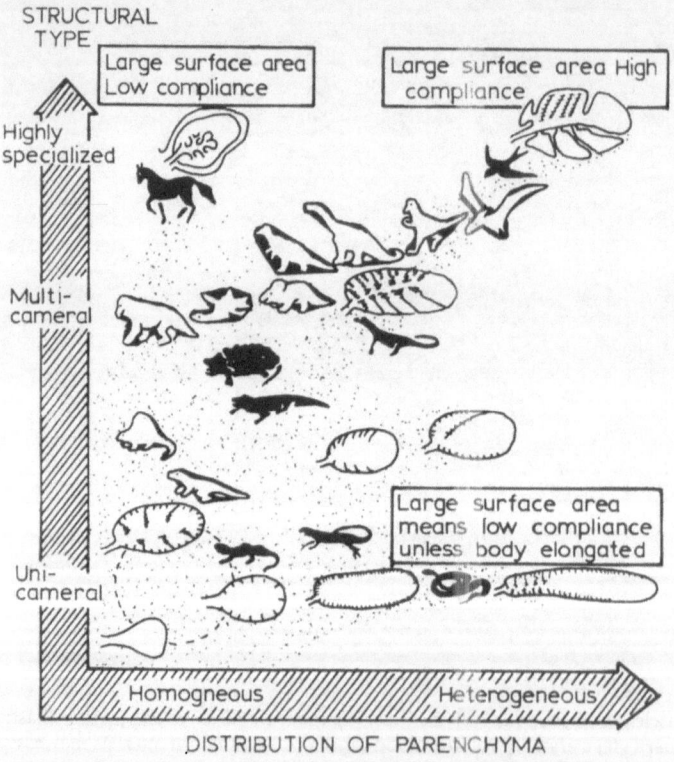

Fig. 81. Scheme of the structural complexities of the vertebrate lungs showing that a parameter such as an extensive respiratory surface area can be attained both in homogenous and heterogenous lungs through different modifications. For example, birds have isolated the ventilatory and the gas exchange parts of the lung and through intense subdivision of the lung achieved an extensive respiratory surface area in a relatively small lung. Pulmonary design corresponds with the metabolic needs and phylogenetic statuses of individual animal groups. Animals shown *on a black background* are now extinct. (Perry 1989)

reptiles, birds, and mammals. In these animals the integument is almost totally impermeable to water, O_2, and CO_2. Among vertebrates, the simplest forms of lungs are the smooth-walled saccular types, e.g., in *Proteus* (Hughes 1970), which are poorly vascularized. Though such lungs develop at the early larval stages, they play very little, if any, role in respiration until the midlarval stage (Goniakowska-Witalinska 1995). In the aquatic or amphibious groups, the lungs play diverse roles such as hydrostatic (Milsom and Johansen 1975; Pohunkova and Hughes 1985a), sexual display, and sensory perception (Duykers and Percy 1978; Schmidt 1982; Ehret et al. 1990). Internal subdivision of the lung imparts a greater respiratory surface area (Maina et al. 1989b; Fig. 82), increases the volume of blood in the lung, provides better exposure of blood to air (Hughes 1978), and ensures mechanical integrity of the organ. In most lungs, the exposure of the pulmonary capillary blood to air is promoted by a construction which entails not only increase in vascularity but also distension of the blood vessels over the epithelial surface, e.g., in the pneumonate gastropods (Maina 1989c; Figs. 59,60) and the mammalian lung (Fig. 87). In the double capillary system of the amphibians (Maina 1989d; Fig. 84) and most reptilian lungs (Maina 1989e; Figs. 83,85), a parallel row of blood capillaries which are exposed to air on only one side occur. A single capillary plan, where a sheet of blood capillaries which are supported by

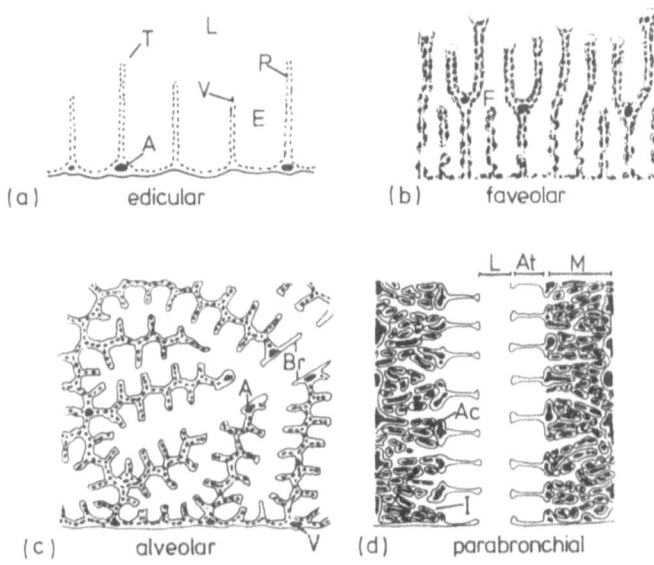

Fig. 82a–d. Subdivisions of the parenchyma of the vertebrate lungs. The simple unicameral lungs (a) have large air spaces, the ediculae; the faveolar lungs have layered air spaces (b), the faveoli; the highly specialized mammalian (c) and avian lungs (d), respectively, have alveoli and air capillaries as the terminal gas exchange components. Evidently, a basic design appears to have given rise to all forms of lungs, the particular design and complexity depending on the level of development and the metabolic needs of different animals. *T* Trabeculae; *A* artery; *V* vein; *E* ediculae; *L* lumen; *P* pneumocytes; *F* faveolus; *A* alveolus; *Br* bronchus; *At* atrium; *M* exchange tissue; *Ac* air capillary; *I* infundibulum. (Perry 1989)

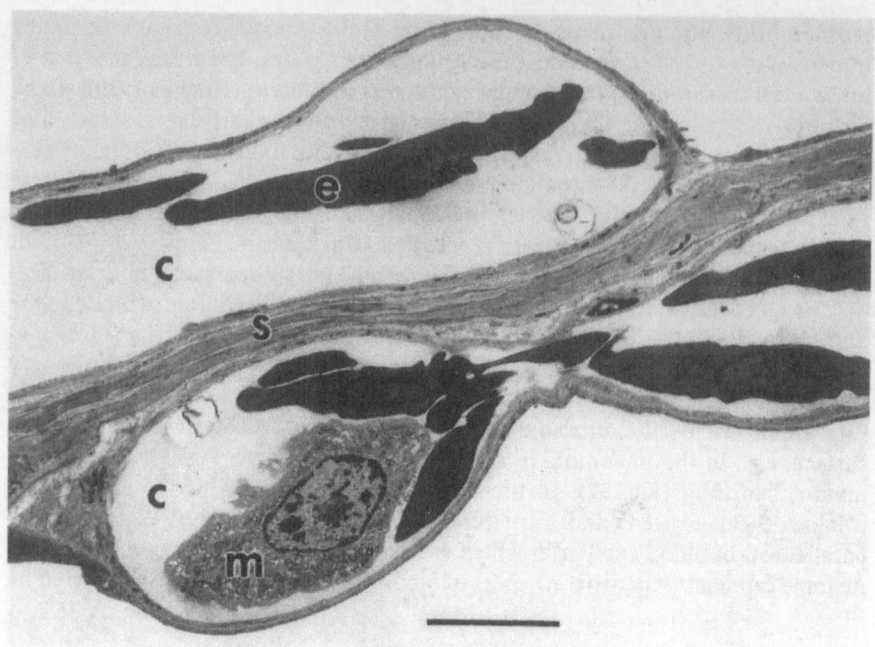

Fig. 83. Double capillary arrangement in the snake lung, *Dendroapis polylepis*. This design is limiting in that only one side of the blood capillary can be used for gas exchange. An attempt is made to increase the available surface area by the blood capillaries protruding over the surface of the lung. *c* Blood capillary; *e* erythrocytes; *s* interfaveolar septum; *m* white blood cell. *Bar* 5 μm. (Maina 1989e)

a thin septum, exists in the highly refined mammalian lungs (e.g., Alcorn et al. 1980; Burri 1984a; Figs. 86,87). In such cases, the capillary blood is essentially suspended in a three-dimensional space. The arrangement provides better exposure of blood to air compared with the double capillary one where the capillary loading, the ratio between the blood volume and the surface area available for gas exchange, is very high (Perry 1983). A notable exception to the general plan in the mammalian lung is that found in the rather placid herbivorous marine mammals, the sirenians, i.e., manatees and dugongs. Abundant connective tissue in the interalveolar septa, large alveoli (Tenney and Remmers 1963), and a well-developed double capillary system (e.g., Belanger 1940; Wislocki and Belanger 1940) occur. The pattern is similar to that found during the embryonic stage of development of the mammalian lung (e.g., Pinkerton et al. 1982). The preponderance of collagen and other supporting tissue elements in the lung (Laurent 1986) may provide the biomechanical support necessary to overcome the hydrostatic forces during deep dives. The exposure of the pulmonary capillary blood to air in the lungs of birds occurs in form of a diffuse arrangement between the air and blood capillaries (Maina 1982a, 1988a; Figs. 88,89) which intimately interdigitate with each other maximizing the respiratory surface area (Dubach 1981; Maina et

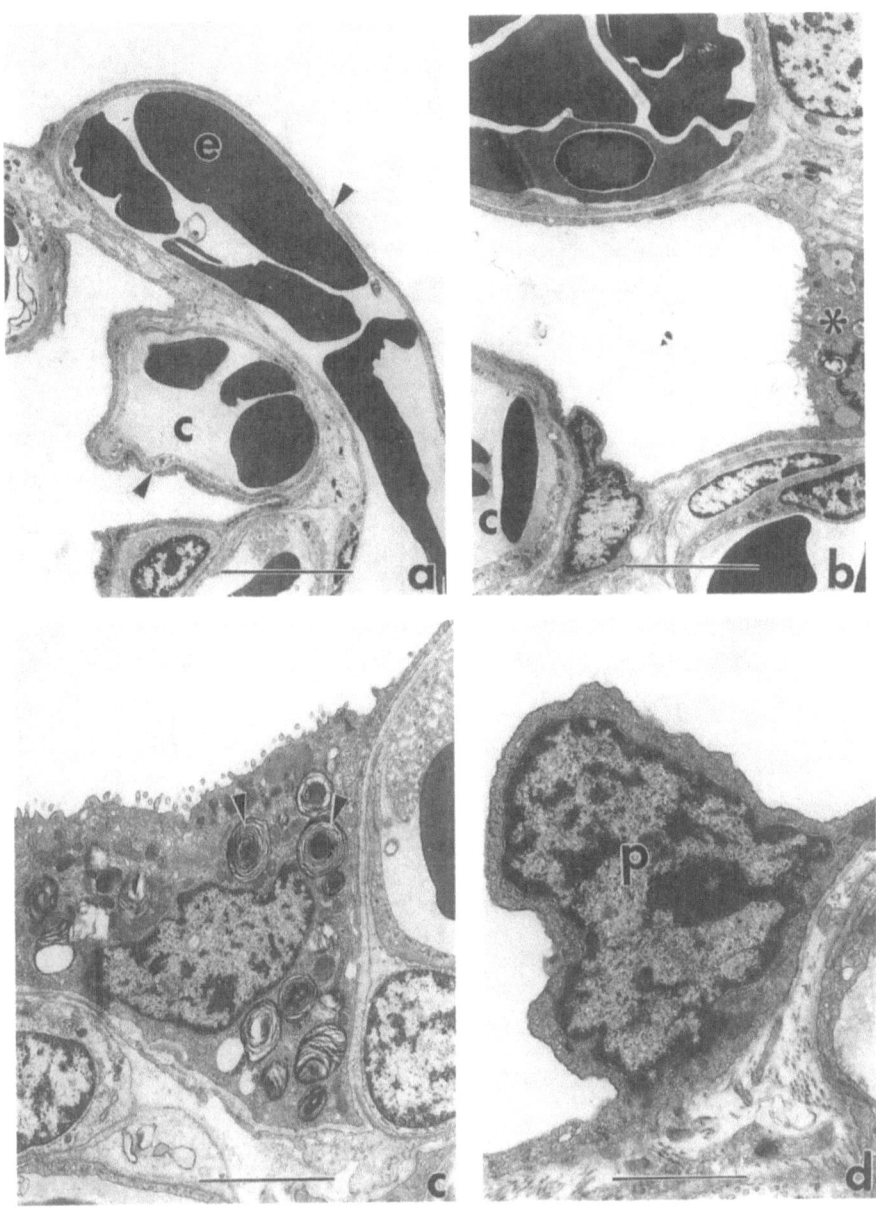

Fig. 84a–d. Structrue of the lung of the tree frog, *Chiromantis petersi*. **a** Blood capillaries, *c*, which contain erythrocytes, *e*, exposed to air across a thin blood-gas barrier, ➤. **b** Blood capillaries, *c*, bulging into an air space; *, granular pneumocyte. **c** Granular pneumocyte containing lamellated osmiophilic bodies, ➤. **d** *p*, nucleus of a Type I pneumocyte overlying a blood capillary. **a** *Bar* 8 μm; **b** 4 μm; **c** 3 μm; **d** 0.2 μm. (Maina 1989d)

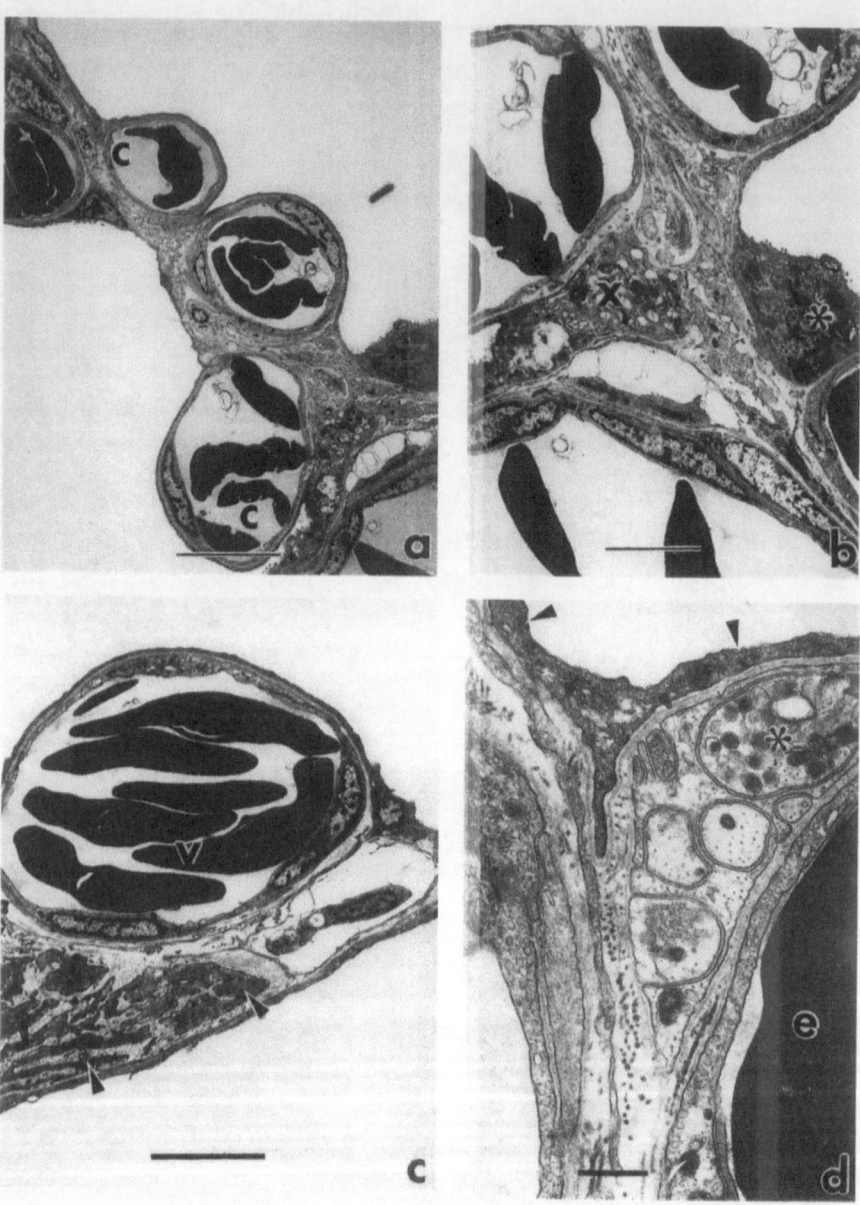

Fig. 85a–d. Structure of the lung of the black mamba, *Dendroapis polylepis*, showing blood capillaries bulging into the air spaces. Reptilian lungs generally manifest a better exposure of blood to air than the amphibian ones. **a** *c* blood capillaries. **b** * granular pneumocyte (type II cell) located between blood capillaries; *x* interstitial macrophange. **c** *v* a blood capillary bulging into an air space; ➤ smooth muscle. **d** * nonmyelinated axons in the lung; ➤ squamous type I cell; *e* erythrocytes. **a Bar** 10 μm; **b** 5 μm; **c** 5 μm; **d** 1 μm. (Maina 1989e)

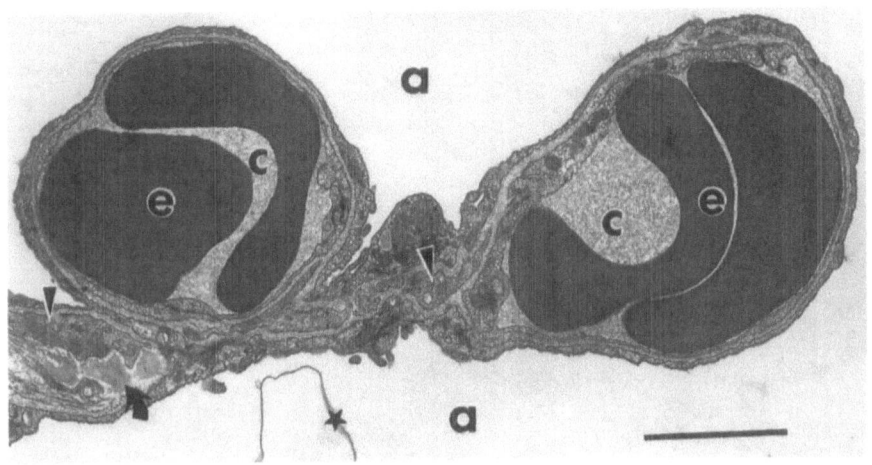

Fig. 86. Lung of a bat, *Epomophorus wahlbergi*, showing alveoli, *a*, and blood capillaries, *c*, containing erythrocytes, *e*. The interalveolar septum contains smooth muscle, ➤, and elastic tissue elements, ➘. The pulmonary capillary blood in the mammalian lung is better exposed to air than in amphibian (cf. Fig. 84) and reptilian (Fig. 85) lungs but thick and thin parts of the blood-gas barrier are evident. ★, detached surfactant lining. *Bar* 1.8 μm. (Maina et al. 1982a)

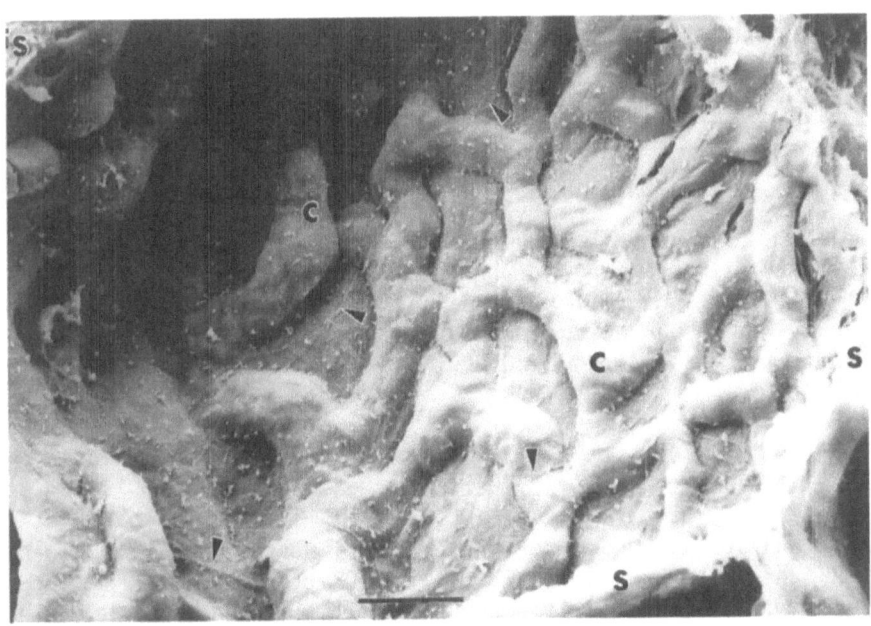

Fig. 87. An alveolus of the lung of the lesser bushbaby, *Galago senegalenis*, showing the blood capillaries, *c*, bulging into the air space. ➤ junctions of type I cells; *s* interalveolar septum. *Bar* 120 μm. (Maina 1990c)

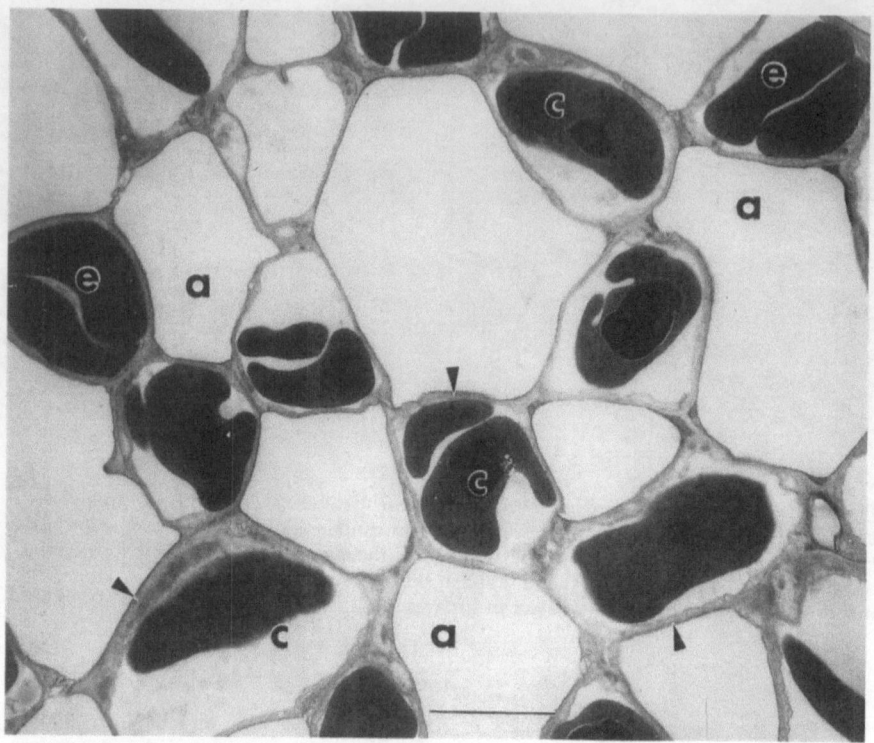

Fig. 88. Exchange tissue of the lung of the house sparrow, *Passer domesticus*, showing the air capillaries, *a*, and blood capillaries, *c*, which interdigitate very profusely. The blood capillaries are exposed to air on all sides giving rise to an extensive respiratory surface area per unit volume of the exchange tissue. ➤ blood-gas barrier; *e* erythrocytes. Bar 3 μm

al. 1989a) and providing an extremely thin blood-gas barrier (Figs. 40a,90). These features generate a remarkably high pulmonary diffusing capacity for O_2 (Maina 1989a, 1993; Maina et al. 1989a). In the lungfish, *Lepidosiren paradoxa*, the pulmonary capillary blood volume constitutes 3.5% of the total lung volume, in the rat lung the value is 14%, and in the bird lung 25% (Maina et al. 1989a).

6.7.2 The Amphibian Lung

The actual evolutionary origin of the amphibians is not well known (e.g., Szarski 1962; Schaeffer 1965b; Løvtrup 1977; Milner 1988). The osteolepiform fish (which are thought to have evolved a primal lung) or the lungfishes (Dipnoi) are conjectured to have given rise to the tetrapods (Romer 1946, 1967, 1972; Pough et al. 1989; Meyer and Dolven 1992). The transition to land by the vertebrates was inaugurated by the amphibians in the Devonian and concluded by the reptiles in

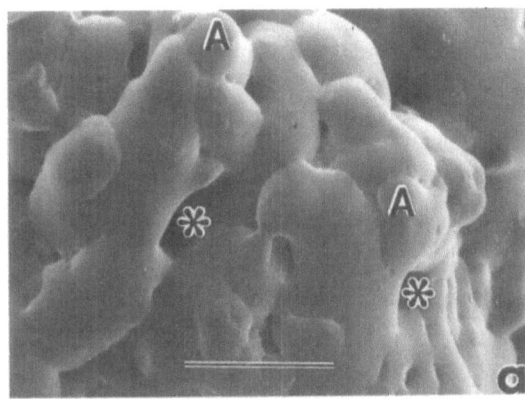

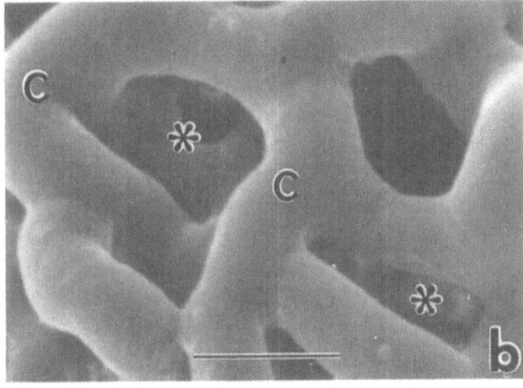

Fig. 89. a Cast preparations of the air capillaries, *A*, and **b** blood capillaries, *C*, of the lung of the domestic fowl, *Gallus domesticus*. **a** ✳ spaces occupied by blood vessels. **b** ✳ spaces occupied by the air capillaries. The two gas exchange units intertwine intimately increasing the surface area available for gas exchange. **a** *Bar* 10 μm; **b** 3 μm. (Maina 1982)

the Paleozoic era. For a good part of their evolution (about 70 million years), the early amphibians were strictly aquatic bimodal breathers (Thompson 1991). The contemporary amphibians occupy a focal point in the study of some fundamental biological processes such as the realization of air breathing and terrestrial colonization (Szarslo 1977; Smits and Flanagin 1994). Dual subsistence in water and land has obligated development of unique physiological and morphological respiratory adaptations in the taxon (Lenfant and Johansen 1967; Duellman and Trueb 1986; Jackson 1987; Burggren 1989). The unique nature of development (metamorphosis from aquatic to air breathing), the diversity of the habitats the taxon occupies, and the multifunctional nature of the lungs (respiration, vocalization, buoyancy control, and defense) explain both the heterogeneity of the lung morphology and the multiple pathways utilized for gas exchange (Guimond and Hutchison 1972, 1973a,b; Rahn and Howell 1976; Burggren and Wood 1981). Amphibians largely live in water and humidic habitats but a few species have adapted to the highly desiccating deserts (e.g., McClanahan et al. 1994). Terrestrial anurans such as *Chiromantis xerampelina*, which have developed an impermeable skin (Stinner and Shoemaker 1987) and ureotelism (Shoemaker et al. 1972), have acquired remarkable tolerance to desiccation. The two species can withstand a water loss in excess of 60% of their body mass (Loveridge 1970). The highly xeric African tree frog, *Chiromantis petersi*, leads a characteristically unamphibian lifestyle: it prefers direct solar insolation and temperatures of 40 to 42 °C (Loveridge 1970, 1976). The need to balance water conservation with gas exchange may explain why there are no large existing amphibians.

During the larval stages, amphibians have transient external and internal gills. Subsequently, the lungs and the bucco-pharngeal cavity become highly vascularized and assume an important respiratory role in air. The skin is a dual respiratory organ, exchanging gases in both water and air. The neotenic amphibians retain the gills throughout life. In the egg, larval caecilians use gills for respiration. The gill are lost soon after hatching (Welsch 1981). Using gills, the larvae of the viviparous caecilian, *Typhlonectes*, exchange gases and nutrients with the lining of the oviduct (Wake 1977): the gills are lost before birth. The neotenic larva of *Amblyostoma* (Axolotl) has perennial external gills. The flow of blood to the various respiratory organs is controlled mainly by ventilatory rate and the levels of ambient O_2 (Wood and Glass 1991). Hypoxia affects the flow of blood to the gills, skin, and the lungs (Shelton et al. 1986; Malvin and Heisler 1988). A drop in the PO_2 in the lung increases the perfusion of the skin (Boutilier et al. 1986) and vice versa. In aerated water, the skin receives 20% of the pulmocutaneous blood flow, the amount decreasing to an insignificant amount when the water is hypoxic (Wood and Glass 1991). In the lungless salamanders (Plethodontidae), gas exchange which occurs entirely across the skin and the buccal cavity (Gatz et al. 1974; Gratz et al. 1974) is diffusion-limited (Piiper et al. 1976). Such animals have acquired long, cylindrical bodies to secure a greater surface-to-volume ratio, a feature which promotes gas exchange across the skin. The lungless salamanders normally live in cold, well-aerated waters. In the aquatic amphibians, locomotion generates passive ventilation of the skin, a process which can increase O_2 uptake markedly (Feder 1985; Full 1985).

Living amphibians exist in three orders, namely the Gymnophiona (= Apoda = caecilians), Salentia (= Anura), and Caudata (= Urodela). Biologically, the caecilians are the least known group (Stiffler et al. 1990; Smit and Flanagin 1994). They are an elusive, vermiform tropical aquatic, semiterrestrial, or subterranean group (Nieden 1913; Taylor 1968; Renous and Gasc 1989). Evolutionarily, the caecilians are a monophyletic distant group (Milner 1988): they have been isolated from the other living amphibian orders for at least 70 million years. As characteristic of other similar animals with thin cylindrical bodies, e.g., the snakes (Ophidia) (e.g., George and Shah 1956), the caecilians possess long, tubular lungs (Fig. 99; Wake 1974; Maina and Maloiy 1988; Renous and Gasc 1989). In some species, the left lung is remarkably reduced or totally missing (e.g., Wiedersheim 1879; Pattle et al. 1977; Maina and Maloiy 1988; Burggren 1989; Fig. 98). In the aquatic *Typhlonectes compressicauda* (Sawaya 1947), however, as many as three lungs occurs (Toews and MacIntyre 1978): the anterior (tracheal) lung is located between the buccal cavity and the heart, while the other two run from the buccal cavity to the cloaca (Fuhrmann 1914; Toews and MacIntyre 1978). The lungs of the caecilians are internally subdivided to form air spaces by a single row of septa which are mechanically reinforced by two diametrically opposite trabeculae (Wiedersheim 1879; Wake 1974; Welsch 1981; Maina and Maloiy 1988; Fig. 99). Marcus (1928, 1937) observed that the structure of the caecilian lung is particularly primitive. The organization of the lung compares with that of the almost limbless, large aquatic salamanders like *Amphiuma* and *Siren* (Bell and Stark-Vancs 1983; Martin and Hutchison 1979). A relatively elaborate lung as in *Boulengerula* (Maina and Maloiy 1988) and *Geotrypetes* (Bennett and Wake 1974) may enable the caecilians to procure adequate amounts of O_2 from the hypoxic fossorial environment in which they live. Adaptively, the hematocrit and the hemoglobin concentration respectively, of the aquatic *Typhlonectes compressicauda* (38% and 11.3 g Hb per 100 cm^3 blood; Toews and MacIntyre 1977, 1978) and the terrestrial *Boulengerula taitanus* (Wood et al. 1975; 40% and 10.3 g Hb per 100 cm^3 blood) are some of the highest values reported in amphibians. The large blood volume (24 to 26% of the body mass; Toews and MacIntyre 1978) and low P_{50} of the blood hemoglobin, e.g., in *Typhlonectes* 3 kPa (Toews and MacIntyre 1978) and 3.7 kPa in *Boulengerula* (Wood et al. 1975), provide an efficient mechanism for uptake and storage of O_2 in the blood. While the caecilians have a lower resting rate of metabolism than the anurans and the urodeles, the aerobic capacity during exercise exceeds that of the two other groups (Smits and Flanagin 1994). The elongated nature of the lung of the caecilians may introduce ventilatory limitations during locomotion as a result of compression of the lung by the trunk muscles. A temporal dissociation between breathing and locomotion has been reported in the running lizard (Carrier 1991), an animal with a similarly long lung and general body form. In many ways, the development of the gas exchangers in the amphibians identifies with that of the lungfishes (Dipnoi) (Wassnetzov 1932; see Fig. 64a,b). Physiological adaptations such as high hemoglobin concentration, small numerous erythrocytes, and large blood volume have been reported in the caecilians which live in hypoxic environments (Wood et al. 1975; Toews and MacIntyre 1978). At 25 °C, O_2 uptake in the fossorial caecilian, *Boulengerula taitanus*, which lives in hypoxic-hypercarbic habitats, is equal to that of other amphibians (Hutchison 1968; Wood et al. 1975). The Caudata, which are mostly aquatic, e.g., the newts, have poorly vascularized lungs with a smooth internal surface (Hightower et al. 1975; Meban 1977; Goniakowska-Witalinska 1980a,b). Such animals mainly use the skin for gas exchange (Noble 1925,1929). The characteristically low metabolic rate newt, *Triturus alpestris*, has smooth-surfaced lungs (e.g., Claussen and Hue 1987) with 569 capillary meshes cm^{-2} while the metabolically more active tree frog, *Hyla arborea* (Goniakowska-Witalinska 1986), has more elaborate lungs with 652 capillary meshes per cm^2 (Czopek 1965). Plethodontidae constitute the largest family among the Caudata (Feder 1976; Ruben et al. 1993; Wake and Marks 1993). They acquire all their O_2 needs from the cold, well-oxygenated water in which they live across their highly vascularized skin. The length of the skin capillaries constitutes 90% of all blood vessels associated with the respiratory surfaces, with the other 10% being in the buccal cavity (Czopek 1965). The epithelial lining of the buccal cavity is very thin (Noble 1931; Czopek 1965). In caudates such as *Salamandra*, *Amphiuma*,

Megalobatrachus, and *Siren*, species which predominantly utilize the lung for gas exchange, the internal surface of the lung is well subdivided (e.g., Goniakowska-Witalinska 1978; Meban 1979; Hashimoto et al. 1983; Matsumura and Setoguti 1984). The skin is poorly vascularized, with the epidermis being very thick (47 to 110 μm) (Czopek 1965). The lungs of most amphibian species such as *Amphiuma* and the toad, *Bufo marinus*, have a preponderance of smooth muscle tissue (Czopek 1962b; Smith and Rapson 1977; Martin and Hutchison 1979; Goldie et al. 1983; Stark-Vancs et al. 1984; Maina and Maloiy 1988; Maina 1989d), which may account for the great compliance of the lungs (Hughes and Vergara 1978). In *Amphiuma*, during expiration, the lung virtually collapses, producing an almost 100% turnover of air (Stark-Vancs et al. 1984). The amphibian lungs are best developed in the Salentia where septa intensely subdivide the lung, converting the large central air space into small stratified air cells (e.g., Okada et al. 1962; Smith and Rapson 1977; Goniakowska-Witalinska 1986). The internal morphology of these relatively elaborate lungs is similar to that of the lungs of the lungfishes (Dipnoi) (e.g., de Groodt et al. 1960; Klika and Lelek 1967; Gannon et al. 1983; Maina 1987a; Fig. 64). The lungs of *Pipa pipa* (Marcus 1937) and *Xenopus laevis* (Goniakowska-Witalinska 1995) are reinforced with septal cartilages to ensure patency of the air passages. In Salentia, the skin contributes very little towards gas exchange. The length of the skin capillaries constitutes only 30% of the total length of the blood capillaries located in the respiratory surfaces (Czopek 1965). However, in two species of Salentia which live in well-oxygenated high mountain lakes, e.g., *Telmatobius* and *Batrachophrynus*, the lungs are very small, the body very well vascularized, and the epidermis very thin (Muratori et al. 1976; Czopek and Szarski 1989). Well-differentiated pneumocytes (Fig. 103) as well as dust cells (free phagoctes; Fig. 102) have occasionally been observed on the surface of the amphibian lung (Welsch 1983; Maina 1989d).

The morphological heterogeneity of the amphibian gas exchangers and the lungs in particular correspond with the remarkable diversity of the environments they occupy, the mode of life they lead, and their property of interrupted development. Though amphibians have multiple options for gas exchange, only one pathway is best developed for optimal performance in a particular environment (Guimond and Hutchison 1976). Pulmonary vascularization correlates with the degree of terrestriality, behavior, and tolerance to dehydration. The skin is the main pathway for gas transfer in the predominantly aquatic species, while in the more terrestrial ones it has been relegated or rendered totally redundant as the lung has assumed a central position in gas exchange. *Necturus* experimentally held in cool, well-aerated water has better-developed gills than those of animals kept in warmer, poorly aerated water (Guimond and Hutchison 1976). Despite its strong reliance on water, the lung of the anuran clawed toad, *Xenopus laevis*, is very well developed for aerial respiration (Smith and Rapson 1977; Pohunkova and Hughes 1985a): 80% of its O_2 needs are transferred across the skin (Emilio and Shelton 1974). The perfusion of the skin is regulated by dilation and constriction of the cutaneous vasculature (Poczopko 1959). In *Bufo marinus* and *Rana catesbeiana* (Hillman 1987a,b), dehydration leads to increased vascular resistance due to a hemoconcentration effect on the blood viscosity. Dehydration in the terrestrial frog, *Eleutherodactylus coqui* (Pough et al. 1983), and the aquatic

Xenopus laevis (Hillman 1987a,b) lowers the capacity to utilize aerobic metabolism during activity. Systemic O_2 transport may be the limiting factor to aerobic capacity in the air-breathing amphibians (Hillman et al. 1985). In the xerophilous anuran, *Phyllomedusa sauvagei* and *Chiromantis xerampelina*, at the resting evaporative water loss (EWL), cutaneous respiration is insignificant, but with increased water loss, fractional cutaneous gas exchange correlates with the cutaneous EWL (Stinner and Shoemaker 1987). Adaptively, the blood of the highly xeric *Chiromantis petersi* has a low O_2 affinity and a temperature-insensitive O_2-hemoglobin binding capacity (Johansen et al. 1980).

Compared with the other air-breathing vertebrates, the amphibians have some of the simplest lungs, generally with low diffusing capacities for O_2 (Glass et al. 1981a). The lungs of *Necturus* and *Cryptobranchus* are thin-walled, transparent, poorly vascularized, and nonsepted (Guimond and Hutchison 1976): hydrostatic control may have the more significant role in such simple lungs. Regarding morphological and morphometric characteristics, generally, the lungs of the anurans and the apodans are more advanced than those of the urodeles (Meban 1980): on average, the arithmetic mean thickness of the blood-gas barrier in the urodeles is $2.59\,\mu m$, $2.35\,\mu m$ in the Apoda, and $1.89\,\mu m$ in the Anura. Some areas of the blood-gas barrier of the lung of the caecilians *Chthonerpoton indistinctum* and *Ichthyophis paucesulcus* (Welsch 1981) may be only $1\,\mu m$ thick, while in the tree frog, *Hyla arborea*, the barrier may be as thin as $0.6\mu m$ (Goniakowska-Witaliniska 1986): *H. arborea* has a relatively high metabolic rate (Goniakowska-Witaliniska 1973) and the skin is richly vascularized (Czopek 1965). The lungs of the terrestrial species, e.g., the toad, *Bufo marinus* (Smith and Rapson 1977), the tree frogs, *Hyra arborea* (Goniakowska-Witalinska 1986) and *Chiromantis petersi* (Maina 1989d; Fig. 84), are eminently elaborate, having a series of hierarchical septa which delineate the air cells, which range in diameter from 1.45 mm in *Rana pipiens* to 2.3 mm in *Bufo marinus* and *Rana catesbeiana* (Tenney and Tenney 1970). The respiratory surface area in the lungs of the more terrestrial species is higher than that in the lungs of the aquatic ones (Tenney and Tenney 1970). Elementary lungs are adequate in amphibians, a group which characteristically has low aerobic metabolism (Whitford and Hutchison 1967; Goniakowska 1973; Goniakowska-Witalinska 1974; Feder 1976; Guimond and Hutchison 1976). Together with the Dipnoi, the amphibians have the largest tissue cells among vertebrates (Wintrobe 1934; Misiek and Szarski 1978; Szarski 1983). *Amphiuma* and *Necturus* exhibit great tolerance to anaerobiosis, withstanding 6h of total anoxia (Rose and Zambernard 1966). When exposed to hypoxia and hypercapnia, the amphibians utilize behavioral hypothermia to reduce their O_2 need (Glass et al. 1983; Riedel and Wood 1988; Wood and Glass 1991).

Though the first vertebrates to invade land some 300 million years ago, by erecting a strong dependence on water for essential processes like aquatic oviparous reproduction, in regard to specific and numerical abundance and ecological distribution, modern amphibians constitute only a rather obscure vertebrate taxon. Their geographical distribution corresponds with freshwater, wet, high humidity, and high rainfall areas, feature which reflect strongly on their physiology. Cutaneous respiration (CR) presents very few possibilities for innovative designs and is generally considered an evolutionary dead end. Gans (1970), however, argued that CR may be a secondary condition in amphibians. This observation was supported by the suggestion by Romer (1972) that the Carboniferous amphibians may have been well scaled and thus had a water-impermeable skin cover. For all it is worth, the water-permeable skin of modern amphibians may represent a specialized secondary condition!

6.7.3 The Reptilian Lung

The reptiles were the first vertebrates to be adequately adapted for terrestrial habitation and pulmonary respiration. By evolving a cleidoic egg and an impermeable surface cover, parameters above those achievable by amphibians from which they evolved (e.g., Olmo 1991), they were able to delink their physiology from water. The Mesozoic era, which lasted for nearly 200 million years, is often called the age of the reptiles, as the taxon dominated the Earth. Reptiles are exclusively lung breathers. Like the amphibians, reptiles display remarkable pulmonary structural diversity which to an extent can be correlated with the diverse habitats occupied and lifestyles led. There is no single model for the reptilian lungs. Based largely on the nature of internal organization (e.g., Milani 1894; Marcus 1937; Baudrimont 1955; Duncker 1978a, 1979; Perry 1983, 1992a; Hlastala et al. 1985), different morphological classifications of the lungs have been attempted. The lungs range in complexity from the profusely compartmentarized (multicameral) ones of the turtles, monitor lizard, crocodiles, and snakes (Perry 1978, 1988; Perry and Duncker 1978, 1980; Maina et al. 1989b; Maina 1989e; see Figs. 92,97) through the less elaborate (paucicameral) ones of the chameleons (Fig. 96) and the iguanids to the simple, saccular, smooth-walled, transparent, (unicameral) ones of, e.g., the teju lizard, *Tupinambis nigropunctatus* (Klemm et al. 1979; Perry 1983). This classification is overly simplistic as transitional forms and gradations occur. The simplest lungs, which correspond in development to the amphibian lung, occur in the Sphenodontia. Such lungs have a central air duct and peripherally situated, shallow air cells, which give a low surface-to-volume ratio.

The brochoalverolar lung of mammals and parabronchial lung of birds are thought to have evolved from transformation of a multicameral lung (e.g., George and Shah 1956, 1965; Duncker 1978a; Klaver 1981; Perry 1983, 1989; Becker et al. 1989). With an elaborate anterior space in which much of the gas exchange occurs, and a simple posterior one analogous to the air sacs of the bird lung, the design of the avian lung-air sac system is more closely related to the reptilian lungs (Brackenbury 1987). While the reptilian lungs fill up from the peripheral walls into the axial air space (centripetal = centralizing = compacting growth), the air conduits constituting the unfilled gaps, the mammalian and avian lungs fill from inside (centrifugal = radiative = diffusive growth), i.e., from outward bifurcation of the central airways. These developmental differences may account for the fact that the volume density of the parenchyma in the reptilian lung is only 25.2% in the tegu lizard, *Tupinambis nigropunctus*, 32.1% in the monitor lizard, *Varanus exanthematicus* (Perry 1981), and 25% in the Nile crocodile, *Crocodylus niloticus* (Perry 1988). In the avian lung the average value is 50% (e.g., Maina et al. 1989a) and in mammals the value is as high as 90% (e.g., Gehr et al. 1981; Maina and King 1984). The faveolar air spaces in the parenchyma constitute 40% of the intrapulmonary air in the crocodile lung (Perry 1988), a value lower than that in the mammalian lung (about 56%; Maina and King 1984) and 53% in the bird lung (Maina et al. 1989a). In the more advanced snakes, e.g., Colubridae, Viperidae, and Elapidae, the left lung is greatly reduced or is totally lacking. In the primitive species, e.g., the boas and the pythons (Cope 1894; Verde 1951; Luchtel and Kardong 1981; Pohunkova and Hughes 1985b; Maina 1989e; Pastor 1995), the left lung occurs. The right lung is atrophied in the Amphisbenia (Gibe 1970). In the order Squamata, single-chambered lungs predominate especially in the families Teiidae (Klemm et al. 1979), Scindae (Gibe 1970), Lacertidae (Meban 1978a), and Gekkonidae (Perry et al. 1989b). Similarly, simple lungs occur in the family Angioidea (Meban 1978b). The land-based chelonians have paucicameral lungs, i.e., lungs with two or three peripheral compartments, which open into a central air space and lack an intrapulmonary bronchus. The marine species have multi-chambered bronchiolated lungs (Solomon and Purton 1984; Pastor et al. 1989). The elongated lungs of the ophidia and the amphisbaenids are divided into two functional zones, an anterior respiratory region which is well vascularized and a posterior one which is saccular and avascular (Kardong 1972; Klemm et al. 1979; Stinner 1982; Maina 1989e; Pastor 1995). In the crocodile lung, most of the parenchyma is located in the anterior two thirds of the lung where the blood makes 38 to 50% of the total volume (Perry 1988). The posterior part of the lung is thought to store air (e.g., Heatwole 1981), serve a hydrostatic role (Graham et al. 1975), and mechanically ventilate the anteriorly located exchange tissue in the manner of air sacs in birds: the arrangement may enhance the efficiency of gas exchange (Gratz et al. 1981; Stinner 1987; Vitalis et al. 1988). In heterogenous lungs, the design should convey a distinct functional advantage since the lungs of the more primitive reptiles are more homogenous (e.g., Luchtel and Kardong 1981). The varanids present the greatest level of pulmonary complexity in the suborder Sauria. *Varanus exanthematicus* and the pancake tortoise, *Malacochersus tornieri*, have multichambered lungs with a bifurcated intrapulmonary bronchi and profuse internal subdivision (Perry and Duncker 1978; Maina et al. 1989b; Fig. 93). The single-chambered lungs with an edicular parenchyma are thought to require low energy for convective ventilation: they occur in animals with low metabolic rates (Pastor 1995). The reptilian lung constitutes 5% of the body weight (Tenney and Tenney 1970). In animals of similar body mass, a reptile has a lung volume which is seven times greater than that of a mammal (Crawford et al. 1976; Glass and Johansen 1981) but the diffusing capacity for O_2 is relatively low (Crawford et al. 1976). The aerobic capacity of reptiles is remarkably lower than that of mammals. At a temperature of 37 °C, a 1-kg lizard consumes $122\,ml\;O_2\;h^{-1}$, a value which constitutes 18% of the O_2 consumption of an equivalent-sized mammal (Bennett and Dawson 1976). The muscle capillary surface per unit muscle mass of a reptile is about 20% the value of a mammal of similar size (Pough 1980). At maximal exercise, anaerobic metabolism provides 86% of the total energy consumption in the water snake, *Natrix rhombifera* (Gratz and Hutchison 1977). At a body temperature of 20 to 23 °C, the physiological diffusing capacity of the reptilian lung (DLo_2p) is an order of magnitude smaller than that of a mammal of the same size (Crawford et al. 1976; Glass and Johansen 1981). The DLo_2p of the lung is similar in reptiles and amphibians (Glass et al. 1981a,b; Gatz et al. 1987; Lutcavage et al. 1987). Compared with mammals (e.g., Baldwin and Winder 1977; Scheuer and Tipton 1977; Dudley et al. 1982), where endurance exercise precipitates changes such as increases in tissue oxidative capacities and O_2 consumption, the adaptive response of lizards, *Amphibolurus nuchalis*, to endurance exercise is different (Gleeson 1979; Garland et al. 1987): trained lizards exhibit decreased heart and muscle masses but increases in liver mass, hematocrit, liver pyruvate kinase, and heart citrate synthetase activities. Interestingly, physical training enhances the swimming performance of the African clawed frog, *Xenopus laevis* (Miller and Camilliere 1981).

The epithelial cells lining the respiratory surface of the reptilian lung, are completely differentiated into types I and II cells (e.g., Okada et al. 1962; Nagaishi et al. 1964; Luchtel and Kardong 1981; Daniels et al. 1990; Maina 1989d,e; Perry et al. 1989b) and type III cells – the brush cells (Gomi 1982). A rare mitochondria-rich cell has been described in the lung of the turtle, *Pseudemys scripta* (Bartels and Welsch 1984). The type I cells are squamous and have remarkably thin, long cytoplasmic extensions. The much smaller, rather cuboidal surfactant-secreting type II cells are scattered between the type I cells (Figs. 84,85,87,90). In general, the pneumocytes in the lungs of the amphibians (Goniakowska-Witalinska 1995) and the lungfishes (Hughes and Weibel 1976; Maina 1987a) are undifferentiated while those in the reptilian, mammalian, and avian ones are. The differentiation of the pneumocytes in the higher vertebrates may be an adaptive strategy for greater functional efficiency (Maina 1994): reducing the numerical density of the more metabolically active surfactant-producing type II cells to a minimum and the type I cells adopting an extremely thin form lowered the overall O_2 consumption by the tissues of the gas exchanger. Furthermore, the design generated a thin blood-gas barrier, an important structural feature which enhanced the diffusing capacity of the gas exchanger for O_2 (Liem 1987a; Maina 1987a). Sporadic attenuation of the endothelium (Figs. 40,90), where extremely thin areas of the blood-gas barrier are generated without jeopardizing the mechanical integrity of the lung (Figs. 29b,40a,90) is a common scheme evoked to maximize gas transfer in the lungs of the higher vertebrates (e.g., Weibel 1973; Maina and King 1982a). Dust cells (macrophages) have been described in some reptilian lungs, e.g. in the turtle, *Testudo graeca* (Pastor et al. 1989). Unlike in the mammalian (Fig. 86) and bird lungs (Figs. 88,90), where owing to a single capillary system arrangement the exposure of the pulmonary capillary blood to air is very efficient, in the reptilian lung, depending on the species and the particular areas of the lung, a double capillary system (Fig. 85) commonly occurs. Reptilian lungs have a prevalence of smooth muscle tissue (Fig. 80). In the tegu and the monitor lizards, respectively, smooth muscle tissue constitutes 7.4 and 1.3% of the nontrabecular tissue (Perry 1981; Perry et al. 1989b,c). The smooth muscle tissue has been associated with intrapulmonary convective movement of air (Klemm et al. 1979; Tenney et al. 1984; Carrier 1988). The compliance of the lung of the garter snake, *Thamnophis sirtalis* of 0.042 ml per cm H_2O per g (Bartlett et al. 1986) is 50 times that of the lung of a mouse, a mammal of about the same body mass (Bennett and Tenney

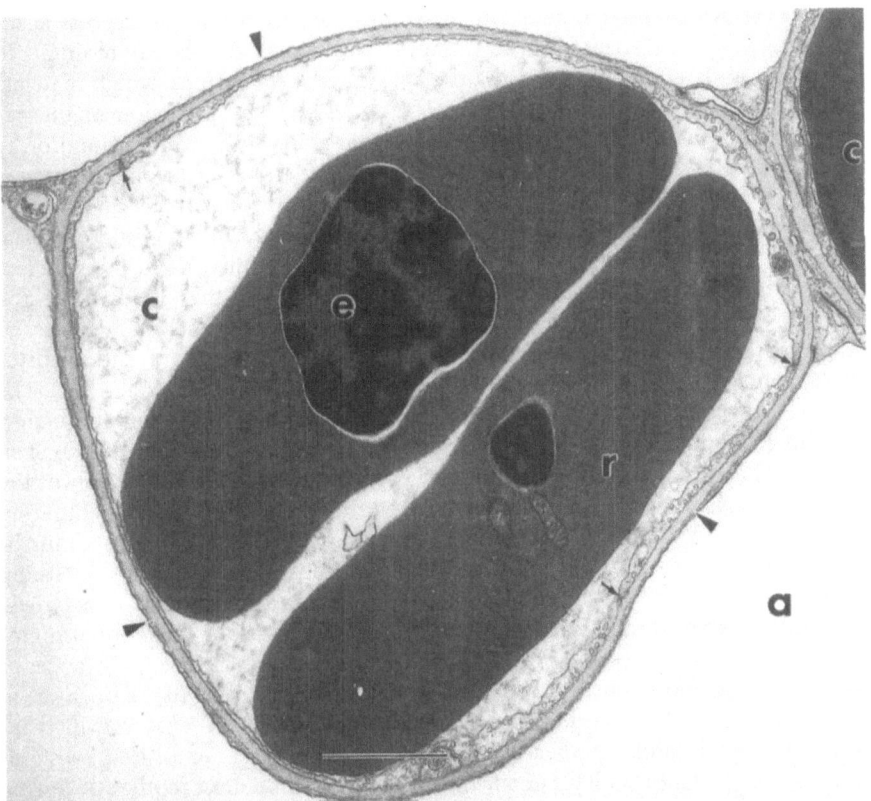

Fig. 90. High power view of the pulmonary blood capillaries, *c*, of the lung of the domestic fowl, *Gallus domesticus*, showing the blood-gas barrier, ➤, which is characterized by sporadic attenuations especially of the endothelial cell. →, endothelial cell junctions; *r* red blood cells; *e* nucleus of the red blood cell; *a* air capillary. *Bar* 0.5 µm

1982). At the peak of an expiratory phase, the residual volume of air $(18\,ml\,kg^{-1})$ in the crocodile lung is only 13% of the maximal lung volume (Perry 1988). The compliance of the lung of the crocodile $(0.7\,ml\,cm^{-1}\,H_2O\,g^{-1})$ is over four times that of the body wall (Perry 1988). The overall compliance of the reptilian lung is determined by the contractile elements of the lung, i.e., smooth muscle and elastic tissue, the saccular nature of the lung (Craig 1975), and the very effective pulmonary surfactant (Perry and Duncker 1978). Coupled with the irregular pattern of breathing, the properties may provide an energy-saving system on the respiratory work (Milsom 1984). The volume-specific lung compliance of the multicameral lung of the crocodile is similar to that of the much simpler lung of the gecko (Perry 1988; Perry et al. 1989b), suggesting that lung compliance in the reptilian lung may be an attribute of the parenchymal structure and not of lung type. The reptilian lungs may serve as air stores during apnea and in the aquatic species may support extended dives (Ackerman and White 1979). In the alligator, the pulmonary O_2 store constitutes 85% of the total lung volume (Andersen 1961).

Turtles can withstand complete anoxia for days or months, in the process accumulating lactic acid levels up to 200 mmol l^{-1}, surviving a decrease in brain pH to 6.4 (Glass and Wood 1983). During the episodic apneic periods, which may last for a few minutes to hours (Glass and Wood 1983), the perfusion of the gas exchanger is reduced through a decrease in heart rate, stroke volume, and/or by the blood being shunted away from the lung. Ventilation-perfusion matching is necessary for efficient gas exchange (Burggren et al. 1977; Wood et al. 1978). In the Chelonia, the difference between the PO_2 in air spaces and that in the arterial blood (a factor determined by central cardiovascular admixture of the systemic venous blood into the systemic arteries and shunting of the blood within the pulmonary circulation; Seymour 1978) may be as high as 6 to 6.7 kPa (Burggren and Shelton 1979), in sea snakes it may exceed 8 kPa (Seymour and Webster 1975), and in resting lizards it ranges from 2.7 to 13 kPa (Mitchell et al. 1981). The respiratory parameters such as ventilation-perfusion ratio, respiratory frequency, and tidal volume are difficult to characterize in the periodic breathers (Glass et al. 1979). It has been hypothesized that structural limitations in the design of the reptilian lungs prevented the reptiles from attaining endothermic-homeothermy (Perry 1992a), relegating their aerobic capacities behind birds and mammals. In the green turtle, *Chelonia mydas*, during swimming, intermittent breathing gives way to a continuous breathing pattern (Butler et al. 1984). Uncharacteristic of reptiles, green turtles can endure sustained exercise: they make long migrations between their feeding grounds and breeding beaches, covering distances of as much as 4800 km, most of the time continuously swimming in the open sea at speeds of 2 m s^{-1} (Carr and Goodman 1970; Carr et al. 1974). They can increase their O_2 consumption by a factor of 3 (Prange 1976; Butler et al. 1984) without resulting in a significant level of anaerobic metabolism as most reptiles do during long bouts of activity (e.g., Bennet 1982). Hypoxia (10% O_2) caused continuous ventilation in resting green turtles and pulmonary blood flow was elevated and sustained (West et al. 1992): during swimming, increased heart rate was accompanied by a sevenfold increase in the ventilatory rate. Interestingly, physiological conditions of periodic ventilatory flow pattern do constitute a significant impairment in the parabronchial gas exchange efficiency in the avian lung (Scheid et al. 1977). This may indicate the remarkable functional reserves inherent in the bird lung (Sect. 6.7.5) where adjustments can be made to overcome certain limitations.

6.7.4 The Mammalian Lung

The beginning of the Tertiary saw the mammals take over and eclipse the reptiles as the dominant terrestrial vertebrate group (Romer 1967). The appearance of the placental mammals (subclass: Eutheria) formed the pinnacle in the development of the taxon. Among vertebrates, the mammalian lung, especially the human one, has been best studied both structurally and functionally. Recent excellent integrative reviews on these aspects include those by Weibel (1984a,b, 1986) and Burri (1984a,b). Owing to the abundance of literature in the area and need for brevity here, in-depth discussion of the mammalian lung (except for that of bats, which are less well known) will not be made. Suffice it to say now that the blind-ending, tidally ventilated mammalian lung falls far short of the level of efficiency attained by other gas exchangers, e.g., the insectan tracheal system and the avian lung-air sac system. With respect to O_2 consumption in standard conditions as well as maximal activity, Dejours (1990) observed that "compared with other mammals, the human exhibits no species-specific characteristics". Among the air-breathing vertebrates, the heterogeneity of the design of the lung decreases from the ecto-therms – the Dipnoi (lungfishes), amphibians, and reptiles to the endotherms – mammals and birds. Morphological homogeneity appears to be an essential at-tribute for optimal design in gas exchangers. Structurally, the lungs of mammals are similar except for fine differences like bronchiolar bifurcation, lobulation, and topographic relationships between the airways and the vasculature. The need to optimize respiratory function must have become critical with changes to lifestyles which called for increased metabolic demands for O_2. Over and above the phylogenetic level of development, the organization complexity of the gas exchangers and respiratory stratagems adopted by an animal are greatly deter-mined by the needs placed on them and the habitat occupied. For example, in the lungs of the nonhuman primates, e.g., the baboon, *Papio anubis* (Maina 1987b), and the vervet monkey, *Cercopithecus aethiops* (Maina 1988b), the lungs of the small, supposedly primitive prosimians, e.g., the lesser bushbaby, *Galago senegalensis* (Maina 1990c), are better from a morphometric perspective. The naked mole rat, *Heterocephalus glaber*, a small, eusocial poikilothermic fossorial bathyergid rodent which lives in thermostable burrows, has remarkably neotenic lungs (Maina et al. 1992). The diffusing capacity of the lung of the Japanese waltzing mouse, *Mus wagneri*, a pathologically hyperactive animal, is 55% greater than that of a normal one (Geelhaar and Weibel 1971). Lungs of high altitude-raised rats have smaller alveoli and a greater respiratory surface area than sea level dwellers (Pearson and Pearson 1976). Experimental lobectomy results in a compensatory overgrowth (Rannels and Rannels 1988), the remaining lobes in young rats attaining the same diffusing capacity as the control animals (Burri and Sehovic 1979).

6.7.4.1 Lungs of Bats

Bats (order: Chiroptera) are unique among mammals by their ability to fly (e.g., Thewissen and Babcock 1992). Flight has enabled remarkable adaptive radiation to occur in bats. Of all known mammalian species, one in five is a bat. About 800 species of bats have been recognized, the number falling between that of Roden-tia (1600) and Insectivora (400). After the human being, *Myotis* (family: Vespertilionidae) is reputed to be the most widely spread naturally occurring mammalian genus on Earth (Yalden and Morris 1975). Despite the abundance and the wide geographical distribution, perhaps due to their elusive nocturnal lifestyle to which they were relegated by the birds, the only other older apparently well-established volant vertebrate, bats remain animals of curiosity, myth, and prejudice. The phylogenetic affinity between bats and the mainstream mammals is not clear (e.g., Novacek 1980, 1982; Scholey 1986). The groups which have evolutionary been closely associated with bats include Scandentia (Tupaiidae or tree shrews), primates, and the Dermoptera (e.g., van Valen 1979; Padian 1982; Pettigrew et al. 1989). Bats comprise the Suborders Megachiroptera and Microchiroptera and are considered by some investigators, e.g. van Valen (1979) and Novacek (1982), to be monophyletic, while others, e.g., Jepsen (1970), Smith (1977), Scholey (1986) and Pettigrew et al. (1989), consider them diphyletic. The earliest reliably known fossil record of a bat is that of *Icaronycteris index* of the Eocene (50 million years ago) which morphologically resembles the modern Microchiroptera (Jepsen 1970): the protobats may have been tree-dwelling omnivores which started to glide between trees while foraging (Norberg 1981, 1986). The general anatomy of bats is mammalian (Yalden and Morris 1975).

Powered flight is defined as capacity to produce lift, accelerate, and maneuver at various speeds (e.g., Pennycuick 1975; Norberg 1976a,b; Rayner 1986). The large number of animals which are said to fly, for example the freshwater butterfly fish, *Pantodon buchholzii* of the West African rivers, the parachuting frog of Borneo, *Rhacophorus dulitensis*, the flying snakes of the jungles of Borneo, *Chrysopelea* sp., the flying squirrel of North America, *Glaucomys volans*, the flying lemur, *Cyanocephalus volans*, and the East Indian gliding lizard, *Draco volans*, are essentially acrobatic passive gliders or parachutists which use part of their body to slow down a fall by using drag and lift (see Scholey 1986; Davenport 1994). Energetically, powered flight is a highly demanding mode of locomotion which has evolved in only a few elite animals (Tucker 1972; Berger and Hart 1974; Carpenter 1975; Thomas 1975), the insects, pterosaurs, birds, and bats – chronologically in that order. The mass-specific aerobic capacities of flying bats are essentially the same as those of forward-flapping birds but are 2.5 to 3 time those of running mammals of the same size (Thomas and Suthers 1972; Carpenter 1975; Thomas 1987). Bats can increase their O_2 consumption during sustained flight by a factor of 20 to 30 times (Bartholomew et al. 1964; Thomas and Suthers 1972). At an ambient temperature of 20 °C, a 12-g bat, *Myotis velifer*, is reported to increase its O_2 consumption by an astounding factor of 130 (Riedesel and Williams 1976). By avian standards, bats are excellent fliers in terms of speed, distance, and maneuverability (Vaughan 1966; Griffin 1970; Norberg 1976a,b; Fenton et al. 1985; Norberg and Rayner 1987). Speeds of $16 \, \mathrm{km \, h^{-1}}$ in *Pipistrellus pipistrellus* (Jones and Rayner 1989), 30 to $50 \, \mathrm{km \, h^{-1}}$ in *Myotis* (Hayward and Davis 1964) and $64 \, \mathrm{km \, h^{-1}}$ in *Eptesicus fuscus* (Petterson and Hardin 1969) have been estimated. Migratory distances of about 1000 km have been reported in *Lasiurus borealis, L. cinereus, Lasionycteris noctivagans, Nyctalus noctula*, and *Tadarida brasiliensis* (Baker 1978; Thomas 1983). As they forage, *Epomophorus wahlbergi* and *Scotophilus viridis* are known to cover distances of about 500 km in a night (Fenton et al. 1985). Flight style and optimum speed in bats depend on, among other factors, choice of food, foraging behavior, and habitat selection (Norberg 1981). The small bats show a greater scope for flight with respect to agility and are even able to momentarily hover (Norberg 1976a,b). However, unlike birds, which commonly use energy-saving modes of flight like gliding and soaring (Rayner 1985), bats only rarely do so. *Pipistrellus pipistrellus* adopts gliding flight for only

13.4% of its flight time and the glides last for only 0.1 to 0.3 s (A.L.R. Thomas et al. 1990).

Despite the basic limitations intrinsic to the design of their characteristically mammalian lungs (Yalden and Morris 1975; Maina 1985), bat lungs have been structurally highly refined (e.g., Maina and Nicholson 1982; Maina et al. 1982a, 1991; Maina and King 1984; Maina 1986). These features have been functionally closely integrated with other anatomical and physiological aspects, enhancing the uptake and transport of the necessary large amounts of O_2 for flight. The strategies necessary to afford flight in bats evolved within the constraints presented by an inferior lung. This is a classic case of the innate plasticity of biological systems. In bats, a typical mammalian lung was exquisitely modified to exchange respiratory gases during flight at rates equal to those of the seemingly better-designed bird lungs. The most important parameters in this portfolio included (1) development of relatively large hearts with a huge cardiac output (Hartman 1963; Snyder 1976; Jürgens et al 1981), (2) high hematocrit, hemoglobin concentration, and O_2 carrying capacity of blood (e.g., Riedesel 1977; Wolk and Bodgdanowicz 1987), and exceptional pulmonary structural parameters (Maina et al. 1982a, 1991; Lechner 1984; Jürgens et al. 1981; Maina and King 1984). *Phyllostomus hastatus* can maintain the same high lung O_2 extraction factor of 20%, a value comparable to that of bird during the metabolic stress of flight (Thomas 1981; Thomas et al. 1984) or when at rest during exposure to a severe hypoxic stress (Farabaugh et al. 1985). The mean resting O_2 extraction factor at thermoneutral range in *Noctilio albiventris* of 18.3% lies between that of a bird (20.8%) and a nonflying mammal (16.6%) of the same body mass (Chappell and Roverud 1990). It increases from 35 to 40% at low ambient temperature. These values surpass those of most birds under similar conditions (Bucher 1985). One of the highest venous hematocrits in the vertebrates (75%) has been reported in a specimen of a 13-g bat, *Tadarida mexicanobrasiliensis* (Black and Wiederhielm 1976). In the five species of bats examined by Jürgens et al. (1981), venous hematocrits ranged from 51 to 63% and hemoglobin concentration and erythrocytes numbers were respectively 24.4 g dl^{-1} and 26.2 million l^{-1}. While high hematocrit may enhance O_2 uptake, the advantages conferred are soon compromised by the increased viscosity of blood (Stone et al. 1968; Hedrick et al. 1986; Hedrick and Duffield 1991). On account of the different effects of hematocrit on blood O_2 capacity and viscosity, where O_2 capacity increases linearly with the hematocrit while viscosity increases exponentially, an optimal O_2 transport level is established (Crosswell and Smith 1967; Shepherd and Riedel 1982; Kiel and Shepherd 1989). Increase in the hematocrit beyond optimal level causes a reduction in cardiac output, maximum O_2 consumption, and aerobic scope (Hillman et al. 1985; Tipton 1986).

In all bats which have been studied, a 1:1 synchronization between wing beat and breathing cycles has been observed (Suthers et al. 1972; Thomas 1981, 1987; Carpenter 1985, 1986): in birds, this occurs in only a few species (e.g., Tomlinson 1963; Berger et al. 1970; Berger and Hart 1974; Butler and Woakes 1980). It has been suggested (e.g., Bramble and Carrier 1983) that locomotory activity provides mechanical assistance to respiratory muscles. In birds, though flight muscles attach on the sternum, there appears to be very little effect on the actual pulmonary respiratory air flow and volume. Bernstein (1987), however, suggests that the disparity between the wing beat to breathing rates may be caused by the complex nature of the air flow in the avian parabronchial lung. In the European starling, *Sternus vulgaris*, the ventilatory volume change associated with wing beat ranges from 3 to 11% at most (Banzett et al. 1992). To a yet undetermined degree, the synchronization of wing beat with breathing cycles must enable bats to ventilate their lungs at a lower cost and probably more efficiently. In the bats *Antrozous pallidus* (Basset and Wiederhielm 1984) and *Myotis daubentoni* (Krátký 1981; Lundberg et al. 1983), the maturation of the O_2 transport system (i.e., blood O_2 capacity) to the adult status closely coincides with the start of flight behavior. Bats have exceptionally large lungs which occupy a large proportion of the coelomic cavity (e.g. Maina and King 1984; Maina et al. 1991). The gastrointestinal system is small and morphologically poorly differentiated (Makanya and Maina 1994; Makanya et al. 1995). Compared with nonflying mammals, bats have relatively large respiratory surface areas. The mass-specific respiratory surface area of $138\,cm^2\,g^{-1}$ reported by Maina et al. (1982a) in the epauletted fruit bat, *Epomophorus wahlbergi*, is the highest value so far reported in a vertebrate. Compared with nonflying mammals, bats have relatively thin blood-gas barriers. The thinnest blood-gas barrier so far reported among mammals is that of $0.1204\,\mu m$ in *Phyllostomus hastatus* (Maina et al. 1991), a bat in which the O_2 extraction factor equals that of an energetic bird of comparable size (Thomas 1987) and which has a venous hematocrit of 60% (Jürgens et al. 1981). Pulmonary respiratory surface area can be increased by an overall enlargement of the lung, as occurred in bats, and/or increased subdivision of the gas exchange tissue, as in birds. In the compliant mammalian lung, increased subdivision of the parenchyma, however, generates smaller alveoli which are not only highly susceptible to collapse (due to large surface tensional forces at the air-tissue interface) but obligate greater amount of energy to inflate.

Unlike birds, which have dispersed widely and penetrated the remote cold regions of the world such as Antarctica, bats are largely tropical and neotropical in distribution (Wimsatt 1970; Yalden and Morris 1975; Carpenter 1985). This may probably be due to the need for reliable food sources in order to procure adequate metabolic substrates for production of the large amounts of energy for flight. Furthermore, due to the relatively poor insulation of the bat wings and presence of skeletal muscles on the wings (compared with the feathered bird wings which do not have muscles on the wing surface), excessive convective heat loss to the cold air at extremely low temperatures may occur in bats. The wing muscles may be cooled to a critical temperature where the proper coordination which is necessary for efficient flight is curtailed. On isolated forearm muscles from a number of species of temperate zone bats, Nelson et al. (1977) observed that the duration of contraction of forearm muscles at temperatures below $8\,°C$ were five to ten times longer than those at $32\,°C$. However, some bats have been reported to fly in ambient temperatures as low as $-5\,°C$ (O'Farrell and Bradely 1977). Temperate-zone bats are known to hibernate or migrate to warmer regions of the world during winter (Kulzer 1965; A.L.R. Thomas et al. 1990). In hibernating *Myotis lucifugus*, O_2 consumption is 1.5% of their respective normothermic resting rates at or about thermoneutrality (Hock 1951): at $5\,°C$, breathing is arrhythmic in character, and apneic phases last for as long as $48\,min$.

In principal, animals use different strategies to attain optimal states (Howell 1983). Due to their remarkably different evolutionary backgrounds, different genetic resources, designs, and strategies were utilized by bats and birds to independently meet the aerodynamic and energetic requirements of flight. In pursuit of optimization of respiration, bats and birds evoked remarkably different paradigms: structural and functional parameters were variably integrated to promote gas exchange efficiency (Maina 1998). From modeling anatomical and physiological data, it appears that in bats, the process occurred through synergism of performances of relatively fewer, highly refined parameters which operate at or close to their maximum capacity. We have called this the narrow-based – high-keyed scheme (NB-HKS). Birds, on the other hand, have incorporated a wide spectrum of parameters in their gas exchange arsenal (and conspicuously conserved some) in what has been termed the broad-based – low-keyed scheme (BB-LKS): similar to this scheme, Heinrich (1983) has described a program of integrated suboptimal parts for a better whole in foraging bumblebees. Inevitably, consequent to the different respiratory contrivances in birds and bats, different functional reserves have evolved. In bats, the NB-HKS leaves a very limited margin of operation while the BB-LKS of birds affords an enormous reserve. This model-based inference (Maina 1998) is supported by the observation made by Chappell and Roverud (1990) that whereas with changing ambient temperature birds and nonflying mammals adjust respiratory frequency and tidal volume to meet changing O_2 demands, O_2 extraction remains fairly constant and in some species actually decreases (Casey et al. 1979; Withers et al. 1979; Bucher 1985; Chappell 1985; Kaiser and Bucher 1985; Chappell and Souza 1988; Bucher et al. 1990), bats (at least the lesser bulldog bat, *Noctilio albiventris*, on which data are available) accommodate varying thermogenetic O_2 consumption by simultaneously changing all three factors, namely respiratory frequency, tidal volume, and O_2 extraction, which may increase by a factor of as much as 2. The margin of operation in bats depends on factors such as the magnitude, nature, and duration of exposure to stress. Some degree of respiratory functional reserve must, however, exist in bats since they are able to absorb changes which call for moderate increases in O_2 demand such as flight during pregnancy, flight after the premigratory weight gain, and thermogenetic heat production associated with low ambient temperatures. For example, a female red bat, *Lasiurus borealis*, reported to weigh 12.9 g and presumed to have a wing loading of $0.09\,\mathrm{g\,cm^{-2}}$ is reported to have carried four young ones whose total mass was 23.4 g, i.e., 181% of her body mass (Staines 1965). In experimental tests to assess weight-lifting capacities, Davies and Cockrum (1964) observed that a female long-eared bat, *Plecotus townsendii*, with a body mass of 10.1 g, could lift 73.3% of her own mass, but *Tadarida brasiliensis* could manage only 9.3%. No bat is truly herbivorous as rodents are. Fermentative microbial digestion of cellulose calls for a voluminous gastrointestinal system (GIT) (e.g., Chivers and Hladik 1980; Warner 1981) and a long transit time for the ingesta (e.g., Balch and Campling 1965). Such an increase in mass for low energy yield per unit weight of the ingesta perhaps would be untenable for flight: bats have simple GITs contained in restrictive abdominal cavities (Forman 1972; Makanya and Maina 1994; Makanya et al. 1995) and the transit times of the ingesta must be relatively short. Bats of the temperate regions are exclusively insectivorous but tropical regions support both fruit-eating and insectivorous bats (Wilson 1973; Yalden and Morris 1975). While there are numerous herbivorous birds, e.g., the hoatzin, *Opisthocomus hoazin* (Dominguez-Bello et al. 1993), such birds are fairly large and have poor, if any, flight capacity.

From differentiation and integration the factors involved in gas exchange (Maina 1998), birds appear to have refined the structural parameters in preference to the physiological ones in contrast to bats, which favored the physiological ones: about 60% of the gas exchange capacity in birds can be attributed to the structural parameters and 40% to functional ones while in bats, 61% of the capacity can be attributed to physiological factors and 39% to the structural ones. Being generally small in size, the heaviest bat, *Pteropus edulis*, being about 1.5 kg (Yalden and Morris 1975; Carpenter 1985; Pough et al. 1989), leading a particularly energetic lifestyle, and operating from a level of a rather inferior lung, bats appear to have had very few choices but to fully refine and maximally exploit practically all the resources available to them. An "optimized" species is threatened with extinction if and when it is faced with increased demands to which it cannot respond (e.g., Minkoff 1983). Appropriate behavior and niche selection help minimize ambient pressures. In bats, we may unconsciously be witnessing a taxon living within very narrow tolerance limits of the environmental factors. The nocturnal lifestyle led by the bats may be explained by their more recent evolution of flight long after the birds had firmly occupied the diurnal niche for about 100 million years. The rather mundane respiratory physiological specializations for flight apparent in bats ruled out any possibility of successfully competing with birds.

6.7.5 The Avian Lung

It has been speculated by, e.g., Duncker (1978b) and Perry and Duncker (1980), that the avian lung-air sac system had its origin in the multicameral reptilian lungs like those found in the monitor lizards. Such lungs have a large ventilatable surface area (Maina et al. 1989a). The close phylogenetic affinity between reptiles and birds (e.g., de Beers 1954; Ostrom 1975; Jones et al. 1993) accounts for the anatomical similarities which abound between the two taxa. The chameleons possesses extensions from the lung similar to the air sacs of birds (Grassé 1970; Patt and Patt 1969; Klaver 1973, 1981; Fig. 96). Having arisen from reptiles much later than mammals, birds are phylogenetically more primitive than mammals (Romer 1967). From the arguments that some degree of endothermy may have developed in the Mesozoic and even Paleozoic reptiles (e.g., Romer 1967; Bakker 1975) and from the presence of a robust pectoral girdle, extensively ossified sternum, and expansive deltopectoral crest of the humerus (Romer 1966) (features which indicate well-developed flight muscles, excellent capacity to fly, and hence a high aerobic capacity), it has been conjectured that the direct progenitors of birds, the pterosaurs (order: Archosauria) of the Jurassic and Cretaceous, e.g., the 250-kg *Quetzalcoatlus northropi* (Langston 1981; Paul 1990, 1991) and *Pteranodon* (Bramwell 1971; Padian 1983) had developed complex, multichambered lungs. We shall, however, never know for certain about the structure and function of the lungs in these interesting now extinct reptiles. Fundamental differences in the design of the lungs with respect to respiratory gas flow pattern and overall efficiency occur between reptiles, birds, and mammals (Scheid and Piiper 1987).

Unique to the other vertebrate air-breathing groups, in birds, the lung has been uncoupled from the compliant bellows-like ventilatory air sacs (e.g., Fedde 1980; Maina 1983, 1996): the avian lungs are compact and virtually nonexpansile (Fig. 91). They are continuously ventilated by a synchronized action of the totally avascular air sacs (Lucas and Denington 1961; Marin-Girón et al. 1975) which play no role in gas exchange (Magnussen et al. 1976). During respiration, the volume of the lungs changes by a mere 1.4% (Jones et al. 1985). Experimental compression of the lung does not result in a substantial collapse of the air capillaries (Macklem et al. 1979). Although birds have smaller lungs per unit body mass than mammals (Maina et al. 1982b, 1983, 1989a; Maina 1989a; Maina and Settle 1982), the virtual rigidity of the avian lung has resulted in a substantial increase in the surface area per unit volume of the lung. The respiratory surface exceeds that of mammals of similar body mass (Maina et al. 1989a). The "insertion" of the bird

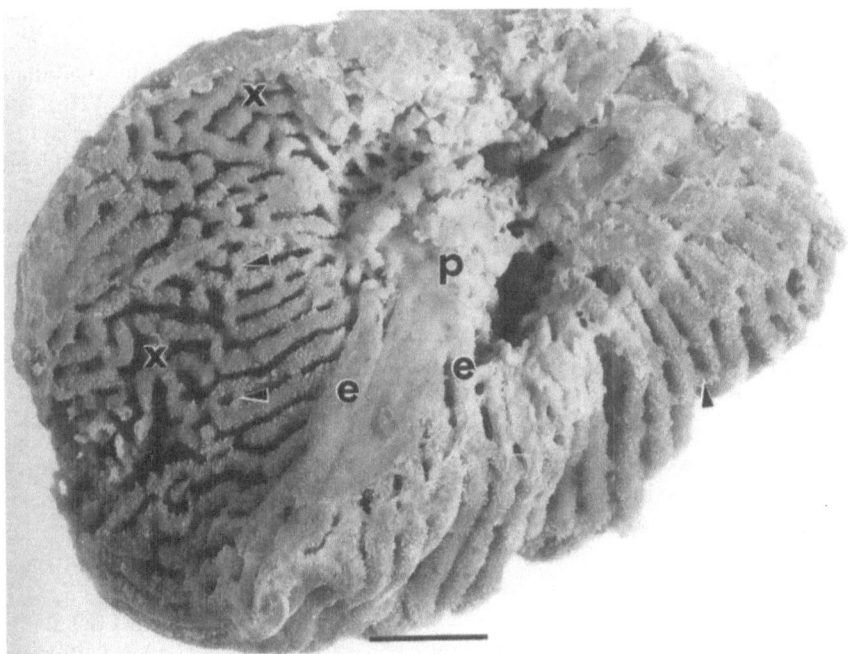

Fig. 91. Cast of the lung of the domestic fowl, *Gallus domesticus*, showing compact nature. The air conduits include the primary bronchus, *p*, the secondary bronchi, *e*, and the parabronchi, *x*. The ➤ show the anastomoses of the parabronchial system. *Bar* 8 mm

lung to the ribs followed by the relegation of the ventilatory compliance to the air sacs meant that surface tension was no longer a restricting factor in the extent of internal subdivision of the parenchyma. This resulted in extremely narrow terminal gas exchange components, the air capillaries, which gave rise to an extensive respiratory surface area and a thin blood-gas barrier (Maina 1981a; Maina and King 1982a). Considering the large number of species of birds (about 9000 species; e.g., Morony et al. 1975; Gruson 1976) and the remarkably diverse habitats they live in, the bird lung is morphologically remarkably homogenous. Subtle differences such as the degree of development of the parabrochi, spatial arrangement of the secondary bronchi, and size and location of air sacs, however, occur. The congruency in the morphological configuration of the bird lung may perhaps arise from the fact that all birds evolved from a common volant lineage (e.g., Cracraft 1986) and hence at one stage, they had to grapple with the demands for flight which some groups abandoned late. A fascinating difference in the morphologies of the bird lung is, however, that of the spatial arrangement, the degree of development, and the location of the tertiary bronchi (parabronchi). Two sets of parabrochial systems, the paleopulmo and neopulmo have been described (e.g., King 1966; Duncker 1974). The main differences between the papeopulmonic and the neopulmonic zones of the lung are: (1) the paleopulmonary parabronchi are located on the dorsocranial region of the lung and constitute about two thirds of the lung volume, while the neopulmonic set is located ventrocaudally and comprises about one third of the lung volume; (2) the paleopulmonic parabronchi are arranged in parallel stacks while the neopulmonic ones are irregularly arranged and anastomose profusely (e.g., López 1995); (3) while the air flow in the paleopulmo is continuous and unidirectional, that in the neopulmo changes with the phase of respiration (e.g., Fedde 1976); and (4) both sets of parabronchi start to develop at different embryonic stages (Romanoff 1960). The lungs of the primitive birds such as the kiwi and the penguin only have the paleopulmonic parabronchi while the neopulmo is variably well developed in the less conserved species, reaching maximal development in the relatively highly evolved passerines (King 1966; Duncker 1974). The substantive functional implication of the presence and absence of the development of the paleo- and neopulmo is still unclear. There are no morphometric differences in the structure of the gas exchange components in the two regions of the lung (Maina 1982b; Maina et al. 1982c). It is plausible that the bidirectionally ventilated neopulmonic parabrochi may provide a site for CO_2 cycling to alleviate occurrence of respiratory alkalosis due to excessive washout of CO_2 across the undirectionally ventilated paleopulmonic parabrochi especially during panting in thermal stressed birds (e.g., Jones 1982). The ostrich can pant continuously for as long as 8 h without experiencing acid-base imbalance (Schmidt-Nielsen et al. 1969). The avian lung is satisfactorily described by the crosscurrent model which considers that blood and parabronchial air interact at right angles (Fig. 32). Unlike in the mammalian alveolar system, under most conditions, the PCO_2 in the arterial blood ($PaCO_2$) and the PCO_2 in the end-expired air ($PECO_2$) are different, with the $PaCO_2$ being as much as 0.8 kPa less than $PECO_2$ (Scheid and Piiper 1970; Meyer et al. 1976; Hastings and Powell 1986). High-frequency respiration reduces the respiratory efficiency of the avian lung (Hastings and Powell 1987).

Flight is the most energetically demanding form of locomotion that has evolved in animals. It places substantial metabolic demands on an animal. Skeletal muscle accounts for 96% of a flying animal's total energy consumption during flight, as it does in a human being at maximum exercise (e.g., Folkow and Neil 1971). In practically all active vertebrate groups, locomotion exerts the greatest demands on the respiratory system (Banzett et al. 1992). A significant metabolic barrier differentiates volant vertebrates from nonvolant ones. The fact that birds and bats are the only extant vertebrates capable of powered flight attests to the extreme selective pressure these taxa have endured during the course of evolution of this mode of locomotion. Wind tunnel experiments show that the energetic demands of flight are beyond those attainable by nonflying vertebrates (e.g., Tucker 1968; Carpenter 1975; Thomas 1975). In turbulent air or when ascending, a bird can increase O_2 consumption for brief periods by about 20 to 30 times whereas even a good human athlete can attain such an increase for only a few minutes (Tucker 1970). The O_2 consumption in the pigeon while running was estimated to be 27.4 ml per min and while flying at 19 m per s was 77.8 ml per min, a factorial difference of 2.8 (Butler et al. 1977; Grubb 1982). Although expensive in its absolute demands for energy, active flight is a highly efficient form of locomotion, as at fast speeds the distance covered per unit of energy expended is less than in most other forms of locomotion (Tucker 1970; Thomas 1975; Rayner 1981). In the bats, *Phyllostomus hastatus* and *Pteropus gouldii*, the energy required to cover a given distance is only one sixth and one fourth, respectively, of that needed by the same-sized nonflying mammal (Thomas 1975). At their optimal speeds, the minimum cost of flying for a 380-g bird is about 30% of the energetic cost of a 380-g mammalian runner (Hainsworth 1981). After evolving independently from reptiles, birds and small mammals acquired aerobic metabolic scopes between resting and maximal rates of exercise or cold-induced thermogenesis which are 4 to 15 times greater than those of their progenitors at the same body temperature (Bennett and Dawson 1976; Bartholomew 1982b; Dawson and Dawson 1982).

Whereas most birds fly at moderate speeds with the smaller and more agile passerines attaining speeds of 15 to 40 kph, the swifts (Apodidae), the loons (Gaviidae), and the pigeons (Columbidae) have been reported to reach speeds of between 90 and 150 kph, while the peregrine falcon (*Falco peregrinus*) has been reported to dive on its prey at a speed in excess of 180 kph (Welty 1964): a yet unsubstantiated speed of 360 kph has been reported in a diving male falcon. When the avian flight speeds, which may appear mediocre by general standards, are normalized with body lengths covered, they turn out to be remarkably fast. The small passeriform birds, e.g., swifts, starlings, and chaffinches, flying at a moderate speed of 40 kph, cover about 100 body lengths per s (Alerstam 1982; Kuethe 1975) compared with only 5 in a highly athletic human being and only 18 body lengths per s in the cheetah, one of the fastest land mammal (Hildebrand 1959, 1961). In its annual migration, the Arctic tern (*Sterna paradisea*) flies from pole to pole, a distance of 35 000 km between breeding seasons (Berger 1961; Salomonsen 1967) while the American golden plover (*Pluvialis dominica*) flies

3300 km nonstop from Aleutian Islands to Hawaii in only 35 h (Johnston and McFarlane 1967). Collision between a vulture with a jet craft at an altitude of 11 km was reported by Laybourne (1974) and a flock of swans (probably whooper, *Cygnus cygnus*) was observed by radar at an altitude of 8.5 km (Stewart 1978; Elkins 1983). Amazingly, the 3-g ruby-throated hummingbird (*Archilochus colubris*) flies nonstop for nearly 1000 km across the Gulf of Mexico from the Eastern United States, a distance which may require about 20 h to cover (Lasiewski 1962, 1963a,b). A 3- to 4-g rufous hummingbird (*Selasphorus rufus*) has about 1 g of flight muscles which consume O_2 at a rate of 82 μmol per minute (Suarez et al. 1990). Many passerine species are known to fly continuously for 50 to 60 h on the Europe to Africa trans-Saharan route (Berger 1961). The swifts (Apodidae) are said to fly continuously, day and night, sleeping, eating, drinking, and mating on the wing (Lockely 1970): they only come to land when nesting or drop down when they die! The same has been said of the wandering albatross (Jameson 1958). The alpine chough (*Pyrrhocorax graculus*), which is reported to nest above 6.5 km (Swan 1961), not only faces hypoxia ($PO_2 = 9$ kPa) but also low temperatures (-27 °C) and a desiccating atmosphere. In premigrating hummingbirds, body mass may increase by up to 60% with fat stores (Odum and Connell 1956; Norris et al. 1957). Carpenter et al. (1983) and Carpenter and Hixon (1988) observed that during the period before migration, the rofous hummingbirds gain weight at an average rate of 0.23 to 0.30 g per day (about 8–10% of lean body mass). Brain hypoxia at altitude inhibits the central nervous stimulus for shivering thermogenesis and may set the limit for high altitude residence (Bernstein 1990). Due to the lower density and viscosity of air at altitude, though the aerodynamic drag is reduced, a bird has to do more work to generate the required thrust for level flight (Tucker 1974).

6.7.5.2 Functional and Structural Respiratory Adaptations for Flight in Birds

The efficiency of the lung-air sac system of birds is remarkable. Its capacity to provide the large amounts of O_2 needed for flight at high speeds, across huge distances, and at high altitude is exceptional by mammalian standards. Unequivocally, the main adaptive feature which enables birds to fly in hypobaric hypoxia is their unrivaled tolerance to hypocapnia (Bouverot et al. 1976; Black et al. 1978; Black and Tenney 1980). The increased ventilatory rate under strenuous activity of flight is not accompanied by an increase in the tidal volume and excessive CO_2 washout (Bernstein 1987). By reducing the PO_2 gradient between the arterial blood and that in the inhaled air (Shams and Scheid 1987) and by evoking a Bohr effect which raises the O_2 content of blood (Grubb et al. 1979), the hyperventilatory response during high altitude hypoxia enhances O_2 uptake. The champion high altitude flyer, the bar-headed goose, *Anser indicus* (Sect. 3.8.2) exhibits unique adaptations for coping with hypoxia. Experimentally, the goose is able to withstand hypoxia to a simulated altitude of 11 km (Black and Tenney 1980) and cerebral O_2 flow is not limiting (Faraci et al. 1984): up to an altitude of 6.1 km, the bird maintains normal gas exchange without hyperventilating, and at about 11 km, when the O_2 concentration is only 1.4 mmol l^{-1}, it extracts adequate amounts of it to necessitate only a minimal increase in ventilation (Black and Tenney 1980). Fedde et al. (1989) observed that muscle blood supply and O_2 loading from the muscle capillaries rather than ventilation or pulmonary gas transfer are the limiting steps in the contraction of the flight muscles of the bar-headed goose under hypoxia. Black and Tenney (1980) observed that the PO_2 in the arterial blood at a simulated altitude of 11.6 km is only 0.13 kPa less than that in the inhaled air, indicating a very high O_2 extraction efficiency of the lung.

The remarkable functional efficiency of the avian lung is a product of a synergism of various structural and functional parameters and processes. The principal ones are: (1) the crosscurrent arrangement between the parabronchial air and the pulmonary venous blood (Figs. 36,41,42); (2) to a yet undetermined extent the countercurrent disposition between the air capillaries and the blood capillaries (Maina 1988a; Fig. 42); (3) a large tidal volume; (4) a large cardiac output; (5) a continuous and highly efficacious unidirectional ventilation of the parabronchial gas exchange tissue (Colacino et al. 1977; Scheid 1979); (6) a short but adequate pulmonary circulatory time (Burton and Smith 1967); and (7) superior morphometric parameters which provide a high diffusing capacity of the lung (Maina 1989a; Maina et al. 1989a). The total volume of the pulmonary system in birds (i.e., the volume of the lungs, air sacs, and pneumatic spaces) on average constitutes about 20% of the total body volume, with the value being as high as 34% in the mute swan, *Cygnus olor* (Duncker and Guntert 1985a,b). The volume of the entire avian respiratory system (lung + air sacs) is three to five times larger than in mammals and two times larger than in reptiles. The total volume of blood in the bird lung comprises as much as 36% of the lung volume, with 58 to 80% of it being held in the capillaries (Duncker and Guntert 1985a,b; Maina et al. 1989a). The pulmonary capillary blood volume in birds is 2.5 to 3 times greater than in the mammalian lung, where only 20% of it is found in the alveolar capillaries (Weibel 1963). The gas exchange tissue (the parenchyma) of the bird lung on average constitutes only about 46% of the lung volume (Maina et al. 1982b) while in the mammalian lung the parameter constitutes a greater proportion (about 90%) of the lung volume (Gehr et al. 1981; Maina and King 1984; Maina et al. 1991). The large surface area per unit volume of the parenchyma in the avian lung (Maina 1989a; Maina et al. 1989a) is achieved through intense subdivision of the parenchyma into the remarkably small terminal gas exchange components, the air capillaries. The air capillaries generally range in diameter from 5 to 10 µm (Maina et al. 1981; Maina 1982a,b; Duncker 1974). In comparison, the smallest alveoli in the mammalian lung are the 30-µm diameter ones of the lung of the shrew (Tenney and Remmers 1963), the smallest extant mammal with the highest mass-specific metabolic rate (Fons and Sicart 1976). The surface density of the blood-gas barrier (the surface area per unit volume of parenchyma) in birds ranges from 172 mm^2 mm^{-3} in the domestic fowl (*Gallus domesticus*) Abdalla et al. 1982) to 389 mm^2 mm^{-3} in the hummingbird, *Colibri coruscans* (Dubach 1981; Duncker and Guntert 1985a,b): in the mammalian lung, the values are about one tenth those of birds (Gehr et al. 1981). It should be empasized that the intense internal subdivision of the gas exchange tissue in the bird lung occurs within the confines of a proportionately much smaller lung which has a smaller parenchymal volume proportion (Maina 1989a; Maina et al. 1989a). This division is so intense that the epithelial surface area of the air capillaries is essentially equal to that of the capillary endothelium. In the process of optimizing the gas transfer surface area, the terminal gas exchange components interdigitate closely with each other, nearly constituting mirror images. The capillary loading, i.e., the ratio of the pulmonary capillary blood volume to the respiratory surface area, a parameter which indicates the degree of exposure of the capillary blood to air, is near unity (Maina 1989a; Maina et al. 1989a). The rigidity of the avian lung (Macklem et al. 1979), of which the change in volume has been estimated not to exceed 1.4% during inspiration (Jones et al. 1985), has facilitated the intense subdivision of the bird lung, providing an extensive and thin blood-gas barrier with a high diffusing capacity for O_2 (Maina and King 1982a; Maina et al. 1989a). While in the mammals the compliance of the respiratory system (excepting the thoracic walls) is determined by the terminal parts of the respiratory tree (mainly the alveolar spaces), in the avian system, compliance is confined to the air sacs (Piiper and Scheid 1989). In the domestic fowl, the maximum compliance of the respiratory system was reported to be $9.6\,ml\,cm^{-1}$ H_2O (Scheid and Piiper 1969), in the duck $30\,ml\,cm^{-1}$ H_2O (Gillespie et al. 1982) and in the anesthetized pigeon $2.8\,ml\,cm^{-1}$ H_2O (Kampe and Crawford 1973).

6.7.5.3 Functional Reserves of the Avian Respiratory System

The lungs of birds maintain a large functional reserve. In the pigeon, *Columba livia*, Bech et al. (1985) and Koteja (1986) observed that cold exposure did not result in any perceivable respiratory distress, and Brackenbury et al. (1989) and Brackenbury and Amaku (1990) noted that experimental isolation of the thoracic and abdominal air sacs, a surgical procedure which renders about 70% of the total air capacity nonfunctional, had no detrimental effect on the gas exchange efficiency of the respiratory system of the domestic fowl: the birds could still run at three times the preexercise metabolic rate. From these experiments, Brackenburry (1991) observed that "there is considerable redundancy within the lung-air sac system in terms of its ability to adapt to removal of functional capacity, both at rest and during exercise". Investigations by other researchers corroborate this conclusion. The lung O_2 extraction was the same in exercising Pekin ducks in both normoxic and hypoxic conditions (Kiley et al. 1985). During flight, at ambient temperatures between 12 and 22 °C, lung O_2 extraction remains unchanged in the crows (Bernstein and Schmidt-Nielsen 1976) and starlings (Torre-Bueno 1978). The mute swan, *Cygnus olor* (Bech and Johansen 1980), has a lung O_2 extraction factor of 33% and the ventilation-perfusion ratio is near unity (Hastings and Powell 1986). Lung O_2 extraction values as high as 60 to 70% have been reported in some species of birds (Brent et al. 1984; Stahel and Nicol 1988). They exceed those of 40 to 45% reported in bats by Chappell and Roverud (1990). In the bar-headed goose, *A. indicus*, neither ventilation nor pulmonary gas transfer were limiting in a bird experimentally exercising in a hypoxic environment (7% oxygen) (Fedde et al. 1989). Hummingbirds, e.g., *Selasphorus platycerus*, can tolerate additional loading up to a maximum of 29.4% (Wells 1993b): a 10% increase in load called for a 5.7% increase in mass-specific O_2 consumption. The rufous-tailed hummingbird, *Amazilia tzacatl* has been reported to lift a mass about 80% of its body mass (Marden 1987). With high operational reserves, figuratively speaking, birds have not found it necessary to refine and/or utilize some of the structural and functional parameters commonly used by other animals. For example, the respiratory rate in birds is generally lower than in mammals (Calder 1968; Lasiewski and Calder 1971; Schmidt-Nielsen 1975) and the hemoglobin concentration and O_2 affinity of blood on average are essentially similar in the two groups (Lutz et al. 1973, 1974; Palomeque et al. 1980). In the mute swan, *Cygnus olor*, the respiratory rate is as low as three times per minute (Bech and Johansen 1980). Shams and Scheid (1989) and Scheid (1990) have suggested that the great endurance of birds to hypoxia is not due to the higher efficiency of the parabronchial lung compared with the broncho-alveolar one but rather in their ability to withstand lower arterial PCO_2 levels than mammals. In the resting domestic fowl, the arterial PCO_2 was estimated to be 4.4 kPa (Kawashiro and Scheid 1975) and in the exercising emu, *Dromaius novaehollandiae*, a value of 4 kPa, which is much lower than the average one of 5.3 kPa in the mammalian lung, was measured (Jones et al. 1983). In the mute swan, a pronounced positive PCO_2 difference between end-tidal gas (5.6 kPa) and mixed venous blood (4.3 kPa) averaging 1.3 kPa was reported by Bech and Johansen (1980). Experimentally, birds have been shown to tolerate arterial PCO_2 below 0.9 kPa (Shams and Scheid 1987; Faraci and Fedde 1986; Scheid 1990). Even after an increase of ventilatory rate by about 78% at a simulated altitude of 6.1 km, house sparrows, *Passer domesticus*, do not incur respiratory alkalosis (Tucker 1968; Lutz and Schmidt-Nielsen 1977). Unlike mammals, birds can withstand the high level of arterial hypoxemia because through a yet unknown mechanism, brain blood flow is not affected by arterial hypocapnia (Grubb et al. 1977; Wolfenson et al. 1982; Faraci and Fedde 1986). Paradoxically, in some species such as pigeons and bar-headed geese, the flow increases during such circumstances (Grubb et al. 1978; Faraci et al. 1984; Pavlov et al. 1987). Grubb et al. (1979) observed that by evoking a Bohr effect, hypocapnic birds reduced arterial PCO_2, raising blood O_2 content for any particular PO_2. This illustrates that in hypoxia, a hypocapnic bird enjoys better cerebral O_2 delivery than a normoxic one! In the human being, a reduction of the arterial PCO_2 to 1.3 kPa results in cerebral vasoconstriction, leading to a reduction of the flow of blood to the brain by about 50% (Wollman et al. 1968). In an apparent reassessment of their earlier view which recognized the superior design of the avian lung as a contributive factor in its exceptional efficiency (e.g., Scheid and Piiper 1987, pp. 123–124), based on small differences of the arterial PO_2 during gas exchange between birds and mammals, Scheid (1979, 1990) and Shams and Scheid (1989) cast doubt on the importance of the structure of the parabronchial lung (compared with the broncho-alveolar one of mammals) during exposure to extreme hypoxia. In Scheid (1990, p. 6), he expresses his sentiments as follows: "we have then to admit that we cannot decide whether the lung structure of birds has evolved out of functional needs or simply out of structural constraints with no significance for the higher efficiency". These sentiments warrant further debate. It is highly unlikely that the evolution of such an intricate and efficient system as the avian-lung air sac system could have been coincidental.

The lungs of the small and the metabolically active species of birds show distinct morphometric specializations (Maina 1993; Duncker and Guntert 1985a,b). The generally small passerine species which have a higher metabolic rate and operate at a higher body temperature (see King and Farner 1969) have lungs superior to those of the nonpasserine ones (Maina 1981b, 1982c,d, 1984). The small highly energetic violet-eared hummingbird, *Colibri coruscans* (Dubach 1981), has remarkably highly refined lungs while the gliding and soaring birds which expend less energy for flight, e.g., gulls (see Norberg 1985), have relatively inferior lungs (Maina and King 1982b; Maina 1987c). Among birds, hummingbirds have the highest hemoglobin content, O_2 carrying capacity, erythrocyte counts, and large hearts (e.g., Carpenter 1975; Johansen et al. 1987). By developing a very small body size, the hummingbirds have occupied an ecological niche used earlier only by the insects (Suarez 1992). Hummingbirds have a heart rate as high as 1300 times min^{-1} during hovering flight (Lasiewski 1964), a wing beat frequency of up to 80 times s^{-1} (Greenewalt 1960), a heart size about twice that of a mammal of equivalent body mass (Schmidt-Nielsen 1984), and whole body circulatory time of about 1 s (Johansen 1987). During hovering, the O_2 consumption is 40 ml O_2 g^{-1} per h (Lasiewski 1963a,b; Epting 1980; Bartholomew and Lighton 1986; Suarez et al. 1990). To support hovering, the most energetical mode of flight when lift is generated both at up- and downstrokes to overcome gravity (Weis-Fogh 1972; Epting 1980), the flight muscles (Pectoralis and supracoracoideus) constitute about 30% of the body mass (Hartman 1961; Suarez et al. 1990). The mitochondrial volume density (the fraction of the muscle fiber volume occupied by the mitochondria) in the flight muscles of hummingbirds is about 35% (Suarez et al. 1991). The highest mass-specific respiratory surface area of about 90 $cm^2 g^{-1}$ has been reported in the small and highly energetic violet-eared hummingbird (Dubach 1981) and the African rock martin, *Hirundo fuligula* (Maina 1984). The value is substantially greater than that of 43 $cm^2 g^{-1}$ in the shrews, *Crocidura flavescens* and *Sorex* sp. (Gehr et al. 1980). The very high value of the repiratory surface area of 800 $cm^2 g^{-1}$ reported in a hummingbird by Stanislaus (1937) must be treated with caution. An extremely thin blood-gas barrier (harmonic mean thickness) of 0.090 μm has been reported in the rock martin (Maina 1984) and the violet-eared humming bird (Dubach 1981): the thickness of the blood-gas barrier in the shrew is 0.334 μm (Gehr et al. 1981). The indolent galliform species, e.g. the domestic fowl, *Gallus gallus* variant *domesticus* (Abdalla et al. 1982), a bird which has been domesticated for over 5000 years (West and Zhou 1988) and the guinea fowl, *Numida meleagris* (Abdalla and Maina 1981), have low pulmonary diffusing capacities. Among birds, the lowest pulmonary morphometric values have been reported in the emu, *Dromaius novaehollandiae* (Maina and King 1989), a huge bird which in nature is exposed to few predators. The Humboldt penguin, *Spheniscus humboldti*, a good diver (Butler and Woakes 1984), has a remarkably thick blood-gas barrier (Maina and King 1987), a feature presumed to enable the lung to resist collapse under hydrostatic pressure (Welsch and Aschauer 1986).

The resemblance between the structure (presence of air sacs and air conduits) and function (unidirectional and continuous air flow) between the avian and insectan respiratory sytems (Sects. 6.7.5 and 6.6.1) is an astonishing example of morphological convergence (Weis-Fogh 1964b; Maina 1989b). The insects evolved at the middle of the Paleozoic era some 400 million years ago (Callahan 1972) and the birds about 150 million years ago (de Beers 1954). The congruent morphology in such evolutionary distant animals presents a historic masterpiece of convergence which occurred as the two groups of animals grappled with the challenges involved in acquiring the large amounts of O_2 needed for volancy. This entailed radical transformations of their ancient gas exchangers. While bats retained but highly refined the structure of the mammalian lung, the insects honed the book lungs into the exceptionally efficient tracheal system, pterodactyls (of which the design of the respiratory system is unknown), and birds modified the ancestral multicameral reptilian lung. Clearly, there is no immutable design of a gas exchanger which is a prerequisite for flight. Subject to other features like aerodynamic configuration of the body (e.g., development of wings) being met, theoretically, at a cost, the mammalian lung can be refined for flight. Except in insects with the highly efficient tracheolar system (Sect. 6.6.1), endothermy appears to be a fundamental requirement for flight. Interestingly, momentary (preflight) endothermy has evolved in many insects (e.g., Heinrich 1992). The different strategies adopted by distant animals to solve common problems of uptake and delivery of O_2 to the tissues during strenuous activities such as flight depend on the different resources available to them to make the necessary adjustments. Those animals which accomplished flight reaped substantial benefits which included occupation of the almost limitless aerial biotope, adoption of a more economical mode of foraging, escape from ground predators, and dispersal to diverse habitats. Such taxa have undergone remarkable adaptive radiation. In the Animal Kingdom, the insects are the most populous taxon with about 750 000 species (Wigglesworth 1972). Among the terrestrial vertebrates, birds constitute the most speciose group with about 9000 species (e.g., Morony et al. 1975; Gruson 1976) compared with 4200 in mammals. Among mammals, the bats, of which there are about 1000 species (Wimsatt 1970; Yalden and Morris 1975), constitute 25% of all mammalian species. The now extinct flying reptiles which included *Pteranodon*, the largest flying animal that ever evolved, displayed exceptional diversity of body form (Norberg 1990).

6.8 The Internal Subdivision of the Lung: the Functional Implications

In the higher vertebrates, an extensive respiratory surface area per unit volume is achieved by internal partitioning of a finite volume of parenchyma, e.g., in reptiles (Figs. 92,93), mammals (Fig. 94), and in birds (Fig. 95). Among the air breathers, except for the avian lung and the insectan tracheal system where the terminal gas exchange components, i.e., the air capillaries and the tracheoles are

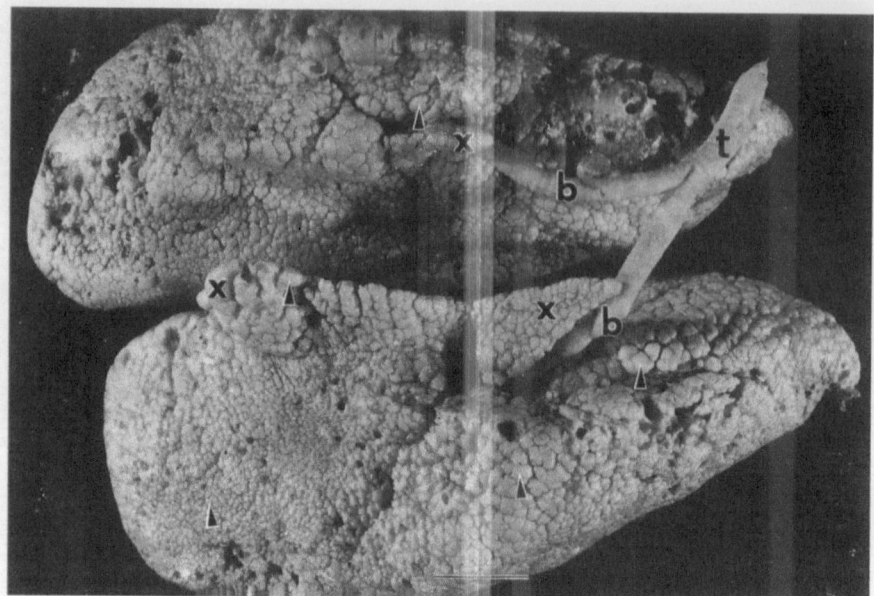

Fig. 92. Cast of the lung of the monitor lizard, *Varanus exanthmaticus*, showing the remarkable differences in the sizes of the air spaces in different regions of the lung, ➤. Unlike the lung of birds (Fig. 91), this lung presents some degree of lobulation, *x*; *t* trachea; *b* principal bronchi. *Bar* 13 mm

tubular structures, the amphibian, reptilian, and mammalian lungs are made up of numerous foam-like air bubbles surrounded by a moist surfactant-lined epithelium. Dilatation of these spaces requires energy. According to Laplace's Law, the pressure required to inflate a sphere is proportional to the surface tension and inversely proportional to the radius of a bubble. In practical terms, it requires more energy to inflate a lung with small alveoli than that with larger ones. It hence follows that while an intensely subdivided lung provides a more extensive respiratory surface area, more energy is needed to ventilate it. There must be an intrinsic compromise between the endeavor to establish an optimal pulmonary design and the cost of operating it. The low metabolism ectotherms have evolved lungs with large air spaces, mammals have developed small alveoli and an efficient ventilatory mechanism, while birds have evolved a rigid, noncompliant lung with remarkably small terminal gas exchange components. Initially, the surfactant developed in the archaic piscine lungs mainly to provide a protective continuous barrier over the epithelium (Liem 1987a). With the increasing internal complexity of the lung, the surfactant was commissioned to serve as a surface tension-reducing agent in order to prevent the collapse of the small terminal gas exchange components (Wilson and Bachofen 1982). In snakes (Ophidia), there is a tendency for a single lung to occur and in chameleons, extrapulmonary saccular extensions are characteristic of the lung (Fig. 96). The general pulmonary design in such reptiles approximates that of birds, where the gas exchange tissue (the

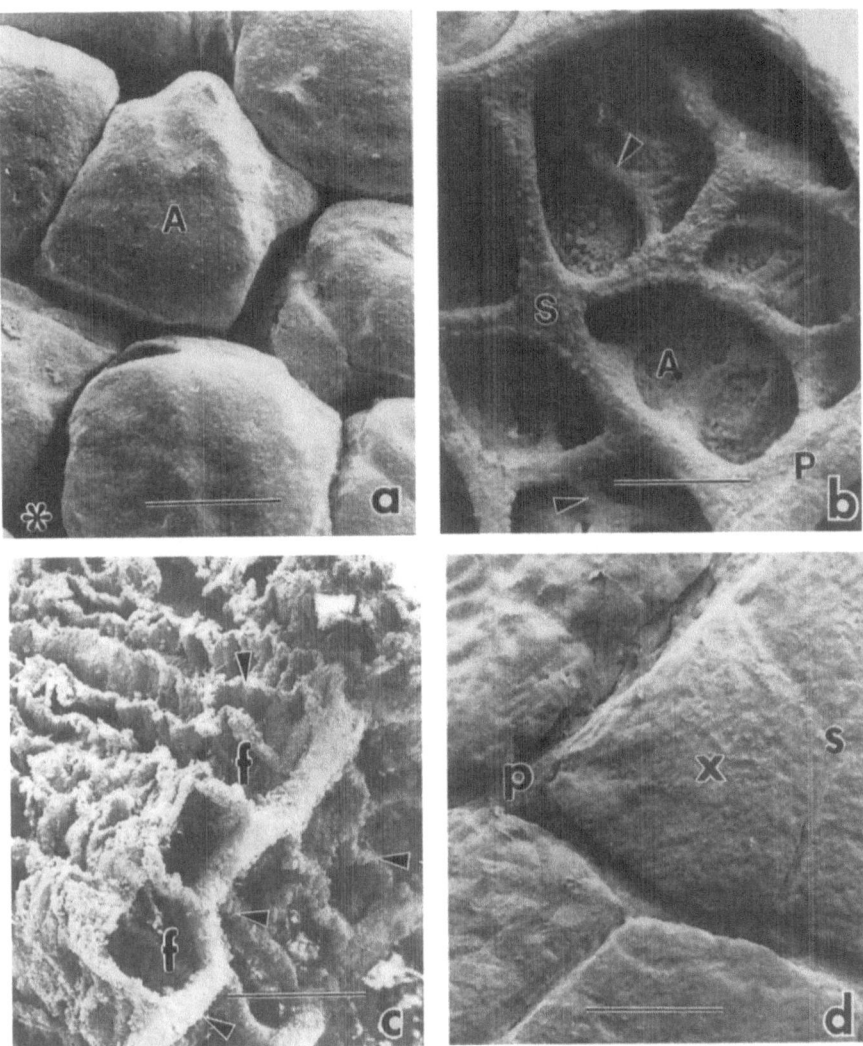

Fig. 93a–d. The respiratory surface area of the lung is increased by internal subdivisions. This process is shown in the reptilian lung. **a** Cast and **b** critical point dried preparation showing spaces, *A*, of the lung of the pancake tortoise, *Malacochersus tornieri*, formed by septation of the lung, ✳. **c** Lung of a snake, *Dendropis polylepi*, showing air spaces, *f*, separated by septae, ➤. **d** Cast preparation of the lung of the monitor lizard, *Varanus exanthematicus*, showing the various levels of septations; *p* primary septum; *s* secondary septum; ➤ tertiary septa; *x* terminal air spaces-definition of symbols also apply to fig. b. **a** *Bar* 300 μm; **b** 400 μm; **c** 500 μm; **d** 360 μm. (**a, b, d** Maina et al. 1989b; **c** Maina 1989e)

respiratory zone) is separated from the avascular (mechanical) region. Gas exchange tissue is located at the cranial part of the lung (Fig. 97) with the caudal saccular part ventilating it. The air passes twice through the gas exchange tissue, i.e., during inspiration and expiration, possibly enhancing O_2 extraction. A

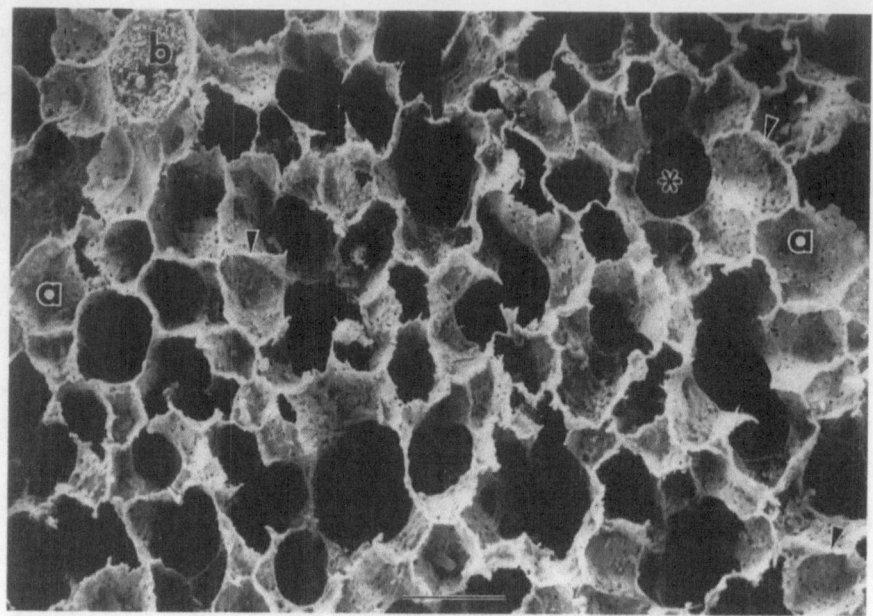

Fig. 94. Lung of the lesser bushbaby, *Galago senegalensis*, showing alveoli, *a*, the terminal gas exchange components. ➤ interalveolar septa; * alveolar duct; *b* blood vessel. *Bar* 133 μm. (Maina 1990c)

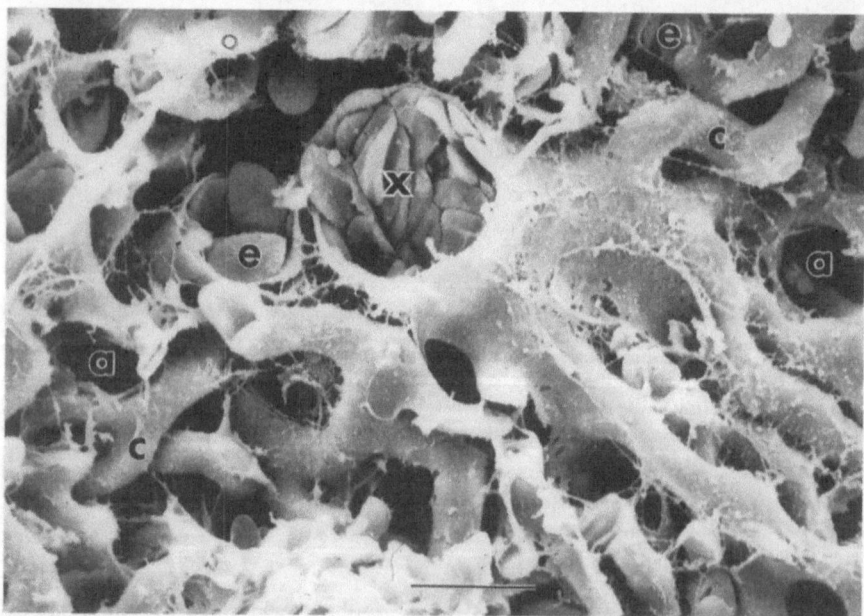

Fig. 95. Exchange region of the lung of the emu, *Dromaius novaehollandiae*, showing an intraparabronchial arteriole, *x*, giving rise to blood capillaries, *c*, which interdigitate with air capillaries, *a*; *e* erythrocytes. *Bar* 2.5 μm. (Maina 1994)

326

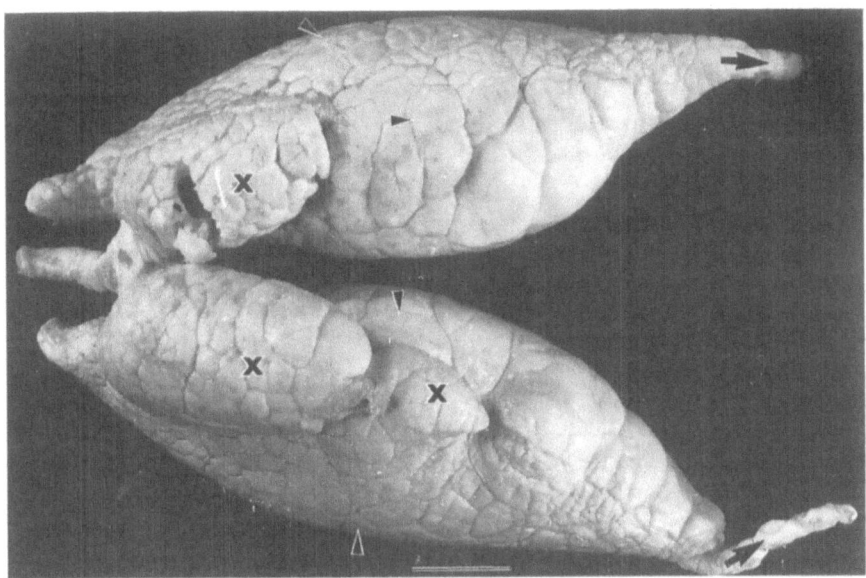

Fig. 96. Cast preparation of the lung of the chameleon, *Chamaeleo chameleo*, showing the notably large terminal gas exchange spaces and the size differences between the dorsal and ventral aspects of the lung, ➤. *Arrows* show the posterior saccular extensions similar to the air sacs of the lungs of birds. Some degree of lobulation of the lung, *x*, is evident. *Bar* 6 mm

similar organizational plan occurs in the lungs of the serpentine caecilians (Apoda), e.g., *Bourengerula taitanus* (Maina and Maloiy 1988). The left lung is very small and the tubular right lung is well subdivided in the cranial region with the caudal part being smooth (Figs. 98,99). Stratified septa provide mechanical support, which is necessary to avoid lung collapse. The smooth muscle and elastic tissue (Stark-Vancs et al. 1984; Goniakowska-Witalinska 1986) impart the tractability essential for pulmonary ventilation in air (Smith and Campbell 1976; Toews and MacIntyre 1978). In two genera of Salentia (Pipidae), the lung is internally supported by cartilaginous plates which are located in the 1st order of the septal walls in *Pipa pipa* (Marcus 1937) while in two Gymnophiona species, *Chthonerpeton indistinctum* and *Ichthyophis paucisulcus*, tiny aggregates of cartilage cells occur in the proximal part of the lung (Welsch 1981).

Internal partitioning of the lung increases its respiratory surface area. The intensity of the internal partitioning positively correlates with factors such as the metabolic rate, the lifestyle, and the environment in which an animal lives. While a sphere of a volume of 1 cm^3 has a surface area of 4.8 cm^2, 1 cm^3 of the lung of the shrew, *Sorex minutus*, has an alveolar surface area of 2100 cm^2 (Gehr et al. 1980). In the human lung, there are about 300 million alveoli of an average diameter of $250 \mu\text{m}$ (Weibel 1963), giving an overall surface area of nearly 150 m^2 and a thickness of the blood-gas barrier of $0.65 \mu\text{m}$ (Gehr et al. 1978). The respiratory surface area is about 100 times the total surface area of the body (Comroe 1974). Such an extensive surface area over which the respiratory media, air and

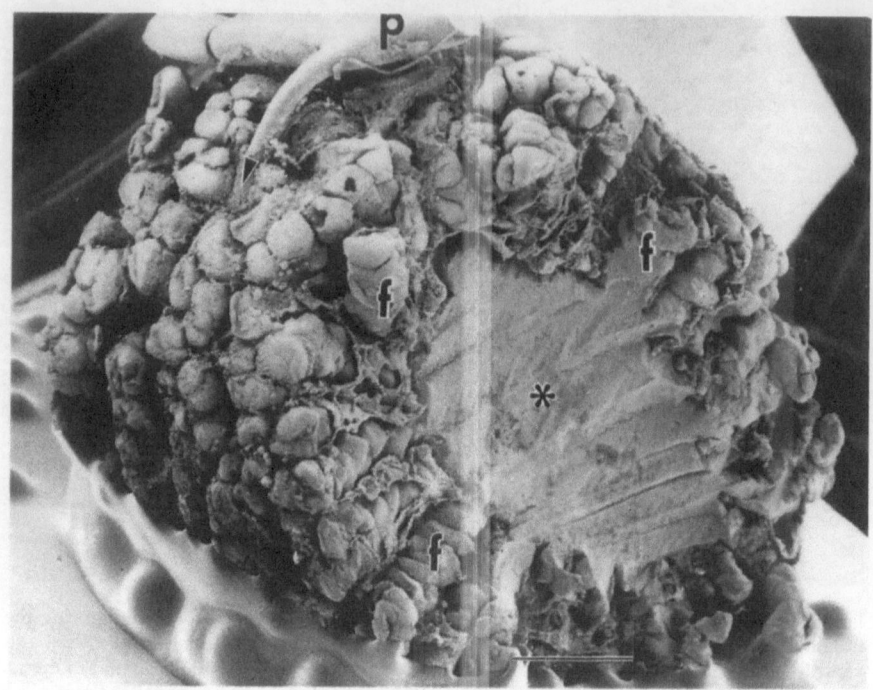

Fig. 97. Semimacerated double cast of the gas exchange region of the lung of the black mamba, *Dendroapis polylepis*, showing the air spaces, *f,* which radiate from the central air duct, *;* *p* dorsal aorta; ➤ pulmonary artery. *Bar* 2 mm

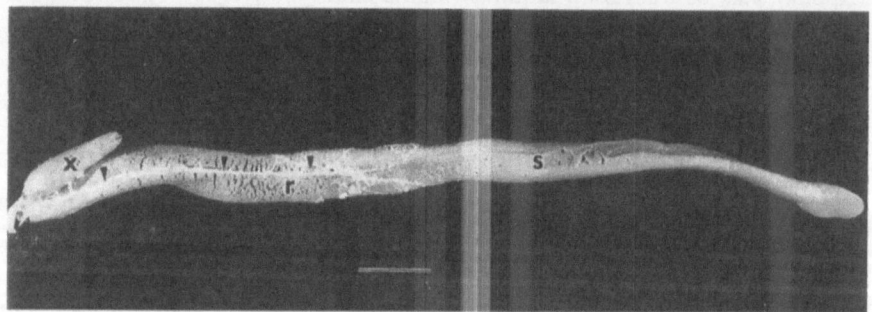

Fig. 98. Cast of a lung of a caecilian, *Boulengerula (Afrocaecicilia) taitanus,* showing the vestigial left lung, *x,* and the pulmonary artery, ➤, which runs three quarters of the length of the right lung. The posterior part of the lung, *s,* is smooth and saccular while the gas exchange region, *r,* is found in the anterior part of the lung. *Bar* 5 mm

blood, are separated by a thin barrier has called for development of various lines of defense to avoid damage by toxic substances and infection by pathogens. These include presence of mucus and ciliated cells on the upper respiratory passages (Figs. 100,101) and alveolar macrophages in the dependent regions (e.g., Brain

328

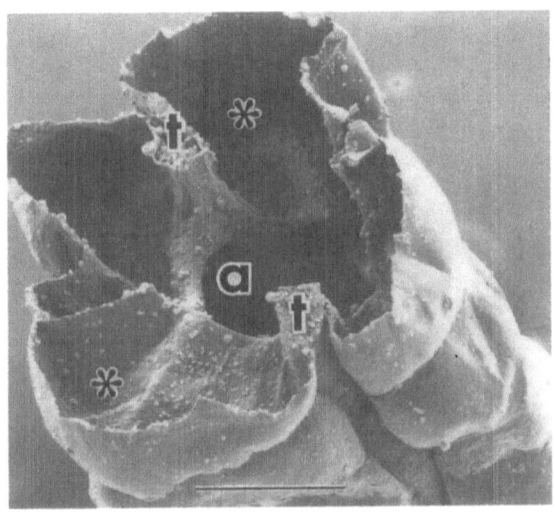

Fig. 99. View of the right lung of the caecilian, *Boulengerula (Afrocaecicilia) taitanus*, showing the central air duct, *a*, and peripheral gas exchange spaces, *∗*. The rather simple lung is supported by two diametrically located trabeculae, *t*. *Bar* 115 µm. (Maina and Maloiy 1988)

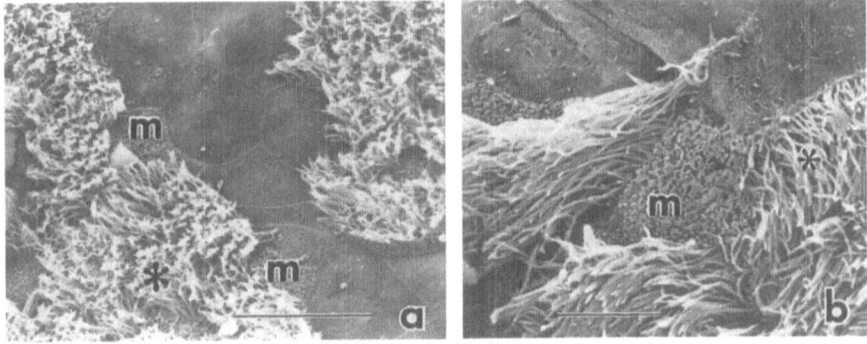

Fig. 100. a, b, Epithelial lining of the trachea of the respiratory system of the lung of the black mamba, *Dendroapis polylepis*, showing ciliated cells, *∗*, and mucus cells, *m*. **a** *Bar* 2 µm; **b** 6 µm

1985; Fig. 102). In spite of the fact that the respiratory mucosa presents the largest interface between the external environment and the internal milieu, in absence of pulmonary disease, below the larynx, the respiratory system is virtually sterile (Skerret 1994; Agostini et al. 1995). In the amphibians (Welsch 1983; Maina 1989d; Goniakowska-Witalinska 1995) and probably in ectothems in general, lung macrophages are rare. The number of pulmonary macrophages appears to

329

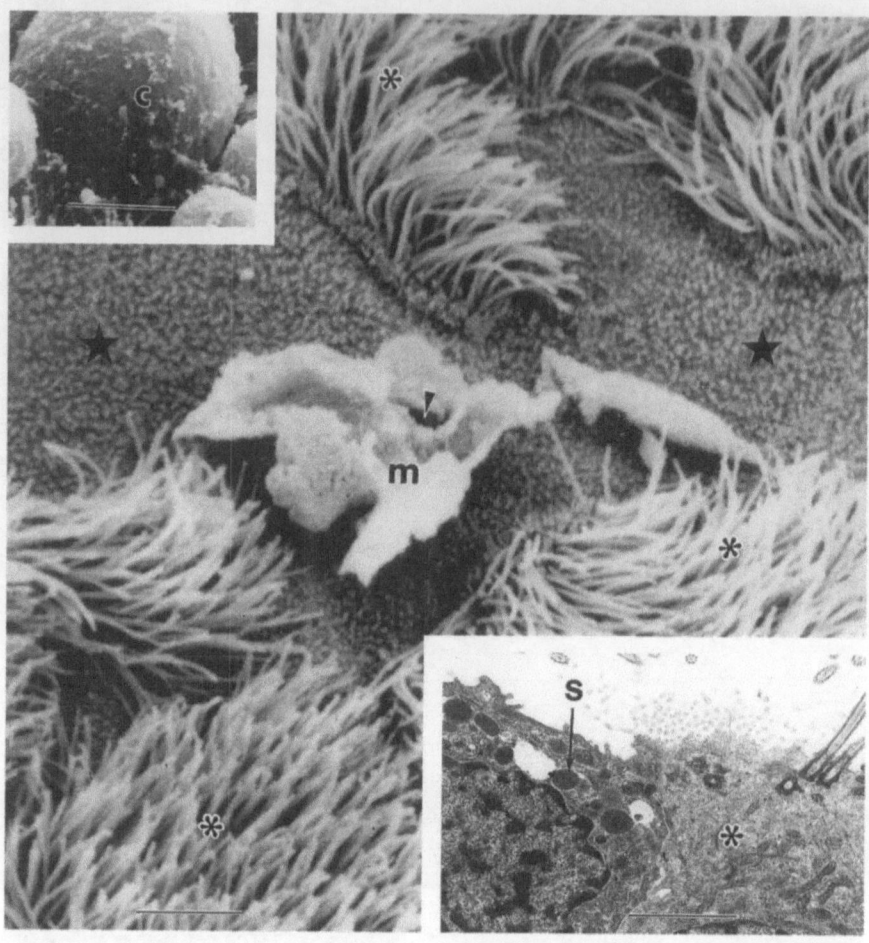

Fig. 101. Epithelial lining of the air passages of the lung of the vervet monkey, *Cercopithecus aethiops*: ∗ (*main figure*) ciliated cells; ★ mucus cells; *m* mucus debri; ➤ secretory pore of a mucus cell. *Top inset* Clara cells, *c*. *Bottom inset* a mucus cell with secretory granules, *s*, next to a ciliated cell, ∗. *Bar* (*main fig.*) 3.6 μm; *top inset* 0.5 μm; *bottom inset* 2 μm. (Maina 1988b)

correlate with the metabolic capacities of animals and the environment in which they live.

6.9 The Surfactant: a Versatile Surface Lining of the Gas Exchangers

While the insectan trachea, which are supported by helical cuticular taenidia, and the fish gills, whose filaments are physically separated by water, have little need for stabilization against surface forces, practically all evolved air-breathing

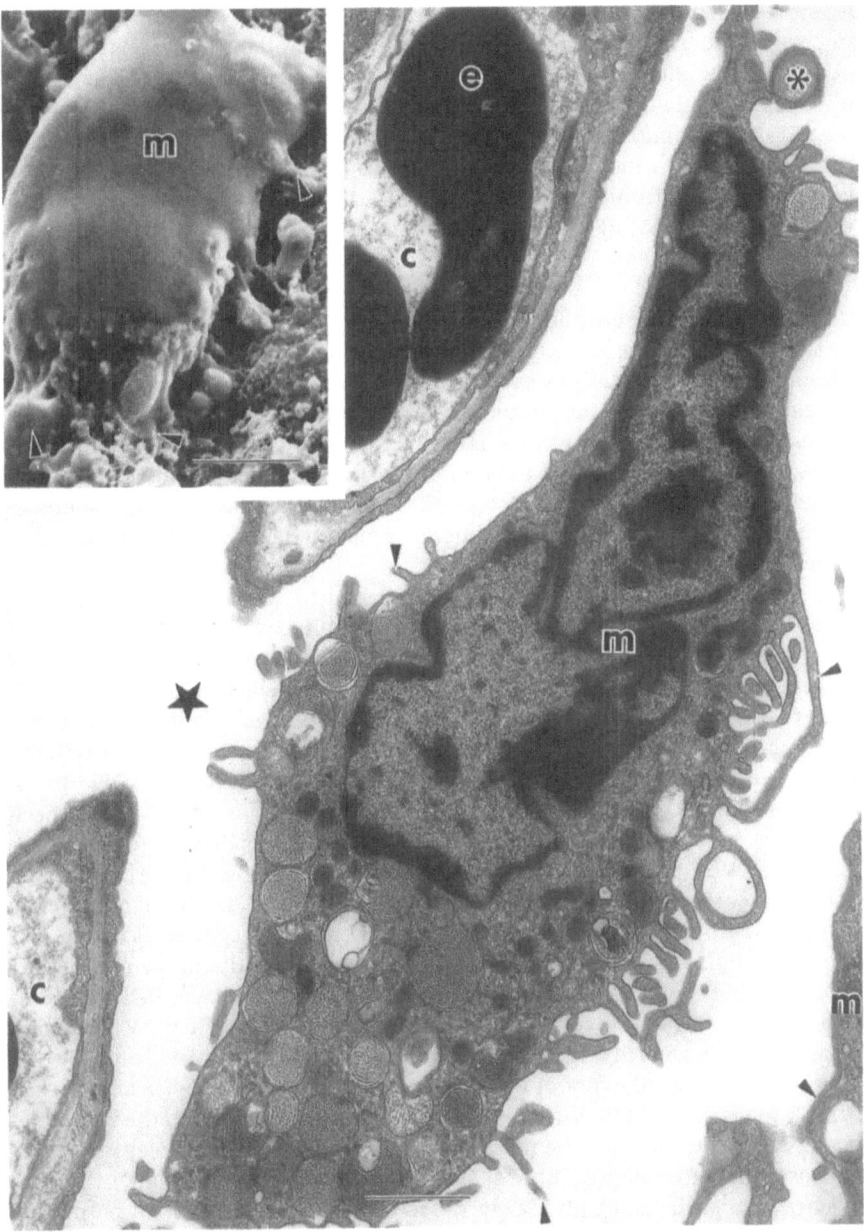

Fig. 102. Alveolar macrophages (*m*). *Main figure* (rat lung), ∗ a particle about to be ingested; ➤ filopodia; ★ interalveolar pore; *c* blood capillary; *e* erythrocytes. The macrophage shown in the *inset* is from a lung of a tree frog, *Chiromantis petersi*: ➤ filopodia. *Bar* 0.6 μm; *inset* 6 μm. (*Inset* Maina 1989d)

331

organs require presence of an active surface film for stability (Stratton 1984). The process of increasing surface area by subdivision of the lung engenders an increase in the surface tension in the terminal respiratory units. The energy necessary to dilate the gas exchanger during inspiration (an important factor in the cost of breathing) rises. High surface tension affects fluid balance across the blood-gas barrier which coupled with the capillary blood pressure enhances filtration of fluids from the capillaries. The surfactant, a phospholipid material (dipalmitoylphosphatidylcholine) which lines the respiratory surfaces in all gas exchangers evolutionary derived from the lung, reduces the deleterious effects of surface tension at the air-water interface by spreading on the respiratory surface as a monomolecular phospholipid film (Cochrane and Revak 1991). By lowering the surface tension, the surfactant, at least in the mammalian lungs, stabilizes the extremely small terminal gas exchange spaces. In human premature newborn, inadequacy of the surfactant due to lack of type II cells leads to the respiratory distress syndrome which may be fatal. In the vertebrate lungs, the surfactant serves to smooth the alveolar air-liquid interface (Bastacky et al. 1995) and to promote the displacement of deposited particles into the aqueous subphase where they are cleared by the pulmonary macrophages and the mucociliary carpet (Schürch et al. 1990).

In various forms and quantities, the surfactant is widely distributed in the vertebrate lungs (e.g., Clements et al. 1970; Hughes 1973; Pattle 1976; Dierichs and Dosche 1982; Hills 1988; McGregor et al. 1993). Lamellar bodies have been described in the lungs of the lungfishes (Hughes and Weibel 1976; Maina 1987a), those of fish such as bichirs, *Polypterus delhezi* and *P. ornatipinnis* (Zaccone et al. 1989), the gar-fish (*Lepisosteus osseous*) and the bowfin, *Amia calva* (Hughes 1973), in the epithelial lining of the fish swim bladder (Copeland 1969; Brooks 1970) and in the distal (respiratory) part of the intestine of the pond roach, *Misgumus fossilis* (Jasinski 1973). In the lungs of the higher vertebrates, it is secreted by the granular pneumocytes, the type II cells (Fig. 103). The composition of the surfactant and probably its function (at least in the human lung) may be regulated by the O_2 tension in the air spaces (Acarregui et al. 1995). The amphibian pulmonary type surfactant may or may not be discharged into the air space as tubular myelin type (Goniakowska-Witalinska 1980a,b; Bell and Stark-Vancs 1983), is mainly composed of phosphatidylcholine (e.g., Vergara and Hughes 1981), occurs in large quantities (Clements et al. 1970), and is not particularly surface-active like that of the higher vertebrates (e.g., Hughes and Vergara 1978; Daniels et al. 1989). The role of the surfactant, especially in the amphibian lung, is enigmatic, as even the most elaborate lungs have wide air cells which are not susceptible to collapse. Pattle et al. (1977) showed that in the lung of *Triturus vulgaris*, the surfactant only partly reduces the surface tension to $0.2\,\mathrm{mN\,m^{-1}}$ compared with the more efficient mammalian one which reduces it to $0.1\,\mathrm{mN\,m^{-1}}$ or less. In addition to the better-known roles of preventing atelectasis, edema, reduction of respiratory work, and stabilization of the terminal gas exchange components (e.g., Fishman 1972; Farrell 1982b), the surfactant plays several roles such as prevention of transendothelial transudation of substances across the blood-gas barrier, immune suppression, chemotaxis of macrophages (e.g., Daniels et al. 1993), and antioxidant function (Brooks 1970). The

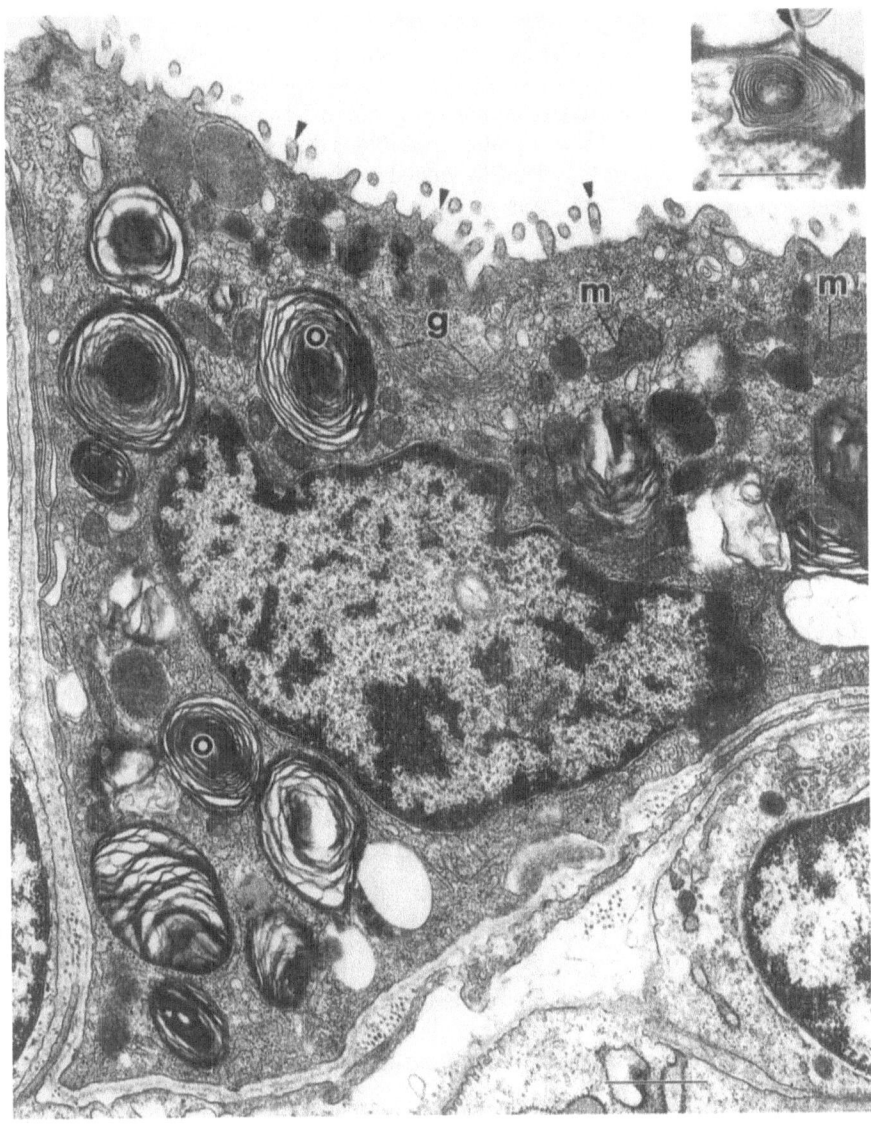

Fig. 103. A Type II cell from the lung of the tree frog, *Chiromantis petersi*, showing mitochondria, *m*, Golgi bodies, *g*, and osmiophilic lamellated bodies, *o*. ➤ microvilli. *Inset* a type II cell from the lung of a caecilian, *Boulengerula (Afrocaecicilia) taitanus*, showing secretion of the surfactant onto the surface of the lung. *Bar* 0.7 µm; *inset* 0.3 µm. (Main figure Maina 1989e; *inset* Maina and Maloiy 1988)

multifunctionality of the surfactant may help explain why the simple lungs, e.g., fish and amphibian lungs which have large air cells, possess the lining. In such animals, surface tension reduction may not be the primary role of the lining. In agamid lizard, *Ctenophorus nuchalis* (Daniels et al. 1990), the salamander,

Amphiuma tridactylum (Stark-Vancs et al. 1984), and *Siren intermedia* (Guimond and Hutchison 1976; Martin and Hutchison 1979), the surfactant-like lipids are envisaged to act as an antiglue preventing epithelial adhesion after near total pulmonary collapse during apnea (Frappell and Daniels 1991; Daniels et al. 1993; McGregor et al. 1993). In aquatic amphibians this occurs during a dive when the hydrostatic pressure may lead to virtual pulmonary collapse (Daniels et al. 1994) and in mammals in cases of alveolar collapse at different points in a ventilatory cycle (Hills 1971; Weibel et al. 1973; Sanderson et al. 1976). In the Weddell seal, *Leptonychotes weddelli*, the alveoli are totally collapsed at depths of 30 to 35 m (Falke et al. 1985) but the lung of the dolphin, *Tursiops truncatus*, does not collapse, at least up to a depth of 70 m (Ridgway and Howard 1979). In the agamid lizard, the lung contains 70 times more surfactant per respiratory surface area than those of a comparable-sized mammal (Daniels et al. 1989). The antiglue role of the surfactant may be important in the lungs of those animals which lack well-configured intrapulmonary conducting airways and a diaphragm (e.g., amphibians, reptiles, and birds), where the lung is mechanically more susceptible to the displacements of the organs in the coelomic cavity (Daniels et al. 1994). The presence of the surfactant on the air capillaries of the bird lung (Fig. 40a), structures which are somewhat fixed in size, is most intriguing. While it may simply be a phylogenetic carryover from a primordial reptilian lung, especially the crocodile one (e.g., Perry 1992b), it may as well have evolved as a means of curtailing transendothelial filtration of fluids from the blood capillaries in order to prevent flooding of the air capillaries. It is interesting that there is no surfactant either in the fish gills or in the accessory respiratory organs of the bimodal breathing fish. The surfactant appears to be an innovation of the lung and its most immediate derivatives. As evident in most facets of respiration, exceptions pervade the field. Some air-breathing fish such as the gourami, which construct bubble nests, secrete a surface-active (surpellic) substance to reduce surface tension at the liquid-air interface (Phleger and Saunders 1978).

6.10 Nonrespiratory Roles of the Gas Exchangers

6.10.1 Trophic, Sensory, and Locomotory Functions

Owing to the fact that the respiratory system shares a common passage, the pharynx, with the digestive system, activities such as change in posture, eating, swallowing, drinking, and rumination affect ventilatory activity, but only momentarily. This results in moderate but noneventful asphyxia. In some animals, however, predominant activities such as feeding (in filter feeders), husbandry of young (e.g., in the mouth brooding fish), and phonation greatly conflict with the gas transfer function. In some cases, respiration may be relegated in preference for a nonrespiratory role. Compared with *Chaetopterus variopedatus*, a filter-feeding marine annelid which ventilates its burrow continuously, the lugworm, *Arenicola marina*, a deposit feeder which swallows the detritus and regurgitates sand after extracting organic matter (Jacobsen 1967) has a ventilatory rate (at

15 °C) about ten times lower (Toulmond 1991): the O_2 extraction coefficient in *C. variopedatus* is 30%, while in *A. marina* the value is 80%. Debate has persisted (e.g., Willmer 1970) as to whether the initial task of the archetype respiratory organ, the gill, was respiratory or trophic. Filter feeders like sponges, ascidians, and lamellibranchs move water over their body surfaces, establishing a PO_2 gradient between the surrounding water and the body cells. In the process, microscopic unicellular organisms are entrapped (Hazelhoff 1939; Jorgensen 1955). Hazelhoff (1939) observed that at the same PO_2 in water, the O_2 extraction coefficient in the filter feeders is lower than that of the nonfilter feeders. This is presumably due to the necessary high flow rate of water needed to procure an adequate quantity of food. Jorgensen (1975) suggested that the rate of water flow over the gills is well synchronized with the O_2 and food concentration in a particular environment. In the polychaete worms, the irrigation of the branchial crown is $70\,\mathrm{ml\,g^{-1}h^{-1}}$ while that of the tube is only $12\,\mathrm{ml\,g^{-1}h^{-1}}$ (Dales 1961): O_2 utilization by the crown is about two times that of the skin. The marine polychaete, *Chaetopterus variopedatus*, a suspension feeder, continuously ventilates its burrow by body movements (Brown 1975, 1977), extracting O_2 and nutrients from the water current (McGinitie 1939). In organisms which use the same site for feeding and respiration, synchronization between gas transfer and feeding is most critical. The larva of the insect *Chironomus plumosus* exhibits cyclic behavior which changes from filter feeding to rest and from rest to respiration (water ventilation). Normally, the rest intervals are brief, with the two other episodes being of about the same duration: ventilation increases in a hypoxic condition at the expense of the feeding interval, the limiting external PO_2 for filter feeding being $2\,\mathrm{kPa}$ (Walshe 1950; Ewer 1959; Rubenstein and Koehl 1977). In aquatic animals such as the sea urchins, the respiratory organs (podia or tube feet) are utilized both for locomotion and sensory perception – in addition to their gas exchange role (Steen 1965). In the sabellid polychaete, *S. spallanzani*, a 60% drop in O_2 uptake was recorded by Fox (1938) after amputation of the branchial crown, a very active ciliary feeding structure and evidently an important respiratory one. In the bivalve mollusks, except for some Protobranchia, the ctenidia are greatly hypertrophied for their added role in extraction and transportation of food (e.g., Atkins 1936). In *Pholas dactylus*, particulate food is filtered from the large volumes of water (about 100 l) which pass through them per day (Knight 1984). The current of water created is far in excess of that needed for respiratory purposes in these rather sluggish, often sessile animals. The cephalopods, a toxon which traces its evolutionary lineage some 500 million years back, have conservatively utilized jet propulsion as the basic locomotor mechanism (e.g., Teichert 1988; Chamberlain 1990; O'Dor and Webber 1991). The ventilatory currents serve a locomotory role (Yonge 1947; Wells 1952). Oxygen extraction in cephalopods is variable, depending on the degree to which the animal uses jet propulsion for locomotion. In *Nautilus*, when in motion, O_2 extraction drops to as low as 4% and while at rest it is on average 5 to 10% (Wells and Wells 1985). *Sepia* and *Octopus*, which can uncouple locomotion from respiration, have O_2 extraction rates which are 35 to 45% or even much higher (O'Dor and Webber 1991). The volume of the water across the gills of swimming squids is so large that O_2 extraction is only 5 to 10% and on exposure to hypoxia and extreme exercise rises to a maximum of 17%

(Wells et al. 1988; Shadwick et al. 1990). The most aerobic muscle fibers in squids, which are located in the inner and outer surfaces of the mantle, contain as much as 50% mitochondria by volume (Bone et al. 1981; Mommsen et al. 1981), a value which compares with that of the insect flight muscle (Elder 1975; Ready 1983).

As a general rule, multifunctionality of a gas exchanger is an indication of primitiveness, especially where the roles greatly conflict, e.g., in designs for such processes as feeding, sensation, and respiration. Such exchangers must inherently be designed to accommodate remarkably different needs which cannot be compromised. Filter feeding occurs in fish such as the paddle fish, *Polyodon spathula*, anchovies, and menhaden (Durbin et al. 1981; Lazzaro 1987; James and Probyn 1989; Burggren and Bemis 1992) which utilize ram ventilation as they swim continuously. In the sabellid, *Schizobranchia insignis*, extirpation of the crown resulted in a 75% drop in O_2 consumption (Vo_2) while those worms which retained the crown but were prohibited from extending it into the water showed a 60% drop in Vo_2 (Dales 1961). The importance of the ciliary organs in respiratory function in polchaetes is remarkable. Assuming that the surgical procedure does not greatly interfere with the metabolic level, the total O_2 uptake by the two halves of a transversely bisected *S. pavolina* is not significantly lower than that of the whole worm but in *Myxicola infundibulum*, a dramatic drop in total O_2 uptake occurs on bisection. The posterior body part gives lower values compared with those of *S. pavolina* (Wells 1952). The differences show that in *S. pavolina*, which irrigates its tube through body movements, the ciliary crown does not contribute to the O_2 requirements of the rest of the well-ventilated body. However, in *Myxicola*, where the tube is not irrigated, dependence on the crown is much greater. The role of the respiratory organs in phonation and the physiological consequences on respiration and acid-base status of blood has been outlined by Dejours et al. (1967) and Bouhuys (1974), while the thermoregulatory aspects have been reported in mammals by Cunningham and O'Riordan (1957) and in birds by Calder and Schmidt-Nielsen (1968) and Lasiewski (1972).

6.10.2 Metabolic and Pharmacokinetic Functions of the Lung

The lung is best known for its role as the oxygenator of the blood and the provider of O_2 to the body tissues (e.g., West 1974; Weibel 1984a). This role is axiomatic from its design, which incorporates a large surface area and thin blood-gas barrier. The facts that the lung is mechanically ventilated with air by the ribs, perfused with blood by the heart, and gas exchange occurs by the passive process of diffusion is suggestive of an inert organ. For these reasons, the biochemical, pharmacological, and metabolic roles of the lungs are not as well recognized as the respiratory one (Heinemann and Fishman 1969; Vane 1969; Tierney 1974). This is despite the fact that as far back as over 70 years ago, Starling and Verney (1925) observed that isolated kidneys perfused with defibrinated blood quickly caused vasoconstriction but if the blood was passed through a heart-lung pump preparation, the vasoconstrictor response was absent. They concluded that the blood was "detoxified" in the lungs. Twenty five years later, Rappaport et al. (1948) identified the serum vasoconstrictor substance as 5-hydroxytryptamine (5-HT) and demonstrated that isolated lung preparations inactivated 5-HT. The processes of substrate handling by the lung was well documented by Eiseman et al. (1964) and the subject reviewed by, among others, Bakhle (1975). Though the complete spectrum of the roles the lung plays in physiological and pathological states remains largely unappreciated, it is now well recognized that the organ performs various important metabolic tasks. It is said to be "a peculiar immuno-logical organ which can operate independently and synchronously with the general immune apparatus" (Agostini et al. 1995). Certain tasks are as important, if not more important, than the respiratory one. Bakhle (1975) asserts that "instead of referring to pharmacokinetics as one of the functions of the lung, we ought to refer to gas exchange as one of the nonpharmacokinetic functions of the lung". In some instances, the lung performs the nonrespiratory functions more efficiently than other organs which are considered more specialized (Table 27). The lung is an important source of biologically active agents, constitutes an important organ for defense, and, paradoxically, may initiate, actively participate in, or even sustain pathological processes such as allergy (Nicolet et al. 1975), emphysema (Kimbel and Weinbaum 1975), shock (Bleyl and Büsing 1971), pulmonary cancer (Cohen 1975), pulmonary necrosis (Reid et al. 1973), and hyaline membrane disease (Avery and Mead 1959). To carry out the roles which range from defense, clearance of mucus, and electrolyte transport, to the pharmacological ones, the lung has a large number of different cell types (e.g., Breeze and Wheeldon 1977; Andrews 1981; Weibel 1984b; Welsh 1987).

Anatomically, the lung is strategically located in the general circulation dividing the venous and the arterial circulations. Owing to its vast vascular transit distance, it contains the highest volume of blood per unit time of any other organ in the body. During the transit of blood through the lung, a fine balance between providing the essential metabolic nutriments to the tissue cells and limiting transport of salts and water into and out of vascular space must be maintained (Welsh 1987). More than in other tissues, the lung must make allowance for unhindered passage of some hormones and biologically active substances from the arterial to the venous side of circulation (Ryan and Ryan 1975). The adjust-

Table 27. N-demethylation rate of several drugs in rabbit microsomes prepared either from lung or liver. (Remmer 1975)

Substrate	Lung	Liver
	nmol HCOH formed mg^{-1} microsomal protein min^{-1}	nmol HCOH formed mg^{-1} microsomal protein min^{-1}
Aminopyrine	5.2	15
Ethylmorphine	2.2	3.5
N-Methyl-alanine	3.0	3.0
N-Methyl-p-chloroaniline	7.2	4.8
(+) – Benzphetamine	8.8	4.7

ments which take place after a substrate passes through the pulmonary circulation can be categorized into intrinsic ones, i.e., those which affect the activity of the substrate itself and the extrinsic ones, i.e., those which entail in situ production of biologically active factors (Bakhle 1975): the latter constitutes an endocrine function. The active factors are handled differently by the endothelial cells of the pulmonary vasculature particularly at the capillary level (e.g., Ryan and Ryan 1975). The endothelial cells of the lung have abundant caveolae cellulares (micropinocytotic vesicles; Figs. 29b,40a), most of which are in direct contact with the circulating blood and are irrefutably the sites of degradation, transformation, interaction, and biosynthesis of the macromolecules for which the lung has affinity and transendothelial transfer of the same to the alveolar surface (e.g., Bignon 1975). The lung has a very high pharmacokinetic specificity for the circulating biologically active enzymatic complexes. For example, the pulmonary circulation acts as a filter, letting adrenaline or histamine pass through but inactivating 5-HT, bradykinin, and synthesizing angiotensin II from angiotensin I (e.g., Bakhle 1975; Ryan and Ryan 1975; Nossaman et al. 1994) (Tables 28,29,30). A well-known example of the endocrine role of the lung is that of appearance of histamine in the lung perfusate when an antigen is passed through the pulmonary circulation of the isolated lungs from sensitized guinea pig (Bartosch et al. 1933).

Table 28. Pulmonary extraction of circulating 5-hydroxytryptamine (5-HT). (Junod 1975)

Species	Concentration of 5-HT	Extraction (%)
Man	Bolus injection of $7.6\,\mu g\,ml^{-1}$	65
Rat	$5-100\,ng\,ml^{-1}$	92
	$0.02\,\mu g\,ml^{-1}$	20–58
Dog	$10-20\,\mu g\,ml^{-1}$	80–98
Rabbit	$250\,ng\,ml^{-1}$	52.6
	$500\,ng\,ml^{-1}$	56
Guinea pig	$1-1000\,ng\,ml^{-1}$	48–60

Table 29. Pulmonary inactivation of bradykinin. (Friedli et al. 1973)

Group	Age of Gestation (Days)	Inactivation (% mean and standard deviation)
Ewes n = 5	Adult	$93.4 \pm SD\ 1.20$
Lambs (n = 5)	Immature (12 Days)	$68.0 \pm SD\ 5.28$
Fetal lambs at term (n = 4)	Term (144 Days)	$46.5 \pm SD\ 4.80$
Premature fetal lambs (n = 4)	Immature (118 Days)	–

It is now well known that different pharmacologically active substances are released during anaphylaxis and that other stimuli are capable of causing the release of active substances from unsensitized lungs (e.g., Piper 1975). The lung secretes numerous active biochemical substances such as histamine (Bakhle 1975), slow reacting substance of anaphylaxis (Brocklehurst 1960), dipeptidyl carboxypeptidase (Roth and Depierre 1975), bradykininase, angiotensin I converting factor (Bakhle 1975), and prostaglandins (e.g., Änggård 1975; Piper 1973) in response to different stimuli which include anaphylaxis, physical deformation, biogenic amines, and peptides. Other than in anaphylaxis, the relevance of these experimental findings on isolated organs of laboratory animals to in vivo situations and particularly extrapolation to the human being is still a matter of speculation (Berry et al. 1971). More recently, however, the use of cardiopulmonary bypass as an investigative procedure (e.g., Parker et al. 1975) has corroborated some of the earlier assumptions. It is now well established that the metabolic and the pharmacokinetic functions of the lung are important for some crucial processes such as blood coagulation and inactivation and destruction of hormones and other biologically active factors. It is plausible that these processes may serve as an alternative safety system should other integral organs like the liver, or the kidneys be rendered ineffective. The nonrespiratory functions of the

Table 30. Effect of passage through the pulmonary circulation of biological activity. (Bakhle 1975)

Class of substance	Activation	Inactivation	No change in activity
Biogenic amines	–	5-Hydroxytryptamine Tryptamine Noradrenaline Acetylcholine	α-Methyl 5- hydroxytryptamine Adrenaline Isoprenaline Histamine
–	Angiotensin I Reno-active peptide	Bradykinin Reno-active peptide	Angiotensin Eledoisin Physalaemine Oxytocin Vasopressin
Prostaglandins	Arachidonic acid Other PG precursors	Prostaglandins of E and F type	Prostaglandins
Nucleotides	–	Adenosine monophosphate (AMP) Adenosine triphosphate (ATP)	–
Basic drugs	–	By absorption: Imipramine chlorocyclizine aphetamine By metabolism: methadone	–

lung are affected by factors such as age (Melmon et al. 1968), pregnancy (Bedwani and Morley 1974), and exposure to gaseous anesthetics (Naito and Gillis 1973).

The diversity of the functions carried out by the lung compares with the well-established multifunctional ones of the fish gills which, in addition to respiration, include osmoregulation, acid-base balance, nitrogen excretion, and modification and conditioning of plasma hormones before perfusion of the systemic circuit (e.g., Olson 1996). The lung is also known to greatly modify biological activity of many substances endogenous and exogenous to the pulmonary circulation (Alabaster and Bakhle 1972; Remmer 1975). Smooth muscle contractility in the femoral arterial blood of cats was much less (after infusion of bradykinin) when the chemical was infused into the right ventricle than when the infusion was made into the aorta just distal to the aortic valve (Alabaster and Bakhle 1972; Levine et al. 1973): an inactivation factor of bradykinin ranging from 75 to 99.9% during the passage from the right ventricle to the heart was determined. The degradation of bradykinin in the lung was shown by Ryan et al. (1970) to be very fast. A peptidase from the pig lung has been shown to split the COOH-terminal dipeptide from bradykinin (Igic et al. 1972; Dorer et al. 1974). The ability of the lungs to inactivate bradykinin, a hypotensive factor, while converting (activating) angiotensin I to angiotensin II, a potent hypertensive agent (Nossaman et al. 1994; Table 30) indicates the central role of the lung in blood pressure regulation. Thomas and Vane (1967) showed that 98% of the biological activity of 5-HT was lost during its passage through the pulmonary circulation (Table 28). The disappearance of 5-HT from the circulation results from its uptake by the lung tissue which is followed by fast oxidative deamination yielding 5-OH indole acetic acid (Junod 1975). The extraction process seems to be: (1) saturable when the concentration of the substrate is increased (Junod 1972; Iwasawa et al. 1973); (2) limited by transport rather than metabolism (Junod 1972); and (3) an active temperature-dependent process (Junod 1975). 5-HT, which is exclusively found in endothelial cells of the pulmonary circulation from large vessels to capillaries (Strum and Junod 1972; Iwasawa et al. 1973) and norepinephrine which, compared with 5-HT, is extracted to a lesser extent (40%) (Hughes et al. 1969b; Iwasawa and Gillis 1974) are the only naturally occurring amines affected to a significant extent by the lung and show similar uptake behavior. Histamine, dopamine, 1-dopa, and norepinephrine are not significantly retained or degraded (Vane 1969). The level of P_{450} in the sheep lung is 0.092 nmol per mg protein (Burns et al. 1975). Besides other possible functions, P_{450} has been implicated with O_2 transfer in the lung (Burns and Gurtner 1973; Burns et al. 1975) and the placenta (Burns and Gurtner 1973; Gurtner et al. 1982).

In summary, the central position of the lung in the circulatory system and the fact that among all organs in the body it contains the largest number of endothelial cells (Junod 1975; Ryan and Ryan 1975) as well as the greatest diversity of constituent cells (Ballard and Ballard 1974), of which as many as 20 types are in place as early as the 14th week of gestation (Avery 1968), to a large extent enables the pulmonary system to protect itself as well as the arterial circulation from the influences of locally produced and exogenous biologically active molecular factors. The large number of different cells provides the necessary receptor sites over which the different functions of the lung are carried out (Conolly and Greenacre 1975). The pulmonary endothelial cells are known to be the active sites of metabolic and biosynthetic functions of the lung (Ryan et al. 1968; Smith and Ryan 1970). It is possible that the number of endothelial cells exposed to the circulating blood can be varied during different states (Ryan and Ryan 1975). Prostaglandins are released from isolated perfused lungs of several species by diverse stimuli which may be immunological, chemical, or mechanical (Saeed and Roy 1972; Änggård 1975), the possible sites of production being the perivascular cells, type II alveolar cells, and macrophages (Smith and Ryan 1973a,b). Under experimental and pathological conditions such as septicemia, improper transfusion of blood, trauma, or poisoning, the lung is highly susceptible to disseminated extraneous intravascular coagulation. Blockage of blood flow may be a secondary cause of death.

6.11 The Implications of Liquid Breathing in Air Breathers

The commonality of the mechanisms of gas exchange by diffusion across the tissue barriers and the ventilatory and circulatory adjustments evoked to meet the metabolic needs independent of the milieu an animal lives in raises a fundamental theoretical and practical question as to whether air breathers can survive by breathing water or any other liquid, especially if such fluid is charged with O_2 (Kylstra 1968, 1969). This would hopefully reveal why the air- and water-breathing organs are structurally so different in spite of the fact that their basic functions are the same. It is now well recognized that in utero, the lungs are not collapsed but are distended by aspirated and secreted fluid measuring at about 30 ml per kg in the human fetal lung. Presence of an appropriate volume of liquid is thought to be necessary for normal intrauterine lung growth and development (Moessinger et al. 1990). Immediately before birth (e.g., Dickson et al. 1986), during labor, and/or soon after (e.g., Brown et al. 1983), pulmonary filtration ceases (as the respiratory epithelium stops secreting Cl^- ions and starts to absorb Na^+ from the lumen of the lung) (e.g., Bland 1990; Chapman et al. 1994). The two process are regulated by levels of circulating catecholamines (e.g., Mortola 1987). Efficient removal of liquid from the air spaces during and after birth is vital for normal switching from placental to pulmonary gas exchange. Liquid is thus not an entirely alien factor to the respiratory surfaces of aerial gas exchangers, at least during the early stages of development. It is of profound interest to know to what extent the structural and functional modifications of the aerial gas exchangers, the lungs, have lost the capacity of utilizing the primordial respiratory medium – water.

In an applied sense, liquid breathing, if and when perfected, will overcome the serious obstacles which human beings would experience for successful survival in deep subaquatic habitats and in the outer space. If a diver breathes a suitable liquid enriched with O_2 instead of air, since the liquid in the lungs would resist the external pressure without significantly changing in volume, it would be possible to descend to depths and ascend to the surface rapidly without the risk of decompression sickness (bends) (Paulev 1965). The ultimate obstacle in the conquest of speed apparently will not be technology but the fragility of the human body to withstand the stresses and strains that have to be endured during such states. Only a few people have "gone supersonic" without the aid of an air craft and survived (Cameron 1990). In future space travels enormous accelerations will be necessary to escape from the gravitational pulls of the much larger celestial bodies. Though constant displacement (speed), irrespective of how fast, produces little stress on the human body, accelerations, decelerations, and sudden maneuvers exert strong forces on the body. Such stresses and strains may not be tolerable to the delicate parts of the human body such as the lungs (e.g., Bullard 1972). The destructive effect could be minimized or even totally eliminated if a whole animal is externally supported by a respirable liquid of the same specific gravity as the body fluids (instead of air). Furthermore, fluid-filled lungs should be able to bear much greater stresses and strains (e.g., Margaria et al. 1958). Absence of gas in the body would equalize the density throughout the body, preventing relative inertial movements of the heart, lungs and other visceral organs while pulmonary arterio-venous shunting would be prevented (Sass et al. 1972). Experimentally, dependent pulmonary atelectasis, arterio-venous shunting and downward displacement of the heart brought about by gravitational-inertial force exposure were prevented in dogs breathing oxygenated liquid fluorocarbon in a whole-body immersed respirator (Sass et al. 1972). Liquid lavage has been used successfully in treating pulmonary conditions such as asthma, bronchiectasis, and mucoviscidosis (Kylstra et al. 1971), in treatment of acute lung injury (e.g., Richman et al. 1993), and management of respiratory failure (Rogers et al. 1972). The pioneering experimentation on breathing O_2-saturated liquids was carried out by, among others, Kylstra and Tissing (1963), Clark and Gollan (1966) and Kylstra et al. (1966).

Due to the remarkably different physicochemical properties of water and air (Tables 4,9), water breathing poses distinct problems to an air breather. The primary limitations are that: (1) water under normal atmospheric pressure contains too little dissolved O_2, and (2) the differences in the ionic composition of water compared with that of blood upsets ionic equilibrium with the body fluids (e.g., Lowe et al. 1979). Owing to its greater viscosity, the maximum flow rate of water from isolated dog lungs is much lower than the maximum expiratory flow rate of air at equal lung volumes (Leith and Mead 1966). Furthermore, the driving pressure affordable for maximum expiratory flow, which is limited by the lung's static recoil pressure in liquid filled lungs (Mead et al. 1967) is less than in an air-filled one (Wood and Bryan 1971). Aquatic organisms have evolved within the physical constraints of their relatively O_2-deficient medium. Most of them possess highly specialized gas exchange systems (gills) which promote extraction of the available O_2. When saline is inhaled into a lung, it physically destroys the delicate terminal gas exchange components (e.g., Curtis et al. 1993), dissolves and mechanically displaces the surfactant (Lewis et al. 1993), osmotically interferes with the composition of the body fluids, causes pathological changes such as interstitial edema, produces intrapulmonary froth and atelectasis (Blenkarn and Hayes 1970), causes loss of macrophages and rupture of the alveolar cell and basement membrane (Huber and Finley 1965), and upsets the integrity of the blood-gas barrier (Reidbord 1967). Direct interference with the surfactant production reduces or even totally eliminates surface tension. This lowers the driving pressure available for maximum expiratory flow to a value no greater than that produced by the elastic recoil of the lung tissue, which is only a small portion of the total static recoil pressure available during air breathing (Kylstra and Schoenfisch 1972). Presence of liquid in the airways increases airway constriction (Yager et al. 1989). With the dissolving of the surfactant and other alveolar hypophase liquid materials, the viscosity and density of the saline increases tremendously, elevating the cost of the ventilatory effort (Blenkarn and Hayes 1970). Addition of exogenous surfactant to liquid ventilated lungs reduces maximal inflatory pressures (Tarczy et al. 1996).

By using isotonic solution supercharged with O_2 to a pressure equivalent to that of air at sea level, it has been possible to keep rats (Pegg et al. 1963), mice (Kylstra 1962), and dogs (Kylstra et al. 1966) alive breathing water for hours. The eventual survival times have depended on factors such as the experimental temperature and the chemical composition of the fluid used. Death during liquid ventilation in a normothermic cat (Clark and Gollan 1966) and the dog (Kylstra et al. 1966) did not result from anoxia but from difficulties in removing CO_2 at the necessary rate. This limitation resulted in acidosis, thought to be secondary to a high arterial PCO_2 (e.g., Shaffer et al. 1976). It was not until mechanically assisted liquid breathing systems were designed (e.g., Saga et al. 1973; Shaffer and Moskowitz 1974) and the effect of CO_2 overloading minimized that the significance of acidosis was fully appreciated as a particular metabolic complication during liquid breathing: unequivocally, O_2 delivery to the tissues during liquid breathing is more than adequate (Shaffer and Moskowitz 1974; Shaffer et al. 1976). The exhaled breath of a mammal on average contains 50 ml of $CO_2 \, l^{-1}$ while at the same temperature and PCO_2, a solution with the same salt concentration as blood contains only about $30 \, ml \, l^{-1}$. This indicates that to remove as much CO_2 through the water as that eliminated via the air, a water-breathing mammal would have to exhale about twice as large a volume of water as of air (Kylstra and Tissing 1963). This is practically made difficult by the narrowness of the bronchial passages and the greater viscosity of water which requires about 36 times more power to move than air. In a state of laminar flow, a process which appeared to occur in saline breathing dogs (Kylstra et al. 1966), a liquid breather would have to expend 60 times the energy required in breathing air. The maximum expiratory flow of a liquid has been estimated to be 40 to 100 times lower than air ventilation (Leith and Mead 1966). Liquid-breathing mice died as a result of exhaustion of the respiratory muscles and accumulation of CO_2 to toxic levels (Kylstra 1962), a complication which has been reported in other animals (e.g., Clark and Gollan 1966; Modell et al. 1970).

To demonstrate the inadequacy of CO_2 elimination during liquid breathing, the survival time of mice was increased to 18 h by addition of tris(hydroxymethyl) amino methane (a substance that minimizes the harmful effects of CO_2 accumulation) to the experimental solution (Kylstra 1962; Nahas 1962). An anesthetized dog cooled to 32 °C was kept alive for 24 min and resuscitated, and normothermic dogs were able to breath liquid for 45 min, 40% of them surviving the exposure well (Kylstra et al. 1966; Mathews et al. 1978): during the experiment, the blood pressure was lower than normal, the heart and respiratory rates were below normal but regular, the arterial blood was fully saturated with O_2 (arterial PO_2, 6.8 to 39 kPa) but the CO_2 content of the arterial blood increased gradually from 5.7 to 10.7 kPa, indicating that the dog's respiratory efforts were not enough to eliminate adequate amounts of CO_2 from the body. West et al. (1965) and Kylstra et al. (1966) observed that the overall pulmonary gas exchange in liquid-filled lungs is diffusion-limited and remarkably large gas tensional gradients occur within the liquid-filled terminal gas exchange components due to the slow rate of diffusion in water. In dogs subjected to lavage with hyperbarically oxygenated saline (PO_2 in the inspired air, 387 kPa), arterial oxygenation was 32 kPa after 15 min, but a severe respiratory acidosis (arterial PCO_2, 9.3 kPa; pH, 7.2) developed.

6.12 Physical Gill and the Plastron: a Unique Underwater Respiratory Strategy

The insectan tracheal system (Sect. 6.6.1) evolved fundamentally as a means for air breathing and hence as an adaptation for terrestrial habitation. A large number of species in the taxon have, however, many times successfully second-arily invaded wet and even aquatic environments. The retention of atmospheric respiration in such cases is a clear manifestation of the advantages derived from air breathing in the taxon and animals in general. In being able to carry air with them in form of gas bubbles (Fig. 104), the insects have retained the major advantage of extracting O_2 from air while subsisting in water. They have gained access to new resources in water and escaped from surface predators. The simplest mode of aerial respiration in aquatic insects is the snorkel one utilized by many larval forms, e.g., the mosquito larva *Culex*, where a breathing tube opening to a spiracle placed on one end of the body is brought into contact with the surface

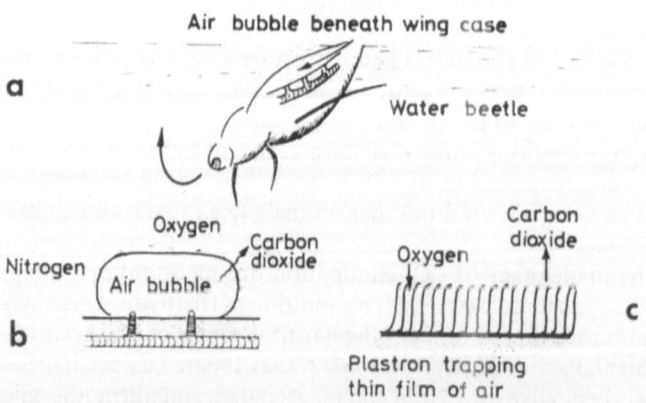

Fig. 104a–c. Underwater air-breathing in insects. a An air bubble entrapped underneath a wing of a beetle. b A physical gill. c Gas exchange in a plastron. (Hughes 1982)

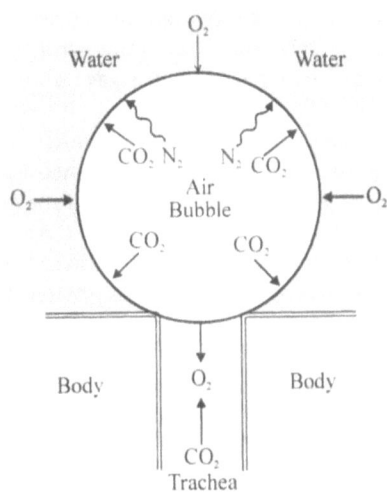

Fig. 105. Schematic diagram of a compressible (physical) gas gill. Oxygen diffuses into the bubble from the surrounding water ( ) to replace the amount consumed in the body. The CO_2 released from the body dissolves into the surrounding water. The partial pressure of nitrogen decreases, leading to eventual collapse of the bubble

of the water. The air is transferred to the body by diffusion, the surface film being utilized for support using hydrofuge hairs. Amazingly, the underwater gas bubble utilized by the insect is comparable to the air held in the lung of a pneumonate gastropod (Hunter 1953), providing a micromileu which facilitates exchange of O_2 and CO_2 across the water-gas interface (Wolvekämp 1955; Rahn and Paganelli 1968).

Two modes of underwater air breathing have evolved in insects especially in the Hemiptera and Coleoptera. In the "compressible gas-gill" (Rahn and Paganelli 1968; Fig. 105), gas pockets or bubbles of air adherent to the body parts (Fig. 104) are utilized as sources of O_2. Groups such as the diving beetles (*Dytiscids*), the black swimmers (*Notonectids*), and water boatman (*Corixa*) (Ege 1918; Crisp 1964) utilize this mode of gas exchange where gas transfer is effected between the trachea, the gas-gill, and the water. In a compressible gas-gill, the pressure of the gases is dependent on the prevailing hydrostatic pressure, i.e., the depth at which the insect operates. Depth shortens the duration of the dive by increasing the rate of O_2 loss into the water due to the increased PO_2 (Rahn and Paganelli 1968): a dive of 2 m increases the PO_2 in the gas-gill to 25 kPa, creating a partial pressure gradient of 4 kPa in favor of loss of O_2 from the gas-gill. Oxygen is extracted by the trachea through the spiracles which open into the bubble. The discharged CO_2 rapidly dissolves into the surrounding water. As the PO_2 in the bubble decreases, in well-oxygenated waters, O_2 diffuses inwards from the surrounding water. Due to the increase of the partial pressure of nitrogen in the bubble, indirectly through loss of CO_2 into the surrounding water and directly through the hydrostatic pressure which increases at a rate of 1 atm per 10 m, N_2 is slowly lost. This results in a steady decrease in the size of the bubble and its eventual collapse at a critical diameter (Rahn and Paganelli 1968; Liew 1970). To increase the rate of diffusion of O_2 into the bubble, some insects actively ventilate the gas-gill by moving water currents around it. The frequency at which an insect makes periodic visits to the surface to pick up fresh air is dependent on: (1) the pressure prevailing in the gas bubble, which in turn is determined by the working depth (the PN_2 between the gill and water increases at a rate $9.9\,kPa\,m^{-1}$); (2) O_2 uptake by the tissues which indirectly leads to an in situ elevation of the concentration of N_2 enhancing its efflux; and (3) the PO_2 in the surrounding water and hence the rate of O_2 recharge of the gas gill from the surrounding water or is even lost from the air bubble when the animal is in hypoxic water. Before the collapse of the gas gill, however, the insect will have extracted a volume of O_2 from the water which is 7 to 13 times greater than that in the bubble at the beginning of the dive. This extends the dive duration 8 to 13 times (Ege 1918; Rahn and Pagannelli 1968). In a *Corixa*, Ege (1918) observed that the insect obtains enough O_2 out of the water by diffusion into the air bubbles covering its body, a supply which extends its underwater stay 10 to 30 times longer than would be possible if it were to rely only on the initial amount of O_2 in the gas bubble(s) at submergence. The spider, *Argyroneta*, which always carries a layer of air around the entire abdomen and part of the thorax, is highly adapted for underwater gas gill respiration, with the air sometimes lasting for several days in summer (Braun 1931). Interestingly, an insect can stay underwater longer when the gas gill is filled with air (in air-saturated water) than when the gill is filled with pure O_2. A backswimmer survives for only 35 min in O_2 saturated water but for as long as 6 h in air-saturated water. When a *Notonecta* was put in O_2-saturated water and also made to breathe from a pure O_2 bubble, the animal became heavier than water in 14 min and succumbed in 35 min. When breathing ordinary air (at the same temperature), however, it could survive for 6 h without surfacing (Ege 1918). This is due to the fact that the presence of N_2 enables the gas gill to function as a physical gill, i.e., it allows O_2 and CO_2 transfer between the water and the gas gill. During winter when the surface of water is covered with ice, a number of species of the Dytiscidae family are highly active, catching air bubbles arising from aquatic plants or from the mud (Krogh 1941). Although such bubbles may contain relatively little O_2, they should replenish the quantity of N_2 in the air bubbles carried by the insect and hence prolong the survival of the gas gill. While the spiracles and hydrophobic hairs may act as barriers, it is intriguing how the insects totally keep the trachea from flooding with water through capillarity. If the trachea are filled with water, the transfer of O_2 would decrease 10 000-fold and the insect would almost certainly drown. This has found a practical application in destruction of mosquito larvae by applying oil on the surface of standing water, a process which eliminates the supporting surface tension. Denney (1993) contemplates that the trachea are lined by a waxy substance which repels water.

In insects which possess "incompressible gas gill" (Fig. 106), a layer of gas is held firmly (over the epicuticle which covers those parts of the body onto which the spiracles open) by stiff hydrophobic hairs or a cuticular meshwork of hairs, the plastron (e.g., Hinton 1966) (Fig. 107). The hairs are about 5 µm long. In the bug, *Aphelocheirus*, the hairs number about $2.5 \times 10^6\,mm^{-2}$, are about 0.5 µm apart, have a diameter of about 0.2 µm, and are strong enough and adequately hydrophobic to withstand 4 to 5 atm of external pressures before the wetting of the cuticle can occur (Thorpe 1950; Crisp 1964). The surface tensional forces

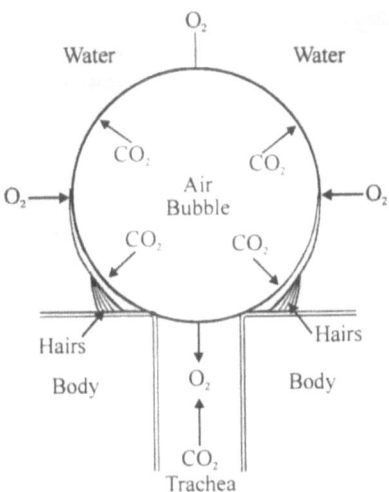

Fig. 106. Schematic diagram of an incompressible gas gill (plastron). The gas air bubble is firmly attached to the body by hydrofuge hairs. As O_2 is consumed and PO_2 drops in the bubble, the gas diffuses inwards from the surrounding water ( ). Carbon dioxide dissolves into the surrounding water. Owing to the firm physical support of the bubble, the pressure of nitrogen in the bubble is not a factor in the longevity of a plastron. Theoretically, the bubble may last indefinitely

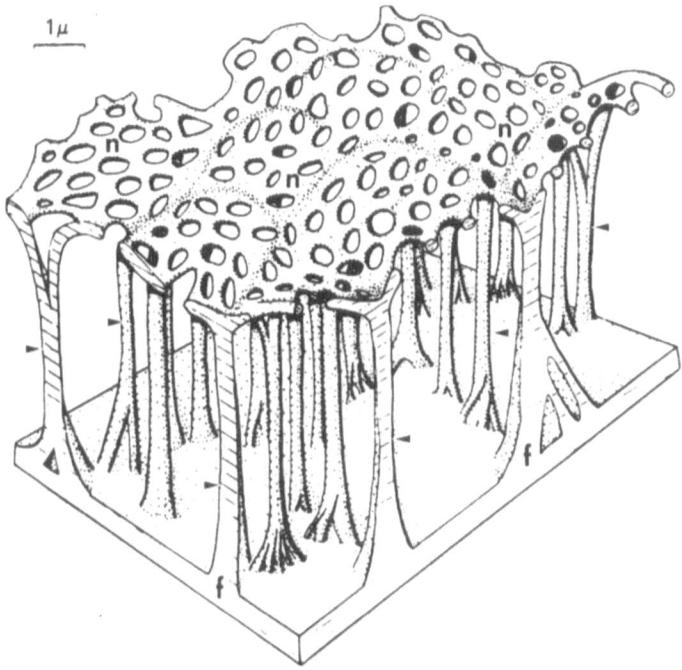

Fig. 107. Schematic drawing of the hairs of a plastron of a cranefly larva, *Dicranomyia*, that consists of narrow spaces delineated by a roof made up of a thin holey cuticle, *n*, and thick basal cuticle, *f*. Vertical struts, ➤, connect the roof to the floor. The air contained in the interstices is not directly subjected to hydrostatic pressures at natural depths. (Hinton 1976)

augmented by the support provided by the hairs counteract the hydrostatic pressure preventing compression of the gas phase, making the volume of the plastron remain fairly constant (Rahn and Paganelli 1968): the PN_2 in the bubble is equal to that in the water and the rate of O_2 uptake by the insects is the same as that diffusing in from the water. Consequently, the loss of the inert gases, mainly N_2, which sets the limit of the gas-gill existence, is avoided. Theoretically, this should allow an insect to remain submerged indefinitely (Thorpe and Crisp 1949; Hinton 1976; Rahn and Paganelli 1968). Strictly, the insects endowed with "compressible" (physical gas gill) or "incompressible" gas bubbles are essentially air breathers as they actually exchange O_2 and CO_2 with a gas phase which, in turn, exchanges gases with the surrounding water. The empiric changes of the gas profiles within the tracheal system and air sacs of an insect utilizing such devices are fundamentally similar to those in an insect breathing free air.

The compressible gas-gill is suitable for insects which operate close to the water surface. In such cases, large energetic costs are not involved for surfacing and diving. Incompressible gas gills, within limits, act well at depths and serve those insects which stay under the water for long periods of time well. While the compressible gas gill requires no definite structural adaptations, plastron respiration demands a greater degree of anatomical specializations in form of hydrophobic hairs and/or other cuticular modifications. The typical environment inhabited by insects with a plastron is that which has well-aerated water and one which dries up at times, exposing the insect to the atmosphere. Plastrons are also quite common on insect eggs which are liable to being covered by water when it rains (Hinton 1953, 1966; Crisp 1964). Such eggs have areas which consist of an intricate meshwork of air channels lined by hydrophobic substances.

A plastron is a highly versatile respiratory organ which functions as an efficient underwater gas exchanger and when the insect is exposed to air provides a satisfactory pathway for gas transfer while minimizing risk of desiccation (Hinton 1966). The insectan plastron corresponds with the fine hydrophobic air channels of many aquatic plants which prevent entrance of water while allowing in air. This makes such plants relatively more buoyant. Of particular interest is the mode of respiration in the larva of the bot-fly, *Gastrophilus intestinalis* which subsists in the virtually anoxic stomach of the horse. In a more or less plastron manner, O_2 is extracted from the air bubbles (which are swallowed during feeding) through a special organ made up of large cells around an array of trachea (Krogh 1941). A totally different mechanism of maintaining a permanent gas-gill has evolved in the volant aquatic elmid beetle: before diving, the beetle carries a large bubble of air which extends from the first femora back over and past the dorsum. In calm water, the beetle has to surface from time to time to replenish the air but in the torrential mountain stream waters, where the currents are faster than $70\,\mathrm{cm\,s^{-1}}$, the Bernoulli-Venturi effect generated by the flow of water across the convex face of the bubble causes the pressure in the bubble to fall below the atmospheric. This effects an inward flux of O_2. The tracheal (air) gills which comprise of an air-filled tracheal system sealed off from water, e.g., the highly tracheated flaps or filamental outgrowths from the thoracic and/or abdominal segments of the body in nymphal Plecoptera and Ephemeroptera and larval Trichoptera constitute an extreme retrogressive transformation which allows permanent aquatic habitation: O_2 diffuses from the surrounding water into the trachea which serve as internal plastrons and CO_2 is discharged and dissolved by the surrounding water.

Remarkable behavioral, morphological, and physiological convergencies in underwater respiration occur. Some high-latitude winter-diving mammals are known to use gas pockets trapped between the water-ice interface (e.g., Hart and Fisher 1964; Mitchell and Reeves 1981; MacArthur 1992). Such air pockets may arise from air entering through natural fissures in the ice, from exhaled gases released by the diving animals (e.g., Harrison et al. 1972), from the hair, or from photosynthetic activity of the aquatic plants. In nature, diving muskrats, *Ondatra zibethicus* (Errington 1963), and experimentally submerged gray seals, *Halichoerus grypus* (Harrison et al. 1972), are reported to breath in air in these gas pockets. Since CO_2 is more soluble (30 times) in water than O_2, the PCO_2 in the air in the bubbles discharged from the lung or from the surface of the body should drop with time. Depending on the PO_2 in the surrounding water, the bubbles trapped under ice may be charged with O_2. A high concentration of O_2 (26.67 to 29.04%) was determined in the air bubbles collected under ice by MacArthur (1992). The investigator, however, attributed it to photosynthetic activity of submerged macrophytes. In addition to the O_2 contained in the intrapulmonary air and the body tissues, the air bubbles in the water/ice interface could provide an important auxiliary source of O_2 during prolonged dives. In the muskrat, *O. zibethicus*, dive duration was increased by 31% when the animals were allowed access to underwater gas pockets (MacArthur 1992).

6.13 The Cleidoic Egg: a Fascinating Gas Exchanger

The evolution of the cleidoic (self-contained) egg in reptiles and birds was a monumental advance in the attainment of terrestriality. For once, the land invaders were able to totally dissociate themselves from water for fundamental activities such as reproduction. In the about 9000 species of birds, in addition to the presence of the ubiquitous feather, oviparity, the mode of reproductive which involves formation of fertilized eggs encased in an egg shell, is a unifying feature. While viviparity has variably evolved in most other vertebrate groups, e.g., amphibians like the caecilian *Typhlonectes compressicauda* (Garlick et al. 1979), reptiles like the lizard, *Sphenomorphorus quoyii* (Grigg and Harlow 1981), and teleost fish like *Zoarces viviparous* (Hartvig and Weber 1984; Weber and Hartvig 1984), *Embiotoca lateralis* (Ingermann and Terwilliger 1981), and *Clinus superciliosus* (Veith 1980), birds constitute the only vertebrate class which is exclusively oviparous. Debate ranges on the reasons for retention of this ancestral reproductive condition (e.g., Blackburn and Evans 1986; Anderson et al. 1987; Lewin 1988): it has been suggested that viviparity in birds would inhibit flight due to the increased pay load, during pregnancy. Such an argument is clearly untenable if it is recalled that bats have evolved both flight and viviparity and of the at least 15 families of birds which have given up flight, none has evolved viviparity.

Simple cost-benefit analysis may explain the case. Being ectothermic, birds can incubate their eggs outside the body – they do not need to retain them in the body for development. Since the eggs are formed at different times (one every 25 h or so), if retained in the oviduct and/or uterus (space allowing), it would mean that embryos would have to be delivered at different times (even when the bird is still laying!) depending on the incubation period and the period of laying (clutch size). The benefits of retaining oviparity perhaps outweigh the costs and risks. Due to its intrinsic self-sufficiency, some authorities (e.g., Bender 1992) consider the egg to be a "cell". In birds, the 1.5-kg ostrich (*Struthio camelus*) egg is hence the largest extant cell while the dinosaur eggs, e.g., those of the sauropod, *Hypselosaurus priscus* (Kerourio 1981), which had a volume of about 2 l (about 40 times the volume of a chicken egg) and the 10-kg ones of the now extinct 500 to 1000-kg elephant bird, *Aepyornis* (Amadon 1947, Rahn et al. 1975; Feducia 1980), are some of the largest cells which ever formed.

Both theoretically and practically, the egg is a marvel of morphogenetic engineering (e.g., Ar et al. 1974). The microcosmos contains all the necessary factors needed for embryonic development such as nutriments, minerals, and water: the only factor missing is O_2, a resource which must be procured from outside. The products of metabolism, CO_2 and water, must be removed while the nongaseous ones (e.g., products of nitrogen metabolism) are stored in the allantois. The shell must be strong enough to mechanically and physically protect the developing chick from trauma during brooding and from toxic, infectious, and parasitic agents: O_2 must be allowed in and CO_2 let out, excessive water loss must be avoided, while the shell must be adequately weak for the chick to break out at the end of incubation (Ar et al. 1979; Schmidt-Nielsen 1984). The eggshell is made up of calcium carbonate ($CaCO_3$) arranged in orderly crystalline arrays embedded in a protein matrix separated by fine air spaces, the pores. The differences in the pattern of deposition of $CaCO_3$ crystals accounts for the disparities in the eggshell conductances of newly laid eggs for O_2 (Tazawa 1987). The conductances of the eggshell of the African parrot, *Enicognathus ferrugineus*, differs by a factor of 7 (Bucher and Barnhart 1984) and in the turkey, *Meleagris gallopavo*, the number of pores in the eggshells changes during a laying cycle, increasing during the late stages of laying (Rahn et al. 1981). The shell of a 60-g chicken egg has a surface area of about 70 cm^2 and about 10 000 pores which are 17 μm in diameter and 0.35 mm in length (Wangensteen et al. 1971; Tazawa 1987). Much as it is a rigid structure, the shell does not completely isolate the developing chick from the external environment. The embryo receives information in form of sound or mechanical movements from the parent(s) and the adjacent eggs, cues used to synchronize hatching in a clutch of eggs (Drent 1975). The need to reconcile remarkably different requirements has led to what appears like near-optimization of the gas transfer capacity of the avian eggshell. Wangensteen and Weibel (1982) observed that the diffusing capacity of the chorioallantois of a 16-day-old egg was equal to the physiological diffusing capacity. Weibel (1984a) attributed the optimal state of the avian eggshell to its transient nature since it does not add immense cost to support and maintain. It is instructive to note that the placenta (Sect. 4.7), an equally ephemeral organ, maintains a substantial functional reserve during most of the gestation period (Karsdorp et al. 1996).

Since the capacity to conduct the respiratory gases is somewhat fixed at the formative stage of the eggshell, i.e., before incubation starts (Wangensteen et al. 1971; Kayar et al. 1981; but see dissenting views e.g., Kutchai and Steen 1971; Lomholt 1976a; Tullet and Board 1976), the design must somewhat adaptively preempt the maximum O_2 and CO_2 flux at the peak of embryonic development. Unique to all gas exchangers, where both CO_2 and O_2 are transported with balanced facility, in the chorioallantois of the bird egg, gases are transported at different rates corresponding with their molecular weights. Unique to all other gas exchangers, each gas has its own diffusion coefficient. On this basis, Rahn and Paganelli (1982) suggested that while the convective transport of the lung can be described as an egalitarian transport, that in the avian egg should be considered an elite transport system. Throughout the incubation period, a time when O_2 consumption increases by a factor of about 800 times (Kutchai and Steen 1971; Rahn et al. 1974; Table 32), the bird embryo is encased in a rigid shell of invariable area and thickness (Table 31). Kutchai and Steen (1971) estimated that O_2 and CO_2

Table 31. Morphometric parameters of the eggs of several species of birds. (Romanoff and Romanoff 1949)

Species	Body wt. (approx – kg)	Egg wt. (g)	Shell thickness (mm)	Mean pore diameter (mm)	Thickness of the membranes (mm)
Aepyornis	500	12000	4.40	–	–
African ostrich	150	1400	1.95	0.035	0.200
Australian swan	17	700	0.69	0.34	0.165
Holland turkey	8	80	0.41	0.040	–
Chickens	3.3	53.7	0.31	0.018	0.065
Pheasant	3.0	32	0.26	0.012	–
Quail	2.0	9	0.13	–	0.067
Finch	0.01	1	0.09	–	0.005
Hummingbird	0.008	0.5	0.06	–	–
Auk	–	–	–	0.041	–
Duck	–	–	–	0.023	–
Gull	–	–	–	0.013	–

Table 32. Day of incubation (DI), cardiac output (CO), oxygen consumption (Vo_2), hemoglobin content (cHb), oxygen capacity (O_2C), transport capacity (TC), and oxygen utilization coefficient (O_2UC) of the chicken during development. (Romanoff 1967; Bartels et al. 1996)

DI	CO	Vo_2	cHb	O_2C	TC	O_2UC
3	20	3.8	0.015	0.0201	0.4	950
5	120	9.4	0.028	0.0375	4.5	210
12	4800	121	0.069	0.0926	445	27
17	6800	305	0.098	0.1310	826	37
17	6800	305	–	0.1100	750	41

permeabilities of the shell and the membranes of a newly laid egg are not adequate to support gas exchange needs during the last stages of the development of the embryo. The megapodes, birds of the Southern-Pacific area, e.g., the Australian malee fowl, *Leipoa ocellata*, and the brush turkey, *Alectura lathami* (Brom and Dekker 1992; Dekker and Brom 1992; Jones and Birks 1992), deposit their eggs deep in the soil among putrefying organic matter and on rare occasions in geothermal heat to provide natural warmth for incubation (e.g., Fleay 1937; Frith 1956). In the egg mound, the eggs must face critical levels of hypoxia and hypercapnia especially during the last stages of incubation. Furthermore, the developing embryos are exposed to a moisture-saturated environment where there is minimal water loss. In most birds, water loss, which may constitute as much as 20% of the initial mass of the egg, is essential for proper development of the eggs (Rahn and Ar 1974; Ar and Rahn 1980; Simkiss 1980). Within a short period, the hatchlings must find their way to the surface or face suffocation. The shells of the eggs of the megapodes are remarkably thin, allowing rapid outward diffusion of metabolic water, out flux of CO_2, and influx of O_2 (Drent 1975). The eggs of the painted turtle, *Chrysemys picta* (Emydidae), present another interesting mode of development: though the eggs hatch in late summer or early autumn, the neonates do not emerge from their subterranean nests until the following spring, i.e., 6 to 9 months after hatching (e.g., Linderman 1991; Constanzo et al. 1995). Though the great mortality of the neonates has been associated with hypothermia (Packard and Packard 1997), hypoxia and hypercapnia (especially when the ground is covered with ice) may be significant contributing factors to the fatalities. Whereas turning of the eggs during incubation is critical for proper embryonic development (e.g., Romanoff 1960; Metcalfe et al. 1979), the eggs of the megapodes develop normally without it (Seymour and Rahn 1978; Seymour and Ackerman 1980). In some reptiles, the sex of the embryo is determined by the ambient temperature (e.g., Ferguson 1992). In fixed incubating eggs, concentration of the albumen occurs arresting the development of the chorioallantoic vessels (Romanoff 1960) and that of the albumin sac (Randles and Romanoff 1950). The normal clutch of about 100 eggs laid at a depth of about 1 m by the green turtle, *Chelonia myda*, face extreme hypoxia and hypercapnia at the end of incubation. The O_2 level falls to 12% and that of CO_2 rises to above 2.24% (Prange and Ackerman 1974). The dinosaurs, e.g., the sauropod, *Hypselosaurus priscus*, are thought to have deposited their eggs in small groups (e.g., Kerourio 1981) instead of about 50 in one clutch, as such a mass of eggs would have consumed O_2 faster than it could diffuse through the walls of the nest (Seymour 1979).

Birds have been able to adapt to a wide variety of nestling conditions by adjusting the eggshell microarchitecture to provide optimal gas exchange for the special needs in their habitats. The incubation period of the bird eggs is inversely related to metabolic rate and the eggshell conductance (Rahn et al. 1974). Pathological conditions which affect the shells of the bird eggs occur frequently especially in cases of malnutrition, e.g., when laying birds experience calcium deficiency. The shells are poorly formed and fragile. A recent environmental problem which has lead to extensive eggshell deformation in birds has resulted from use of pesticides such as DDT (Cooke 1976; Fox 1976; Risenbrough 1986).

During the 21-day incubation period, a 60-g chicken egg will take 6l (8.6 g) of O_2, give off 4.5 l (8.8 g) of CO_2, lose 9 g in weight (=11 l of water vapor) and by the end of incubation, about 30 kcal of the initial 100 kcal energy in the egg will have been consumed to form the 39 g of the chick (e.g., Carey et al. 1980; Rahn and Ar 1980). In the species which nest at high elevations (see Rahn 1977), the respiratory demands of the eggs are daunting. Not only do the eggs of such species have to cope with the prevailing hypoxia but must also experience low ambient temperatures and excessive water loss due to the low vapor pressure prevalent at elevation (Rahn et al. 1976). Hatchability of chicken and turkey eggs decreases with altitude (Weiss 1978), the mortality rate being particularly more pronounced during the second week of incubation. Incubated under the same conditions, the eggs of species from wet habitats lose mass at a higher rate than those from drier habitats (Lomholt 1976b). Eggs laid by chickens acclimatized to an altitude of 3.8 km showed a reduction in the total pore area (Wangensteen et al. 1974; Packard et al. 1977; Carey 1980a,b), a feature which should curtail water loss (Tazawa 1987). In eggs that are laid at higher altitude, the number of pores in the shell is less, the shells are thicker, and the water content is higher (Rahn et al. 1977; Carey 1980a,b). The eggs of the birds which nest over water, e.g., the pied-billed grebe, *Podilymbus podiceps*, have a high density of pores which allows them to lose water to the relatively humid air around them (Drent 1975). The hatchability of the eggs laid by high-altitude (3.8 km) hens improves with the generations of residence at elevation (e.g., Carey et al. 1982). The total effective pore surface area of eggshells decreases with the barometric pressure (Packard et al. 1977). Increased diffusivity of gases through the eggshell due to enhanced gas-phase diffusion compensates for the prevailing hypoxia but results in hypocapnia due to increased CO_2 efflux (Rahn and Ar 1974). The PCO_2 in the air cell drops from 4.3 to 2.9 kPa when sea level eggs are exposed to 0.5 atm of a gas mixture containing 40% O_2 in N_2, to prevent hypoxia. Hyperoxia has a damaging effect on the development of the chick embryo. Exposure of O_2 at 5 atm for 3 h on 72-h-old chick causes a more than 50% mortality and 20 to 30% of those chicks which hatch have deformities of the brain, eyes, upper jaw, legs, feet, and heart (Pizarello and Shircliffe 1967). Due to the gas diffusion changes which occur at altitude and the fact that the measure of egg shell conductance is somewhat fixed once an egg is laid, fertile eggs laid at sea level but incubated at altitude require specific changes in the composition of gases such as enrichment with O_2 and CO_2 in the incubator for normal hatching to occur (Visschedjik and Rahn 1981). Incubation of eggs laid at sea level at an altitude of between 3.1 and 3.8 km results in reduced metabolic rates, prolonged incubation periods, low body mass at hatching, and unpredictable hatchability (Beattie and Smith 1975). To achieve the same conductance, the shells of the large eggs are relatively thicker than those of small eggs but are much more porous (e.g., Ar et al. 1974). The pores occupy 0.02% of the area of a chicken's egg shell but 0.2% of the area of that of an ostrich.

Gas exchange in the avian egg from the ambient air to the chorioallantoic capillary blood takes place by diffusion in the gas phase. The pathway through which O_2 diffuses from outside comprises of: (1) a variably thick cuticle (a noncellular mucinous coat closely applied on the surface of the eggshell); (2) a shell (a hard calcaneous and porous structure); (3) outer and inner shell membranes; and

(4) allantoic capillary blood (Fig. 108). The thickness of the eggshell (0.3 mm in the domestic fowl) varies among species (Table 31), depending on factors such as body and egg size and nutrition and nutritional status (Romanoff and Romanoff 1949). The inner shell membrane surrounds the albumin and the outer one is cemented to the shell. While the inner shell membrane has been presumed to be "wet" and hence presents significant resistance to O_2 transfer (e.g., Kutchai and Steen 1971; Wangsteen 1972; Tullet and Board 1976; Lomholt 1976a), morphometric determination of the diffusing capacity of the egg by Wangensteen and Weibel (1982) indicated that the membrane is "dry" and offers negligible resistance to O_2 flux. Of the three components of the chorioallantoic blood-gas barrier, the blood-gas barrier itself of which the harmonic mean thickness is 0.40 µm offers 10%, the plasma layer about 2%, and the O_2 binding rate 88% of the total resistance to O_2 flow (Wangensteen and Weibel 1982). Feeding acidic salts such as ammonium chloride (Hunt and Aitken 1962) and exposure to high concentrations of CO_2 for half a day for about 2 days (Helbacka et al. 1963; Hunt and Simkiss 1967), factors which lead to metabolic acidosis, decrease shell thickness. Pore geometry and the diffusion coefficient of a particular gas determine the permeability of the eggshell. With the onset of incubation when the embryonic heart starts beating, the circulation of the blood through the capillaries of the embryo

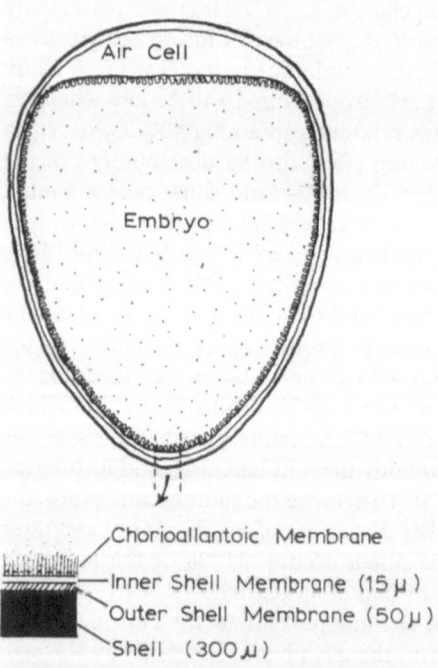

Chorioallantoic Membrane

Inner Shell Membrane (15 µ)

Outer Shell Membrane (50 µ)

Shell (300 µ)

Fig. 108. Cross-section of an avian egg at an advanced stage of development. The dimensions of the main components which form the gas exchange pathway are shown. (Wangensteen et al. 1971)

add a convective process to the earlier entirely diffusive one. The formation of blood starts about 25h after the start of incubation and hemoglobin containing erythrocytes appear by the 2nd day (Romanoff 1960). As occurs in the lungs of birds (Maina et al. 1989a), the greatest impedance to O_2 uptake in the egg lies in the reaction between O_2 and the hemoglobin. The process contributes 88% of the overall resistance, the plasma only 2%, and the remarkably thin blood-gas barrier 10% (Wangensteen and Weibel 1982). The concentration of the hemoglobin does not change significantly between the 12th day of incubation and hatching, when 90% of it is converted to the adult form (Sandreuter 1951). Similarly, over the greater part of the period, the O_2-carrying capacity of blood does not change considerably (Bartels et al. 1966). Erythropoiesis in the developing chick is suppressed by hyperoxia and enhanced by hypoxia (Jalavisto et al. 1965). In complete contrast to the mammalian placenta (Sect. 4.7) which adjusts its gas exchange capacity with gestation, i.e., as the metabolic requirements of the fetus increase, the gas exchange machinery of the egg especially regarding pore number, size, and geometry are somewhat fixed despite the gradual increases in O_2 demands with incubation (Wangensteen et al. 1971). Romijn (1950) and Kutchai and Steen (1971), however, observed that eggshell permeability increases with incubation and estimated that at day 20 of incubation, 575 ml O_2 diffuse into the egg per day. Like the mammalian embryo, the avian one subsists under perpetual hypoxia. Towards the end of incubation, high PCO_2 and low PO_2 in the air space (Romijn 1948) may prompt the chick to break through the shell. A change in the O_2 dissociation curve of the blood of 20-day-old embryos was observed by Bartels et al. (1966): the P_{50} was only 4kPa in the embryonic blood compared with 6.7kPa after a few days of hatching. In the egg, the PO_2 and PCO_2 of the air cell is set by the O_2 uptake and CO_2 release (metabolism) of the embryo and the conductances of the egg shell (Wangensteen and Rahn 1970; Paganelli 1980). Throughout incubation, CO_2 permeability of the eggshell and membranes is two to three times higher than that of O_2, showing that diffusion occurs mainly through air-filled rather than fluid-filled pores (Kutchai and Steen 1971). Given the relatively low diffusion coefficient of O_2 in water, eggs with solid shells would theoretically have to be very small and possess a very high density of pores and low rate of O_2 consumption to survive in water. Such eggs occur in skates and rays (Denney 1993). In a water-saturated environment, the embryos in the larger avian eggs soon drown. The chorioallantoic membrane, a very well-vascularized organ which is closely applied to the inner shell membrane, is the only adaptable structural factor in the egg. It develops steadily and only completely underlies the whole shell towards the beginning of last half of the incubation period (Wangensteen and Rahn 1970; Wangensteen 1972). The degree of vascularization of the outer allantoic surface is increased when the eggs are incubated in a lower PO_2 (Remotti 1933).

The chorioallantoic membrane is considered to be a homolog of the mammalian placenta (Metcalfe and Stock 1993). The diffusing capacity of the chorioallantoic membrane for O_2 in chicks increases sixfold between days 10 and 18 of incubation. During late incubation, the mass-specific diffusing capacity and the capillary blood volume are similar to those of the human lung (Tazawa and Mochizuki 1977). Compared with the viviparous species, where the change from embryonic respiration is abrupt on the breaking of the umbilical cord, in the oviparous species, the change is a gradual process. Mainly owing to the fact that the metabolic rate of the developing embryo increases steadily while the permeability of the shell arguably remains almost constant (Kutchai and Steen 1971; Rahn et al. 1974), in bird eggs, a critical hypoxic phase occurs towards the end of embryogenesis (Freeman and Misson 1970; Girard and Muffat-Joly 1971; Wangensteen 1972). As incubation progresses, the PO_2 in the air cell decreases from 18.7 to 13.3 kPa and that of CO_2 increases from 0.67 to 5.3 kPa (Wangensteen 1972; Table 33). At about day 19, the concentration of O_2 is about 10 to 12% and that of CO_2 lies between 6 and 8% (Tazawa et al. 1983b). As the embryo uses the O_2 inside the shell's air space (Fig. 108), the PO_2 decreases. This increases the partial pressure gradient of O_2 between the ambient air and that inside the egg, enhancing the O_2 influx. Until the 8th day of incubation, the partial pressure gradient of O_2 between the outside of the shell and the air space is about 1.3 kPa, at the 16th day it rises to 6.0, and rises to 10 kPa by the 18th day (Metcalfe 1967). This indicates that as hatching advances, the PO_2 in the allantoic arterial blood must fall to very low levels, increasing the danger of the chick succumbing to hypoxia. Bartels et al. (1966) observed that after day 17, compared with the state after hatching, the chick embryo showed signs of respiratory acidosis: the CO_2 content of blood at a PCO_2 of 5.3 kPa is 54 ml per 100 ml of blood at the 17th day and decreases to about 36 ml after hatching. The autoregulating process of the PO_2 in and outside the developing egg ascertains that the embryo is well supplied with O_2 by diffusion. In regulating O_2 flux, the embryo is not a passive participant. The hypoxia which occurs during the late embryogenesis induces increase in the level of the catecholamines which improve the blood gas status (Wittman and Prechtl 1991). The permeability of the eggshell must adaptively be designed to provide the maximum amount of O_2 required by the embryo, regulating it around the

Table 33. Measured air cell gas tensions (mmHg)[a] of an incubated chicken egg. (Wangensteen and Rahn 1970)

Embryo age (days)	PAO_2	$PACO_2$
7	141.6	13.6
8	140.0	12.7
9	138.1	16.2
10	137.5	16.0
11	134.7	18.3
12	131.3	19.4
13	129.2	21.2
14	124.5	24.8
15	117.9	30.8
16	111.9	36.6
17	106.4	41.8
18	109.9	38.7
19	102.3	43.3

[a] To convert to kPa multiply by 0.013.

critical minimum arterial PO$_2$ needed for survival. Reciprocatively, the O$_2$ consumption of an embryo is determined by the conductance of the shell and the lowest tolerable arterial PO$_2$ (Wangensteen and Rahn 1970). The acid-base status of the embryo is regulated by increases in plasma concentration of the HCO$_3^-$ ions in direct proportion with the changes in the PCO$_2$. This maintains a constant pH of about 7.45 (Erasmus et al. 1970/71). During incubation, the arterial PCO$_2$ reaches a value of about 5.3 kPa, which is about the same level as that in the lungs of the hatched chicks. Wangensteen and Rahn (1970) considered this to be a preparation for smooth transition to air breathing when the gas exchange process changes from diffusion respiration of the chorioallantois to the convective one of the lungs. The changeover from total chollioallantoic respiration to pulmonary respiration takes 22 to 30 h in the chicken (Visschedijk 1968a; Vince 1973). The time of pipping corresponds with increased PCO$_2$ or decreased PO$_2$ in the egg air space (Vince et al. 1975), the stimulating effect of CO$_2$ being two times more potent than that of O$_2$ (Visschedijk 1968b).

6.14 The Bottom Line

In nearly all evolved life forms, throughout their existence, molecular O$_2$ to varying extents has been in constant demand. Respiration has hence been integral in molding the complex morphologies in biology. As a terminal hydrogen acceptor in the ubiquitous scheme of intracellular oxidation by electron transfer, O$_2$ is involved in the vital energy-producing processes which sustain life. The gas exchangers have evolved and adapted in tandem with the respiratory requirements of whole organisms in different states, environments, and habitats. Once they were genomically inaugurated, the designs were continually fashioned by needs and circumstances. The shared morphological contrivances evident in the constructional plans of the gas exchangers indicate that similar pressures have inspired the congruous designs. The aphorism that necessity is the mother of invention is as relevant to evolution and adaptation to novel designs in biology as it is to human activities and needs in the contemporary world. Change in size and activity and subsistence in unique habitats has called for appropriate adjustments for efficient O$_2$ uptake, transfer, and utilization. Among many animals, solutions to these needs have differed only in details, and not in generality. For example, sheet flow of blood at the gas exchanger, closed circulation, double circulation, presence of respiratory pigments in blood and muscle tissue, etc., are some of such prevalent devices. There are no rules in the construction and working of the gas exchangers: means justify ends! Contingent upon the available resources and the specific requirements for survival, the final solutions are arrived at independently. For instance, in insects, the lengthy and costly developments that would have entailed formation of, e.g., a circulatory system, blood, carrier pigments, and blood cells were ingeniously circumvented simply by elaboration of a tracheal system which delivered O$_2$ directly to the tissue cells from the atmosphere. An animal's success in life is determined by the arsenal of vital adaptations it has creatively appropriated, devised, and harnessed along the way. In the formidable continuous forward momentum enforced by natural selection, befitting improvements in the respiratory processes are pivotal for survival. Our response to the question asked by Weibel (1983b) – is the lung built reasonably? – is yes, and to that by Dempsey (1986) – is the lung built for exercise? – is no. We hasten to qualify the latter response as follows: exercise is an extreme costly state of operation which in virtually all animals is endured only momentarily. Gas exchangers are designed to operate at the most economical levels but have inbuilt plasticity of making functional adjustments in response to moderate pressures.

As established in the laws of thermodynamics, a fundamental aspect integral to life's existence on Earth is that it survives on finite, diminishing resources. This rather fatalistic state of affairs calls for efficient means of procurement and economical utilization of the limited resources. It obliges optimal designs of biological systems. Strategies for thrifty use and maximization of the energy reserves and supply developed very early in evolution (e.g., Priede 1977; Szarski 1983). Through painstaking cost-benefit analysis of causes and effects and rationalization and appraisal of the utilities of the attributes that they are endowed with, animals are innately engineering and carefully crafted by natural selection to a high fidelity state from which they are able to ameliorate or overcome the external pressures which continually besiege them. The practical consequence of evolution and adaptation has been to fashion optimal systems where form and function associate harmonically. Nowhere else has this need been of more paramount importance than in the design of the gas exchangers. On a low-energy life-style animal, the snail, Ramsay (1968) observed that "the more one looks into the snail's way of life (plant food is abundant in near vicinity and when threatened it takes refuge in its shell) the more one sees that it has nothing to gain by stepping up its activity, and for its low level of activity a respiratory organ and a circulatory system of moderate efficiency are adequate". In extreme circumstances and when need has justified it, animals have adopted extreme measures of reducing their metabolic requirements, e.g., by going into programmed low energy retreats like estivation and cryptobiosis (e.g., Storey 1988, 1989; Storey and Storey 1990). When warranted, some animals have resolutely deconstructed organs and organ systems and, where need has justified it, readopted past simpler designs and more economic strategies of operation. In many cases, if not in most, when all else has failed, two or more animals (and even animals and plants!) have abandoned their genetically programmed evolutionary trajectories and adopted beneficial symbiotic (= cooperative) associations (e.g., Margulis 1979; Trench 1979; May 1981; Childress et al. 1989; Fenchel and Finlay 1991a,b; McFall-Ngai and Ruby 1991; Finlay and Fenchel 1993; Rennie 1992; Vogel 1997). By pooling their capabilities, the partnership ensured access to resources which otherwise would be unreachable to the individuals alone. Identifying and reconciling the factors that enforce the different or similar morphologies and lifestyles on organisms remains an interesting and challenging aspect in biology. It calls for a vast synthesis of different disciplines of science. Cognizance of the effects past events have engendered on present life is both theoretically engaging and practically useful. It is hopefully by understanding how we got here that we may be able to better anticipate where we are going. The very survival of the humankind may rest squarely on understanding and considerate management of our single most important resource – biodiversity. The opinion offered by Thomas Thomson in 1802 is as valid now as it was some two centuries ago. It is an appropriate closing introspection.

"As soon as man begins to think and to reason, the different objects which surround him on all sides naturally engage his attention. He cannot fail to be struck with their number, diversity, and beauty; and naturally feels a desire to be better acquainted with their properties and uses. If he reflect also, that he himself is altogether dependent upon these objects, not merely for his pleasures and comforts, but for his very existence, this desire must become irresistible. Hence (it is) that curiosity, that eager thirst for knowledge, which animates and distinguishes generous minds." A system of chemistry, by Thomas Thomson, 1802 – quoted in Holmyard and Palmer (1952)

References

Abdalla MA, Maina JN (1981) Quantitative analysis of the exchange tissue of the avian lung (Galliformes). J Anat 134:677–680

Abdalla MA, Maina JN, King AS, King DZ, Henry J (1982) Morphometrics of the avian lung. 1. The domestic fowl (*Gallus gallus* variant *domesticus*). Respir Physiol 47:267–278

Abel DC, Koenig SS, Davis WP (1987) Immersion in the mangrove forest fish, *Rivulus marmoratus*: a unique response to hydrogen sulfide. Envion Biol Fishes 18:67–72

Able KW, Twichell DC, Grimes CB, Jones RS (1987) Tile fishes of the genus *Caulolatilus* construct burrows in the sea floor. Bull Mar Sci 40:1–10

Abzug MJ (1994) Characterization of the placental barrier to murine enterovirus. Placenta 16:207–219

Acarregui MJ, Brown JJ, Mallampalli RK (1995) Oxygen modulates surfactant protein mRNA expression and phospholipid production in human foetal lung in vitro. Am J Physiol 268:L818–L825

Accurso FJ, Alpert B, Wilkening RB, Petersen RG, Meschia G (1986) Time dependent response of foetal pulmonary blood flow to an increase in foetal oxygen tension. Respir Physiol 63:43–52

Ackerman RA, White FN (1979) Cyclic carbon dioxide exchange in the turtle *Pseusemys scripta*. Physiol Zool 52:378–389

Adams D (1979) The hitchhiker's guide to the galaxy. Pocket Books, New York

Adams FH, Yanagisawa M, Kuzela D, Martinek H (1971) The disappearance of foetal lung fluid following birth. J Paediatr 63:881–888

Adams JM, Cory S (1991) Transgenic models for haemopoietic malignancies. Biochim Biophys Acta 1072:9–31

Adelman IR, Smith LL (1970) Effects of oxygen on growth and food conversion efficiency of northern pike. Prog Fish Cult 32:93–96

Adelman R, Saul RL, Ames BN (1988) Oxidative damage to DNA: relation to species metabolic rate and life span. Proc Natl Acad Sci USA 85:2706–2708

Adler K, Stauffer-Holden W, Repine J (1990) Oxygen metabolites stimulate release of high molecular weight glycoconjugates by cell and organ cultures of rodent respiratory epithelium via an arachidonic acid-dependent mechanism. J Clin Invest 85:75–85

Adolph EF (1943) On the appearance of vascular filaments on the pectoral fin of *Lepidosiren paradoxa*. Anat Anz 33:27–30

Agar WE (1908) On the appearance of vascular filaments on the pectoral fin of *Lepidosiren paradoxa*. Anat Anz 33:27–30

Agostini C, Cipriani A, Cadrobbi P, Semenzato G (1995) The pulmonary immune system and HIV infection. Eur Respir Monogr 2:89–124

Aherne W, Dunnill MS (1966) Morphometry of the human placenta. Br Med Bull 22:5–8

Ahlgren JA, Cheng CC, Shrag JD, De Vries AL (1988) Freezing avoidance and the distribution of antifreeze glycoproteins in body fluids and tissues of Antarctic fish. J Exp Biol 137:549–563

Akester AR (1974) Deformation of red blood cells in avian blood capillaries. J Anat 117:658

Alabaster VA, Bakhle YS (1972) The inactivation of bradykinin in the pulmonary circulation of isolated lungs. Br J Pharmacol 45:299–309

Alberch P (1980) Ontogenesis and morphological diversification. Am Zool 20:653–667

Alberch P (1985) Problems with the interpretation of developmental sequences. Syst Zool 34:46–58

Alcorn D, Alexander IGS, Adamson TM, Maloney JE, Ritchie BC, Robinson PM (1977) Morphological effects of chronic tracheal ligation and drainage in the foetal lamb lungs. J Anat 123:649–660

Alcorn D, Adamson TM, Maloney JE, Robinson PM (1980) Morphological effects of either chronic bilateral phrenectomy or vagotomy in the foetal lamb lung. J Anat 130:683–695

Aldrich JC, McMullan PM (1979) Observations on non-locomotory manifestations of biological rhythmns and exitement in the oxygen consumption rates of crabs. Comp Biochem Physiol 62:707–709

Aleksiuk M, Frohlinger A (1971) Seasonal metabolic organization in the muskrat (Ondatza zibethica). I. Changes in growth, thyroid activity, brown adipose tissue, and organ weights in nature. Can J Zool 49:1143–1154

Alerstam T (1982) Fagelflyttning (bird migration). Signum, Lund

Alexander RMcN (1967) Functional design in fishes. Hutchison, London

Alexander RMcN (1981) Factors of safety in the structure of animals. Sci Prog Oxf 67:109–130

Alexander RMcN (1982a) The energetics of vertical migration by fishes. Symp Soc Exp Biol 26:273–294

Alexander RMcN (1982b) Optima for animals. Edward Amold, London

Alexander RMcN (1985) The ideal and the feasible: physical constraints on evolution. Biol J Linn Soc 26:345–358

Alexander RMcN (1990) Size speed and buoyancy adaptations in aquatic animals. Am Zool 30:189–196

Alexander RMcN (1993) Buoyancy. In: Evans DH (ed) The physiology of fishes. CRC Press, Boca Raton, pp 75–97

Alexander RMcN (1996) Optima for animals. Princeton University Press, Princeton

Alford CA, Neva FA, Weller TH (1964) Virologic and serologic studies on human products of conception after maternal rubella. N Engl J Med 271:1275–1281

Allegre CJ, Schneider SH (1994) The evolution of the earth. Sci Am 271:144–151

Alsterberg G (1922) Die respiratorischen Mechanismen der Tubificiden. Acta Univ Lund NF Avd 2 Bd 18:1–176

Al-Tinawi A, Madden JA, Dawson CA, Linehan JH, Harder DR et al. (1991) Distensibility of small arteries of the dog lung. J Appl Physiol 71:1714–1722

Altman PL, Dittmer DS (1961) Blood and other body fluids. Fed Am Soc Exp Biol, Washington, DC

Al-Wassia AH, Innes AJ, Whiteley NH, Taylor EW (1989) Aerial and aquatic respiration in the ghost crab Ocypode saratan. I. Fine structure of respiratory surfaces, their ventilation and perfusion, oxygen consumption, and carbon dioxide production. Comp Biochem Physiol 94A:755–764

Amadon D (1947) An estimated weight of the largest known bird. Condor 49:159–164

Amaladoss ASP, Burton GJ (1985) Organ culture of human placental villi in hypoxic and hyperoxic conditions: a morphometric study. J Dev Physiol 7:113–118

Ames BN, Shigenaga MK, Hagen TM (1993) Oxidants, antioxidants, and the degenerative diseases of aging. Proc Natl Acad Sci USA 90:7915–7922

Amoore JE (1961) Dependence of mitosis and respiration in roots upon O_2 tension. Proc R Soc Lond 154B:109–129

Ancilotto F, Chiarotti GL, Scandolo S, Tosatti E (1997) Dissociation of methane into hydrocarbons at extreme planetary pressure and temperature. Science 275:1288–1290

Andersen HT (1961) Physiological adjustments to prolonged diving in the American alligator, Alligator mississippiensis. Acta Physiol Scand 53:23–45

Andersen HT (1966) Physiological adaptations in diving vertebrates. Physiol Rev 46:212–243

Anderson AE, Felbeck H, Childress JJ (1990) Aerobic metabolism is maintained in animal tissue during rapid sulfide oxidation in the symbiont-containing clam Solemya reidi. J Exp Zool 256.130–134

Anderson DJ, Stoyan NC, Ricklefs RE (1987) Why are there no vivaparous birds? A comment. Am Nat 130:941–947

Anderson JF (1970) Metabolic rates of spiders. Comp Biochem Physiol 33:51–72

Anderson JF, Prestwich KN (1982) Respiratory gas exchange in spiders. Physiol Zool 55:72–90

Anderson M (1978) Optimal foraging area: size and allocation of search effort. Theor Popul Biol 13:397–409

Anderson SA, Bankier AI, Barrell BG, de Brujin MHL et al. (1981) sequence and organization of the human mitochondrial genome. Nature (Lond) 290:457–465

Andrews EA (1955) Some minute movements in protoplasm. Biol Bull 108:121–124

Andrews EB, Taylor PM (1988) Fine structure, mechanism of heart function and hemodynamics in the prosobranch gastropod mollusc *Littorina littorea* (L.). J Comp Physiol B 158:247–262

Andrews P (1981) Characteristic surface topographies of cells lining the respiratory tract. Biomed Res 2:281–288

Andriashev AP (1962) A general view of the Antarctic fish fauna. In: Oye V, Mieghem J (eds) Biogeography and ecology in Antarctica. Dr W Junk, The Hague, pp 491–550

Angersbach D (1978) Oxygen transport in the blood of the tarantula *Eurypelma californicum*: PO_2 and pH during rest, activity and recovery. J Comp Physiol 123:113–125

Angersbach D, Decker (1978) Oxygen transport in crayfish blood: effect of thermal acclimation and short-term fluctuations related to ventilation and cardiac performance. J Comp Physiol 123B:105–112

Änggård E (1975) Biosynthesis and metabolism of prostaglandins in the lung. In: Junod AE, Haller R (eds) Lung metabolism. Academic Press, New York, pp 301–311

Anker GC, Dullemeijer P (1996) Transformation morphology on structures in the head of cichlid fishes. In: Datta-Munshi JSD, Dutta HM (eds) Fish morphology: horizon of new research. Science Publishers, Lebanon, New Hampshire, pp 1–20

Anonymous (1993) Systematics agenda 2000: charting the biosphere. Department of Ornithology, American Museum of Natural History, New York, 10pp

Antonini E (1967) Haemoglobin and its reaction with ligands. Science 158:1417–1425

Antonini E, Rossi-Farelli A, Caputo A (1962) Studies on chlorocruorin. I. The oxygen equilibrium of *Spirographis chlorocruorin*. Arch Biochem Biophys 97:336–342

Antony EH (1961) Survival of the goldfish in presence of carbon monoxide. J Exp Biol 38:109–125

Ar A (1987) Physiological adaptations to underground life in mammals. In: Dejours P (ed) Comparative physiology of environmental adaptations, vol 2. Karger, Basel, pp 208–221

Ar A, Rahn H (1980) Water in the avian egg: overall budget of incubation. Am Zool 20:373–395

Ar A, Paganelli CV, Reeves RB, Greene DG, Rahn H (1974) The avian egg: water vapour conductance, shell thickness, and functional pore area. Condor 76:153–158

Ar A, Rahn H, Paganelli CV (1979) The avian egg: mass and strength. Condor 81:331–337

Arieli R (1979) The atmospheric environment of the fossorial mole rat (*Spalax ehrenbergi*): effects of season, soil texture, rain, temperature and activity. Comp Biochem Physiol 63A:569–575

Arieli R, Ar A (1979) Ventilation of a fossorial mammal *Spalax ehrenbergi* in hypoxic and hypercapnic conditions. J Appl Physiol 47:1011–1017

Arieli R, Ar A, Shkolnik A (1977) Metabolic responses of fossorial rodent (*Spalax ehrenbergi*) to simulated burrow conditions. Physiol Zool 50:61–75

Armstrong HG (1952) Boiling of body fluids. In: Dill DB, Adolph EF, Wilber CG (eds) Principles and practices of aviation medicine, 3rd edn. Williams and Wilkins, Baltimore

Armstrong JD, Priede IG, Lucas MC (1992) The link between respiratory capacity and changing metabolic demands during growth of northern pike, *Esox lucius*. J Fish Biol 41:65–75

Armstrong W (1970) Rhizosphere oxidation in rice and other species: a mathematical model based on the oxygen flux component. Physiol Plant 23:623–630

Arp AJ (1991) The role of heme compounds in sulfide tolerance in the echurian worm *Urechis caupo*. In: Vinogradov D, Kapp C (eds) Structure and function of invertebrate oxygen proteins. Springer, Berlin Heidelberg New York, pp 337–346

Arrigo KR, Worthen DL, Lizotte MP, Dixon P, Dieckmann G (1997) Primary production in Antarctic ice. Science 276:394–397

Arthur WB (1997) Wonders: how fast is technology evolving? Sci Am 276:89–91

Arudpragasm KD, Naylor E (1964a) Gill ventilation and the role of reversed respiratory currents in *Carcinus maenas* (L.). J Exp Biol 41:299–307

Arudpragasm KD, Naylor E (1964b) Gill ventilation volumes, oxygen consumption and respiratory rhythms in *Carcinus maenas* (L.). J Exp Biol 41:2307–2321

Aschoff J, Pohl H (1970) Rhythmic variations in energy metabolism. Fed Proc Am Soc Exp Biol 29:1541–1552

Assali NS, Kirschbaum TH, Dilts PV (1968) The effects of hyperbaric oxygen on uteroplacental and foetal circulation. Circ Res 22:573–588

Atchley WR, Hall BK (1991) A model for development and evolution of complex morphological structures. Biol Rev 66:101–157

Atkins D (1936) On the ciliary mechanism and interrelationship of Lamellibranches. Part I. New observations on sorting mechanisms. Q J Microsc Sci 79:181–309

Atkinson RJA, Taylor AC (1988) Physiological ecology of burrowing decapods. In: Fincham A, Rainbow PS (eds) Aspects of the biology of decapod Crustacea. Oxford University Press, Oxford, pp 201–226

Atkinson RJA, Pelster B, Bridges CR, Taylor AC, Morris S (1987) Behavioural and physiological adaptations to a burrowing life-style in the snake blenny, *Lumpenus lampretaeformis*, and the red band fish, *Cepola rubescens*. J Fish Biol 31:639–659

Augee ML, Elsner RW, Gooden BA, Wilson PR (1970/71) Respiratory and cardiac responses of a burrowing animal, the echidna. Respir Physiol 11:327–334

Averof , Akam M (1995) Insect-crustacean relationships: insights from comparative developmental and molecular studies. Philos Trans R Soc Lond B 347:293–303

Avery ME (1968) The lung and its disorders in the newborn infant, 1st edn. WB Saunders, Philadelphia

Avery ME, Mead J (1959) Surface properties in relation to atelectiasis and hyaline membrane disease. Am J Dis Child 97:517–523

Axelsson M, Farrell AP, Nilsson S (1990) Effects of hypoxia and drugs on the cardiovascular dynamics of the Antlantic hagfish *Myxine glutinosa*. J Exp Biol 151:297–316

Axelsson M, Davison W, Forster ME, Farrell AP (1992) Cardiovascular responses of the red-blooded Antarctic fishes, *Pagothenia bernachii* and *P. borchgrevinki*. J Exp Biol 167:179–201

Azar C, Rodh N (1997) Targets for stabilization of atmospheric CO_2. Science 276:1818–1819

Bacon BJ, Gilbert RG, Kaufmann P (1984) Placental anatomy and diffusing capacity in guinea pigs following long term maternal hypoxia. Placenta 5:475–488

Bader R (1937) Bau, Entwicklung und Funktion des akzessorischen Atmungsorgans der Labyrinthfische. Z Wiss Zool 149:323–401

Bailey L (1954) The respiratory currents in the tracheal system of the adult bee. J Exp Biol 31:589–595

Baird B, Cook SF (1962) Hypoxia and reproduction in Swiss mice. Am J Physiol 202:611–615

Bainerd EL (1994) The evolution of lung-gill bimodal breathing and the homology of vertebrate respiratory pumps. Am Zool 34:289–299

Baker CL (1949) Comparative anatomy of the aortic arches of the urodeles and their relation to respiration and degree of metamorphosis. J Tenn Acad Sci 24:12–40

Baker MA (1982) Brain cooling in endotherms in heat and exercise. Annu Rev Physiol 44:85–96

Baker RR (1978) The evolutionary ecology of animal migration. Hodder, London

Baker VR, Strom RG, Gulick VC, Kargel JS, Komatsu G, Kale VS (1991) Ancient oceans, ice sheets and hydrological cycles on Mars. Nature (Lond) 352:589–594

Bakhle YS (1975) Pharmocokinetic function of the lung. In: Junod AF, Haller R (eds) Lung metabolism. Academic Press, New York, pp 293–299

Bakker RT (1975) Dinosaur renaissance. Sci Am 232:58–78

Balasch J, Musqueras S, Palacios L et al. (1976) Comparative haematology of some Falconiformes. Condor 78:258–273

Balch CC, Campling RC (1965) Rate of passage of digesta through the ruminant digestive tract. In: Dougherty RW (ed) Physiology of digestion in the ruminant. Butterworths, Washington, DC, pp 108–146

Baldauf SL, Palmer JD (1993) Animals and fungi are each other's closest relatives: congruent evidence from multiple proteins. Proc Natl Acad Sci USA 90:11558–11562

Baldridge GW, Gerard RW (1933) Extra respiration of phagocytosis. Am J Physiol 103:235–236

Baldwin KM, Winder WW (1977) Adaptive responses in different types of muscle fibers to endurance exercise. Ann NY Acad Sci 301:411–423

Ballard PL, Ballard RA (1974) Cytoplasmic receptor for glucocoticoids in lung of the human foetus and neonate. J Clin Invest 53:477–486

Ballew C, Haas JD (1986) Hematologic evidence of fetal hypoxia among newborn infants at high altitude in Bolivia. Am J Obstet Gynecol 155:166–169

Ballingand JL, Kobzik L, Han X, Kaye DM et al. (1995) Nitric oxide dependent parasymphathetic signaling is due to activation of constitutive endothelial (type III) nitric oxide synthetase in cardiac myocytes. J Biol Chem 270:14582–14586

Ballintijn CM (1972) Efficiency, mechanics, and motor control of fish respiration. Respir Physiol 14:124–141

Ballintijn CM (1982) Neural control of respiration in fishes and mammals. In: Adink ADF, Spronk N (eds) 3rd congress of ESCPB, vol I. Pergamon Press, Oxford, pp 127–140

Balon TW, Nadler JL (1997) Evidence that nitric oxide increases glucose transport in skeletal muscle. J Appl Physiol 82:359–363

Balter M (1996) Looking for clues to the mystery of life on Earth. Science 273:870–872

Banzett RB, Butler PJ, Nations CS, Barnas GM, Lehr JL, Jones JH (1987) Inspiratory aerodynamic valving in goose lungs depends on gas density and velocity. Respir Physiol 70:287–300

Banzett RB, Nations CS, Wang N, Fredberg JJ, Butler JP (1991) Pressure profiles show features essential to aerodynamic valving in geese. Respir Physiol 84:295–309

Banzett RB, Nations CS, Wang N, Butler PJ, Lehr JL (1992) Mechanical interdependence of wing beat and breathing in starlings. Respir Physiol 89:27–36

Barcroft J, Barron DH (1946) Observations upon the form and relations of the maternal and foetal vessels in the placenta of sheep. Anat Rec 94:569–595

Bard H, Fouton JC, Robillard JE, Cornet A, Soukini MA (1978) Red cell oxygen affinity in foetal sheep: role of 2,3-DPG and adult haemoglobin. J Appl Physiol 45:7–10

Bardack D (1991) First fossil hagfish (Myxinoidea): a record from the Pennsylvanian of Illinois. Science 254:701–703

Barham EG (1966) Deep-scattering layer migration and composition: observations from a diving saucer. Science 151:1399–1416

Barel CDN (1993) Concepts of an architectonic approach to transformation morphology. Acta Biotheor 41:345–381

Barel CDN, Anker GC, Witte F, Hoogerhoud RJC, Goldschmidt T (1989) Constructional constraint and its ecomorphological implications. Acta Morphol Neerl Scand 27:83–109

Barman SA, McCloud LL, Catravas JD, Ehrhart IC (1996) Measurement of pulmonary blood flow by fractal analysis of flow heterogeneity in isolated canine lungs. J Appl Physiol 81:2039–2045

Barnard KH (1943) Revision of the indigenous fresh water of the S.W. Cape region. Ann S Afr Mus 36:101–262

Barnes P (1990) Reactive oxygen species and airway inflammation. Free Radical Biol Med 9:235–243

Barnhart MC (1986a) Respiratory gas tensions and gas exchange in active and dormant land snails, *Otala lactea*. Physiol Zool 59:733–745

Barnhart MC (1986b) Control of acid base status in active and dormant land snail, *Otala lactea* (Pulmonata, Helicidae). J Comp Physiol B156:347–354

Barnhart MC (1986c) Hemocyanin function in active and dormant land snails, *Otala lactea*. Physiol Zool 59:725–732

Barnola JM, Raynaud D, Korotkevich YS, Lorius C (1987) Vostok ice core provides 160 000-year record of atmospheric CO_2. Nature (Lond) 329:408–414

Barnsley MF, Massopust P, Strickland H, Sloan AD (1987) Fractal modelling of biological structure and fuction. Ann NY Acad Sci 504:179–194

Bar-Nun A, Shaviv A (1975) Dynamics of the chemical evolution of the Earth's primitive atmosphere. Icarus 24:197–210

Barrell J (1916) Influence of Silurian-Devonian climates on the rise of air breathing vertebrates. Bull Geol Soc Am 27:387–436

Barrell BG, Bankier AT, Drouin J (1979) A different genetic code in human mitochondria. Nature (Lond) 282:189–194

Barron DH, Meschia G (1954) A comparative study of the exchange of the respiratory gases across the placenta. Cold Spring Harbour symposium on quantitative biology, vol XIX, the mammalian fetus: physiological aspects of development. Long Island Biological Association, New York, pp 93–101

Barron DH, Metcalfe J, Meschia G, Huckabee W, Hellegers A, Prystowsky H (1964) Adaptations of pregnant ewes and their fetuses to high altitude. In: Weihe WH (ed) The physiological effects of high altitude. Pergamon Press, Oxford, pp 115–129

Barron ET, Thompson SL, Schneider SH (1981) An ice-free Cretaceous? results from climate model simulations. Science 212:501–514

Bartels H (1970) The diffusion capacity of the placenta. In: Neuberger A, Tatum EL (eds) Frontiers of biology, vol 17, prenatal respiration. North Holland, London, pp 61–67

Bartels H, Metcalfe J (1965) Some aspects of the comparative physiology of placental gas exchange. Int Union Physiol Sci 4:34–52

Bartels H, Welsch U (1983) Freeze-fracture study of the turtle lung. 1. Intercellular junctions in the air-blood barrier of *Pseudemys scripta*. Cell Tissue Res 231:157–172

Bartels H, Welsch U (1984) Freeze-fracture study of the turtle lung. 2. Rod-shaped particles in the plasma membrane of a mitochondria-rich pneumocyte in *Pseudemys (Chrysemys) scripta*. Cell Tissue Res 236:453–467

Bartels H, Haller G, Reinhardt W (1966) Oxygen affinity of chicken blood before and after hatching. Respir Physiol 1:345–356

Barthelemy L (1987) Oxygen poisoning. In: Dejours P (ed) Comparative physiology of environmental physiology, vol 2. Karger, Basel, pp 152–162

Bartholomew GA (1982a) Scientific innovation and creativity: a zoologists point of view. Am Zool 22:227–235

Bartholomew GA (1982b) Energy metabolism. In: Gordon MS (ed) Animal physiology: principles and adaptations. Macmillan, New York, pp 57–110

Bartholomew GA (1988) Interspecific comparison as a tool for ecological physiology. In: Feder ME, Bennet AF, Burggren WW, Huey RB (eds) New directions in ecological physiology. Cambridge University Press, Cambridge, pp 11–37

Bartholomew GA, Barnhart CM (1984) Tracheal gases, respiratory gas exchange, body temperature and flight in some tropical cicadas. J Exp Biol 111:131–144

Bartholomew GA, Lighton JRB (1986) Oxygen consumption during hover-feeding in free-ranging Anna hummingbirds. J Exp Biol 123:191–199

Bartholomew GA, Leitner P, Nelson JE (1964) Body temperature, oxygen consumption, and heart rate in three species of Australian flying foxes. Physiol Zool 37:179–198

Bartlett D, Mortola JP, Doll EJ (1986) Respiratory mechanics and control of the ventilatory cycle in the garter snake. Respir Physiol 64:13–27

Barton M, Elkins K (1988) Significance of aquatic surface respiration in the comparative adaptation of two species of fishes (*Notropis chrysocephalus* and *Fundulus catenus*) to headwater environments. Trans Ky Acad Sci 49:69–73

Bartosch R, Feldberg W, Nagel E (1933) Weitere Versuche über Freiwerden eines histaminähnlichen Stoffes aus der durchströmten Lunge sensibilisierter Meerschweinchen beim Auslösen einer anaphylaktischen Lungenstarre. Pfluegers Archiv Gesamte Physiol Henschen Tiere 231:616–629

Basalla G (1989) The evolution of technology. Cambridge University Press, New York

Baskin DG, Detmers PA (1976) Electron microscopic study of the gill bars of *Amphioxus* (*Brachiostoma californiense*) with special reference to neurocilliary control. Cell Tissue Res 166:167–178

Basset JE, Wiederhielm CA (1984) Postnatal changes in hematology of the bat *Antrozous pallidus*. Comp Biochem Physiol 78A:737–742

Bassingthwaighte JB (1988) Physiological heterogeneity: fractals link determinism and randomness in structures and functions. News Physiol Sci 3:5–9

Bassingthwaighte JB, King RB, Roger SA (1989) Fractal nature of regional myocardial blood flow heterogeneity. Circ Res 65:578–590

Bassingthwaighte JB, Liebovitch LS, West BJ (1994) Fractal physiology. Oxford University Press, New York

Bast A, Haenen G, Doelman C (1991) Oxidants and antioxidants: state of the art. Am J Med 91:S2–S13

Bastacky J, Hook GR, Finch GL, Goerke J et al. (1987) Low temperature scanning electron microscopy of frozen hydrated mouse lung. Scanning 9:57–70

Bastacky J, Goerke J, Lee CYC, Yager D et al. (1993) Alveolar lining liquid layer is thin and continuous: low temperature SEM of normal rat lung. Am Rev Respir Dis 147:148

Bastacky J, Lee CYC, Goerke J, Koushafar H et al. (1995) Alveolar lining layer is thin and continuous: low temperature scanning electron microscopy of rat lung. J Appl Physiol 79:1615–1628

Bates JHT (1993) Stochastic model of the pulmonary airway tree and its implications for bronchial responsiveness. J Appl Physiol 75:2493–2499

Batterton CV, Cameron JN (1978) Characteristics of resting ventilation and response to hypoxia, hypercapnia and emersion in the blue crab *Callinectes sapidus* (Rathbun). J Exp Zool 203:403–418

Baudrimont A (1955) Organization générale du poumon et structure des alvéoles pulmonaires des vertébrés (amphibiéns, reptiliens, mammiféres) considérés dans leurs rapports avec la mécanique respiratoire, la circulation pulmonaire fonctionelle et l'activité métabolique de ces animaux. Arch Anat Histol Embryol 38:99–136

Bauer C (1974) On the respiratory function of haemoglobin. Rev Physiol Biochem Pharmacol 70:1–31

Baum DA, Larson A (1991) Adaptation reviewed: a phylogenetic methodology for studying character macroevolution. Syst Zool 40:1–18

Baumann FH, Baumann R (1977) A comparative study of the respiratory properties of bird blood. Respir Physiol 31:333–343

Baumgarten-Schumann D, Piiper J (1968) Gas exchange in the gills of resting unanaethetized dogfish, *Scyliorhinus*. Respir Physiol 5:317–325

Bazzaz FA, Fajer ED (1992) Plant life in a CO_2-rich world. Sci Am 266:18–24

Beachy CK, Bruce RC (1992) Lunglessness in plethodontid salamanders is consistent with the hypothesis of a mountain stream origin: a response to Ruben and Boucot. Am Nat 139:839–847

Beadle LC (1932) Scientific results of the Cambridge Expedition to the East African Lakes 1930–31. 4. The waters of some East African Lakes in relation to their fauna and flora. J Linn Soc (Zool) 38:157–211

Beadle LC (1957) Respiration in the African swampworm *Alma emini* Mich. J Exp Biol 34:1–10

Beadle LC (1974) The inland waters of tropical Africa: an introduction to tropical limnology. Longman, London

Beament JWL (1960) Physical models in biology. In: Beament JWL (ed) Models and analogues in biology. Symp Soc Exp Biol No 14. Academic Press, New York, pp 66–123

Beattie J, Smith AH (1975) Metabolic adapatations of the chick embryo to chronic hypoxia. Am J Physiol 228:1346–1350

Bech C, Johansen K (1980) Ventilation and gas exchange in the mute swan, *Cygnus olor*. Respir Physiol 39:285–295

Bech C, Rautenberg W, May B (1985) Ventilatory O_2 extraction during cold exposure in the pigeon (*Columba livia*). J Exp Biol 116:499–502

Becker HO, Bohme W, Perry SF (1989) The lung morphology of lizards (Reptilia:Varaniidae) and its taxonomic-phylogenetic meaning. Bonn Zool Beitr 40:27–56

Becker V (1963) Funktionelle Morphologie der Placenta. Arch Gynaekol 198:3–28

Bedwani JR, Morley PB (1974) Increased inactivation of prostaglandin E_2 by the rabbit lung during pregnancy. Br J Pharmacol 50:459–460P

Behrensmeyer AK, Kidwell SM (1985) Taphonomy and paleobiology. Paleobiology 11:105–119

Beitinger TL, Pettit MJ (1984) Comparison of low oxygen avoidance in a bimodal breather, *Erpetoichthys calabaricus* and an obligate water breather, *Percina caprodes*. Environ Biol Fishes 11:235–240

Belanger LF (1940) A study of the histological structure of the respiratory portion of the lungs of aquatic mammals. Am J Anat 67:437–461

Belkin DA (1964) Variation in heart rate during voluntary diving in the turtle *Pseudemys concinna*. Copeia 2:321–330

Belkin DA (1968) Aquatic respiration and underwater survival of two fresh water turtle species. Respir Physiol 4:1–14

Belkin S, Douglas CN, Jannasch HW (1986) Symbiotic assimilation of CO_2 in two hydrothermal vent animals, the mussel *Bathymodiolus thermophilus* and the tube worm *Riftia pachyptila*. Biol Bull 170:110–121

Bell PB, Stark-Vancs VI (1983) SEM study of the microarchitecture of the lung of the giant salamnder *Amphiuma tridactylum*. Scanning Electron Microsc 1983:449–456

Bellairs R, Griffiths I, Bellairs AA (1955) Placentation in the adder, *Vipera berus*. Nature (Lond) 176:657–658

Belman BW (1975) Some aspects of the circulatory physiology of the spiny lobster *Panulirus interruptus*. Mar Biol 29:295–305

Bemis WE, Lauder GV (1986) Morphology and function of the feeding apparatus of the lungfish, *Lepidosiren paradoxa* (Dipnoi). J Morphol 187:81–108

Ben-Avraham Z (1981) The movement of the continents. Am Sci 69:291–299

Bender HG (1974) Placenta-Insuffizienz. Morphometrische Untersuchungen am Modell der Rhesus-Placenta. Arch Gynaekol 216:289–300

Bender L (1992) The human body: its mysteries and marvels. Bramley Books, Godalming, Surrey

Bender ML (1984) On the relationship between ocean chemistry and atmospheric PCO$_2$ during the Cenozoic. In: Climate processes and climate sensitivity. Geophys Monogr 29:352–359

Benedict FG (1938) Vital energetics: a study in comparative basal metabolism. Carnegie Institute, Washington, DC, Publ No 503

Bennett AF (1978) Activity metabolism in the lower vertebrates. Annu Rev Physiol 400:447–469

Bennett AF (1982) The energetics of reptilian activity. In: Gans C, Pough FH (eds) Biology of the Reptilia, physiological ecology. Academic Press, New York, pp 155–199

Bennett AF (1988) Structural and functional determinates of metabolic rate. Am Zool 28:699–708

Bennett AF, Dawson WR (1976) Metabolism. In: Gans C, Dawson WR (eds) Biology of Reptilia, vol 5. Academic Press, New York, pp 127–223

Bennett AF, Ruben JA (1979) Endothermy and activity in vertebrates. Science 206:649–655

Bennett AF, Tenney SM (1982) Comparative mechanics of mammalian respiratory system. Respir Physiol 49:131–140

Bennett AF, Wake MH (1974) Metabolic correlates of activity in the caecilian, *Geotrypetes seraphini*. Copeia 1874:764–769

Bennett MB (1988) Morphometric analysis of the gills of the European eel, *Anguilla anguilla*. J Zool (Lond) 215:549–560

Bentley PJ, Shield JW (1973) Ventilation of toad lungs in absence of the buccopharyngeal pump. Nature (Lond) 243:538–539

Benton MJ (1993) Late Triassic extinctions and the origin of the dinosaurs. Science 260:769–770

Benton MJ (1995) Diversification and extinction in the history of life. Science 268:52–58

Berg K (1951) On the respiration of some molluscs from running and stagnant water. Ann Biol 33:561–567

Berg T, Steen JB (1965) Physiological mechanisms for aerial respiration in the eel. Comp Biochem Physiol 15:469–484

Berger AJ (1961) Bird study, 1st edn. Dover Publications, New York

Berger M, Hart JS (1974) Physiology and energetics of flight. In: Farner DS, King JR (eds) Avian biology, vol 4. Academic Press, New York, pp 415–477

Berger M, Roy OZ, Hart JS (1970) The coordination between respiration and wing beats in birds. Z Vergl Physiol 66:190–200

Berger PJ, Walker AM, Horne R, Brodecky V, Wikinson MH et al. (1986) Phasic respiratory activity in the foetal lamb during late gestation and labour. Respir Physiol 65:55–68

Berger WH, Spitzy A (1988) History of atmospheric CO$_2$: constraints from the deep-sea records. Paleoceanography 3:401–411

Berker LV, Marshall LC (1965) The origin and rise of oxygen concentration in the Earth's atmosphere. J Atmos Sci 22:225–261

Berkner LV, Marshall LC (1965) On the origin and rise of oxygen concentration in the Earth's atmosphere. J Atmos Sci 22:225–261

Berner RA (1963) Electrodes studies of hydrogen sulfide in marine sediments. Geochim Cosmochim Acta 27:563–575

Berner RA (1991) Model for atmospheric CO$_2$ over the Phanerozoic. Am J Sci 291:339–376

Berner RA (1997) The rise of plants and their effect on weathering and atmospheric CO$_2$. Science 276:544–545

Berner RA, Canfield DE (1989) A new model for atmospheric oxygen over Phanerozoic time. Am J Sci 289:333–361

Bernhardt R (1995) Cytochrome P_{450}: structure, function, and generation of reactive oxygen species. Rev Physiol Biochem Pharmacol 127:137–221

Bernstein MH (1987) Respiration in flying birds. In: Seller TJ (ed) Bird respiration, vol II. CRC Press, Boca Raton, Florida, pp 43–73

Bernstein MH (1989) Temperature and oxygen supply in the avian brain. In: Wood SC (ed) Comparative pulmonary physiology: current concepts. Marcel Dekker, New York, pp 343–368

Bernstein MH (1990) Avian respiration and high altitude tolerance. In: Sutton JR, Coates GC, Remmers JE (eds) Hypoxia: the adaptations. BC Decker, Burlington, Ontario, pp 30–40

Bernstein MH, Schmidt-Nielsen K (1976) Vascular responses and foot temperature in pigeons. Am J Physiol 226:1350–1355

Bernstein MH, Duran HL, Pinshow B (1984) Extrapulmonary gas exchange enhances brain oxygen in pigeons. Science 226:564–566

Berry EM, Edmonds JF, Wyllie JH (1971) Release of prostaglandin E_2 and unidentified factors from ventilated lungs. Br J Surg 58:189–192

Berry WBN, Wilde P (1978) Progressive ventilation of the oceans: an explanation for the distribution of the lower Paleozoic black shales. Am J Sci 278:257–275

Berschick P, Bridges CR, Grieshaber MK (1987) The influence of hyperoxia, hypoxia and temperature on the respiratory physiology of the intertidal rockpool fish, *Gobius cobitis pallas*. J Exp Biol 130:369–387

Bertin L (1958) Organes de la respiration aérienne. Grassé PG (ed) Traite de zoologie, vol 13, pt 2. Massonet, Paris, pp 1363–1398

Betke K (1958) Hämatologie der ersten Lebenszeit. Ergeb Inn Med Kinderheilkd 9:437–509

Bettex-Galland M, Hughes GM (1973) Contractile filamentous material in the pillar cells of fish gills. J Cell Sci 13:359–366

Betticher DC, Geiser J, Tempini A (1991) Lung diffusing capacity and red blood cell volume. Respir Physiol 85:271–278

Betts RA, Cox PM, Lee SE, Woodward FI (1997) Contrasting physiological and structural vegetation feedbacks in climate change simulations. Nature (Lond) 387:196–799

Bevelander G (1934) The gills of *Amia calva* specialized for respiration in an oxygen deficient habitat. Copeia 3:123–127

Bidani A, Crandall ED (1988) Velocity of CO_2 exchanges in the lungs. Annu Rev Physiol 50:639–652

Bignon J (1975) Several plasma proteins demonstrated in lung tissue by immuno-electron microscopy using the peroxidase coupling technique. In: Junod AF, Haller R (eds) Lung metabolism. Academic Press, New York, pp 425–431

Billet FS, Courtenay TH (1973) A stereoscan study of the origin of ciliated cells in the embryonic epidermis of *Amblyostoma mexicanum*. J Embryol Exp Morphol 29:549–558

Birks EK, Mathieu-Costello O, Fu Z, Tyler WS, West JB (1997) Very high pressures are required to cause stress failure of pulmonary capillaries in thoroughbred racehorses. J Appl Physiol 82:1584–1592

Bishop SC (1943) Handbook of salamanders. Comstock Publications, Ithaca

Bishop WW, Trendall AF (1967) Erosion surfaces, tectonics, and volcanic activity in Uganda. Q J Geol Soc Lond 122:385–420

Black CP, Tenney SM (1980) Oxygen transport during progressive hypoxia in bar-headed geese (*Anser indicus*) acclimatized to sea-level and 5600 m. In: Piiper J (eds) Respiratory function in birds, adult and embryonic . Springer, Berlin Heidelberg New York, pp 79–83

Black CP, Tenney SM (1980) Oxygen transport during progressive hypoxia in high altitude and sea level water-fowl. Respir Physiol 39:217–239

Black CP, Tenney SM, Kroonenburg MV (1978) Oxygen transport during progressive hypoxia in bar-headed geese (*Anser anser*) acclimated to sea level and 5600 m. In: Piiper J (ed) Respiratory function in birds adult and embryonic. Springer, Berlin Heidelberg New York, pp 79–83

Black LL, Wiederhielm CA (1976) Plasma oncotic pressures and hematocrit in the intact, anaesthetized bat. Microvasc Res 12:55–58

369

Blackburn DG (1982) Evolutionary origins of viviparity in the Reptitlia. 1. Sauria. Amphibia-Reptilia 3:185–205

Blackburn DG (1993) Chorioallantoic placentation in squamate reptiles: structure, function, development, and evolution. J Exp Zool 266:414–430

Blackburn DG, Evans HE (1986) Why are there no viviparous birds? Am Nat 128:165–190

Blair HS (1994) Molecular evidence for the origin of birds. Proc Natl Acad Sci USA 91:2621–2624

Bland RD (1990) Lung epithelial ion transport and fluid movement during the perinatal period. Am J Physiol 259:L30–L37

Bland RD, McMillan DD, Bressack MA, Dong L (1980) Clearance of liquid from lungs of newborn rabbits. J Appl Physiol 49:171–177

Bland RD, Hansen TN, Haberkern CM, Bresack MA et al. (1982) Lung fluid balance in lambs before and after birth. J Appl Physiol 53:992–1004

Blatchford JG (1971) Hemodynamics of *Carcinus maenas* (L.). Comp Biochem Physiol 9A:193–202

Blenkarn GD, Hayes JA (1970) Bilateral lung lavage with hyperbarically oxygenated saline in dogs. J Appl Physiol 29:786–793

Bleyl U, Büsing CM (1971) Perpetuation des Shocks durch die Schocklunge. Z Prakt Anaesth Wiederbeleb 6:249–262

Bliss DE (1979) From sea to tree: saga of a land crab. Am Zool 19:385–410

Bliss DE, Mantel LH (1968) Adaptations of crustaceans to land: a summary analysis of new findings. Am Zool 8:673–685

Bloch EH (1962) A quantitative study of the hemodynamics in the living microvascular system. J Anat 110:125–145

Block BA (1987) Strategies for regulating brain and eye temperatures: a thermogenic tissue in fish. In: Dejours P, Bolis L, Taylor CR, Weibel ER (eds) Comparative physiology: life in water and on land. Fidia Research Series, vol 9. Liviana Press, New York, pp 401–420

Block BA (1991a) Endothermy in fish: thermogenesis, ecology and evolution. In: Hochachka PW, Mommsen T (eds) Elsevier, Amsterdam, pp 269–298

Block BA (1991b) Evolutionary novelties: how fish have built a heater out of muscle. Am Zool 31:726–742

Block ER, Patel JM, Angelides KJ, Sherdan NP, Garg C (1986) Hyperoxia reduces plasma membrane fluidity: a mechanism for endothelial cell dysfunction. J Appl Physiol 60:826–835

Blomqvist CG, Stone HL (1983) Cardiovascular adjustments to gravitational stress. In: Fenn WO, Rahn H (eds) Handbook of physiology: the cardiovascular system, peripheral circulation and organ blood flow, sect 2, vol III, Part 2. Am Physiol Soc, Bethesda pp 1025–1063

Blum HF (1955) Time's arrow and evolution. Princeton University Press, Princeton

Bock WJ, von Wahlert G (1965) Adaptation and the form-function complex. Evolution 19: 269–299

Bøe F (1954) Vascular morphology of the human placenta – cold Spring Harbour Symposium. Q J Biol 19:29–35

Boggs DF, Kilgore DL, Birchard GF (1984) Respiratory physiology of burrowing mammals and birds. Comp Biochem Physiol 77A:1–7

Böhlke JE, Chaplin CCG (1968) Fishes of the Bahamas and adjacent tropical waters. Academy of Natural Sciences, Philadelphia

Bohr C (1909) Ueber die spezifische Tätigkeit der Lungen bei der respiratorischen Gasaufnahme. Scand Arch Physiol 22:221–280

Boland EJ, Olson KR (1979) Vascular organization of the catfish gill filament. Cell Tissue Res 198:487–500

Bolin B, Houghton J, Filho LGM (1994) Radiative forcing of climatic change: the 1994 report of the Scientific Assessment Working Group of Intergovernmental Panel on Climatic Change (IPCC). WHO and UNEP, Meteorological Office Marketing Cummunications Studio, Washington, DC, 48pp

Bond AN (1960) An analysis of the response of salamander gills to changes in the oxygen concentration of the medium. Dev Biol 2:1–20

Bond CF, Gilbert PW (1958) Comparative study of blood volume in representative aquatic and nonaquatic birds. Am J Physiol 194:519–521

Bone Q, Pulsford A, Chubb AD (1981) Squid mantle muscle. J Mar Biol Assoc UK 61:327–342

Bone Q, Brown ER, Travers G (1994) On the respiratory flow in the cuttlefish, *Sepia officinalis*. J Exp Biol 194:153–165

Bonner JT (1988) The evolution of complexity. Princeton University Press, Princeton

Booth DT (1995) Oxygen availability and embryonic development in sand snail (*Polinices sordidus*) egg masses. J Exp Biol 198:241–247

Booth JH (1978) The distribution of blood flow in the gills of fish: application of a new technique to rainbow trout (*Salmo gairdnen*). J Exp Biol 73:119–129

Booth JH (1979) The effect of oxygen supply, epinephrine and acetycholine on the distribution of blood flow in trout gills. J Exp Biol 83:31–39

Borden MA (1931) A study of the respiration and the function of haemoglobin in *Planorbis corneus* and *Arenicola marina*. J Mar Biol Assoc UK 17:709–738

Borell U, Fernstrom I (1962) The shape of the foetal chest during its passage through the birth canal: a radiolographic study. Acta Obstet Gynecol Scand 41:213–222

Borradaille LA, Potts FA, Eastham LES, Saunders JT (1963) Invertebrata, 4th edn (Revised by GA Kerkut). Cambridge University Press, Cambridge

Borteux S (1993) Properties and biological functions of the NTH and FPG proteins of *Escherichia coli*: two DNA glycosylases that repair oxidative damage in DNA. Photochem Photobiol B 19:87–96

Bouchet JY, Truchot JP (1985) Effects of hypoxia and L-lactate on the hemocyanin oxygen affinity of the lobster, *Homarus vulgaris*. Comp Biochem Physiol 80A:69–73

Boucot AJ, Gray J (1982) Paleozoic data of climatological significance and their use for interpreting Silurian-Devonian climate. Geophysics Study Commitee, National Academy Press, Washington, DC, pp 189–198

Bouhuys A (1974) Breathing: physiology, environment and lung disease. Grune and Stratton, New York

Boulière F (1975) Mammals, small and large: the ecological implication of size. In: Golley FB, Petrusewicz K (eds) Small mammals: their productivity and population dynamics. Cambridge University Press, Cambridge, pp 1–8

Bourne GB, Redmond JR (1977) Hemodynamics in the pink albarone, *Haliotis corrugata*. I. Pressure relations and pressure gradients in intact animals. J Exp Zool 200:9–16

Boussaad S, Tazi A, Leblac RM (1997) Chlorophyll *a* dimer: a possible primary electron donor for the photosynthesis II. Proc Natl Acad Sci USA 94:3404–3506

Boutilier RG (1990) Respiratory gas tensions in the environment. In: Boutilier RG (ed) Advances in comparative and environmental physiology, vol 6: Vertebrate gas exchange from environment to cell. Springer, Berlin Heidelberg New York, pp 1–13

Boutilier RG, Toews DP (1981) Respiratory, circulatory and acid-base, adjustments to hypercapnia in a strictly aquatic and predominantly skin breathing urodele *Cryptobranchus alleganiensis*. Respir Physiol 46:177–192

Boutilier RG, Randall DJ, Shelton G, Toews DP (1979a) Acid-base relationships in the blood of the toad *Bufo marinus*. I. The efects of environmental CO_2. J Exp Biol 84:289–302

Boutilier RG, Randall DJ, Shelton G, Toews DP (1979b) Acid-base relationships in the blood of the toad *Bufo marinus*. III. The effects of burrowing. J Exp Biol 82:357–365

Boutilier RG, Heming TA, Iwama GK (1984) Physicochemical parameters for use in fish respiratory physiology. In: Hoar WS, Randall DJ (eds) Fish physiology, vol 10A. Academic Press, New York, pp 403–456

Boutilier RG, Glass ML, Heisler N (1986) The relative distribution of pulmocutaneous blood flow in *Rana catesbeiana*: effects of pulmonary or cutaneous hypoxia. J Exp Biol 126:33–39

Boutilier RG, West TG, Pogson GH, Mesa KA, Wells J, Wells MJ (1996) Nautilus and the art of metabolic maintenance. Nature (Lond) 382:535

Bouverot P (1985) Adaptation to altitude-hypoxia in vertebrates. Springer, Berlin Heidelberg New York

Bouverot P, Hildewein G, Oulhen P (1976) Ventilatory and circulatory O_2 convecion at 400 m in pigeons at neutral or cold temperature. Respir Physiol 28:371–385

Bozinovic F (1993) Scaling basal and maximal metabolic rate in rodents and the aerobic capacity model for the evolution of endothermy. Physiol Zool 65:921–932

Bradbury M (1979) The concept of blood-brain barrier. John Wiley, New York

Bradford SM, Taylor AC (1982) The respiration of *Cancer pagurus* under normoxic and hypoxic conditions. J Exp Biol 97:273–288

Brafield AE (1964) The oxygen content of interstitial water in sandy shores. J Anim Ecol 33:97–116

Brain JD (1985) Macrophages in the respiratory tract. In: Fishman AP, Fisher AB (eds) Handbook of physiology, vol 1. Circulation and nonrespiratory functions. American Physiological Society, Bethesda, pp 447–471

Brainerd EL (1994) Lung ventilation in fishes and amphibians: the evolution of vertebrate airbreahing mechanisms. Am Zool 34:289–299

Brainerd EL, Ditelberg JS, Bramble DM (1993) Lung ventilation in salamanders and the evolution of vertebrate air-breathing mechanisms. Biol J Linn Soc 49:163–183

Brakenbury JH (1984) Physiological responses of birds to flight and running. Biol Rev 59:559–575

Brakenbury JH (1987) Ventilation of the lung-air sac system. In: Seller TJ (ed) Bird respiration vol. I. CRC Press, Boca Raton, pp 39–69

Brakenbury JH (1991) Ventilation, gas exchange and oxygen delivery in flying and flightless bird. In: Woakes AJ, Grieshaber MK, Bridges CR (eds) Physiological strategies for gas exchange and metabolism. Cambridge University Press, Cambridge, pp 125–147

Brakenbury JH, Amaku J (1990) Effects of combined abdominal and thoracic air-sac occlusion on respiration in domestic fowl. J Exp Biol 152:93–100

Brakenbury JH, Avery P (1980) Energy consumption and ventilatory mechanisms in the exercising fowl. Comp Biochem Physiol 66A:439–445

Brakenbury JH, Avery P, Gleeson M (1981) Respiration in exercising fowl. I. Oxygen consumption, respiratory rate, and respired gases. J Exp Biol 93:317–325

Brakenbury JH, Darby C, El-Sayed MS (1989) Respiratory function in exercising fowl following occlusion of the thoracic air sacs. J Exp Biol 145:227–237

Bramble DM, Carrier DR (1983) Running and breathing in mammals. Science 219:251–256

Bramwell CD (1971) Aerodynamics of *Pteranodon*. J Linn Soc Biol 3:313–328

Braulin EA, Wahler GM, Swayze CR, Lucas RV, Fox IJ (1986) Myoglobin facilitated oxygen diffusion maintains mechanical function of mammalian cardiac muscle. Cardiovasc Res 20:627–636

Braun F (1931) Beiträge zur Biologie und Atmungsphysiologie der *Argyroneta aquatica*. Zool Jahrb Abt Syst 62:175–262

Bray AA (1985) The evolution of the terrestrial vertebrates: environmental and physiological considerations. Philos Trans R Soc Lond 309B:289–322

Breeze RG, Wheeldon EB (1977) The cells of the pulmonary airways. Am Rev Respir Dis 116:705–777

Brent R, Pedersen PF, Bech C, Johansen K (1984) Lung ventilation and temperature regulation in the European coot (*Fulica atra*). Physiol Zool 57:19–25

Brett JR (1972) The metabolic demand for oxygen in fish, particularly salmonids and a comparison with other vertebrates. Respir Physiol 14:151–174

Brett RA (1986) The ecology and behaviour of the naked mole rat (*Heterocephalus glaber* Ruppell) (Rodentia: Bathyergidae). PhD Diss, University of London, London

Brigdes CR (1986) A comparative study of the respiratory properties and physiological function of haemocyanin in two burrowing and two non-burrowing crustaceans. Comp Biochem Physiol 83A:261–270

Brigdes CR (1987) Environmental extremes – the respiratory physiology of intertidal rockpool fish and sublittoral burrowing fish. Zool Beitr 30:65–84

Bridges CR (1988) Respiratory adaptations in intertidal fish. Am Zool 18:79–96

Bridges CR, Morris S (1986) Modulation of hemocyanin oxygen affinity by L-lactate; a role for other cofactors. In: Lizen B (ed) Invertebrate oxygen carriers. Springer, Berlin Heidelberg New York, pp 341–352

Bridges CR, Kester P, Scheid P (1980) Tracheal volume in the pupa of the saturniid moth *Hyalophora cecropia* determined with inert gases. Respir Physiol 40:281–291

Briggs J (1992) Fractals: The pattern of chaos. Simon and Schuster, New York

Britton JC (1970) The Lucinidae (Mollusca: Bivalvia) of the western Antlantic Ocean. PhD Diss, The George Washington University, Washington, DC

Brix O (1982) The adaptive significance of the reversed Bohr and Root shifts in blood from the marine gastropod, *Buccinum undulum*. J Exp Zool 221:27–36

Brix O, Bardgard A, Cau A, Colosimo A, Condo SG, Giardina B (1989) Oxygen-binding properties of cephalopod blood with special reference to environmental temperatures and ecological distribution. J Exp Zool 252:34–42

Brocher F (1920) Étude expérimentale sur le fonctionnement du vaisseau dorsal et sur la circulation du sang chez le lnsectes. III. Le *Sphinx convolvuli*. Arch Zool Exp Gén 60:1–45

Brocher F (1931) Le mécanisme de la respiration et celui de la circulation du sang chez les insectes. Arch Zool Exp Gén 74:25–32

Brocklehurst WE (1960) The release of histamine and formation of slow-reacting substance (SRS-A) during anaphylactic shock. J Physiol (Lond) 151:416–425

Broecker WS (1982) Glacial to interglacial changes in ocean chemistry. Prog Oceanogr 11:151–197

Brom TG, Dekker RWRJ (1992) Current studies of megapode phylogeny. Zool Verh (Leiden) 278:7–17

Bromhall C (1987) Spider heart rates and locomotion. J Comp Physiol B 157:451–460

Brooks D (1994) How to perform reduction. Philos Phenomen Res 54:84–99

Brooks RE (1970) Ultrastructure of the physosostomatous swim bladder of rainbow trout *Salmo gairdneri*. Z Zellforsch Mikrosk Anat 106:473–483

Browman MW, Kramer DL (1985) *Pangasius sutchi* (Pangassidae), an air-breathing catfish that uses the swim bladder as an accessory respiratory organ. Copeia 1985:994–998

Brown AC (1993) Variability in biological systems. S Afr J Sci 89:308–309

Brown AC (1994) Is biology science? Trans R Soc S Afr 49:141–146

Brown AC, Terwilliger NB (1992) Developmental changes in ionic and osmotic regulation in the Dungeness crab, *Cancer magister*. Biol Bull Mar Biol Lab Woods Hole 182:270–277

Brown J (1995) Macroecology. University of Chicago Press, Chicago

Brown JH (1981) Two decades of homage to Santa Rosalia: toward a general theory of diversity. Am Zool 21:877–888

Brown JH, Marquet PA, Taper ML (1993) Evolution of body size: consequences of an energetic definition of fitness. Am Nat 142:573–584

Brown JR, Doolittle WF (1995) Root of universal tree of life based on ancient aminoacyl-tRNA synthetase gene duplications. Proc Natl Acad Sci USA 92:2441–2445

Brown ME, Hill RE (1996) Discovery of an extended atmosphere around Europa. Nature (Lond) 380:229–231

Brown MJ, Olver RE, Ramsden CA, Strang LB, Walters DV (1983) Effects of adrenaline and of spontaneous labour on the secretion and absorption of lung liquid in foetal lamb. J Physiol (Lond) 344:137–152

Brown SC (1975) Biomechanics of water pumping by *Chaetopterus variopedatus* Renier: Skeletomusculature and kinematics. Biol Bull 149:136–150

Brown SC (1977) Biomechanics of water pumping by *Chaetopterus variopedatus* Renier: kinematics and hydrodynamics. Biol Bull 153:121–132

Brown SC, McGee-Russel S (1971) *Chaetopterus* tubes: ultrastructural architecture. Tissue Cell 3:65–70

Brown-Borg HM, Borg KE, Meliska CJ, Bartke A (1996) Dwarf mice and ageing process. Nature (Lond) 384:33

Broyles RH (1981) Changes in the blood during amphibian metamorphosis. In: Gilbert Ll, Frieden E (eds) Metamorphosis: a problem in developmental biology, 2nd edn. Plenum Press, New York, pp 461–490

Bruna EM, Fisher RN, Case TJ (1996) Morphological and genetic evolution appear decoupled in Pacific skinks (Squamata: Scincidae: Emoia). Proc R Soc Lond 263 B:681–688

Brunori M (1975) Molecular adaptation to physiological requirements: the hemoblobin systems of trout. Curr Top Cell Regul 9:1–39

Bruton MN (1979) The breeding biology and early development of *Clarias gariepinus* (Pisces: Clariidae) in Lake Sibaya, South Africa, with a review of breeding in species of subgenus *Clarias*. Trans Zool Soc Lond 35:1–45

Bruton MN, Cabral AJP, Fricke H (1992) First capture of a coelacanth, *Latimeria chalumnae* (Pisces, latimeriidae) off Mozambique. S Afr J Sci 48:225–227

Bryan JD, Hill LG, Neill WH (1984) Interdependence of acute temperature preference and respiration in the plains minnows. Trans Am Fish Soc 113:557–562

Bshouty Z, Younes M (1990) Distensibility and pressure-flow relationship of the pulmonary circulation. II. Multibranched model. J Appl Physiol 68:1514–1527

Bucher TL (1985) Ventilation and oxygen consumption in *Amazona viridigenalis*: a reappraisal of resting respiratory parameters in birds. J Comp Physiol B 155:269–276

Bucher TL, Barnhart MC (1984) Varied egg gas conductance, air cell gas tensions and development in *Agapornis roseicollis*. Respir Physiol 55:277–294

Bucher TL, Chappell MA (1989) Energy metabolism and patterns of ventilation in euthermic and torpid hummingbirds. In: Bech C, Reinertsen RE (eds) Physiology of cold adaptation in birds. Plenum Press, New York, pp 187–195

Bucher TL, Chappell MA, Morgan KR (1990) The ontogeny of oxygen consumption and ventilation in the Adélie penguin, *Pygoscelis adelie*. Respir Physiol 82:269–388

Buck JB (1948) The anatomy and physiology of the light organs in fire flies. Ann NY Acad Sci 49:397–482

Buck JB (1962) Some physical aspects of insect respiration. Annu Rev Entomol 7:27–56

Buck JB, Keister M (1955) Further studies of gas filling in the insect tracheal system. J Exp Biol 32:681–691

Budgett JS (1900) Observations on *Polypterus* and *Protopterus*. Proc Cambr Philos Soc 10:236–240

Bugge J (1960) The heart of the African lungfish, *Protopterus*. Vidensk Medd Dan Naturhist Foren 123:193–210

Buick FJ, Gledhill N, Froese AB, Spriet L, Meyers EC (1984) Effect of induced erythrocythemia on aerobic work capacity. J Appl Physiol 48:636–642

Bullard RW (1972) Physiological problems of space travel. Annu Rev Physiol 34:205–234

Bünning E (1973) The physiological clock, 3rd edn. Springer, Berlin Heidelberg New York

Burda H (1993) Evolution of eusociality in Bathyergidae: the case of the giant mole rats (*Cryptomys mechowi*). Naturwissenschaften 80:235–237

Burg M, Green N (1977) Bicarbonate transport by isolated perfused rabbit proximal convoluted tubules. Am J Physiol 233:F307–F314

Burger PJ, Bradley SE (1951) The general form of the circulation of the dogfish (*Squalus acanthias*). J Cell Comp Physiol 37:389–402

Burger RE, Meyer M, Werner G, Scheid P (1979) Gas exchange in the parabronchial lung of birds: experiments in unidirectionally ventilated ducks. Respir Physiol 36:19–37

Burggren WW (1975) A quantitative analysis of ventilation tachycardia and its control in two chelonians *Pseudemys scripta* and *Testudo graeca*. J Exp Biol 63:367–380

Burggren WW (1977) Circulation during intermittent lung ventilation in the garter snake, *Thamnophis*. Can J Zool 55:1720–1725

Burggren WW (1979) Bimodal gas exchange during variation in environmental oxygen and carbon dioxide in the air-breathing fish *Trichogaster trichopterus*. J Exp Biol 82:197–213

Burggren WW (1982a) "Air gulping" improves blood oxygen transport during aquatic hypoxia in the goldfish *Carassius auratus*. Physiol Zool 55:327–334

Burggren WW (1982b) Pulmonary blood plasma filtration in reptiles: a wet vertebrate lung. Science 215:77–78

Burggren WW (1989) Lung structure and function: amphibians. In: Wood SC (ed) Comparative pulmonary physiology: current concepts. Marcel Dekker, New York, pp 153–192

Burggren WW (1991) Does comparative respiratory physiology have a role in evolutionary biology (and vice versa)? In: Woakes AJ, Grieshaber MK, Bridges CR (eds) Physiological strategies for gas exchange and metabolism. Cambridge University Press, Cambridge, pp 1–13

Burggren WW, Demis WE (1992) Metabolism and ram gill ventilation in juvenile paddlefish, *Polyodon spathula* (Chondrostei: Polodontidae). Physiol Zool 65:515–539

Burggren WW, Doyle M (1986) Ontogeny of regulation of gill and lung ventilation in the bullfrog, *Rana catesbeiana*. Respir Physiol 66:279–291

Burggren WW, Feder ME (1985) Skin breathing in vertebrates. Sci Am 253:106–118

Burggren WW, Infantino RL (1994) The respiratory transition from water- to air breathing during amphibian metamoprphosis. Am Zool 34:238–246

374

Burggren WW, Johansen K (1982) Ventricular hemodynamics in the monitor lizard *Varanus exanthematicus*: pulmonary and systemic pressure separation. J Exp Biol 96:343–354

Burggren WW, Johansen K (1986) Circulation and respiration in lungfishes (Dipnoi). J Morphol Suppl 1:217–236

Burggren WW, Johansen K (1987) Circulation and respiration in lungfishes (Dipnoi). In: Bemis WE, Burggren WW, Kemp NE (eds) The biology and evolution of lungfishes. Alan R Liss, New York, pp 217–236

Burggren WW, McMahon BR (1981) Hemolymph oxygen transport, acid-base status, and hydromineral regulation during dehydration in three crabs, *Cardisoma*, *Birgus* and *Coenobita*. J Exp Zool 218:53–64

Burggren WW, McMahon BR (1988a) Circulation. In: Burggren BB, McMahon BR (eds) Biology of the crabs. Cambridge University Press, Cambridge, pp 298–332

Burggren WW, McMahon BR (1988b) Biology of the land crabs: an introduction. In: Burggren BB, McMahon BR (eds) Biology of the crabs. Cambridge University Press, Cambridge, pp 1–5

Burggren WW, Mwalukoma A (1983) Respiration during chronic hypoxia and hyperoxia in larval and adult bullfrogs (*Rana catesbeiana*). I. Morphological responses of lungs, skin and gills. J Exp Biol 105:191–203

Burggren WW, Pinder AW (1991) Ontogeny of cardiovascular and respiratory physiology in lower vertebrates. Annu Rev Physiol 53:107–135

Burggren WW, Roberts J (1991) Respiration and metabolism. In: Prosser CL (ed) Environmental and metabolic animal physiology. Wiley/Liss, New York, pp 353–435

Burggren WW, Shelton G (1979) Gas exchange and transport during intermittent breathing in chelonian reptiles. J Exp Biol 82:75–92

Burggren WW, West NH (1982) Changing importance of gills, lungs and skin during metamorphosis in the bullfrog *Rana catesbeiana*. Respir Physiol 47:151–164

Burggren WW, Wood SC (1981) Respiration and acid-base balance in the salamander, *Ambystoma tigrum*: influence of temperature acclimation and metamorphosis. J Comp Physiol 144:241–246

Burggren WW, McMahon BR, Costerton JW (1974) Branchial water- and blood-flow patterns and the structure of the gills of the crayfish *Procambarus clarkii*. Can J Zool 52:1511–1518

Burggren WW, Glass ML, Johansen K (1977) Pulmonary ventilation/perfusion relationships in terrestrial and aquatic chelonian reptiles. Can J Zool 55:2024–2034

Burggren WW, Dunn J, Barnard K (1979) Branchial circulation and gill morphometrics in the sturgeon, *Acipenser transmontanus*. Can J Zool 57:2160–2170

Burggren WW, Feder ME, Pinder AW (1983) Temperature and the balance between aerial and aquatic respiration in larvae of *Rana berlandieri* and *Rana catesbeiana*. Physiol Zool 56:263–273

Burggren WW, Pinder AW, McMahon BR, Wheatly M, Doyle M (1985a) Ventilation, circulation and their interactions in the land crabs, *Cardisoma guanhumi*. J Exp Biol 117:133–154

Burggren WW, Johansen K, McMahon B (1985b) Respiration in phyletically ancient fish. In: Foreman RE, Gorbman A, Dodd JM, Olson R (eds) Evolutionary biology of primitive fishes. Plenum Press, New York, pp 217–252

Burggren WW, Infantino L, Townsend DS (1990) Developmental changes in cardiac and metabolic physiology of the direct developing tropical frog. *Eleutherodactylus coqui*. J Exp Biol 152:129–147

Burggren WW, McMahon BR, Powers D (1991) Respiratory functions of blood. In: Prosser CL (ed) Environmental and metabolic animal physiology. Wiley/Liss, New York, pp 445–508

Burggren WW, Bicudo JE, Glass ML, Abe AS (1992) Development of blood pressure and cardiac reflexes in the frog, *Pseudis paradoxus*. Am J Physiol 263:R602–R608

Burleson ML, Milsom WK (1990) Propranolol inhibits oxygen sensitive chemoreceptor activity in trout gills. Am J Physiol 258:R1089–R1091

Burnett BR (1972) Aspects of the circulatory system of *Pollicipes polymerus* (Cirripedia: Thoracica). J Morphol 136:79–108

Burnett LE, Bridges CR (1981) The physiological properties and function of ventilatory pauses in the crab, *Cancer paguras*. J Comp Physiol 145:81–88

Burnett LE, McMahon B (1987) Gas exchange hemolymph acid-base status and the role of branchial water stores during air exposure in three littoral crab species. Physiol Zool 60:27–36

Burnett LE, DeFur PL, Jorgesen DD (1981) Application of the thermal dilution technique measuring cardiac output and assessing stroke volume in crabs. J Exp Zool 218:165–173

Burns B, Gurtner GH (1973) A specific carrier for O_2 and CO_2 in the lung and placenta. Drug Metab Dispos 1:374–379

Burns B, Gurtner GH, Peavy H, Cha YN (1975) A specific carrier for oxygen and carbon dioxide. In: Junod AF, Haller R (eds) Lung metabolism. Academic Press, New York, pp 159–184

Burns B, Young-Nam C, Purcell JM (1976) A specific carrier for O_2 and CO_2 in the lung: effects of volatile anaesthetics on gas transfer and drug metabolism. Chest 69:316–321

Burrell MH (1985) Endoscopic and virological observations of respiratory disease in a group of young thoroughbred horses in training. Equine Vet J 17:99–103

Burri PH (1984a) Foetal and postnatal development of the lung. Annu Rev Physiol 46:617–628

Burri PH (1984b) Lung development and histogenesis. In: Fishman AP, Fischer AB (eds) handbook of physiology; respiration, vol 4. American Physiological Society Press, New York, pp 1–46

Burri PH (1985) Morphology and respiratory function of the alveolar unit. Int Arch Allergy Appl Immunol 76:2–12

Burrin PH, Sehovic S (1979) The adaptive responses of the rat lung after bilobectomy. Am Rev Respir Dis 119:769–777

Bursell E (1970) An introduction to insect physiology. Academic Press, London

Burton GJ, Mayhew TM, Robertson LA (1989) Stereologic re-examination of the effects of varying oxygen tensions on human placental villi maintained in organ culture for up to 12 hr. Placenta 10:263–273

Burton RR, Smith AH (1967) Blood and air volumes in the avian lung. Poult Sci 47:85–91

Burwell CS, Metacalfe J (1958) Heart disease and pregnancy; physiology and management. Little Brown and Company, Boston

Butler J, Woakes AJ (1980) Heart rate, respiratory frequency and wing beat frequency of free-flying barnacle geese, *Branta leucopsis*. J Exp Biol 85:213–226

Butler J, Woakes AJ (1984) Heart rate and earobic metabolism in Humboldt penguins, *Spheniscus humboldti*, during voluntary dives. J Exp Biol 108:419–428

Butler PJ (1991a) Respiratory adaptations to limited oxygen supply during diving in birds and mammals. In: Woakes AJ, Grieshaber MK, Bridges CR (eds) Physiological strategies for gas exchange and metabolism. Cambridge University Press, Cambridge, pp 235–257

Butler PJ (1991b) Exercise in birds. J Exp Biol 160:233–262

Butler PJ, Metcalfe JD (1983) Control of respiration and circulation. In: Rankin JC, Pitcher TJ, Duggan R (eds) Control processes in fish physiology. Croom Helm, Beckenham, pp 41–65

Butler PJ, West NH, Jones DR (1977) Respiratory and cardiovascular responses of the pigeon to sustained level flight in a wind tunnel. J Exp Biol 71:7–26

Butler PJ, Taylor EW, McMahon BR (1978) Respiratory and circulatory changes in the lobster (*Homarus vulgaris*) during long-term exposure to moderate hypoxia. J Exp Biol 73:131–146

Butler PJ, Milsom WK, Woakes AJ (1984) Respiratory cardiovascular and metabolic adjustments during steady state swimming in the green trutle, *Chelonis mydas*. J Comp Physiol 154B:167–174

Butler PJ, Banzett RB, Fredberg JJ (1988) Inspiratory valving in avian bronchi: aerodynamic consideration. Respir Physiol 72:241–256

Butlin RK, Tregenza T (1997) Is speciation no accident?. Nature (Lond) 387:551–553

Cala P (1987) Aerial respiration in the catfish, *Eremophilus mutisii* (Trichomycteridae, Siluriformes), in the Rio Bogota Basin, Colombia. J Fish Biol 31:301–303

Calder WA (1968) Respiratory and heart rates of birds at rest. Condor 70:358–365

Calder WA (1984) Size, function, and life history. Harvard University Press, Cambridge

Calder WA (1987) Scaling energetics of homeothermic vertebrates. An operational allometry. Annu Rev Physiol 49:107–120

Calder WA, King JR (1974) Thermal and caloric relations of birds. In: Farner DS, King JR (eds) Avian biology, vol 4. Academic Press, London, pp 259–413

Calder WA, Schmidt-Nielsen K (1968) Panting and blood carbon dioxide in birds. Am J Physiol 215:477–482

Callahan PS (1972) The evolution of insects. Holliday House, New York

Callender GS (1940) Variations of the amounts of CO_2 in different air currents. Q J R Soc 66:395–400

Calvin M (1956) Chemical evolution and the origin of life. Am Sci 44:248–263

Cameron E (ed) (1990) Flight: How things work. The Time Inc Book Company, Alexandria, Virginia

Cameron JC (1989) The respiratory physiology of animals. Oxford University Press, Oxford

Cameron JN (1978) Regulation of blood pH in teleost fish. Respir Physiol 33:129–144

Cameron JN (1979) Excretion of CO_2 in water breathing animals: a short review. Mar Biol Lett 1:3–13

Cameron JN (1981) Brief introduction to the land crabs of the Palau Islands: stages in the transition to air breathing. J Exp Zool 218:1–5

Cameron JN, Cech JJ (1970) Notes on the energy cost of gill ventilation in teleosts. Comp Biochem Physiol 34:447–455

Cameron JN, Iwama GK (1987) Compensation of progressive hypercapnia in channel catfish and blue crabs. J Exp Biol 133:183–197

Cameron JN, Mecklenburg TA (1973) Aerial gas exchange in the coconut crab, *Birgus latro*, with some notes on *Gecarcoidea Ialandii*. Respir Physiol 19:245–261

Cameron JN, Randall DJ, Davis JC (1977) Regulation of the ventilation-perfusion ratio in gills of *Dasyatis sabina and Squalus suckleyi*. Comp Biochem Physiol 39A:505–519

Cane MA (1983) Oceanographic events during El Niño. Science 222:1189–1195

Canfield DE, Teske A (1996) Late Proterozoic rise in atmospheric oxygen concentration inferred from phylogenetic and sulphur-isotope studies. Nature (Lond) 382:127–132

Cannon WB (1939) The wisdom of the body. Norton, New York

Cantin AM, North SL, Hubbard RC, Crystal RC (1987) Normal alveolar epithelial lining fluid contains high levels of glutathione. J Appl Physiol 63:152–157

Capone DG, Zehr JP, Paerl HW, Bergman B, Carpenter EJ (1997) *Trichodesmium*, a globally significant marine cyanobacterium. Science 276:1221–1229

Cappellen PV, Ingall ED (1996) Redox stabilization of the atmosphere and oceans by phosphorus-limited marine productivity. Science 271:493–496

Carey C (1980a) Physiology of the avian egg. Am Zool 20:325–489

Carey C (1980b) Adaptation of the avian egg to high altitude. Am Zool 20:449–464

Carey C, Rahn H, Parisi P (1980) Calories, water, lipid and yolk in avian eggs. Condor 82:335–345

Carey FG (1973) Fishes with warm bodies. Sci Am 228:36–54

Carey FG, Lawson KD (1973) Temperature regulation in free-swimming bluefin tuna. Comp Biochem Physiol 44A:375–386

Carey FG, Robinson BH (1981) Daily patterns in the activities of swordfish, *Xiphius gladius*, observed by acoustic telemetry. Fish Bull US 79:277–289

Carey FG, Teal JM (1966) Heat conservation in tuna fish muscle. Proc Natl Acad Sci (USA) 56:1464–1469

Carey FG, Teal JM (1969) Mako and porbeagle: warm-bodied sharks. Comp Biochem Physiol 28:199–204

Carey FG, Teal JM, Kanwisher JW, Lawson KD, Beckett JS (1971) Warm bodied fish. Am Zool 11:137–158

Carey FG, Kanwisher JW, Brazier O, Gabrielson G, Casey JG, Pratt HW (1982) Temperature and activities of a white shark. Copeia 1982:254–260

Carey SW (1976) The expanding Earth. Elsevier, Amsterdam

Carlson CW (1960) Aortic rupture. Turkey Producer. January Issue

Carney JM, Starke-Reed PE, Oliver CN, Landum RW, Cheng MS, Wu JF, Floyd RA (1991) Reversal of age-related increase in brain protein oxidation, decrease in enzyme activity, and loss in temporal and spatial memory by chronic administration of the spin-trapping compound N-tert-butyl-alpha-phenylnitrone. Proc natl Acad Sci USA 88:3633–3636

Caro CG, Saffman PG (1965) Extensibility of blood vessels in isolated rabbit lungs. J Physiol (Lond) 178:193–210

Carpenter FL (1975) Bird hematocrits: effects of high altitude and strength of flight. Comp Biochem Physiol 50A:415–417

Carpenter FL, Hixon MA (1988) A new function for torpor: fat conservation in a wild migrant hummingbird. Condor 90:373–378

Carpenter FL, Paton DC, Hixon MA (1983) Weight gain and adjustment of feeding territory size in migrant Rufous hummingbirds. Proc Natl Acad Sci USA 80:7263

Carpenter RE (1975) Flight metabolism in flying foxes. In: Wu TYT, Brokaw CJ, Brennen C (eds) Swimming and flying in nature, vol 2. Plenum Press, New York, pp 883–890

Carpenter RE (1985) Flight physiology of flying foxes, *Pteropus poliocephalus*. J Exp Biol 114:619–747

Carpenter RE (1986) Flight physiology of intermediate sized fruit-bats (Family: Pteropodidae). J Exp Biol 120:79–103

Carr AF (1952) Handbook of turtles. Cornell University, Ithaca

Carr AF, Goodman D (1970) Ecological implications of size and growth in *Chelonia*. Copeia 1970:783–786

Carr AF, Ross P, Carr S (1974) Interesting behaviour of the green turtle, *Chelonia mydas*, at a mid-ocean island breeding colony ground. Copeia 1974:702–706

Carr MH (1996) Water on Mars. Oxford University Press, Oxford

Carrier DR (1984) Lung ventilation in running lizards. Am Zool 24:84A

Carrier DR (1987a) The evolution of locomotor stamina in tetrapods: circumventing a mechanical contraint. Paleobiology 13:326–341

Carrier DR (1987b) Lung ventilation during walking and running in four species of lizards. Exp Biol 47:33–42

Carrier DR (1988) Ventilation in *iguana iguana*: an action of tonic muscle. Am Zool 28:197A

Carrier DR (1991) Conflict in the hypaxial musculoskeletal system: documenting an evolutionary constraint. Am Zool 31:644–654

Carroll RL (1970) Quantitative aspects of the amphibian-reptilian transition. Forma Functio 3:165–178

Carroll RL (1988) Vertebrate palaeontology and evolution. WH Freeman, New York

Carslaw HS, Jaeger JC (1959) Conduction of heat in solids. Oxford University Press, Oxford

Cartee GD, Farrar RP (1987) Muscle respiratory capacity and VO_{2max} in identically trained young and old rats. J Appl Physiol 63:257–261

Carter GS (1935) The fresh waters of the rain forest areas of British Guiana. J Linn Soc Lond Zool 39:147–193

Carter GS (1955) The papyrus swamps of Uganda. W Heffer, Cambridge

Carter GS (1957) Air breathing. In: Brown ME (ed) The physiology of fishes. Academic Press, London, pp 65–79

Carter GS (1967) Structure and habitat in vertebrate evolution. Washington University Press, Seattle

Carter GS, Beadle LC (1930) Notes on the habits and development of *Lepidosiren paradoxa*. J Linn Soc Zool 37:327–368

Carter GS, Beadle LC (1931) The fauna of the swamps of the Paraguayan Chaco in relation to its environment: respiratory adaptations in fishes. J Linn Soc Zool 37:205–258

Caruthers SD, Harris TR (1994) Effects of pulmonary blood flow on the fractal nature of flow heterogeneity in sheep lungs. J Appl Physiol 77:1474–1479

Casey TM, Withers PC, Casey KK (1979) Matabolic and respiratory responses of Arctic mammals to ambient temperature during summer. Comp Biochem Physiol 155B:751–758

Cassin S, Dawes GS, Mott JC, Ross BB, Strang LB (1964) The vascular resistance of the foetal and newly ventilated lung of the lamb. J Physiol (Lond) 171:61–79

Cassini A, Favero M, Albergoni V (1993) Comparative studies in red-blooded and white-blooded Antarctic teleost fish, *Pagothenia bemacchii* and *Chionodraco hamatus*. Comp Biochem Physiol 106C:333–336

Cech JJ, Rowell DM, Glasgow JS (1977) Cardiovascular responses of the winter flounder *Pseudopleuronectes americanus* to hypoxia. Comp Biochem Physiol A 53:123–125

Chabotarev DF, Korlsushko OV, Ivanov LA (1974) Mechanisms of hypoxia in the elderly. J Gerontol 29:393–400

Chaline J, Laurine B (1986) Phyletic gradualism in a European Plio-Pleistocene *Mimomys* lineage (Arvicolidae, Rodentia). Paleobiology 12:203–216

Chamberlain JA (1990) Jet propulsion of Nautilus: a surviving example of early Paleozoic cephalopod locomotor design. Can J Zool 68:806–814

Chance B, Sies H, Boveris A (1979) Hydroperoxide metabolism in mammalian organs. Physiol Rev 59:527–605

Chang DM, Mark R, Miller SL, Strathearn GE (1983) Prebiotic organic syntheses and the origin of life. In: Schopf JW (ed) Earth's earliest biosphere: its origin and evolution. Princeton University Press, Princeton, pp 53–92

Chapin FS, Zimov SA, Shaver GR, Hobble SE (1996) CO_2 fluctuations at high latitudes. Nature (Lond) 383:585–586

Chapman CB, Jensen D, Wildenthal K (1963) On circulatory control mechanisms in the Pacific hagfish. Circ Res 12:427–440

Chapman DJ, Ragan MA (1980) Evolution of biochemical pathways: evidence from comparative biochemistry. Annu Rev Plant Physiol 31:639–678

Chapman DJ, Schopf JW (1983) Biological and biochemical effects of the development of an aerobic environment. In: Schopf JW (ed) Earth's earliest atmosphere: its origin and evolution. Princeton University Press, Princeton, pp 302–320

Chapman DL, Carlton DP, Nielson DW, Cummings JJ, Poulain FR, Bland RD (1994) Changes in lung liquid during spontaneous labour in foetal sheep. J Appl Physiol 76:523–530

Chapman RC, Bennett AF (1975) Physiological correlates of burrowing in rodents. Comp Biochem Physiol 51A:599–603

Chappell MA (1985) Effects of ambient temperature and altitude on ventilation and gas exchange in deer mice (*Peromyscus maniculatus*). J Comp Physiol 155B:751–758

Chappell MA, Roverud RC (1990) Temperature effects on metabolism, ventilation and oxygen extraction in a Neotropical bat. Respir Physiol 81:401–412

Chappell MA, Souza SL (1988) Thermoregulation, gas exchange and ventilation in Adélie penguins (*Pygoscelis adelie*). J Comp Physiol 157B:783–790

Chappellaz J, Barnola JM, Raynaud D, Korotkevich YS, Lorius C (1992) Ice-core record of atmospheric methane over the past 160 000 years. Nature (Lond) 345:127–131

Charnov EI (1993) Life history invariants, Oxford University Press, Oxford

Charnov EI (1997) Trade-off-invariant rules for evolutionary stable state life histories. Nature (Lond) 387:393–394

Cheatum EP (1934) Limnological investigations on respiration, annual migratory cycle and other related phenomena in fresh water pulmonate snails. Trans Am Microsc Soc 53:348–407

Chen C, Rabourdin B, Hammen CS (1987) The effect of hydrogen sulfide on the metabolism of *Solemya velum* and enzymes of sulfide oxidation in gill tissue. Comp Biochem Physiol 88B:949–952

Chien S (1970) Shear dependence of effective cell volume as a determinant of blood viscosity. Science 168:977–979

Chien S (1985) Role of blood cells in microcirculatory regulation. Microvasc Res 29:129–151

Chien S, Usami S, Dellenback RJ, Bryant CA (1971) Comparative hemorheology – hematological implications of species differences in blood viscosity. Biorheology 8:35–57

Childress JJ, Mickel TJ (1982) Oxygen and sulfide consumption rates of the vent cram *Calyptogena pacifica*. Mar Biol Lett 3:73–79

Childress JJ, Felbeck H, Somero GN (1989) Symbiosis in the deep sea. Sci Am 256:107–112

Chinard FP (1992) Quantitative assessment of epithelial fluid in the lung. Am J Physiol 263:L617–L618

Chivers DJ, Hladik CM (1980) Morphology of the gastrointestinal tract in primates: comparison with other animals in relation to diet. J Morphol 166:337–386

Chyba CF (1997) Life on other planets. Nature (Lond) 385:201

Clark A, Clark PAA, Connett RJ, Gayaski TEJ, Honig CR (1987) How large is the drop in PO_2 between the cytosol and mitochondrion? Am J Physiol 252:C583–587

Clark JM, Lambertsen CJ (1971) Pulmonary oxygen toxicity: a review. Pharmacol Rev 23:37–98

Clark LC, Gollan F (1966) Survival of mammals breathing organic liquids equilibrated with oxygen at atmospheric pressure. Science 152:1755–1756

Clarke FW (1911) The data of geochemistry. US Geol Surv Bull 491, pp 43–158

Clarke MR (1962) Respiratory and swimming movements in the cephalopod, *Crachia scabra*. Nature (Lond) 196:351–352

Clarke J, Nicol S (1993) Blood viscosity of little penguin, *Eudyptula minor*, and the Adélie penguin, *Pygoscelis adeliae*: effect of temperature and shear rate. Physiol Zool 66:720–731

Clausen G, Ersland A (1968) The respiratory properties of the blood of two diving rodents, the beaver and the water vole. Respir Physiol 5:350–359

Claussen CP, Hue A (1987) Light- and electronmicroscopic studies of the lung of *Triturus alpestris* (Laurenti) (Amphibia). Zool Anz 218:115–128

Clayton DA, Vaughan TC (1986) Territorial acquisition in the mudskipper *Boleophthalmus boddarti* (Teleosti, Gobiidae) on the mudflats of Kuwaiti. J Zool (Lond) 209:501–519

Clements JA, Nellenbogen J, Trahan HJ (1970) Pulmonary surfactant and evolution of the lungs. Science 169:603–604

Clench M (1978) Tracheal elongation in birds-of-paradise. Condor 80:423–430

Cloud P (1973) Paleoecological significance of the banded iron-formation. Econ Geol 68:1135–1143

Cloud P (1974) Evolution of ecosystems. Am Sci 62:54–66

Cloud P (1988) Oasis in space. WW Norton, New York

Cloud P (1983a) The biosphere. Sci Am 249:176–189

Cloud P (1983b) Early biogeologic history: the emergence of a paradigm. In: Schopf JW (ed) Earth's earliest biosphere: its origin and evolution. Princeton University Press, Princeton, pp 14–31

Cloudsley-Thompson JL (1977) The water and temperature relations of woodlice. Meadowfield Press, Durham, England

Cloutier R, Forey PL (1991) Diversity of extinct and living actinistian fishes (Sarcopterygii). Environ Biol Fishes 32:59–74

Cochrane CG (1991) Cellular injury by oxidants. Am J Med 91:23S–30S

Cochrane CG, Revak SD (1991) Pulmonary surfactant protein B (SP-B):structure-function relationships. Science 254:566–568

Codispoti LA, Christensen JP (1985) Nitrification, denitrification and nitrous oxide cycling in the eastern tropical south Pacific Ocean. Mar Chem 16:277–300

Cody ML (1974) Optimization in ecology. Science 183:1156–1164

Cohen GM (1975) Pulmonary metabolism of inhaled substances and possible relationship with carcinogenicity and toxicity. In: Junod AF, Haller R (eds) Lung metabolism. Academic Press, New York, pp 185–200

Cohen J (1995) Getting all tunned around over the origins of life on Earth. Science 267:1265–1266

Cohen JE (1964) Age and the pulmonary diffusing capacity. In: Cander L, Moyer J (eds) Aging of the lung. Grune and Stratton, New York, pp 163–172

Cohen WD (1978) On erythrocyte morphology. Blood Cells 4:449–451

Cohn LE, Kinnula VL, Adler KB (1994) Antioxidant properties of guinea pig tracheal epithelial cells in vitro. Am J Physiol 266:L397–L404

Coin JT, Olson JS (1979) The rate of oxygen uptake by human erythrocytes. J Biol Chem 254:1178–1190

Colacino JM, Kraus DW (1984) Haemoglobin-containing cells of *Neodasys* (Gastrotricha, Chaetonotida). II. Respiratory significance. Comp Biochem Physiol 79A:363–369

Colacino JM, Hector DH, Schmidt-Nielsen K (1977) Respiratory responses of ducks to simulated altitude. Respir Physiol 29:265–281

Colbert EH, Morales M (1991) Evolution of the vertebrates, 4th edn. Wiley-Liss, New York

Coles GC (1970) Some biochemical adaptations of the swampworm *Alma emini* to low oxygen levels in tropical swamps. Comp Biochem Physiol 34:481–489

Collazo A (1993) Evolutionary correlations between early development and life history in plethodontid salamanders and teleost fishes. Am Zool 33:60A

Collins-George N (1959) The physical environment of soil animals. Ecology 40:550–557

Colwell RK, Coddington JA (1994) Estimating terrestrial biodiversity through extrapolation. Philos Trans R Soc Lond 345B:101–118

Comroe JH (1966) The lung. Sci Am 57–68

Comroe JH (1974) Physiology of respiration: an introductory text. Year Book Medical Publishers, Chicago

Comroe JH, Dripps RD, Dumke PR, Deming M (1945) Oxygen toxicity: the effect of inhalation of high concentrations of oxygen for 24 hours on normal men at sea level and at simulated altitude of 18 000 ft. J Am Med Assoc 128:710–725

Conley KE, Kayar SR, Rösler K, Hoppeler H, Weibel ER, Taylor CR (1987) Adaptive variation in the mammalian respiratory system in relation to energetic demand: IV. Capillaries and their relationship to oxidative capacity. Respir Physiol 69:47–64

Connett RJ, Gayeski TEJ, Honig CR (1985) An upper bound on the minimum PO_2 for O_2 consumption in red muscle. In: Kreuzer F, Cain SM, Turek Z, Goldstick TK (eds) Oxygen transport to tissue, vol 7. Plenum Publishing Corporation, New York, pp 291–300

Connett RJ, Honig CR, Gayeski TEJ, Brooks GA (1990) Defining hypoxia: a systems view of O_2, glycolysis, energetics, and intracellular PO_2. J Appl Physiol 68:833–842

Conolly ME, Greenacre JK (1975) Hormone receptors in the lung. In: Junod AF, Haller R (eds) Lung metabolism. Academic Press, New York, pp 233–249

Constantinopol M, Jones JH, Weibel ER, Hoppelar H, Lidholm A, Karas RH (1989) Oxygen transport during exercise in large mammals. II. Oxygen uptake by the pulmonary gas exchanger. J Appl Physiol 67:871–878

Constanzo JP, Iverson JB, Wright MF, Lee RE (1995) Cold hardiness and overwintering strategies of hatchilings in an assemblage of northern turtles. Ecology 76:1772–1785

Conway-Morris S (1993) The fossil record and the early evolution of Metazoa. Nature (Lond) 361:219–225

Cook RH, Boyd S (1965) The avoidance by Gammarus aceanicus Segerstrale (Amphiopoda: Crustacea) of anoxic regions. Can J Zool 43:971–975

Cooke AS (1976) Changes in egg shell characteristics of the sparrow hawk (Accipeter nisus) and peregrine (Falco peregrinus) associated with exposure to pollutants during recent decades. J Zool (Lond) 187:245–263

Cooke IRC, Berger PJ (1990) Precursor of respiratory pattern in the early gestation mammalian fetus. Brain Res 522:330–336

Coope A, Penny D (1997) Mass survival across the Cretaceous-Tertiary boundary: molecular evidence. Science 275:1109–1113

Cope ED (1885) The retrograde metamorphosis of Siren. Am Nat 19:1226–1227

Cope ED (1894) On the lungs of Ophidia. Am Philos Soc 33:217–224

Cope ED (1896) The primary factor of organic evolution. Open Court, Chicago

Copeland DE (1969) Fine structural study of gas secretion in the physoclistous swim bladder of Fundulus heteroclitus and Gadus callarias and in the euphysoclistous swim bladder of Opsanus tau. Z Zellforsch 93:305–331

Cornfield DN, Stevens T, McMurtry IF, Abman SH, Rodman DM (1994) Acute hypoxia causes membrane depolarization and calcium influx in fetal pulmonary artery smooth muscle cells. Am J Physiol 266:L469–L475

Cory S, Adams JM (1988) Transgenic mice and oncogenesis. Annu Rev Immunol 6:25–48

Cosgrove WB, Schwartz JB (1965) The properties and function of the blood pigment of the earthworm, Lumbricus terrestris. Physiol Zool 38:206–212

Cossins AR (1991) Cold facts and naked truth. Nature (Lond) 353:699

Costa HH (1967) Responses of Gammarus pulex (L) to modified environment. III. Reactions to low oxygen tensions. Crustaceana 13:175–189

Costello ML, Mathieu-Costello O, West JB (1992) Stress failure of alveolar epithelial cells studied by scanning electron microscopy. Am Rev Respir Dis 145:1446–1455

Cotton CU, Lawson EE, Boucher RC, Gatzy JT (1983) Bioelectric properties and ion transport of airways excised from adult and fetal sheep. J Appl Physiol 55:1542–1549

Coulson JM, Richardson JF (1965) Chemical engineering. Pergamon Press, Oxford

Cournand A, Richards DW, Bader RA, Bader ME, Fishman AP (1954) The oxygen cost of breathing. Trans Assoc Phys 67:162–173

Coutant CC (1987) Thermal preference: when does an asset become a liability? Env Biol 18:161–172

Cracraft JA (1983) The significance of phylogenetic classification for systematic and evolutionary biology. In: Felsenstein J (ed) Numerical taxonomy. Springer Heidelberg, Berlin New York, pp 1–17

Cracraft JA (1986) The origin and early diversification of birds. Paleobiology 12:383–399

Cracraft JA (1994) Species diversity, biogeography, and the evolution of biotas. Am Zool 34:33–47

Craig P (1975) Respiration and body weight in the reptilian genus *Lacerta*: a physiological, anatomical and morphometric study. PhD Thesis, University of Bristol, England

Crandall RR, Smith AH (1952) Tissue metabolism in growing birds. Proc Soc Exp Biol Med 79:345–346

Crapo JD (1986) Morphologic changes in pulmonary oxygen toxicity. Annu Rev Physiol 48:721–731

Crapo JD (1987) Hyperoxia: lung injury and localization of antioxidant defenses. In: Dejours P (ed) Comparative physiology of environmental adaptations, vol 2. Karger, Basel, pp 163–176

Crapo JD, Crapo RO (1983) Comparison of total lung diffusion capacity and the membrane component of diffusion capacity as determined by physiologic and morphometric techniques. Respir Physiol 51:221–280

Crapo JD, McCord JM (1976) Oxygen-induced changes in pulmonary superoxide dismutase, assayed by antibody titrations. Am J Physiol 231:1196–1203

Crapo JD, Tierney DF (1974) Superoxide dismutase and pulmonary oxygen toxicity. Am J Physiol 226:1401–1407

Crapo JD, Barry BE, Chang LY, Mercer RR (1984) Alterations in lung structure by inhalation of oxidants. J Toxicol Environ Health 13:301–321

Crapo JD, Crapo RO, Jensen RL, Mercer RR, Weibel ER (1988) Evaluation of lung diffusing capacity by physiological and morphometric techniques. J Appl Physiol 64:2083–2091

Crapo RO, Crapo JD, Weibel ER (1986) Comparison of physiologic and morphometric estimations of the components of CO diffusing capacity. Prog Respir Res 21:88–91

Crawford EC, Schultetus RR (1970) Cutaneous gas exchange in the lizard, *Sauromalus obesus*. Copeia 1970:179–180

Crawford EC, Gatz RN, Magnussen H, Perry SF, Piiper J (1976) Lung volumes, pulmonary blood flow and carbon monoxide diffusing capacity of turtles. J Comp Physiol 107:169–178

Crawford RMM, Hendry GAF, Goodman BA (eds) (1994) Oxygen and environmental stress in plants. Proc R Soc Edinb 102B:1–549

Crezee M (1976) Solenofilomorphidae (Acoela), major component of a new turbellarian association in sulfide system. Int Rev Gesamten Hydrobiol Hydrogr 61:105–129

Crews D, Grassman M, Garstka WR, Halpert A, Camazine B (1987) Sex and seasonal differences in metabolism in the red-sided garter snake, *Thamnophis sirtalis parietalis*. Can J Zool 65:2362–2368

Crick FH (1966) Codon-anticodon pairing: the wobble hypothesis. J Mol Biol 38:367–379

Crisp DJ (1964) Plastron respiration. In: Danielli JF, Pankhurst KGA, Riddiford AC (eds) Recent advances in surface science. Academic Press, London, pp 377–425

Croll RP (1985) Sensory control of respiratory pumping in *Aplysia californica*. J Exp Biol 117:15–27

Crompton AW, Taylor CR, Jagger JA (1978) Evolution in homeothermy in mammals. Nature (Lond) 272:333–336

Crone C, Lassen NA (1970) Capillary permeability. The transfer of molecules and ions between capillary blood and tissue. Munksgaard, Copenhagen

Crossfill ML, Widdicombe JG (1961) Physical characteristics of the chest and lungs and the work of breathing in different mammalian species. J Physiol (Lond) 158:1–14

Crosswell JW, Smith EE (1967) Determination of optimal hematocrit. J Appl Physiol 22:930–933

Csillag C, Aldhous P (1992) Signs of damage by radicals. Science 258:1875–1876

Culotta E, Koshland DE (1992) Molecule of the year: no news is good news. Science 258:1862–1865

Cunningham DJ, O'Riordan JLH (1957) The effect of rise in temperature of the body on the respiratory response to carbon dioxide. Q J Exp Physiol 42:329–345

Cunningham JT, Reid DM (1932) Experimental researches on the emission of oxygen by the pelvic filaments of the male *Lepidosiren* with some experiments on *Synbranchus marmoratus*. Proc R Soc Lond 110B:234-248

Currie JA (1962) The importance of aeration in providing the right conditions for plant growth. J Sci Food Agric 13:380-385

Currie JA (1984) Gas diffusion through soil crumbs: the effects of compaction and wetting. J Soil Sci 35:1-10

Currey JD (1967) The failure of exoskeletons and endoskeletons. J Morphol 123:1-16

Currey JD (1984) The mechanical adaptations of bones. Princeton University Press, Princeton

Curtis SE, Peek JT, Kelly DR (1993) Partial liquid breathing with perfluocarbon improves arterial oxygenation in acute canine lung injury. J Appl Physiol 75:2696-2702

Cutler A (1995) Notes from the underground. The Sciences (Public NY Acad Sci) Jan/Feb Issue: 36-40

Czietrich HM, Bridges CR, Grieshaber MK (1987) Purine metabolism of the crayfish *Astacus leptodactylus*. Verh Dtsch Zool Ges 80:207

Czopek J (1962a) Vascularization of respiratory surfaces in some caudata. Copeia 1962:576-587

Czopek J (1962b) Smooth muscle in the lungs of some urodeles. Nature (Lond) 193:798-811

Czopek J (1965) Quantitative studies of the morphology of respiratory surfaces in amphibians. Acta Anat 62:296-323

Czopek J, Szarski H (1989) Morphological adaptations to water movements in the skin of anuran amphibians. Acta Biol Cracov 31:81-96

Dahr E (1924) Die Atmungsbewegungen der Landpulmonaten. Lund Univ Aarsskr NF Avd 20:1-19

Dahr E (1927) Studien uber die Respiration der Landpulmonaten. Lund Univ Aarsskr NF Avd 23:1-118

Dales RP (1961) Observations on the respiration of the sabellid polychaete *Schizobranchia insignis*. Biol Bull Mar Biol Lab Woods Hole 121:82-91

Dales RP (1969) Respiration and energy metabolism in annelids. In: Florkin M, Stoltz EH (eds) Chemical zoology, vol 4. Academic Press, New York, pp 93-109

Dandy JWT (1970) Activity response to oxygen in brook trout, *Salvelinus fontinalis* (Mitchill). Can J Zool 48:1067-1078

Daniel TL (1981) Fish mucus. In situ measurements of polymer drag reduction. Biol Bull 60:376-382

Daniel MJ, Boyden CR (1975) Diurnal variations in physicochemical conditions within rockpools. Field Stud 4:161-176

Daniels CB, Barr HA, Nicholas TE (1989) A comparison of the surfactant associated lipids derived from reptilian and mammalian lungs. Respir Physiol 75:335-348

Daniels CB, Barr HA, Power JHT, Nicholas TE (1990) Body temperature alters the lipid composition of pulmonary surfactant in the lizard *Ctenophorus nuchalis*. Exp Lung Res 16:435-449

Daniels CB, Eskandari-Marandi BD, Nicholas TE (1993) The role of surfactant in the static lung mechanics of the lizard *Ctenophorus nuchalis*. Respir Physiol 94:11-23

Daniels CB, Orgeig S, Wilsen J, Nicholas TE (1994) Pulmonary type surfactants in the lungs of terrestrial and aquatic amphibians. Respir Physiol 95:249-258

D'Aoust G (1970) The role of lactic acid in gas secretion in the teleost swim bladder. Comp Biochem Physiol 32:637-668

Darden TR (1970) Respiratory adaptations of a fossorial mammal, the pocket gopher (*Thomomys bottae*). PhD Diss, University of California, Davis

Darnell RM (1949) The aortic arches and associated arteries of caudate amphibians. Copeia 1949:18-31

Darwin CR (1851) A monograph of the subclass Cirripedia. The Lepadidae. John Murray, London

Darwin CR (1859) On the origin of species by means of natural selection, or the preservation of favoured races in the struggle for life. John Murray, London

Darwin CR, Wallace AR (1958) On the tendency of species to form varieties and on the perpetuation of varieties and species by natural means of selection. J Linn Soc Zool 3:45-62

Das BK (1927) The bionomics of certain air-breathing fishes of India together with an account of the development of the air-breathing organs. Philos Trans R Soc Lond 216B:183-219

Das BK (1934) The habits and structure of *Pseudapocryptes lanceolatus*, a fish in the first stages of structural adaptation to aerial respiration. Proc R Soc Lond 115B:422-435

Das BK (1940) Nature and causes of evolution and adaptation of the air-breathing fishes. Proc 27th Indian Sci Congr 2:215-260

Davenport HW (1974) The ABC of acid-base chemistry, 6th edn. The University of Chicago Press, Chicago

Davenport J (1985) Environmental stress and behavioural adaptation. Croom Helm, London

Davenport J (1994) How and why do flying fish fly? Rev Fish Biol Fisheries 4:184-214

Davenport J, Woolmington AD (1981) Behavioural responses of some rocky shore fish exposed to adverse environmental conditions. Mar Behav Physiol 8:1-12

David A, Rao NGS, Ray P (1974) Tank fishery resources of Karnataka. Bull Cent Ini Fish Res Inst Barrackpore 20:1-87

Davie P, Farrell AP (1991) Cardiac performance of an isolated heart preparation from the dogfish (*Squalus acanthias*): the effects of hypoxia and coronary artery perfusion. Can J Zool (Lond) 69:1822-1828

Davies JC (1975) Minimal dissolved oxygen requirements of aquatic life with emphasis on Canadian species: a review. J Fish Res Board Can 32:2295-2395

Davies P (ed) (1989) The new physica. Cambridge University Press, Cambridge

Davies R, Cockrum EL (1964) Experimentally determined weight-lifting capacity of five species of western bats. J Mammal 45:643-644

Davis GE, Foster J, Warren CE, Daudoroff P (1963) The influence of oxygen concentration on the swimming performance of juvenile Pacific salmon at various temperatures. Trans Am Fish Soc 92:111-124

Dawes GS (1965) Oxygen supply and consumption in late foetal life and the onset of breathing at birth. In: Fenn WO, Rahn H (eds) Handbook of physiology, sect 3, respiration, vol II, American Physiolgical Society, Washington, DC, pp 1313-1328

Dawes GS, Mott JC (1962) The vascular tone of the foetal lung. J Physiol (Lond) 164:465-477

Dawes GS, Mott JC, Widdicombe JG (1953) Changes in the lungs of the newborn lamb. J Physiol (Lond) 121:141-162

Dawes GS, Mott JC, Widdicombe JG (1954) The foetal circulation in the lamb. J Physiol (Lond) 126:563-587

Dawson TJ, Dawson WR (1982) Metabolic scope and conductance in response to cold of some dasyurid marsupials and Australian rodents. Comp Biochem Physiol 71A:59-64

Dean AM, Golding GB (1997) Protein engineering reveals ancient adaptive replacements in isocitrate dehydrogenase. Proc Natl Acad Sci USA 94:3104-3109

DeAngelis DL, Godbout L, Shuter BJ (1991) An individual based approach to predicting density dependent dynamics in smallmouth bass populations. Ecol Model 57:91-115

de Beer M (1989) Light measurements in Lake Victoria, Tanzania. Ann Mus R Afr Cent Ser Quarto Zool 257:57-60

de Beers G (1951) Embryos and ancestors. Oxford University Press, Oxford

de Beers G (1954) *Archeopteryx lithographica*. Br Mus Nat Hist Lond

De Belder AJ, Radomski MW, Why HJF, Richardson CA et al. (1993) Nitric oxide synthase activities in human myocardium. Lancet 341:84-85

DeFur PL, Mangum CP (1979) The effects of environmental variables on the heart rate of invertebrates. Comp Biochem Physiol A 62:283-294

DeFur PL, McMahon BR (1978) Respiratory responses of *Cancer productus* to air exposure. Am Zool 18:605A

DeFur PL, Pease AL (1988) Metabolic and Respiratory compensation during long term hypoxia in blue crabs, *Callinectes sapidus*. Understanding the Estuary, Advances in Chesapeake Bay Research, Chesapeake Res Consortium, Publication No 129

Degan H, Kristensen B (1981) Low sensitivity of *Tubifex* sp. respiration to hydrogen sulfide inhibitors. Comp Biochem Physiol 69B:809-817

De Groodt M, Lagasse A, Sebruyns M (1960) Electronenmikroskopische Morphologie der lungenalveolen des *Protopterus* und *Amblystoma*. Proc Int Congr Electron Microscopy, Springer, Berlin Heidelberg New York, 418 pp

Dehadrai PV (1962) Respiratory function of the swim bladder of *Notopterus* (Lacepede). Nature (Lond) 185:929

Dehadrai PV, Tripathi SD (1976) Environment and ecology of fresh water air-breathing teleosts. In: Hughes GM (ed) Respiration of amphibious vertebrates. Academic Press, London, pp 39–72

De Jong HJ (1962) Activity of the body wall musculature of the African clawed toad, *Xenopus laevis* (Daudin) during diving and respiration. Zool Med 47:135–144

De Jong HJ, Gans C (1969) On the mechanism of respiration in the bullfrog, *Rana catesbeiana*: reassessment. J Morphol 127:259–290

Dejours P (1973) Problems of control of breathing in fishes. In: Bolis L, Schmidt-Nielsen K, Madrell SHP (eds) Comparative physiology. Elsevier/North Holland, Amsterdam, pp 117–133

Dejours P (1975) Principles of comparative respiratory physiology, 1st edn. Elsevier-North Holland, Amsterdam

Dejours P (1981) Principles of comparative respiratory physiology, 2nd edn. Elsevier-North Holland, Amsterdam

Dejours P (1982) Mount Everest and beyond: breathing air. In: Taylor CR, Johansen K, Bolis L (eds) A companion to animal physiology. Cambridge University Press, Cambridge, pp 17–27

Dejours P (1988) Respiration in water and air: adaptations, regulation and evolution. Elsevier, Amsterdam

Dejours P (1989) From comparative physiology of respiration to several problems of environmental adoptations and to evolution. J Physiol (Lond) 410:1–19

Dejours P (1990) Comparative aspects of maximal oxygen consumption. Respir Physiol 80:155–162

Dejours P (1994) Environmental factors as determinats in bimodal breathing: an introductory overview. Am Zool 34:178–183

Dejours P, Truchot JP (1988) Respiration of the immersed shore crab at variable ambient oxygenation. J Comp Physiol 158B:387–391

Dejours P, Wagner S, Dejager M, Vichon MJ (1967) Ventilation et gaz alvéolaire pendant langage parlé. J Physiol (Paris) 59:386

Dejours P, Garey WF, Rahn H (1970) Comparison of ventilatory and circulatory flow rates between animals in various physiological conditions. Respir Physiol 9:108–117

Dejours P, Toulmond A, Truchot JP (1977) The effect of hyperoxia on breathing of marine fishes. Comp Biochem Physiol 58A:409–411

Dekker RWRJ, Brom TG (1992) Megapode phylogeny and the interpretation of incubation strategies. Zool Verh Leiden 278:9–31

DeLaney RG, Fishman AP (1977) Analysis of lung ventilation in the aestivating lungfish *Protopterus aethiopicus*. Am J Physiol 233:R181–187

DeLaney RG, Lahiri GS, Fishman AP (1974) Aestivation of the African lungfish *Protopterus aethiopicus*: cardiovascular and respiratory functions. J Exp Biol 61:111–118

DeLaney RG, Laurent P, Galante R, Pack AI, Fishman AP (1983) Pulmonary mechanoreceptors in the dipnoan lungfish *Protopterus* and *Lepidosiren*. Am J Physiol 244:R418–R428

Delmas RJ, Ascencio JM, Legrand M (1980) Polar ice evidence that the atmospheric CO_2 20 000 yr BP was 50% of the present. Nature (Lond) 284:155–157

Demoll R (1927) Untersuchungen über die Atmung der Insekten. Z Biol 87:8–22

Demple B, Harrison L (1994) Repair of oxidative damage to DNA enzymology and biology. Annu Rev Biochem 63:915–948

Dempsey EW (1960) Histophysical considerations: In: Villee CA (ed) The placenta and foetal membranes. Williams and Wilkins, Baltimore, pp 29–35

Dempsey JA (1986) Is the lung built for exercise? Med Sci Sports Exercise 18:143–155

Dence WA (1933) Notes on large bowfin (*Amia calva*) living in a mud puddle. Copeia, Ichtyol Notes 1:35

Denison RH (1941) The soft anatomy of *Bothriolepis*. J Paleontol 15:553–561

Denney MW (1993) Air and water: the biology and physics of life's media. Princeton University Press, Princeton

Densmore LD, Owen RD (1989) Molecular systematics of the order Crocodilia. Am Zool 29:181A

Denton EJ, Shaw TI, Gilpin-Brown JB (1958) Bathyscaphoid squid. Nature (Lond) 182:1810

Denton EJ, Liddicoat JD, Taylor DW (1970) Impermeable "silvery" layers in fishes. J Physiol (Lond) 207:64P

De Queiroz K (1985) The ontogenetic method for determining character polarity and its relevance to phylogenetic systematics. Syst Zool 34:280–299

De Saint-Aubain ML, Wingstrand K (1981) A sphincter in the pulmonary artery of the frog *Rana temporaria* and its influence on blood flow in skin and lungs. Acta Zool (Stockh) 60:163–172

Desbruyères D, Laubier B (1986) Les Alvineliidae, une famille nouvelle d'annélides polychètes inféodées aux sources hydrothermales sous-marines: systématique, biologie et écologie. Can J Zool 64:2227–2245

Desbruyères D, Crassous P, Grassle J, Khripounoff A, Reyss D, Rio M, van Praet M (1982) Données écologiques sur un nouveau site d'hydrothermalisme actif de la ride du Pacifique oriental. C R Hebd Séances Acad Sci Paris Ser III 295:489–494

Des Marais DJ, Strauss H, Simmons RE, Hayes JM (1992) Carbon isotope evidence for the stepwise oxidation of the Proterozoic environment. Nature (Lond) 359:605–609

De Villiers CJ, Hodgson AN (1987) The structure of the secondary gills of *Siphonaria capensis* (Gastropoda; Pulmonata). J Molluscan Stud 53:129–138

De Vries AL (1971) Glycoproteins as biological antifreeze agents in Antarctic fishes. Science 172:1152–1155

De Vries R, De Jager (1984) The gill in the spiny dogfish, *Squalus acanthias*: respiratory and nonrespiratory function. Am J Anat 169:1–29

De Vries TJ, Ortlieb L, Diaz A (1997) Determining the early history of El Niño. Science 276:965–966

D'Hondt S, Arthur MA (1997) Late Cretaceous oceans and the cool tropical paradox. Science 271:1838–1841

Diamond JM (1992) The red flag of optimality. Nature (Lond) 355:204–205

Diamond JM (1998) Evolution of biological safety factors: a cost benefit analysis. In: Weibel ER, Taylor CR, Bolis L (eds) Principles of animal design: the optimization and symmorphosis debate. Cambridge University Press, London, pp 21–27

Diamond JM, Hammond KA (1992) The matches, achieved by natural selection, between biological capacities and their natural loads. Experientia 48:551–557

Diamond JM, Karasov WH, Phan D, Carpenter FL (1986) Digestive physiology is a determinant of foraging bout frequency in hummingbirds. Nature (Lond) 320:62–63

Diaz H, Rodriguez G (1977) The branchial chamber in terrestrial crabs: a comparative study. Biol Bull 153:485–504

Dickson KA, Graham JB (1986) Adaptations to hypoxic environments in the erythrinid fish, *Hoplias microlepis*. Environ Biol Fish 15:301–308

Dickson KA, Maloney JE, Berger PJ (1986) Decline in liquid volume before labour in foetal lambs. J Appl Physiol 61:2266–2272

Dierichs R, Dosche C (1982) The alveolar lining layer in the lung of the axolotol, *Ambryostoma mexicanum*. Cell Tissue Res 222:677–686

Dill DB, Graybill A, Hurtado A, Taguini AC (1963) Gaseous exchange in the lungs in old age. J Am Geriatr Soc 11:1063–1076

DiMagno L, Chan CK, Jia Y, Lang MJ, Newman JR, Mets L, Fleming GR, Haselkorn R (1995) Energy transfer and trapping in photosystem reaction centers from cyanobacteria. Proc Natl Acad Sci USA 92:2715–2719

Dios-Escolar J, Gallego B, Tejero C, Escolar MA (1994) Changes occurring with increasing age in the rat lung: morphometrical study. Anat Rec 239:287–306

Di Prampero PE (1985) Metabolic and circulatory limitations to Vo$_{2max}$ at the whole animal level. J Exp Biol 115:319–331

Dixon B (1994) Power unseen: how microbes rule the world. WH Freeman, New York

Dixon D (1987) The planet Earth. World Book, Chicago

Dobzhansky T (1973) Nothing in biology makes sense except in the right of evolution. Am Biol Teach 35:125–129

Dock DS, Kraus WL, McGuire LB, Hyland JW, Haynes FW, Dexter L (1961) The pulmonary blood volume in man. J Clin Invest 40:317–328

Doebeli M (1993) The evolutionary advantage of controlled chaos. Proc R Soc Lond 254B:281–285

Doebeli M, Blarer A, Ackerman M (1997) Population dynamics, demographic stochasticity, and the evolution of cooperation. Proc Natl Acad Sci USA 94:5167–5171

Doeller JE, Kraus DW, Colacino JM, Wittenberg JB (1988) Gill haemoglobin may deliver sulfide to bacterial symbionts of *Solemya velum* (Bivalvia, Mollusca). Biol Bull 175:388–396

Donahue TM (1966) The problem of atomic hydrogen. Ann Geophys 22:175–188

Dominguez-Bello MG, Ruiz MC, Michelangeli F (1993) Evolutionary significance of fore gut fermentation in the hoatzin (*Opisthocomus hoazin*) – Aves:Opisthocomidae. J Comp Physiol 163:594–601

Dorer FE, Kahn JR, Lentz KE, Levine M, Skeggs WT (1974) Hydrolysis of bradykinin by angiotensin-converting enzyme. Circ Res 34:824–827

Dorsey NE (1940) Properties of ordinary water substance. Reinhold, New York

Doudoroff P, Shumway DL (1970) Dissolved oxygen requirements of freshwater fishes. FAO Fish Tech Pap 86:291

Douglas EL, Peterson KS, Gysi JR, Chapman DJ (1985) Myoglobin in the heart tissue of fishes lacking haemoglobin. Comp Biochem Physiol 81:885–888

Drabkin DL (1950) The distribution of the chromoproteins, myoglobin, and cytochrome *c*, in the tissues of different species, and the relationship of the content of each chromoprotein to body mass. J Biol Chem 182:317–333

Dratisch L (1925) Über das Leben der Salamandra-Larven bei hohem und niedrigem Sauerstoffpartialdruck. Z Vergl Physiol 2:632–657

Drent R (1975) Incubation. In: Farner DS, King JR (eds) Avian biology, vol 5. Academic Press, New York, pp 333–419

Dresco-Derouet L (1974) Étude des mygales II. Premier résultats sur la biologie et le métabolisme respiratoire de différentes espèces tropicales en captivité. Bull Mus Natl Hist Nat 42:1054–1062

Driedzic WR (1988) Matching cardiac oxygen delivery and fuel supply to energy demand in teleosts and cephalopods. Can J Zool 66:1078–1083

Driedzic WR, Gesser H (1994) Energy metabolism and contractility in ectothermic vertebrate hearts: hypoxia, acidosis, and low temperature. Physiol Rev 74:221–258

Driedzic WR, Stewart JM (1982) Myoglobin content and the activities of enzymes of energy metabolism in red and white fish hearts. J Comp Physiol 149:67–73

Driedzic WR, Phleger CF, Fields JH, French C (1978) Alterations in energy metabolism associated with transitions from water to air breathing fish. Can J Zool 56:730–735

Dubach M (1981) Quantitative analysis of the respiratory system of the house sparrow, budgerigar, and violet-eared hummingbird. Respir Physiol 46:43–60

Dubale MS (1959) A comparative study of the oxygen carrying capacity of the blood in water and air-breathing teleosts. J Anim Morphol Physiol 6:48–54

Dube SC (1972) Investigations on the functional capacity of respiratory organs of certain fresh water teleostean fishes. PhD Thesis, Banaras University, India

Dudley GA, Abraham WM, Terjung RL (1982) Influence of exercise intensity and duration on biochemical adaptations in skeletal muscle. J Appl Physiol 53:844–850

Duellman WE, Trueb L (1986) Biology of amphibians. McGraw-Hill, New York

Duling BR, Kuschinsky W, Wahl M (1979) Measurements of the perivascular PO_2 in the vicinity of the pial vessels of the cat. Pfluegers Arch 383:29–34

Dullemeijer P (1974) Concepts and approaches in animal morphology. Van Corkum, Assen

Duncker H-R (1974) Structure of the avian respiratory tract. Respir Physiol 22:1–34

Duncker H-R (1978a) General morphological principles of amniotic lungs. In: Piiper J (ed) Respiratory function in birds, adult and embryonic. Springer, Berlin Heidelberg New York, pp 1–15

Duncker H-R (1978b) Development of the avian respiratory system. In: Piiper J (ed) Respiration in birds, adult and embryonic. Springer, Berlin Heidelberg New York, pp 260–273

Duncker H-R (1979) General morphological principles of amniotic lungs. In: Piiper J (ed) Respiratory function in birds, adult and embryonic. Springer, Berlin Heidelberg New York, pp 1–15

Duncker H-R (1985) The present situation of morphology and its importance for biological and medical sciences. In: Duncker H, Fisher W (eds) Vertebrate morphology. Fischer, Stuttgart, p 9

Duncker H-R (1991) The evolutionary biology of homoithermic vertebrates: the analysis of complexity as a specific task of morphology. Verh Dtsch Zool Ges 84:39–60

Duncker HR, Guntert M (1985a) The quantitative design of the avian respiratory system: from hummingbird to the mute swan. In: Nachtigall W (ed) BIONA report No 3. Gustav Fischer, Stuttgart, pp 361–378

Duncker HR, Guntert M (1985b) Morphometric analysis of the avian respiratory system. In: Duncker HR, Fleischer G (eds) Vertebrate morphology. Gustav Fischer, Stuttgart, pp 383–387

Dunel-Erb S, Laurent P (1980a) Ultrastructure of marine teleost gill epithelia: SEM and TEM study of the chloride cell apical membrane. J Morphol 165:175–186

Dunel-Erb S, Laurent P (1980b) Functional organization of the gill vasculature in different classes of fish. In: Lahlou B (ed) Epithelial transport in the lower vertebrates. Cambridge University Press, Cambridge, pp 37–58

Dunham AE (1993) Population responses to environmental change: physiologically structured models, operative environments, and population dynamics. In: Karieva PM, Kingsolver JS, Huey RB (eds) Biotic interactions and global change. Sinauer, Sunderland, pp 196–275

Dupré RK, Wood SC (1988) Behavioural temperature regulation by aquatic ectotherms during hypoxia. Can J Zool 66:2649–2652

Dupré RK, Burggren WW, Vitalis TZ (1991) Dehydration decreases cutaneous gas diffusing capacity in the toad, *Bufo woodhouseii*. Am Zool 31:75A

Durbin AG, Durbin G, Verity PG, Smayda TJ (1981) Voluntary swimming speeds and respiration rates of a filter-feeding planktivore, the Atlantic menhaden, *Brevoortia tyranuus* (Pisces: Clupeidae). Fish Bull 78:877–886

Durmowicz AG, Hofmeister S, Kadyraliev TK, Aldashev AA, Stenmark KR (1993) Functional and structural adaptation of the yak pulmonary circulation to residence at high altitude. J Appl Physiol 74:2276–2285

Duval A (1983) Heartbeat and blood pressure in terrestrial slugs. Can J Zool 61:987–992

Duykers LRB, Percy JL (1978) Lung resonance characteristics of submerged mammals. J Acoust Soc Am 64:S97

Dyer BD, Ober RA (1994) Tracing the history of the eukryotic cells. Columbia University Press, New York

Dyer MF, Uglow RF (1978) Gill chamber ventilation and scaphognathite movements in *Crangon crangon* (L). J Exp Mar Biol Ecol 31:195–207

Eastman JT (1991) Evolution and diversification of Antarctic notothenioid fishes. Am Zool 31:93–109

Eastman JT (1993) Antarctic fish biology: evolution in a unique environment. Academic Press, San Diego

Eaton R, Arp AJ (1990) The effect of sulfide on oxygen consumption rate of *Urechis caupo*. Am Zool 30:69A

Eaton R, Arp AJ (1993) Aerobic respiration during sulfide exposure in the marine echiuran worm *Urechis caupo*. Physiol Zool 66:1–19

Ebeling AW, Bernal P, Zuleta A (1970) Emersion of the amphibious Chilean clingfish *Sicyases sanguineus*. Biol Bull Mar Biol Lab woods Hole 139:115–137

Eberly LB, Kanz JE, Taylor C, Pinsker H (1981) Environmental modulation of a central pattern generator in freely behaving *Aplysia*. Behav Neural Biol 32:21–34

Economos AC (1979) Gravity, metabolic rate, and body size of mammals. Physiologist 22:S71

Eddy FB, Bamford OS, Maloiy GMO (1981) Na^+ and Cl^- effluxes and ionic regulation in *Tilapia grahami*, a fish living in conditions of extreme alkalinity. J Exp Biol 91:349–353

Edney EB (1960) Terrestrial adaptations. In: Waterman TH (ed) The physiology of Crustacea, vol I. Academic Press, New York, pp 367–393

Edney EB, Spencer J (1955) Cutaneous respiration in woodlice. J Exp Biol 32:256–269

Edsall JT (1972) Blood and haemoglobin: the evolution of knowledge of functional adaptation in a biochemical system. J Hist Biol 5:205–257

Edwards GA, Ruska H, Harven de E (1958) The fine structure of insect tracheoblasts, tracheae and tracheoles. Arch Biol 69:351–369

Edwards RRC (1971) An assessment of the energy cost of gill ventilation in the plaice (*Pleuronectes platessa* L). Comp Biochem Physiol 40:391–398

Effros RM, Mason GR (1983) Measurements of pulmonary epithelial permeability in vivo. Am Rev Respir Dis 127:S59–S65

Effros RM, Murphy C, Ozker K, Hacker A (1992) Kinetics of urea exchange in air- filled and fluid-filled lungs. Am J Physiol 263:L619–L626

Egan EA, Olver RE, Strand LB (1975) Changes in non-electrolyte permeability of alveoli and the absorption of lung liquid at the start of breathing in the lamb. J Physiol (Lond) 244:161–179

Ege R (1916) Less known respiratory media. Vidensk Medd Dan Naturhist Foren 67:14–16

Ege R (1918) On the respiratory function of the air stores carried by some aquatic insects (Corixidae, Dytiscidae, and Notonecta). Z Vergl Physiol 17:81–124

Eger WH (1971) Ecological and physiological adaptations of intertidal clingfishes (Teleostei; Gobisocidae) in the Gulf of California. PhD Dissertation, Department of Biological Sciences, University of Arizona, Tucson

Ehrenfeld J, Garcia-Romeu F (1977) Active hydrogen excretion and sodium absorption through isolated frog skin. Am J Physiol 233:F46–F54

Ehret G, Tautz J, Schmitz B (1990) Hearing through the lungs: lung-eardrum transmission of sound in the frog Eleutherodactylus coqui. Naturwissenschaften 77:192–194

Ehrlich HL (1996) Geomicrobiology, 3rd edn. Marcel Dekker, New York

Eigen M, Schuster P (1979) The hypercycle. Springer, Berlin Heidelberg New York

Eiseman B, Bryant B, Waltuch T (1964) Metabolism of vasomotor agents by the isolated per-fused lung. J Thorac Cardiovasc Surg 48:798–806

Ekblom B, Wilson G, Åstrand PO (1975) Central circulation during exercise after venesection and reinfusion of red blood cells. J Appl Physiol 40:379–383

Elder HY (1975) Muscle structure. In: Usherwood PNR (ed) Insect muscle. Academic Press, New York, pp 1–74

Eldredge N (1993) History, function, and evolutionary biology. Evol Biol 27:33–50

Eldredge N, Gould SJ (1972) Punctuated equilibria: an alternative to phyletic gradualism. In: Schopf TJM (ed) Models in paleobiology. Freeman Cooper, San Francisco, pp 82–115

El Haj AJ, Innes AJ, Taylor EW (1986) Ultrastructure of the pulmonary, cutaneous and branchial gas exchange organs of the Trinidad mountain crab. J Physiol (Lond) 373:84P

Elkins N (1983) Weather and bird behaviour. T and AD Poyser, Calton, Stoke on Trent

Ellington CP (1981) The aerodynamics of flapping animal flight. Am Zool 24:95–105

Ellington CP (1984) The aerodynamics of hovering insect flight. VI. Life and power require-ments. Philos Trans R Soc Lond 305B:145–181

Ellis CG, Potter RF, Groom AC (1983) The Krogh cylinder geometry is not appropriate for modelling O_2 transport in contracted skeletal muscle. Adv Exp Med Biol 159:253–268

Else PL, Hubert AJ (1981) Comparison of the "mammal machine" and the "reptilian machine" energy production. Am J Physiol 240:R3–R9

Else PL, Hubert AJ (1983) A comparative study of the metabolic capacity of hearts from reptiles and mammals. Comp Biochem Physiol 76A:553–557

Else PL, Hubert AJ (1985) Mammals: an allometric study of metabolism at tissue and mitochon-drial level. Am J Physiol 248:R415–R421

Embley MT, Hirt RP, Williams DM (1994) Biodiversity at the molecular level: the domains, kingdoms and phyla of life. Philos Trans R Soc Lond 345B:21–33

Emery SH, Szczepanski A (1986) Gill dimensions in pelagic elasmobranch fishes. Biol Bull 171:441–449

Emilio MG, Shelton G (1974) Gas exchange and its effect on blood-gas concentrations in the amphibian, Xenopus laevis. J Exp Biol 60:567–579

Emilio MG, Machado MM, Menano HP (1970) The production of hydrogen ion gradient across the isolated frog skin: quantitative aspects and the effects of acetazolamide. Biochem Biophys Acta 203:394–409

Emmett B, Hochachka PW (1981) Scaling of oxidative and glycolytic enzymes in mammals. Respir Physiol 45:261–272

Engel LA (1991) Effect of microgravity on the respiratory system. J Appl Physiol 70:1907–1911

Engelhard EK, Kam-Morgan LNW, Washburn JO, Volkman LE (1994) The insect tracheal system: a conduit for the systemic spread of Autographa californica nuclear polyhedrosis virus. Proc Natl Acad Sci USA 91:3224–3227

Enns T, Scholander PF, Bradstreet ED (1965) Effect of hydrostatic pressure on gases dissolved in water. J Phys Chem 69:389–391

Epe B (1995) DNA damage profiles induced by oxidizing agents. Rev Physiol Biochem Pharmacol 127:223–249

Epting RJ (1980) Functional dependence of the power for hovering on wing disc loading in hummingbirds. Physiol Zool 53:347–352

Erasmus B, Howell B, Rahn H (1970/71) The ontogeny of acid-base balance in the bullfrog and chicken. Respir Physiol 11:46–53

Eriksen CH, Brown RJ (1980a) Comparative respiratory physiology and ecology of phyllopod Crustacea. I. Conchostraca. Crustaceana 39:1–10

Eriksen CH, Brown RJ (1980b) Comparative respiratory physiology and ecology of phyllopod Crustacea. II. Anostraca. Crustaceana 39:11–21

Errington PL (1963) Muskrat populations. Iowa State University, Iowa

Erskine RLA, Ritchie JWK (1985) Umbilical artery blood flow characteristics in normal and growth retarded fetuses. Br J Obstet Gynecol 5:605–610

Erwin DH (1993) The great Paleozoic crisis: life and death in the Permian. Columbia University Press, New York

Erwin DH (1994) The Permo-Triassic extinction. Nature (Lond) 367:231–236

Erwin DH (1996) The mother of mass extinctions. Sci Am 275:56–62

Escourrou P, Qi X, Weiss M, Mazmanian GM, Gaultier C, Hervé P (1993) Influence of pulmonary blood flow on gas exchange in piglets. J Appl Physiol 75:2478–2483

Evans HM (1929) Some notes on the anatomy of the electric eel, *Gymnotus electrophorus*, with special reference to a mouth-breathing organ and the swim bladder. Proc Zool Soc Lond 57:17–23

Evans HM, Damant GCC (1928) Observations on the physiology of the swim bladder of cyprinoid fishes. J Exp Biol 6:42–55

Everson I, Ralph R (1968) Blood analyses of some Antarctic fish. Br Antarct Surv Bull 15:59–62

Ewer DW (1941) The blood systems of *Sabella* and *Spirographis*. Q J Microsc Soc 82:587–620

Ewer DW (1959) A toad (*Xenopus leavis*) without haemoglobin. Nature (Lond) 183:271

Faber JJ (1993) Diffusion permeability of the immature placenta of the rabbit embryo in inert hydrophilic molecules. J Appl Physiol 265:H1804–H1808

Faber JJ, Rahn H (1970) Gas exchange between air and water and the ventilation pattern of the electric eel. Respir Physiol 9:151–161

Faber JJ, Thornburg KL, Binder NC (1992) Physiology of placental transfer in mammals. Am Zool 32:343–354

Fahlén G (1971) The functional morphology of the gas bladder of the genus *Salmo*. Acta Anat 78:161–184

Fairbridge RW (1966) The encyclopedia of oceanography. Reinhold Publishing, New York

Falke KJ, Hill RD, Schneider RC, Guppy M, Liggins GC et al. (1985) Seal lungs collapse during free diving: evidence from arterial nitrogen tensions. Science 229:556–567

Falkowski PG (1997) Evolution of the nitrogen cycle and its influence on the biological sequestration of CO_2 in the ocean. Nature (Lond) 387:272–275

Fanelli GM, Goldstein L (1964) Ammonia excretion in the neotenous newt *Necturus maculosus* (Raf). Comp Biochem Physiol 13:193–204

Fänge R (1953) The mechanisms of gas transport in the euphysoclist swim bladder. Acta Physiol Scand 30:1–133

Fänge R (1966) Physiology of the swim bladder. Physiol Rev 46:299–322

Fänge R (1976) Gas exchange in the swim bladder. In: Hughes GM (ed) Respiration in amphibious vertebrates. Academic Press, London, pp 189–211

Fänge R (1983) Gas exchange in the fish swim bladder. Rev Physiol Biochem Pharmacol 97:111–158

Farabaugh AT, Thomas DB, Thomas SP (1985) Ventilation and body temperature of the bat *Phyllostomus hastatus* during flight at different air temperatures. Physiologist 28:272

Faraci MF (1990) Cerebral circulation during hypoxia: is a bird brain better? In: Sutton JR, Coates G, Remmers JE (eds) Hypoxia: the adaptations. BC Decker, Burlington, Ontario, pp 26–29

Faraci MF, Fedde MR (1986) Regional circulatory responses to hypocapnia and hypercapnia in bar-headed geese. Am J Physiol 250:R499–R504

Faraci MF, Kilgore DL, Fedde MR (1984) Oxygen delivery to the heart and brain during hypoxia: Pekin duck vs bar-headed geese. Am J Physiol 247:R69–R75

Farhi L (1964) Gas sores of the body. In: Fenn WO, Rahn H (eds) Handbook of physiology, sect 3. respiration, vol I. Am Physiol Soc, Washington, DC, pp 873–924

Farhi L, Rahn H (1955) Gas stores of the body and the steady state. J Appl Physiol 7:472–484

Farley RD (1990) Regulation of air and blood flow through the book lungs of the desert scorpion, *Paruroctonus mesaensis*. Tissue Cell 22:547–569

Farley RD, Case JF (1968) Perception of external oxygen by the burrowing shrimp, *Callianassa californiensis* Dan and *C. affinis* Dana. Biol Bull Mar Biol Lab Woods Hole 134:261–265

Farrell AP (1982a) Cardiovascular changes in the unanethetized lingcod (*Ophiodon elongatus*) during short-term, progressive hypoxia and spontaneous activity. Can J Zool 60:933–941

Farrell AP (1982b) Lung development: biological and clinical perspectives. Academic Press, London

Farrell AP (1984) A review of cardiac performance in the teleost heart: intrinsic and humoral regulation. Can J Zool 62:523–536

Farrell AP (1991a) Circulation of body fluids. In: Prosser CL (ed) Environmental and metabolic animal physiology. Wiley/Liss, New York, pp 509–558

Farrell AP (1991b) Cardiac scope in lower vertebrates. Can J Zool 69:1981–1984

Farrell AP (1993) Cardiovascular system. In: Evans DH (ed) The physiology of fishes. CRC Press, Boca Raton, pp 219–250

Farrell AP, Jones DR (1992) The heart. In: Hoar WS, Randall DJ, Farrell AP (eds) Fish physiology: cardiovascular systems, vol 12. Academic Press, New York, pp 1–87

Farrell AP, Randall DJ (1978) Air-breathing mechanisms in two Amazonian teleosts, *Arapaima gigas* and *Hoplerythrinus unitaeniatus*. Can J Zool 56:939–945

Farrell AP, Daxboeck C, Randall DJ (1979) The effect of input pressure and flow on the pattern and resistance to flow in isolated perfused gills. J Comp Physiol 133:233–248

Farrell AP, Sobin SS, Randall DJ, Crosby S (1980) Sheet blood flow in the secondary lamellae of teleost gills. Am J Physiol 239:R428–436

Farrelly CA, Greenaway P (1987) The morphology of and vasculature of the lungs and gills of the soldier crab *Mictyris longicarpus*. J Morphol 193:285–304

Farrelly CA, Greenaway P (1992) Morphology and ultrastructure of the gills of terrestrial crabs (Crustacae, Gecarcinidae and Grapsidae): adaptations for air-breathing. Zoomorphology 112:39–49

Farris R (1976) Systematics and ecology of Gnathostomulida from North Carolina and Bermuda. PhD Diss, University of North Carolina, Chapel Hill

Fauvel P (1927) Polychaètes sédentaires. In: Office Central de Faunistique (ed) Faune de France, vol 16. Lechevalier, Paris, pp 1–494

Fay P (1965) Heterotrophy and nitrogen fixation in *Chlorogloea fritschii*. J Gen Microbiol 39:11–20

Fedde MR (1976) Respiration. In: Sturkie PD (ed) Avian physiology, 3rd edn. Springer, Berlin Heidelberg New York, pp 122–145

Fedde MR (1980) The structure and gas-flow pattern in the avian respiratory system. Poult Sci 59:2642–2653

Fedde MR, Kuhlmann WD (1978) Intrapulmonary carbon dioxide sensitive receptors: amphibians to mammals. In: Piiper J (ed) Respiratory function in birds, adult and embryonic. Springer, Berlin Heidelberg New York, pp 33–50

Fedde MR, Faraci FM, Kilgore DL, Cardinet GH, Chatterjee A (1985) Cardiopulmonary adaptations in birds for exercise at high altitude. In: Giles R (ed) Circulation, respiration, and metabolism. Springer, Berlin Heidelberg New York, pp 149–163

Fedde MR, Orr JA, Shams H, Scheid P (1989) Cardiopulmonary function in exercising bar-headed geese during normoxia and hypoxia. Respir Physiol 77:239–262

Feder ME (1976) Lungless, body size, and metabolic rate in salamanders. Physiol Zool 49:398–418

Feder ME (1985) Effects of thermal acclimation on locomotor energetics and locomotor performance in a lungless salamander. Physiologist 28:342

Feder ME, Burggren WW (1985a) Cutaneous gas exchange in vertebrates: design, patterns, control, and implications. Biol Rev 60:1–45

Feder ME, Burggren WW (1985b) The regulation of cutaneous gas exchange in vertebrates. In: Giles R (ed) Circulation, respiration, and metabolism. Springer, Berlin Heidelberg New York, pp 101–112

Feder ME, Pinder AW (1988) Ventilation and its effect on "infinite pool" exchangers. Am Zool 28:973–983

Feder ME, Lynch JF, Shaffer HB, Wake DB (1982) Field body temperatures of tropical and temperate zone salamanders. Smithson Herp Inf Service 52:1–23

Federspiel WJ (1989) Pulmonary diffusing capacity: implications of two-phase blood flow in capillaries. Respir Physiol 77:119–134

Federspiel WJ, Popel AS (1986) A theoretical analysis of the effect of the particulate nature of blood on oxygen release in capillaries. Microvasc Res 32:164–189

Feducia A (1980) The age of birds. Harvard University Press, Cambridge

Felbeck H (1983) Sulfide oxidation and carbon fixation by the gutless clam Solemya reidi: an animal-bacteria symbiosis. J Comp Physiol 152:3–11

Felbeck H, Childress JJ, Somero GN (1981) Calvin-Benson cycle and sulphide oxidation enzymes in animals from sulphide-rich habitats. Nature (Lond) 293:291–293

Felder DL (1979) Respiratory adaptations of the estuarine mud shrimp, Callianassa jamaicense (Schmitt 1935) (Crustacea, Decapoda, Thalassinidea). Biol Bull Mar Biol Lab Woods Hole 157:125–138

Felgenhauer BE, Abele LG (1983) Branchial water movement in the grapsid crab Sesarma reticulatum Say. J Crustacean Biol 3:187–195

Fellenius ACB, Idstrom JP, Holm S (1984) Muscle respiration during exercise. Am Rev Respir Dis 129:S10–S12

Feller G, Gerdy C (1987) Metabolic pattern of the heart of a haemoglobin-free Antarctic fish Channichthys rhinoceratus. Polar Biol 7:225–229

Fenchel T (1992) What can ecologists learn from microbes: life beneath a square centimetre of sediment surface. Funct Ecol 6:499–507

Fenchel T (1993) There are more small than large species? Oikos 68:375–378

Fenchel T, Finlay BJ (1990a) Anaerobic free-living protozoa: growth efficiencies and the structure of anaerobic communities. FEMS Microbiol Ecol 74:269–276

Fenchel T, Finlay BJ (1990b) Oxygen toxicity, respiration and behavioural responses to oxygen in free-living anaerobic ciliates. J Gen Microbiol 136:1953–1959

Fenchel T, Finlay BJ (1991a) The biology of free-living anaerobic ciliates. Eur J Prostist 26:201–215

Fenchel T, Finlay BJ (1991b) Endosymbiotic metanogenic bacteria in anerobic ciliates: significance of the growth efficiency of the host. J Protoz 38:18–32

Fenchel T, Finlay BJ (1994) The evolution of life without oxygen. Am Nat 82:22–29

Fenchel T, Riedl R (1970) The sulfide system: a new biotic community underneath the oxidised layer of marine sand bottoms. Mar Biol 7:255–268

Fenton MB, Bringham RM, Mills AM, Rautenbach IL (1985) The roosting and foraging areas of Epomophorus wahlbergi (Pteropodida) and Scotophilus viridis (Vespertilionide) in Kruger National Park, South Africa. J Mammal 66:461–468

Ferguson MWJ (1992) Temperature-dependent sex determination in alligators: phenomenon, mechanism and evolution. Biol J Linn Soc 46:49–58

Fernandes MN, Rantin FT (1985) Gill morphometry of the teleost Hoplias malabaricus (Bloch). Bol Fisiol Anim Univ São Paulo 9:57–65

Fernandes MN, Rantin FT (1989) Respiratory responses of Oreochromis niloticus (Pisces, Cichlidae) to environmental hypoxia under different thermal conditions. J Fish Biol 35:509–519

Fernandes MN, Rantin FT, Losa LL (1984) Estudo morfo-functional comparativo das brânquias de três espécies fa família Erythrinidae (Pisces; Teleosti): Hoplias malabaricus, Hoplias lacerdae e Hoplererythrinus unitaeniatus – implicaçōoes ecológicas. Cie Cult (Suppl) 36:608–609

Fielder DR (1970) The feeding behaviour of the sand crab Scopimera inflata (Decapoda, Ocypodidae). J Zool (Lond) 160:35–49

Fige FHJ (1936) The differential reaction of the blood vessels of a branchial arch of Amblystoma tigrum (Colorado Axolotl). I. The reaction of adrenaline, oxygen and carbon dioxide. Physiol Zool 9:79–101

Fincke T, Paul R (1989) Booklung function in arachnids. III. The function and control of the spiracles. J Comp Physiol 159B:433–441

Finlay BJ, Fenchel T (1993) Methanogens and other bacteria as symbionts of free-living anaerobic ciliates. Symbiosis 14:375–390

Finley TN, Pratt SA, Ladman AJ, Brewer L, McKay MB (1968) Morphological and lipid analysis of the alveolar lining material in dog lung. J Lipid Res 9:351–365

Firth JA, Farr A (1977) Structural features and quantitative age-dependent changes in the intervascular barrier of the guinea pig haemochorial placenta. Cell Tissue Res 184:507–516

Fischkoff S, Vanderkooi JM (1975) Oxygen diffusion in the biological and artificial membrane determined by fluorochrome pyrene. J Gen Physiol 65:663–676

Fish GR (1956) Soma aspects of the respiration of six species of fish from Uganda. J Exp Biol 33:186–195

Fisher CR (1990) Chemoautotrophic and methanotrophic symbioses in marine invertebrates. Rev Aquatic Sci 2:399–436

Fisher MR, Hand SC (1985) Chemoautotrophic symbionts in the bivalve *Lucida floridana* from seagrass beds. Biol Bull 167:445–459

Fishman AP (1972) Pulmonary edema: the water-exchange function of the lung. Circulation XLVI:390–408

Fishman AP (1983) Comparative biology of the lung. Am Rev Respir Dis 128:S90–S91

Fishman AP, Becker EL, Fritts HW, Heinemann HO (1957) Apparent volumes of distribution of water, electrolytes and hemoglobin within the lung. Am J Physiol 188:95–101

Fishman AP, Galante RJ, Pack AI (1989) Diving physiology: lungfish. In: Wood SC (ed) Comparative pulmonary physiology: current concepts, vol 39: Lung biology in health and disease. Marcel Dekker, New York, pp 645–676

Fitch NA, Johnston IA, Wood RE (1984) Skeletal muscle capillary supply in fish that lacks respiratory pigments. Respir Physiol 57:201–211

Flauenfelder H, Shigar SG, Wolynes PG (1991) The energy landscapes and motion of proteins. Science 254:1598–1602

Fleay DH (1937) Nesting habits of the brush turkey. Emu 36:153–163

Fletcher BD, Sachs BF, Kotas RV (1970) Radiologic demonstration of postnatal liquid in the lungs of newborn lambs. Paediatrics 46:252–258

Floyd RA (1991) Oxidative damage to behaviour during aging. Science 254:1597

Foelix RF (1982) Biology of spiders. Harvard University Press, Cambridge

Foley JA, Kutzbach JE, Coe MT, Levis S (1994) Feedbacks between climate and boreal forests during the Halocene epoch. Nature (Lond) 371:52–54

Folkow B, Neil E (1971) Circulation. Oxford University Press, New York

Fons R, Sicart R (1976) Contribution à la connaissance du métabolisme énergétique chez deux Crocidurinae: *Suncus etruscus* (Savi 1822) et *Crocidura russula* (Herman 1780), Insectivora, Soricidae. mammalia. Mammalia 40:229–311

Fontana W, Buss LW (1993) The arrival of the fitness: toward a theory of biological organization. Technical Report No SFI 93-09-055, Santa Fe Institute, Santa Fe, pp 1–71

Forbes WA (1882) On the convoluted trachea of two species of manucode (*Manucodia atra*) and (*Phonygama gould*); with remarks on similar structures on other birds. Proc Zool Soc Lond 1882:347–353

Forey P, Janvier P (1994) Evolution of the early vertebrates. Am Sci 82:554–565

Forgue J, Massabuau JC, Truchot JP (1992a) When are resting water breathers lacking O_2? Arterial PO_2 at the anaerobic threshold in crab. Respir Physiol 88:247–256

Forgue J, Truchot JP, Massabuau JC (1992b) Low arterial PO_2 in resting crustaceans is independent of blood oxygen-affinity. J Exp Biol 170:257–264

Forman GL (1972) Comparative morphological and histochemical studies of stomachs of selected American bats. Univ Kans Sci Bull 49:591–729

Forman HJ, Fisher AB (1981) Antioxidant defense. In: Glber DL (ed) Oxygen and living processes: an interdisplinary approach. Springer, Berlin Heidelberg New York, pp 235–249

Forster ME, Davie PS, Davison W, Satchell GH, Wells RMG (1988) Blood pressures and heart rates in swimming hagfish. Comp Biochem Physiol 89:247–250

Forster ME, Axelsson M, Farrell AP, Nilsson S (1991) Cardiac function and circulation in hagfishes. Can J Zool 69:1985–1992

Forster RE (1964) Diffusion of gases. In: Fenn WO, Rahn H (eds) Handbook of physiology, sect 3. respiration, vol I. Amer Physiol Soc, Washington, DC, pp 839–872

Forster RE (1973) Some principles governing maternal-foetal transfer in the placenta. In: Comline KS, Cross KW, Dawes GS, Nathanielz PW (eds) Foetal and neonatal physiology, Proc of the Sir Henry Barcroft Centenary Symposium. Cambridge University Press, Cambridge, pp 223–237

Fox GA (1976) Eggshell: its ecological and physiological significance in a DDT contaminated tern population. Wilson Bull 88:459–477

Fox H (1964a) The pattern of villous variability in the normal placenta. J Obstet Gynecol 71:749–758

Fox H (1964b) The villous trophoblast as an index of placental ischaemia. J Obstet Gynecol 71:885–983

Fox H (1986) Pathology of the placenta. Clinical Obstet Gynecol Lond 13:501–509

Fox HM (1921) Methods of studying the respiratory exchange in small aquatic organisms with particular reference to the use of flagellates as an indicator for oxygen consumption. J Gen Physiol 3:565–573

Fox HM (1938) On blood circulation and metabolism of sabellids. Proc R Soc Lond 125B:554–569

Fox HM (1955) The effect of oxygen on the concentration of haem in invertebrates. Proc R Soc Lond 143B:203–214

Fox HM, Taylor AER (1954) The tolerance of oxygen by aquatic invertebrates. Proc R Soc Lond 143B:214–225

Fox HM, Simmonds BG, Washbourn R (1935) Metabolic rates of ephemerid nymphs from swiftly flowing and from still waters. J Exp Biol 12:179–184

Fox GE, Stackebrandt E, Hespell RB, Gibson J, Maniloff J et al. (1980) The phylogeny of procaryocytes. Science 209:457–463

Foxon GEH (1964) Blood and respiration. In: Moore JA (ed) Physiology of Amphibia. Academic Press, New York, pp 151–209

Fraenkel G (1932) Der Atmungsmechanismus des Skorpions. Z Vergl Physiol 11:656–661

Fraenkel G, Herford GVB (1938) The respiration of insects through the skin. J Exp Biol 15:266–280

Fraga CG, Shigenaga MK, Park JW, Degan P, Ames BN (1990) Oxidative damage to DNA during aging: 8-hydroxy-2′-deoxyguanosine in rat organ DNA and urine. Proc Natl Acad Sci USA 87:4533–4537

Frakes LA (1979) Climates throughout geologic time. Elsevier, Amsterdam

Frappell P, Daniels CB (1991) Temperature effects on ventilation and metabolism in the lizard *Ctenophorus nuchalis*. Respir Physiol 86:257–270

Frazzetta TH (1975) Complex adaptations in evolving populations. Sinauer, Sunderland

Freadman MA (1981) Swimming energetics of striped bass (*Morone saxatilis*) and bluefish (*Pomatomus saltatrix*): hydrodynamic correlates of locomotion and gill ventilation. J Exp Biol 90:253–265

Frederickson JK, Onstott TC (1996) Microbes deep inside the Earth. Sci Am 275:42–47

Fredericq L (1878) Recherches sur la physiologie de poulpe commun. Arch Zool Exp Gen 7:535–583

Freedman LS, Samuels I, Fish SA, Schwartz B, Lange M, Morgano L (1980) Sparing of the brain in neonatal undernutrition: amino acid transport and incorporation into brain and muscle. Science 207:902–904

Freeman BA, Crapo JD (1982) Free radicals and tissue injury. Lab Invest 47:412–426

Freeman BM, Misson BH (1970) pH, PO_2 and PCO_2 of blood from foetus and neonate of *Gallus domesticus*. Comp Biochem Physiol 33:763–772

Fricke H (1988) Coelacanths, the fish that time forgot. Natl Geogr 173:824–838

Fridovich I (1975) Superoxide dismutases. Am Rev Biochem 44:147–159

Fridovich I (1976) Oxygen radicals, hydrogen peroxide, and oxygen toxicity. In: Pryor WG (ed) Free radicals in biology, vol 1. Academic Press, New York, pp 239–277

Fridovich I (1978) Biology of oxygen radicals. Science 201:875–879

Friedli B, Kent G, Olley P (1973) Inactivation of bradykinin in the pulmonary vascular bed of newborn and foetal lambs. Circ Res 33:421–427

Friedmann EI (1982) Endolithic microorganisms in the Antarctic cold desert. Science 215:1045–1053

Friedmann EI, Ocampo R (1976) Endolithic blue algae in the dry valleys: primary producers in the Antarctic desert ecosystems. Science 193:1247–1249

Frils-Christensen E, Lassen K (1991) Length of the solar cycle: an indicator of solar activity closely associated with climate. Science 254:698–700

Frith HJ (1956) Breeding habits in the family Magapodiidae. Ibis 98:620–640

Fritsche R (1990) Effects of hypoxia on blood pressure and heart rate in three marine teleosts. Fish Physiol Biochem 8:85–92

Fritsche R, Thomas S, Perry SF (1993) Effects of serotonin on circulation and respiration in the rainbow trout Oncorhynchus maykiss. J Exp Biol 178:191–204

Fry M, Jenkins DC (1984) Nematoda: aerobic respiratory pathways of adult parasitic species. Exp Parasitol 57:86–92

Full RJ (1985) Exercising without lungs: energetics and endurance in a lungless salamander, Plethodon jordani. Physiologist 28:342

Fung YB (1993) Biomechanics: mechanical properties of living tissues, 2nd edn. Springer, Berlin Heidelberg New York

Fung YB, Sobin SS (1969) Theory of sheet flow in lung alveoli. J Appl Physiol 26:472–488

Fuhrmann O (1914) Le genre Typhlonectes. Neuchatel Mem Soc Sci Nat 5:112–123

Fustec A, Desbruyères D, Juniper SK (1987) Deep-sea hydrothermal vent communities at 13°N on the East Pacific rise: microdistributions and temporal variations. Biol Oceanogr 4:121–164

Futuyuma DJ (1986) Evolutionary biology, 2nd edn. Sinauer, Sunderland

Gaehtgens P (1990) Avian versus mammalian blood: experimental observations and physiologic relevance. In: Sutton JR, Coates GC, Remmers JE (eds) Hypoxia: the adaptations. BC Decker, Toronto, pp 20–25

Gaehtgens P, Will G, Schmidt F (1981) Comparative rheology of nucleated and non-nucleated red blood cells. II. Rheological properties of avian red cell suspensions in narrow capillaries. Pfluegers Arch Ges Physiol 390:283–287

Gahlenbeck H, Bartels H (1970) Blood gas transport properties in gill and lung forms of the axolotol (Ambryostoma mexicanum). Respir Physiol 9:175–182

Gahlenbeck H, Frerking H, Rathschlag-Schaefer AM, Bartels H (1968) Oxygen and carbon dioxide exchange across the cow placenta during the second part of pregnancy. Respir Physiol 4:119–131

Galla HJ (1993) Nitric oxide: an intracellular messanger. Angew Chem Int Ed Engl 32:378–380

Gamble JC (1971) The responses of the marine amphipods Corphium arenarium and C. volutator to gradients and to choices of different oxygen concentrations. J Exp Biol 54:275–290

Gameson ALH, Robertson KG (1955) The solubility of oxygen in pure water and sea water. J Appl Chem 5:502–523

Gennon BJ, Campbell G, Randall DJ (1973) Scanning electron microscopy of the vascular casts for the study of vessel connections in a complex vascular bed – the trout gill. Proc Elect Microsc Soc Am 31:442–443

Gannon BJ, Randall DJ, Browning J, Lester RJG, Rogers LJ (1983) The microvascular organization of the gas exchange organs of the Australian lungfish, Neoceratodus forsteri (Krefft). Aust J Zool 31:651–673

Ganote CE (1983) Contraction band necrosis and irrevasible myocardial injury. J Mol Cell Cardiol 15:67–73

Gans C (1970) Respiration in the early tetrapods: the frog is a red herring. Evolution 24:740–751

Gans C (1971) Strategy and sequence in the evolution of the external gas exchangers of ectothermal vertebrates. Forma Functio 3:66–104

Gans C (1976) Ventilatory mechanisms and problems in some amphibious aspiration breathers (Chelydra, Caiman- Reptilia). In: Hughes GM (ed) Respiration of amphibious vertebrates. Academic Press, London, pp 357–374

Gans C (1979) Momentarily excessive construction as the basis for protoadaptation. Evolution 331:227–233

Gans C (1983) On the fallacy of perfection. In: Fay RR, Gourevitch G (eds) Perspectives on modern auditory research: papers in honour of EG Wever. Amphora Press, Groton, pp 101–114

Gans C (1985) Vertebrate morphology: tale of a phoenix. Am Zool 25:689–694

Gans C (1988) Adaptation and the form-function relation. Am Zool 28:681–697

Gans C, Clark B (1976) Studies on ventilation of *Caiman crocodilus* (Crocodilia, Reptilia). Respir Physiol 26:285–301

Gans C, Clark B (1978) Air flow in reptilian ventilation. Comp Biochem Physiol 60A;453–457

Gans C, Hughes GM (1967) The mechanism of lung ventilation in the tortoise *Testudo graeca* Linne. J Exp Biol 47:1–20

Garcia AGP, Basso NGD, Fonseca MEF, Zuardi JAT, Outanni HN (1991) Enterovirus associated placental morphology: a light virological, electron microscopic and immunologcic study. Placenta 12:533–547

Gardner BG (1980) Tetrapod ancestry: a reappraisal. In: Panchen AL (ed) The terrestrial environment and the origin of land vertebrates. Academic Press, London, pp 135–207

Garey WF, Rahn H (1970) Normal arterial gas tensions and pH and the breathing frequency of the electric eel. Respir Physiol 9:141–150

Garland RJ, Milson WK (1994) End tidal gas composition is not correlated with episodic breathing in hibernating ground squirrels. Can J Zool 72:1141–1148

Garland T, Huey RB (1987) Testing symmorphosis: does structure match functional requirements? Evolution 41:1404–1409

Garland T, Else PL, Hulbert AJ, Tap P (1987) Effects of endurance training on activity metabolism of lizards. Am J Physiol 21:R450–R456

Garlick RL, Davis BJ, Farmer M, Fyhn UEH, Noble RW et al. (1979) A foetal maternal shift in the oxygen equilibrium of haemoglobin from the viviparous caecilian, *Typhlonectes compressicauda*. Comp Biochem Physiol 62A:239–244

Garrow JS, Hawes SF (1971) The relationship of the size and composition of the human placenta to its functional capacity. J Obstet Gynecol Br Commw 78:22–28

Gatz RN, Crawford EC, Piiper J (1974) Respiratory properties of the blood of lungless and gill-less salamander, *Desmognathus fuscus*. Respir Physiol 20:33–41

Gatz RN, Glass ML, Wood SC (1987) Pulmonary function of the green sea turtle, *Chelonia mydas*. J Appl Physiol 62:459–463

Gatzy JT (1975) Ion transport across the excised bullfrog lung. Am J Physiol 228:1162–1171

Gaunt AS, Gans C (1969) Mechanics of resiration in the snapping turtle, *Chelydra serpentina* (Linne). J Morphol 128:195–228

Gaunt AS, Gaunt SLL, Prange HD, Wasser JS (1987) The effects of tracheal coiling on the vocalization of the cranes (Aves: Gruidae), J Comp Physiol 161A:43–58

Gaustad JE, Vogel SN (1982) High energy solar radiation and the origin of life. Origins Life 12:3–8

Gee JH (1976) Buoyancy and aerial respiration: factors influencing the evolution of reduced swim bladder volume of some Central American catfishes (Trichomycteridae, Callichthyidae, Loricariidae, Astroblepidae). Can J Zool 54:1030–1037

Gee JH (1981) Coordination of respiratory and hydrostatic functions of the swim bladder in the Central American mudminnow, *Umbra limi*. J Exp Biol 92:37–53

Gee JH, Gee PA (1995) Aquatic surface respiration, buoyancy control and the evolution of air-breathing gobies (Gobiidae: Pisces). J Exp Biol 198:79–89

Gee JH, Graham JP (1978) Respiratory and hydrostatic functions of the intestine of the catfishes *Hoplosternum thoracatum* and *Brochis splendens* (Callichthydidae). J Exp Biol 74:1–16

Geelhaar A, Weibel ER (1971) Morphometric estimation of pulmonary diffusing capacity. III. The effect of increased oxygen consumption in Japanese waltzing mice. Respir Physiol 11:354–366

Gehr P, Erni H (1980) Morphometric estimation of pulmonary diffusion capacity in two horse lungs. Respir Physiol 41:199–210

Gehr P, Bachofen M, Weibel ER (1978) The normal human lung: ultrastructure and morphometric estimation of diffusion capacity. Respir Physiol 32:121–140

Gehr P, Sehovic S, Burri PH, Claasen H, Weibel ER (1980) The lung of shrews: morphometric estimation of diffusion capacity. Respir Physiol 44:61–86

Gehr P, Mwangi DK, Amman A, Maloiy GMO, Taylor CR, Weibel ER (1981) Design of the mammalian respiratory system: V. Scaling morphometric diffusing capacity to body mass: wild and domesic animals. Respir Physiol 44:61–86

Gehrke PC, Fielder DR (1988) Effects of emperature and dissolved oxygen on heart rate, ventilation rate and oxygen consumption of spangled perch, *Leiopotherapon unicolor* (Günther 1859) (Percoidei, Teraponidae). J Comp Physiol 157:771–782

Geiger R (1965) Das Klima der bodennahen Luftschicht. F Viewig, Brunswick, Germany. Translated as The climate near the ground. Harvard University Press, Cambridge

Geiser J, Betticher DC (1989) Gas transfer in isolated lungs perfused with red cell suspension or haemoglobin solution. Respir Physiol 77:31–40

George JC, Shah RV (1956) Comparative morphology of the lung in snakes with remarks on the evolution of the lung in reptiles. J Anim Morphol Physiol 3:1–7

George JC, Shah RV (1965) Evolution of air sacs in Sauropodia. J Anim Morphol Physiol (Bombay) 12:255–263

Gerth WA, Hemmingsen FA (1982) Limits of gas secretion by salting-out effect in the fish swim bladder rete. J Comp Physiol 146B:129–136

Ghiretti F (1966) Respiration. In: Wilbur KM, Yonge CM (eds) Physiology of Mollusca. Academic Press, London, pp 175–298

Ghiretti F, Ghiretti M (1975) Respiration. In: Fretter VV, Peake J (eds) Pulmonates, vol V. Academic Press, London, pp 33–52

Giaver J, Keese CR (1989) Fractal motion of mammalian cells. Physica D 38:128–133

Gibe J (1970) L'appareil respiratoire. In: Grasse PP (ed) Traité de zoologie, tome XIV, fascicule III. Masson, Paris, pp 499–520

Giebisch GH, Granger JP, Greenleaf JE, Lydie RB, Mitchell RH et al. (1990) What's past is prologue. Physiologist 33:161–180

Gibert CR (1993) Evolution and phylogeny. In: Evans DH (ed) The physiology of fishes. CRC Press, Boca Raton, pp 1–45

Gilbert LI (1988) Developmental biology. Sinauer, Sutherland

Gilbert RD, Cummings LA, Juchau MR, Longo LD (1979) Placental diffusing capacity and foetal development in exercising or hypoxic guinea pigs. J Appl Physiol 46:828–834

Gilchrist BM (1954) Haemoglobin in *Artemia*. Proc R Soc Lond 143B:136–146

Gillen RC, Riggs A (1973) Structure and function of the isolated haemoglobins of the American eel, *Anguilla rostrata*. J Biochem 248:1961–1969

Gilles R, Pequeux A (1985) Ion transport in crustacean gills: physiological and ultrastructural approaches. In: Gilles R, Gilles-Baillien M (eds) Transport process, iono- and ultrastructural approaches. Springer, Berlin Heidelberg New York, pp 136–158

Gillespie JR, Sagot JC, Gendner JP, Bouverot P (1982) Impedance of the lower respiratory system in ducks under four conditions: pressure breathing, anaesthesia, paralysis or breathing CO_2-enriched gas. Respir Physiol 47:177–191

Gillis AM (1991) Can organisms direct their evolution? BioScience 41:202–205

Giorgio P (1990) Adaptation to water and the evolution of echolocation in the cetacea. Ethol Ecol Evol 2:135–163

Giovane A, Greco G, Maresca A, Tota B (1980) Myoglobin in the heart ventricle of tuna and other fishes. Experientia 36:219–220

Girard H, Muffat-Joly M (1971) Evolution de la pression partielle d'oxygène et du pH sanguins chez l'embryon de poulet au cours de la croissance. Pfluegers Arch Ges Physiol 328:21–35

Glass ML, Johansen K (1981) Pulmonary diffusing capacity in reptiles (relations to temperature and oxygen uptake). J Comp Physiol 107:169–178

Glass ML, Wood SC (1983) Gas exchange and control of breathing in reptiles. Physiol Rev 63:232–260

Glass ML, Hicks JW, Riedesel ML (1976) Respiratory responses to long-term temperature exposure in the box turtle, *Terrapene ornata*. J Comp Physiol 131:353–359

Glass ML, Wood SC, Hoyt RW, Johansen K (1979) Chemical control of breathing in the lizard *Varanus exanthematicus*. Comp Biochem Physiol 62A:999–1003

Glass ML, Abe AS, Johansen K (1981a) Pulmonary diffusing capacity in reptiles: relations to temperature and O_2-uptake. J Comp Physiol 142:509–514

Glass ML, Burggren WW, Johansen K (1981b) Pulmonary diffusing capacity of the bullfrog (*Rana catesbeiana*). Acta Physiol Scand 113:485–490

Glass ML, Boutilier RG, Heisler N (1983) Ventilatory control of arterial PO_2 in the turtle, *Chrysemys picta bellii*: effects of temperature and hypoxia. J Comp Physiol 151:145–153

Glass ML, Ishmatsu A, Johansen K (1986) Responses of aerial ventilation to hypoxia and hypercapnia in *Channa argus*, an air-breathing fish. J Comp Physiol B 156:425–430

Glazier JB, Hughes JMB, Maloney JE, West JB (1967) Vertical gradient of alveolar size in lungs and of dogs frozen intact. J Appl Physiol 23:694–705

Gleeson TT (1979) The effects of training and captivity on the metabolic capacity of the lizard *Sceloporus occidentalis*. J Comp Physiol 129:123–128

Gleeson TT, Mitchell GS, Bennett AF (1980) Cardiovascular responses to graded activity in the lizards *Varanus* and *Iguana*. Am J Physiol 239:R174–R179

Gleick J (1987) Chaos. Sphere Books, New York

Glenny RW (1992) Spatial correlation of regional pulmonary perfusion. J Appl Physiol 72:2378–2386

Glenny RW, Robertson HT (1990) Fractal properties of pulmonary blood flow characterization of spatial heterogeneity. J Appl Physiol 69:532–545

Glenny RW, Robertson HT (1991a) Fractal modeling of pulmonary blood flow heterogeneity. J Appl Physiol 70:1024–1030

Glenny RW, Robertson HT (1991b) Gravity is a minor determinant of pulmonary blood flow distribution. J Appl Physiol 71:620–629

Glenny RW, Robertson HT, Yamashiro S, Bassinthwaighte JB (1991) Applications of fractal analysis to physiology. J Appl Physiol 70:2351–2367

Gnaiger E (1983) In situ measurement of oxygen profiles in lakes: microstratifications, oscillations, and the limits of comparison with chemical methods. In: Gnaiger E, Forstner H (eds) Polarographic oxygen sensors. Springer, Berlin Heidelberg New York, pp 245–264

Gnaiger E (1991) Animal energetics at very low oxygen: information from calorimetry and respirometry. In: Waokes AJ, Grieshaber MK, Bridges CR (eds) Physiological strategies for gas exchange and metabolism. Cambridge University Press, Cambridge, pp 149–171

Gnaiger E, Forstner H (1983) Polarographic oxygen sensors. Springer, Berlin Heidelberg New York

Godfray HCJ (1997) Making life simpler. Nature (Lond) 387:351–352

Godoy MP (1975) Familia Erythrinidae. In: Godoy MP (ed) Peixes do Brasil, suborden Characoidei, vol 3. Editora Fransiscana, Piracicaba, SP, pp 23–57

Goldberger AL (1991) Is the normal heart beat chaotic or homeostatic? News Physiol Sci 6:87–91

Goldberger AL, West BJ (1987) Fractals in physiology and medicine. Yale J Biol Med 60:421–435

Goldberger AL, Bhargava V, West BJ, Mandell AJ (1985) On a mechanism of cardial electrical stability. Biophys J 48:525–528

Goldberger AL, Rigney DR, West BJ (1990) Chaos and fractal in human physiology. Sci Am 262:34–41

Golde LMG, Batenburg JJ, Robertson B (1994) The pulmonary surfactant system. News Physiol Sci 9:13–20

Goldie RG, Bertram JF, Warton A, Papadimitriou JM, Paterson JW (1983) Pharmacological and ultrastructural study of alveolar contractile tissue in toad (*Bufo marinus*) lung. Comp Biochem Physiol 75:343–349

Goldspink G (1985) Malleability of the motor system: a comparative approach. J Exp Biol 115:375–391

Goldstein L (1982) Gill nitrogen excretion. In: Houlihan DF, Rankin JC, Shuttleworth HR (eds) Gills. Cambridge University Press. Cambridge, pp 93–206

Gomi T (1982) Electron microscopic studies of the alveolar brush cell of the striped snake (*Elaphe quadrivirgata*). J Med Soc Toho Univ 29:481–102

Goniakowska-Witalinska L (1973) Metabolism, resistance to hypotonic solutions and ultrastructure of erythrocytes of five amphibian species. Acta Biol Cracov 16:114–123

Goniakowska-Witalinska L (1974) Respiration, resistance to hypotonic solutions and ultrastructure of erythrocytes of *Salamandra salamandra*. Bull Acad Pol Sci 22:59–75

Goniakowska-Witalinska L (1978) Ultrastructure and morphometric study of the lung of the European salamander, *Salamandra salamandra* L. Cell Tissue Res 191:343–356

Goniakowska-Witalinska L (1980a) Ultrastructural and morphometric changes in the lung of newt, *Triturus cistatus camifex* (Laur.) during ontogeny. J Anat 130:571–583

Goniakowska-Witalinska L (1980b) Scanning and transmission electron microscopic study of the lung of the newt, *Triturus alpestris* Laur. Cell Tissue Res 205:133–145

Goniakowska-Witalinska L (1982) Development of the larval lung of *Salamandra salamandra* L. Anat Embryol 164:113–137

Goniakowska-Witalinska L (1986) Lung of the tree frog, *Hyla arborea*: a scanning and transmission electron microscope study. Anat Embryol 174:379–389

Goniakowska-Witalinska L (1995) The histology and ultrastructure of the amphibian lung. In: Pastor LM (ed) Histology, ultrastructure and immunohistochemistry of the respiratory organs in non-mammalian vertebrates. Publicaciones Universidad de Murcia 1995, Murcia, Spain, pp 77–112

Gonzalez NC, Clancy RL, Wagner PD (1993) Determinants of maximal oxygen uptake in rats acclimated to simulated altitude. J Appl Physiol 75:1608–1614

Gonzalez RJ, McDonald DG (1992) The relationships between oxygen consumption and ion loss in fresh water fish. J Exp Biol 163:317–326

Gonzalez-Grussi F, Boston RW (1972) The absorptive function of the neonatal lung. Ultrastructural study of horseradish peroxidase uptake at the onset of ventilation. Lab Invest 26:114–121

Goodrich ES (1930) Air bladder and lungs. In: Studies on the structure and development of vertebrates (Chap XI). Macmillan, London

Gordon JE (1978) Structure; or, why things don't fall down. Plenum Press, New York

Gordon MS, Boetius I, Evans DH, McCarthey R, Oglesby LC (1969) Aspects of the physiology of terrestrial life in amphibious fishes. I. The mudskipper *Periophthalmus sobrinus*. J Exp Biol 50:141–149

Gordon MS, Ng WW, Yip AY (1978) Aspects of terrestrial life in amphibious fishes. III. The Chinese mudskipper, *Periophthalmus cantonensis*. J Exp Biol 72:57–75

Gordon MS, Gabaldon DJ, Yip AY (1985) Exploratory observations on microhabitat selection within the intertidal zone by the chinese mudskipper fish *Periophthalmus cantonensis*. Mar Biol 85:209–215

Gordon JW, Scangos GA, Plotkin DJ, Barbosa JA, Ruddle FH (1980) Genetic transformation of mouse embryos by microinjecion of purified DNA. Proc Natl Acad Sci USA 77:7380–7384

Gorge G, Chatelain P, Schaper J, Lerch R (1991) Effects of increasing degrees of ischemic injury on myocardial oxidative metabolism early after reperfusion in isolated rat hearts. Circ Res 68:1681–1692

Gorin AB, Steward PA (1979) Differential permeability of endothelial and epithelial barriers to albumin flux. J Appl Physiol 47:1315–1324

Gorr T, Kleinschmidt T, Fricke H (1991) Close tetrapod relationships of the coelacanth *Latimeria* indicated by heamoglobin sequences. Nature (Lond) 351:394–397

Gosline JM, Steeves JD, Harman AD, deMont ME (1983) Patterns of circular and radial muscle activity in respiration and jetting of the squid *Loligo opalescens*. J Exp Biol 104:97–109

Gould SJ (1966) Allometry and size in ontogeny and phylogeny. Biol Rev 41:587–640

Gould SJ (1989) Wonderful world. Penguin Books, London

Gould SJ (1994) The evolution of life on Earth. Sci Am 271:63–69

Gould SJ, Eldridge N (1977) Punctuated equilibria: the tempo and mode of evolution reconsidered. Paleobiology 3:115–151

Gould SJ, Lewontin RC (1979) The spandrels of San Macro and Panglossian paradigm: a critique of adaptationist programme. Proc R Soc Lond 205B:581–598

Grace J, Lloyd J, McIntyre J, Miranda AC et al. (1995) Carbon dioxide uptake by undisturbed tropical rain forest in southwest Amazonia, 1992 to 1993. Nature 375:811–819

Graham JB (1970) Preliminary studies on the biology of the amphibious clinid *Mnierpes macrocephalus*. Mar Biol 5:136–140

Graham JB (1973) Terrestrial life of the amphibious fish *Mnierpes macrocephalus*. Mar Biol XXX 83–91

Graham JB (1974) Aquatic respiration of the seasnake, *Pelamis platurus*. Respir Physiol 21:1–7

Graham JB (1976) Respiratory adaptations of marine air-breathing fishes. In: Hughes GM (ed) Respiration of amphibious vertebrates. Academic Press, New York, pp 165–187

Graham JB (1990) Ecological, evolutionary, and physical factors influencing aquatic animal respiration. Am Zool 30:137–146

Graham JB (1994) An evolutionary perspective for biomdal respiration: a biological synthesis of fish air breathing. Am Zool 34:229–237

Graham JB, Gee JH, Robinson FS (1975) Hydrostatic and gas exchange functions of the lung of the sea snake, *Pelamis platurus*. Comp Biochem Physiol 50:477–482

Graham JB, Kramer DL, Pineda E (1977) Respiration of the air-breathing fish *Piabucina festae*. J Comp Physiol 122B:295–310

Graham JB, Kramer DL, Pineda E (1978a) Comparative respiration of an air-breathing and non-air-breathing characoid fish and the evolution of aerial respiration in characins. Physiol Zool 51:279–288

Graham JB, Rosenblatt RH, Gans C (1978b) Vertebrate air breathing arose in fresh waters and not in the ocean. Evolution 32:459–463

Graham JB, Baird TA, Stockmann W (1987) The transition to air-breathing in fishes. IV. Impact of branchial specializations for air-breathing on the aquatic respiratory mechanisms and ventilatory costs of the swamp eel *Synbranchus marmoratus*. J Exp Biol 129:83–106

Graham JB, Dudley R, Agullar NM, Gans C (1995) Implications of the late Paleozoic oxygen pulse of physiology and evolution. Nature (Lond) 375:117–120

Graham JM (1949) Some effects of temperature and oxygen pressure on the metabolism and activity of the speckled trout, *Salvelunus fontinalis*. Can J Res 27:270–288

Graiger E, Steinlechner P, Maran R, Méndez G, Eberl T, Margreiter R (1995) Control of mitochondrial and cellular respiration by oxygen. J Bioenerg Biomembr 27:583–596

Grassé PP (1970) Traité de zoologie, vol XIV, Reptilia. Masson, Paris

Grassle JF (1985) Hydrothermal vent animals: distribution and biology. Science 229:713–717

Gratz RK, Ar A, Geiser J (1981) Gas tension profile of the lung of the viper, *Vipera xanthina palestinae*. Respir Physiol 44:165–171

Gratz RK, Hutchison H (1977) Energetics for activity in the diamondback water snake, *Natrix rhombifera*. Physiol Zool 50:99–114

Gratz RK, Crawford EC, Piiper J (1974) Metabolic and heart rate response of the plethodontid salamander *Desmognathus fuscus* to hypoxia. Respir Physiol 20:43–49

Graur D (1993) Molecular phylogeny and the higher classification of eutherian mammals. Trends Ecol Evol 8:141–147

Gray IE (1954) Comparative study of the gill area of marine fish. Biol Bull Mar Biol Lab Woods Hole 107:219–225

Gray IE (1957) Comparative study of the gill area of crabs. Biol Bull Mar Biol Lab Woods Hole 112:34–42

Gray J (1968) Animal locomotion. Weidenfeld and Nicolson, London

Gray J (1985a) The microfossil record of early land plants: advances in understanding of early terrestriality, 1970–1984. Philos Trans R Soc Lond 309B:167–195

Gray J (1985b) Ordovician-Silurian land plants: the interdependence of ecology and evolution. Spec Pap Palaeontol 32:281–295

Gray MW (1992) The endosymbiont hypothesis revisited. Int Rev Cytol 141:233–357

Green J, Corbet SA, Betney E (1973) Ecological studies on crater lakes in West Cameron. The blood of endemic cichlids in Barombi Mbo in relation to stratification and vertical distribution of the zooplankton. J Zool (Lond) 170:30–67

Greenaway P, Taylor HH (1976) Aerial gas exchange in Australian arid-zone crab, *Parathelphusa transversa*, von Mertens. Nature (Lond) 262:711–713

Greenaway P, Taylor HH, Bonaventura J (1983) Aquatic gas exchange in Australian fresh water/land crabs of the genus *Holthuisana*. J Exp Biol 103:237–251

Greenaway P, Morris S, MacMahon BR (1988) Adaptations to a terrestrial existence by the robber crab *Birgus latro*. II. *In vivo* respiratory gas exchange and transport. J Exp Biol 140:493–509

Greenberg MJ (1985) Ex Bouillabaisse Lux: the charm of comparative physiology and biochemistry. Am Zool 25:737–749

400

Greenewalt CH (1960) Hummingbirds. Doubleday, New York

Greenewalt CH (1968) Bird song: acoustics and physiology. Smithsonian Institution Press, Washington, DC

Greenwood PH (1961) A revision of the genus *Dinotopterus* Blgr (Pisces, Clariidae) with notes on the comparative anatomy of the suprabranchial organs in the Clariidae. Bull Br Mus Nat Hist 7:217–241

Greer AC, Parker F (1979) On the density of the New Guinea scincid lizard *Lygosoma fragile* Macleay 1877, with notes on its natural history. J Herpetol 13:221–225

Gregersen MI, Rawson RA (1959) Blood volume. Physiol Rev 39:307–342

Gregory RB (1977) Synthesis and total excretion of waste nitrogen by fish of the *Periophthalmus* (mudskipper) and Scartelaos families. Comp Biochem Physiol 57A:33–36

Gregory EM, Fridovich I (1973) Induction of superoxide dismutase by molecular oxygen. J Bacteriol 114:543–548

Gresson R (1927) On the structure of the branchiae of the gilled Oligochaete *Alma nilotica*. Ann Mag Nat Hist 19:348

Grieshaber MK, Harding I, Kreutzer U, Pörtner HO (1994) Physiological and metabolic responses to hypoxia in invertebrates. Rev Physiol Biochem Pharmacol 125:44–144

Griffin DR (1970) Migrations and homing of bats. In: Wimsatt WA (ed) Biology of bats, vol 1. Academic Press, London, pp 233–264

Grigg GC (1965) Studies on the Queensland lungfish, *Neoceratodus forsteri* (Krefft). Aust J Zool 13:243–257

Grigg GC (1969) The failure of oxygen transport in a fish at low levels of ambient oxygen. Comp Biochem Physiol 29:1253–1257

Grigg GC, Harlow P (1981) A foetal-maternal shift of blood oxygen affinity in an Australian viviparous lizard, *Sphenomorphus quoyii* (Reptilia, Scincidae). J Comp Physiol B 142:495–499

Grimes CB, Able KW, Jones RS (1986) Tilefish, *Lopholatilus chamaeoleonticeps*, habitat, behaviour and community structure in mid-Atlantic and southern New England waters. Environ Biol Fishes 15:273–292

Grimme JD, Lane SM, Maron MB (1997) Alveolar liquid clearance in multiple nonperfused canine lung lobes. J Appl Physiol 82:348–353

Groebe K, Thews G (1987) Time courses of erythrocytic oxygenation in capillaries of the lung: lower and upper bounds on red cell transit times. Adv Exp Med Biol 215:165–169

Groom AC (1987) The microcirculatory society Eugene M. Landis award lecture. Microcirculation of the spleen: new concepts, new challenges. Microvasc Res 34:269–289

Groom AC, Ellis CG, Potter RF (1984a) Microvascular architecture and red cell perfusion in skeletal muscle. Prog Appl Microcirc 5:64–83

Groom AC, Ellis CG, Potter RF (1984b) Microvascular geometry in relation to modeling oxygen transport in contracted skeletal muscle. Am Rev Respir Dis 129:S6–S9

Gros G (1991) The role of carbonic anhydrase within the tissues, with special reference to mammalian striated muscle. In: Woakes AJ, Grieshaber MK, Bridges CR (eds) Physiological strategies for gas exchange and metabolism. Cambridge University Press, Cambridge, pp 35–54

Gross MG (1990) Oceanography: a view of the Earth, 5th edn. Prentice Hall, Englewood Cliffs

Grote J (1967) Die Sauerstoffdiffusionkonstanten in Lungengewebe und Wasser und ihre Temperaturabhängigkeit. Pfuegers Arch Gesamte Physiol Menschen Tiere 295:245–254

Grubb BR (1982) Cardiac output and stroke volume in exercising ducks and pigeons. J Appl Physiol 53:203–211

Grubb BR (1983) Allometric relations of cardiovascular function in birds. Am J Physiol 14:H567–H572

Grubb BR, Mills CD, Colacino JM, Schmidt-Nielsen K (1977) Effect of arterial carbon dioxide on cerebral blood flow in ducks. Am J Physiol 232:H596–H601

Grubb BR, Colacino JM, Schmidt-Nielsen K (1978) Cerebral blood flow in birds: effects of hypoxia. Am J Physiol 234:H230–H243

Grubb BR, Jones JH, Schmidt-Nielsen K (1979) Avian cerebral blood flow: influence of the Bohr effect on oxygen supply. Am J Physiol 236:H744–H753

Gruson ES (1976) Checklist of birds of the world. William Collins, London

Guard CL, Murrish DE (1975) Effects of temperature on the viscous behaviour of blood from Antarctic birds and mammals. Comp Biochem Physiol 52A:287–290

Guillette LJ (1982) The evolution of viviparity and placentation in the high elevation Mexican lizard, *Sceloporus aeneus*. Herpetologia 38:94–103

Guillette LJ (1989) The evolution of vertebrate viviparity: morphological modifications and endocrine control. In: Wake D, Roth G (eds) Complex organismal functions: integration and evolution in vertebrates. J Wiley (Dahlem Workshop Report), Chichester, pp 219–233

Guillette LJ (1991) The evolution of viviparity in amniote vertebrates: new insights, new questions. J Zool (Lond) 521–526

Guillette LJ (1993) The evolution of viviparity in lizards. BioScience 43:742–750

Guillette LJ, Hotton N (1986) The evolution of mammalian reproductive characteristics in therapsid reptiles. In: Hotton N, MacLean PD, Roth JJ, Roth EC (eds) The ecology and biology of mammal-like reptiles. Smithsonian Institution, Washington, DC, pp 239–250

Guillette LJ, Jones RE (1985) Ovarian, oviductal and placental morphology of the reproductively bimodal lizard species, *Sceloporus aeneus*. J Morphol 84:85–98

Guimond RW, Hutchison HV (1972) Pulmonary branchial and cutaneous gas exchange in the mudpuppy, *Necturus maculosus maculosus* (Ranfinesque). Comp Biochem Physiol 42A:367–393

Guimond RW, Hutchison HV (1973a) Trimodal gas exchange in the large aquatic salamander *Siren lacertina*. Comp Biochem Physiol 46A:249–268

Guimond RW, Hutchison HV (1973b) Aquatic respiration: an unusual strategy in the hellbender, *Cryptobranchus alleganiensis alleganiensis* (Dudin). Science 182:1263–1265

Guimond RW, Hutchison VH (1976) Gas exchange of the giant salamanders of North America. In: Hughes GM (ed) Respiration of amphibious vertebrates. Academic Press, New York, pp 313–338

Gunsalus IC, Sligar SG (1978) Oxidation reduction by the P_{450} monoxygenase system. Adv Enzymol 47:1–44

Guntheroth WG, Luchtel DL, Kawabori I (1982) Pulmonary microcirculation: tubules rather than sheet and post. J Appl Physiol 153:510–515

Gupta KP, van Golen KL, Randerath E, Randerath K (1990) Age dependent covalent DNA alterations (I-compounds) in rat mitochondrial DNA. Mutat Res 237:17–27

Gumett DA, Kurth WS, Roux A, Bolton SJ, Kennel CF (1996) Evidence for a magnetosphere at Ganymede from plasma-wave observations by Galileo spacecraft. Nature (Lond) 384:535–537

Gurtner GH, Traystman RJ, Bums B (1982) Interactions between placental O_2 and CO transfer. J Appl Physiol 52:479–487

Gutman WF (1977) Phylogenetic reconstruction: theory, methodology, and application to chordate evolution. In: Hecht MK, Goody PC, Hecht BM (eds) Major patterns of vertebrate evolution. NATO Adv Study Inst Series, vol 14. Plenum Press, New York, pp 45–96

Gutman WF, Bonik K (1981) Kritische Evolutionstheorie. Gerstenberg, Hildesheim

Gutteridge JMC (1993) Free radicals in disease processes: a complication of cause and consequences. Free Radic Res Commun 19:141–158

Guyer C, Slowinski JB (1993) Adaptive radiation and the tolology of large phylogenies. Evolution 47:253–263

Haas JD (1976) Prenatal and infant growth and development. In: Baker PT, Little MA (ed) Man in the Andes: a multidisplinary study of high altitude Quechua. Hutchinson Ross, Stroudsburg, Pennsylvania, pp 161–179

Haas JD, Frongillo EA, Stepick CD (1980) Altitude, ethnic and sex differences in birth weight in Bolivia. Hum Biol 52:459–477

Hackstein JHP, Stumm CK (1994) Methane production in terrestrial arthropods. Proc Natl Acad Sci USA 91:5441–5445

Hadley NF (1980) Surface waxes and integumentary permeability. Am Sci 68:546–553

Hagerman L, Uglow RF (1985) Effects of hypoxia on the respiratory and circulatory regulation of *Nephros norvegicus*. Mar Biol 87:273–278

Hainsworth FW (1981) Locomotion. In: Hainsworth FW (ed) Animal physiology: adaptation in function. Addison-Wesley, Reading, pp 259–292

Hainsworth R (1986) Vascular capacitance: its control and importance. Rev Physiol Biochem Pharmacol 105:101–173

Hakim A, Munshi JSD, Hughes GM (1978) Morphometrics of the respiratory organs of the Indian green snakehead fish, *Channa punctata*. J Zool (Lond) 184:519–543

Haldane JS (1922) Respiration. Yale University Press, New Haven

Hall VE (1931) The muscular activity and oxygen consumption of *Urechis caupo*. Biol Bull 61:400–416

Hallam A (1987) End-Cretaceous mass extinction event: argument for terrestrial causation. Science 238:1237–1242

Hallam JF, Dawson TJ, Holland RAB (1989) Gas exchange in the lung of a dasyurid marsupial: morphometric estimation of diffusion capacity and blood oxygen uptake kinetics. Respir Physiol 77:309–322

Halliwell B (1978) Biochemical mechanisms accounting for the toxic action of oxygen on living organisms. The key role of superoxide dismutase. Cell Biol Int Rep 2:113–129

Halliwell B (1994) Free radicals, antioxidants and human disease: curiosity, cause or consequence? Lancet 344:721–724

Halliwell B, Gutteridge JMC (1985) Free radicals in biology and medicine. Clarendon, Oxford

Hamlett WC (1986) Prenatal nutruent absorptive structures in selachains. In: Uyeno T, Arai T, Taniuchi T, Matsuura K (eds) Indo-Pacific fish biology. Ichtyol Soc Japan, Tokyo, pp 333–344

Hamlett WC (1987) Comparative morphology of the elasmobranch placental barrier. Arch Biol Brux 98:135–162

Hamlett WC (1989) Evolution and morphogenesis of the placenta in sharks. J Exp Zool (Suppl) 2:35–52

Hamlin RL, Kondrich RM (1969) Hypertension, regulation of heart rate, and possible mechanism contributing to aortic rupture in turkeys. Proc Fed Am Soc Exp Biol 28:451–456

Hammen CS (1969) Metabolism of the oyster, *Crassostrea virginica*. Am Zool 9:309–318

Hammen CS (1976) Respiratory adaptations: invertebrates. In: Wiley M (ed) Estuarine processes. Academic Press, New York, pp 347–355

Hammersley JR, Olson DE (1992) Physical models of the smaller pulmonary airways. J Appl Physiol 72:2402–2414

Hammond P (1992) Species inventory. In: Groombridge B (ed) Global biodiversity: status of the Earth's living resources. Chapman and Hall, London, pp 17–39

Handerson LJ (1913) The fitness of the environment. Macmillan, New York

Handy RD (1989) The ionic composition of rainbow trout body mucus. Comp Biochem Physiol 93A:571–575

Hansell DA, Bates NR, Carson CA (1997) Predominance of vertical loss of carbon from the surface waters of the equatorial Pacific Ocean. Nature (Lond) 386:59–61

Hansen CA, Sidell BD (1983) Atlantic hagfish cardiac muscle: metabolic basis of tolerance to anoxia. Am J Physiol 244:R356–R362

Hansen HJ (1893) Organs and characters in different orders of Arachnida. Entomol Medd 4:135–144

Hansen VK, Wingstrand KG (1960) Further studies on the nonnucleated erythrocytes of *Maurolieus mulleri* and comparison with blood cells of related fish. AF Host, Dana Report No 54, Copenhagen

Hanson D, Johansen K (1970) Relationship of gill ventilation and perfusion in dogfish, *Squalus suckleyi*. J Fish Res Board Can 27:551–564

Haq BU, Hardenbol J, Vail PR (1987) Chronology of fluctuating sea levels since the Triassic. Science 235:1156–1167

Hardisty MW (1979) Biology of cyclostomes. Chapman and Hall, London

Harms JW (1932) Die Realisation von Genen und die konsekutive Adaptation. II. *Birgus latro* L. als Landkrebs und seine Beziehungen zu den Coenabiten. Z Wiss Zool 140:167–190

Harnisch O (1937) Primäre und sekundäre Oxybiose wirbelloser Tiere. Verh Dtsch Zool Ges 1937:129–136

Harrison JF, Fewell JH, Roberts SP, Hall HG (1996) Achievement of thermal stability by varying metabolic heat production in flying honeybees. Science 274:88–90

Harrison P, Zummo G, Farina F, Tota B, Johnson IA (1991) Gross anatomy, myoarchitecture, and ultrastructure of the heart ventricle in hemoblobinless icefish *Chaenocephalus aceratus*. Can J Zool 69:1339–1347

Harrison RJ, Ridway SH, Joyce PL (1972) Telemetry of heart rate in diving seals. Nature (Lond) 238:280

Hart JS, Fisher HD (1964) The question of adaptations to polar environments in marine mammals. Fed Proc 23:1207–1214

Hartman FA (1954) Cardiac and pectoral muscles of trochlids. Auk 71:467–469

Hartman FA (1955) Heart weight in birds. Condor 57:221–238

Hartman FA (1961) Locomotor mechanisms of birds. Smithson Misc Collect 143:1–91

Hartman FA (1963) Some flight mechanisms of bats. Ohio J Sci 63:59–65

Hartnoll RG (1988) Evolution, systematics, and geographical distribution. In: Burggren WW, McMahon BR (eds) Biology of the land crabs. Cambridge University Press, Cambridge, pp 6–54

Hartvig M, Weber RE (1984) Blood adaptations for maternal-foetal oxygen transfer in the viviparous teleost, *Zoarces viviparous* L. In: Seymour RS (ed) Respiration and metabolism of embryonic vertebrates. Junk, Dordrecht, pp 17–30

Harvey EN (1928) The oxygen consumption of luminous bacteria. J Comp Physiol 11:469–475

Harvey HW (1957) The chemistry and fertility of sea waters. Cambridge University Press, London

Harvey PH (1993) The ecology of evolutionary succession. Curr Biol 3:106–108

Hashimoto K, Yamaguchi Y, Matsura F (1960) Comparative studies on the haemoglobins of salmon. IV. Oxygen dissociation curve. Bull Jpn Soc Sci Fish 26:827–834

Hashimoto T, Gomi T, Kimura A, Tsuchiya H (1983) Light electron microscopic study of the lung of the giant salamander *Megalobateracus japonicus davidanus*. J Med Soc Toho Univ 29:52–69

Hastings RH, Powell FL (1986) Physiological dead space and effective parabronchial ventilation in ducks. J Appl Physiol 60:85–91

Hastings RH, Powell FL (1987) High frequency ventilation of ducks and geese. J Appl Physiol 63:413–417

Haswell MS, Randall DJ (1978) The pattern of carbon dioxide excretion in the rainbow trout *Salmo gairdneri*. J Exp Biol 72:17–24

Haughton TM, Kerkut GA, Munday KA (1958) The oxygen dissociation and alkaline denaturation of haemoglobin from two species of earthworms. J Exp Biol 35:360–368

Hawking S (1993) Black holes and baby universes and other essays. Bantam Books, New York

Hawkins AJS, Jones MB (1982) Gill area and ventilation in two mud crabs, *Helice crassa* Dana (Grapsidae) and *Macrophthalmus hirtipes* (Jacquinot) (Ocypodidae). J Exp Mar Biol Ecol 60:103–118

Hawksworth DL (1991) The fungal dimension of biodiversity: magnitude, signficance, and conservation. Mycol Res 95:641–655

Hayes B (1994) Nature's algorithms. Am Sci 82:206–210

Hayes JD, Imbrie J, Shackleton NJ (1976) Variations in the Earth's orbit: pacemaker of the ice ages. Science 194:1121–1131

Hayes K, Gibas H (1971) Placental cytomegalovirus infection without foetal involvement following primary infection in pregnancy. J Paediatr 79:401–405

Hayward B, Davis R (1964) Flight speeds in western bats. J Mammal 45:236–242

Hazel JR (1993) Thermal biology. In: Evans DH (ed) The physiology of fishes. CRC Press, Boca Raton, pp 427–467

Hazelhoff EH (1939) Über die Ausnützung des Sauerstoffs bei verschiedenen Wassertieren. Z Vergl Physiol 26:306–327

Hearse DJ, Humprey SM, Chain EB (1973) Abrupt reoxygenation of the anoxic potassium-arrested perfused rat heart: a study of myocardial enzyme release. J Mol Cell Cardiol 5:395–407

Heath D, Williams DR (1981) Man at high altitude: the pathophysiology of acclimatization and adaptation. Churchill Livingstone, Edinburgh

Heath D, Williams DR (1989) High altitude pulmonary oedema. High altitude medicine and pathology. Butterworths, London, pp 132–148

Heath D, Williams DR, Dickson J (1984) The pulmonary arteries of the yak. Cardiovasc Res 18:133–139

Heatwole H (1981) Role of the saccular lung in the diving of the sea krait, *Laticuda colubrina* (Serpentes: Laticaudidae). Aust J Herpetol 1:11–16

Hedges SB, Sibley CG (1994) Molecules vs. morphology in avian evolution: the case of the "pelecaniform" birds. Proc Natl Acad Sci USA 91:9861–9865

Hedrick MS, Duffield DA (1986) Blood viscosity and optimal hematocrit in deep-diving mammal, the northern elephant seal (*Mirounga angustirostris*). Can J Zool 64:2081–2085

Hedrick MS, Duffield DA (1991) Hematological and rheological characteristics of blood in seven marine mammal species: physiological implications for diving behaviour. J Zool (Lond) 225:273–283

Hedrick MS, Jones DR (1993) The effects of altered aquatic and aerial respiratory gas concentrations on air-breathing patterns in a primitive fish *Amia calva*. J Exp Biol 181:81–94

Hedrick MS, Duffield DA, Cornell LH (1986) Blood viscosity and optimal hematocrit in a deep-diving mammal, the northern elephant seal (*Mirounga angustirostris*). Can J Zool 64:2081–2085

Hedrick MS, Burleson ML, Jones DR, Milsom WK (1991) An examination of central chemosensitivity in an air-breathing fish (*Amia calva*). J Exp Biol 155:165–174

Heijden FL (1981) Compensation mechanisms for experimental reduction of the functional capacity of the guinea pig placenta. Acta Anat 111:359–366

Heinemann HO, Fishman AP (1969) Nonrespiratory functions of mammalian lung. Physiol Rev 49:1–61

Heinrich B (1983) Do bumblebees forage optimally, and does it matter? Am Zool 23:273–281

Heinrich B (1992) The hot-blooded insects. Harvard University Press, Cambridge

Heisler N (1982a) Intracellular and extracellular acid-base regulation in the tropical fresh water teleosts *Synbranchus marmoratus* in response to the transition from water-breathing to air-breathing. J Exp Biol 99:9–28

Heisler N (1982b) Transepithelial ion transfer processes as mechanisms for fish acid-base regulation in hypercapnia and lactacidosis. Can J Zool 60:1108–1122

Heisler N (1984) Acid-base regulation in fishes. In: Hoar WS, Randall DJ (eds) Fish physiology vol XA. Academic Press, New York, pp 315–401

Heisler N (1989) Interactions between gas exchange, metabolism and ion transport in animals: an overview. Can J Zool 67:2923–2935

Heisler N, Forcht G, Ultsch GR, Anderson JF (1982) Acid-base regulation in response to environmental hypercapnia in two aquatic salamanders, *Siren lacertina* and *Amphiuma means*. Respir Physiol 49:141–158

Helbacka NV, Caserline JL, Smith CJ (1963) The effect of high CO_2 atmosphere on the laying hen. Poult Sci 42:1082–1097

Hellems HK, Haynes FW, Dexter L (1949) Pulmonary capillary pressure in man. J Appl Physiol 2:24–52

Hellin B, Chardon M (1981) Observations sur le trajet de l'air durant la respiration aérienne chez *Clarias lazera* Cuvier et Valenciennes, 1840. Ann Soc R Zool Belg 113:97–106

Hellman LM, Kobayashi M, Tolles WE, Cromb E (1970) Ultrasonic studies of the volumetric growth of the human placenta. Am J Obstet Gynecol 108:740–750

Hemmingsen EA (1963) Enhancement of oxygen transport by myoglobin. Comp Biochem Physiol 10:239–244

Hemmingsen EA (1965) Transfer of oxygen through solutions of heme pigments. Acta Physiol Scand Suppl 246

Hemmingsen EA, Douglas EL (1970) Respiratory characteristics of the haemoglobin-free fish, *Chaenocephalus aceratus*. Comp Biochem Physiol 32:733–744

Hemmingsen EA, Douglas EL, Johansen K, Millard RW (1972) Aortic blood flow and cardiac output in the haemoglobin-free fish, *Chaenocephalus aceratus*. Comp Biochem Physiol A 43:1045–1051

Hendry GAF (1993) Oxygen, free radicals processes and seed longevity. Seed Sci Res 3:141–153

Henk WG, Haldiman JT (1990) Microanatomy of the lung of the bowhead whale, *Balaena mysticetus*. Anat Rec 226:187–197

Henry JD, Fedde MR (1970) Pulmonary circulatory time in the chicken. Poult Sci 49:1286–1293

Henry RP (1994) Morphological, behavioural, and physiological characterization of bimodal breathing crustaceans. Am Zool 34:205–215

Henry RP, Perry HM, Trigg CB, Handley HL, Krarup A (1990) Physiology of two species of deep water crabs, *Chaceon fenneri* and *C. quiquedens*: gill morphology, and haemolymph ionic and nitrogen concentrations. J Crust Biol 10:375–381

Henze M (1910) Ueber den Einfluss des Sauerstoffdrucks auf den Gaswechsel einiger Meerrestiere. Biochem Z 26:255–278

Herbert CV, Jackson DC (1985) Temperature effects on the responses to prolonged submergence in the turtle *Chrysemys picta belii*. II. Metabolic rate, blood acid-base and ionic changes, and cardiovascular function in aerated and anoxic water. Physiol Zool 58:670–681

Herman AB, Spicer RA (1996) Paleobotanical evidence for a warm Cretaceous Arctic ocean. Nature (Lond) 380:330–333

Herreid CF (1980) Hypoxia in invertebrates. Comp Biochem Physiol 67A:311–320

Herreid CF, Full RJ (1988) Energetics of locomotion. In: Burggren WW, McMahon BR (eds) Biology of the land crabs. Cambridge University Press, Cambridge, pp 333–377

Herreid CF, Lee LW, Spampata R (1981) How do spiders breathe? Am Zool 21:917

Hers MJ (1943) Relation entre respiration et circulation chez *Anodonta cygnea* L. Ann Soc R Zool Belg 74:45–54

Hershenson MB, Abe MK, Kelleher MD, Naureckas ET, Garland A et al. (1994) Recovery of airway structure and function after hyperoxic exposure in immature rats. Am J Respir Crit Care Med 149:1663–1669

Heth G, Frankenberg E, Nevo E (1985) Adaptive optimal sound for vocal communication in tunnels of a subterranean mammal (*Spalax ehrenbergi*): a physcial analysis. Experientia 42:1287–1289

Heusner AA (1983) Body size, energy metabolism and the lung. J Appl Physiol 54:867–873

Hewitt JA, Kilmartin JV, Ten Eyck LF, Perutz MF (1972) Noncooperativty of the αβ-dimer in the reaction of haemoglobin with oxygen. Proc Natl Acad Sci USA 69:302–307

Heymann MA, Hoffman JIE (1984) Pulmonary hypertension in the newborn. In: Weir EK, Reeves JT (eds) Pulmonary hypertension. Futura, New York, pp 45–71

Hickman CS (1984) Form, function, and evolution of gastropod filters. Am Malacol Bull 3:95

Hickman CS (1987) Analysis of form and function in fossils. Am Zool 28:775–793

Hicks JW, Wood SC (1985) Temperature regulation in lizards: effects of hypoxia. Am J Physiol 248:R595–R600

Hiebl I, Braunitzer G, Schneeganss D (1987) The primary structures of the major and minor haemoglobin-components of adult Andean goose (*Chleophaga melanoptera*, anatidae) the mutation α-Leu-Ser in position 55 of the β-chains. Biol Chem 368:1559–1569

Hightower A, Burke JD, Haar JL (1975) A light microscopic study of the respiratory epithelium of the adult newt, *Notophthalmus viridescens*. Can J Zool 53:465–478

Hikida RS, Staron RS, Hagerman FC, Sherman WM, Costill DL (1983) Muscle fiber necrosis associated with human marathon runners. J Neurol Sci 59:185–203

Hildebrand M (1959) Motions of running cheetah and horse. J Mammal 40:481–495

Hildebrand M (1961) Further studies on the motion of the cheetah. J Mammal 42:84–91

Hill AV (1950) The dimensions of animals and their muscular dynamics. Sci Prog 38:209–230

Hill B (1981) Respiratory adaptations of three species of *Upogebia* (Thalassinidea, Crustacea) with reference to low tide periods. Biol Bull Mar Biol Lab Woods Hole 160:272–279

Hill EP, Power GG, Longo LD (1973) Mathematical simulation of pulmonary O_2 and CO_2 exchange. Am J Physiol 224:909–917

Hill HAO (1978) The superoxide iron and the toxicity of molecular oxygen. In: Williams RJP, Da Silva JRRF (eds) New trends in bioinorganic chemistry. Academic Press, New York, pp 173–208

Hill JE (1944) Rodent miners. Nat Hist 53:21

Hillman SS (1987a) Dehydrational effects of cardiovascular and metabolic capacity of two amphibians. Physiol Zool 60:608–613

Hillman SS (1987b) The roles of oxygen delivery and electrolyte levels in the dehydrational death of *Xenopus laevis*. J Comp Physiol 128:169–178

Hillman SS, Withers PC, Hedrick MS, Kimmel PB (1985) The effects of erythrocythemia on blood viscosity, maximal systemic oxygen transport capacity and maximal rates of oxygen consumption in an amphibian. J Comp Physiol 155B:577–581

Hills BA (1971) Geometric irreversibility and compliance hysteresis in the lung. Respir Physiol 13:50–61

Hills BA (1972) Diffusion and convection in lungs and gills. Respir Physiol 14:105–114

Hills BA (1988) The biology of the surfactant. Cambridge University Press, Cambridge

Hills BA (1996) Effects of gill dimensions on respiration. In: Munshi JSD, Dutta HM (eds) Fish morphology: horizon of new reseach. Science Publishers, Lebanon NH, pp 235–247

Hills BA, Hughes (1970) A dimensional analysis of oxygen transfer in the fish gill. Respir Physiol 9:126–140

Hilpert P, Fleischmann RG, Kempe D, Bartels H (1963) The Bohr effect related to blood and erythrocytes pH. Am J Physiol 205:337–340

Hinds DS, Calder WA (1971) Tracheal dead space in the respiration of birds. Evolution 25:429–440

Hinton HE (1953) Some adaptations of insects to environments that are alternately dry and flooded, with some notes on the habits of the Stratiomyidae. Trans Soc Br Entomol 11:209–227

Hinton HE (1966) Plastron respiration. Nature (Lond) 209:220–221

Hinton HE (1971) Reversible suspension of metabolism. In: Marois M (ed) From theoretical physics to biology. Editions du Centre National de la Recherche Scientifique, Paris, pp 332–357

Hinton HE (1976) Plastron respiration in bugs and beetles. J Insect Physiol 22:1529–1550

Hitschold T, Beck T, Muntefering H, Berle P (1992) Plazentamorphometrie bei diastolischem Null- und Negativeflow der Nabelarterien. Geburtsh Frauenheilkd 52:270–274

Hitzig BM, Jackson DC (1978) Central chemical control of ventilation in the unanaesthetized turtle. Am J Physiol 235:R257–R264

Hlastala MP, Standaert TA, Pierson DJ, Luchtel DL (1985) The matching of ventilation and perfusion in the lung of the tegu, Tupinambis nigropunctus. Respir Physiol 60:277–294

Hlastala MP, Bernard SL, Erickson H et al. (1996) Pulmonary blood flow distribution in standing horses is not dominated by gravity. J Appl Physiol 81:1051–1061

Ho YS (1994) Transgenic models for the study of lung biology and disease. Am J Physiol 266:L319–L353

Ho YS, Dey MS, Crapo JD (1996) Antioxidant enzyme expression in rat lungs during hyperoxia. Am J Physiol 270:L810–L818

Hoagland M, Dodson B (1995) The way life works. Ebury Press, London

Hochachka PW (1979) Cell metabolism, air breathing, and the origins of endothermy. In: Wood SC, Lenfant C (eds) Evolution of respiratory processes: a comparative approach. Marcel Dekker, New York, pp 253–288

Hochachka PW (1987) Limits: how fast and how slow muscle metabolism can go. In: Benzi G (ed) Advances in myochemistry, vol 1. John Libbey, Eurotext, New York, pp 3–12

Hochachka PW (1988) Metabolic suppression and O_2 availability. Can J Zool 66:152–158

Hochachka PW, Guppy M (1987) Metabolic arrest and the control of biological time. Harvard University Press, Cambridge

Hochachka PW, Fields J, Mustafa T (1973) Animal life without oxygen: basic biochemical mechanisms. Am Zool 13:543–555

Hochachka PW, Guppy M, Goderley HE, Storey KB, Hulbert WC (1978) Metabolic biochemistry of water- vs. air-breathing fishes: muscle enzymes and ultrastructure. Can J Zool 56:736–750

Hochachka PW, Emmett B, Suarez RK (1988) Limits and contraints in the scaling of oxidative and glycolytic enzymes in homeotherms. Can J Zool 66:1128–1138

Hock RJ (1951) The metabolic rates an body temperatures of bats. Biol Bull Mar Biol Lab Woods Hole 101:289–299

Hock RJ (1964) Animals in high altitudes: reptiles and amphibians. In: Dill DB, Adolph EF, Wilber CG (eds) Handbook of physiology, sect 4, adaptation to environment. American Physiological Society, Washington DC, pp 841–842

Hoeger U, Mommsen TP (1985) Role of free amino acids in the oxidative metabolism of cephalopod hearts. Circulation, respiration, and metabolism. Springer, Berlin Heidelberg New York, pp 367–376

Hoese B (1983) Struktur und Entwicklung der Lungen der Tylidae (Crustacea, Isopoda, Oniscoidea). Zool Jb Anat 109:487–501

Hoffman LH (1970) Placentation in the garter snake, *Thamnophis sirtalis*. J Morphol 131:57–88

Hogg JC, Nepszy S (1969) Regional lung volume and pleural pressure gradient estimated from lung density in dogs. J Appl Physiol 27:198–203

Hogg JC, McLean T, Martin BA, Wiggs B (1988) Erythrocyte transit and neutrophil concentration in the dog lung. J Appl Physiol 65:1217–1225

Holden T (1993) Giants in the Earth. Velikovskian 1(4):7–17

Holeton GF (1970) Oxygen uptake and circulation by a haemoglobinless Antarctic fish (*Chaenocephalus aceratus*, Lonnberg) compared with three red-blooded Antarctic fish. Comp Biochem Physiol 34:457–471

Holeton GF (1974) Metabolic cold adaptation of polar fish: fact or artifact? Physiol Zool 47:137–154

Holeton GF, Heisler N (1983) Contribution of net ion transfer mechanisms to acid-base regulation after exhausting activity in the larger spotted dogfish (*Scyliorhinus stellaria*). J Exp Biol 103:31–46

Holland HD (1978) The chemistry of the atmosphere and the oceans. Wiley, New York

Holland JB (1984) The chemical evolution of the atmosphere and oceans. Princeton University Press, Princeton

Holland RAB, Forster RE (1966) The effect of size of red cells on the kinetics of their oxygen uptake. J Gen Physiol 49:727–742

Holle JP, Meyer M, Scheid P (1977) Oxygen affinity of duck blood determined by in vivo and in vitro techniques. Respir Physiol 29:355–361

Hollinger DY, Kellihaer FM, Schulze ED, Köstner MM (1994) Coupling of tree transpiration to turbulence. Nature (Lond) 371:60–62

Holm-Hansen O (1968) Ecology, physiology, and biochemistry of blue-green algae. Annu Rev Microbiol 22:47–70

Holmyard EJ, Palmer WG (1952) A higher school inorganic chemistry. JM Dent, London

Homberger DG (1988) Models and tests in functional morphology: the significance of description and integration. Am Zool 28:217–229

Hong SK, Rahn H (1967) The diving women of Korea and Japan. Sci Am 216:34–43

Hook GR, Bastacky J, Conhaim RL, Staub NC, Hayes TL (1987) A new method for pulmonary edema research: scanning electron microscopy of frozen hydrated edematous lung. Scanning 9:71–79

Hoppeler H, Kayar SR, Claassen H, Uhlman E, Karas RH (1987) Adaptive variation in the mammalian respiratory system in relation to energetic demand: III. Skeletal muscles: setting the demand for oxygen. Respir Physiol 69:27–46

Horn MH, Gibson RN (1989) Intertidal fishes. In: Gould JL, Gould CG (eds) Life at the edge: readings from Scientific American. WH Freeman, New York, pp 59–67

Horn MH, Messer KS (1992) Fish guts as chemical reactors: a model of the alimentary canals of marine herbivorous fishes. Mar Biol 113:527–535

Hora SL (1935) Physiology, bionomics and evolution of the air-breathing fishes of India. Trans Natl Inst Sci India 1:1–16

Horsfield K (1981) The science of branching systems. In: Scadding JG, Cumming G (eds) Scientific foundation of respiratory medicine. Heinemann, London, pp 45–54

Horsfield K, Thurlbeck A (1981) Relation between diameter and and flow in the bronchial tree. Bull Math Biol 43:681–691

Horsfield K, Woldenberg MJ (1986) Branching ratio and growth of tree-like structures. Respir Physiol 63:97–107

Horvath SM, Borgia JF (1984) Cardiopulmonary gas transport and aging. Am Rev Respir Dis 129:S68–S71

Hosler P (1977) Castrophic chemical events in the history of the oceans. Nature (Lond) 267:403–408

Hoss LM (1973) A study of the benthos of an anoxic marine basin and factors affecting its distribution. MSc Thesis, Dalhouse Univ, Nova Scotia

Hossler FE (1980) Gill arch of the mullet, *Mugil cephalus*. III. Rate of response to salinity change. Am J Physiol Regul Integr Comp Physiol 238:379–398

Hossler FE, Harpole JH, King JA (1986) The gill arch of the striped bass, *Morone saxatilis*. I. Surface ultrastructure. J Submicrosc Cytol 18:519–528

Hou PCL, Burggren WW (1991) Hemodynamic development in anuran larvae *Xenopus laevis*. Am Zool 31:76A

Houde P (1986) Ostrich ancestors found in the Northern Hemisphere suggest new hypothesis of ratite origin. Nature (Lond) 324:563–565

Houlihan DF, Innes AJ, Wells MJ, Wells J (1982) Oxygen consumption and blood gases. J Comp Physiol B 148:35–40

Houston CS, Sutton JR, Cymerman A, Reeves JT (1987) Operation Everest II. Man at extreme attitude. J Appl Physiol 63:877–882

Howell BJ (1969) Acid-base balance in the transition from water-breathing to air-breathing. In: Farhi LE, Rahn H (eds) Studies in pulmonary physiology, mechanics, chemistry, and circulation. USAF School of Aerospace Medicine, Aerospace Medical Division (AFSC), Brooks Airforce Base, Texas, pp 270–275

Howell BJ (1970) Acid-base balance in transition from water to air breathing. Fed Proc 29:1130–1134

Howell BJ, Baumgardner FW, Bondi K, Rahn H (1970) Acid-base balance in cold-blooded vertebrates as a function of body temperature. Am J Physiol 218:600–606

Howell DJ (1983) Optimization of behaviour: introduction and overview. Am Zool 23:257–260

Howlett R (1966) Root cause of fish buoyancy. Nature (Lond) 380:203

Huang W, Ten RT, McLaurie ·M, Bledsoe G (1996) Morphometry of the human pulmonary vasculature. J Appl Physiol 81:2123–2133

Huber GL, Finley TN (1965) Effect of isotonic saline on the alveolar architecture. Anaesthesiology 26:252–253

Huchzermeyer FW (1986) Causes and prevention of broiler ascites. SAPA Poult Bull 346

Huchzermeyer FW, de Ruyk AMC (1986) Pulmonary hypertension syndrome associated with ascites in broilers. Vet Rec 119:94

Huchzermeyer FW, de Ruyk AMC, van Ark H (1988) Broiler pulmonary hypertension syndrome. III. Commercial broiler strains differ in their susceptibility. Onderstepoort J Vet Res 55:5–9

Huebner E, Chee G (1978) Histological and ultrastructural specialization of the digestive tract of the intestinal air breather *Hoplostemum thoracatum* (Teleost). J Morphoe 157:301–325

Huelsenbeck JP, Rannala B (1997) Phylogenetic methods come of age: testing hypotheses in an evolutionary context. Science 276:227–232

Huey RB (1987) Phylogeny, history and comparative method. In: Burggren WW, Huey RB (eds) New directions in ecological physiology. Cambridge University Press, Cambridge, pp 77–98

Huey RB, Kingsolver JG (1993) Evolution of resistance to high temperature in ectotherms. Am Nat 142:S21–S46

Hughes DM, Wimpenny JWT (1969) Oxygen metabolism by microorganisms. In: Rose AH, Wilkinson JF (eds) Advances in microbial physiology, vol 3. Academic Press, London, pp 197–231

Hughes GM (1963) Comparative physiology of vertebrate respiration, 1st edn. Heineman, London

Hughes GM (1965) Comparative physiology of vertebrate respiration, 2nd edn. Heineman, London

Hughes GM (1966) Evolution between air and water. In: de Reuck AVS, Porter R (eds) Development of the lung. Churchill, London, pp 64–80

Hughes GM (1967) Experiments on the respiration of the trigger fish (*Balistes capriscus*). Experientia 23:1077

Hughes GM (1970) Ultrastructure of the air-breathing organs of some lower vertebrates. 7th Int Congr Electron Microsc, Grenoble 7:599–600

Hughes GM (1972a) Distribution of oxygen tension in the blood and water along the secondary lamella of the icefish gill. J Exp Biol 56:481–492

Hughes GM (1972b) Morphometrics of the fish gills. Respir Physiol 14:1–25

Hughes GM (1973) Ultrastructure of the lung of *Neoceratodus* and *Lepidosiren* in relation to the lungs of other vertebrates. Folia Morphol (Pragne) 2:155–161

Hughes GM (1976) Fish respiratory physiology. In: Spencer-Davies P (ed) Perspectives in environmental biology. Pergamon, Oxford, pp 235–245

Hughes GM (1978) Some features of gas transfer in fish. Bull Inst Math Appl 14:39–43

Hughes GM (1979) Scanning electron microscopy of the respiratory surfaces of trout gills. J Zool Lond 188:553–453

Hughes GM (1980) Functional morphology of fish gills. In: Lahlou B (ed) Epithelial transport in lower vertebrates. Cambridge University Press, Cambridge, pp 15–36

Hughes GM (1981) The effects of low oxygen and pollution on the respiratory systems of fish. In: Pickering AD (ed) Stress in fish. Academic Press, New York, pp 121–146

Hughes GM (1982) An introduction to the study of gills. In: Houlihan DF, Rankin JC, Shuttleworth TJ (eds) Gills. Cambridge University Press, Cambridge, pp 1–24

Hughes GM (1984) General anatomy of the gills. In: Randall DJ (ed) Fish physiology, vol XA. Academic Press, London, pp 1–72

Hughes GM (1995) The gills of the coelacanth, *Latimeria chalumnae*, a study in relation to body size. Philos Trans R Soc Lond 347B:427–438

Hughes GM, Al-Kadhomiy NK (1986) Gill morphometry of the mudskipper, *Boleophthalmus boddarti*. J Mar Biol Assoc UK 66:671–682

Hughes GM, Iwai T (1978) A morphometric study of the gills in some Pacific deep-sea fishes. J Zool (Lond) 184:155–170

Hughes GM, Kikuchi Y (1984) Effects of in vivo and in vitro changes in PO_2 on the deformability of erythrocytes of rainbow trout (*Salmo gairdnen*). J Exp Biol 111:253–257

Hughes GM, Mondolfino RM (1983) Scanning electron microscopy of the gills of *Trachurus mediterraneus*. Experientia 39:518–519

Hughes GM, Morgan M (1973) The structure of fish gills in relation to their respiratory function. Biol Rev 48:419–475

Hughes GM, Munshi JSD (1968) Fine structure of the respiratory surfaces of an air-breathing fish, the climbing perch, *Anabas testudineus*. Nature (Lond) 219:1382–1384

Hughes GM, Munshi JSD (1979) Fine structure of the gills of some Indian air-breathing fishes. J Morphol 160:169–194

Hughes GM, Munshi JSD (1986) Scanning electron microscopy of the accessory respiratory organs the snake-head fish, *Channa strata* (Bloch) (Channidae, Channiformes). J Zool (Lond) 209:305–317

Hughes GM, Pohunkova H (1980) Scanning and transmission electron microscopy of the lungs of *Polypterus senegalensis*. Folia Morphol 110:112–123

Hughes GM, Shelton G (1958) The mechanism of gill ventilation in three fresh water teleosts. J Exp Biol 35:807–823

Hughes GM, Singh BN (1970a) Respiration in air-breathing fish, the climbing perch, *Anabas testudineus* II. Respiratory patterns and the control of breathing. J Exp Biol 53:281–298

Hughes GM, Singh BN (1970b) Respiration in air-breathing fish, the climbing perch, *Anabas testudineus* Bloch. I. Oxygen uptake and carbon dioxide release into air and water. J Exp Biol 53:265–280

Hughes GM, Singh BN (1971) Gas exchange with air and water in an air-breathing catfish, *Saccobranchus (Heteropneustes) fossilis*. J Exp Biol 55:667–682

Hughes GM, Umezawa SI (1983) Gill structure of the yellowtail and frogfish. Jpn J Ichthyol 30:176–183

Hughes GM, Vergara GA (1978) Static pressure-volume curves for the lung of the frog (*Rana pipiens*). J Exp Biol 76:140–165

Hughes GM, Weibel ER (1976) Morphometry of fish lungs. In: Hughes GM (ed) Respiration of amphibious vertebrates. Academic Press, London, pp 213–232

Hughes GM, Wright DE (1970) A comparative study of the ultrastructure of the water-blood pathways in the secondary lamellae of teleost and elasmobranch fishes – benthic forms. Z Zellforsch 104:478–493

Hughes GM, Knight B, Scammel CA (1969a) The distribution of PO₂ and hydrostatic pressure changes within the branchial chambers in relation to gill ventilation in the shore crab *Carcinus maenas* L. J Exp Biol 51:203–220

Hughes J, Gillis CN, Bloom FE (1969b) The uptake and disposition of dl-norepinephrine in perfused rat lung. J Pharmacol Exp Ther 169:237–248

Hughes GM, Dube SC, Munshi JSD (1973) Surface area of the respiratory organs of the climbing perch, *Anabas testudineus* (Pisces: Anabantidae). J Zool (Lond) 170:227–243

Hughes GM, Singh BR, Guha G, Dube SC, Munshi JSD (1974) Respiratory surface areas of an air-breathing siluroid fish *Saccobranchus (Heteropneustes) fossilis* in relation to body size. J Zool (Lond) 172:215–232

Hughes GM, Horimoto M, Kikuchi Y, Kakiuchi Y, Koyama T (1981) Blood flow velocity in microvessels of the gill filaments of goldfish (*Carassius auratus* L.). J Exp Biol 90:327–331

Hughes GM, Kikuchi Y, Watari H (1982) A study of the deformability of erythrocytes of a teleost fish, the yellow tail (*Seriola quinqueradiata*) and a comparison with human erythrocytes. J Exp Biol 96:209–220

Hughes GM, Munshi JSD, Ojha J (1986a) Post-embryonic development of water- and air-breathing organs of *Anabas testudineus* (Bloch). J Fish Biol 29:443–450

Hughes GM, Perry SF, Piiper J (1986b) Morphometry of the gills of the elasmobranch *Scyliohinus stellaris* in relation to body size. J Exp Biol 121:27–42

Hughes GM, Roy PK, Munshi JSD (1992) Morphometric estimation of oxygen-diffusing capacity for the air sac in the catfish *Heteropneustes fossilis*. J Zool (Lond) 227:193–209

Hulbert AJ, Else PL (1983) Oxygen demand of organ systems. Proc Physiol Soc NZ 3:95–101

Humburger V (1980) Embryology and the modern synthesis in evolutionary theory. In: Mayr E, Provone WB (eds) The evolutionary synthesis: perspectives on the unification of biology. Harvard University Press, Cambridge, pp 97–111

Humpreys PW, Normand ICS, Reynolds EOR, Strang LB (1967) Pulmonary lymph flow and the uptake of liquid from the lungs of the lamb at the start of breathing. J Physiol (Lond) 193:1–29

Hunt BG (1979) The effects of past variations of the Earth's rotation rate on climate. Nature (Lond) 281:188–191

Hunt JR, Aitken JR (1962) The effect of ammonium and chloride ions in the diet of hens on egg shell quality. Poult Sci 41:343–357

Hunt JR, Simkiss K (1967) Acute respiratory acidosis in the domestic fowl. Comp Biochem Physiol 21:223–237

Hunten DM (1973) The escape of light gases from planetary atmospheres. J Atmos Sci 30:1481–1494

Hunten DM, Donahue TM (1976) Hydrogen loss from the terrestrial planets. Annu Rev Earth Planet Sci 4:265–292

Hunten DM, Strobel DF (1974) Production and escape of terrestrial hydrogen. J Atmos Sci 31:305–317

Hunter WR (1953) The condition of mantle cavity in two pulmonate snails living in Loch Lomond. Proc R Soc Edinb 65B:143–165

Hurst LD, McVean G (1996) Evolutionary genetics and scandalous symbionts. Nature (Lond) 381:650–651

Hutchison AA, McNicol KJ, Loughlin GM (1985) The effect of age on pulmonary epithelial permeability in unanaethetized lambs. Pediatr Pulmonol 1:53–58

Hutchison GE (1959) Homage to Santa Rosalia, or why are there so many kinds of animals? Am Nat 93:145–159

Hutchison GE (1975) A treatise on limnology, vol I: geography, physics and chemistry. John Wiley, London

Hutchison VH (1968) Relation of body size and surface area to gas exchange in anurans. Physiol Zool 41:65–85

Hutchison VH, Haines HB, Engbretson G (1976) Aquatic life at high altitude: respiratory adaptations in the Lake Titicaca frog, *Telmatobius culeus*. Respir Physiol 27:115–129

Huxley VH, Kutchai H (1983) Effect of diffusion boundary layers on the initial uptake of O₂ by red cells. Theory versus experiment. Microvasc Res 26:89–107

Hyde DM, Robinson NE, Gillespie JR, Tyler WS (1977) Morphometry of the distal air spaces in lungs of aging dogs. J Appl Physiol 43:86–91

Hyden P, Lindberg R (1970) Hypoxia induced torpor in pocket mice (genus *Perognathus*). Comp Biochem Physiol 33A:167–179

Hyman LH (1951) The invertebrates: Platyhelminthes and Rynchocoela. The Acoelomate Bilateria, vol II. McGraw-Hill, New York

Hyman LH (1955) The invertebrates, vol IV. Echinodermata, water vascular system. McGraw-Hill, London

Hyman LH (1967) The invertebrates VI: Mollusca I. McGraw-Hill, New York

Igic R, Erdös EG, Yeh HSJ, Sorrells K, Nakajima Y (1972) Angiotensin I converting enzyme of the lung. Circ Res 30 & 31, Suppl II:51–61

Imbrie J, McIntyre A, Mix A (1989) Oceanic response to orbital forcing in the late Quaternary: observational and experimental strategies. In: Berger A (ed) Climate and geosciences. Yale University Press, New Haven, pp 121–164

Infantino RL, Burggren WW, Twonsend DS (1988) Physiology of direct development in the Puerto Rican frog *Eleutherodactylus coqui*. Am Zool 28:23A

Inger RF (1957) Ecological aspects of the origin of the tetrapods. Evolution 11:373–376

Ingermann RL, Terwilliger RC (1981) Oxygen affinities of foetal and maternal haemoglobins of the viviparous seaperch, *Embiotoca lateralis*. J Comp Physiol 142:523–531

Innes AJ (1985) Aerobic scope for activity of burrowing mantis shrimp, *Heterosquilla tricarinata*, at low environmental oxygen tensions. Comp Biochem Physiol 81:827–832

Innes AJ, El Haj AJ, Gobin JF (1986) Scaling of the respiratory, cardiovascular and skeletal muscle systems of the fresh water/terrestrial mountain crab, *Pseudothelphusa garhami garhami*. J Zool (Lond) 209:595–606

Innes AJ, Mardsen ID, Wong PPS (1984) Bimodal respiration of intertidal pulmonates. Comp Biochem Physiol 77A:441–445

Innes AJ, Taylor EW (1986a) The evolution of air breathing in crustaceans: a functional analysis of branchial, cutaneous and pulmonary gas exchange. Comp Biochem Physiol 85A:621–637

Innes AJ, Taylor EW (1986b) Air breathing crabs of Trinidad: adaptive radiation into the terrestrial environment. I. Aerobic metabolism and habitat. Comp Biochem Physiol 85A:373–382

Ishii Y, Hasegawas S, Uchiyama Y (1989) Twenty four-hour variations in subcellular structures of the rat type II alveolar epithelial cell: a morphometric study at the electron microscopic level. Cell Tissue Res 256:347–353

Ishmatsu A, Itazawa Y (1981) Ventilation of the air-breathing organ in the snakehead *Channa argus*. Jpn J Ichthyol 28:276–282

Ishmatsu A, Itazawa Y (1983) Difference in blood oxygen levels in the outflow vessels of the heart of an air-breathing fish, *Channa argus*. Do separate blood streams exist in a teleost heart? J Comp Physiol B 149:435–440

Ishmatsu A, Itazawa Y, Takeda T (1979) On the circulatory systems of the snakeheads *Channa maculata* and *C. argus* with reference to bimodal breathing. Jpn J Ichthyol 26:167–180

Isozaki Y (1997) Permo-Triassic boundary superanoxia and stratified superocean: records from lost deep sea. Science 276:235–238

Itazawa Y, Oikawa S (1983) Metabolic rates in excised tissues of the carp. Experientia 39:160–161

Ito T (1953) The permeability of the integument to oxygen and carbon dioxide in vivo. Biol Bull 105:308–315

Iwai T, Nakamura I (1964) Branchial skeleton of the bluefin tuna, with special reference to the gill rays. Bull Misaki Mar Biol Inst Kyoto Univ 6:21–25

Iwasawa Y, Gillis CN (1974) Pharmacological analysis of norepinephrine and 5-hydroxytryptamine removal from the pulmonary circulation: differentiation of uptake sites for each amine. J Pharmacol Exp Ther 188:386–393

Iwasawa Y, Gillis CN, Aghajanian G (1973) Hypothermic inhibition of 5-hydroxytryptamine and norepinephrine uptake by lung: cellular location of amines after uptake. J Pharmacol Exp Ther 186:498–507

Jablonski D (1997) Body size evolution in Cretaceous molluscs and the status of Cope's rule. Nature (Lond) 385:250–252

Jackson DC (1978) Respiratory control in air-breathing ectotherms. In: Davies DG, Barnes CD (eds) Regulation of ventilation and gas exchange. Academic Press, London, pp 93–130

Jackson DC (1986) Acid-base regulation of reptiles. In: Heisler N (ed) Acid-base regulation in animals. Elsevier, Amsterdam, pp 235–263

Jackson DC (1987) How do amphibians breathe both water and air. In: Dejours P, Bolis L, Taylor CR, Weibel ER (eds) Comparative physiology: life in water and on land. Springer, Berlin Heidelbery New York, pp 49–58

Jackson DC, Schmidt-Nielsen K (1964) Counter-current heat exchange in the respiratory gases. Proc Natl Acad Sci USA 51:1192–1197

Jackson DC, Allen J, Strupp PK (1976) The contribution of the nonrespiratory surfaces to CO_2 loss in six species of turtle at 20 °C. Comp Biochem Physiol 55A:243–246

Jackson MR, Joy CF, Mayhew TM, Haas JD (1985) Stereological studies on the true thickness of the villous membrane in human term placentae: a study of placentae from high altitude pregnancies. Placenta 6:249–258

Jackson MR, Mayhew TM, Haas JD (1988a) On the factors which contribute to the thinning of the villous membrane in human placentae at high altitude. I. Thinning and regional variation in thickness of trophoblast. Placenta 9:1–8

Jackson MR, Mayhew TM, Haas JD (1988b) On the factors which contribute to thinning of the villous membrane in human placentae at high altitude. II. An increase in the degree of peripherization of foetal capillaries. Placenta 9:9–18

Jackson MR, Mayhew TM, Boyd PA (1992) Quantitative description of the elaboration and maturation of villi from ten weeks of gestation to term. Placenta 13:357–370

Jackson MR, Walsh AJ, Morrow RJ, Brendan J, Mullen M et al. (1995) Reduced placental villous tree elaboration in small-for-gestational-age pregnancies: relationship with umbilical artery Doppler waveforms. Am J Obstet Gynecol 172:518–525

Jacobs W (1938) Untersuchungen zur Physiologie der Schwimmblase der Fische. IV. Die erste Gasfüllung der Schwimmblase bei jungen Seepferdchen. Z Vergl Physiol 25:379–388

Jacobsen VH (1967) The feeding of the lugworm (*Arenicola marina*) (L.). Quntitative studies. Ophelia 4:91–109

Jahnke L, Klein HP (1979) Oxygen as a factor in eukaryocyte evolution: some effects of low levels of oxygen on *Saccharomyces cerevisiae*. Origins Life 9:329–334

Jakubowski M, Byczkowska-Smyk W, Mikhalev Y (1969) Vascularization and size of the respiratory surfaces in the antarctic white-blooded fish *Chaenichthys rugosus* Regan (Percoidei, Chaenichthydea). Zool Pol 19:303–317

Jalavisto E, Kuorinka J, Kyllästinen M (1965) Responsiveness of the erythron to variations of oxygen tension in the chick embryo and young chicken. Acta Physiol Scand 53:479–486

James AG, Probyn T (1989) The relationship between respiration rate, swimming speed and feeding behaviour in the Cape anchovy *Engraulis capensis* Gilchrist. J Exp Mar Biol Ecol 131:81–100

Jameson W (1958) The wandering albatross. Hart-Davis, London

Janis C (1993) Victory by default: the mammalian succession. In: Gould SJ (ed) The book of life. Ebury-Hutchison, London, pp 169–217

Janke A, Feldmaier-Fuchs G, Thomas WK, Haeseler AV, Paabo S (1994) The marsupial mitochondrial genome and the evolution of placental mammals. Genetics 137:243–256

Jansen RG, Randall DJ (1975) The effects of changes in pH and PCO_2 in blood and water on breathing in rainbow trout, *Salmo gairdneri*. Respir Physiol 25:235–245

Jarman C (1970) Evolution of life. Bantam Books, Toronto

Jarvik E (1968) Aspects of vertebrate phylogeny. In: Orvig T (ed) Current problems of lower vertebrate phylogeny. Interscience, New York, pp 497–527

Jarvik E (1980) Basic structure and evolution of vertebrates, vol 2. Academic Press, London

Jarvis JUM, Bennet NC (1990) The evolutionary history, population biology and social structure of African mole rats: family Bathyergidae. In: Nevo E, Reig OA (eds) Evolution of subterranean mammals at the organismal and molecular levels. Wiley-Liss, New York, pp 97–128

Jasinski A (1973) Air-blood barrier in the respiratory intestine of the pond roach, *Misgumus fossilis* L. Acta Anat 86:376–391

Jasper JP, Hayes JM (1990) A carbon isotope record of CO_2 levels during the late Quartenary. Nature (Lond) 347:462–464

Jell PA (1978) Trilobite respiration and genal caeca. Alcheringa 2:251–260

Jenkins RJF (1991) The early environment. In: Bryant CH (ed) Metazoan life without oxygen. Chapman and Hall, London, pp 38–64

Jenkyns HC (1980) Cretaceous anoxic events: from continents to oceans. J Geol Soc (Lond) 137:171–188

Jensen D (1966) The hagfish. Sci Am 214:82–90

Jensen FB (1991) Multiple strategies in oxygen and carbon dioxide transport by haemoglobin. In: Woakes AJ, Grieshaber MK, Bridges CR (eds) Physiological strategies for gas exchange and metabolism. Cambridge University Press, Cambridge, pp 55–78

Jensen FB, Weber RE (1985) Kinetics of the acclimational responses of tench to combined hypoxia and hypercapnia. I. Respiratory responses. J Comp Physiol B 156:197–203

Jepsen GL (1970) Bat origins and evolution. In: Wimsatt WA (ed) The biology of bats, vol I. Academic Press, New York, pp 1–64

Jervis P (1995) Water: vital resource in home for agriculture and for industry. Franklin Watts, London

Jesse MJ, Shub C, Fishman AP (1967) Lung and gill ventilation of the African lungfish. Respir Physiol 3:267–287

Jessop NM (1995) General zoology, 6th edn. McGraw-Hill, New York

Jeuken M (1957) A study of the respiration of *Misgurnus fossilis* (L.), the pond loach. Thesis, University of Leiden, Leiden

Jia L, Bonaventura C, Bonaventura J, Stamler J (1996) S-Nitrosohaemoglobin: a dynamic activity of blood involved in vascular control. Nature (Lond) 380:221–226

Joenje H (1989) Genetic toxicology of oxygen. Mutat Res 219:193–208

Johansen K (1960) Circulation in the hagfish, *Myxine glutinosa* L. Biol Bull 118:289–295

Johansen K (1966) Air-breathing in the teleost *Synbranchus marmoratus*. Comp Biochem Physiol 18:383–395

Johansen K (1968) Air-breathing fishes. Sci Am 219:102–111

Johansen K (1970) Air-breathing in fish. In: Hoar WS, Randall DJ (eds) Fish physiology, vol 4. Academic Press, London, pp 361–411

Johansen K (1971) Comparative physiology: gas exchange and circulation in fishes. Annu Rev Physiol 33:569–612

Johansen K (1972) Heart and circulation in gill, skin and lung breathing. Respir Physiol 14:193–210

Johansen K (1979) Cardiovascular support of metabolic functions invertebrates. In: Wood SC, Lenfant C (eds) Evolution of respiratory processes: a comparative approach. Marcel Dekker, New York, pp 107–192

Johansen K (1982) Respiratory gas exchange of vertebrate gills. In: Houlihan DF, Rankin JC, Shuttleworth TJ (eds) Gills. Cambridge University Press, Cambridge, pp 99–128

Johansen K (1987) The world as a laboratory: physiological insights from Nature's experiments. In: McLennan H, Ledsome JR, McIntosh CHS, Jones DR (ed) Advances in physiological research. Plenum Press, New York, pp 377–396

Johansen K, Burggren WW (1980) Cardiovascular function in lower vertebrates. In: Bourne GH (ed) Hearts and heart-like organs. Academic Press, New York, pp 61–117

Johansen K, Hanson D (1968) Functional anatomy of the hearts of lungfishes and amphibians. Am Zool 8:191–210

Johansen K, Hol R (1968) A radiological study of the central circulation in the lungfish *Protopterus aethiopicus*. J Morphoe 126:333–348

Johansen K, Lenfant C (1966) Gas exchange in the Cephalopod, *Octopus dofleini*. Am J Physiol 210:910–918

Johansen K, Lenfant C (1967) Respiratory function in the South American lungfish. J Exp Biol 46:205–218

Johansen K, Lenfant C (1968) Respiration in the African lungfish, *Protopterus aethiopicus*. II. Control of breathing. J Exp Biol 49:453–468

Johansen K, Lenfant C (1972) A comparative approach to the adaptability of O_2-Hb affinity. In: Roth M, Astrup P (eds) Oxygen affinity of haemoglobin and red cell acid-base status. Munksgaard, Copenhagen, pp 750–780

Johansen K, Martin AW (1966) Circulation in the giant earthworm, *Glossoscolex giganteus*. I. Contractile processes and pressure gradients in the large blood vessels. J Exp Biol 43:333–347

Johansen K, Reite OB (1968) Effects of acetylcholine and biogenic amines on pulmonary smooth muscle in the African lungfish *Protopterus aethiopicus*. Acta Physiol Scand 71:465–471

Johansen K, Lenfant C, Grigg GF (1967) Respiratory control in the lungfish, *Neoceratodus forsteri* (Krefft). Comp Biochem Physiol 20:835–854

Johansen K, Lenfant C, Hanson D (1968a) Cardiovascular dynamics in the lungfish. Z Vergl Physiol 59:157–186

Johansen K, Lenfant C, Schmidt-Nielsen K, Petersen JA (1968b) Gas exchange and control of breathing in the electric eel *Electrophorus electricus*. Z Vergl Physiol 61:137–163

Johansen K, Lenfant C, Hanson D (1970a) Respiration in a primitive air-breather, *Amia calva*. Respir Physiol 9:162–174

Johansen K, Lenfant C, Hanson D (1970b) Phylogenetic development of pulmonary circulation. Fed Proc 29:1135–1140

Johansen K, Maloiy GMO, Lykkeboe G (1975) A fish in extreme alkalinity. Respir Physiol 24:159–162

Johansen K, Lomholt JP, Maloiy GMO (1976) Importance of air and water breathing in relation to size of the African lungfish *Protopterus amphibius*, Peters. J Exp Biol 65:395–399

Johansen K, Burggren WW, Glass M (1977) Pulmonary stretch receptors regulate heart rate and pulmonary blood flow in the turtle, *Pseudemys scripta*. Comp Biochem Physiol 58:185–191

Johansen K, Mangum CP, Lykkeboe G (1978) Respiratory properties of the blood of Amazon fishes. Can J Zool 56:898–906

Johansen K, Lykkeboe G, Kornerup S, Maloiy GMO (1980) Temperature-insensitive oxygen binding in blood of the tree frog *Chiromantis petersi*. J Comp Physiol 136:71–76

Johansen K, Berger M, Bicudo JEPW, Ruschi A, De Almeida PJ (1987) Respiratory properties of blood and myoglobin in hummingbirds. Physiol Zool 60:269–278

Johnson LF (1964) The effects of decreased barometric pressure on maximum pressure-volume relationships of the human respiratory system. J Aerosp Med 35:637–642

Johnson L, Rees CJ (1988) Oxygen consumption all gill surface area in relation to habitat and lifestyle of four crab species. Comp Biochem Physiol 89A:243–246

Johnson ML (1942) The respiratory function of the haemoglobin of the earthworm. J Exp Biol 18:266–277

Johnson RE, Liu M (1996) The loss of atmosphere from mass. Science 274:1932

Johnson SL (1988) The effects of the 1983 El Niño on Oregon's coho (*Oncorhynchus kitutch*) and chinook (*O. tshawytscha*) salmon. Fish Res 6:105–123

Johnston AI, Harrison P (1985) Contractile and metabolic characteristics of muscle fibers from Antarctic fish. J Exp Biol 116:223–237

Johnston AI, Fitch N, Zummo G, Wood RE, Harrison P, Tota B (1983) Morphometric and ultrastructural features of the ventricular mrocardium of the haemoglobin-less icefish *Chaenocephalus aceratus*. Comp Biochem Physiol 76:475–480

Johnston JD, Mcforlane RW (1967) Migration and bioenergeties of flight in the Pacific golden prover. Condor 69:156–168

Johnston KS, Beehler CL, Sakamoto-Arnold CM, Childress JJ (1986) In situ measurements of chemical distributions in a deep-sea hydrothermal vent field. Science 231:1139–1141

Johnston KS, Childress JJ, Beehler CL (1988) Short term temperature variability in the Rose Garden hydrothermal vent field: an unstable deep-sea environment. Deep Sea Res 35:1711–1721

Jokumsen A, Fyhn HJ (1982) The influence of aerial exposure upon respiratory and osmotic properties of haemolymph from intertidal mussels, *Mytilus edulis* L and *Modiolus modiolus*. J Exp Mar Biol Ecol 61:189–203

Jollie WP, Jollie LG (1967) Electron microscopic observations on accommodations to pregnancy in the uterus of the spiny dogfish, *Squalus acanthias*. J Ultrastruct Res 20:161–178

Jones C, Fox H (1977) Syncytial knots and intervillous bridges in the human placenta: an ultrastructural study. J Anat 124:275–286

Jones D (1996) Hydrogen for ever. Nature (Lond) 383:766

Jones D, Birks S (1992) Megapodes: recent ideas on origins, adaptations and reproduction. Trends Ecol Evol 7:88–91

Jones DP (1986) Intracellular diffusion gradients of O_2 and ATP. Am J Physiol 250:C663–675

Jones DR, Johansen K (1972) The blood vascular system of birds. In: Farner DS, King JR (eds) Avian biology, vol II. Academic Press, New York, pp 157–285

Jones DR, Randall DJ (1978) The respiratory and circulatory systems during exercise. In: Hoar WS, Randall DJ (eds) Fish physiology, vol 7. Academic Press, New York, pp 425–501

Jones DR, Schwarzfeld T (1974) The oxygen cost to the metabolism and efficiency of breathing in trout (*Salmo gairdnen*). Respir Physiol 21:1992–1999

Jones DR, Shelton G (1972) Factors affecting diastolic blood pressures in the systemic and pulmocutaneous arches of anuran amphibia. J Exp Biol 57:789–803

Jones G, Rayner JMV (1989) Optimal flight speed pipistrelle bats (*Pipistrellus pipistrellus*). In: Hanak V, Horaceck I, Gaisler J (eds) European bat research 1987: Proc 4th Europ Bat Res Symp. Charles University Press, Praha, pp 87–103

Jones HD (1983) Circulatory systems of gastropods and bivalves. In: Saleuddin ASM, Wilbur MW (eds) The Mollusca, vol 5, physiology, part 2. Academic Press, London, pp 189–238

Jones JD (1961) Aspects of respiration in *Planorbis corneus* L. and *Lymnaea stagnalis* L. (Gastropoda: Pulmonata). Comp Biochem Physiol 4:1–29

Jones JD (1964) Respiratory gas exchange in the aquatic pulmonate *Biompalaria sudanica*. Comp Biochem Physiol 12:297–310

Jones JD (1972) Comparative physiology of respiration. Edward Amold, London

Jones JG, Roystom D, Minty BD (1983) Changes in alveolar-capillary barrier function in animals and humans. Am Rev Respir Dis 127:S51–S59

Jones JH (1982) Pulmonary blood flow in distribution in panting ostriches. J Appl Physiol 53:1411–1417

Jones JH, Effmann EL, Schmidt-Nielsen K (1981) Control of air flow in bird lungs: radiographic studies. Respir Physiol 45:121–131

Jones JH, Grubb B, Schmidt-Nielsen K (1983) Panting in the emu causes arterial hypoxemia. Respir Physiol 54:189–195

Jones JH, Effmann EL, Schmidt-Nielsen K (1985) Lung volume changes in during respiration in ducks. Respir Physiol 59:15–25

Jones JH, Longworth KE, Lindholm A, Conley KE, Karas RH et al. (1989) Oxygen transport during exercise in large mammals. I. Adaptive variation in oxygen demand. J Appl Physiol 67:862–870

Jones JRE (1952) The reaction of fish to water of low oxygen concentration. J Exp Biol 29:403–415

Jones RE, Smith HM, Bock CE (1993) Reptilian and avian ovarian cycles and evolutionary origin of volant birds. J Comp Physiol 163:594–601

Jorgensen CB (1952) On the relationship between water transport and food requirements in some marine filter feeding invertebrates. Biol Bull Mar Biol Lab Woods Hole 103:356–363

Jorgensen CB (1955) Quantitative aspects of filter feeding in invertebrates. Biol Rev 30:391–453

Jorgensen CB (1975) Comparative physiology of suspension feeding. Annu Rev Physiol 37:57–70

Joseph JH (1967) Diurnal and solar variations of neutral hydrogen in the atmosphere. Ann Geophys 23:365–373

Joss JM, Cramp N, Baverstock PR, Johnson AM (1991) A phylogenetic comparison of 18S ribosomal RNA sequences of lungfish with those of other chordates. Aust J Zool 39:509–518

Jouin C, Toulmond A (1989) The ultrastructure of the gill of the lugworm, *Arenicola marina*. Acta zool 70:121–129

Jouve-Duhamel A, Truchot JP (1983) Ventilation in the shore crab *Carcinus maenas* (L) as a function of ambient oxygen and carbon dioxide field and laboratory studies. J Exp Mar Biol Ecol 70:281–296

Joyce GF (1992) Directed molecular evolution. Sci Am 267:48–55

Joyce GF (1997) Evolutionary chemistry: getting from there from here. Science 276:1658–1657

Jukes TH (1985) A change in the genetic code in *Mycoplasma capricolum*. J Mol Evol 22:361–362

Jukes TH, Osawa S (1993) Evolutionary changes in the genetic code. Comp Biochem Physiol 106B:489–494

Julian D, Arp AJ (1992) Sulfide perrmeability in the marine invertebrates *Urechis caupo*. J Comp Physiol 162B:59–67

Julian D, Menon JG, Arp AJ (1991) Structural and functional adaptations to hypoxia and sulfide in *Urechis caupo*. Am Zool 31:73A

Julian RJ (1987) The effects of increased sodium in the drinking water on right ventricular failure and ascites in broiler chickens. Avian Pathol 16:61–71

Julian RJ, Wilson JB (1986) Right ventricular failure as a cause of ascites in broiler and roaster chickens. Proc IVth Int Symp Vet Lab Diag 1986:608–611

Julian RJ, Moran ET, Revinton W, Hunter DB (1984) Acute hypertensive angiopathy as a cause of sudden death in turkeys. Proc Am Vet Med Assoc 117

Junod AF (1972) Uptake, metabolism and efflux of ^{14}C-5-hydroxytryptamine in isolated, perfused rat lungs. J Pharmacol Exp Ther 183:431–355

Junod AF (1975) Mechanism of uptake of biogenic amines in the pulmonary circulation. In: Junod AF, Haller R (eds) Lung metabolism. Academic Press, New York, pp 387–396

Junqueira IC, Steen JB, Tincoce RM (1967) The respiratory area of the fishes of teleosts from Rio Negro and Rio Branco area. Research Papers from the Alpha Helix Amazon Expedition, pp B20–21

Jürgens JD, Bartels H, Bartels R (1981) Blood oxygen transport and organ weight of small bats and small nonflying mammals. Respir Physiol 45:243–260

Just JJ, Gatz RN, Crawford EC (1973) Changes in respiratory functions during metamorphosis of the bullfrog, *Rana catesbeiana*. Respir Physiol 17:276–282

Kaestner A (1929) Bau und Funktion der Fächertracheen einiger Spinnen. Z Morphol Ökol Tiere 13:463–558

Kaestner A (1970) Invertebrate zoology, vol 3 (transl by Levi HW and Levi LR). Wiley, New York

Kagan VE, Day BW, Elsayed NM, Gorbunov NV (1996) Dynamics of haemoglobin. Nature (Lond) 383:30–31

Kaiser T, Bucher TL (1985) The consequences of reverse sexual size dimorphism for oxygen consumption, ventilation and water loss in relation to ambient temperature in the prairie falcon, *Falco mexicanus*. Physiol Zool 58:748–758

Kamau JMZ, Maina JN, Maloiy GMO (1984) The design and the role of the nasal passages in temperature regulation in the dik-dik antelopes, *Rhynchotragus kirkii* with observations on the carotid rete. Respir Physiol 56:183–194

Kammer AE, Heinrich B (1978) Insect flight muscle metabolism. Adv Insect Physiol 13:133–228

Kampe G, Crawford EC (1973) Oscillatory mechanics of the respiratory system of pigeons. Respir Physiol 18:188–193

Kämpfe L (1980) Evolution und Stammesgeschichte der Organismen. Fischer, Stuttgart

Kandel ER (1979) Behavioural biology of *Aplysia*. WH Freeman, San Francisco

Kanwisher JW (1966) Tracheal gas dynamics in pupae of the *Cecropia* silkworm. Biol Bull 130:96–105

Kanwisher JW, Ebeling A (1957) Composition of the swim bladder gas in bathypelagic fishes. Deep Sea Res 2:211–223

Kanz JE, Quast WD (1990) Respiratory pumping seizure: a newly discovered spontaneous stereotyped behaviour pattern in the opisthobranch mollusc *Aplysia califomica*. J Comp Physiol 166A:619–627

Kanz JE, Quast WD (1992) Respiratory pumping behaviour in the marine snail *Aplysia califomica* as a function of ambient hypoxia. Physiol Zool 65:35–54

Karas RH, Taylor CR, Jones JH, Linstedt SL, Reeves RB, Weibel ER (1987a) Adaptive variation in the mammalian respiratory system in relation to energetic demand. VII. Flow of oxygen across the pulmonary gas exchanger. Respir Physiol 69:101–115

Karas RH, Taylor CR, Rösler K, Hoppeler H (1987b) Adaptive variation in the mammalian respiratory system in relation to energetic demand: V. Limits to oxygen transport by circulation. Respir Physiol 69:65–79

Karasov WH, Diamond JH (1985) Digestive adaptations for fuelling the cost of endothermy. Science 228:202–204

Karasov WH, Phan D, Diamond JH, Carpenter FL (1986) Food passage and intestinal nutrient absorption in hummingbirds. Auk 103:453–464

Kardong KV (1972) Morphology of the respiratory system and its musculature in diffrerent snake genera (part I), *Crotalus* and *Elaphe*. Gegenbaurs Morphol Jahrb 117:285–302

Kardong KV (1995) Vertebrate: comparative anatomy, function and evolution. Wm C Brown, Dubuque, Iowa

Kargel JS, Strom RG (1996) Global climatic change on Mars. Sci Am 275:60–68

Karlberg P, Adams FH, Geubelle F, Wallgren G (1962) Alterations of the infant's thorax during vaginal delivery. Acta Obstet Gynecol Scand 41:223–229

Karlsson L (1983) Gill morphology in the zebra fish, *Brachydanio rerio* (Hamilton-Buchanan). J Fish Biol 23:511–524

Karsdorp VHM, van Vugt JMG, van Geijn HP, Kostense PJ, Arduini D et al. (1994) Clinical significance of absent or reversed end diastolic velocity waveforms in umbilical artery. Lancet 344:1664–1668

Karsdorp VHM, Dirks BK, van der Linden JC, van Vugt JMG, Baak JPA, van Geijn HP (1996) Placenta morphology and absent or reversed end diastolic flow velocities in the umbilical artery: a clinical and morphometrical study. Placenta 17:393–399

Kaschke M, Russell MJ (1994) [FeS/FeS$_2$]: a redox system for the origin of life. Origins Life Evol Biosphere 24:43–56

Kasting JF (1997) Warming early Earth and Mars. Science 276:1213–1215

Kasting JF, Walker JCG (1981) Limits on oxygen concentration in prebiological atmosphere and the rate of abiotic fixation of nitrogen. J Geophys Res 86:1147–1158

Kasting JF, Liu SC, Donahue TM (1979) Oxygen levels in the prebiological atmosphere. J Geophys Res 84:3097–3107

Katušic ZS, Cosentino F (1994) Nitric oxide synthetase: from molecular biology to cerebrovascular physiology. News Physiol Sci 9:64–66

Kaufmann P (1972) Untersuchungen uber die Langhanszellen in der menschlichen Placenta. Z Zellforsch 128:283–302

Kaufmann P, Davidoff M (1977) The guinea-pig placenta. Adv Anat Embryol Cell Biol 53:1–90

Kaufmann P, Gentzen DM, Davidoff M (1977) Die Utrastruktur von Langhanszellen in pathologischen menschlichen Placenten. Arch Gynecol 22:319–332

Kawamoto N (1927) The anatomy of *Caudina chilensis* (J Müller) with special reference to the perivisceral cavity, the blood and the water vascular systems in their relation to the blood circulation. Sci Rep Tôhoku Univ IV Biol 2:239–264

Kawasaki M (1993) Independently evolved jamming avoidance responses employ identical computational algorithms: a behavioural study of the African electric fish, *Gymnarchus niloticus*. J Comp Physiol A 173:9–22

Kawashiro T, Scheid P (1975) Arterial blood gases in undisturbed gases in undisturbed resting brids: measurements in chicken and duck. Respir Physiol 23:337–342

Kayar SR, Snyder GK, Birchard GF, Black CP (1981) Oxygen permeability of the shell and membranes of chicken eggs during development. Respir Physiol 11:16–34

Keeling CD, Chin JFS, Whorf TP (1996) Increased activity of northern vegetation inferred from atmospheric CO$_2$ measurements. Nature (Lond) 382:146–149

Keevil T, Mason HS (1978) Molecular oxygen in biological oxidations – an overview. In: Fleischer S, Paker L (eds) Biomembranes: methods in enzymology, vol 3. Academic Press, New York, pp 3–40

Keilin D (1924) On appearance of gas in the trachea of insects. Proc Camb Philos Soc Biol Sci 1:63–70

Kendal MW, Dale JE (1979) Scanning and transmission electron microscopic observations of rainbow trout (*Salmo gairdnen*) gill. J Fish Res Board Can 36:1072–1079

Kennedy B (1979) Blood circulation in polychaete gills. Am Zool 19:868

Kennerly TE (1964) Microenvironmental conditions of the pocket gopher burrow. Tex J Sci 16:395–441

Kerourio P (1981) Nouvelles observations sur le mode de nidification et de ponte chez les dinosauriens du Cretace terminal du Midi de la France. C R Somm Séances Soc Geol Fr 1:25–28

Kerr RA (1988) Was there a prelude to the dinosaurs demise? Science 239:729–730

Kerr RA (1996a) Ancient life on Mars? Science 273:864–866

Kerr RA (1996b) Did a plate tectonic surge flood earth? Science 274:1611

Kerr RA (1997) An ocean emerges on Europa. Science 276:355

Kerr RJ (1898) The dry-season habits of *Lepidosiren*. Proc Zool Soc Lond 1898:41–44

Khalaf El Duweini A (1957) On the gills and respiration in *Alma nilotica* Grube. Publ 2nd Sci Arab Congr (1955) Cairo 833–838

Kiceniuk JW, Jones DR (1977) The oxygen transport system in trout (*Salmo gairdnen*) during sustained exercise. J Exp Biol 69:247–260

Kiel JW, Shepherd AP (1989) Optimal hematocrit for canine gastric oxygenation. Am J Physiol 256:H472–H477

Kikkawa Y (1970) Morphology of alveolar lining layer. Anat Rec 167:389–400

Kikuchi S (1992) Histological and fine structural evidence of the ion-transporting role of the gill and the neck organ of a fresh water brachiopod, *Branchinella kugenumaensis*. Annu Rep Iwate Med U Sch Lib Arts Sci 27:13–28

Kiley JP, Faraci FM, Fedde MR (1985) Gas exchange during exercise in hypoxic ducks. Respir Physiol 59:105–115

Kilgore DL, Boggs DF, Birchard GF (1979) Role of the rete mirabile ophthalmicum in maintaining body-to-brain temperature difference in pigeons. J Comp Physiol 129:119–122

Kimball RE, Reddy K, Peirce TH, Schwartz LW, Mustafa MG, Cross CE (1976) Oxygen toxicity: augmentation of antioxidant defense mechanisms in rat lung. Am J Physiol 230:1425–1431

Kimbel P, Weinbaum G (1975) Role of leucoproteases in the genesis of emphysema. In: Junod AF, Haller R (eds) Lung metabolism. Academic Press, New York, pp 25–37

Kimoto T, Fujinaga T (1990) Non-biotic synthesis of organic polymers on H_2S-rich sea floor: a possible reaction in the origin of life. Mar Chem 30:179–192

Kimura M (1983) The neutral theory of molecular evolution. Cambridge University Press, Cambridge

Kimura A, Gomi T, Kikuchi Y, Hashimoto T (1987) Anatomical studies of the lung of air breathing fish. I. Gross anatomical and light microscopic observations of the lungs of the African lungfish *Protopterus aethiopicus*. J Med Soc Toho Univ 34:1–18

King AS (1966) Structural and functional aspects of the avian lung and its air sacs. Int Rev Gen Exp Zool 2:171–267

King AS, McLelland J (1975) Outlines of avian anatomy. Bailliéle Tindall, London

King MC, Wilson AC (1975) Evolution at two levels in human and chimpanzees. Science 188:107–116

King JR, Farner DA (1969) Energy metabolism, thermoregulation, and body temperature. In: Marshall AJ (ed) Biology and comparative physiology of birds, vol 2. Academic Press, London, pp 215–279

Kinney JL, White FN (1977) Oxidative ventilation in a turtle, *Pseudemys floridana*. Respir Physiol 31:327–332

Kinnula VL, Adler K, Akley N, Crapo J (1992) Release of reactive oxygen species by guinea pig tracheal epithelial cells *in vitro*. Am J Physiol 262:L708–L712

Kirsch R, Nonnotte G (1977) Cutaneous respiration in there fresh water teleosts. Respir Physiol 29:339–354

Kirschner LB (1982) Physical basis of solute and water transfer across gills. In: Houlihan DF, Rankin JC, Shuttleworth TJ (eds) Gills. Cambridge University Press, Cambridge, pp 63–76

Kirschner LB (1993) The energetics of osmotic regulation in ureotelic and hypoosmotic fishes. J Exp Zool 267:19–26

Kirschner RP (1994) The Earth's elements. Sci Am 271:37–43

Kitcher P (1984) 1953 and all that: a tale of two sciences. Philos Rev 93:102–124

Kitterman JA, Ballard PL, Clements JA, Mescher EJ Tooley WH (1979) Tracheal fluid in fetal lambs: spontaneous decrease prior to birth. J Appl Physiol 47:985–989

Klaver CJJ (1973) Lung anatomy: aid in chameleon taxonomy. Beaufortia 20:155–177

Klaver CJJ (1981) Lung morphology in the chamaeleonidae (Sauria) and its bearing upon phylogeny, systematics and Zoogeography. Z Zool Syst Evolutionsforsch 19:36–58

Kleiber M (1965) Respiratory exchange and metabolic rate. In: Fenn WO, Rahn H (eds) handbook of physiology, sect 3, respiration, vol II. American Physiological Society, Washington, DC, pp 927–938

Kleinschmidt T, Sgouros JG (1987) Haemoglobin sequences. Biol Chem Hoppe-Seyler 368:579–615

Klemm RD, Gatz RN, Westfall JA, Fedde MR (1979) Microanatomy of the lung parenchyma of a tegu lizard *Tupinambis nigropunctatus*. J Morphol 161:257–280

Klika E, Lelek A (1967) A contribution to the study of the lungs of *Protopterus annectens* and *Polypterus segegalensis*. Folia Morphol 15:168–175

Klite PD (1965) Intestinal bacterial flora and transit time in three neotropical bat species. J Bacteriol 90:375–379

Knight DR, Schaffartzik W, Poole DC, Hogan MC, Bebout DE, Wagner PD (1992) Hyperoxia improves leg VO_{2max}. FASEB J 139:A1466

Knight J (1984) Studies on the biology and biochemistry of *Pholas dactylus* L., PhD Thesis, University of London, London

Knight J, Knight R (1986) The blood vascular system of the gills of *Pholas dactylus* L. (Mollusca, Bivalvia, Eulamellibranchia). Philos Trans R Soc Lond 313B:509–523

Knoll AH (1979) Archean photoautotrophy: some alternatives and limits. Origins Life 9:313–327

Knoll AH (1991) The end of the Proterozoic Eon. Science 265:64–73

Knoll AH (1996) Breathing room for early animals. Nature (Lond) 382:111–112

Knutton S, Jackson D, Graham JM, Micklem KJ, Pastemak CA (1976) Microvilli and cell swelling. Nature (Lond) 262:52–54

Kobayashi H, Pelster B, Scheid P (1989a) Solute back-diffusion in counter-current flow. Respir Physiol 78:59–71

Kobayashi H, Pelster B, Scheid P (1989b) Water and lactate movement in the swim bladder of the eel, *Anguilla anguilla*. Respir Physiol 78:45–57

Kobayashi H, Pelster B, Scheid P (1990) Carbon dioxide back-diffusion in the rete aids O_2 secretion in the swim bladder of the eel. Respir Physiol 79:231–242

Koch H (1938) The absorption of chloride ions by the anal papillae of diptera larvae. J Exp Biol 15:152–160

Koch R, Seymour S, Bartholomew GA, Barnhart MC (1983) Respiration and heat production by inflorescence of *Philodendron selloum*. Planta 157:336–343

Koch R, Seymour S, Barnhart MC, Bartholomew GA (1984) Respiratory gas exchange during thermogenesis in *Philodendron selloum*. Planta 161:229–232

Kokko JP, Tisher CC (1976) Water movement across nephron segments involved with the counter-current multiplication system. Kidney Int 10:64–81

Kon K, Maeda N, Sekiya M, Shiga T, Suda T (1980) A method for studying oxygen diffusion barrier in erythrocytes: effects of haemoglobin content and membrane cholestrol. J Physiol (Lond) 309:569–590

Kon K, Maeda N, Shiga T (1983) The influence of deformation of transformed erythrocytes during flow on the rate of oxygen release. J Physiol (Lond) 339:573–584

König MF, Lucocq JM, Weibel ER (1993) Demonstration of pulmonary vascular perfusion by electron and light microscopy. J Appl Physiol 75:1877–1883

Konings WN, van Driel R, van Bruggen EF, Gruber M (1969) Structure and properties of hemocyanins. V. Binding oxygen and copper in *Helix pomatia* hemocyanin. Biochim Biophys Acta 194:55–66

Kooyman GL (1985) Physiology without restraint in diving mammals. Mar Mammal Sci 1:166–178

Kooyman GL, Cornell LH (1981) Flow properties of expiration and insiration in a trained bottlenosed porpoise. Physiol Zool 54:55–61

Kooyman GL, Sinnett EE (1979) Pulmonary shunts in harbour seals and sea lions during simulated dives to depth. Physiol Zool 55:105–111

Koshland DE (1992) The molecule of the year. Science 258:1861

Kostelecka-Mycha A (1987) Respiratory function of a unit of blood volume in the little auk (*Plautus alle*) and the Arctic tern (*Stema prardisaea*). Comp Biochem Physiol 86A:117–120

Koteja P (1986) Maximum cold-induced oxygen consumption in the house sparrow, *Passer domesticus* (L.). Physiol Zool 59:43–48

Kozlowski J (1993) Measuring fitness in life history studies. Trends Ecol Evol 8:84–85

Kramer DL (1978) Ventilation of the respiratory gas bladder in *Hoplerythrinus unitaeniatus* (Pisces, Characoidei, Erythrinidae). Can J Zool 56:931–938

Kramer DL (1980) A comparative test of the adaptive significance of aquatic surface respiration in fishes. Am Zool 20:742

Kramer DL (1983) The evolutionary ecology of respiratory mode in fishes: an analysis based on the costs of breathing. Environ Biol Fish 9:145–158

Kramer DL (1987) Dissolved oxygen and fish behaviour. Environ Biol Fish 18:91–92

Kramer DL (1988) The behavioural ecology of air breathing by aquatic animals. Can J Zool 66:89–94

Kramer DL, Graham JB (1976) Synchronus air breathing, a social component of respiration in fishes. Copeia 1976:689–697

Kramer DL, McClure M (1981) The transit cost of aerial respiration in the catfish, *Corydoras aeneus* (Callichthyidae). Physiol Zool 54:189–194

Kramer DL, McClure M (1982) Aquatic surface respiration, widespread adaption to hypoxia in tropical fresh water fishes. Environ Biol Fish 7:47–55

Kramer DL, Lindsey CC, Moodie GEE, Stevens ED (1978) The fishes and the aquatic environment of the Central Amazon Basin, with particular reference to respiratory patterns. Can J Zool 56:717–729

Kràtký J (1981) Postnatale Entwicklung der Wasserfledermaus, *Myotis daubentoni* Kuhl, 1981 und bisherige Kenntnis dieser Problematik im Rahmen der Unterordnung Microchiroptera (Mammalia: Chiroptera). Fol Mus Rer Natur Bohem Occident Zool 16:1–34

Krebs JR, Harvey PH (1986) Busy doing nothing – efficiently. Nature (Lond) 320:18–19

Krenz GS, Linehan JH, Dawson CA (1992) A fractal continuum model of the pulmonary arterial tree. J Appl Physiol 72:2225–2237

Kreuzer F (1970) Facilitated diffusion of oxygen and its possible significance: a review. Respir Physiol 9:1–30

Krogh A (1910) On the mechanism of the gas exchange in the lungs. Skand Arch Physiol 23:248–278

Krogh A (1913) On the composition of the air in the tracheal system of some insects. Skand Arch Physiol 29:29–36

Krogh A (1920a) Studien über Tracheen Respiration. II. Über Gasdiffusion in den Tracheen. Pfluegers Arch Gesamte Physiol Menschen Tiere 179:95–112

Krogh A (1920b) Studien über Tracheen Respiration. III. Die Kombination von mechanischer Ventilation mit Gasdiffusion nach Versuchen an Dytiscuslarven. Pfluegers Arch Gesamte Physiol Menschen Tiere 179:113–120

Krogh A (1941) The comparative physiology of respiratory mechanisms. University of Pennsylvania Press, Philadelphia

Kruhoffer M, ML, Abe AS, Johansen K (1987) Control of breathing in an amphibian *Bufo paracnemius*: effects of temperature and hypoxia. Respir Physiol 69:267–275

Kuethe AM (1975) Prototypes in nature. The carry-over into technology. TechniUM, Spring Issue 1975:3–20

Kuethe DO (1988) Fluid mechanical valving of air flow in bird lungs. J Exp Biol 136:1–12

Kuhn W, Ramel A, Kuhn H, Marti E (1963) The filling mechanism of the swim bladder. Generation of high gas pressures through hairpin counter-current multiplication. Experientia 19:497–511

Kulzer E (1965) Temperaturregulation bei Fledermäusen (Chiroptera) aus verschiedenen Klimazonen. Z Vergl Physiol 50:1–34

Kurland CG (1992) Evolution of mitochondria genomes and the genetic code. BioAssays 14:709–714

Kutchai H, Steen JB (1971) Permeability of the shell and shell membranes of hen's eggs during development. Respir Physiol 11:265–278

Kutty MN, Saunders RL (1973) Swimming performance of young Atlantic salmon (*Salmon salar*) as affected by reduced ambient oxygen concentration. J Fish Res Board Can 30:223–227

Kylstra JA (1962) Drowning: the role of salts in the drowning fluid. Acta Physiol Pharmacol Neerl 10:327–334

Kylstra JA (1968) Experiments in water-breathing. Sci Am 219:66–74

Kylstra JA (1969) The feasibility of liquid breathing and artificial gills. In: Bennet PB, Elliot DH (eds) The physiology and medicine of diving and compressed air work. Bailleire, London, pp 193–212

Kylstra JA, Schoenfisch WH (1972) Alveolar surface tension in fluorocarbon-filled lungs. J Appl Physiol 33:32–35

Kylstra JA, Tissing MO (1963) Fluid breathing. In: Clinical applications of hyperbaric oxygen. Proc 1st Int Congr, Amsterdam, pp 371–379

Kylstra JA, Paganelli CV, Lanphier EH (1966) Pulmonary gas exchange in dogs ventilated with hyperbarically oxygenated liquid. J Appl Physiol 21:177–184

Kylstra JA, Rausch DC, Hall KD, Spock A (1971) Volume-controlled lung lavage in the treatment of asthma, bronchiectasis, and mucoviscidosis. Am Rev Respir Dis 103:651–665

Labra MA, Rosenmann M (1994) Energy metabolism and evaporative water loss of *Pristidactylus* lizards. Comp Biochem Physiol A 109:369–376

Laburn HP, Goelst K, Mitchell D (1994) Body temperatures of lambs and their mothers measured by radio-telemetry during parturition. Experientia 50:708–711

Laga EM, Driscoll SG, Munro HN (1974) Human placental structure: relationship to foetal nutrition. In: Josmovich JB, Reynods M, Cobo E (eds) Problems of human reproduction, vol 2, lactogenic hormones, foetal nutrition and nutrition. John Wiley, London, pp 143–181

Lahiri S (1975) Blood oxygen affinity and alveolar ventilation in relation to body weight in mammals. Am J Physiol 229:529–536

Lahiri S, Szidon JP, Fishman AP (1970) Potential respiratory and circulatory adjustments to hypoxia in the African lungfish. Fed Proc 29:1141–1148

Lai NC, Graham JB, Burmett L (1990) Blood respiratory properties and the effect of swimming on blood gas transport in the leopard shark *Triakis semifasciata*. J Exp Biol 151:161–173

Laitman TJ, Reidenberg JS, Marquez S, Gannon PJ (1996) What the nose knows: new understandings of the Neanderthal upper respiratory tract. Proc Natl Acad Sci USA 93:10543–10545

Lallier F, Truchot JP (1989) Hemolymph oxygen transport during environmental hypoxia in the shore crab, *Carcinus maenas*. Respir Physiol 77:323–336

LaManna JC, Vendel LM, Farrell RM (1992) Brain adaptation to chronic hypobaric hypoxia in rats. J Appl Physiol 72:2238–2243

Lambertsen CJ (1961) Respiration. In: Bard P (ed) Medical physiology. Mosby, St Louis, pp 32–98

Lamy J, Truchot JP, Giles R (1985) (eds) Respiratory pigments in animals. Springer, Berlin Heidelberg New York

Landes RR, Leohardt KO, Duruman N (1964) A clinical study of the oxygen tension of the urine and renal structures. II. J Urol 92:171–178

Landis GP, Snee LW (1991) 40Ar/39r systematics and argon diffusion in amber: implictions for ancient Earth atmosphere. Paleogeogr Paleoclimatol Paleoecol 97:63–67

Lang BF, Burger G, O'Kelly CJ, Cedergren R et al. (1997) An ancestral mitochondrial DNA resembling a eubacterial genome in miniature. Nature (Lond) 387:493–496

Langille BL, Jones (1975) Central cardiovascular dynamics of ducks. Am J Physiol 228:1856–1861

Langston W (1981) Pterosaurs. Sci Am 244:92–102

Lanphier EJ (1969) Pulmonary function. In: Bennett PB, Elliot DH (eds) The physiology and medicine of diving and compressed-air work. Bailere, Tindall and Cassell, London, pp 58–112

Lapennas GN, Schimdt-Nielsen K (1977) Swim bladder permeability to oxygen. J Exp Biol 67:175–196

Larimer JL, Schimdt-Nielsen K (1960) A comparison of blood carbonic anhydrase of various mammals. Comp Biochem Physiol 1:19–23

Larson A, Chippindale P (1993) Molecular approaches to the evolutionary biology of plethodontid salamanders. Herpetologia 49:204–215

Lasiewski RC (1962) The energetics of migrating hummingbirds. Condor 64:324

Lasiewski RC (1963a) Oxygen consumption of torpid, resting, active, and evaporative water loss in hummingbirds. Physiol Zool 36:122–140

Lasiewski RC (1963b) The energetic cost of small size hummingbirds. Proc XIII Int Congr 1095–1103

Lasiewski RC (1964) Body temperatures, heart and breathing rate, and evaporative water loss in hummingbirds. Physiol Zool 37:212–223

Lasiewski RC (1972) Respiration function in birds. In: Famer DS, King JR, Parkes KC (eds) Avain biology, vol II. Academic Press, New York, pp 271–342

Lasiewski RC, Calder WR (1971) A preliminary allometric analysis of respiratory variables in resting birds. Respir Physiol 11:152–166

Lasiewski RC, Weathers WW, Bernstein MV (1967) Physiological responses of the giant hummingbird, *Patagona gigas*. Comp Biochem Physiol 23:797–813

Lauder GV (1980) Evolution of feeding mechanisms in primitive actinopterygian fishes: A functional anatomical analysis of *Polypterus, Lepisosteus* and *Amia*. J Morphol 163:283–317

Lauder GV (1981) Form and function: structural analysis in evolutionary morphology. Paleobiology 7:430–442

Lauder GV, Liem KF (1983) The evolution and interrelationships of actinopterygian fishes. Bull Mus Comp Zool 150:95–197

Lauder GV, Liem KF (1989) The role of historical factors in the evolution of complex organismal functions. In: Wake DB, Roth G (eds) Complex organismal functions: integration and evolution in vertebrates. John Wiley, London, pp 63–78

Laurent GJ (1986) Lung collagen: more than scaffoding. Thorax 41:418–428

Laurent P (1982) Structure of vertebrate gills. In: Houlihan DF, Rankin JC, Shuttleworth TJ (eds) Gills. Cambridge University Press, Cambridge, pp 25–43

Laurent P (1984) Gill internal morphology. In: Hoar WS, Randall DJ (eds) Fish physiology, vol A. Academic Press, New York, pp 73–183

Laurent P (1985) Organization and control of the respiratory vasculature in lower vertebrates: are there anatomical gill shunts? In: Johansen K, Burrgren WW (eds) Cardiovascular shunts. Alfred Benzon Symposium 21, Munksgaard, Copenhagen, pp 57–70

Laurent P (1996) Vascular organization of lungfish, a landmark in ontogeny and phylogeny of air-breathers. In: Munshi JSD, Dutta HM (eds) Fish morphology: horizon of new research. Science Publishers, Lebanon, New Hampshire, pp 47–58

Laurent P, Dunel-Erb S (1980) Morphology of gill epithelia. Am J Physiol 238:R147–R159

Laurent P, Hebibi (1989) Gill morphometry and fish osmoregulation. Can J Zool 67:3055–3063

Laurent P, Perry SF (1990) Effects of cortisol on gill chloride cell morphology and ionic uptake in the fresh water salmonid fish. Cell Tissue Res 259:429–442

Laurent P, Perry SF (1991) Environmental effects on gill morphology. Physiol Zool 69:2–25

Laurent P, Delaney RG, Fishman AP (1978) The vasculature of the gills in the aquatic and aestivating lungfish (*Protopterus aethiopicus*). J Morphol 156:173–208

Laurent P, Maina JN, Bergman HL, Narahara A, Walsh PJ, Wood CM (1995) Gill structure of a fish from an alkaline lake: effect of short-term exposure to neutral conditions. Can J Zool 73:1170–1181

Laverack MS (1963) The physiology of earthworms. Pergamon Press, Oxford

Laybourne RC (1974) Collision between a vulture and an aircraft at an altitude of 37 000 ft. Wilson Bull 86:461–462

Lazzaro X (1987) A review of planktivorous fishes: their evolution, feeding behaviours, selectivities, and impacts. Hydrobiologia 146:97–167

Leatherland JF, Hyder M, Ensor DM (1974) Regulation of Na^+ and K^+ concentrations in five African species of *Tilapia* fishes. Comp Biochem Physiol 48A:699–710

Lechner AJ (1984) Pulmonary design in microchiropteran lung (*Pipistrellus subflavus*) during hibernation. Respir Physiol 59:301–312

Lee DH, Granja JR, Martinez JA, Severin K, Ghadiri MR (1996) A self-replicating peptide. Nature (Lond) 382:525–531

Lee R, Mayhew TM (1995) Volumes of villi and intervillous pores in placentae from low and high altitude pregnancies. J Anat 186:349–355

Lefevre J (1983) Teleonomical optimization of a fractal model of the pulmonary arterial bed. J Theor Biol 102:225–248

Leigh GJ (1997) Biological nitrogen fixation and model chemistry. Science 275:1442

Leith DE, Mead J (1966) Maximum expiratory flow in liquid filled lungs. Fed Proc 25:506

Leming TD, Stuntz WE (1984) Zones of coastal hypoxia revealed by satellite scanning have implications for strategic fishing. Nature (Lond) 310:136–138

Lenfant C (1973) High altitude adaptation in mammals. Am Zool 13:447–456

Lenfant C, Johansen K (1967) Respiratory adaptations in selected amphibians. Respir Physiol 2:247–260

Lenfant C, Johansen K (1968) Respiration in an African lungfish, *Protopterus aethiopicus*: respiratory properties of blood and normal patterns of breathing and gas exchange. J Exp Biol 49:437–452

Lenfant C, Johansen K (1972) Gas exchange in gill, skin and lung breathing. Respir Physiol 14:211–218

Lenfant C, Johansen K, Grigg GC (1966) Respiratory properties of blood and pattern of gas exchange in the lungfish *Neocertodus forsteri* (Kreffti). Respir Physiol 2:1–21

Lenfant C, Elsner R, Kooyman GL, Drabek CM (1969) Respiratory function of blood of adult and fetus Weddell seal *Leptonychotes weddelli*. Am J Physiol 216:1595–1597

Lenfant C, Johansen K, Torrance JD (1970a) Gas transport and oxygen storage capacity in some pinnipeds and the sea otter. Respir Physiol 9:277–286

Lenfant C, Johansen K, Hanson D (1970b) Bimodal gas exchange and ventilation-perfusion relationship in lower vertebrates. Fed Proc 29:1124–1129

Lennard R, Huddart H (1989) Electrophysiology of the flounder heart (*Platichthys flesus*) – the effect of agents which modify transmembrane ion transport. Comp Biochem Physiol 93C:499–509

Leopold LB, Davies KS (1968) Water. Time-Life Books, Amsterdam

Lessard J, Val AL, Aota S, Randall DJ (1995) Why is there no carbonic anhydrase activity available to fish plasma? J Exp Biol 1995:31–38

Levi HW (1967) Adaptations of respiratory systems of spiders. Evolution 21:571

Levine BW, Talamo RC, Kazemi H (1973) Action and metabolism of bradykinin in dog lung. J Appl Physiol 34:821–826

Levine OR, Mellins RB, Fishman AP (1965) Quantitative assessment of pulmonary oedema. Circ Res 17:414–423

Levine S (1976) Competitive interactions in ecosystems. Am Nat 110:903–910

Levins R, Lewontin R (1985) The dialetical biologist. Harvard University Press, Cambridge

Levinton JS (1992) The big bang of animal evolution. Sci Am 267:52–59

Levy M, Achituv Y, Susswein AJ (1989) Relationship between respiratory pumping and oxygen consumption in *Aplysia depilans* and *A. fasciata*. J Exp Biol 141:389–405

Lewin R (1988) Egg laying in birds remains a hot issue. Science 233:465

Lewis AB, Heymann MA, Rudolph AM (1976) Gestational changes in pulmonary vascular responses in foetal lambs *in utero*. Circ Res 39:536–541

Lewis JF, Tabor B, lkegami M, Jobe AH, Joseph M, Absolom D (1993) Lung function and surfactant distribution in saline-lavaged sheep given instilled vs. nebulized surfactant. J Appl Physiol 74:1256–1264

Lewis RW (1970) The densities of three classes of marine lipids in relation to their possible role as hydrostatic agents. Lipids 5:151–165

Lewis SV (1980) Respiration in lampreys. Can J Fish Aquatic Sci 37:1711–1722

Lewis SV, Potter IC (1982) A light and electron microscope study of the gills of the lampreys (*Geotria australis*) with particular reference to the water-blood pathway. J Zool (Lond) 198:157–176

Lewontin RG (1979) Fitness, survival and optimality. In: Horn DH, Mitchell R, Stairs GR (eds) Analysis of ecological systems. Ohio State University, Columbus, pp 3–21

Liem KF (1961) Tetrapod parallelisms and other features in the functional morphology of the blood vascular system of *Fluta alba* Zuiew (Pisces: Teleostei). J Morphol 108:131–143

Liem KF (1980) Air ventilation in advanced teleosts: biomechanical and evolutionary aspects. In: Ali MA (ed) Environmental physiology of fishes. Plenum Press, New York, pp 57–91

Liem KF (1981) Larvae of air-breathing fishes as counter-current flow devices in hypoxic environments. Science 211:1177–1179

Liem KF (1984) The muscular basis of aquatic and aerial ventilation in the air breathing teleost fish *Channa*. J Exp Biol 113:1–18

Liem KF (1985) Ventilation. In: Hildebrand M, Bramble DM, Liem KF, Wake DB (eds) Functional vertebrate morphology. Harvard University Press, Cambridge, pp 185–209

Liem KF (1987a) Form and function of lungs: the evolution of air-breathing mechanisms. Am Zool 28:739–759

Liem KF (1987b) Functional design of the air ventilation apparatus and overland excursions by teleosts. Fieldiana Zool New Ser 37:1–29

Liem KF (1989) Respiratory gas bladders in teleosts: functional conservatism and morphological diversity. Am Zool 29:333-352

Liem KF (1991) Towards a new morphology: pluralism in research and education. Am Zool 31:759-767

Liem KF, Wake DB (1985) Morphology: current approaches and concepts. In: Hildebrand M (ed) Functional vertebrate morphology. Harvard University Press, Cambridge, pp 269-305

Lien DC, Wagner WW, Capen C, Haslett WL et al. (1987) Physiological neutrophil sequestration in the lung: visual evidence for localization in capillaries. J Appl Physiol 62:1236-1243

Liew HD (1970) Maintenance of underwater gas bubbles: how sea weed solves the problem. In: Farhi LE, Rahn H (eds) Studies in pulmonary physiology: mechanics, chemistry, and circulation of the lung. USAF School of Aerospace Medicine, Aerospace Medical Division (AFSC), Brooks Air Force Base, Houston Texas, pp 237-253

Liggins GC, Vilos GA, Campos GA, Kitterman JA, Lee CH (1981) The effect of spinal cord transection on lung development in foetal sheep. J Dev Physiol 3:267-274

Lilja C (1997) Oxygen consumption and vital organ masses in young growing quail *Coturnix coturnix japonica*. Acta Physiol Scand 160:113-114

Lillegraven JA, Thompson SD, McNab BK, Patton JL (1987) The origin of the Eutherian mammals. Biol J Linn Soc 32:281-336

Lindal T (1990) Repair of intrisic DNA lesions, Mut at Res 238:305-311

Linderman PV (1991) Survivorship of overwintering painted turtles, *Chrysemys picta* in Northern Idaho. Can Field Nat 105:263-266

Lindroth A (1938a) Studien über die respiratorischen Mechanismen von *Nereis virens* Sars. Zool Bidr Upps 17:367-497

Lindroth A (1938b) Gibt es bei den Polychäten intestinale Atmung per anum? Z Vergl Physiol 25:283-292

Lindroth A (1939) Beobachtungen an Capitelliden, besonders hinsichtlich ihrer Respiration. Zool Anz 127:285-297

Lindsay HA, Friex ED, Hoversland AS et al. (1971) Respiration and circulation. In: Altiman PL, Dittmer DS (eds) Handbook of physiology: respiration and circulation. Am Soc Exp Biol, Bethesda, pp 438-543

Lindstedt SL (1984) Pulmonary transit time and diffusing capacity in mammals. Am J Physiol 246:R384-R388

Lindstedt SL, Jones JH (1988) Symmorphosis: The concept of optimal design. In: Feder ME, Bennet AF, Burggren WW, Huey RB (eds) New directions in ecological physiology. Cambridge University Press, Cambridge, pp 290-310

Lindstedt SL, Hokanson JF, Wells DJ, Swain SD, Hoppelar H, Navarro V (1991) Running energetics in the pronghom antelope. Nature (Lond) 353:748

Linzen B, Gallowitz P (1975) Enzyme activity patterns in muscles of the lycosid spider, *Cupiennius salei*. J Comp Physiol 96:101-109

Little C (1983) The colonization of land. Origins and adaptations of terrestrial animals. Cambrigde University Press, Cambridge

Little C (1990) The terrestrial invasion: an ecophysiological approach to the origins of land animals. Cambridge University Press, Cambridge

Liversmore RA, Smith AG, Briden JC (1985) Palaeomagnetic constraints on the distribution of continents in the rate Silurian and early Devonian. Philos Trans R Soc Lond 309B:29-56

Lloyd D, Rossi EL (1993) Biological rhythms as organization and information. Biol Rev 68:563-577

Locke MJ (1958a) The structure of insect tracheae. Q J Microsc Sci 98:487-492

Locke MJ (1958b) Coordination of growth in the tracheal system of insects. Q J Microsc Sci 99:373-391

Locke MJ (1958c) The formation of trachea and tracheoles in *Rhodinius prolixus*. Q J Microsc Sci 99:29-46

Lockely Rm (1970) The most aerial bird in the world. Animals 13:4-7

Lockwood APM (1968) Aspects of physiology of Crustacea. Oliver Boyd, Edinburgh

Loesch H (1960) Sporadic mass shoreward migration of demersal fish and crustaceans in Moblie Bay, Alabama. Ecology 41:291-298

Logan GA, Hayes JM, Hieshima GB, Summon RE (1995) Terminal Proterozoic reorganization of biogeochemical cycles. Nature (Lond) 376:53–56

Lomholt JP (1976a) The development and of the oxygen permeability of the avian egg shell during incubation. J Exp Zool 198:177–203

Lomholt JP (1976b) Relationship of weight loss to ambient humidity of birds eggs during incubation. J Comp Physiol 105:189–196

Lomholt JP, Johansen K (1976) Gas exchange in the amphibious fish, *Amphipnous cuchia*. J Comp Physiol 107:141–157

Lomholt JP, Johansen K, Maloiy GMO (1975) Is aestivating lungfish the first vertebrate with suctional breathing? Nature (Lond) 257:787–788

Londraville RL, Sidell BD (1990) Ultrastructure of aerobic muscles in Antarctic fishes may contribute to maintenance of diffusive fluxes. J Exp Biol 150:205–220

Longmuir IS (1976) Search for new tissue oxygen carriers. In: Oxygen and physiological function. Professional Information Library, Dallas

Longmuir IS, Bourke A (1959) Application of Warburg's equation to tissue slices. Nature (Lond) 184:634–635

Longmuir IS, Bourke A (1960) The measurement of the diffusion of oxygen through respiring tissue. Biochem J 76:225–229

Longmuir IS, Sun S (1970) A hypothetical tissue oxygen carrier. Microvasc Res 2:287–293

Longo LD, Power GG, Forster RE (1967) Respiratory function in the placenta as determined by carbon monoxide in sheep and dogs. J Clin Invest 46P:812–828

Longo LD, Power GG, Forster RE (1969) Placental diffusing capacity for carbon monoxide at varying aprtial pressures of oxygen. J Appl Physiol 26:360–370

Longo LD, Hill EP, Power GG (1973) Factors affecting placental oxygen transfer. Chemical Engineering in Medicine. Adv Chem Ser 118:88–129

Longworth KE, Jones JH, Bicudo JEP, Taylor CR, Weibel ER (1989) High rate of O_2 consumption in exercising foxes: large PO_2 difference drives diffusion across the lung. Respir Physiol 77:263–276

López J (1995) Anatomy and histology of the lung and air sacs of birds. In: Pastor LM (ed) Histology, ultrastructure and immunohistochemistry of the respiratory organs in non-mammalian vertebrates. Publicaciones de la Universitatd de Murcia, University of Murcia, Murcia, Spain, pp 179–233

Lorenz RD, McKay CP, Lunine JI (1997) Photochemically driven collapse of Titan's atmosphere. Science 275:642–643

Losa GA, Baumann G, Nonnenmacher TF (1992) Fractal dimension of pericellular membranes in human lymphocytes and lymphoblastic leukemia cells. Pathol Res Pract 188:680–686

Lotshaw DP (1977) Temperature adaptation and effects of thermal acclimation in *Rana sylvatica* and *Rana catesbeiana*. Comp Biochem Physiol 56A:287–294

Loudon C (1989) Tracheal hypertrophy in mealworms: design and plasticity in oxygen supply systems. J Exp Biol 147:217–235

Lovejoy TE (1994) The quantification of biodiversity: an esoteric quest or a vital component of sustainable development? Philos Trans R Soc Lond 345B:81–87

Loveridge JP (1970) Observations on nitrogenous excretion and water relations of *Chiromantis xerapherina* (Amphibia, Anura). Arnoldia 5:1–6

Loveridge JP (1976) Strategies of water conservation in the Southern Africa frogs. Zool Afr 11:319–333

Loveridge JP (1980) Cuticle water relations. In: Miller TA (ed) Cuticle techniques in arthropods. Springer, Berlin Heidelberg New York, pp 301–366

Løvtrup S (1977) The phylogeny of vertebrata. Wiley, London

Low FN (1953) The pulmonary alveolar epithelium of laboratory animals and man. Anat Rec 117:241–263

Low WP, Lane DJ, Ip YK (1988) A comparative study of terrestrial adaptations of the gills in three mudskippers – *Periophthalmus chrysospolos*, *Boleophthalmus boddaerti*, and *Periophthalmus schlosseri*. Biol Bullmar Biol Woods Hole 175:334–438

Low WP, Ip YK, Lane DJ (1990) A comparative study of the gill morphology of the mudskippers, *Periophthalmus chrysospilos*, *Boleophthalmus boddaerti* and *Periophthalmus schlosseri*. Zool Sci 7:29–38

Lowe CA, Tuma RF, Sivieri EM, Shaffer TH (1979) Liquid ventilation: cardiovascular adjustments with secondary hyperlactatemia and acidosis. J Appl Physiol 47:1051–1057

Lucas AM, Denington EM (1961) A brief report on anatomy, histology and nomenclature of air sacs in the fowl. Avian Dis 5:460–481

Luchtel DL, Kardong KV (1981) Ultrastructure of the lung of the rattlesnake, *Crotalus viridis oreganus*. J Morphol 169:29–47

Luckett WP (1976) Ontogeny of amniote foetal membranes and their applications to phylogeny. In: Hecht MK, Goody PC, Hecht BM (eds) The major patterns in vertebrate evolution. Plenum Press, London, pp 439–516

Luckett WP (1993) Uses and limitations of mammalian fetal membranes and placenta for phylogenetic reconstruction. J Exp Zool 266:514–527

Luckett WP, Hartenberger JL (1993) Monophyly or polyphyly of the order rodentia: possible conflict between morphological and molecular interpretations. J Mammal Evol 1:127–147

Ludwig KS (1965) Zur Feinstruktur der materno-foetalen Verbindung im Placenta des Schafes (*Ovis ovis*). Experientia 18:212–213

Luft UC (1965) Aviation physiology. In: Fenn WO, Rahn H (eds) Handbook of physiology, sect 3, respiration vol II. American Physiological Society, Washinton, DC, pp 1099–1147

Lüling KH (1964) Über die Atmung des *Hoploerythrinus unitaeniatus* (Pisces, Erythrinidae). Bonn Zool Beitr 15:90–102

Lundberg K, Almgren B, Odelberg C (1983) Nagot om vattenfladdermusens (*Myotis dubenonoi*) ekologi. Fauna Flora (Stockh) 78:237–242

Lundberg S, Persson L (1993) Optimal body size and resource density. J Theor Biol 164:163–180

Lutcavage ME, Lutz PL, Baier H (1987) Gas exchange in the loggerhead sea turtle, *Caretta*. J Exp Biol 131:365–372

Lutz BR (1930) The effect of low oxygen tension on the pulsations of the isolated holothurian cloaca. Biol Bull 58:74–84

Lutz PL, Schmidt-Nielsen K (1977) Effects of simulated altitude on blood gas transport in the pigeon. Respir Physiol 30:383–395

Lutz PL, Longmuir IS, Turtle JV, Schmidt-Nielsen K (1973) Dissociation curve of bird blood and effect of red cell oxygen consumption. Respir Physiol 17:269–275

Lutz PL, Longmuir IS, Schmidt-Nielsen K (1974) Oxygen affinity of bird blood. Respir Physiol 20:325–330

Lykkeboe G, Johansen K (1975) Functional properties of haemoglobins in the teleost, *Tilapia graphami*. J Comp Physiol 104:1–11

Lykkeboe G, Weber RE (1978) Changes in the respiratory properties of the blood in the carp, *Cyprinus carpio*, induced by diurnal variation in ambient oxygen tension. J Comp Physiol 128:117–125

Lynch JJ, King JE, Chamberlain TK, Smith AL (1947) Effects of aquatic weed infestations on the fish and wildlife of the Gulf States. US Dept Int Spec Sci Rep 39:1–71

Maarek JMI, Chang HK (1991) Pulsatile pulmonary microvascular pressure measured with vascular occlusion techniques. J Appl Physiol 70:998–1005

Maarek JMI, Grimbert F (1994) Segmental pulmonary vascular resistances during oleic acid lung injury in rabbits. Respir Physiol 98:179–191

MacArthur RA (1992) Gas bubble release by muskrats diving under ice: lost gas or potential oxygen pool? J Zool Lond 226:151–164

MacArthur RH, Levine R (1967) The limiting similarity, convergence, and divergence of coexisting species. Am Nat 101:377–395

Macchin J (1974) Water relations. In: Fretter VV, Peake J (eds) Pulmonates, vol V. Academic Press, London, pp 105–163

MacFadden BJ (1985) Patterns of phylogeny and rates of evolution in fossil horses: Hipparions from the Miocene and Pliocene of North America. Paleobiology 11:245–257

Mackay AR, Gelperin A (1972) Pharmacology and reflex responsiveness of the heart in the giant garden slug *Limax maximus*. Comp Biochem Physiol 43A:877–896

Macklem PT, Bouverot P, Scheid P (1979) Measurement of the distensibility of the parabronchi in duck lungs. Respir Physiol 38:23–35

Madan JJ, Wells MJ (1996) Why squid can breathe easy. Nature (Lond) 380:590

Madigan MT, Marrs BL (1997) Extremophiles. Sci Am 276:66–71

Maeda N, Shiga T (1994) Velocity of oxygen transfer and erythrocytes rheology. News Physiol Sci 9:22–27

Magnussen H, Willmer H, Scheid P (1976) Gas exchange in the air sacs: contribution to respiratory gas exchange in ducks. Respir Physiol 26:129–146

Maina JN (1981a) Morphometric study of the blood-gas barrier of the avian lung. J Anat 130

Maina JN (1981b) Morphometry of the passerine and non-passerine lungs. Zentralbl Veterinaermed Reihe C Anat Histol Embryol 9:366–367

Maina JN (1982a) A scanning electron microscopic study of the air and blood capillaries of the lung of the domestic fowl (*Gallus domesticus*). Experientia 35:614–616

Maina JN (1982b) A stereological analysis of the paleopulmo and neopulmo respiratory regions of the avian lung (*Streptopelia decaocto*). Int Res Comm Syst (Biochem) 10:328

Maina JN (1982c) Relationship between pulmonary morphometric characteristics and energetic requirements in four avian orders. J Anat 135:825

Maina JN (1982d) A morphometric comparison of the lungs of two species of birds of different exercise capacities. J Anat 134:604–605

Maina JN (1983) The bird lung: how is it made and how does it work? East African Natural History Bulletin, Nairobi, Kenya, July/August Issue, pp 51–54

Maina JN (1984) Morphometrics of the avian lung. 3. The structural design of the passerine lung. Respir Physiol 55:291–309

Maina JN (1985) A scanning and transmission electron microscopic study of the bat lung. J Zool (Lond) 205B:19–27

Maina JN (1986) The structural design of the bat lung. Myotis 23:71–77

Maina JN (1987a) The morphology of the lung of the African lungfish, *Protopterus aethiopicus*: a scanning electron microscopic study. Cell Tissue Res 250:191–196

Maina JN (1987b) The morphology and morphometry of the adult normal baboon lung, *Papio anubis*. J Anat 150:229–245

Maina JN (1987c) Morphometrics of the avian lung. 4. The structural design of the charadriiform lung. Respir Physiol 68:99–119

Maina JN (1988a) Scanning electron microscopic study of the spatial organization of the air- and blood conducting components of the avian lung. Anat Rec 222:145–153

Maina JN (1988b) The morphology and morphometry of the normal lung of the adult vervet monkey *Cercopithecus aethiops*. Am J Anat 183:258–267

Maina JN (1989a) Morphometrics of the avian lung. In: King AS, McLelland J (eds) Form and function in birds, vol 4. Academic Press, London, pp 307–368

Maina JN (1989b) A scanning and transmission electron microscopic study of the tracheal airsac system in a grasshopper (*Chrotogonus senegalensis*, Kraus) – (Orthoptera: Acrididae: Pygomorphinae). Anat Rec 223:393–405

Maina JN (1989c) The morphology of the lung of a tropical terrestrial slug, *Trichotoxon copleyi* (Mollusca: Gastropoda; Pulmonata): a scanning and transmission electron microscopic study. J Zool (Lond) 217:335–366

Maina JN (1989d) The morphology of the lung of the East African tree frog *Chiromantis petersi* with observations on the skin and the buccal cavity as secondary gas exchange organs: a TEM and SEM study. J Anat 165:29–43

Maina JN (1989e) The morphology of the lung of the black mamba *Dendroaspis polylepis* (Reptilia: Ophidia: Elapidae): a scanning and transmission electron microscopic study. J Anat 167:31–46

Maina JN (1990a) A study of the morphology of of the gills of an extreme alkalinity and hyperosmotic adapted teleost *Oreochromis alcalicus grahami* (Boulenger) with particular emphasis on the ultrastructure of the chloride cells and their modifications with water dilution. Anat Embryol 181:83–98

Maina JN (1990b) The morphology of the gills of the African fresh water crab *Potamon niloticus* (Ortmann-Crustacea-Brachyura-Potamonidae): a scanning and transmission electron microscopic study. J Zool (Lond) 221:499–515

Maina JN (1990c) The morphology and morphometry of the lung of the lesser bushbaby *Galago senegalensis*. J Anat 172:129–144

Maina JN (1991) A morphometric analysis of chloride cells in the gills of the teleosts *Oreochromis alcalicus* and *Oreochromis niloticus* and a description of presumptive urea excreting cells in *Oreochromis alcalicus*. J Anat 175:131–145

Maina JN (1993) Morphometries of the avian lung: the structural-functional correlations in the design of the lungs of birds. Comp Biochem Physiol 105:397–410

Maina JN (1994) Comparative pulmonary morphology and morphometry: The functional design of respiratory systems. In: Gilles R (ed) Advances in comparative and environmental physiology, vol 20. Springer, Berlin Heidelberg New York, pp 111–232

Maina JN (1996) Perspectives on the structure and function in birds. In: Rosskoff E (ed) Diseases of cage and aviary birds. Williams and Wilkins, Baltimore, pp 163–256

Maina JN (1998) The lungs of the volant vertebrates – birds and bats: how are they relatively structurally optimized for this elite mode of locomotion? In: Weibel ER, Taylor CR, Bolis L (eds) Principles of animal design: the optimization and symmorphosis debate. Cambridge University Press, London, pp 177–185

Maina JN, King AS (1982a) The thickness of the avian blood-gas barrier: qualitative and quantitative observations. J Anat 134:553–562

Maina JN, King AS (1982b) Morphometrics of the avian lung. 2. The wild mallard (*Anas platyrhynchos*) and the greylag goose (*Anser anser*). Respir Physiol 50:299–313

Maina JN, King AS (1984) The structural functional correlation in the design of the bat lung. A morphometric study. J Exp Biol 111:43–63

Maina JN, King AS (1987) A morphometric study of the lung of the Humboldt penguin (*Spheniscus humboldti*). Zentralbl Veterinaermed Reihe C Anat Histol Embryol 16:293–297

Maina JN, King AS (1989) The lung of the emu, *Dromaius novaehollandiae*: a microscopic and morphometric study. J Anat 163:67–74

Maina JN, Maloiy GMO (1985) The morphometry of the lung of the lungfish (*Protopterus aethiopicus*): its structural-functional correlations. Proc R Soc Lond 244B:399–420

Maina JN, Maloiy GMO (1986) The morphology of the respiratory organs of the African air-breathing catfish (*Clarias mossambicus*): a light, and electron microscopic study, with morphometric observations. J Zool (Lond) 209:421–445

Maina JN, Maloiy GMO (1988) A scanning and transmission electron microscopic study of the lung of a caecilian *Boulengerula taitanus*. J Zool (Lond) 215:739–751

Maina JN, Nicholson T (1982) The morphometric diffusing capacity of a bat *Epomophorus wahlbergi*. J Physiol (Lond) 325:36–37

Maina JN, Settle JG (1982) Allometric comparison of some morphometric parameters of avian and mammalian lungs. J Physiol (Lond) 330:28P

Maina JN, Howard CV, Scales L (1981) The determination of the length densities and size distribution of blood and and capillaries in the avian lung involving log normal fitting procedure. Stereol Yugosl 3:673–678

Maina JN, King AS, King DZ (1982a) A morphometric analysis of the lungs of a species of bat. Respir Physiol 50:1–11

Maina JN, Abdalla MA, King AS (1982b) Light microscopic morphometry of the lungs of 19 avian species. Acta Anat 112:264–270

Maina JN, Howard CV, Scales L (1982c) The length densities and size distributions of the air and blood capillaries of the paleopulmo and neopulmo regions of the avian lung. Acta Stereol 2:101–107

Maina JN, King AS, King DZ (1983) Lung volume-body weight correlation in birds and mammals. Zentralbl Veterinaermed Reihe C Anat Histol Embryol 11:362

Maina JN, King AS, Settle G (1989a) An allometric study of the pulmonary morphometric parameters in birds, with mammalian comparison. Philos Trans R Soc Lond 326B:1–57

Maina JN, Maloiy GMO, Warui CN, Njogu EK, Kokwaro ED (1989b) A scanning electron microscope study of the reptilian lungs: the savanna monitor lizard (*Varanus exanthematicus*) and the pancake tortoise (*Malacochersus tornieri*). Anat Rec 224:514–522

Maina JN, Thomas SP, Dallas DM (1991) A morphometric study of bats of different size: correlations between structure and function of the chiropteran lung. Philos Trans R Soc Lond 333B:31–50

429

Maina JN, Maloiy GMO, Makanya AN (1992) Morphology and morphometry of the lungs of two East African mole rats, *Tachyoryctes splendens* and *Heterocephalus glaber* (Mammalia, Rodentia). Zoomorphology 112:167–179

Maina JN, Kisia SM, Wood CM, Narahara AB, Bergman HL, Laurent P, Walsh PJ (1996a) A comparative allometric study of the morphometry of the gills of an alkalinity-adapted cichlid fish, *Oreochromis alcalicus grahami*, of Lake Magadi, Kenya. Int J Salt Lake Res 5:131–156

Maina JN, Wood CM, Narahara A, Bergman HL, Laurent P, Walsh P (1996b) Morphology of the swim (air) bladder of a cichlid teleost: *Oreochromis alcalicus grahami* (Trewavas, 1983), a fish adapted to a hyperosmotic, alkaline and hypoxic environment: a brief outline of the structure and function of the swim bladder. In: Dutta HH, Munshi JSD (eds) Fish morphology: horizon of new research. Science Publishers, Lebanon, New Hampshire, pp 179–192

Maina JN, Maloiy GMO, Wood CM (1998) Respiratory stratagems, mechanisms, and morphology of the "lung" up a tropical swampworm, *Alma emini* Mich. (Oligochaeta: Glossoscolecidae): A transmission and scanning electron microscopic study, with field and laboratory observations. J Zool Lond, in press

Maitland DP (1986) Crabs that breathe air with their legs – *Scopimera and Dotilla*. Nature (Lond) 319:493–495

Maitland DP (1987) A highly complex invertebrate lung: the gill chambers of the soldier crab *Mictyris longicarpus*. Naturwissenschaften 74:293–295

Maitland DP (1990a) Aerial respiration in the semiphore crab, *Holoecius cordiformis*, with or without branchial water. Comp Biochem Physiol 95A:267–274

Maitland DP (1990b) Carapace and branchial water circulation, and water-related behaviours in the semiphore crab *Heloecius cordiformis* (Decapoda: Brachyura: Ocypode). Mar Biol 105:275–286

Makanya AN, Maina JN (1994) Comparative morphology of the gastrointestinal tract of fruit and insect-eating bats. Afr J Ecol 32:158–168

Makanya AN, Mayhew TM, Maina JN (1995) Morphometry of the gastrointestinal system of insectivorous and frugivorous bats: analysis of an anisotropic tissue. J Anat 187:361–368

Malan A (1982) Respiration and acid-base state in hibernation. In: Layman CP, Willis JS, Malan A, Wang LCH (eds) Hibernation and torpor in mammals and birds. Academic Press, New York, pp 237–282

Mallat J, Paulsen C (1986) Gill ultrastructure of the Pacific hagfish *Eptatretus stouti*. Am J Anat 177:243–269

Maloney JE (1984) The development of the respiratory system in placental mammals. In: Seymour R (ed) Respiration and metabolism in embryonic vertebrates. Dr W Junk, Dordrecht, pp 57–109

Maloney JE, Rooholamini SA, Wexler L (1970) Pressure-diameter relations of small blood vessels in isolated dog lung. Microvasc Res 2:1–12

Maloney JE, Drian-Smith C, Takahashi V, Limpus CJ (1990) The environment for development of the embryonic loggerhead turtle (*Caretta caretta*) in Queensland. Copeia 1990:2

Malvin GM (1988) Microvascular regulation of cutaneous gas exchange in amphibians. Am Zool 28:999–1007

Malvin GM (1989) Gill structure and function: amphibian larvae. In: Wood SC (ed) Comparative pulmonary physiology: current concepts, vol 39. Lung biology in health and disease. Marcel Dekker, New York, pp 121–151

Malvin GM, Heisler N (1988) Blood flow patterns in the salamander, *Ambryostoma tigrinum*, before, during and after metamorphosis. J Exp Biol 137:53–74

Malvin GM, Hlastala MP (1986) Regulation of cutaneous gas exchange by environmental O_2 and CO_2 in the frog. Respir Physiol 65:99–111

Mandelbrot BB (1977) Form, chance, and dimension. Freeman, New York

Mandelbrot BB (1983) The fractal geometry of nature. Freeman, New York

Mangum CP (1963) Oxygen consumption in different species of polchaete worms. Comp Biochem Physiol 10:335–349

Mangum CP (1964) Activity patterns in metabolism and ecology of polychaetes. Comp Biochem Physiol 11:239–250

Mangum CP (1976a) Primitive adaptations. In: Newell RC (ed) Adaptations to the environment: essays on the physiology of marine animals. Butterworth, London, pp 191–278

Mangum CP (1976b) The oxygenation of haemoglobin in lugworms. Physiol Zool 49:85–99

Mangum CP (1980a) Respiratory function of the hemocyanins. Am Zool 20:19–38

Mangum CP (1980b) Distribution of the respiratory pigments and the role of anaerobic metabolism in the lamellibranch molluscs. In: Gilles R (ed) Animals and environmental fitness. Pergamon, Oxford, pp 171–184

Mangum CP (1982a) The functions of gills in several groups of invertebrate animals. In: Houlihan DF, Rankin JC, Shuttleworth TJ (eds) Gills. Cambridge University Press, Cambridge, pp 77–97

Mangum CP (1982b) On the relationship between P_{50} and the mode of gas exchange in tropical crustaceans. Pac Sci 36:403–410

Mangum CP (1983) Oxygen transport in blood. In: Mantel LH (ed) The physiology of Crustacea, vol 5. Academic Press, London, pp 373–429

Mangum CP (1985) Oxygen transport in invertebrates. Am J Physiol 248:R505–R514

Mangum CP (1990) Gas transport in the blood. In: Gilbert DL, Adelman WJ, Arnold JM (eds) Squid as experimental animals. Plenum Press, New York, pp 443–468

Mangum CP (1992) Physiological adaptation of crustacean hemocyanins: an extended investigation of the blue crab, Callinectes sapidus. In: Wood SC, Weber RE, Hargens AR, Millard RW (eds) Physiological adaptations in vertebrates: respiration, circulation, and metabolism. Marcel Dekker, New York, pp 279–293

Mangum CP (1994) Multiple sites of gas exchange. Am Zool 34:184–193

Mangum CP, Lykkeboe G (1979) The influence of inorganic ions and pH on oxygenation properties of the blood in the gastropod mollusc Busycon canaliculatum. J Exp Zool 207:417–430

Mangum CP, Lykkeboe G, Johansen K (1975) Oxygen uptake and the role of haemoglobin in the East African swampworm Alma emini. Comp Biochem Physiol 52A:477–482

Manier G, Moinard J, Téchoueyres P, Varène N, Guénard H (1991) Pulmonary diffusion limitation after strenous exercise. Respir Physiol 83:143–154

Mann KH (1982) Ecology of coastal waters: a system approach. Blackwell, Oxford

Manning AM, Trotman CNA, Tate WP (1990) Evolution of a polymeric globin in the brine shrimp Artemia. Science 248:653–656

Manwell C (1958) The oxygen respiratory pigment equilibrium of the hemocyanin and myoglobin of the amphineuron mollusc Cryptochiton stelleri. J Comp Physiol 52:341–353

Manwell C (1960) Histological specifity of respiratory pigments. I. Comparisons of the coelom and muscle haemoglobins of the polychaete worm Travisia pupa and the echiuroid worm Arhynchite pugettensis. Comp Biochem Physiol 1:267–276

Manwell C (1963) The chemistry and biology of haemoglobin in some marine clams. I. Distribution of the pigment and the properties of the oxygen equilibrium. Comp Biochem Physiol 8:209–218

Marcus H (1928) Lungenstudien III und IV. Gegenbaurs Morphol Jahrb 59:287–342

Marcus H (1937) Lungen. In: Bolk L, Goppert E, Kallius E, Lubosch W (eds) Handbuch der vergleichenden Anatomie der Wirbeltiere, III. Urban and Schwarzenberg, Berlin, pp 909–1018

Marcus E, Marcus C (1960) On Siphonaria hispida. Bull Fac Filos Ciencias e letras, Univ de Sao Paulo 23:107–140

Marden JH (1987) Maximum lift production during take-off in flying animals. J Exp Biol 130:235–258

Marden JH, Kramer MG (1994) Surface-skimming stoneflies: a possible intermediate stage in insect flight evolution. Science 266:427–430

Mardsen ID (1976) Effect of temperature on the microdistribution of the isopod Sphaeroma rugicauda from a salt marsh habitat. Mar Biol 38:117–128

Margaria R (1976) Biomechanics and energetics of muscular exercise. Clarendon Press, Oxford

Margaria R, Gualtierotti T, Spinelli D (1958) Protection against acceleration forces in animals by immersion in water. J Aviat Med 29:433–437

Margaria R, Caproresi E, Aghemo P, Sassi G (1972) The effect of O_2 breathing on maximal aerobic power. Pfluegers Archir Gesamte Physiol Menschen Tiere 336:225–235

Margulis L (1970) Origin of eukaryotic cells. Yale University Press, New Haven

Margulis L (1979) Symbiosis and evolution. In: Life – origin and evolution: readings from Scientific American, WH Freeman, San Francisco, pp 101–110

Margulis L (1981) Symbiosis in cell evolution. WH Freeman, New York

Marin-Girón F, Cedres T, Otero A (1975) Contribución al estudio microscópico de la histología de los sacos aéreos. Bol R Soc Española Hist Nat (Biol) 73:275–296

Marsh BA, Branch GM (1979) Circadian and circatidal rhythms of oxygen consumption in sandy-beach isopod *Tylos granulatus* Kraus. J Exp Mar Biol Ecol 37:77–89

Marshall C, Schultze HP (1992) Relative importance of molecular, neontological, and paleontological data in understanding the biology of the vertebrate invasion of land. J Mol Evol 35:93–101

Marshall DJ, McQuaid CD (1992) Comparative aerial metabolism and water relations of the intertidal limpets, *Patella granularis* L. (Mollusca: Prosobranchia) and *Siphonaria oculus* Kr. (Mollusca: Pulmonata. Physiol Zool 65:1040–1056

Marshall NB (1960) Swim bladder structure of deep sea fishes in relation to their systematics and biology. Discovery Rep 31:1–122

Martin AP, Palumbi SR (1993) Body size, metabolic rate, generation time, and the molecular clock. Proc Natl Acad Sci USA 90:4097–4091

Martin KM, Hutchison VH (1979) Ventilatory activity in *Amphiuma tridactylum* and *Siren lacertina* (Amphibia, Caudata). J Herpatol 13:427–434

Martin TE (1996) Fitness cost of resource overlap among coexisting bird species. Nature (Lond) 380:338–340

Masahiko K, Paul JL, Thurlbeck WM (1984) The effect of age on lung structure in male BALB/cNNia. Am J Anat 170:1–21

Mason DK, Collins AE, Watkins KL (1983) Exercise induced pulmonary haemorrhage in horses. In: Snow DH, Persson SGB, Rose RJ (eds) Equine exercise physiology. Granta Editions, Cambridge, pp 57–63

Mass JA (1939) Über die Atmung von *Helix pomatia*, Z Vergl Physiol 26:605–610

Massabuau JC, Dejours P, Sakakibara Y (1984) Ventilatory CO_2 drive in the crayfish: influence of oxygen consumption level and water oxygenation. J Comp Physiol 154B:65–72

Massabuau JC, Burtin B, Wheatly M (1991) How is O_2 consumption maintained independent of ambient oxygen in mussel *Anodonta cygnea*? Respir Physiol 83:103–114

Matalon S, Egan EA (1981) Effects of 100% oxygen breathing on permeability of alveolar epithelium to solute. J Appl Physiol 50:859–863

Mathieu-Costello O (1990) Histology of flight: tissue and muscle gas exchange. In: Sutton JR, Coates G, Remmers JE (eds) Hypoxia: the adaptations. BC Dekker, Toronto, pp 13–19

Mathieu-Costello O, Szewczak JM, Logemann RB, Agey PJ (1992) Geometry of blood-tissue exchange in bat flight muscle compared with bat hindlimb and rat soleus muscle. Am J Physiol 262:R955–R965

Matsui T, Abe Y (1986) Evolution of an impact-induced atmosphere and magma ocean on the accreting Earth. Nature (Lond) 319:303–305

Matsumura H, Setoguti T (1984) Electron microscopic studies of the lung of the salamander, *Hynobius nebulosus*. I. A scanning and transmission microscopic observation. Okajimas Folia Anat Jpn 61:15–25

Matthews WH, Balzer RH, Shelburne JD, Pratt PC, Kylstra JA (1978) Steady-state gas exchange in normothermic, anaethetized, liquid-ventilated dogs. Undersea Biomed Res 5:341–354

Maury W, Potts BJ, Rabson AB (1989) HIV 1 infection of first trimester and term human placental tissue: a possible mode of maternal-foetal transmission. J Infect Dis 160:15–23

Maxwell MH, Robertson GW, Spence S (1986a) Studies on an ascitic syndrome in young broilers. I. Haematology and pathology. Avian Pathol 1986:15:511–524

Maxwell MH, Robertson GW, Spence S (1986b) Studies on an ascitic syndrome in young broilers. II. Ultrastructure. Avian Pathol 1986:15:525–538

May EB (1973) Extensive oxygen depletion in Mobile Bay, Alabama. Limnol Oceanogr 18:353–366

May RM (1981) The evolution of cooperation. Nature (Lond) 292:291–292

May RM (1988) How many species are there on Earth? Science 241:1441–1449

May RM (1990) How many species? Philos Trans R Soc Lond 330B:293–316

May RM (1992) How many species inhabit Earth? Sci Am October Issue:18–24

Mayhew TM (1991) Scaling placental oxygen diffusion to birthweight: studies on placentae from low- and high-altitude pregnancies. J Anat 175:187–194

Mayhew TM (1992) The structural basis of oxygen diffusion in the human placenta. In: Egginton S, Ross HF (eds) Oxygen transport in biological tissue systems: modelling pathways from environment to cell. Cambridge University Press, Cambridge, pp 79–101

Mayhew TM, Simpson RA (1994) Quantitative evidence for the spatial dispersal of trophoblast nuclei in human placental villi during gestation. Placenta 15:837–844

Mayhew TM, Wadrop E (1994) Placental morphogenesis and the star volumes of villous trees and intervillous pores. Placenta 15:209–217

Mayhew TM, Joy CF, Haas JD (1984) Structure-function correlation in the human placenta: the morphometric diffusing capacity for oxygen at term. J Anat 139:691–708

Mayhew TM, Jackson MR, Haas JD (1986) Microscopical morphology of the human placenta and its effects on oxygen diffusion: a morphometric model. Placenta 7:121–131

Mayhew TM, Jackson MR, Haas JD (1990) Oxygen diffusive conductances of human placentae from term pregnancies at low and high altitudes. Placenta 11:493–503

Maynard-Smith J (1968) Mathematical ideas in biology. Cambridge University Press, Cambridge

Maynard-Smith J (1978) Optimization theory in evolution. Annu Rev Ecol Syst 9:31–56

Maynard-Smith J (1996) The games lizards play. Nature (Lond) 380:198–199

Maynard-Smith J, Szathmary E (1996) On the likelihood of habitable worlds. Nature (Lond) 384:107

Mayr E (1942) Systematics and the origin of species from the viewpoint of a zoologist. Columbia University Press, New York

Mayr E (1960) The emergence of evolutionary novelties. In: Tax S (ed) Evolution after Darwin, vol 2. University of Chicago Press, Chicago, pp 349–380

McClanahan LL, Rodolfo R, Shoemaker VH (1994) Frogs and toads in deserts. Sci Am 273:82–88

McClary A (1964) Surface inspiration and cilliary feeding in *Pomacea paludosa* (Prosobranchia: Mesogastropoda: Ampullaridae). Malacologia 2:87–104

McCord JM (1988) Free radicals and myocardial ischaemia: overview and outlook. Free Radic Biol Med 4:9–14

McCutcheon FH (1936) Haemoglobin function during the life history of the bullfrog. J Cellular Comp Physiol 8:63–81

McCutcheon FH (1954) Phylogenetic aspects of respiratory function. Evolution 8:181–191

McCutcheon FH (1964) Organ systems in adaptation: the respiratory system. In: Dill DB, Adolph EF, Wilber CG (eds) Handbook of physiology, sect 4, adaptation to the environment. American Physiological Society, Washington, DC, pp 167–191

McDonald DG, McMahon BR, Wood CM (1977) Patterns of heart and scaphognathite activity in the crab *Cancer magister*. J Exp Zool 202:33–44

McDonald DG, Caudek V, Ellis R (1991) Gill design in fresh water fishes: interrelationships among gas exchange, ion regulation, and acid base regulation. Physiol Zool 64:103–123

McDougall JDB, McCobe M (1967) Diffusion coeffcent of oxygen through tissues. Nature (Lond) 215:1173–1174

McElhinny MW, Taylor SR, Stevenson DJ (1978) Limits to the expansion of the Earth, Moon, Mars and Mercury and to changes in the gravitational constant. Nature (Lond) 271:316–321

McElroy MB (1983) Marine biological controls on atmospheric CO_2 and climate. Nature (Lond) 302:328–329

McFall-Ngai MJ, Ruby EG (1991) Symbiont recognition and subsequent morphology as early events in animal-bacterial mutualism. Science 254:1491–1494

McFarland RA, Evans JN (1939) Alterations in dark adaptation under reduced oxygen tensions. Am J Physiol 127:37–50

McGhee GR (1989) Frasian-Famennian extinction event. In: Briggs DEG, Crowther PR (eds) Paleobiology – a synthesis. Blackwell, Oxford, pp 97–176

McGinitie GE (1939) The method of feeding in *Chaetopterus*. Biol Bull 77:115–118

McGrath MW, Thomson ML (1959) The use of helium and sulphur hexafluoride for assessing diffusive mixing in the lung. J Physiol (Lond) 148:72–73

McGregor LK, Daniels CB, Nicholas TE (1993) Lung ultrastructure and surfactant-like system of the central Australian netted dragon, *Ctenophoruos nuchalis*. Copeia 1993:326–333

McKay C (1997) Organic synthesis in experimental impact shocks. Science 276:390–392

McKay CP, Pollack JB, Courtin R (1991) The greenhouse and antigreenhouse effects on Titan. Science 253:1118–1121

McKay DS, Gibson EK, Thomas-Kerpta KL, Vali H, Romanek CS et al. (1996) Search for pas tlife on Mars: possible relic biogenic activity in martian meteorite ALH 84001. Science 273:924–930

McLaughlin PA (1983) Internal anatomy. In: Mantel LH (ed) The biology of Crustacea, vol 5. Academic Press, New York, pp 1–53

McLean DM (1978) Land floras: the major late Proterozoic atmospheric carbon dioxide/oxygen control. Science 200:1060–1062

McLelland J (1989) Larynx and trachea. In: King AS, McLelland J (eds) Form and function in birds, vol 4. Academic Press, London, pp 69–103

McMahon BR (1969) A functional anlysis of the aquatic and aerial respiratory movements of an African lungfish, *Protopterus aethiopicus*, with refrence to the evolution of the lung ventilation mechanism in vertebrates. J Exp Biol 51:407–430

McMahon BR (1988) Physiological responses to periodic emergency in intertidal molluscs. Am Zool 28:97–114

McMahon BR, Burggren WW (1979) Respiration and adaptation to the terrestrial habitat in the land hermit crab, *Coenobita clypeatus*. J Exp Biol 79:265–281

McMahon BR, Burggren WW (1987) Respiratory physiology of intestinal air breathing in the teleost fish, *Misgurnus anguillicaudatus*. J Exp Biol 133:371–393

McMahon BR, Burggren WW (1988) Respiration. In: Burggren WW, McMahon BR (eds) Biology of the land crabs. Cambridge University Press, Cambridge, pp 249–297

McMahon BR, Wilkens JL (1977) Periodic respiratory and circulatory performance in the red rock crab, *Cancer productus*. J Exp Zool 202:363–374

McMahon BR, Wilkens JL (1983) Ventilation, perfusion and oxygen uptake. In: Mantel LH (ed) The physiology of Crustacea, vol 5. Academic Press, London, pp 289–372

McMahon BR, Wilkes PRH (1983) Emergence response and aerial ventilation in normoxic and hypoxic crayfish, *Orconectes rusticus*. Physiol Zool 51:133–141

McMahon RF (1983) Physiological ecology of fresh water pulmonates. In: Russell-Hunter WD (ed) The Mollusca, vol 6. Academic Press, Orlando, pp 359–430

McMahon RF (1985) Functions and functioning of crustacean hemocyanin. In: Lamy J, Truchot JP, Gilles R (eds) Respiratory pigments in animals. Springer, Berlin Heidelberg New York, pp 35–58

McMahon TA, Bonner JT (1983) On size and life. Scientific American Library, New York

McNab B (1966) The metabolism of fossorial rodents: a study of convergence. Ecology 47:712–733

McNamara JM, Houston AI (1996) State-dependent life histories. Nature (Lond) 380:215–221

McShea DW (1991) Complexity and evolution: what everbody knows. Biol Philos 6:303–324

Mead J, Turner JM, Macklem PT, Little JB (1967) Significance of the relationship between lung recoil and maximum expiratory flow. J Appl Physiol 22:95–108

Meban C (1977) Ultrastructure of the respiratory epithelium in the lungs of the newt *Triturus cristatus*. Acta Zool Stockh 58:151–167

Meban C (1978a) Functional anatomy of the lungs of the green turtle, *Lacerta viridis*. J Anat 125:421–436

Meban C (1978b) The respiratory epithelium in the lungs of the slow-worm, *Anguilis fragilis*. Cell Tissue Res 190:337–354

Meban C (1979) An electron microscopy study of the respiratory epithelium in the lungs of the fire salamander (*Salamandra salamandra*). J Anat 128.215–221

Meban C (1980) Thicknesses of the air-blood barriers in vertebrate lungs. J Anat 131:299–307

Meduna JL (1950) Carbon dioxide therapy. CC Thomas, Springfield, Illinois

Meilin S, Rogatsky GG, Thoms R, Zarchin N et al. (1996) Effect of carbon monoxide on brain may be mediated by nitric oxide. J Appl Physiol 81:1078–1083

Melchior FM, Srinivasan RS, Charles JB (1992) Mathematical modeling of human cardiovascular system for simulation of orthostatic response. Am J Physiol 262:H1920–H1933

Melmon KL, Cline MJ, Hughes T, Nies AS (1968) Kinins; possible mediators of neonatal circulatory changes in man. J Clin Invest 47:1279–1302

Melsom MN, Johansen JK, Flatebø T, Müller C (1997) Distribution of pulmonary ventilation and perfusion measured simultaneously in awake goats. Acta Physiol Scand 159:199–220

Meng H, Bentley TB, Pittman RN (1992) Oxygen diffusion in hamster striated muscle: comparison of in vitro and near in vitro conditions. Am J Physiol 263:H35–H39

Menon JG, Arp AJ (1992a) Morphological adaptations of the respiratory hind gut of a marine echiurian worm. J Morphol 214:131–138

Menon JG, Arp AJ (1992b) Symbiotic bacteria may prevent sulfide poisoning of the body wall of *Urechis caupo*. Am Zool 32:59A

Meredith JL (ed) (1985) Hydrothermal vents of the eastern Pacific: An overview. Bull Biol Soc Washington, December Issue

Merrill EW (1969) Rheology of blood. Physiol Rev 19:863–888

Mertens R (1960) The world of amphibians and reptiles. McGraw-Hill, New York

Mertens S, Noll T, Spahr R, Krützfeldt A, Piper HM (1990) Energetic response of coronary endothelial cells to hypoxia. Am J Physiol 258:H689–H694

Metcalfe J, Stock MK (1993) Oxygen uptake in the chorioallantoic membrane, avian homologue of the mammalian placenta. Placenta 14:605–614

Metcalfe JD, Butler PJ (1984) On the nervous regulation of gill blood flow in the dogfish (*Scyliorhinus canicula*). J Exp Biol 113:253–267

Metcalfe JD, Butler PJ (1986) The functional anatomy of the gills of the dogfish (*Scyliorhinus canicula*). J Zool (Lond) 208:519–630

Metcalfe J, Meschia G, Hellegers A, Prystowsky H, Huckabee W, Barron DH (1962) Observations on the placental exchange of the respiratory gases in pregnant ewe at high altitude. Q J Exp Physiol 47:74–92

Metcalfe J, Bartels H, Moll W (1967) Gas exchange in the pregnant uterus. Physiol Rev 47:782–838

Metcalfe J, Bissonnette JM, Bowles RE, Matsumoto JA, Dunham SJ (1979) Hen's eggs with retarded gas exchange. I. Chorioallantoic capillary growth. Respir Physiol 36:97–108

Metcalfe R (1967) The oxygen supply of the foetus. In: Reuck AVS, Porter R (eds) Development of the lung. Churchill, London, pp 37–63

Meyer A, Dolven SI (1992) Molecules, fossils, and the origin of tetrapods. J Mol Evol 35:102–113

Meyer M, Worth H, Scheid P (1976) Gas-blood CO_2 equilibration in parabronchial lungs of birds. J App. Physiol 41:302–309

Meyer WV (1988) The role of form and function in the collegiate biology curriculum. Am Zool 28:619–664

Michels DB, West JB (1978) Distribution of pulmonary ventilation and perfusion during short periods of weightlessness. J Appl Physiol 45:987–998

Michels MD, Friedman PJ, West JB (1979) Radiographic comparison of human lung shape during normal gravity and weightlessness. J Appl Physiol 47:851–885

Midtgärd U (1983) Scaling of the brain and eye cooling system in birds: a morphometric analysis of the rete ophthalmicum. J Exp Zool 225:197–207

Midtgärd U (1984) Blood vessels and the occurrence of arteriovenous anastomoses in the cephalic heat loss areas of the mallards, *Anas platyhynchos* (Aves). Zoomorphology 104:323–335

Milani A (1894) Beiträge zur Kenntnis der Reptilienlunge. II. Zool Jahrb Abt Anat Ontog Tiere 7:545–592

Milburn TR, Beadle LC (1960) The determination of total carbon dioxide. J Exp Biol 37:444–460

Milhorn HT, Benton R, Ross R, Guyton AC (1965) A mathematical model of the human respiratory control system. Biophys J 5:27–46

Milic-Emili J (1991) Work of breathing. In: West JB, Crystal RG (eds) The lung: scientific foundations. Raven Press, New York, pp 1065–1075

Milic-Emili J, Henderson JAM, Kaneko K (1966) Regional distribution of inspired gas in the lung. J Appl Physiol 21:749–759

Miller AH (1949) Some ecological and morphologic considerations in the evolution of higher taxonomic categories. In: Mayr E, Schuz E (eds) Ornithologie als biologische Wissenschaft. Carl Winter Universitätsverlag, Heidelberg, pp 84–88

Miller K, Camilliere JJ (1981) Physical training improves swimming performance of the African clawed frog, *Xenopus laevis*. Herpetologica 37:1–10

Miller MA (1948) Seasonal trends in burrowing of pocket gophers (*Thomomys*). J Mammal 29:38–44

Miller PL (1960) Respiration in the desert locust. III. Ventilation and spiracles during flight. J Exp Biol 37:264–278

Miller PL (1966) The supply of oxygen to the active flight muscles of some large beetles. J Exp Biol 45:285–304

Miller PL (1974) Respiration – aerial gas transport. In: Rockstein M (ed) The physiology of insects, 2nd edn. Academic Press, New York, pp 346–402

Miller SL, Orgel LE (1974) The origins of life on Earth. Prentice-Hall, Englewood Cliffs, New Jersey

Mills RM, Brinster RL (1967) Oxygen consumption of preimplantation mouse embryos. Exp Cell Res 47:337–344

Milner AR (1988) The relationships and origins of living amphibians. In: Benton MJ (ed) The phylogeny and classification of tetrapods, vol 1, Amphibians, reptiles and birds. Clarendon Press, Oxford, pp 59–102

Milner WR (1980) Pulmonary circulation. In: Mountcastle VB (ed) Medical physiology, vol 2. Mosby, St Louis, pp 1108–1117

Milnor WR (1982) Hemodynamics. Williams and Williams, Baltimore

Milsom WK (1984) The interrelationship between pulmonary mechanics and spontaneous breathing pattern in the Tokay lizard *Gekko gecko*. J Exp Biol 113:203–214

Milsom WK (1988) Control of arrythmic breathing in aerial breathers. Can J Zool 66:99–108

Milsom WK (1989) Mechanisms of ventilation in lower vertebrates: adaptations to respiratory and non-respiratory constraints. Can J Zool 67:2943–2963

Milsom WK (1990) Control and co-ordination of gas exchange in air breathers. In: Boutilier RG (ed) Advances in comparative environmnetal physiology, vol 6. Vertebrate gas exchange from environment to cell. Springer, Berlin Heidelberg New York, pp 374–400

Milsom WK (1991) Intermittent breathing in vertebrates. Annu Rev Physiol 53:87–105

Milsom WK, Johansen K (1975) The effect of buoyancy-induced lung volume changes on respiratory frequency in a chelonian (*Caretta caretta*). J Comp Physiol 98:157–160

Milsom WK, Jones DR (1979) The role of pulmonary afferent information and hypercapnia in the control of the breathing pattern in chelonia. Respir Physiol 37:101–107

Milsom WK, Jones DR (1985) Characteristics of mechanoreceptors in the air breathing organ of the holostean fish, *Amia calva*. J Exp Biol 117:389–399

Milton P (1971) Oxygen consumption and osmoregulation in the shanny, *Blennius pholis*. J Mar Biol Assoc UK 51:247–265

Minkoff EE (1983) Evolutionary biology. Addison-Wesley, Reading, Massachusetts

Misiek L, Szaski H (1978) Dimensions of cells in some tissues of six amphibian species. Acta Biol Cracov Ser Zool 21:127–136

Mitchell E, Reeves RR (1981) Catch history and cumulative catch estimates of initial population size of cetaceans in the eastern Canadian Arctic. Rep Int Whale Comm 31:645–682

Mitchell GS, Gleeson TT, Bennett AF (1981) Pulmonary oxygen transport during activity in lizards. Respir Physiol 43:365–375

Mitchell HA (1964) Investigation of the cave atmosphere of a Mexican bat colony. J Mammal 45:568–577

Mix AC (1989) Influence of productivity variations on long-term atmospheric CO_2. Nature (Lond) 337:541–544

Moalli R, Meyers RS, Jackson DC, Millard RW (1980) Skin circulation of the frog, *Rana catesbeiana*: distribution and dynamics. Respir Physiol 40:137–148

Modell JH, Gollan F, Giammona ST, Parker D (1970) Effect of fluorocarbon liquid on surface tension properties of pulmonary surfactant. Chest 57:263–265

Moessinger AC, Harding R, Adamson TM, Singh M, Kiu GT (1990) Role of lung fluid volume in growth and maturation of the foetal sheep lung. J Clin Invest 86:1270–1277

Mojzsis SJ, Arrhenius G, McKeegan KD, Harrison TM, Nutman AP, Friends CRL (1996) Evidence of life on Earth before 3800 million years ago. Nature (Lond) 384:55–59

Moll W (1966) The diffusion coefficient of haemoglobin. Respir Physiol 1:357–365

Mommsen TP, Ballantyne J, MacDonald D, Gosline J, Hochachka PW (1981) Analogous of red and white muscle in squid mantle. Proc Natl Acad Sci USA 78:3274–3278

Moncada S, Palmer RMJ, Higgs EA (1991) Nitric oxide: physiology, pathophysiology, and pharmacology. Pharmacol Rev 43:109–142

Monge MC, Monge CC (1968) Adaptation to high altitude. In: Hafez ESE (ed) Adaptation of domestic animals. Lea and Febiger, Philadelphia, pp 194–201

Moore JA (1990a) Science as a way of knowing form VII: a conceptual framework for biology part III. Am Zool 30:1–123

Moore JA (1990b) The ability to live on dry land, rather than in water, required major adjustments in structure and physiology. Am Zool 30:847–849

Moore LD, Jahnigen D, Rounds SS, Reeves JT, Grover RF (1982) Maternal hyperventilation helps preserve arterial oxygenation during high altitude pregnancy. J Appl Physiol 52:690–694

Moore SJ (1976) Some spider organs as seen by the scanning electron microscope, with special reference to the booklung. Bull Br Arachnol Soc 3:177–187

Morgan N (1995) Chemistry in action: the molecules of everyday life. Oxford University Press, New York

Morin FC, Egan EA (1992) Pulmonary hemodynamics in fetal lambs duuring development at normal and increased oxygen tension. J Appl Physiol 73:213–218

Moritz AR (1944) Chemical methods for the determination of drowning. Physiol Rev 24:70–88

Morkin E, Collins JA, Goldman HS, Fishman P (1965) Pattern of blood flow in the pulmonary veins of the dog. J Appl Physiol 20:1118–1128

Morony JJ, Bock WJ, Farrand J (1975) Reference list of the birds of the world. Department of Ornithology, American Museum of Natural History, New York

Morris S (1991) Respiratory gas exchange and transport in crustaceans: ecological determinats. Mem Queense Mus 31:241–261

Morris S, Bridges CR (1986) Novel non-lactate cofactors of hemocyanin oxygen affinity in crustacean. In: Linzen B (ed) Invertebrate oxygen carriers. Springer Berlin Heidelberg New York, pp 353–356

Morris S, Bridges CR (1994) Properties of respiratory pigments in bimodal breathing animals: air- and water-breathing by fish and crustaceans. Am Zool 34:216–228

Morris S, Greenaway P (1990) Adaptations to a terrestrial existence in the robber crab, *Birgus latro* L.V. Preliminary investigations of carbonic anhydrase activities. J comp Physiol 160B:217–221

Morris S, Taylor AC (1985) The respiratory response of the intertidal prawn *Palaemon elegans* (Rathke) to hypoxia and hyperoxia. Comp Biochem Physiol 81A:633–639

Morrison DW (1980) Efficiency of food utilization by fruit bats. Oecologia 45:270–273

Morrison PR, Ryser FA, Dawe AR (1959) Studies on the physiology of the masked shrew, *Sorex cinereus*. Physiol Zool 32:256–271

Mortola JP (1987) Dynamics of breathing in newborn mammals. Physiol Rev 67:187–243

Morton JE (1979) Molluscs. Hutchison, London

Moshiri GA, Goldman CR, Godshalk GL, Mull DR (1970) The effects of variations in oxygen tension on certain aspects of respiratory metabolism in *Pacifastacus leniusculus* (Dana) (Crustacea: Decapoda). Physiol Zool 43:23–29

Moss ML (1962) The functional matrix. In: Kraus B, Reidel R (eds) Vistas in orthodontics. Lea and Febiger, Philadelphia, pp 85–98

Mossiman HW (1987) Vertebrate foetal membranes. Rutgers University Press, New Brunswick, New Jersey

Muir BS, Buckley RM (1967) Gill ventilation in *Remora remora*. Copeia 1967:581–586

Muir BS, Kendall JI (1968) Structural modifications in the gills of tunas and some other oceanic fishes. Copeia 1968:388–398

Mukai H, Koike I (1984) Behaviour and respiration of burrowing shrimps *Upogebia major* (de Haan) and *Callianassa japonica* (de Haan). J Crust Biol 4:191–200

Müller W (1950) Die Mündung des Luftganges beim Messerfisch und ihre konstruktive Gestaltung. Zool Anz Ergänz 145:635–642

Munro HN, Downie ED (1964) Relationship of liver composition to intensity of protein metabolism in different mammals. Nature (Lond) 203:603–604

Munshi JSD (1968) The accessory respiratory organs of *Anabis testudineus* (Bloch) (Anabantidae, Pisces). Proc Linn Soc Lond 179:107–126

Munshi JSD (1976) Gross and fine structure of the respiratory organs of air-breathing fishes. In: Hughes GM (ed) Respiration of amphibious vertebrates. Academic Press, London, pp 73–102

Munshi JSD, Hughes GM (1986) Scanning electron microscopy of the respiratory organs of juvenile and adult climbing perch, *Anabas testudineus.* Jpn J Ichthyol 33:39–45

Munshi JSD, Hughes GM (1991) Structure of the respiratory islets of accessory respiratory organs and their relationship with the gills in the climbing perch. *Anabas testudineus* (Teleostei, Perciformes). J Morphol 209:241–256

Munshi JSD, Hughes GM (1992) Air breathing fishes of India: their structure, function and life history. AA Balkema Uitgevers BV, Rotterdam

Munshi JSD, Singh BN (1968) On the respiratory organs of *Aphipnous cuchia* (Ham. Buch). J Morphol 124:423–444

Munshi JSD, Srivastava MP (1988) Natural history of fishes and systematics of fresh water fishes of India. Narendra Publishing House, New Delhi

Munshi JSD, Olson KR, Ojha J, Ghosh TK (1986a) Morphology and vascular anatomy of the accessory respiratory organs of air-breathing climbing perch, *Anabas testudineus* (Bloch). Am J Anat 176:321–331

Munshi JSD, Weibel ER, Gehr P, Hughes GM (1986b) Structure of the respiratory air sac of *Heteropneustes fossilis* (Bloch) (Heteropneustidae, Pisces) – an electron microscope study. Proc Indian Nat Sci Acad 52B:703–713

Munshi JSD, Hughes GM, Gehr P, Weibel ER (1989) Structure of the air-breathing organs of the swamp mud eel, *Monopterus cuchia.* Jpn J Ichthyol 35:453–465

Munshi JSD, Olson KR, Ghosh TK(1990) Vasculature of the head and respiratory organs in an obligate air-breathing fish, the swamp eel *Monopterus* (=*Amphipnous*) *cuchia.* J Morphol 203:181–201

Munshi JSD, Roy PK, Ghosh TK, Olson KR (1994) Cephalic circulation in the air-breathing snakehead fish, *Channa punctata, C. gashua,* and *C. marulius* (Ophiocephalidae, Ophiocephaliformes). Anat Rec 238:77–91

Muratori RA, Falugi C, Colosi R (1976) Osservazioni su alcuni aspetti dell'anatomia del *Telmatobius culeus* (Gorman 1875) visti come addattamento al particolare ambiente del lago Titicaca. Atti Accad Naz Lincei 61:508–519

Murdock GR, Currey JD (1978) Strength and design of the two ecologically distinct barnacles, *Balanus balanus* and *Semibalanus balanoides.* Biol Bull 155:169–192

Murray JW, Barber RT, Roman MR, Bacon MD, Feely RA (1994) Physical and biological controls on carbon cycling in the equatorial Pacific Ocean. Science 266:58–65

Murry CD (1926) The physiological principle of minimum work. I. The vascular system and the cost of blood volume. Proc Natl Acad Sic USA 12:207–214

Musacchia XK, Volkert WA (1971) Blood gases in hibernating and active ground squirrels: HbO_2 affinity at 6 and 38 °C. Am J Physiol 221:128–130

Mutsaddi KB, Bal DV (1969) Some observations on habits and habitat of *Boeophthalmus dussumierei* (Cuv. and Val.). J Univ Bombay 65:33–41

Myers AC (1972) Tube worm sediment relationships of *Diopatra cuprea* (Polychaeta: Onuphidae). Mar Biol 17:350–356

Nachtigall W (1991) Flight. In: Witt R, Lieckfeld CP (eds) Bionics: nature's patents. Pro Futura, Munich, pp 7–31

Nagaishi Ch, Okada Y, Ishiko S, Daido S (1964) Electron microscopic observations of the pulmonary alveoli. Exp Med Surg 22:81–97

Nagel E (1961) The structure of science. Routledge and Kegan Paul, London

Nagy JA, Odell DK, Seymour RS (1972) Temperature regulation by the inflorescence of Philodendron. Science 178:1195–1197

Nahas GG (1962) The pharmacology of *tris* (hydroxymethyl) methane (THAM). Pharmacol Rev 14:447–472

Naito H, Gillis CN (1973) Effects of halothane and nitrous oxide on removal of norepinephrine from the pulmonary circulation. Anaesthesiology 30:575–580

Nakao T (1974) The fine structure and innervation of gill lamellae in Anodonta. Cell Tissue Res 157:239–254

Narahara AB, Bergman HL, Maina JN, Lauren P, Walsh PJ, Wood CM (1996) Respiratory physiology of the Lake Magadi Tilapia (*Oreochromis alcalicus grahami*), a fish adapted to a hot, alkaline, and frequently hypoxic environment. Physiol Zool. 69(5):1114–1136

Nassar SAK, Munshi JSD (1971) Studies on the macrophytic biomass production and fish population in an abandoned pond at Bhagalpur, Bihar. J Sci Bihar University, India IV:8–16

Nathan C (1992) Nitric oxide as a secretory product of mammalian cells. FASEB J 6:3051–3064

Neckvasil NP, Olson KR (1986) Extraction and metabolism of circulating catecholamines by the trout gill. Am J Physiol 19:R5276–5287

Neftel A, Oeschger H, Schwander J, Stauffer B, Zunbrunn R (1982) Ice core sample meaurements give atmospheric CO_2 content during the past 40 000 yrs. Nature (Lond) 295:220–223

Nelson G (1978) Ontogeny, phylogeny, paleontology and biogenic law. Syst Zool 27:324–345

Nelson JS (1976) Fishes of the world. Wiley, New York

Nelson TR, Manchester DK (1988) Modeling of lung morphogenesis using fractal geometries. IEEE Trans Med Imaging 7:321–327

Nelson TR, West BJ, Goldberger AL (1990) The fractal lung: universal and species related scaling patterns. Experientia 46:251–254

Nelson ZC, Hirshfield D, Schreiweis DO, O'Farrell MJ (1977) Flight muscle contraction in relation to ambient temperature in some species of desert bats. Comp Biochem Physiol 56A:31–36

Nevo E (1979) Adaptive convergence and divergence of subterranean mammals. Annu Rev Ecol Syst 10:269–308

Nevo E, Guttman R, Haber M, Erez E (1982) Activity patterns of evolving mole rats. J Mammal 63:453–463

Newell RC, Courtney WAM (1965) Respiratory movements in *Holothuria forskali*. J Exp Biol 42:45–57

Newman RA (1992) Adaptive plasticity in amphibian metamorphosis. BioScience 42:671–678

Newman MJ, Rood RT (1977) Implications of solar evolution for Earth's atmosphere. Science 198:1035–1037

Nguyen Phu D, Yamaguchi K, Scheid P, Piiper J (1986) Kinetics of oxygen uptake and release by erythrocytes of the chicken. J Exp Biol 125:15–27

Nicloux M (1923) Action de l'oxyde de carbone sur les poissons et capacité respiratoire du sang de ces animaux. Cr Blanc Soc Biol 89:1328–1331

Nicol JAC (1960) The biology of marine animals. Interscience, New York

Nicolet J, Bannerman ESN, Haller R (1975) Mycotic proteolytic enzymes. In: Junod AF, Haller R (eds) Lung metabolism. Academic Press, New York, pp 57–65

Nicoll PA (1954) The anatomy and behaviour of the vascular systems in *Nereis virens* and *Nereis limbata*. Biol Bull Mar Lab Woods Hole 106:69–82

Nieden F (1913) Gymnophiona (Amphibia: Apoda). Tierreich 37:1–31

Nielsen EG, Gargas E (1984) Oxygen nutrients and primary production in the open Danish waters. Limnologica 15:303–310

Nikinmaa M (1990) Vertebrate erythrocytes: adaptation of function to respiratory requirements. Springer, Berlin Heidelberg New York

Nikinmaa M, Huestis WH (1984) Shape changes in goose erythrocytes. Biochim Biophys Acta 773:317–320

Niles E (1993) History, function, and evolutionary biology. Evol Biol 27:33–50

Nilsson GR, Löfman CO, Block M (1995) Extensive erythrocytes deformation in fish gills observed by in vivo microscopy: apparent adaptations for enhancing oxygen uptake. J Exp Biol 198:1151–1156

Nilsson S (1985) Filament position in fish gills is influenced by a smooth muscle enervated by adrenergic nerves. J Exp Biol 118:433–437

Nilsson S (1986) Control of gill blood flow. In: Nilsson S, Holmgren S (eds) Fish physiology: recent advances. Croom Helm, Dover HH, pp 86–101

Nisbet EG (1988) The young Earth. Allen and Unwin, London

Nisbet UG, Cann JR, Dover van CL (1995) Origins of photosynthesis. Nature (Lond) 373:479–480

Noble GK (1925) Integumentary, pulmonary, and cardiac modifications correlated with increased cutaneous respiration in the amphibian: a solution to the "hairy frog" problem, J Morphol Physiol 40:341–416

Noble GK (1929) The adaptive modifications of the arboreal tadpoles of *Hoplophryne* and torrent tadpoles of *Staurois*. Bull Am Mus Nat Hist 58:291–334

Noble GK (1931) The biology of Amphibia. McGraw-Hill, New York

Nonnenmacher TF (1987) A scaling model for dichotomous branching processes. Biol Cybern 56:155–157

Nonnenmacher TF (1988) Fractal shapes of cell membranes and pattern formation by dichotomous branching processes. In: Lamprecht I, Zotin Al (eds) Thermodynamics and pattern formation in biology. Walter de Gruyter, Berlin, pp 371–394

Nonnenmacher TF (1989) Fractal scaling mechanisms in biomembranes: oscillations in the lateral diffusion coefficient. Eur Biophys J 16:375–379

Norberg UM (1976a) Aerodynamics of hovering flight in long-eared bat *Plecotus auritus*. J Exp Biol 65:459–470

Norberg UM (1976b) Kinematics, aerodynamics, and energetics of horizontal flapping flight in the long-eared bat *Plecotus auritus*. J Exp Biol 65:179–212

Norberg UM (1981) Why foraging birds in trees should climb and hop upwards rather than downwards. Ibis 123:281–288

Norberg UM (1985) Flying, gliding, and soaring. In: Hildebrand M (ed) Functional vertebrate morphology. Harvard University Press, Cambridge, pp 366–377

Norberg UM (1986) On the evolution of flight and wing forms in bats. In: Nachtigall W (ed) Bat flight. BIONA report 5, Gustav Fischer, Stuttgart, pp 13–26

Norberg UM (1990) Vertebrate flight: mechanics, physiology, morphology, ecology and evolution. Springer, Berlin Heidelberg New York

Norberg UM, Rayner JMV (1987) Ecological morphology and flight in bats (Mammalia: Chiroptera): wing adaptations, flight performance, foraging strategy and echolocation. Philos Trans R Soc Lond 316B:335–427

Norell MA (1989) The higher level relationships of the extant crocodilia. J Herpetol 23:325–335

Normand ICS, Olver RE, Reynolds EOR, Strang LB (1971) Permeability of lung capillaries and alveoli to nonelectrolytes in the foetal lamb. J Physiol Lond 219:303–330

Norris RA, Connell CE, Johnston DW (1957) Notes on fall plumages, weights and fat condition in the ruby-throated hummingbird. Wilson Bull 69:155–163

Norsk P, Foldager N, Bonde-Petersen F, Elmann-Larsen B, Johansen TS (1987) Central venous pressure in humans during short periods of weightlessness. J Appl Physiol 63:2433–2437

Northcutt RG (1990) Ontogeny and phylogeny: a reevaluation of conceptual relationships and some applications. Brain Behav Evol 36:116–140

Nossaman BD, Feng CJ, Wang J, Kadowitz PJ (1994) Analysis of angiotensins I, II, III in pulmonary vascular bed of the rat. Am J Physiol 266:L389–L396

Novacek MJ (1980) Cranioskeletal features in tupaiids and selected *Eutheria* as phylogenetic evidence. In: Luckett WP (ed) Comparative biology and evolutionary relationships of tree shrews. Plenum, New York, pp 35–67

Novacek MJ (1982) Information for molecular studies from anatomical and fossil evidence on higher eutherian phylogeny. In: Goodman M (ed) Macromolecular sequences in systematic and evolutionary biology. Plenum Press, New York, pp 3–41

Novy MJ, Parer JT (1969) Absence of high blood oxygen affinity in the foetal cat blood. Respir Physiol 6:144–151

Nursall JR (1959) Oxygen as a prerequisite to the origin of the Metazoa. Nature (Lond) 183:1170–1172

Nurse P (1997) The ends of understanding. Nature (Lond) 387:657

Nyberg R, West B (1957) The influence of oxygen tension and some drugs on human blood vessels. Acta Physiol Scand 39:216–227

O'Dor RK (1982) Respiratory metabolism and swimming performance of the squid, *Loligo opalescens*. Can J Fish Aquat Sci 39:580–587

O'Dor RK, Webber DM (1991) Invertebrate athletes: trade-offs between transport efficiency and power density in cephalopod evolution. J Exp Biol 160:93–112

Odum EP, Connell CE (1956) Lipid levels in migrating birds. Science 123:892–894

O'Farrell MJ, Bradley WG (1977) Comparative thermal relationships of flight for some bats in southwestern United States. Comp Biochem Physiol 58A:223–227

Officer CB, Biggs RB, Taft JL, Cronin LE, Tyler MA, Boynton WR (1984) Chesapeake Bay anoxia: origin, development and significance. Science 223:22–27

Ogden E (1945) Respiratory flow in *Mustelus*. Am J Physiol 145:134–139

Ohkuwa T, Sato Y, Naoi M Chang LY (1997) Glutathione status and reactive oxygen generation in tissues of young and old exercised rats. Acta Physiol Scand 159:237–244

Oikawa S, Itazawa Y (1993) Allometric relationship between tissue respiration and body mass in a marine teleost, pogy *Pagrus major*. Comp Biochem Physiol 195:129–133

Ojha J, Singh SK (1986) Scanning electron microscopy of the gills of a hill-stream fish, *Danio dangila* (Ham.). Arch Biol Bruxelles 97:455–467

Okada O, Presson RG, Kirk KR, Godbey PS, Capen RL, Wagner WW (1992) Capillary perfusion patterns in single alveolar walls. J Appl Physiol 72:1838–1844

Okada Y, Ishiko S, Daido S, Kim J, Ikeda S (1962) Comparative morphology of the lung with special reference to the alveolar epithelial cells. I. Lungs of Amphibia. Acta Tuberc Jpn 11:63–87

Olmo E (1991) Genome variations in the transition from amphibians to reptiles. J Mol Evol 33:68–75

Olsen CR, Hale FC, Elsner R (1969) Mechanics of ventilation in the pilot whale. Respir Physiol 7:137–149

Olsen GJ, Woese CR (1993) Ribosomal RNA: a key to phylogeny. FASEB J 7:113–123

Olsen LF, Degn H (1985) Chaos in biological systems. Q Rev Biophys 18:165–225

Olson E, Miller R (1958) Morphological integration. Univ Chicago Press, Chicago

Olson KR (1991) Vasculature of the fish gills: anatomical correlates of physiological function. J Electron Microsc Tech 2:217–228

Olson KR (1994) Circulation in bimodally breathing fish. Am Zool 34:280–288

Olson KR (1996) Scanning electron microscopiy of the fish gill. In: Munshi JSD, Dutta HM (eds) Fish morphology: horizon of new research. Science Publishers, Lebanon, New Hampshire, pp 31–45

Olson KR, Fromm PO (1973) A scanning electron microscope study of secondary lamellae and chloride cells of rainbow trout (*Salmo gairdnen*). Z Zellforsch Mikrosk Anat 143:439–449

Olson KR, Kullman DK, Narkates AJ, Oparil S (1986a) Angiotensin extraction by trout tissues *in vivo* and metabolism by perfused gill. Am J Physiol 250:R532–541

Olson KR, Munshi JSD, Ghosh TK, Ojha J (1986b) Gill microcirculation of the air-breathing climbing perch, *Anabas testudineus* (Bloch): relationships with the accessory respiratory and systemic circulation. Am J Anat 176:305–320

Olson KR, Terwilliger N, Capuzzo MJ (1988) Structure of hemocyanin in larval and adult American lobsters. Am Zool 28:47A

Olson KR, Munshi JSD, Ghosh TK (1990a) Vascular organization of the head and respiratory organs of the air-breathing catfish, *Heteropneustes fossilis*. J Morphol 203:165–179

Olson KR, Taylor A, Capuzzo MJ (1990b) Correlation between hemocyanin structure and function in American lobsters. Am Zool 30:94A

Olson KR, Roy PK, Ghosh TK, Munshi JSD (1994) Microcirculation of gills and accessory respiratory organs from the air-breathing snakehead fish, *Channa punctata*, *C. gachua*, and *C. marulius*. Anat Rec 238:92–107

Olver RE, Strang LB (1974) Ion fluxes across the pulmonary epithelium and the secretion of lung liquid in the fetal lamb. J Physiol (Lond) 241:327–357

O'Mahoney P, Full RJ (1984) Respiration of crabs in air and water. Comp Biochem Physiol 79A:275–282

Onimaru H, Homma I (1987) Respiratory rhythm generator neurons in medulla of brain stem-spinal cord prepartion from new born rat. Brain Res 403:380–384

Oparin AI (1938) The origin of life. Macmillan, New York

Oparin AI (1953) The origin of life. Dover Publications, London

Opell BD (1987) The influence of web monitoring tactics of the tracheal systems of spiders in the family Uloboridae (Arachnida, Areneida). Zoomorphology 107:255–259

Orgel LE (1994) The origin of life on the earth. Sci Am 271:53–61

Orr WC, Sohal RS (1994) Extension of lifespan by overexpression of superoxide dismutase and catalase in *Drosophila melanogaster*. Science 263:1128–1130

Osaki S (1996) Spider silk as mechanical lifeline. Nature (Lond) 384:419

441

Osawa S, Jukes TH, Watanabe K, Muto A (1992) Recent evidence for evolution of the genetic code. Microbiol Rev 56:229–264

Oseid D, Smith L (1974) Chronic toxicity of hydrogen sulfide to *Gammarus pseudolimnaeus*. Trans Am Fish Soc 103:819–822

Oshino N, Sugano T, Oshino R, Chance B (1974) Mitochondrial function under hypoxic conditions: The steady states of cytochrome a + a^3 and their relation to mitochondrial energy state. Biochim Biophys Acta 368:298–310

Osterberg R (1974) Origins of metal ions in biology. Nature (Lond) 249:382–383

Östlund E, Fänge R (1962) Dasodilation by adrenaline and noradrenaline and the effects of some other substances on perfused fish gills. Comp Biochem Physiol 97:292–303

Ostrom JH (1975) The origin of birds. Annu Rev Earth Planet Sci 3:55–77

Owens T, Cess RD, Ramanathan V (1979) Enhanced carbon dioxide greehouse to compensate for reduced solar luminosity on early Earth. Nature (Lond) 277:640–641

Pace N, Smith AH (1981) Gravity, metabolic rate effects in mammals. Physiologist 24:S37–40

Pace NR, Stahl DA, Lane DJ, Olsen GJ (1985) Analyzing natural microbial populations by rRNA sequences. Am Soc Microbiol News 51:4–12

Pack AI, Galante R, Fishman AP (1984) Breuer-Hering reflexes in the African lungfish (*Protopreus annectens*). Fed Proc 43:A433

Pack AI, Galante R, Fishman AP (1992) Role of lung inflation in control of air breath duration in African lungfish (*Protopterus annectens*). Am J Physiol 262:879–884

Packard A (1972) Cephalopods and fish: the limits of convergence. Biol Rev 47:241–307

Packard CC, Sotherland PR, Packard MJ (1977) Adaptive reduction in permeability of avian eggshells to water vapour at high altitudes. Nature (Lond) 266:252

Packard GC (1974) The evolution of air-breathing in Paleozoic gnathostome fishes. Evolution 28:320–325

Packard GC, Packard MJ (1997) Type of soil affects survival by overwintering hatchilings of the painted turtle. J Therm Biol 22:53–58

Packard GC, Sotherland PR, Packard MJ (1997) Adaptive reduction in permeability of avian egg shells to water vapour at high altitudes. Nature (Lond) 266:255–256

Packard GC, Elinson RP, Gavaud J, Guillette L, Lombardi J et al. (1989) How are reproductive systems integrated and how has viviparity evolved? In: Wake D, Roth G (eds) Complex organismal functions: integration and evolution in vertebrates. John Wiley (Dahlem Workshop Report), Chichester, pp 281–293

Padian K (1982) Macroevolution and the origin of major adaptations: vertebrate flight as a paradigm for the analysis of pattern. Proc 3rd N Am Paleontol Convec, pp 387–392

Padian K (1983) A functional analysis of flying and walking in pterosaurs. Paleobiology 9:218–239

Paganelli CV (1980) The physics of gas exchange across the avian egg shell. Am Zool 20:329–338

Page TL (1994) Time is the essence: molecular analysis of the biological clock. Science 263:1570–1572

Palmer JD (1997) The mitochondrion time forgot. Nature (Lond) 387:454–455

Palmer MF (1968) Aspects of the respiratory Physiology of *Tubifex tubifex* in relation to its ecology. J Zool (Lond) 154:463–473

Palmer RMJ, Rerrige AC, Moncada S (1987) Nitric oxide release accounts for the biological activity oe endothelium derived relaxing factor. Nature (Lond) 327:524–526

Palomeque J, Rodriguez JD, Placios L, Planas J (1980) Blood respiratory properties of swifts. Comp Biochem Physiol 67A:91–95

Panchen AL, Smithson TR (1988) The relationships of the earliest tetrapods. In: Benton MJ (ed) The phylogeny and classification of the tetrapods, vol 1: Amphibians, reptiles birds. Clarendon Press, Oxford, pp 1–32

Parker DJ, Cook S, Warwick MT (1975) Serum complement studies during and following cardiopulmonary bypass. In: Junod AF, Haller R (eds) Lung metabolism. Academic Press, New York, pp 481–487

Parker HW (1940) The Percy Sladen Trust Expedition to Lake Titicaca, Amphibia ser 3. Trans Linn Soc Lond 12:203–216

Parker WN (1892) On the anatomy and physiology of *Protopterus annectens*. Trans R Irish Acad 30:111–230

Part P, Tuurala H, Nikinmaa M, Kiessling A (1984) Evidence for nonrespiratory intralamellar shunt in perfused rainbow trout gills. Comp Biochem Physiol 79A:29–34

Pastor LM (1995) The histology of the reptilian lungs. In: Pastor LM (ed) Histology, ultrastructure and immunohistochemistry of respiratory organs in non-mammalian vertebrates. Secretariado de Oublicaciones de la Universidad de Murcia. Spain, pp 131–153

Pastor LM, Ballesta J, Castells MT, Perez-Tomas R, Marin JA, Madrid JF (1989) A microscopic study of the lung of Testudo graeca. J Anat 162:19–33

Patel S, Spencer (1963) Studies on the haemoglobin of Arenicola marina. Comp Biochem Physiol 8:65–82

Patt DI, Patt GR (1969) The respiratory system. In: Comparative vertebrate histology. Harper and Row, New York

Pattinson RC, Odendall HJ, Kirsten G (1993) The relationship between absent end-diastolic velocities of the umbilical artery and perinatal mortality and morbidity. Early Human Dev 33:61–69

Pattle RE (1976) The lung surfactant in the evolutionary tree. In: Hughes GM (ed) Respiration of amphibious vertebrates. Academic Press, London, pp 233–255

Pattle RE, Schock C, Creasey JM, Hughes GM (1977) Surpellic films, lung surfactant, and their cellular origin in newt, caecilian and frog. J Zool (Lond) 182:125–136

Paul GS (1990) A re-evaluation of the mass of and flight of giant pterosaurs. J Vertebr Paleontol 10:37

Paul GS (1991) The many myths, some old, some new, of dinosaurology. Mod Geol 16:69–99

Paul R, Fincke T, Linzen B (1987) Respiration in the tarantula Eurypelma califormicum: evidence of diffusion lungs. J Comp Physiol B 157:209–217

Paul RJ (1989) Smooth muscle energetics. Annu Rev Physiol 51:331-349

Paul RJ (1992) Gas exchange, circulation and energy metabolism in spiders. In: Wood SC, Weber RE, Hargens AR, Millard RW (eds) Physiological adaptations in vertebrates: respiration, circulation, and metabolism. Marcel Dekker, New York, pp 169–197

Paulev P (1965) Decompression sickness following repeated breathhold dives. J Appl Physiol 20:1028–1031

Pavlov NA, Krivchenko AT, Cherepivskaya EN, Zagvazdin YS, Zayat ND (1987) Reactivity of cerebral vessels in the pigeon, Columba livia. J Evol Biochem Physiol 23:447–451

Pearson OP, Pearson AK (1976) A streological analysis of the ultrastructure of the lungs of wild mice living at low and high altitude. J Morphol 150:359–368

Peebles PJE, Schramm DN, Turner EL, Kron RG (1994) The evolution of the Universe. Sci Am 271:29–33

Pegg JH, Horner TL, Wahrenbrock EA (1963) Breathing of pressure-oxygenated liquids. Proc 2nd Symp Underwater Physiol, Natl Acad Sci Natl Res Council Publ 1191, pp 166–170

Peitgen HO, Richter PH (1986) The beauty of fractals: images of complex dynamical systems. Springer, Berlin Heidelberg New York

Pelseneer P (1935) Essai d'éthologie zoologique. Acad R Belgique Classe des Sciences, Publ Found Brussels, Agathon Potter, Paris

Pelster B (1985) Mechanismen der Anpassung an das Leben in extremen Biotopen: vergleichede Studien zur Atmungsphysiologie bei Lumpenus lampretaeformis und Bleennius pholis. PhD Thesis, University of Düsseldorf

Pelster B, Scheid P (1991) Activities of enzymes for glucose catabolism in the swim bladder of the European eel Anguilla anguilla. J Exp Biol 156:207–213

Pelster B, Scheid P (1992a) Counter current concentration and gas secretion in the fish swim bladder. Physiol Zool 65:1–16

Pelster B, Scheid P (1992b) Metabolism of the swim bladder epithelium and the single concentrating effect. Comp Biochem Physiol 105A:383–388

Pelster B, Scheid P (1992c) The influence of gas gland metabolism and blood flow on gas deposition into the swim bladder of the European eel Anguilla anguilla. J Exp Biol 173:205–216

Pelster B, Scheid P (1993) Glucose metabolism of the swim bladder tissue of the European eel, Anguilla anguilla. J Exp Biol 185:169–178

Pelster B, Bridges CR, Grieshaber MK (1988a) Respiratory adaptations of the burrowing marine teleost *Lumpenus lampretaeformis*. (Walbaum). II. Metabolic adaptations. J Exp Mar Biol Ecol 124:43–55

Pelster B, Bridges CR, Taylor AC, Morris S, Artikinson RJA (1988b) Respiratory adaptations of the burrowing marine teleost *Lumpenius lampretaeformis* (Walbaum). I. O_2 and CO_2 transport, acid-base balance: a comparison with *Depola rubescens*. J Exp Mar Biol Ecol 124:31–42

Pelster B, Kobayashi H, Scheid P (1988c) Solubility of nitrogen and argon in eel whole blood and its relation to pH. J Exp Biol 135:243–252

Pelster B, Kobayashi H, Scheid P (1989) Metabolism of the perfused swim bladder of the European eel: oxygen, carbon dioxide, glucose and lactate balance. J Exp Biol 144:495–506

Pelzenberger M, Pohla H (1992) Gill surface area of water and air breathing fish. Rev Fish Biol Fisheries 2:187–216

Pendergast DR, Olszowska AJ, Rokitka MA, Farhi LE (1987) Gravitational force and the cardio-vascular system. In: Dejours P (ed) Comparative physiology of environmental adaptations, vol 2. Karger, Basel, pp 15–26

Penman HL (1940a) Gas and vapour movements in the soil. I. The diffusion of vapours through porous solids. J Agric Sci 30:437–462

Penman HL (1940b) Gas and vapour movements in the soil. II. The diffusion of carbon dioxide through porous solids. J Agric Sci 30:570–581

Penney DG (1977) Effects of prolonged diving anoxia on the turtle, *Pseudemys scripta elegans*. Comp Biochem Physiol A 47:933–941

Penny DG, Hasegawa M (1997) The platypus put in its place. Nature (Lond) 387:549–550

Pennycuick CJ (1975) Mechanics of flight. In: Famer DS, King JR (eds) Avian biology, vol 5. Academic Press, New York, pp 1–75

Pennycuick CJ (1992) Newtonian rules in biology. Oxford University Press, New York

Pennycuick CJ, Rezende MA (1984) The specific power output of aerobic muscle, related to the power density of mitochondria. J Exp Biol 108:377–392

Penry DL, Jumars PA (1987) Modelling animal guts as chemical reactors. Am Nat 129:69–96

Perkins JR (1964) Historical development of respiratory physiology. In: Handbook of physiology. Respiration, sect 3, vol I, Am Physiol Soc, Washington, DC, pp 1–62

Perlo S, Jalowayski AA, Durand CM, West JB (1975) Distribution of red and white blood cells in alveolar walls. J Appl Physiol 38:117–124

Perry SF (1978) Quantitative anatomy of the lungs of the red-eared turtle. *Pseudemys scripta elegans*. Respir Physiol 35:245–262

Perry SF (1981) Morphometric analysis of pulmonary structure: methods for evaluation of unicameral lungs. Microscopie 38:278–293

Perry SF (1983) Reptilian lungs: functional anatomy and evolution. Adv Anat Embryol Cell Biol 79:1–81

Perry SF (1988) Functional morphology of the lungs of the Nile crocodile *Crocodylus niloticus*: non-respiratory parameters. J Exp Biol 143:99–117

Perry SF (1989) Mainstreams in the evolution of vertebrate respiratory structures. In: King AS, McLelland J (eds) Form and function in birds, vol V, Academic Press, London, pp 1–67

Perry SF (1992a) Evolution of the lung and its diffusing capacity. In: Bicudo JPW (ed) Vertebrate gas transport cascade adaptations to environment and mode of life. CRC press, Boca Raton, pp 142–153

Perry SF (1992b) Gas exchange strategies in reptiles and the origin of the avian lung. In: Wood SC, Weber RE, Hargens AR, Millard RW (eds) Physiological adaptations in vertebrates: respiration, circulation, and metabolism. Marcel Dekker, New York, pp 149–167

Perry SF, Duncker HR (1978) Lung architecture, volume and static mechanics in five species of lizards. Respir Physiol 34:61–81

Perry SF, Duncker HR (1980) Interrelationship of static mechanical factors and anatomical structure in lung ventilation. J Comp Physiol 138:321–334

Perry SF, Laurent P (1990) The role of carbonic anhydrase in carbon dioxide excretion, acid base balance and ionic regulation in aquatic gill breathers. In: Truchot JP, Lahlou B (eds) Transport, respiration and excretion: comparative and environmental aspects. Karger, Basel, pp 39–67

Perry SF, McDonald G (1993) Gas exchange. In: Evans DH (ed) The physiology of fishes. CRC Press, Boca Raton, pp 251–278

Perry SF, Darian-Smith C, Alston D, Limpus CJ, Maloney JE (1989a) Histological structure of the lungs of the loggerhead turtle, *Caretta caretta*, before and after hatching. Copeia 1989:1000–1010

Perry SF, Bauer AM, Russell AP, Alston JT, Maloney JE (1989b) Lungs of the geccko *Rhacodactylus leachianus* (Reptilia: Gekkonidae): a correlative gross anatomical and light and electron microscopic study. J Morphol 199:23–46

Perry SF, Auman U, Maloney JE (1989c) Intrinsic lung musculature and associated ganglion cells in a teiid lizard, *Tupinambis nigropunctatus spix*. Herpetologica 45:217–227

Perutz MF (1970) Stereochemistry of cooperative effects in haemoglobin. Haem-haem interaction and problem of allostery. The Bohr effect and combination with organic phosphates. Nature (Lond) 228:726–733

Perutz MF (1979) Regulation of oxygen affinity of haemoglobin: influence of structure of the globin on the haeme iron. Annu Rev Biochem 48:327–386

Perutz MF (1983) Species adaptation in a protein molecule. Mol Biol Evol 1:1–28

Perutz MF (1990a) Molecular inventiveness. Nature (Lond) 348:583–584

Perutz MF (1990b) Mechanisms regulating the reactions of human haemoglobin with oxygen and carbon monoxide. Annu Rev Physiol 52:1–25

Perutz MF (1996) Taking the pressure off. Nature (Lond) 380:205–206

Peters HM (1978) On the mechanism of air ventilation in anabantoids (Pisces, Teleostei). Zoomorphologie 89:93–123

Petersen JA, Fyhn HJ, Johansen K (1974) Ecophysiological studies of an intertidal crustacean, *Pollicipes polymerus* (Cirripeda, Lepadomorpha): Aquatic and aerial respiration. J Exp Biol 61:309–320

Petit-Maire N (1991) Paléoenvironments du Sahara. CNRS, Paris

Petroski H (1985) To engineer is human. St Martin's Press, New York

Petschow D, Würdinger, I, Baumann R, Duhm J, Braunitzer G, Bauer C (1977) Causes of high blood oxygen affinity of animals living at high altitude. J Appl Physiol 42:139–143

Petterson A, Hardin J (1969) Flight speeds of five species of vespertilionid bats. J Mammal 50:152–153

Pettersson K, Nilsson S (1979) Nervous control of the branchial vascular resistance of the Atlantic cod, *Gadus morhua*. J Comp Physiol 129:179–183

Pettigrew JD, Jamieson BGM, Robson SK, Hall LS, McNally KI, Cooper HM (1989) Phylogenetic relations between microbats, megabats and primates (Mammalia: Chiroptera and Primates). Philos Trans R Soc Lond 325B:489–559

Philander SGH (1983) El Niño southern oscillation phenomena. Nature (Lond) 302:295–301

Phillips CG, Kaye SR, Schroter RC (1994) A diameter-based reconstruction of the branching pattern of the human bronchial tree part I. Description and application. Respir Physiol 98:193–217

Phillipson J (1981) Bioenergetic options and phylogeny. In: Townsend CR, Calow P (eds) Physiological ecology: an evolutionary approach to resource use. Blackwell, London, pp 20–45

Phleger CF, Saunders BS (1978) Swim bladder surfactants of Amazon air breathing fishes. Can J Zool 56:946–952

Phleger CF, Smith DG, Macintyre DH, Saunders BS (1978) Alveolar and saccular lung phospholipids of the anaconda, *Eunectes murinus*. Can J Zool 56:1009–1013

Pickard WF (1974) Transition regime diffusion and the structure of the insect tracheolar system. J Insect Physiol 20:947–956

Pierce VA, Crawford DL (1997) Phylogenetic analysis of glycolytic enzyme expression. Science 276:256–259

Piiper J, Scheid P (1972) Maximum gas transfer efficacy of models for fish gills, avian lungs and mammalian lungs. Respir Physiol 14:115–124

Piiper J, Scheid P (1975) Gas transfer efficacy of gills, lungs and skin: theory and experimental data. Respir Physiol 23:209–221

Piiper J, Scheid P (1980) Blood-gas equilibration in lungs. In: West JB (ed) Pulmonary gas exchange, vol I. Academic Press, New York, pp 121–171

Piiper J, Scheid P (1981) Model for capillary-alveolar equilibration with special reference to O_2 uptake in hypoxia. Respir Physiol 46:193–205

Piiper J, Scheid P (1989) Respiratory mechanics and air flow in birds. In: King AS, McLelland J. Form and function in birds, vol. 4. Academic Press, London, pp 369–391

Piiper J, Scheid p (1992) Modeling of gas exchange in vertebrate lungs, gills, and skin. In: Wood SC, Weber RE, Hargens AR, Millard RW (eds) Physiological adaptations in vertebrates: respiration, circulation, and metabolism. Marcel Dekker, New York, pp 69–95

Piiper J, Humphrey HT, Rahn H (1962) Gas composition of pressurized, perfused gas pockets and the fish swim bladder. J Appl Physiol 17:275–282

Piiper J, Dejours P, Haab P, Rahn H (1971) Concepts and basic quantities in gas exchange physiology. Respir Physiol 13:292–304

Piiper J, Gatz RN, Crawford EC (1976) Gas transport characteristics in an exclusively skin breathing salamander, Desmognathus fuscus (Plethodontidae). In: Hughes GM (ed) Respiration in amphibious vertebrates. Academic Press, London, pp 339–356

Piiper J, Meyer M, Worth H, Willmer H (1977) Respiration and circulation during swimming activity in the dogfish, Scyliorhinus stellaris. Respir Physiol 30:338–349

Piiper J, Tazawa H, Ar A, Rahn H (1980) Analysis of chorioallantoic gas exchange in the chick embryo. Respir Physiol 39:273–287

Piiper J, Scheid P, Perry SF, Hughes GM (1986) Effective and morphometric oxygen diffusing capacity of the gills of the elasmobranch Scyliorhinus stellaris. J Exp Biol 123:27–41

Pilgrim M (1966) The anatomy and histology of the blood system of the maldanid polychaetes Clymenella torquata and Euclymene oerstedi. J Zool (Lond) 149:261

Pilson ME (1965) Variation of hemocyanin concentration in the blood of four species of haliotis. Biol Bull 128:459–472

Pinkerton KE, Barry BE, O'Neil JJ, Raus JA, Pratt PC, Crapo JD (1982) Morphologic changes in the lung during the lifespan of Fischer 344 rats. Am J Anat 164:155–174

Pinshaw B, Bernstein MH, Arad Z (1985) Effects of temperature and PCO_2 on O_2 affinity of pigeon blood: implications for brain O_2 supply. Am J Physiol 249:R759–R764

Piper HM, Noll T, Siegmund B (1994) Mitochondrial function in the oxygen depleted and reoxygenated myocardial cell. Cardiovasc Res 28:1–15

Piper PJ (1973) Distribution and metabolism. In: Cuthbert MF (ed) The prostaglandins. Pharmacological and therapeutic aspects. Heineman, London, pp 125–150

Piper PJ (1975) Conditions of release of prostaglandins from the lung. In: Junod Af, Haller R (eds) Lung metabolism. Academic Press, New York, pp 315–319

Pizarro B, Salas A, Paredes J (1970) Mal de altura en aves. Inst Vet Invest Trop Altura Cuarto Boletin Extraord 1970:147–151

Pizarello DJ, Shircliffe AC (1967) Hyperbaric oxygen: toxic effects in chick embryos. Am Surg 33:958–957

Platt T, Irwin B (1972) Phytoplankton productivity and nutrient measurements in Petpeswick Inlet, 1971–1972. Fish Res Board Can Tech Rep 314

Plattner W (1941) Etudes sur la fonction hydrostatique de la vessie natatoire des poissons. Rev Suisse Zool 48:201–338

Poczopko P (1959) Changes in blood circulation in Rana escutenta L. while diving. Zool Pol 10:29–43

Pohunkova H (1967) The ultrastructure of the lung of the snail Helix pomatia. Folia Morphol 15:250–257

Pohunkova H (1969) Lung ultrastructure of the Arachnida-Arachinoidea. Folia Morphol Prague 17:309–361

Pohunkova H, Hughes GM (1985a) Structure of the lung of the clawed toad (Xenopus laevis Daudin). Folia Morphol XXXIII:385–390

Pohunkova H, Hughes GM (1985b) Ultrastructure of the lungs of the garter snake. Folia Morphol Prague 23:254–258

Polanyi M (1968) Life's irreducible structure. Science 160:1308–1312

Policard A (1929) Les nouvelles idées sur la disposition de la surface respiratoire pulmonaire. Presse Med 80:1–20

Polimanti O (1912) Über den Beginn der Atmung bei Embryonen von Scyllium. Z Biol 57:237–272

Polimanti O (1913) Sui rapporti fra peso del corpo e ritmo respiratoria di *Octopus vulgaris* Lam. Z Allg Physiol 15:449–455

Popper K (1968) The logic of scientific discovery. Hutchinson, London

Popper K (1969) Conjectures and refutations. Routledge and Kegan Paul, London

Portier P (1933) Locomotion aérienne et respiration des lépidoptères, un nouveau rôle physiologique des ailes et des écailles. Trav V Congr Int Ent Paris 2:25–31

Portier P, Duval M (1929) Recherches physiologiques sur la teneur en gaz carbonique de l'atmosphére interne des fourmilieres. CR Soc Biol 102:906–908

Postgate JR (1987) Nitrogen fixation. Edward Arnold, London

Potter EL, Bohlender GP (1941) Intrauterine respiration in relation to development of foetal lung with report of 2 unusual anomalies of respiratory system. Am J Ogstet Gynecol 42:14–22

Potter GE (1927) Respiratory function of swim bladder in *Lepidosteus*. J Exp Zool 49:45–52

Pough FH (1980) Blood oxygen transport and delivery in reptiles. Am Zool 31:455–456

Pough FH, Taigen TL, Stawart MM, Brussard PF (1983) Behavioural modification of evaporative water loss by a Puerto Rican frog. Ecology 60:608–613

Pough FH, Heiser JB, McFarland WN (1989) Vertebrate life, 3rd edn. Macmillan, New York

Powell CS (1993) Livable planets. Sci Am 268:7–8

Powell FL (1982) Diffusion in avian lungs. Fed Proc 41:53–55

Powell FL (1990) Acclimatization to high altitude. In: Sutton JR, Coates GC, Remmers JE (eds) Hypoxia: the adaptations. BC Decker, Toronto, pp 41–44

Powell FL, Scheid P (1989) Physiology of gas exchange in the avian respiratory system. In: King AS, McLelland J (eds) Form and function in birds, vol 4. Academic Press, London, pp 393–437

Powell FL, Wagner PD (1982a) Ventilation-perfusion inequality in the avian lungs. Respir Physiol 38:233–241

Powell FL, Wagner PD (1982b) Measurement of continuous distributions of ventilation-perfusion in non-alveolar lungs. Respir Physiol 48:219–232

Powell MA, Arp AJ (1989) Hydrogen sulfide oxidation by abundant nonhaemoglobin heme compounds to marine invertebrates from sulfide rich habitats. J Exp Zool 249:121–132

Powell MA, Somero GN (1985) Sulfide oxidation occurs in the animal tissue of the gutless clam, *Solemya reidi*. Biol Bull 169:164–181

Powell MA, Crenshaw MA, Rieger RN (1979) Adaptations to sulfide in the meiofauna of the sulfide system. l[35] S-sulfide accumulation and the presence of a sulfide detoxification system. J Exp Mar Biol Ecol 37:57–76

Powers DA (1972) Haemoglobin adaptation for fast and slow water habitats in sympatric catostomid fishes. Science 177:360–362

Powers LW, Bliss DE (1983) Terrestrial adaptations. In: Vernberg FJ, Vernberg WB (eds) Environmental adaptations: biology of Crustacea, vol 8. Academic Press, New York, pp 271–333

Powers DA, Fyhn HJ, Fyhn UEH, Martin JP, Garlick RL, Wood SC (1979) A comparative study of the oxygen equilibria of blood from 40 genera of Amazonian fishes. Comp Biochem Physiol 62A:67–86

Prange HD (1976) Energetics of swimming in a sea turtle. J Exp Biol 64:1–12

Prange HD, Ackerman RA (1974) Oxygen consumption and mechanisms of gas exchange of green turtle (*Chelonia mydas*) eggs and hatchlings. Copeia 3:758–763

Prange HD, Wasser JS, Gaunt AS, Gaunt SLL (1985) Respiratory responses to acute heat stress in cranes (Gruidae): the effects of tracheal coiling. Respir Physiol 62:95–103

Prankerd TAJ (1961) The red cell: an account of its chemical physiology and pathology. Blackwell, Oxford

Prasad MS (1988) Morphometrics of gills during growth and development of air-breathing habit in *Colisa fasciatus* (Bloch and Schneider). J Fish Biol 32:367–381

Precht H (1939) Die Lungenatmung der Süsswasserpulmonaten. Z Vergl Physiol 26:696–738

Prestwich KN (1983) The roles of aerobic and anaerobic metabolism in active spiders. Physiol Zool 56:122–132

Priede IG (1977) Natural selection for energetic efficiency and relationship between activity level and mortality. Nature (Lond) 267:610

Prigogine I, Stengers I (1984) Order of chaos: man's new dialogue with nature. Heineman, London

Pringle JWS (1983) Insect flight. Carolina Biology Readers, Burlington, North Carolina

Prinzinger R, Hinninger CH (1992) Endogenous? Diurnal rhythmn in the energy metabolism of pigeon embryos. Naturwissenschaften 79:278–279

Prior DJ, Hume M, Varga D, Hess SD (1983) Physiological and behavioural aspects of water balance and respiratory function in the terrestrial slug, *Limax maximus*. J Exp Biol 104:111–127

Prisk GK, Guy HJB, Eliot AB, Deutschman RA, West JB (1993) Pulmonary diffusing capacity, capillary blood volume and cardiac output during sustained microgravity. J Appl Physiol 75:15–26

Pritchard A, White GN (1981) Metabolism and oxygen transport in the innkeeper worm *Urechis caupo*. Physiol Zool 54:44–54

Proctor DF, Caldini P, Permutt S (1968) The pressure surrounding the lungs. Respir Physiol 5:130–144

Prosser CL (1958) The nature of physiological adaptations. In: Prosser DL (ed) Physiological adaptations. American Physiological Society, Washington, DC, pp 167–180

Prosser CL (1961) Oxygen: respiration and metabolism. In: Prosser CL, Brown FA (eds) Comparative animal physiology. Saunders, Philadelphia, pp 198–287

Prosser CL (1973) Comparative animal physiology, 3rd edn. Saunders, Philadelphia

Prosser CL (1986) Adaptational biology: molecules to organisms. John Wiley, New York

Prosser CL, Brown FA (1962) Comparative animal physiology, 2nd edn. WB Saunders, London

Prothero J (1986) Scaling of energy metabolism in unicellular organisms: a re-analysis. Comp Biochem Physiol 83A:2243–248

Pryor WA (1986) Oxy-radicals and related species: their formation, life times, and reactions. Annu Rev Physiol 48:657–667

Pugh LGCE (1962) Physiological and medical aspects of the Himalayan scientific and mountaineering expedition, 1960–61, Br Med J 2:621–626

Pullin RSV, Morris DJ, Bridges CR, Atkinson RJA (1980) Aspects of the respiratory physiology of the burrowing fish *Cepola rubescens* L. Comp Biochem Physiol 66A:35–42

Qasim SZ, Qayyum A, Garg KK (1960) The measurement of carbon dioxide produced by air-breathing fishes and evidence of respiratory function of accessory respiratory organs. Proc Indian Acad Sci 52:19–26

Quinlan MC, Hadley NF (1993) Gas exchange, ventilatory patterns, and water loss in two lubber grasshoppers: quantifying cuticular water and respiratory transpiration. Physiol Zool 66:628–642

Quist J, Hill RD, Schneider RC, Falke KJ, Liggins GC et al. (1986) Hemoglobin concentrations and blood gas tensions of free-diving Weddell seals. J Appl Physiol 61:1560–1569

Rabalais NN, Cameron JN (1985) Physiological and morphological adaptations of adult *Uca subcylindrica* to semi-arid environments. Biol Bull 168:135–146

Rabinowitch E, Govindjee I (1965) The role of chlorophyll in photosynthesis. Sci Am 213:74–82

Rahn H (1966) Aquatic gas exchange: theory. Respir Physiol 1:1–12

Rahn H (1967) Gas transport from the external environment to the cell. In: de Reuck AVS, Porter R (eds) Development of the lung: a CIBA Foundation Symposium. Churchill, London, pp 3–23

Rahn H (1974) PCO_2, pH and body temperature. In: Nahas G, Smith KE (eds) Carbon dioxide and metabolic regulation. Springer, Berlin Heidelberg New York, pp 752–761

Rahn H (1977) Adaptations of the avian embryo to high altitude. In: Paintal AS, Gill-Kumar P (eds) Respiratory adaptations, capillary exchange and reflex mechanisms. Proc Krogh Centenary Symp, New Delhi University, India, pp 94–105

Rahn H, Ar A (1974) The avian egg: incubation time and water loss. Condor 76:147–152

Rahn H, Ar A (1980) Gas exchange of the avian egg: time, structure and function. Am Zool 20:477–484

Rahn H, Howell BJ (1976) Bimodal gas exchange. In: Hughes GM (ed) Respiration of amphibious vertebrates. Academic Press, London, pp 271–285

448

Rahn H, Paganelli CV (1968) Gas exchange in gas gills of diving insects. Respir Physiol 5:145–164

Rahn H, Paganelli CV (1982) Role of diffusion in gas exchange of the avian egg. Fed Proc 41:2134–2146

Rahn H, Rahn KB, Howell BJ, Gans C, Tenney SM (1971) Air-breathing of the garfish, *Lepisosteus osseus*. Respir Physiol 11:285–307

Rahn H, Paganelli CV, Ar A (1974) The avian egg: air-cell gas tension, metabolism and incubation time. Respir Physiol 22:297–309

Rahn H, Paganelli CV, Ar A (1975) Relation of avian egg weight to body weight. Auk 92:750–765

Rahn H, Paganelli CV, Nisbet ICT, Whittow GC (1976) Regulation of incubation water loss in eggs of seven species of terns. Physiol Zool 49:245–259

Rahn H, Carey C, Balmas K, Bhatia B, Paganelli C (1977) Reduction of pore area of the avian eggshell as an adaptation to altitude. Proc Natl Acad Sci USA 74:3095–3098

Rahn H, Christensen VL, Edens FW (1981) Changes in shell conductans, pores, and physical dimensions of egg and shell during the first breeding cycle of turkey hens. Poult Sci 60:2536–2557

Ralph R, Everson I (1968) The respiratory metabolism of some Antarctic fish. Comp Biochem Physiol 27:299–307

Ramsay JA (1968) Physiological approach to the lower animals, 2nd edn. Cambridge University Press, Cambridge

Rancour-Laferriére D (1985) Signs of the fresh: an essay on the evolution of hominid sexuality. Mouton de Gluyter, Berlin

Rand DA, Wilson HB (1993) Evolutionary catastrophes, punctuated equilibria and gradualism in ecosystem evolution. Proc R Soc Lond 253B:137–141

Randall DJ (1970) Gas exchange in fish. In: Hoar WS, Randall DJ (eds) Fish physiology, vol 4. Academic Press, London, pp 252–292

Randall DJ (1972) Respiration. In: Hardisty MW, Potter IC (eds) The biology of lampreys, vol 2. Academic Press, London, pp 287–306

Randall DJ (1982) The control of respiration and circulation in fish during exercise and hypoxia. J Exp Biol 100:275–288

Randall DJ, Cameron JN (1973) Respiratory control of arterial pH as temperature changes in rainbow trout, *Salmo gairdneri*. Am J Physiol 225:997–1002

Randall DJ, Daxboeck C (1984) Oxygen and carbon dioxide transfer across fish gills. In: Hoar WS, Randall DJ (eds) Fish physiology, vol 10, part A. Academic Press, New York, London, pp 263–314

Randall DJ, Jones DR (1973) The effect of deafferentation of the pseudobranch on the respiratory response to hypoxia and hyperoxia in the trout (*Salmo gairdneri*). Respir Physiol 17:291–301

Randall DJ, Shelton G (1963) The effects of changes in the environmental gas concentrations on the breathing and heart rate of a teleost fish. Comp Biochem Physiol 9:229–239

Randall DJ, Holeton GF, Stevens ED (1967) The exchange of oxygen and carbon dioxide across the gills of the rainbow trout. J Exp Biol 46:339–348

Randall DJ, Baumgarten D, Malyusz M (1972) The relationship between gas and ion transfer across the gills of fishes. Comp Biochem Physiol 41A:629–637

Randall DJ, Farrell AP, Haswell MS (1978a) Carbon dioxide excretion in the jeju, *Hoplerythrinus unitaeniatus*, a facultative air-breathing teleost. Can J Zool 56:970–973

Randall DJ, Farrell AP, Haswell MS (1978b) Carbon dioxide excretion in the piracucu (*Arapiama gigas*), an obligate air-breathing fish. Can J Zool 56:977–982

Randall DJ, Burggren WW, Farrell AP, Haswell MS (1981) The evolution of air-breathing in vertebrates. Cambridge University Press, Cambridge

Randall DJ, Wood CM, Perry SF, Bergman H, Maloiy GMO, Mommsen TP, Wright PA (1989) Ureotelism in a completely aquatic teleost fish: a strategy for survival in an extremely alkaline environment. Nature (Lond) 337:165–166

Randles CA, Romanoff AL (1950) Some physical aspects of the amnion and allantois of the developing chick embryo. J Exp Zool 114:87–105

Rannels DE, Rannels SR (1988) Compensatory growth of the lung following partial pneumonectomy. Exp Lung Res 14:157–182

Rantin FT, Johansen K (1984) Responses of the teleost *Hoplias malabaricus* to hypoxia. Environ Biol Fish 11:221–228

Rantin FT, Kalinin AL, Glass ML, Fernandes MN (1992) Respiratory responses to hypoxia in relation to mode of life of two erythrinid species (*Hoplias malabaricus* and *Hoplias lacerdae*). J Fish Biol 41:805–812

Rantin FT, Glass ML, Kalinin AL, Verzola RMM, Fernandes MN (1993) Cardio-respiratory responses in two ecologically distinct erythrinids (*Hoplias malabaricus* and *Hoplias lacerdae*) exposed to graded environmental hypoxia. Environ Biol Fish 36:93–97

Rappaport MM, Greene AA, Page IH (1948) Serum vasomotor (serotonin). J Biol Chem 176:1243–1251

Rashevsky N (1960) Mathematical biophysics: physico-mathematical foundations of biology. Dover, New York

Raup DM, Jablonski D (1993) Geography of end-Cretaceous marine bivalve extinctions. Science 260:971–973

Raup DM, Sepkoski JJ (1984) Periodicity of extinctions in the geologic past. Proc Natl Acad Sci USA 81:801–805

Raven PH, Wilson EO (1992) A fifty-year plan for biodiversity surveys. Science 258:1099–1100

Raymo M, Ruddiman WF (1992) Tectonic forcing of late Cenozoic climate. Nature (Lond) 359:117–122

Raynaud D, Jouzel J, Barnola JM, Chapellaz J, Delmas RJ, Lorius C (1993) The ice record of greenhouse gases. Science 259:926–934

Rayner JWV (1981) Flight adaptations in vertebrates. Symp Zool Soc (Lond) 48:137–172

Rayner JMV (1985) Bounding and undulating flight in birds. J Theor Biol 117:47–77

Rayner JMV (1986) Vertebrate flapping flight mechanics and aerodynamics, and the evolution of flight in bats. In: Nachtigall W (ed) BIONA report No 5, Bat flight – Fledermausflug. Gustav Fischer, Stuttgart, pp 27–74

Reader J (1986) The rise of life: the first 3.5 billion years. Alfred A Knopf, New York

Ready NE (1983) Wing development in hemimetabolous insects. PhD Thesis, University of California, Irvine

Redfield AC (1958) The biological control of chemical factors in the environment. Sci Am 46:205–221

Redfield AC, Florkin M (1931) The respiratory function of the blood of *Urechis carpo*. Biol Bull 61:185–210

Reeve HK, Westneat DF, Noon WA, Sherman PW, Aquadro F (1990) DNA fingerprinting reveals high levels of inbreeding in colonies of the eusocial naked mole-rat. Proc Natl Acad Sci USA 87:2496–2500

Reeves JT, Leathers JE (1964) Circulatory changes following birth of the calf and the effect of hypoxia. Circ Res 15:343–354

Reeves RB (1977) The interaction of body temperatures and acid-base balance in ectothermic vertebrates. Annu Rev Physiol 40:559–586

Reid WD, Ilett KF, Glick JM, Krishna G (1973) Metabolism and binding of aromatic hydrocarbons in the lung: relationship to experimental bronchiolar necrosis. Am Rev Respir Dis 107:539–551

Reid RC, Sherwood TK (1966) The properties of gases and liquids, 2nd edn. McGraw-Hill, New York

Reidbord HE (1967) An electronmicroscopic study of the alveolar capillary wall following intratracheal administration of saline and water. Am J Pathol 50:275–283

Reimers CE, Fischer KM, Merewether R, Smith KL, Jahnke RA (1986) Oxygen microporfiles measured in situ in deep ocean sediments. Nature (Lond) 320:741–744

Reite OB, Maloiy GMO, Aasenhaug B (1974) pH, salinity and temperature tolerance of Lake Magadi, Kenya. Nature (Lond) 247:315

Remane A, Storch V, Welsch U (1980) Systematische Zoologie. Fischer, Stuttgart

Remmer H (1975) Pulmonary drug-metabolizing enzymes. In: Junod AF, Haller R (eds) Lung metabolism. Academic Press, New York, pp 133–158

Remotti E (1933) Development of allantoic circulation in response to external variations in gas content. Bull Mus Lab Zool Anat Comp, Univ Genova 13:1–19

450

Rennard SL, Basset G, Lecossier K, O'Donnell KM et al. (1986) Estimation of epithelial lining fluid recovered by lavage using urea as a marker for dilution. J Appl Physiol 60:532–538

Rennie J (1992) Living together. Sci Am 266:104–113

Renous S, Gasc JP (1989) Body and vertebral proportions in Gymnophiona (Amphibia): diversity of morphological types. Copeia 1989:837–847

Repetski JE (1978) A fish from Upper Cambrian of North America. Science 200:529

Revelle R (1982) Carbon dioxide and world climate. Sci Am 247:33–41

Revsbech NP, Jørgensen BB, Blackburn PH (1980a) Oxygen in the sea bottom measured with microelectrode. Science 207:1355–1356

Revsbech NP, Sørensen J, Blakburn TH, Lomholt JP (1980b) Distribution of oxygen in marine sediments measured with microelectrodes. Limnol Oceanogr 25:403–411

Reynolds WW, McCauley RW, Casterlin ME, Crawsha LI (1976) Body temperature of behaviourally thermoregulating largemouth bass, *Micropterus salmoides*. Comp Biochem Physiol 59A:461–475

Reznick DN, Shaw FH, Rodd FH, Shaw RG (1997) Evaluation of the rate of evolution in natural populations of guppies (*Poecilia reticulata*). Science 275:1934–1936

Rhoads DC, Morse PW (1971) Evolutionary and ecologic singificance of oxygen-deficient marine basins. Letharia 4:413–428

Rice SA (1980) Hydrodynamic and diffusion considerations of rapid-mix experiments with erythrocytes. Biophys J 29:65–78

Richards AB (1957) Studies on arthropod cuticle. XIII. The penetration of dissolved oxygen and electrolytes in relation to the multiple barriers of the epicuticle. J Insect Physiol 1:23–29

Richards AG, Korda FH (1950) Studies on arthropod cuticle. IV. An electron microscope survey of the intima of arthropod tracheae. Ann Entomol Soc Am 43:49–71

Richardson J (1976) Autumnal migration over Puerto Rico and the Western Atlantic: a radar study. Ibis 118:309–332

Richman PS, Wolfson MR, Shaffer TH (1993) Lung lavage with oxygenated perfluorochemical liquid in acute lung injury. Crit Care Med 21:768–774

Richter C, Park JW, Ames BN (1988) Normal oxidative damage to mitochondrial and nuclear DNA is extensive. Proc Natl Acad Sci USA 85:6465–6467

Riddle WA (1983) Physiological ecology of land snails and slugs. In: Russell-Hunter WD (ed) The Mollusca. Academic Press, New York, pp 431–448

Ridgway SH, Howard R (1979) Dolphin lung collapse and intramuscular circulation during free diving: evidence from nitrogen washout. Science 206:1182–1183

Ridgway SH, Johnston DG (1966) Blood oxygen and ecology of porpoises of three genera. Science 151:456–467

Riedel C, Wood SC (1988) Effects of hypercapnia and hypoxia on temperature selection of the toad, *Bufo marinus*. Fed Proc 2:500A

Riedesel ML (1977) Blood physiology. In: Wimsatt WA (ed) Biology of bats, vol II. Academic Press, New York, pp 485–517

Riedesel ML, Williams BA (1976) Continuous 24-hr oxygen consumption studies of *Myotis velifer*. Comp Biochem Physiol 54A:95–99

Riedl R (1978) Order in living organisms. John Wiley, New York

Rieppel O (1993) The conceptual relation of ontogeny, phylogeny, and classification: the taxic approach. Evol Biol 27:1–32

Riggs A (1976) Factors in the evolution of haemoglobin function. Fed Proc 35:2115–2118

Riggs A (1979) Studies of the haemoglobins of Amazonian fishes: A review. Comp Biochem Physiol 62A:257–271

Riggs DS (1963) The mathematical approach to physiological problems. Williams and Wilkins, Baltimore

Riley JP, Skirrow G (1975) Chemical oceanography, 2nd edn, vol 2. Academic Press, London, pp 134–198

Risenbrough RW (1986) Pesticides and bird population. In: Johnston RF (ed) Current ornithology, vol 3. Plenum Press, New York, pp 397–427

Roberts JL (1975) Respiratory adaptations of aquatic animals. In: Vernberg FJ (ed) Physiological adaptation to the environment. Intext Educational Publishers, New York pp 395–435

Roberts JL, Rowell DM (1988) Periodic respiration of gill breathing fishes. Can J Zool 66:182–192

Robertson JI (1913) The development of the heart and vascular system of *Lepidosiren paradoxa*. Q J Microsc Sci Lond 59:53–132

Robertson RJ, Gilcher R, Metz KF et al. (1982) Effect of induced erythrocythemia on hypoxia tolerance during physical exercise. J Appl Physiol 53:490–495

Robin ED, Bromberg PA, Cross CE (1969) Some aspects of the evolution of vertebrate acid-base regulation. Yale J Biol Med 42:448–476

Robinson JM (1991) Global planetary change. Paleobiology 97:51–62

Rogers PJ, Stewart PR (1973) Respiratory development in *Saccharomyces cerevisiae* grown at controlled oxygen tension. J Bacteriol 115:88–97

Rogers RM, Braunstein MS, Shuman JF (1972) Role of bronchopulmonary lavage in the treatment of respiratory failure: a review. Chest 62:95S–105S

Rohmer M, Bouvier P, Ourisson G (1979) Molecular evolution of biomembranes: structural equivalents and phylogenetic precusors of sterols. Proc Natl Acad USA 76:847–851

Rolschau J (1978) A prospective study of the placental weight and content of protein, RNA and DNA. Acta Obstet Gynecol Scand (Suppl) 72:28–43

Romanoff AL (1960) The avian embryo. Macmillan, New York

Romanoff AL (1967) Biochemistry of the avian embryo. John Wiley, New York

Romanoff AL, Romanoff AJ (1949) The avian egg. John Wiley, New York

Romer AS (1946) The early evolution of fishes. Q Rev Biol 21:33–69

Romer AS (1966) Vertebrate paleontology, 3rd edn. University of Chicago Press, Chicago

Romer AS (1967) Major steps in vertebrate evolution. Science 158:1629–1637

Romer AS (1972) Skin breathing – primary or secondary? Respir Physiol 14:183–192

Romijn C (1948) Respiratory movements of the chicken during the parafoetal period. Physiol Comp Oecol 1:24–48

Romijn C (1950) Foetal respiration in the hen: gas diffusion through the egg shell. Poult Sci 29:42–51

Ronan CA (1991) The natural history of the Universe. Doubleday, London

Roper CFE (1969) Systematics and zoogeography of the worldwide bathypelagic squid *Bathyteuthis* (Cephalopoda: Oegopsida). Smithsonian Inst Press, Washington, DC, pp 1–210

Rose FL, Zambernard J (1966) Cardiac glycogen depletion in *Amphiuma means* during induced anoxia. J Morphol 120:391–396

Rose KD, Bown TM (1984) Gradual phyletic evolution at the generic level in early Eocene omomyid primates. Nature (Lond) 309:250–252

Rosen R (1967) Optimality principles in biology. Plenum Press, New York

Rosen DE, Forey PL, Gardiner BG, Petterson C (1981) Lungfishes, tetrapods paleontology and plesiomorphology. Bull Am Mus Nat Hist 167:159–276

Rosen P, Stier A (1973) Kinetics of CO_2 and O_2 complexes of rabbit liver microsomal cytochrome P_{450}. Biochem Biophys Res Commun 51:603–611

Rossi-Fanelli A, Antonini E (1957) A new type of myoglobin isolated and crystallized from muscles of Aplysiae. Biochemistry (USSR) 22:312–321

Rossitti S, Löfgren J (1993a) Vascular dimensions of the cerebral arteries follow the principle of minimum work. Stroke 24:371–377

Rossitti S, Löfgren J (1993b) Optimality principles and flow oderliness at the branching points of cerebral arteries. Stroke 24:1029–1032

Rossitti S, Stephensen H (1994) Temporal heterogeneity of the blood flow velocity at the middle cerebral artery in the normal human characterized by fractal analysis. Acta Physiol Scand 1511:191–198

Roth M, Depierre D (1975) Dipeptidyl carboxypeptidase in lung and blood plasma. In: Junod AE, Haller R (ed) Lung metabolism. Academic Press, New York, pp 337–345

Roughton FJW (1945) The average time spent by the blood in the human lung capillary and its relation to the rate of CO_2 uptake and elimination. Am J Physiol 143:621

Roughton FJW, Forster RE (1957) Relative importance of diffusion and chemical reaction rates in determining rate of O_2 exchange of pulmonary membrane and volume of blood in lung capillaries. J Appl Physiol 11:290–302

Royer WE, Love WE, Fenderson FF (1985) Cooperative dimeric and tetrameric clam haemoglobins are novel assemblages of myoglobin folds. Nature (Lond) 316:277–280

Ruben JA, Reagan NL, Verrell PA, Boucot AJ (1993) Plethodontid salamabder origins: a response to Beachy and Bruce. Am Nat 142:1038–1051

Rubenstein DL, Koehl MAR (1977) The mechanism of filter-feeding: some theoretical consideration. Am Nat 111:981–994

Ruckhäberle KE, Franke J, Viehweg B, Gerl D (1977) Quantitative Veränderungen an Resorptionszotten normaler menschlicher Plazenten im Verlauf der Gestation. Zentralbl Gynaekol 99:1313–1322

Rudwick MJS (1964) The inference of function from structure in fossils. Br J Philos Sci 15:27–40

Runham NW, Hunter PJ (1970) Terrestrial slugs. Hutchison, London

Runnegar B (1982) Oxygen requirements, biology and phylogenetic significance of the late Precamrian worm Dickinsonia, and the evolution of the burrowing habit. Alcheringa 6:223–239

Runnegar B (1992) Evolution of the earliest animals. In: Schopf JW (ed) Major events of the history of life. Johns and Bartlett, Boston, pp 65–95

Rurak DW, Gruber NC (1983) Increased oxygen consumption associated with breathing activity in foetal lambs. J Appl Physiol 54:701–707

Rushner RF (1965) General characteristics of the cardiovascular system. In: Ruch TC, Patton HD (eds) Physiology and biophysics, 19th edn. Saunders, Philadelphia, pp 543–549

Russell CW, Evans BK (1989) Cardiovascular anatomy and physiology of the black-lip abalone, *Haliotis ruber*. J Exp Zool 252:105–117

Russell MJ, Daniel RM (1992) Emergence of life via catalytic hydrothermal colloidal iron sulphide membranes. Ann Geophys Suppl III 10:506

Russell MJ, Daniel RM, Hall AJ, Sherringham JA (1994) A hydrothermally precipitated catalytic iron sulphide membrane as a first step toward life. J Mol Evol 39:231–243

Ruthen R (1993) Adapting to complexity. Sci Am 268:130–140

Rutten MG (1970) The history of atmospheric oxygen. Space Life Sci 2:5–17

Ruud JT (1954) Vertebrates without erythrocytes and blood pigment. Nature (Lond) 173:848–850

Ruud JT (1965) The ice fish. Sci Am 213:108–114

Ryan JW, Ryan US (1975) Metabolic activities of plasma membrane and caveolae of pulmonary endothelial cells, with a note on pulmonary prostaglandin synthetase. In: Junod AE, Haller R (eds) Lung metabolism. New York, Academic Press, pp 399–424

Ryan JW, Roblero J, Stewart JM (1968) Inactivation of bradykinin in the pulmonary circulation. Biochem J 110:795–797

Ryan JW, Roblero J, Stewart JM (1970) Inactivation of bradykinin in rat lung. Adv Exp Med Biol 8:263–274

Rzoska J (1974) The Upper Nile swamps: a tropical wetland study. Freshwater Biol 4:1–30

Sacca R, Burggren WW (1982) Oxygen uptake in water and air in the air-breathing reedfish *Calamoichthys calabaricus* role of skin, gills and lungs. J Exp Biol 97:179–186

Sachs G (1977) Ion pumps in the renal tubule. Am J Physiol 233:F359–365

Saeed SA, Roy AC (1972) Purification of 15-hydroxyprostaglandin dehydrogenase from bovine lung. Biochem Res Commun 47:96–107

Safford-Black V (1944) Gas exchange in the simbladder of the mudminnow, *Umbra limi* (Kirtland). Proc Nova Scotia Inst Sci 21:1–22

Saga S, Modell HJ, Calderwood HW, Lucas AJ, Tham ML, Swenson EW (1973) Pulmonary function after ventilation with fluorocarbon liquid P-12F (caroxin-F). J Appl Physiol 34:160–164

Sagan C (1994) The search for extraterrestrial life. Sci Am 271:71–77

Sagan C, Chyba C (1997) The early faint sun paradox: organic shielding of ultraviolet-labile greenhouse gases. Science 276:1217–1221

Salomonsen F (1967) Migratory movements of the Arctic tern (*Sterna paradisea pontoppidan*) in the Southern Ocean. Det Kgl Dan Videns Selsk Biol Med 24:1–37

Saltin B (1985) Malleability of the system in overcoming limitations: functional elements. J Exp Biol 115:345–354

Saltin B, Gollnick PD (1983) Skeletal muscle adaptability: singificance for metabolism and performance. In: Peachy LD, Adrian RH, Geiger SR (eds) Handbook of physiology: skeletal muscle. Williams and Wilkinson, Baltimore, pp 555–631

Samuelson RE, Hanel RA, Kunde VG, Maguine WC (1981) Mean molecular weight and hydrogen abundance of Titan's atmosphere. Nature (Lond) 292:688–698

Sanderson RJ, Paul CW, Vatter AE, Filley GF (1976) Morphological and physical basis for lung surfactant action. Respir Physiol 27:379–392

Sandreuter A (1951) The structure and function of the avian egg. Acta Anat II, Suppl 14, pp 1–72. Quoted from Romanoff (1960)

Santos EA, Baldisseroto B, Biachini A, Colares EP, Nery LEM, Manzoni GC (1987) Respiratory mechanisms and metabolic adaptations of an intertidal crab, *Chasmagnathus graulata* (Dana, 1851). Comp Biochem Physiol 88A:21–25

Sarnthein M, Winn K, Duplessy JC, Fontugne MR (1988) Global variations of surface ocean primary productivity in low and mid latitudes: influence on CO_2 reservoirs of the deep ocean and atmosphere during the last 21 000 years. Paleoceanography 3:361–399

Sass DJ, Ritman EL, Caskey PE, Banchero N, Wood EH (1972) Liquid breathing: prevention of pulmonary arterial-venous shunting during accelaration. J Appl Physiol 32:451–455

Sassaman C, Mangum CP (1972) Adaptations to environmental oxygen levels in infaunal and epifaunal sea anemones. Biol Bull 143:657–678

Sassaman C, Mangum CP (1973) Adaptations to environmental oxygen levels in infaunal sea anemones. Biol Bull Mar Biol Lab Woods Hole 143:657–678

Sassaman C, Mangum CP (1974) Gas exchange in a cerianthid. J Exp Zool 188:297–305

Sassone-Corsi P (1996) Same clock, different works. Nature (Lond) 384:613–614

Satchell GH (1971) Circulation in fishes. Cambridge University Press, Cambridge

Satchell GH (1976) The circulatory system of air-breathing fish. In: Hughes GM (ed) Respiration of amphibious vertebrates. Academic Press, New York, pp 105–124

Satchell GH (1984) Respiratory toxicology of fishes. In: Weber LJ (ed) Aquatic toxicology, vol 2. Raven Press, New York, pp 1–35

Satchell GH (1992) The venous system. In: Hoar WS, Randall DJ, Farrell AP (eds) Fish physiology, vol 12. Academic Press, New York, pp 141–234

Saunders RL (1953) The swim bladder gas content of some fresh water fish with particular reference to the physostomes. Can J Zool 31:547–560

Saunders RL, Sutterlin AM (1971) Cardiac and respiratory responses to hypoxia in the sea raven, *Hemitripterus americanus*, and an investigation of possible control mechanisms. J Fish Res Board Can 28:491–503

Savage RM (1935) The ecology of young tadpoles, with special reference to some adaptations to the habitat of mass spawning in *Rana temporaria* L. Proc Zool Soc (Lond) 605–610

Sawaya P (1947) Metabolisms respiratoria de Amphibio Gymnophiona, *Typhylonectes compressicauda* (Dum et Bibr). Biol Fac Fil Cien Let U Sao Paulo 1947:51–56

Saxena DB (1960) On the asphyxiation and influence of CO_2 on respiration of air-breathing fish, *Heteropneustes fossilis* and *Clarias batrachus*. J Zool (Lond) 12:114–124

Saxena DB (1962) Studies on the physiology of respiration in fishes: V. Comparative study of the gill area in the fresh water fishes *Labeo rohita, Ophicephalus* (= *Channa*) *striatus* and *Anabas testudineus*. Ichthyologica 1:59–70

Saxena DB (1963) A review of ecological studies and their importance in the physiology of air breathing fish. Ichthyologica 2:116–128

Sayer MDJ, Davenport J (1991) Amphibious fish: why do they leave water? Rev Fish Biol Fisheries 1:159–181

Scammell CA, Hughes GM (1981) Comparative study of the functional anatomy of the gills and ventilatory currents in some British decapod crustaceans. Biol Bull 156:35–47

Schaeffer B (1965a) The role of experimentation in the origin of higher levels of organization. Syst Zool 14:318–336

Schaeffer B (1965b) The rhipidistian amphibian transition. Am Zool 5:267–276

Schaffer WM, Kot M (1986) Chaos in ecological systems: the coals that Newcastle forgot. Trends Ecol Evol 1:58–63

Scheid P (1978) Analysis of gas exchange between air capillaries and blood capillaries in the avian lung. Respir Physiol 32:27–49

Scheid P (1979) Mechanisms of gas exchange in bird lungs. Rev Physiol Biochem Pharmacol 86:137–186

Scheid P (1985) Significance of lung structure for performance at high altitude. In: Ilyicher VD, Gavrilov VM (eds) Acta XVIII Int Congr of Ornithology, vol III. Nauka, Moscow, 976pp

Scheid P (1987) Cost of breathing in water- and air-breathers. In: Dejours P, Taylor CR, Weibel ER (eds) Comparative physiology: life on land and water, Fidia Res Series, vol 9. Liviana Press, Padova, pp 83–92

Scheid P (1990) Avian respiratory system and gas exchange. In: Sutton JR, Coates G, Remmers JE (eds) Hypoxia: the adaptations. BC Decker, Burlington, Ontario, pp 4–7

Scheid P, Kawashiro T (1975) Metabolic changes in avian blood and their effects on determination of blood gases and pH. Respir Physiol 23:291–300

Scheid P, Piiper J (1969) Volume, ventilation and compliance of the respiratory system in the domestic fowl. Respir Physiol 6:298–308

Scheid P, Piiper J (1970) Analysis of gas exchange in the avian lung: theory and experiments in the domestic fowl. Respir Physiol 9:246–262

Scheid P, Piiper J (1972) Cross-currrent gas exchange in the avian lungs: effects of reversed parabronchial air flow in ducks. Respir Physiol 16:304–312

Scheid P, Piiper J (1976) Quantitative functional analysis of branchial gas transfer: theory and application to *Scyliorhinus stellaris* (Elasmobrachii). In: Hughes GM (ed) Respiration of amphibious vertebrates. Academic Press, New York, pp 17–38

Scheid P, Piiper J (1987) Gas exchange and transport. In: Seller TJ (ed) Bird respiration, vol I. CRC Press, Boca Raton, pp 97–129

Scheid P, Piiper J (1989) Respiratory mechanics and air flow in birds. In: King AS, McLelland J (eds) Form and function in birds, vol 4. Academic Press, London, pp 369–391

Scheid P, Worth H, Holle JP, Meyer M (1977) Effects of oscillating and intermittent ventilatory flow on efficacy of pulmonary O_2 transfer in the duck. Respir Physiol 31:251–258

Scheuer J, Tipton CM (1977) Cardiovascular adaptations to physical training. Annu Rev Physiol 39:221–251

Schick JM (1991) A functional biology of sea anemones. Chapman and Hall, London

Schidlowski M (1975) Archean atmosphere and evolution of the terrestrial O_2 budget. In: Windley BF (ed) The early history of the Earth. John Wiley, London, pp 125–201

Schmalhausen II (1968) The origin of terrestrial vertebrates. Academic Press, Lonon

Schmid-Schönbein H (1975) Erythrocytes rheology and the optimization of mass transport in the circulation. Blood Cells 1:285–306

Schmid-Schönbein H (1988) Conceptual proposition for a specific microcirculatory problem: maternal blood flow in hemochorial multivillous placentae as percolation of a "porous medium". Trophoblast Res 3:17–38

Schmidt H, Kamp G (1996) The "Pasteur effect" in facultative anaerobic metazoa. Experientia 52:440–448

Schmidt-Nielsen K (1975) Recent advances in avian respiration. In: Peaker M (ed) Avian physiology. Academic Press, London, pp 33–47

Schmidt-Nielsen K (1984) Scaling: why is animal size so important? Cambridge University Press, Cambridge

Schmidt-Nielsen K (1990) Animal physiology: adaptation and environment, 4th edn. Cambridge University Press, Cambridge

Schmidt-Nielsen K, Larimer JL (1958) Oxygen dissociation curves of mammalian blood in relation to body size. Am J Physiol 195:424–428

Schmidt-Nielsen K, Taylor CR (1968) Red blood cells: why or why not? Science 162:274–275

Schmidt-Nielsen K, Schmidt-Nielsen B, Jarnum SA, Houpt TR (1957) Body temperature of the camel and its relation to water economy. Am J Physiol 188:103–112

Schmidt-Nielsen K, Kanwisher J, Lasiewski RC, Cohn JE, Bretz WL (1969) Temperature regulation and respiration in the ostrich. Condor 71:341–352

Schmidt RS (1982) Possible importance of lung inflation related sensory input to frog calling circuits. Copeia 1982:196–198

Schneiderman HA (1960) Discontinuous respiration in insects; role of the spiracles. Biol Bull Mar Biol Lab Woods Hole 119:494–528

Schoene RB, Swenson ER, Pizzo CJ, Hackett PH, Roach RC et al. (1988) The lung at high altitude: bronchoalveolar lavage in acute mountain sickness, and pulmonary oedema. J Appl Physiol 64:2605–2613

Scholander PF (1954) Secretion of gases against high pressures in the swim bladder of deep sea fishes. II. The rete mirabile. Biol Bull Mar Biol Lab Woods Hole 107:247–259

Scholander PF (1958) Counter-current exchange. A principle in biology. Hvalrådets Skr 44:1–24

Scholander PF (1960) Oxygen transport through haemoglobin solutions. How does the presence of haemoglobin in a wet membrane mediate an eightfold increase in oxygen passage? Science 131:585–590

Scholander PF, van Dam L (1953) Composition of the swim bladder gas in deep sea fishes. Biol Bull 104:75–97

Scholey K (1986) The evolution of flight in bats. In: Nachtigall W (ed) BIONA report no 5. Gustav Fischer, Stuttgart, pp 1–12

Schömig A, Fischer S, Kurz Th, Richardt G, Schömig E (1987) Nonexcytotic release of endogenous noradrenaline in the ischemic and anoxic heart: mechanism and metabolic requirements. Circ Res 60:194–205

Schopf JW (1978) The evolution of the earliest cells. Sci Am 239:110–138

Schopf JW (1980) Paleoceanography. Harvard University Press, Cambridge

Schopf JW (1983) Earth's earliest biosphere: its origins and evolution. Princeton University Press, Princeton

Schopf JW (1984) Rates of evolution and notion of living fossils. Annu Rev Earth Planet Sci 12:245–292

Schopf JW (1989) The evolution of the earliest cells. In: Gould JL, Gould CG (eds) Life at the edge: readings from the Scientific American Magazine. WH Freeman, New York, pp 7–23

Schopf JW (1993) Microfossils of the early Archean Apex chert: new evidence of antiquity of life. Science 260:640–646

Schopf JW, Oehler DZ (1976) How old are the eukaryotes? Science 193:47–49

Schopf JW, Walter MR (1983) Archean microfossils: new evidence of ancient microbes. In: Schopf JW (ed) Earth's earliest biosphere: its origin and evolution. Princeton University Press, Princeton, pp 214–239

Schopf JW, Hayes JM, Walter MR (1983) Evolution of Earth's earliest ecosystems: recent progress and unresolved problems. In: Schopf JW (ed) Earth's earliest biosphere: its origin and evolution. Princeton University Press, Princeton, pp 361–384

Schöttle E (1932) Morphologie und Physiologie der Atmung bei wasserschlamm- und landlebenden Gobiiformes. Z Wiss Zool 140:1–114

Schöttler U, Wienhausen G, Zebe E (1983) The mode of energy production in the lugworm Arenicola marina at different oxygen concentrations. J Comp Physiol 149:547–555

Schöttler U, Wienhausen G, Werterman J (1984) Anaerobic metabolism in the lugworm Arenicola marina L.: the transition from aerobic to anaerobic metabolism. Comp Biochem Physiol 79B:93–103

Schumann D, Piiper J (1966) Der Sauerstoffbedarf der Atmung bei Fischen nach Messungen an der narkotisierten Schleie (Tinca tinca). Arch Gesamte Physiol Mens Tiere (Pfluegers) 288:14–26

Schürch S, Bachofen H, Weibel ER (1985) Alveolar surface tension in exercised rabbit lungs: effect of temperature. Respir Physiol 62:31–45

Schürch S, Gehr P, Hof VI, Geiser M, Green F (1990) Surfactant displaces particles toward the epithelium in airways and alveoli. Respir Physiol 80:17–32

Schurmann H, Steffensen JF (1992) Lethal oxygen levels at different temperature and the preferred temperature during hypoxia in the Atlantic cod, Gadus morhua L. J Fish Biol 41:927–934

Schurmann H, Steffensen JF, Lomholt JP (1991) The influence of hypoxia on the preferred temperature of rainbow trout Oncorhynchus mykiss. J Exp Biol 157:75–86

Schwartz JH (1976) H^+ current response to CO_2 and carbonic anhydrase inhibition in turtle bladder. Am J Physiol 231:565–572

Schwerdtfeger WK (1979) Morphometrical studies of the ultrastructure of the epidermis of the guppy, Poecilia reticulata Peters, following adaptation to sea water and treatment with prolactin. Gen Comp Endocrinol 38:476–483

Scrutton CT (1978) Periodic growth features in fossil organisms and the length of the day and month. In: Brosche P, Sunderman (eds) Tidal friction and the Earth's rotation. Springer Berlin Heidelberg, New York, pp 87–169

Secomb TW (1991) Erythrocytes mechanics and capillary blood rheology. Cell Biophys 18:231–251

Seed R (1983) Structural organization, adaptive radiation and classification of molluscs. In: Hochachka PW (ed) The Mollusca, vol I. Academic Press, New York, pp 1–54

Seeherman HJ, Taylor CR, Maloiy GMO, Armstrong RB (1981) Design of the mammalian respiratory system. II. Measuring maximal aerobic capacity. Respir Physiol 44:11–23

Seibel BA, Childress JJ (1996) Deep sea breathing cephalopods? Nature (Lond) 384:421

Seifert R, Schultz C (1991) The superoxide-forming NADPH oxidase of phagocytes: an enzyme system regulated by multiple mechanisms. Rev Physiol Biochem Pharmacol 117:1–338

Seifriz W (1943) Protoplasmic streaming. Biol Rev 9:49–123

Selden P, Edwards D (1989) Colonization of the land. In: Allen KC, Briggs DEG (eds) Evolution and ecology. Pinter, London, pp 67–127

Seliger HH, Boggs JA, Biggley WH (1985) Catastrophic anoxia in the Chesapaeke Bay in 1984. Science 228:70–73

Semlitsch RD, Wilbur HM (1989) Artificial selection for paedomorphosis in the salamander, *Ambystoma talpoideum*. Evolution 43:105–112

Sen CK (1995) Oxidants and antioxidants in excercise. J Appl Physiol 79:675–686

Serfaty A, Gueutal J (1943) La résistance de la grenouille à l'asphyxie lors d'une immersion prolongée. CR Soc Biol 137:154–156

Service RF (1997) Microbiologists explore life's rich, hidden kingdoms. Science 275:1740–1742

Setnikar IE, Agostini E, Taglietti A (1959) The foetal lung, a source of amniotic fluid. Proc Soc Exp Biol Med 101:842–845

Severinghaus JW (1971) Transarterial leakage: a possible mechanism of high altitude pulmonary oedema. In: Porter R, Knight J (eds) High altitude physiology: cardiac and respiratory aspects. Churchill-Livingstone, Edinburgh, pp 61–77

Seymour RS (1978) Gas tensions and blood distribution in sea snakes at surface pressure and at simulated depth. Physiol Zool 51:388–407

Seymour RS (1979) Dinosaur eggs: gas conductance through the shell, water loss during incubation and clutch size. Paleobiology 5:1–11

Seymour RS (1991) Analysis of heat production in a thermogenic arum lily, Philodendron *selloum* by three calorimetric methods. Thermochim Acta 193:91–97

Seymour RS (1997) Plants that warm themselves. Sci Am 276:90–95

Seymour RS, Ackerman RA (1980) Adaptations to underground nesting in birds and reptiles. Am Zool 20:437–457

Seymour RS, Rahn H (1978) Gas conductance in the eggshell of mound building brush turkey. In: Piiper J (ed) Respiratory function in birds, adult and embryonic. Springer, Berlin Heidelberg New York, pp 243–262

Seymour RS, Schultze-Motel P (1996) Thermoregulating lotus flowers. Nature (Lond) 383:305

Seymour RS, Webster MED (1975) Gas transport and blood acid-base balance in diving sea snakes. J Exp Zool 191:169–181

Seymour RS, Spragg RG, Hartman MT (1981) Distribution of ventilation and perfusion in the sea snake, *Pelamis platurus*. J Comp Physiol 60A;145:109–115

Shackleton NJ (1993) The climate system in the recent geological past. Philos Trans R Soc Lond 341B:209–213

Shackleton NJ, Pisias NG (1985) Atmospheric carbon dioxide, orbital forcing and climate. In: Sundquist ET, Broecker WS (eds) The carbon cycle and atmospheric CO_2: natural variations Archean to present. Geophys Monogr 32:303–317

Shadwick RE, O'Dor RK, Gosline JM (1990) Respiratory and cardiac function during exercise in squid. Can J Zool 68:792–798

Shaffer G (1990) A non-linear climate oscillator controlled by biochemical cycling in the ocean: an alternative model for Quatenary ice ages cycles. Clim Dyn 4:127–143

Shaffer TH, Moskowitz GD (1974) Demand-controlled liquid ventilation of the lungs. J Appl Physiol 36:208–215

Shaffer TH, Rubenstein D, Moskowitz GD, Delivoria-Papadopoulos M (1976) Gaseous exchange and acid-base balance in premature lambs during liquid ventilation since birth. Pediatr Res 10:227–231

Shams H, Scheid P (1987) Respiration and blood gases in the duck exposed to normocapnic and hypercapnic hypoxia. Respir Physiol 67:1–12

Shams H, Scheid P (1989) Efficiency of the parabronchial gas exchange in deep hypoxia measurements in the resting duck. Respir Physiol 77:135–146

Shannon P, Kramer DL (1988) Water depth alters respiratory behaviour of *Xenopus laevis*. J Exp Biol 137:597–602

Sheehan PM, Fastovsky DE, Hoffman RG, Berghaus CB, Gabriel DL (1991) Sudden extinction of the dinosaurs: latest Cretaceous, Upper Great Plains, USA. Science 254–835–838

Sheldon RE, Peeters LLH, Jones MD, Makowski EL, Meschia G (1978) Redistribution of the cardiac output and oxygen delivery in the hypoxemic foetal lamb. Am J Obstet Gynecol 135:1071–1078

Shelton G (1970) The effect of lung ventilation on blood flow to the lungs and body of the amphibian, *Xenopus laevis*. Respir Physiol 9:183–196

Shelton G (1976) Gas exchange, pulmonary blood supply and the partially divided amphibian heart. In: Spencer-Davies P (ed) Perspectives in environmental biology. Pergamon Press, Oxford, pp 247–259

Shelton G (1992) Model applications in respiratory physiology. In: Egginton S, Ross HF (eds) Oxygen transport in biological systems: modelling of pathways. Cambridge University Press, Cambridge, pp 1–44

Shelton G, Burggren WW (1976) Cardiovascular dynamics of the chelonia during apnea and lung ventilation. J Exp Biol 64:323–343

Shelton G, Croghan PC (1988) Gas exchange and its control in non-steady state systems: the consequences of evolution from water to air breathing in vertebrates. Can J Zool 66:109–123

Sheltor G, Jones DR (1965) Central blood pressure and heart output in surfaced and submerged frogs. J Exp Biol 42:339–357

Shelton G, Jones DR, Milsom WK (1986) Control of breathing in ectothermic vertebrates. In: Cherniack NS, Widdicombe JG (eds) Handbook of physiology, sect 3. Respiratory system, vol 2. Control of breathing. American Physiological Society, Bethesda, pp 857–909

Shepherd AP, Riedel GL (1982) Optimal hematocrit for oxygenation of canine intestine. Circ Res 51:233–240

Shepherd SA, Thomas IM (1989) Marine invertebrates of Southern Australia. South Australian Govt Press, Adelaide

Sheridan MA (1994) Regulation of lipid metabolism in poikilothermic vertebrates. Comp Biochem Physiol 107B:495–508

Sherman DR, Guinn B, Perdok MM, Goldberg DE (1992) Components of sterol biosynthesis assembled on the oxygen-avid haemoglobin of *Ascaris*. Science 258:1930–1932

Shield JW, Bentley PJ (1973a) Respiration of some urodele and anuran Amphibia in water. I. Role of the skin and the gills. Comp Biochem Physiol 46A:17–28

Shield JW, Bentley PJ (1973b) Respiration of some urodele and anuran Amphibia in air. II. Role of the skin and lungs. Comp Biochem Physiol 46A:29–38

Shiga T (1994) Oxygen transport in microcirculation. Jpn J Physiol 44:19–34

Shigenaga MK, Gimeno CJ, Ames BN (1989) Urinary 8-Hydroxy-2'-deoxyguanosine as a biological marker of in vivo oxidative DNA damage. Proc Natl Acad Sci USA 86:9697–9701

Shiklomanov IA (1993) Water in crisis: a guide to the worlds fresh water resources. In: Gleick PH (ed) Land and water use. Oxford University Press, New York, pp 13–24

Shine R (1983) Reptilian viviparity in cold climates: testing the assumptions of an evolutionary hypothesis. Oecologia 57:397–405

Shine R (1985) The evolution of viviparity in reptiles: an ecological analysis. In: Gans C, Billett F (eds) Biology of Reptilia, vol 15. John Wiley, New York, pp 604–694

Shine R (1989) Ecological influences on the evolution of vertebrate viviparity. In: Wake D, Roth G (eds) Complex organismal functions: integration and evolution in vertebrates. (Dahlem Workshop Report), John Wiley, Chichester, pp 263–278

Shine R, Guillette LJ (1988) The evolution of viviparity in reptiles: a physiological model and its ecological consequences. J Theor Biol 132:43–50

Shlaifer A, Breder CM (1940) Social and respiratory behaviour of small tarpon. Zoologica 25:493–512

Shoemaker VH, Balding D, Raubal R, McClanahan LL (1972) Uricotelism and low evaporative water loss in a South American frog. Science 175:1018–1020

Sibley CG, Ahlquist JE (1990) Phylogeny and classification of birds: a study in molecular evolution. Yale University Press, New Haven

Sidell BD, Driedzic WR, Stowe DB, Johnston IA (1987) Biochemical correlations of power development and metabolic fuel preferanda in fish hearts. Physiol Zool 60:221–232

Sidell BD, Vayda ME, Small DJ, Moylan TJ, Londraville RL et al. (1997) Variable expression of myoglobin among the haemoglobinless Antarctic icefishes. Proc Natl Acad Sci USA 94:3420–3424

Sieker HO, Hickam JB (1956) Carbon dioxide intoxication: the clinical syndrome, its aetiology and management with particular reference to the use of mechanical respirators. Medicine 35:389–408

Sies H (1991) Oxidative stress: oxidants and antioxidants. Academic Press, Orlando

Sies H, Cadenas E (1985) Oxidative stress: damage to intact cells and organs. Philos Trans R Soc Lond 311B:617–631

Siever R (1979) The Earth – readings from Scientific American: life – origin and evolution. WH Freeman, San Francisco, pp 25–31

Sikand RS, Magnussen H, Scheid P, Piiper J (1976) Convection and diffusive gas mixing in human lungs: Experiments and model analysis. J Appl Physiol 40:362–371

Simchon S, Jan KM, Chien S (1987) Influences of reduced red cell deformability on regional blood flow. Am J Physiol 253:H898–H903

Simkiss K (1980) Eggshell porosity and the water metabolism of the chick embryo. J Zool (Lond) 192:1–18

Simons GH, Sussana PG (1886) Aquatic respiration in soft-shelled turtles: a contribution to the physiology of respiration in vertebrates. Am Nat 20:233–236

Simpson GG (1953) The major features of evolution. Columbia University Press, New York

Simpson RA, Mayhew TM, Barnes PR (1992) From 13 weeks to term, the trophoblast of human placenta grows by the continuous recruitment of new proliferative units: a study of nuclear number using the disector. Placenta 13:501–512

Sinclair JD (1987) Respiratory drive in hypoxia: carotid body and other mechanisms compared. News Physiol Sci 2:57–69

Singh BN (1976) Balance between aquatic and aerial respiration. In: Hughes GM (ed) Respiration of amphibious vertebrates. Academic Press, New York, pp 125–164

Singh BN, Hughes GM (1973) Respiration of an air-breathing catfish Clarias batrachus (Linn). J Exp Biol 55:421–434

Singh BN, Munshi JSD (1968) On the respiratory organs and mechanics of breathing in Periophthlmus vulgaris. Zool Anz 183:92–110

Singh BR, Mishra AP (1980) Development of air-breathing organs in Anabas testudineus (Bloch). Zool Anz 205:359–370

Singh BR, Mishra AP, Singh RP (1982) Development of the air-breathing organ in the snake-headed fish, Channa punctatus. Zool Anz 208:428–439

Singh BR, Mishra AP, Singh I (1984) Development of the air-breathing organ in the mud-eel, Amphipnous cuchia (Ham.). Zool Anz 213:395–407

Singh MP, Khetarpal K, Sharan MA (1980) A theoretical model for studying the rate of oxygenation of blood in pulmonary capillaries. J Math Biol 9:305–330

Sinha AK, Gleed RD, Hakim TS, Dobson A, Shannon J (1996) Pulmonary capillary pressure during exercise in horses. J Appl Physiol 80:1792–1798

Siskin M, Katritzky AR (1991) Reactivity of organic compounds in hot water: geochemical and technological implications. Science 254:231–327

Skalak R, Branemark PI (1969) Deformation of erythrocytes in capillaries. Science 164:717–719

Skerret SJ (1994) Host defenses against respiratory infection. Med Clin North Am 78:941–965

Slater TF (1984) Free radical mechanisms in tissue injury. Biochem J 222:1–15

Slonim NB, Hamilton LH (1971) Respiratory physiology, 2nd edn. CV Mosby Company, St Louis

Smatresk NJ (1979) Scaphognathite activity, heart rate and acid base balance after exercise in the coconut crab Birgus latro. Alpha Helix Rep, Scripps Oceanogr Inst, La Jolla, pp 108–119

Smatresk NJ (1988) Control of the respiratory mode of air-breathing fishes. Can J Zool 148:88–152

Smatresk NJ (1990) Chemoreceptor modulation of the endogenous respiratory rhythm in vertebrates. Am J Physiol 259:R887–897

Smatresk NJ (1994) Respiratory control in the transition from water breathing to air breathing in vertebrates. Am Zool 34:264–279

Smatresk NJ, Azizi SQ (1987) Characteristics of lung mechanoreceptors in spotted gar, a bimodal breather, *Lepisosteus oculatus*. Am J Physiol 252:R1066–R1072

Smatresk NJ, Cameron JN (1982a) Respiration and acid-base physiology of the spotted gar, a bimodal breather. I. Normal values and the response to severe hypoxia. J Exp Biol 96:263–280

Smatresk NJ, Cameron JN (1982b) Respiration and acid-base physiology of the spotted gar, a bimodal breather. II. Responses to temperature change and hypercapnia. J Exp Biol 96:281–293

Smatresk NJ, Burleson ML, Azizi SQ (1986) Chemoreflexive responses to hypoxia and NaCN in longnose gar: evidence for two chemoreceptor loci. Am J Physiol 251:R116–R125

Smil V (1997) Cycles of life: civilization and the biosphere. Scientific American Library. WH Freeman, New York

Smits AW, Flanagin JI (1994) Bimodal respiration in aquatic and terrestrial apodan amphibians. Am Zool 34:247–263

Smith AH (1976) Physiological changes associated with-long term increases in acceleration. In: Sneath PHA (ed) COSPAR: life sciences and space research 14. Akademie-Verlag, Berlin, pp 91–100

Smith AH (1978) The role of body mass and gravity in determining the energy requirements of homoitherms. In: Holmquist R, Stickland AC (eds) COSPAR: life sciences and space research 16. Pergamon Press, Oxford, pp 83–88

Smith DG, Campbell G (1976) The anatomy of the pulmonary vascular bed in the toad, *Bufo marinus* and *Xenopus laevis*. Cell Tissue Res 178:1–14

Smith DG, Gannon BJ (1978) Selective control of branchial arch perfusion in an air-breathing Amazonian fish *Hoplerythrinus unitaeniatus*. Can J Zool 56:959–964

Smith DG, Rapson L (1977) Differences in pulmonary microvascular anatomy between *Bufo marinus* and *Xenopus laevis*. Cell Tissue Res 178:1–15

Smith FGW (1957) Rivers in the sea. Smithson Inst Annu Rep 1956:431–441

Smith FM, Jones DR (1982) The effect of changes in blood oxygen carrying capacity on ventilation volume in the rainbow trout (*Salmo gairdneri*). J Exp Biol 97:325–334

Smith HW (1929) The excretion of ammonia and urea by the gills of fish. J Biol Chem 81:727–742

Smith JH (1985) Breeders must respond to market trends. Poultry-Misset Int 34: January Issue

Smith JLB (1956) Old four legs. Longman, London

Smith JD (1977) Comments on flight and the evolution of bats. In: Hecht MK, Goody PC, Hecht M (eds) Major problems in vertebrate evolution. Plenum Press, New York, pp 427–437

Smith JC, Ellenberger HH, Ballanyi K, Richter DW, Feldman JL (1991) Pre-Botzinger complex: a brainstem region that may generate respiratory rhythm in mammals. Science 254:726–729

Smith L, Oseid D, Olson L (1976) Acute and chronic toxicity of hydrogen sulfide to the fathead minnow *Pimephales promelas*. Environ Sci Technol 10:565–568

Smith RE (1952) Cyprinodont fishes from a sulphur-producing lake in Cyrenaica. Ann Mag Nat Hist Ser 125:888–892

Smith RE (1956) Quantitative relations between liver mitochondria metabolism and total body weight in mammals. Ann NY Acad Sci 62:403–422

Smith S (1957) Early development and hatching. In: Brown ME (ed) The physiology of fishes, vol 1. Academic Press, New York, pp 1–57

Smith TB, Wayne RK, Girman DJ, Bruford MW (1997) A role for ecotones in generating rainforest biodiversity. Science 276:1855–1857

Smith TG, Marks WB, Lange GD, Sheriff WH, Neale EA (1989) A fractal analysis of cell images. J Neurosci Methods 27:173–180

Smith U, Ryan JW (1970) An electron Microscopic study of the vascular endothelium as a site for bradykinin and ATP inactivation in the rat lung. Adv Exp Med Biol 8:249–262

Smith U, Ryan JW (1973a) Electron microscopy of endothelial cells collected on cellulose acetate paper. Tissue Cell 5:333–336

Smith U, Ryan JW (1973b) Electron microscopy of endothelial and epithelial components of the lungs: correlations of structure and function. Fed Proc 32:1957–1966

Snider GL, Kleinerman J, Thurlbeck WM, Bengali ZH (1985) The definition of emphysema. Report of a National Heart, Lung, and Blood Institute, division of lung disease workshop. Am Rev Respir Dis 132:182–185

Snow DH (1985) The horde and the dog, elite athletes – why and how. Proc Nutr Soc 44:267–272

Snyder GK (1973) Erythrocyte evolution: the significance of the Fahraeus-Lindqvist phenomenon. Respir Physiol 19:271–278

Snyder GK (1976) Respiratory characteristics of whole blood and selected aspects of circulatory physiology in the common short-nosed fruit bat, *Cynopterus brachyotes*. Respir Physiol 28:239–247

Snyder GK (1977) Blood corpuscles and blood haemoglobin: a possible example of coevolution. Science 195:412–421

Snyder GK (1983) Respiratory adaptations in diving mammals. Respir Physiol 54:269–294

Sohal RS, Weindruch R (1996) Oxidative stress, calorific restriction, and aging. Science 273:59–63

Soivio A, Hughes GM (1978) Circulatory changes in secondary lamellae of *Salmo gairdneri*. Ann Zool Fenn 15:221–225

Soivio A, Nikinmaa M (1981) The swelling of erythrocytes in relation to the oxygen affinity of the blood of the rainbow trout, *Salmo gairdneri* Richardson. In: Pickering AD (ed) Stress and fish. Academic Press, London, pp 103–119

Soivio A, Tuurala H (1981) Structural and circulatory responses to hypoxia in the secondary lamellae of *Salmo gairdneri* gills at two temperatures. J Comp Physiol 145:37–43

Solem A (1985) Origin and diversification of pulmonate land snails. In: Trueman ER, Clarke MR (eds) The Mollusca, vol 10, Evolution. Academic Press, London, pp 269–293

Solomon SE, Purton M (1984) The respiratory epithelium of the lung in the green turtle (*Chelonia mydas* L). J Anat 139:353–361

Somero GN (1991) Biochemical mechanisms of cold adaptation and stenothermality in Antarctic fish. In: di Prisco G, Maresca B, Tota B (eds) Biology of Antarctic fish. Springer, Berlin Heidelberg New York, pp 231–287

Somero GN (1992) Adaptations to high hydrostatic pressure. Annu Rev Physiol 54:557–577

Somero GN, Childress JJ, Anderson AE (1989) Transport, metabolism, and detoxification of hydrogen sulfide in animals from sulfide-rich marine environments. CRC Crit Rev Mar Aquat Sci 1:591–614

Sorbini CH, Grassi V, Solinas E, Muiesan G (1968) Arterial oxygen tesion in relation to age in healthy subjects. Respiration 25:3–13

Sotavalta O (1947) The flight tone (wing stroke frequency) of insects. Acta Entomol Fenn 4:1–117

Sparti A (1992) Thermogenic capacity of shrews (Mammalia, Soricidae) and its relationship with basal rate of metabolism. Physiol Zool 65:77–96

Speckmann EW, Ringer RK (1963) The cardiac output and carotid and tibial blood pressure of the turkey. Can J Biochem Physiol 41:2337–2354

Sperry DG, Wassersug RJ (1976) A proposd function for microridges on epithelial cells. Anat Rec 185:253–258

Squires RW, Buskirk ER (1982) Aerobic capacity during acute exposure to simulated altitude, 914 to 2286 m. Med Sci Sports Exercise 14:36–40

Stadtman ER (1992) Protein oxidation and aging. Science 257:1220–1224

Stahel CD, Nicol SC (1988) Ventilation and oxygen extraction in the little penguin (*Eudyptula minor*), at different temperatures in air and water. Respir Physiol 71:387–398

Staines HJ (1965) Female red bat carrying four young. J Mammal 46:333

Stamler JS, Singel DJ, Loscalzo J (1992) Biochemistry of nitric oxide and its redox-activated forms. Science 258:1898–1902

Standaert T, Johansen K (1974) Cutaneous gas exchange in snakes. J Comp Physiol 89:313–320

Stanislaus M (1937) Untersuchungen an der Kolibrilunge. Z Morphol Oekol Tiere 33:261–289

Stanley SM (1979) Macroevolution; pattern and process. WH Freeman, San Francisco

Stanley SM (1987) Extinction: a scientific American book. WH Freeman, San Francisco

Starck D (1959) Ontogenie und Entwicklungsphysiologie der Säugetiere. Handbuch der Zoologie, 8(1). De Gruyter, Berlin

Stark-Vancs V, Bell PB, Hutchison VH (1984) Morphological and pharmacological basis for pulmonary ventilation in *Amphiuma tridactylum*: an ultrastructural study. Cell Tissue Res 238:1–12

Starling EH, Verney EB (1925) The secretion of urine as studied on the isolated kidney. Proc R Soc Lond 97B:321–363

Staub NC (1974) Pulmonary oedema. Physiol Rev 54:678–811

Staub NC, Nagano H, Paerce NL (1967) Pulmonary oedema in dogs, especially the sequence of fluid accumulation in lungs. J Appl Physiol 22:227–240

Stearns SC (1982) The role of development in the evolution of life histories. In: Bonner JT (ed) Evolution and development. Springer, Berlin Heidelberg New York, pp 237–258

Stebbins GL (1984) [Quoted in) Fremontia. January 1984, p 16

Steen JB (1965) Comparative aspects of the respiratory gas exchange of sea urchins. Acta Physiol Scand 63:164–170

Steen JB (1971) Comparative physiology of respiratory mechanisms. Academic Press, London

Steen JB, Berg T (1966) The gills of two species of haemoglobin-free fishes compared to those of other teleosts, with a note on severe anemia in an eel. Comp Biochem Physiol 18:517–526

Steen JB, Kruysse A (1964) The respiratory function of teleostan gills. Comp Biochem Physiol 12:127–142

Steffensen JF, Lomholt JP (1983) Energetic cost of active branchial ventilation in the sharksucker *Echeneis naucrates*. J Exp Biol 103:185–192

Stein JC, Ellsworth ML (1993) Capillary oxygen transport during severe hypoxia: role of haemoglobin oxygen affinity. J Appl Physiol 75:1601–1607

Steinacker A (1975) Perfusion of the central nervous system of decapod crustaceans. Comp Biochem Physiol 52A:103–104

Stephens RH, Benjamin AR, Walters DV (1996) Volume and protein concentration of epithelial lining liquid in perfused in situ postnatal sheep lungs. J Appl Physiol 80:1911–1920

Stephenson J (1930) The Oligochaeta. Clarendon Press, Oxford

Sterrer W, Rieger R (1974) Retronectidae: a new cosmopolitan marine family of Catenulida (Turbellaria). In: Riser N, Morse M (eds) The biology of Turbellaria. McGraw-Hill, New York, pp 108–147

Stevens ED, Carey FG (1981) On why of warm-blooded fish. Am J Physiol 240:R151–R175

Stevens ED, Holeton GF (1978a) The aprtitionin of oxygen uptake from air and from water by erythrinids. Can J Zool 56:965–969

Stevens ED, Holeton GF (1978b) The partitioning of oxygen uptake from air and from water by the large obligate air breathing teleost, pirarucu (*Arapaima gigas*). Can J Zool 56:974–976

Stevens ED, Randall DJ (1967) Changes in gas concentrations in blood and water during moderate swimming activity in rainbow trout. J Exp Biol 46:329–337

Stevenson DJ (1996) When Galileo met Ganymede. Nature (Lond) 384:511–512

Stewart AG (1978) Swans flying at 8000 m. Brit. Birds 71:459–460

Stewart CB, Schilling JW, Wilson AC (1987) Adaptive evolution in the stomach lysosomes of foregut fermenters. Nature (Lond) 330:401–404

Stewart I (1990) Does God play dice?: the mathematics of chaos. Blackwell, Cambridge

Stewart JR, Thompson MB (1994) Placental structure of the Australian lizard, *Niveoscincus metallicus* (Squamata: Scincidae). J Morphol 220:223–236

Stiffler DF, DeRuyter ML, Talbot CR (1990) Osmotic and ionic regulation in the aquatic caecilian *Typhlonectes compressicauda* and the terrestrial caecial *Ichthyophis kohtaoensis*. Physiol Zool 63:649–668

Stinner JN (1982) Functional anatomy of the lung of the snake, *Pituophis melanoleucus*. Am J Physiol 243:R251–257

Stinner JN (1987) Gas exchange and air flow in the lung of the snake, *Pituophis melanoleucus*. J Comp Physiol 157:307–314

Stinner JN, Shoemaker VH (1987) Cutaneous gas exchange and low evaporative water loss in the frogs *Phyllomedusa sauvagei* and *Chiromantis xeraphelina*. J Comp Physiol 157B:423–427

Stoll HM, Schrag DP (1996) Evidence for glacial control of rapid sea level changes in early Cretaceous Arctic Ocean. Science 272:1771–1774

Stone HO, Thomson HK, Schmidt-Nielsen K (1968) Influence of erythrocytes on blood viscosity. Am J Physiol 214:913–918

Stong CL (1979) The amateur scientist. Life: origin and evolution: readings from Scientific American. WH Freeman, San Francisco, pp 57–62

Stonier T (1990) Information and the internal structure of the Universe. Springer, Berlin Heidelberg New York

Storey KB (1988) Suspended animation: the molecular basis of metabolic depression. Can J Zool (Lond) 66:124–132

Storey KB (1989) Integrated control of metabolic rate depression via reversible phosphorylation of enzymes in hibernating mammals. In: Malan A, Canguilhem B (eds) Living in cold II. John Libbey Eurotext, Montrouge, pp 309–319

Storey KB, Storey JM (1986) Freeze-tolerant frogs: cryoprotectants and tissue metabolism during freeze-thaw cycles. Can J Zool 64:49–56

Storey KB, Storey JM (1988) Freeze tolerance in animals. Physiol Zool 68:27–84

Storey KB, Storey JM (1990) Facultitative matabolic rate depression: molecular regulation and biomedical adaptation in anaerobiosis, hibernation, and estivation. Q Rev Biol 65:145–174

Stoz F, Schuhmann RA, Schebesta B (1988) The development of the placental villus during normal pregnancy: morphometric data base. Arch Gynecol Obstet 244:23–32

Strang LB (1967) Uptake of liquid from the lungs at the start of breathing. In: de Reuck AVS, Porter R (eds) Development of the lung. Little Brown, Boston, pp 348–361

Strang LB (1977) Growth and development of the lung: foetal and postnatal. Annu Rev Physiol 39:253–276

Strathmann RR (1963) The behaviour of myxine and other Myxinoids. In: Brodal A, Fänge R (eds) The biology of myxine. Universitesforlaget, Oslo, pp 22–32

Strathmann RR (1990) Why life histories evolve differently in the sea. Am Zool 30:197–207

Stratton CJ (1984) Morphology of surfactant producing cells and of the alveolar lining layer. In: Robertson B, van Golde LMG, Batenburg JJ (eds) Pulmonary surfactant. Elsevier, Amsterdam, pp 67–118

Strazny F, Perry SF (1984) Morphometric diffusion capacity and functional anatomy of the book lungs in the spider, Tegenaria sp. (Agelenidae). J Morphol 182:339–354

Strazny F, Perry SF (1987) Respiratory system: structure and function. In: Netwig W (ed) Ecophysiology of the spiders. Springer, Berlin Heidelberg New York, pp 78–94

Street FA, Grove AT (1976) Late Quartenary lake level fluctuations in Africa: environmental and climatic implications. Nature (Lond) 261:385–390

Strum J, Junod AF (1972) Autoradiographic demonstration of ^{3}H-5-hydroxytryptamine uptake by pulmonary endothelial cells. J Cell Biol 54:456–467

Stuart AJ (1991) Mammalian extinctions in the late Pleistocene of Northern Eurasia and North America. Biol Rev 66:453–562

Stulc J (1989) Extracellular transport pathways in the haemochorial placenta. Placenta 10:113–119

Sturgess JM (1979) Mucous secretions in the respiratory tract. Pediatr Clin N Am 26:481–501

Sturkie PD (1954) Avian physiology. Comstock, Ithaca

Suarez RK (1992) Hummingbird flight: sustaining the highest mass-specific metabolic rates among vertebrates. Experientia 48:565–570

Suarez PK, Lighton JRB, Moyes CD, Brown GS et al. (1990) Fuel selection in hummingbirds: ecological implications of metabolic biochemistry. Proc Natl Acad Sci USA 87:9207–9210

Suarez RK, Lighton JRB, Brown GS, Mathieu-Costello O (1991) Mitochondrial respiration in hummingbird flight muscles. Proc Natl Acad Sci USA 88:4870–4873

Sugano T, Oshino N, Chance B (1974) Mitochondrial functions under hypoxic conditions. The steady states of cytochrome c reduction and energy metabolism. Biochim Biophys Acta 347:340–358

Sullivan B, Riggs A (1967) Structure, function, and evolution of turtle haemoglobins III. Oxygenation properties. Comp Biochem Physiol 23:459–474

Sundin L, Nilsson GE, Block M, Löfman CO (1995) Control of gill filament flow by serotonin in the rainbow trout, Oncorhynchus mykiss. Am J Physiol 268:R1224–R1229

Susan ET, Chadwick OA, Amundson R (1996) Rapid exchange between soil carbon and atmospheric carbon dioxide driven by temperature change. Science 272:393–396

Suthers RA, Thomas SP, Suthers BJ (1972) Respiration, wing-beat and ultrasonic pulse emission in an echolocating bat. J Exp Biol 56:37–48

Sverdrup HU, Johnson MW, Fleming RH (1949) The oceans: their physics, chemistry and general biology. Prentice-Hall, New York

Swain R, Marker PF, Richardson AMM (1987) Respiratory responses to hypoxia in stream-dwelling (*Astacopsis franklinii*) and burrowing (*Parastacoides tasmanicus*) parastacid crayfish. Comp Biochem Physiol 87A:813–817

Swan LW (1961) The ecology of the high Himalayas. Sic Am 205:67–78

Swan LW (1970) Goose of the Himalayas. Nat Hist 79:68–75

Sweeney BM (1987) Rhythmic phenomena in plants. Academic Press, San Diego

Swenson ER (1990) Kinetics of oxygen and carbon dioxide exchange. In: Boutilier RG (ed) Advances in comparative and environmental physiology, vol 6: Vertebrate gas exchange from environment to cell. Springer, Berlin Heidelberg New York, pp 163–210

Szarski H (1962) The origin of Amphibia. Q Rev Biol 37:189–241

Szarski H (1983) Cell size and the concept of wasteful and frugal evolutionary strategies. J Theor Biol 105:201–243

Szarslo H (1977) Sarcopterygii and the origin of the tetrapods. In: Hecht MK, Goody PC, Hecht BM (eds) Major patterns of vertebrate evolution. Plenum Press, New York, pp 517–544

Szathmàry E (1997) The first two billion years. Nature (Lond) 387:662–663

Szewczak JM, Jackson DC (1992) Apneic oxygen uptake in the torpid bat, *Eptesicus fuscus*. J Exp Biol 173:217–227

Tait JS (1956) Nitrogen and argon in salmonoid swim bladders. Can J Zool 34:58–62

Taketo M, Schroeder AC, Mobraaten LE, Gunning KB, Hanten G et al. (1991) FVB/N: an inbred mouse strain preferable for transgenic analyses. Proc Natl Acad Sci USA 88:2065–2069

Takeuchi H, Uyeda S, Kanamori H (1970) Debate about the Earth. Freeman, Cooper, San Francisco

Talbot CR, Feder ME (1992) Relationships among cutaneous surface area, cutaneous mass and body mass in frogs: a reappraisal. Physiol Zool 65:1135–1147

Talling JF (1957) The longitudinal succession of water characteristics in the White Nile. Hydrobiologia 11:73–89

Tamura M, Hazeki O, Nioka S, Chance B (1989) In vivo study of tissue oxygen metabolism using optical and nuclear magnetic resonance spectroscopies. Annu Rev Physiol 51:813–834

Tamura O, Moriyama T (1976) On the morphological feature of the gill of amphibious and air-breathing fishes. Bull Fac Fish, Nagasaki Univ 41:1–8

Tamura O, Morii H, Yazuriha M (1976) Respiration of amphibious fishes *Periophthalmus cantonensis* and *Boleophthalmus chinensis* in water and land. J Exp Biol 65:97–107

Tan AL, De Young A, Noble RW (1972) The pH dependence of the affinity, kinetics and cooperativity of ligand binding to carp haemoglobin, *Cyprinus carpio*. J Biol Chem 247:2493–2498

Tappan H (1974) Molecular evolution. In: Hayaishi O (ed) Molecular oxygen in biology. Elsevier-North Holland, Amsterdam, pp 81–135

Tarczy-Hornoch P, Hildebrandt J, Mates EA, Standaert TA, Lamm WJE et al. (1996) Effects of exogenous surfactant on lung pressure-volume characteristics during liquid ventilation. J Appl Physiol 80:1764–1777

Tarsitano SF, Frey E, Riess J (1989) The evolution of the crocodilia: a conflict between morphological and biochemical data. Am Zool 29:843–856

Taylor AC (1984) Branchial ventilation in the burrowing crab, *Atelecyclus rotundatus*. J Mar Biol Assoc UK 64:7–20

Taylor AC, Atkinson RJA (1991) Respiratory adaptations of aquatic decapod crustaceans and fish to a burrowing mode of life. In: Woakes AJ, Grieshaber MK, Bridges CR (eds) Physiological strategies for gas exchange and metabolism. Cambridge University Press, Cambridge, pp 211–234

Taylor CR (1977) Why large animals? Cornell Vet 67:155–175

Taylor CR (1987) Structural and functional limits to oxidative metabolism: insights from scaling. Annu Rev Physiol 49:135–146

Taylor CR, Weibel ER (1981) Design of the mammalian respiratory system. Respir Physiol 11:1–10

Taylor CR, Maloiy GMO, Weibel ER, Langman VA, Kamau JMZ et al. (1981) Design of the mammalian respiratory system. III. Scaling maximum aerobic capacity to body mass: wild and domestic mammals. Respir Physiol 44:25–37

Taylor CR, Karas RH, Weibel ER, Hoppeler H (1987a) Adaptive variation in the mammalian respiratory system in relation to energetic demand: II. Reaching limits to oxygen flow. Respir Physiol 69:7–26

Taylor CR, Weibel ER, Karas RH, Hoppelar H (1987b) Adaptive variation in the mammalian respiratory system in relation to energetic demand: VIII. Structural and functional design principles determining the limits to oxidative metabolism. Respir Physiol 69:117–127

Taylor CR, Weibel ER, Karas RH, Hoppelar H (1989) Matching structures and functions in the respiratory system: allometric and adaptive variations in energy demand. In: Wood SC (ed) Comparative pulmonary physiology: current concents. Marcel Dekker, New York, pp 27–65

Taylor DJ, Mathews PM, Radda GK (1986) Myoglobin-dependent oxidative metabolism in hypoxic rat heart. Respir Physiol 63:275–283

Taylor EH (1968) The caecilians of the world. University of Kansas Press, Lawrence

Taylor EW (1982) Control and co-ordination of ventilation and circulation in crustaceans: responses to hypoxia and exercise. J Exp Biol 100:289–319

Taylor EW, Butler PJ (1973) The behaviour and physiological responses of the shore crab *Carcinus maenus* during changes in environmental oxygen tension. Neth J Sea Res 7:496–505

Taylor EW, Butler PJ (1978) Aquatic and aerial respiration in the shore crab, *Carcinus maenas* (L.), acclimated to 15 °C. J Comp Physiol 127:315–323

Taylor EW, Innes AJ (1988) A functional analysis of the shift from gill- to lung breathing during the evolution of land crabs (Crustacea, Decapoda). Biol J Linn Soc 34:309–316

Taylor EW, Wheatly MG (1980) Ventilation, heart rate and respiratory gas exchange in the crayfish *Austropotamobius pallipes* (Lereboullet) submerged in normoxic water and following 3-hour exposure in air at 15 °C. J Comp Physiol 138:67–78

Taylor EW, Butler PJ, Sherlock PJ (1973) The respiratory and cardiovascular changes associated with the emersion response of *Carcinus maenas* (L.) during environmental hypoxia, at three different temperatures. J Comp Physiol 86:95–116

Taylor HH, Greenaway P (1979) The structure of the gills and lungs of the arid zone crab, *Holthuisana (Austrothelphusa) transversa* Morgens (Sundathelphusidae: Brachyura) including observations on arterial vessels within the gills. J Zool (Lond) 189:359–384

Taylor HH, Greenaway P (1984) The role of the gills and branchiostegites in gas exchange in a bimodally breathing crab, *Holthuisana transversa*: evidence for a facultative change in the distribution of respiratory circulation. J Exp Biol 11:103–122

Taylor HH, Wheatly MG (1979) The behaviour and respiratory physiology of the shore crab, *Carcinus maenus* (L.) at moderately high temperatures. J Comp Physiol 130:309–316

Tazawa H (1987) Embryonic respiration. In: Seller TJ (ed) Bird respiration. CRC Press, Boca Raton, pp 3–41

Tazawa H, Mochizuki M (1976) Estimation of contact time and diffusing capacity of oxygen in the chollioallantoic vascular plexus. Respir Physiol 28:119–128

Tazawa H, Mochizuki M (1977) Oxygen analyses of chicken embryo blood. Respir Physiol 31:203–216

Tazawa H, Visschedijk AHJ, Piiper J (1983a) Blood gases and acid-base status in chicken embryos with naturally varying egg shell conductance. Respir Physiol 53:137–158

Tazawa H, Visschedijk AHJ, Wittmann J, Piiper J (1983b) Gas exchange, blood gases and acid base status in the chick before, during and after hatching. Respir Physiol 53:173–185

Teal JM, Kanwisher JW (1966) Gas transport in the marsh grass, *Spartina alterniflora*. J Exp Bot 17:355–361

Teasdale F (1978) Functional significance of the zonal morphologic differences in the normal human placenta: a morphometric study. Am J Obstet Gynecol 130:773–781

Teasdale F (1980) Gestational changes in the functional structure of the human placenta in relation to foetal growth: a morphometric study. J Obstet Gynecol 137:560–568

Teasdale F, Jean-Jacques G (1986) Morphometry of the microvillous membrane of the human placenta in maternal diabetes mellitus. Placenta 7:81–88

Teichert C (1988) Main features of cephalopod evolution. In: Clarke MR, Trueman ER (eds) The Mollusca, vol 12, Paleontology and neontology of cephalopods. Academic Press, London, pp 11–79

Teitel DF, Iwamoto HS, Rudolph AM (1987) Effects of birth-related events on central blood flow patterns. Pediatr Res 22:557–566

Ten Eyck LF (1972) Stereochemistry of haemoglobin. In: Rorth M, Astrup P (eds) Oxygen affinity of haemoglobin and red cell acid-status. Alfred Benzon Symposium IV, Munksgaard, Copenhagen, pp 19–31

Tenney SM (1979) A synopsis of breathing mechanisms. In: Wood SC, Lenfant C (eds) Evolution of respiratory processes. A comparative approach. Marcel Dekker, New York, pp 51–106

Tenney SM (1980) Avian physiology and performance at altitude. In: Sutton JR, Coates GC, Remmers JE (eds) Hypoxia: the adaptations. BC Decker, Toronto, pp 1–3

Tenney SM, Boggs DF (1985) Comparative mammalian respiratory control. In: Fishman AP (ed) Handbook of physiology: the respiratory system II, sect 3. American Physiological Society, Bethesda, pp 833–855

Tenney SM, Remmers JE (1963) Comparative morphology of the lung: diffusing area. Nature (Lond) 197:54–56

Tenney SM, Tenney JB (1970) Quantitative morphology of cold blooded lungs: Amphibia and Reptilia. Respir Physiol 9:197–215

Tenney SM, Bartlett D, Farber JP, Remmers JE (1984) Mechanics of the respiratory cycle in the green turtle (*Chelonia mydas*). Respir Physiol 22:361–368

Terroine EF, Roche J (1925) Production calorique et respiration des tissus *in vitro* chez les homèothermes. CR Acad Sci 180:225–227

Terwilliger NB, Brown C (1993) Ontogeny of hemocyanin function in the Dungeness crab, *Cancer magister*: the interactive effects of developmental stage and divalent cations on hemocyanin oxygenation properties. J Exp Biol 183:1–13

Terwilliger NB, Terwilliger RC (1984) Haemoglobin from "Pompeii worm", *Alvinella pompejana*, an annelid from deep sea hot hydrothermal vent environment. Mar Biol Lett 5:191–201

Theede J, Ponat A, Hiroki K, Schlieper C (1969) Studies on the resistance of marine bottom invertebrates to oxygen deficiency and hydrogen sulfide. Mar Biol 2:325–337

Thewissen JMG, Babcock SK (1992) The origin of flight in bats: to go where no mammal has gone before. BioScience 42:340–345

Thomas ALR, Jones G, Rayner JMV, Hughes PM (1990) Intermittent gliding flight in the pipistrelle bat (*Pipistrellus pipistrellus*) (Chiroptera: Vespertilionidae). J Exp Biol 149:407–416

Thomas DP, Vane JR (1967) 5-hydroxytryptamine in the circulation of the dog. Nature (Lond) 216:335–338

Thomas DW (1983) The annual migrations of three species of West African fruit bats (Chiroptera: Pteropodidae). Can J Zool 61:2266–2272

Thomas DW, Cloutier D, Gagne D (1990) Arrythmic breathing, apnea and non-steady-state oxygen uptake in hibernating little brown bats (*Myotis lucifugus*). J Exp Biol 149:395–406

Thomas HJ (1954) The oxygen uptake of the lobster (*Homarus vulgaris* Edw). J Exp Biol 31:228–251

Thomas S, Fievet B, Barthelemy L, Peyraud C (1983) Comparisons of exogenous and endogenous hypercapnia on ventilation and oxygen uptake in the rainbow trout (*Salmo gairdneri* R). J Comp Physiol 151B:185–190

Thomas SP (1975) Metabolism during flight in two species of bats, *Phyllostomus hastatus* and *Pteropus gouldii*. J Exp Biol 63:273–293

Thomas SP (1981) Ventilation and oxygen extraction in the bat, *Pteropus gouldii*, during rest and during steady flight. J Exp Biol 94:231–250

Thomas SP (1987) The physiology of bat flight. In: Fenton MB, Racey P, Rayner JMV (eds) Recent advances in the study of bats. Cambridge University Press, Cambridge, pp 75–99

Thomas SP, Suthers R (1972) The physiology and energetics of bat flight. J Exp Biol 57:317–335

Thomas SP, Lust MR, van Riper HJ (1984) Ventilation and oxygen extraction in the bat *Phyllostomus hastatus* during rest and steady flight. Physiol Zool 57:237–250

Thompson AJ (1911) Introduction to science. Henry Holt, New York

Thompson D'AW (1959) On growth and form, 2nd edn. Cambridge University Press, Cambridge

Thompson KS (1969) The environment and distribution of Paleozoic sarcopterygian fishes. Am J Sci 267:457–464

Thompson KS (1971) The adaptation and evolution of early fishes. Q Rev Biol 46:139–166

Thompson KS (1980) The ecology of the Devonian lobe-finned fish. In: Panchen AL (ed) The terrestrial environment and the origin of land vertebrates, Systematics Association, Spec Vol - 15. Academic Press, London, pp 45–156

Thompson KS (1991) Where did tetrapods come from? Sci Am 79:488–490

Thompson DJ (1995) The seasons, global temperature and procession. Science 268:59–68

Thomson KS (1961) Water miracle of nature. Collier-Macmillan, London

Thomson KS (1986) Marginalia: a fishy story. Sci Am 74:169–171

Thorpe WH (1930) The biology, post-embryonic development, and economic importance of *Cryptochaetum iceryae* (Diptera, Agromyzidae) parasitic on *Icerya purchasi* (Coccidae, Monophlebini). Proc Zool Soc Lond 60:929–971

Thorpe WH (1932) Experiments upon respiration in the larvae of certain parasitic Hymenoptera. Proc R Soc Lond 109B:450–471

Thorpe WH (1950) Plastron respiration in aquatic insects. Biol Rev 25:344–391

Thorpe WH, Crisp DJ (1941) Studies on plastron respiration. II. The respiratory efficiency of the plastron in *Amphelocheirus*. J Exp Biol 24:270–303

Thorpe WH, Crisp DJ (1949) Studies on plastron respiration. IV. Plastron respiration in the Coleoptera. J Exp Biol 26:219–260

Thurlbeck WM (1980) The effect of age on the lung. In: Dietz AA (ed) Aging - its chemistry. American Association for Clinical Chemistry, Washington, DC, pp 114–131

Tierney DF (1974) Lung metabolism and biochemistry. Annu Rev Physiol 36:209–234

Tilley SG, Bernado J (1993) Life history evolution in plethodontid salamanders. Herpetologia 49:154–163

Timiras PS, Krum AA, Pace N (1957) Body and organ weights of rats during acclimatization to an altitude of 12,470 feet. Am J Physiol 191:598–604

Timwood KT, Julian LM (1983) Early lung growth in the turkey. Proc 32 Western Poult Dis Conf, Davis, pp 21–23

Timwood KT, Hyde DM, Plopper CG (1987) Lung growth of the turkey, *Meleagris gallopavo*: II. Comparison of two genetic lines. Am J Anat 178:158–169

Tipton CM (1986) Determinants of VO_{2max}: insights gained from non-human species. Acta Physiol Scand 128:33–43

Toda N, Okamura T (1991) Role of nitric oxide in neurally induced cerebroarterial relaxation. J Pharmacol Exp Ther 258:1027–1032

Todd ES (1971) Respiratory control in the longjaw, *Gillichthys mirabilis*. Comp Biochem Physiol 39A:147–164

Todd ES (1973) Positive buoyancy and air-breathing: a new piscine gas bladder function. Copeia (1973):461–464

Todd ES, Ebeling AW (1966) Aerial respiration in the longjaw mudsucker *Gillichthys mirabilis* (Teleostei: Gobiidae). Biol Bull Mar Biol Lab Woods Hole 130:265–288

Todd GT (1980) Evolution of the lung and the origin of bony fishes - a casual relationship? Am Zool 20:757A

Toews DP (1971) Factors affecting the onset and termination of respiration in the salamander, *Amphiuma tridactylum*. Can J Zool 49:1231–1237

Toews DP, Boutilier RG (1986) Acid-base regulation in the amphibia. In: Heisler N (ed) Acid-base regulation in animals. Elsevier, Amsterdam, pp 266–308

Toews DP, MacIntyre D (1977) Blood respiratory properties of a viviparous amphibian. Nature (Lond) 266:464–465

Toews DP, MacIntyre D (1978) Respirattion and circulation in an apodan amphibian. Can J Zool 56:199–214

Toews DP, Boutilier RG, Todd L, Fuller N (1978) Carbonic anhydrase in the amphibia. Comp Biochem Physiol 59A:211–213

Toloza EM, Lam M, Diamond JM (1991) Nutrient extraction by cold exposed mice: a test of digestive safety margins. Am J Physiol 261:G608–G620

Tominaga T, Page EW (1966) Accommodation of the human placenta to hypoxia. J Obstet Gynecol 94:679–691

Tomlinson JT (1963) Breathing of birds in flight. Condor 514–523

Torre-Bueno JR (1978) Evaporative cooling and water balance during flight in birds. J Exp Biol 75:231–236

Torre-Bueno JR (1985) The energetics of avian flight at altitude. In: Nachtigall W (ed) Bird flight, BIONA report 3. Gustav Fischer, Stuttgart, pp 45–87

Tota B, Hamlett WC (1989) Epilogue: evolutionary and contemporary biology of elasmo-branchs. J Exp Zool 2:193–196

Tota B, Aciero R, Agnisola C (1991) Mechanical performance of the isolated and perfused heart of the haemoglobinless Antarctic icefish *Chionodraco hamatus*. Philos Trans R Soc Lond 332B:191–198

Toulmond A (1975) Blood oxygen transport and metabolism of the confined *Iugworm Arenicola marina* (L). J Exp Biol 63:647–660

Toulmond A (1985) Circulating respiratory pigments in marine animals. In: Laverack MS (ed) Physiological adaptations of marine animals. Cambridge University Press, Cambridge, pp 163–206

Toulmond A (1991) Respiratory and metabolic adaptations of aquatic annelids to low environ-mental oxygen tensions. In: Woakes AJ, Grieshaber MK, Bridges CR (eds) Physiological strategies for gas exchage and metabolism. Cambridge University Press, Cambridge, pp 191–210

Toulmond A, Tchernigovtzeff C (1984) Ventilation and respiratory gas exchanges of the lug-worm *Arenicola marina* (L.) as functions of ambient PO_2 (20–700 torr). Respir Physiol 57:349–363

Toulmond A, Tchernigovtzeff C, Greber P, Jouin C (1984) Epidermal sensitivity to hypoxia in the lugworm. Experientia 40:541–543

Toulmond A, de Frescheville J, Frisch MH, Jouin C (1988) Les pigments respiratoires de la faune inféodée à l'hydrothermalisme profond. Oceanol Acta 8:195–201

Towe KM (1970) O_2 – collagen priority and early Metazoan fossil record. Proc Natl Acad Sci USA 65:781–788

Townsend LP, Earnest D (1940) The effects of low oxygen and other extreme conditions on salmonod fish. Proc 6th Pax Sci Congr 3:345–351

Trench RK (1979) The cell-biology of plant-animal symbiosis. Annu Rev Plant Physiol 30:485–531

Trivers R (1985) Social evolution. Benjamin/Cummings, Menlo Park

Truchot JP (1975) Blood acid-base changes during experimental emersion and reimmersion of the intertidal crab *Carcinus maenas* (L). Respir Physiol 23:351–360

Truchot JP (1980) Lactate increases the oxygen affinity of crab hemocyanin. J Exp Zool 214:205–208

Truchot JP (1987) Comparative aspects of extracellular acid-base balance. Springer, Berlin Heidelberg New York

Truchot JP (1990) Respiratory and ionic regulation in the invertebrates exposed to both water and air. Annu Rev Physiol 52:61–76

Truchot JP, Jouve-Duhamel A (1980) Oxygen and carbon dioxide in the marine intertidal environment: diurnal and tidal changes in rockpools. Respir Physiol 39:241–254

Trudinger BJ, Cook CM, Giles WB, Fong E, Connelley A, Wilcox W (1991) Foetal umbilical artery velocity waveforms and subsequent neonatal outcome. Br J Obstet Gynecol 98:378–384

Tsonis AA, Tsonis PA (1987) Fractals: a new look biological shape and patterning. Perspect Biol Med 30:355–361

Tsukimoto K, Mathieu-Costello O, Prediletto R, Elliot AR, West JB (1991) Ultrastructural appearances of pulmonary capillaries at high transmural pressures. J Appl Physiol 71:573–582

Tucker V (1968) Respiratory physiology of house sparrows in relation to high altitude flight. J Exp Biol 48:55–66

Tucker V (1970) Energetic cost of locomotion in mammals. Comp Biochem Physiol 34:841–846

Tucker VA (1972) Respiration during flight in birds. Respir Physiol 14:75–82

Tucker VA (1974) Energetics of natural avian flight. In: Paynter RA (ed.) Avian energetics. Nuttal Ornithological Club, Cambridge, pp 298–333

Tucker CE, James WE, Berry MA, Johnstone CJ, Glover RF (1976) Depressed myocardial function in the goat at high altitude. J Appl Physiol 41:356–361

Tulkki P (1965) Dissapearance of the benthic fauna from the basin of Bronholm (Southern Baltic) due to oxygen deficiency. Can Biol Mar 6:455–463

Tullett SG (1981) Theorerical and practical aspects of eggshell porosity. Turkeys 29:24–58

Tullett SG, Board RG (1976) Oxygen flux across the integument of the avian egg during incubation. Br Poult Sci 17:441–450

Tullett SG, Deeming DC (1982) The relationship between eggshell porosity and oxygen consumption of the embryo in the domestic fowl. Comp Biochem Physiol 72A:529–540

Turrens JF, Crapo JD, Freeman BA (1984) Protection against oxygen toxicity by intravenous injection of liposome-entrapped catalase and superoxide dismutase. J Clin Invest 73:87–95

Turrens JF, Freeman BA, Crapo JD (1982a) Hyperoxia increases H_2O_2 release by lung mitochondria and microsomes. Arch Biochem Biophys 217:411–421

Turrens JF, Freeman BA, Levitt JG, Crapo JD (1982b) The effect of hyperoxia on superoxide production by lung submitochondrial particles. Arch Biochem Biophys 217:401–410

Turrens JF, Beconi M, Barilla J, Chavez UB, McCord JM (1991) Mitochondrial generation of oxygen radicals during reoxygenation of ischaemic tissues. Free Radic Res Commun 12–13:681–689

Tuurala H, Part P, Nikinmaa M, Soivio A (1984) The basal channels of secondary lamellae in Salmo gairdneri gills – a nonrespiratory shunt. Comp Biochem Physiol 79A:35–39

Tytler P, Vaughan TC (1983) Thermal ecology of the mudskippers, Periophthalmus koelreuteri (Pallas) and Boleophthalmus boddarti (Pallas), of Kuwaiti bay. J Fish Biol 23:327–337

Uchiyama M, Yoshizawa H, Wakasugi C, Oguro C (1990) Structure of the internal gills in tadpoles of the crab-eating frog, Rana cancrivora. Biol Sci 7:623–630

Ulrich SPH, Bartels H (1963) Über die Atmungsfunktion des Blutes von Spitzmäusen, weissen Mäusen und syrischen Goldhamstern. Pfluegers Archiv Gesamte Physiol Menschen Tiere 277:150–165

Ultsch GR (1973) The effects of water hyacinth (Eichhornia crassipes) on the microenvironment of aquatic communities. Arch Hydrobiol 4:460–473

Ultsch GR (1976) Eco-physiological studies of some metabolic and respiratory adaptations of sirenid salamanders. In: Hughes GM (ed) Respiration of amphibious vertebrates. Academic Press, New York, pp 287–312

Ultsch GR, Gros G (1979) Mucus, a diffusion barrier to oxygen: possible role in O_2 uptake at low pH in carp (Cyprinus carpio) gills. Comp Biochem Physiol 62A:685–689

Ultsch GR, Jackson DC (1982) Longterm submergence at 3 °C of the turtle, Chrysemys picta belii, in normoxic and severely hypoxic water. I. Survival, gas exchange and acid-base status. J Exp Biol 96:11–28

Ulvedal F, Morgan TE, Cutler RG, Welch BE (1963) Respiratory function studies during prolonged exposure to simulated altitude without hypoxia. USAF School of Aerospace Medicine, Rept No SAM-TDR-63-31, Brooks AFB, Texas

Untersee P, Gil J, Weibel ER (1971) Visualization of extracellular lining layer of lung alveoli by freeze etching. Respir Physiol 13:171–185

Urey HC (1959) The atmosphere of the planets, Pt II, The Earth's atmosphere. In: Flügge S (ed) Encyclopedia of physics, LII, Astrophysics III, The solar system. Springer, Berlin Heidelberg New York, pp 366–383

Usry JL, Turner LW, Stahly TS, Bridges TC, Gates RS (1991) GI tract simulation model of the growing pig. Trans Am Soc Agric Eng 34:1879–1890

Vaïda P, Kays CK, Rivière D, Tèchoueyres P, Lachaud JL (1997) Pulmonary diffusing capacity and pulmonary capillary blood volume during parabolic flights. J Appl Physiol 82:1091–1097

Val AL, Fonseca VM, Affonso EG (1990) Adaptive features of Amazon fishes: haemoglobins, hematology, intraerythrocytic phosphates and whole blood Bohr effect of Pterygoplichthys multiradiatus (Siluriformes). Comp Biochem Physiol 97B:435–440

Valentine JW, Moores EM (1976) Plate tectonics and history of life in the oceans. In: Wilson IT (ed) Continents adrift and continents aground. WH Freeman, San Fransisco, pp 196–205 van Beek JHGM, Roger SA, Bassingthwaighte JB (1989) Regional myocardial flow heterogenenity explained with fractal networks. Am J Physiol 257:H1670–1680

van Dam L (1935) On the utilization of oxygen by *Mya arenaria*. J Exp Biol 12:86–94

van Dam L (1938) On the utilization of oxygen and regulation of breathing in some aquatic animals. Diss, Groningen

van Dam L (1954) On the respiration in scallops (Lamellibranchia). Biol Bull Mar Biol Lab Woods Hole 107:192–202

van der Burgh J, Visscher H, Dilcher DL, Kurschner WM (1993) Paleoatmospheric signatures on Neogene fossil leaves. Science 260:1788–1790

Vandergriff KD, Olson JS (1984) Morphological and physiological factors affecting oxygen uptake and release by red blood cells. J Biol Chem 259:12619–12627

Vane JR (1969) The release and fate of vasoactive hormones in pulmonary circulation. Br J Pharmacol 35:209–242

van Holde KE, Miller KI (1982) Hemocyanins. Q Rev Biophys 15:1–129

Vannier G (1983) The importance of ecophysiology for both biotic and abiotic studies of take in the rainbow trout (Sads in soil biology). Ottignies-Louvain-La-Neuve, Dieu-Brichart, pp 289–314

van Oijen MJP, Witte F, Witte-Mass ELM (1981) Introduction to ecology and taxonomic investigations on the haplochromine cichlids from the Mwanza Gulf of Lake Victoria. Neth J Zool 31:149–174

van Valen L (1971) The history and stability of atmospheric oxygen. Science 171:439–443

van Valen L (1979) The evolution of bats. Evol Theory 4:103–121

van Vass O, Vass KF (1960) Appraisal of inland fishery resources. Biological appraisal. In: Lectures presented at the 3rd Int Training Congr. FAO, Bogor, Indonesia (1955). 1:199

van Zalen MM, van Vugt JMG, Colebrander GJ, van Geijn HP (1994) First-trimester uteroplacental and foetal blood flow velocity waveforms in normally developing fetuses: a longitudinal study. Ultrasound Obstet Gynecol 4:284–288

Varansi U, Markey D (1978) Uptake and release of lead and cadmium in skin and mucus of coho salmon (*Oncorhynchus kitutch*). Comp Biochem Physiol 60C:187–191

Varansi U, Robisch PA, Malins DC (1975) Structural alterations in fish epidermal mucus produced by water-borne lead and mercury. Nature (Lond) 258:431–432

Vaughan TA (1966) Morphology and flight characteristics of mollosid bats. J Mammal 47:75–82

Veerannan KM (1974) Respiratory metabolism of crabs from marine and estuarine habitats: an interspecific comparison. Mar Biol 26:35–43

Veith WJ (1980) Viviparity and embryonic adaptations in the teleost *Clinus superciliosus*. Can J Zool 58:1–12

Verbanck S, Linnarsson D, Prisk GM, Paiva M (1996) Specific ventilation distribution in microgravity. J Appl Physiol 180:1458–1465

Verde MR (1951) The morphology and histology of the lung in snakes. J Univ Bombay 19:79–89

Verdier B (1975) Etude de l'atmosphère du sol. Eléments de comparaison et signification écologique de l'atmosphère d'un sol brun calcaire et d'on sol lessivé podzolique. Rev Ecol Biol Sol 12:591–626

Vergara AG, Hughes GM (1981) Phospholipids in washings from the lung of the frog (*Rana pipiens*). J Comp Physiol 27:117–120

Verma A, Hirsch DJ, Glatt CE, Ronnett GV, Snyder SH (1993) Carbon monoxide: a putative neural messanger. Science 259:381–384

Vernberg FJ (1954) The respiratory metabolism of tissues of tissues of marine teleosts in relation to activity and body size. Biol Bull 106:360–370

Vetter RD, Wells ME, Kurtsman AL, Somero GN (1987) Sulfide detoxification by the hydrothermal vent crab *Bythograea thermydron* and other decapod crustaceans. Physiol Zool 60:121–137

Vidal RA, Bahr D, Baragiola RA, Peters M (1997) Oxygen on Ganymede: laboratory studies. Science 276:1839–1842

Vidyadaran MK, King AS, Kassim H (1987) Deficient anatomical capacity for oxygen uptake of the developing lung of the female domestic fowl when compared with red-jungle fowl. Schweiz Arch Tierheilkd 129:225–237

470

Vidyadaran MK, King AS, Kassim H (1988) Quantitative studies of the lung of the domestic fowl, *Gallus gallus* var *domesticus*. Partanika 11:229–238

Vidyadaran MK, King AS, Kassim H (1990) Quantitative comparisons of lung structure of adult domestic fowl and the red-jungle fowl, with reference to broiler ascites. Avian Pathol 19:51–58

Vince M (1973) Effects of external stimulation on the onset of lung ventilation and the time of hatching in the fowl, duck and goose. Br Poult Sci 14:389–405

Vince M, Misson BH, Freeman BM (1975) Blood gas partial pressures and the onset of lung ventilation in the chick embryo. Comp Biochem Physiol 51A:457–469

Viscor G, Fuentes J, Palomeque J (1984) Blood rheology in the pigeon (*Columba livia*), hen (*Gallus gallus domesticus*), and black-headed gull (*Larus ridibundus*). Can J Zool 62:2150–2156

Visschedijk AHJ (1968a) The air space and embryonic respiration. II. The times of pipping and hatching as influenced by artificially changed permeability of the shell over the air space. Br Poult Sci 9:185–196

Visschedijk AHJ (1968b) The air space and embryonic respiration. III. The balance between oxygen and and carbon dioxide in the air space of incubating chicken egg and its role in stimulating pipping. Br Poult Sci 9:197–213

Visschedijk AHJ, Rahn H (1981) Incubation of chicken eggs at altitude: theoretical consideration of optimal gas composition. Br Poult Sci 22:451–460

Visschedijk AHJ, Tazawa H, Piiper J (1985) Variability of shell conductance and gas exchange of chicken eggs. Respir Physiol 59:339–354

Visser SA (1963) Gas production in the decomposition of *Cyperus papyrus*. J Water Pollut Control Fed 35:973–788

Vitalis TZ, Furilla RA, Burggren WW (1988) Ventilation and gas exchange in the snake, *Thamnophis elegans*. Am Zool 28:47A

Vleck D (1979) The energy cost of burrowing by the pocket gopher *Thomomys bottae*. Physiol Zool 52:122–125

Vleck D (1981) Burrow structure and foraging costs in the fossorial rodent, *Thomomys bottae*. Oecologia (Berl) 49:391–396

Vogel G (1997) Parasite shed light on cellular evolution. Science 275:1422

Vogel GM (1980) Oxygen uptake and transport in the sabellid polychaete *Eudistylia vancouveri* (Kinburg). MA Thesis, College of William and Mary, Williamsburg, Virginia

Vogel S (1977) Flows in organisms induced by movement of the external medium. In: Pedley TJ (ed) Scale effects in animal locomotion. Academic Press, London, pp 285–297

Vogel S (1988) Life's devices: the physical world of animals and plants. Princeton University Press, Princeton

Vogel S, Wainright SA (1969) A functional bestiary. Addison-Wesley, Reading

Vollrath F (1992) Spider webs and silks. Sci Am 266:70–76

Von Bertalanffy L, Estwick RR (1953) Tissue respiration of musculature in relation to body size. Am J Physiol 173:58–60

Von Sonntag C (1987) The chemical basis of radiation biology. Taylor and Francis, London

Voss RF (1988) Fractals in nature: from characterization to simulation. In: Petgen HO, Saupe D (eds) The science of fractal images. Springer, New York, pp 21–770

Vyas AB, Laliwala SM (1976) Anatomical studies on the book lungs of the scorpion *Buthus tamulus* with a note on the respiratory mechanism. J Anim Morphol Physiol 23:3–7

Waarde V, Thillart VD, Kesbeke F (1983) Anaerobic energy metabolism of the European eel, *Anguilla anguilla* L. J Comp Physiol 149:469–475

Wache S, Terwilliger NB, Terwilliger RC (1988) Hemocyanin structure changes during early development of the crab *Cancer productus*. J Exp Biol 247:23–32

Wagner F, Below R, Klerk PM, Dilcher DL, Joosten H, Kürschner WM, Visscher H (1996) A natural experiment on plant acclimation: lifetime stomatal atmospheric carbon dioxide increase. Pro Natl Acad Sci USA 93:11705–11708

Wagner GP (1989) The origin of morphologicalcharacters and the biological basis of homology. Evolution 43:1157–1171

Wagner JR, Hu CC, Ames BN (1992) Endogenous oxidative damage of deoxycytidine in DNA. Proc Natl Acad Sci USA 89:3380–3384

Wagner PD (1977) Diffusion and chemical reaction of pulmonary gas exchange. Physiol Rev 57:257–312

Wagner PD (1993) Algebraic analysis of the determinants of VO_{2max}. Respir Physiol 93:221–237

Wagner PD, West JB (1972) Effects of diffusion impairment on O_2 and CO time courses in pulmonary capillaries. J Appl Physiol 33:62–71

Wagner PD, Gale GE, Moon RE, Torre-Bueno JR, Stolp BW, Saltman HA (1986) Pulmonary gas exchange in humans exercising at sea level and simulated altitude. J Appl Physiol 61:260–270

Wagner S, Castel M, Gainer H, Yarom Y (1997) GABA in the mammalian suprachiasmic nucleus and its role in diurnal rhythmicity. Nature (Lond) 387:598–603

Wainright SA (1988) Form and function in organisms. Am Zool 28:671–680

Wainright SA, Biggs WD, Currey JD, Gosline JM (1976) Mechanical design in organisms. Wiley, New York

Wake MH (1974) The comparative morphology of the caecilian lung. Anat Rec 178:483

Wake MH (1977) The reproductive biology of caecilians: an evolutionary perspective. In: Taylor DH, Guttman SI (eds) The reproductive biology of amphibians. Plenum Press, New York, pp 73–101

Wake MH (1989) Phylogenesis of direct development and viviparity in vertebrates. In: Wake D, Roth G (eds) Complex organismal functions: integration and evolution in vertebrates. (Dahlem Workshop Report), John Wiley Chichester, pp 235–250

Wake MH (1990) The evolution of integration of biological systems: an evolutionary perspective through studies on cells, tissues, and organs. Am Zool 30:897–906

Wake MH (1993) Evolution of oviductal gestation in amphibians. J Exp Zool 266:394–413

Wake MH, Marks SB (1993) Development and evolution of plethodontid slalmanders: a review of prior studies and a prospectus for future research. Herpetologia 49:194–203

Wake MH, Roth G (1989) Paedomorphosis: new evidence for its importance in salamander evolution. Am Zool 29:134A

Walker BR, Voelkel NF, McMurtry IF, Adams EM (1982) Evidence for diminished sensitivity of the hamster pulmonary vasculature to hypoxia. J Appl Physiol 52:1571–1574

Walker BR, Berend N, Voelkel NF (1984) Comparison of muscular pulmonary arteries in low and high altitude hamsters and rats. Respir Physiol 56:45–50

Walker JCG (1974) Stability of the atmospheric oxygen. Am J Sci 274:193–214

Walker JCG (1977) Evolution of the atmosphere. Macmillan, New York

Walker JCG (1978) Oxygen and hydrogen in the primitive atmosphere. Pure Appl Geophys 116:222–231

Walker JCG (1983) Possible limits on the composition of the Archean Ocean. Nature (Lond) 302:518–520

Walker JCG (1985) Carbon dioxide on the early earth. Origin Life Evol Biosphere 16:117–127

Walker JCG (1987) Was the Archean biosphere upside down? Nature (Lond) 329:710–712

Walker JCG, Klein C, Schidlowski M, Schopf JW, Stevenson DJ, Walter MR (1983) Environmental evolution of the Archean-early Paleozoic Earth. In: Schopf W (ed) The Earth's earliest biosphere: its origin and evolution. Princeton University Press, Princeton, pp 260–290

Walker JG (1970) Oxygen poisoning in the annelid *Tubifex tubifex*. I. Response to oxygen exposure. Biol Bull 138:235–244

Walker RM, Johansen PH (1977) Anaerobic metabolism in the goldfish (*Carassius auratus*). Can J Zool 55:1304–1331

Walkinshaw LH (1945) Aortic rupture in field sparrow due to fright. Auk 62:141

Wallengren H (1914) Physiolog.-Biolog. Studien über die Atmung bei den Arthropoden. III. Die Atmung der Aeschnalarven. Lunds Univ Aasskr NF Avd 10:1–28

Walsby AE (1972) Gas-filled structures providing buoyancy in photosynthetic organisms. In: Soc exp biol symp No 26, The effects of pressure on organisms. The Company of Biologists, Cambridge, pp 233–250

Walshe BM (1948) The oxygen requirements and thermal resistance of chironomid larvae from flowing and from still waters. J Exp Biol 25:35–44

Walshe BM (1950) The function of haemoglobin in *Chironomus plumosus* under natural conditions. J Exp Biol 27:73–95

Walters DV, Olver RE (1978) The role of catecholamines in lung liquid absorption at birth. Pediatr Res 12:239–242

Wang J, Welkowitz TW, Kostis J, Semmlow J (1989) Incremental network analogue model of the coronary artery. Med Biol Eng Comput 27:416–422

Wang LCH (1978) Energetic and field aspects of mammalian torpor: the Richardson's ground squirrel. In: Wang LCH, Hudson JW (eds) Strategies in cold: natural torpidity and thermogenesis. Academic Press, New York, pp 109–145

Wang N, Banzett RB, Butler JP, Fredberg JJ (1988) Bird lung models show that convective inertia effects inspiratory aerodynamic valving. Respir Physiol 73:109–124

Wang N, Banzett RB, Nations CS, Jenkins F (1992) An aerodynamic valve in the avian primary bronchus. J Exp Zool 262:441–445

Wangensteen OD (1972) Gas exchange by a bird's embryo. Respir Physiol 14:64–74

Wangensteen OD, Rahn H (1970) Respiratory gas exchange by the avian embryo. Respir Physiol 11:31–45

Wangensteen OD, Weibel ER (1982) Morphometric evaluation of chorioallantoic oxygen transport in the chick embryo. Respir Physiol 47:1–20

Wangensteen OD, Wilson D, Rahn H (1971) Diffusion across the shell of the hen's egg. Respir Physiol 11:16–30

Wangensteen OD, Rahn H, Burton RR, Smith AH (1974) Respiratory gas exchange of high altitude-adapted chick embryos. Respir Physiol 21:61–70

Ward DM, Weller R, Bateson MM (1990) 16S rRNA sequences reveal numerous uncultured inhabitants in a natural community. Nature (Lond) 345:63–65

Warneck P (1988) Chemistry of the natural atmosphere. Academic Press, San Diego

Warner ACI (1981) Rate of passage of digesta through the gut of mammals and birds. Nutr Abstr Rev 51B:789–820

Wasawo DPS, Visser SA (1959) Swamp worms and tussock mounds in the swamps of Teso, Uganda. E Afr Agric J 25:86–90

Wasserman K (1994) Coupling of external to cellular respiration during exercise: the wisdom of body revisited. Am J Physiol 29:E159–E539

Wasserman K, Butler J, Kessel VA (1966) Factors affecting the capillary blood blood flow pulse in amn. J Appl Physiol 21:890–900

Wasserzug RJ, Paul RD, Feder ME (1981) Cardiorespiratory synchrony in anuran larvae (*Xenopus laevis*, *Pachymedusa dacnicolor*, and *Rana berlandien*). Comp Biochem Physiol 70A:329–334

Wassnetzov W (1932) Über die Morphologie der Schwimmblase. Zool Jahrb Abt Anat Ont Tiere 56:1–36

Waugh RE, Evans EA (1976) Viscoelastic properties of erythrocyte membranes of different vertebrate animals. Micovasc Res 12:291–304

Weathers WW (1976) Influence of temperature on the optimal hematocrit of the bullfrog, *Rana catesbeiana*. J Comp Physiol 105:173–184

Webb CL, Milsom WK (1994) Ventilatory responses to acute and chronic hypoxic hypercapnia in ground squirrel. Respir Physiol 98:137–152

Webber PJ (ed) (1979) High altitude geoecology. Westview, Boulder

Weber RE (1978) Respiration. In: Mill JP (ed) Physiology of annelids. Academic Press, London, pp 369–392

Weber RE (1992) Molecular strategies in the adaptation of vertebrate haemoglobin function. In: Wood SC, Weber RE, Hargens AR, Millard RW (eds) Physiological adaptations in vertebrates: respiration, circulation, and metabolism. Marcel Dekker, New York, pp 257–277

Weber RE, Hartvig M (1984) Specific foetal haemoglobin underlies the foetal-maternal shift in blood oxygen affinity in a viviparous teleost. Mol Physiol 6:27–32

Weber RE, Jesnen FB (1988) Functional adaptations in haemoglobins from ectothermic vertebrates. Annu Rev Physiol 50:161–179

Weber RE, Lykkeboe G (1978) Respiratory adaptations in carp blood. Influences of hypoxia, red cell organic phosphates, divalent cations and CO_2 on haemoglobin-oxygen affinity. J Comp Physiol 128:127–137

Weber RE, Pauptit E (1972) Molecular and functional heterogeneity in myoglobin from the polchaete *Arenicola marina* (L). Arch Biochem Biophys 148:322–324

Weber RE, Wood SC, Lomholt JP (1976) Temperature acclimation and oxygen binding properties of blood and multiple haemoglobins of rainbow trout. J Exp Biol 65:333–345

Weber RE, Wood SC, Davis BJ (1979) Acclimation to hypoxic water in facultative air-breathing fish: blood oxygen affinity and allosteric effectors. Comp Biochem Physiol 62A:125–129

Weber RE, Jesnen TH, Malte H, Tame J (1993) Mutant haemoglobin (α-Ala and β-Ser): functions related to high-altitude respiration in geese. J Appl Physiol 75:2646–2655

Wedler FC (1987) Determination of molecular heat stability. In: Henle KJ (ed) Thermotolerance, vol II: Mechanisms of heat tolerance. CRC Press, Boca Raton, pp 1–18

Weekes HC (1935) Review of placentation among reptiles with particular regard to the function and evolution of the placenta. Proc Zool Soc (Lond) 2:625–645

Weibel ER (1963) Morphometry of the human lung. Springer, Berlin Heidelberg New York

Weiber ER (1970/71) Morphometric estimation of pulmonary diffusion capacity. I. Model and method. Respir Physiol 11:54–75

Weibel ER (1973) Morphological basis of the alveolar-capillary gas exchange. Physiol Rev 53:419–495

Weibel ER (1979) Oxygen demand and size of respiratory structures in mammals. In: Wood SC, Lenfant C (eds) Evolution of respiratory processes: a comparative approach. Marcel Dekker, New York, pp 289–346

Weibel ER (1982) The pathway for oxygen: lung to mitochondria. In: Taylor CR, Johansen K, Bolis L (eds) A companion to animal physiology. Cambridge University Press, Cambridge, pp 31–48

Weibel ER (1983a) How does lung structure affect gas exchange? Chest 83:657–665

Weibel ER (1983b) Is the lung built reasonably? Am J Respir Dis 128:752–760

Weibel ER (1984a) The pathways for oxygen. Harvard University Press, Harvard

Weibel ER (1984b) Lung cell biology. In: Fishman AP, Fisher AB (eds) Handbook of physiology: respiration, vol 4. American Physiological Society, Washington, DC, pp 47–91

Weibel ER (1985a) Design performance of muscular systems: an overview. J Exp Biol 115:405–412

Weibel ER (1985b) Lung cell biology. In: Fishman AP (ed) Handbook of physiology, vol III, sect 2. American Physiological Society, Bethesda, pp 47–91

Weibel ER (1986) Functional morphology of lung parenchyma. In: Handbook of physiology. The respiratory system. Mechanics of breathing, sect 3, vol III, chapt 8. Am Physiol Soc, Bethesda, pp 89–111

Weibel ER (1987) Scaling of structural and functional variables in the respiratory system. Ann Rev Physiol 49:147–159

Weibel ER (1989) Lung morphometry and models in respiratory physiology. In: Chang HK, Paiva M (eds) Respiratory physiology: an analytical approach. Marcel Dekker, New York, pp 1–56

Weibel ER (1990) Morphometry: stereological theory and practical methods. In: Gill J (ed) Models of lung disease: microscopy and structural methods. Marcel Dekker, New York, pp 199–251

Weibel ER (1991) Fractal geometry: a design principle for living organisms. Am J Physiol 261:L361–L369

Weibel ER (1994) Design of biological organisms and fractal geometry. In: Nonnenmacher T, Losa GA, Weibel ER (eds) Fractals in biology and medicine. Birkhäuser, Basel, pp 68–85

Weibel ER (1996) The structural basis of lung function. In: West JB (ed) Respiratory physiology: people and ideas. Oxford University Press, New York, pp 3–46

Weibel ER, Gil J (1968) Electron microscopic demonstration of an extracellular duplex lining layer of alveoli. Respir Physiol 4:42–57

Weibel ER, Taylor CR (1986) Morphometric modelling of pulmonary diffusing capacity. Prog Respir Res 21:52–55

Weibel ER, Untersee P, Gil J, Zulauf M (1973) Morphometric estimation of pulmonary diffusion capacity. VI. Effect of varying positive pressure inflation of air spaces. Respir Physiol 18:285–308

Weibel ER, Taylor CR, O'Neil JJ, Leith DE, Gehr P, Hoppeler H, Langman V, Baudinette RV (1983) Maximal oxygen consumption and pulmonary diffusing capacity: a direct comparison of physiologic and morphometric measurements in canids. Respir Physiol 54:173–188

Weibel ER, Taylor CR, Hoppeler H, Karas RH (1987a) Adaptive variation in the mammalian respiratory system in relation to energetic demand: I. Introduction to problem and strategy. Respir Physiol 68:1–6

Weibel ER, Marques LB, Constantinopol F, Doffey F, Gehr P, Taylor CR (1987b) Adaptive variation in the mammalian respiratory system in relation to energetic demand: VI. The pulmonary gas exchanger. Respir Physiol 69:81–100

Weibel ER, Taylor CR, Hoppeler H (1991) The concept of symmorphosis: A testable hypothesis of structure-function relationship. Proc Natl Acad Sci 88:10357–10361

Weibel ER, Federspiel WJ, Doffey FF, Hsia CCW, König M, Navarro VS, Vock R (1993) Morphometric model for pulmonary diffusing capacity. I. Membrane diffusing capacity. Respir Physiol 93:125–149

Weibel ER, Taylor CR, Bolis L (eds) (1998) Principles of animal design: the optimization and symmorphosis debate. Cambridge University Press, London, pp 1–314

Weidenschilling SJ, Marzari F (1996) Gravitational scattering as a possible origin for giant planets at small stellar distances. Nature (Lond) 384:619–620

Weinberg S (1994) Life in the Universe. Sci Am 271:22–27

Weingarden M, Mizukami H, Rice SA (1982) Factors defining the rate of oxygen uptake by the erythrocytes. Bull Math Biol 44:135–147

Weinstein Y, Bernstein MH, Bickler PE, Gonzales DV, Samaniego FC, Escobedo MA (1985) Blood respiratory properties in pigeons at high altitudes: effects of acclimation. Am J Physiol 249:R765–R776

Weis-Fogh T (1964a) Diffusion in insect flight muscle, the most active tissue known. J Exp Biol 41:229–256

Weis-Fogh T (1964b) Functional design of the tracheal system of flying insects as compared with the avian lung. J Exp Biol 41:207–228

Weis-Fogh T (1967) Respiration and tracheal ventilation in locusts and other flying insects. J Exp Biol 47:561–587

Weis-Fogh T (1972) Energetics of hovering flight in hummingbirds and in *Drosophila*. J Exp Biol 56:79–104

Weis-Fogh T (1973) Quick estimates of flight fitness in hovering animals, including novel mechanisms for lift production. J Exp Biol 59:169–230

Weiss HS (1978) The role of shell diffusion area in incubating eggs at simulated altitude. J Appl Physiol 45:551–556

Welch HG (1987) Effects of hypoxia and hyperoxia on human performance. Exercise Sport Sci Rev 15:191–221

Welch HG, Pedersen PK (1981) Measurement of metabolic rate in hyperoxia. J Appl Physiol 51:725–731

Welch PS (1952) Limnology. McGraw Hill, New York

Wells DJ (1993a) Muscle performance in hovering hummingbirds. J Exp Biol 178:39–57

Wells DJ (1993b) Ecological correlates of hovering flight of hummingbirds. J Exp Biol 178:59–70

Wells GP (1949) Respiratory movements of *Arenicola marina* L. Intermittent irrigation of the tube and intermittent aerial respiration. J Mar Biol Assoc UK 28:447–464

Wells GP (1952) The respiratory significance of the crown in the polychaete worms *Sabella* and *Myxicola*. Proc R Soc Lond 138B:278–299

Wells GP (1966) The lugworm (*Arenicola*): a study in adaptation. Neth J Sea Res 3:294–313

Wells MJ (1962) Brain and behaviour in cephalopods. Heineman, London

Wells MJ (1983) Circulation in the cephalopods. In: Saleuddin ASM, Wilbur KM (eds) The Mollusca, vol 5: Physiology, part 2. Academic Press, New York, pp 239–290

Wells MJ, Wells J (1984) The effects of reducing gill area on the capacity to regulate oxygen uptake and on metabolic scope in a cephalopod. J Exp Biol 108:393–401

Wells MJ, Wells J (1985) Ventilation and oxygen uptake by *Nautilus*. J Exp Biol 118:297–312

Wells MJ, O'Dor RK, Mangold K, Wells J (1983) Oxygen consumption in movement by *Octopus*. Mar Behav Physiol 9:289–303

Wells MJ, Hanlon RT, Lee PG, DiMarco FP (1988) Respiratory and cardiac performance in *Loliguncula brevis* Blainville, 1823 (Cephalopoda: Myopsida); the effects of activity, temperature and hypoxia. J Exp Biol 138:17–36

Wells NA, Dorr JA (1985) From and function in the fish *Bothriopepis* (Devonian: Placodermi, Antiarchi): the first terrestrial vertebrate? Mich Acad 17:157–173

Wells RMG (1990) Haemoglobin physiology in vertebrate animals: a cautionary approach to adaptationists thinking. In: Boutilier RG (ed) Advances in comparative and environmental physiology, vol 6: Vertebrate gas exchange from environment to cell. Springer, Berlin Heidelberg New York, pp 143–162

Wells RMG, Dales RP (1975) Haemoglobin function in *Terebella lapidaria* L., an intertidal terebellid polychaete. J Mar Biol Assoc UK 55:419–495

Wells RMG, Weber RE (1982) The Bohr effect of the hemocyanin-containing blood from the terrestrial slug, *Arion ater*. Mol Physiol 2:149–159

Wells RMG, Jarvis PJ, Shumway SE (1980) Oxygen uptake, the circulatory system, and haemoglobin function in the intertidal polychaete *Terebella haplochaeta* (Ehlers). J Exp Mar Biol Ecol 46:255–277

Welsch U (1981) Fine structure and enzyme histochemical observations on the respiratory epithelium of the caecelian lungs and gills. A contribution to the understanding of the evolution of the vertebrate respiratory epithelium. Arch Histol Jpn Okayama 44:117–133

Welsch U (1983) Phagocytosis in the amphibian lung. Anat Anz 154:323–327

Welsch U, Aschauer B (1986) Ultrastructural observations on the lung of the Emperor penguin (*Apternodytes forsteri*). Cell Tissue Res 243:137–144

Welsh MJ (1987) Electrolyte transport by airway epithelia. Physiol Rev 67:1143–1184

Welty JC (1964) The life of birds, 1st edn. Constable, London

Welty JC (1979) The life of birds, 2nd edn. Saunders, Philadelphia

Wendelaar-Bonga SE, Meis S (1981) Effects of external osmolarity, calcium and prolactin on growth and differentiation of the epidermal cells of the cichlid teleost *Sarotheradon mossambicus*. Cell Tissue Res 221:109–123

Went FW (1968) The size of man. Sci Am 56:400–413

West B, Zhou BX (1988) Did chickens go north? New evidence for domestication. J Archaeol Sci 15:515–533

West BJ (1985) An essay on the importance of being nonlinear – lecture notes in biomaterials, 62. Springer, Berlin Heidelberg New York

West BJ (1987) Fractals, intermittency and morphogenesis. In: Degn H, Holden AV, Olsen LF (eds) Chaos in biological systems. Plenum Press, New York, pp 305–314

West BJ (1990) Fractal physiology and chaos in medicine. World Scientific, Singapore

West BJ, Goldberger AL (1987) Physiology in fractal dimensions. Am Sci 75:354–365

West BJ, Bhargava V, Goldberger AL (1986) Beyond the principle of similitude in the bronchial tree. J Appl Physiol 60:1089–1097

West BJ, Brown JH, Enquist BJ (1997) A general model for the origin of allometric scaling laws in biology. Science 276:122–126

West JB (1974) Respiratory physiology: the essentials. Williams and Wilkins, Baltimore

West JB (1977a) Ventilation/blood flow and gas exchange, 3rd edn. Blackwell, Oxford

West JB (1977b) Regional differences in the lung. Academic Press, New York

West JB (1983) Climbing Mt Everest without oxygen: an analysis of maximal exercise during extreme hypoxia. Respir Physiol 52:265–274

West JB (1991) High altitude. In: Crystal RG, West JB (eds) The lung: scientific foundations. Raven Press, New York, pp 2093–2107

West JB, Dollery CT (1965) Distribution of blood flow and ventilation-perfusion ratio in the lung, measured with radioactive CO_2. J Appl Physiol 15:405–410

West JB, Jones NL (1965) Effects of changes in topographical distribution of lung blood flow on gas exchange. J Appl Physiol 20:825–835

West JB, Matthews FL (1978) Stresses, strains, and surface pressures in the lung caused by its weight. J Appl Physiol 32:332–345

West JB, Wagner PD (1980) Predicted gas exchange on the summit of Mt Everest. Respir Physiol 42:1–11

West JB, Dollery CT, Naimark A (1964) Distribution of blood flow in isolated lung: relation to vascular and alveolar pressures. J Appl Physiol 19:713–724

West JB, Dollery CT, Matthews CME, Zardini P (1965) Effects of aerolized artificial surfactant on repeated oleic acid injury in sheep. Am Rev Respir Dis 141:1014–1019

West JB, Lahiri S, Maret KH, Peters RM, Pizzo CJ (1983) Barometric pressure at extreme altutudes on Mt Everest: Physiological significance. J Appl Physiol 54:1188–1194

West JB, Tsukimoto K, Mathieu-Costello O, Prediletto R (1991) Stress failure in pulmonary capillaries. J Appl Physiol 70:1731–1742

West JB, Mathieu-Costello O, Jones JH, Birks EK, Logerman RB, Pascoe JR, Tyler WS (1993) Stress failure in pulmonary capillaries in race horses with excercise-induced pulmonary haemorrhage. J Appl Physiol 75:1097–1109

West NH, Burggren WW (1982) Gill and lung ventilatory responses to steady-state aquatic hypoxia and hyperoxia in the bullfrog tadpole. Respir Physiol 47:165–176

West NH, Burggren WW (1983) Reflex interactions between aerial and aquatic gas exchange organs in larval bullfrogs. Am J Physiol 244:R770–R777

West NH, Burggren WW (1984) Control of pulmonary and cutaneous blood flow in the toad, Bufo marinus. Am J Physiol 247:R884–R894

West NH, Jones DR (1975) Breathing movements in the frog Rana pipiens. I. The mechanical events associated with lung and buccal ventilation. Can J Zool 53:332–344

West NH, Butler PJ, Bevan RM (1992) Pulmonary blood flow at rest and during swimming in the green turtle, Chelonia mydas. Physiol Zool 65:287–310

Weymouth FW, Crimson JM, Hall VE, Belding HS, Field II (1944) Total and tissue respiration in relation to body weight. A comparison of the kelp crab with other crustaceans and with mammals. Physiol Zool 17:50–71

Wheatly MG, Taylor EW (1979) Oxygen levels, acid-base status and heart rate during emersion of the shore crab, Carcinus maenas (L) into air. J Comp Physiol 132B:305–311

Wheatly MG, Taylor EW (1981) The effect of progressive hypoxia on heart rates, ventilation, gas exchange and acid-base status in the crayfish, Austropotamobius pallipes. J Exp Biol 92:125–141

White FN (1978) Comparative aspects of vertebrate cardiorespiratory physiology. Annu Rev Physiol 40:471–499

White FN, Bickler PE (1987) Cardiopulmonary gas exchange in the turtle: a model analysis. Am Zool 27:31–40

White FN, Ross G (1965) Blood flow in turtles. Nature (Lond) 208:759–760

White FN, Ross G (1966) Circulatory changes during experimental diving in the turtle. Am J Physiol 211:15–18

White FN, Kinney J, Siegfried WR, Kemp AC (1984) Thermal and gaseous conditions of hornbill nests. Natl Geogr Res Rep 17:931–936

White RE, Coon MJ (1980) Oxygen activation by cytochrome P_{450}. Annu Rev Biochem 49:315–356

Whitford RW, Hutchison VH (1967) Body size and metabolic rate in salamanders. Physiol Zool 40:127–133

Whiting HP, Bone Q (1980) Ciliary cells in the epidermis of the larval Australian dipnoan, Neoceratodus. J Linn Soc Zool Lond 68:125–137

Whitmore CM, Warren CE, Doudoroff P (1960) Avoidance reactions of salmonid and centrachid fishes to low oxygen concentrations. Trans Am Fish Soc 89:17–26

Wickler SJ, Marsh RL (1981) Effects of nestling age and burrow depth on CO_2 and O_2 concentrations in the burrows of bank swallows (Riparia riparia). Physiol Zool 54:132–136

Wicksten M (1994) Sytematics agenda 2000 (letters to the editor). Am Sci 82:205

Widdicombe J (1997) Airway alveolar permeability and surface liquid thickness: theory. J Appl Physiol 183:3–12

Widmer HR, Hoppeler H, Nevo A, Taylor CR, Weibel ER (1997) Working underground: respiratory adaptations in the blind mole rat. Proc Natl Acad Sci USA 94:2062–2067

Wiebe AH (1933) The effect of high concentrations of dissolved oxygen on several species of pond fishes. Ohio J Sci 33:110–126

Wiedersheim R (1879) Die Anatomie der Gymnophionen. Fisher, Jena

Wiener F, Morkin E, Skalak R, Fishman P (1966) Wave propagation in the pulmonary circulation. Circ Res 19:834–850

Wierenga PJ, Nielsen DR, Hagan RM (1969) Thermal properties of a soil based upon field and laboratory measurements. Soil Soc Am Proc 33:354–360

Wigglesworth VB (1950) The principles of insect physiology. Methuen, London

Wigglesworth VB (1953) Surface forces in the tracheal system of insects. Q J Microsc Sci 94:507–522

Wigglesworth VB (1965) The principles of insect physiology, 6th edn. Methuen, London

Wigglesworth VB (1972) The principles of insect physiology, 7th edn. Chapman and Hall, London

Wigglesworth VB, Lee WM (1982) The supply of oxygen to the flight muscles of insects: a theory of tracheole physiology. Tissue Cell 14:501–518

Wignall PB, Twitchett RJ (1996) Oceanic anoxia and the end-Permian mass extinction. Science 272:1155–1158

Wilkie DR (1977) Metabolism and body size. In: Pedley TJ (ed) Scale effects in animal locomotion. Academic Press, New York, pp 23–36

Wilkin PJ, Williams MH (1993) Comparison of the aerodynamic forces on a flying sphingid moth with those predicted by quasi-steady theory. Physiol Zool 66:1015–1044

Williams DD, Rausch RL (1973) Seasonal carbon dioxide and oxygen concentrations in the dens of hibernating mammals (Sciuridae). Comp Biochem Physiol 44A:1227–1235

Williams MH, Wesseldine S, Somma T, Schuster R (1981) The effect of induced erythrocythemia upon 5-mile treadmill run time. Med Sci Sports Exercise 13:169–175

Williams RJP, Da Silva JJRF (1978) High redox potential chemicals in biological systems. In Williams RJP, Da Silva JJRF (eds) New trends in bio-inorganic chemistry. Academic Press, London, pp 121–171

Willmer EN (1934) Some observations on the respiration of certain tropical fresh water fish. J Exp Biol 11:283–306

Willmer EN (1970) Cytology and evolution, 2nd edn. Academic Press, London

Wilson DE (1973) Bat faunas: a trophic comparison. Syst Zool 22:14–29

Wilson EO (1992) The diversity of life. Belknap Press, Cambridge

Wilson KJ, Kilgore DL (1978) The effects of location and design on the diffusion of respiratory gases in mammal burrows. J Theor Biol 71:73–101

Wilson TA (1981) Relations among recoil pressure, surface area, and surface tension in the lung. J Appl Physiol 50:921–926

Wilson TA, Bachofen H (1982) A model for mechanical structure of the alveolar duct. J Appl Physiol 52:1064–1070

Wilson TA, Beck KC (1992) Contributions of ventilation and perfusion inhomogeneities to the VA/Q distribution. J Appl Physiol 72:2298–2304

Wimsatt WA (1970) Biology of bats. Academic Press, London

Winick M, Coscia A, Nobble A (1967) Cellular growth in human placenta. I. Normal cellular growth. Pediatrics 39:248–251

Winker S, Woese CR (1991) A definition of the domains Archea, Bacteria and Eucarya in terms of small subunit ribosomal RNA characteristics. Syst Appl Microbiol 14:305–310

Winterstein H (1908) Beiträge zur Kenntnis der Fischatmung. Pfluegers Arch Gesamte Pysiol Menschen Tiere 125:73–98

Winterstein H (1925) Über die chemische Regulierung der Atmung bei den Cephalopoden. Z Vergl Physiol 2:315–328

Wintrobe M (1934) Variations in size and haemoglobin content of erythrocytes in the blood of various vertebrates. Folia Hematol 51:32–47

Wislocki GB, Belanger LF (1940) The lungs of the larger cetacea compared to those of smaller species. Ibid 78:289–297

Withers PC (1992) Comparative animal physiology. Saunders, New York

Withers PC, Casey TM, Casey KK (1979) Allometry of the respiratory and hematological parameters of Arctic mammals. Comp Biochem Physiol 64A:343–350

Withers PC, Hillman SS, Simmons LA, Zygmut AC (1988) Cardiovascular adjustments to enforced activity in the anuran amphibian, Bufo marinus. Comp Biochem Physiol A 89:45–49

Wit F (1932) Über den Einfluss der Luftfeuchtigkeit auf die Grösse der Atemöffnung bei Landpulmonaten. Z Vergl Physiol 18:116–124

Witt R, Lieckfeld CP (eds) (1991) Bionics: nature's patents. Pro Futura Verlag, Munich

Wittenberg BA, Wittenberg JB (1985) Oxygen pressure gradients in isolated cardiac myocytes. J Biol Chem 260:6548–6554

Wittenberg JB (1965) The secretion of oxygen into the swim bladder of fish. J Gen Physiol 44:521–526

Wittenberg JB (1976) Facilitation of oxygen diffusion by intracellular haemoglobin: oxygen and physiological function. Professional Information Library, Dallas

Wittenberg JB, Wittenberg BA (1962) Active secretion of O_2 in the eye of fish. Nature (Lond) 42:214–232

Wittenberg JB, Wittenberg BA (1987) Myoglobin mediated oxygen delivery to mitochondria of isolated cardiac myocytes. Proc Natl Acad Sci USA 84:7503–7507

Wittenberg JB, Wittenberg BA (1989) Transport of oxygen in muscle. Annu Rev Physiol 51:857–878

Wittmann J, Prechtl J (1991) Respiratory function of catecholamines during the late period of avian development. Respir Physiol 83:375–386

Wolfenson D, Frei YF, Berman A (1982) Blood flow distribution during artificially induced respiratory hypocapnic alkalosis in the fowl. Respir Physiol 50:87–92

Wolk E, Bogdanowicz W (1987) Hematology of the hibernating bat, *Myotis daubentoni*. Comp Biochem Physiol 88A:637–639

Wollman H, Smith TC, Stephen GW, Colton ET, Gleaton HE, Alexander SC (1968) Effects of extremes of respiratory and metabolic alkalosis on cerebral blood flow in man. J Appl Physiol 24:60–65

Wolvekämp HP (1955) Die physikalische Kieme der Wasserinsekten. Experientia 11:294–301

Wolvekämp HP, Waterman TH (1960) Respiration. In: Waterman TH (ed) The physiology of Crustacea, vol I. Academic Press, New York, pp 35–100

Wolvekämp HP, Baerends GP, Kok B, Mommaerts WFHM (1942) O_2 and CO_2 binding properties of the blood of the catfish (*Sepia officinalis*) and the common squid (*Loligo vulgaris*). Arch Neerl Physiol 26:203–218

Wood CM, Cadwell FH (1978) Renal regulation of acid-base balance in a fresh water fish. J Exp Biol 205:301–317

Wood CM, Randall DJ (1973) The influence of swimming activity on sodium balance in the rainbow trout (*Salmo gairdnen*). J Comp Physiol 82:207–233

Wood CM, Randall DJ (1981) Oxygen and carbon dioxide exchange during exercise in the land crab (*Cardisoma carnifex*). J Exp Zool 218:7–22

Wood CM, McMahon BR, McDonald DG (1970) Respiratory gas exchange in the resting starry flounder, *Platichthys stellatus*: a comparison with other teleosts. J Exp Biol 78:167–183

Wood CM, Perry SF, Randall DJ, Wood CM, Bergman HL (1989) Ammonia and urea dynamics in the Lake Magadi tilapia, a ureotelic fish adapted to an extremely alkaline environment. Respir Physiol 77:1–20

Wood CM, Bergman HL, Laurent P, Maina JN, Narahara A, Walsh PJ (1994) Urea production, acid base regulation and their interactions in the Lake Magadi tilapia, a unique teleost adapted to highly alkaline environment. J Exp Biol 189:13–36

Wood LDH, Bryan AC (1971) Mechanical limitations of exercise ventilation at increased ambient pressure. In: Lambertsen CJ (ed) Underwater physiology. Academic Press, New York, pp 125–205

Wood SC (1971) Effects of metamorphosis on blood respiratory properties and erythrocytes adenosine trophosphate level of the salamander, *Dicamptodon ensatus*. Respir Physiol 12:53–65

Wood SC, Glass ML (1991) Respiration and thermoregulation of amphibians and reptiles. In: Woakes AJ, Grieshaber MK, Bridges CR (eds) Physiological strategies for gas exchange and metabolism. Cambridge University Press, Cambridge, pp 107–124

Wood SC, Johansen K (1974) Respiratory adaptations to diving in the Nile monitor lizard, *Varanus niloticus*. J Comp Physiol 89:145–158

Wood SC, Lenfant CMJ (1976) Physiology of fish lungs. In: Hughes GM (ed) Respiration of amphibious vertebrates. Academic Press, London, pp 257–270

Wood SC, Lenfant CMJ (1979) Oxygen transport and oxygen delivery. In: Wood SC, Lenfant C (eds) Evolution of respiratory processes: a comparative approach. Lung biology in health and disease, vol 13. Marcel Dekker, New York, pp 193–223

Wood SC, Moberly WR (1970) The influence of temperature on the respiratory properties of iguana blood. Respir Physiol 10:20–29

Wood SC, Weber RE, Maloiy GMO, Johansen K (1975) Oxygne uptake and blood respiratory properties of the caecilian *Boulengerula taitanus*. Respir Physiol 24:355–363

Wood SC, Johansen K, Gatz RN (1978) Pulmonary blood flow, ventilation-perfusion ratio, and oxygen transport in a varanid lizard. Am J Physiol 233:R89–R93

Woodbury LA (1942) A sudden mortality of fishes accompanying a supersaturation of oxygen in Lake Waubesa, Wisconsin. Trans Am Fish Soc 71:112–117

Woodson RD (1984) Hemoglobin concentration and exercise capacity. Am Rev Respir Dis 129:S72–S75

Wourms JP (1993) Maximization of evolutionary trends for placental viviparity in the spadenose shark, *Scoliodon laticaudus*. Environ Biol Fish 38:269–294

Wourms JP, Callad IP (eds) (1992) Evolution of viviparity in vertebrates. Am Zool 32:249–354

Wourms JP, Groove BD, Lombardi J (1988) The maternal embryonic relationship in vivaparous fishes. In: Hoar WS, Randall DJ (eds) Fish physiology, II: The physiology of developing fish. Academic Press, New York, pp 1–134

Wray GA, Levinton JS, Shapiro LH (1996) Molecular evidence for deep Precambrian divergences among metazoan phyla. Science 274:568–573

Wu ER (1993) The development and evolution of a key morphological innovation: air breathing organs in the anabantoidei. Am Zool 33:14A

Wu HW, Chang HW (1947) On the arterial system of the gills and the suprabranchial cavities in *Ophiocephalus argus*, with special reference to the correlation with bionomics of the fish. Sinensia 17:1–15

Wüst G, Brogmus W, Noodt E (1954) Die zonale Verteilung von Salzgehalt, Niederschlag, Verdunstung, Temperatur und Dichte an der Oberfläche der Ozeane. Kiel Meeresforsch 10:137–161

Xu L, Mortola JP (1989) Effects of hypoxia on the lung of the chick embryo. Can J Physiol Pharmacol 67:515–519

Yager D, Butler JP, Bastacky J, Israel E, Smith G, Drazen JM (1989) Amplification of airway constriction due to liquid filling of airway interstices. J Appl Physiol 66:2873–2884

Yalden DW, Morris PA (1975) The lives of bats. Quadrangle – The New York Times Book Co, New York

Yamaguchi KD, Nguyen-Phu D, Scheid P, Piiper J (1985) Kinetics of O_2 uptake and release by human erythrocytes studies by stopped-flow technique. J Appl Physiol 58:1215–1224

Yamaguchi KD, Jürgens H, Bartels H, Scheid P, Piiper J (1988) Dependence of O_2 transfer conductance of red blood cells on cellular dimensions. In: Mochizuki M, Honig CR, Koyama T, Goldstick TK, Bruley DF (eds) Oxygen transport to tissue, vol. X. Plenum Press, New York, pp 571–578

Yamao F, Muto A, Kawauchi Y, Iwami M, Iwagami S et al. (1985) UGA is read as trytophan in mycoplasma capricolum. Proc Natl Acad Sci USA 82:2306–2309

Yang D, Oyaizu Y, Oyaizu H, Olsen GF, Woese CR (1985) Mitochondrial origins. Proc Natl Acad Sci USA 82:4443–4447

Yeliseev AA, Krueger KE, Kaplan S (1997) A mammalian mitochondrial drug receptor functions as a bacterial "oxygen" sensor. Proc Natl Acad Sci USA 94:5101–5106

Yen MRT (1989a) Elastic properties of pulmonary blood vessels. In: Chang HK, Paiva M (eds) Respiratory physiology: an analytical approach. Marcel Dekker, New York, pp 533–559

Yen MRT (1989b) Elasticity of microvessels in postmortem human lungs. In: Lee JS, Skalak TC (eds) Microvascular mechanics. Springer, Berlin Heidelberg New York, pp 175–190

Yoder MC, Checkley LL, Giger U, Hanson WL, Kirk RL, Capen RL, Wagner WW (1990) Pulmonary microcirculatory kinetics of neutrophils deficient in leukocyte adhesion-promoting glycoproteins. J Appl Physiol 69:207–213

Yonge CM (1947) The pallial organs in the aspidobranch Gastropoda and their evolution throughout the Mollusca. Philos Trans R Soc Lond 232B:443–518

Yonge CM (1952) The mantle cavity in *Siphonaria alternata* Say. Proc Malacol Soc Lond 29:190–199

Yonge CM (1958) Observations on the pulmonate limpet *Trimusculus* (*Gadinia*) *reticulatus* (Sowerby). Proc Malacol Soc Lond 33:31–37

Youlson JH, Freeman PA (1976) Morphology of the gills of larval and parasitic adult sea lamprey, *Petromyzon marinus* L. J Morphol 149:73–104

480

Young BA (1992) Trachea diverticula in snakes: possible functions and evolution. J Zool (Lond) 227:567–583

Young JS (1973) A marine kill in New Jersey coastal waters. Mar Pollut Bull 4:70P

Young RE (1972) The physiological function of hemocyanin in some selected crabs. II. The characteristics of haemocyanin in relation to terrestrialness. J Exp Mar Biol Ecol 10:193–206

Young RE (1978) Correlated activities in the cardioregulatory nerves and ventilatory system in the Norwegian lobster, *Nephrops norvegicus* (L). Comp Biochem Physiol 61A:387–394

Young RE, Coyer PE (1979) Phase co-ordination in the cardiac and ventilatory rhythms of the lobster *Momarus americanus*. J Exp Biol 62:53–74

Young FN, Zimmerman JR (1956) Variations in temperature in small aquatic situations. Ecology 37:609–611

Youvan D, Marrs B (1987) Molecular mechanisms of photosynthesis. Sci Am 256:42–50

Ysseling MA (1930) Über die Atmung der Weinbergschnecke (*Helix pomatia*). Z Vergl Physiol 13:1–60

Zaccone G, Fasulo S, Ainis L (1995) Gross anatomy, histology and immunohistochemistry of respiratory organs of air-breathing and teleost fishes. In: Pastor LM (ed) Histology, ultra-structure and immunohistochemistry of respiratory organs in non-mammalian vertebrates. Servicio de Publicaciones de la Universidad de Murcia, Murcia, Spain, pp 15–33

Zaccone G, Goniakowska-Witalinska L, Lauweryns JM, Fasulo S, Tagliafierro G (1989) Fine structure and serotonin immunohistochemistry of the neuroendocrine cells in the lungs of the bichirs *Polpterus delhezi* and *P. ornatipinnis*. Bas Appl Histochem 33:277–294

Zadunaisky JA (1984) The chloride cell: the active transport of chloride and the paracellular pathways. In: Hoar WS, Randall DJ (eds) Fish physiology, vol XB. Academic Press, London, pp 129–176

Zabzoule M, Marc-Vergnes JP (1986) A global mathematical model of the cerebral circulation in man. J Biomech 19:1015–1022

Zander R, Schmid-Schörbein H (1973) Intracellular mechanisms of oxygen transport in flowing blood. Respir Physiol 19:279–289

Zapol WM, Liggins GC, Schneider RC, Qvist J, Snider MT, Creasy RK, Hochachka PW (1979) Regional blood flow during simulated diving in the conscious Weddell seal. J Appl Physiol 47:968–973

Zeuthen E (1953) Oxygen uptake as related to body size in organisms. Q Rev Biol 28:1–12

Zeuthen E (1970) Rate of living as related to body size in organisms. Pol Arch Hydrobiol 17:21–30

Zhao-Xian W, Ning-Zhen S, Wei-Ping M, Jie-Ping C, Gong-Qing H (1991) The breathing pattern and heart rates of *Alligator sinensis*. Comp Biochem Physiol 98A:77–87

Zhuang FY, Fung YC, Yen RT (1983) Analysis of blood flow in cat's lung with detailed anatomi-cal and elasticity data. J Appl Physiol 55:1341–1348

Ziebis W, Forster S, Huettel M, Jørgensen BB (1996) Complex burrows of the mud shrimp, *Callianassa truncata*, and their geochemical impact in the sea bed. Nature (Lond) 382:619–622

Zimmer G, Beyersdorf F, Fuchs J (1985) Decay of structure and function of heart mitochondria during hypoxia and related stress and its treatment. Mol Physiol 8:495–513

Zinkler D (1966) Comparative metabolism of invertebrates. Z Vergl Physiol 52:99–144

Zoond A (1931) Studies on the localization of respiratory exchange in invertebrates. III. The book lungs. J Exp Biol 8:263–266

Subject Index

Amphibians   20, 35, 38–39, 51, 60, 91, 95–96,
    102, 109, 122, 125, 127–128, 149, 156
    174–175, 184–191, 193, 204, 219, 225, 227,
    233, 236, 239–243, 254, 262 271, 275–277,
    289, 292–293, 298–304, 306, 309, 324, 329,
    332–334, 349
Amphibious fish   219, 227–228, 261
Amphibolurus nuchalis   56, 306
Amphioxus   97, 184
Amphipnous cuchia   201, 222, 244, 253, 257
Amphipods   99
Amphiuma means   233
– *tridactylum*   271, 334
Amyda mutica   201
Anabas testudineus   102, 165, 197, 203, 228,
    231, 243, 253, 257
Anaerobic fermentation   28
Anaerobic glycolysis (*see* Glycolysis)
Anaerobic habitats   45, 153
– metabolism   12, 166, 241, 279, 285, 305, 308
– microorganisms   30
– state   84
Anal gills   188
Anas platyhynchos   176
Ancistrus anisitsi   253
Ancylus fluvialis   97
Andean goose   176
Angiosperms   35, 227
Angiotensin   17, 183, 338–340
Anguilla anguilla   51, 60, 197, 266–267, 269
– *vulgaris*   253
Annelids   92, 95, 98, 102, 106, 172, 185, 188, 200
Anomura   248–249
Anoxia   29, 46, 48, 116, 156, 165, 172–173,
    198, 213, 303, 308, 343
Anser anser   176
– *indicus*   128, 176, 318
Antarctic icefish (*see* Icefish)
Antioxidants   35
Antrozous pallidus   312
Anura   51, 115, 191, 193, 204, 300–303
Aorta   197, 242, 267, 269, 340
Aortic arch   69, 187
Aphrodite aculeata   200
Aplacophora   187
Aplysia californica   189, 198
– *depilans*   110
– *fasciata*   189
Apneustic breathing   240
Apoda   190, 300, 303, 327
Aquatic breathers   49, 109, 160, 219, 234, 271
– burrowers   172
– gas exchanger   127, 183, 223–224
– habitat   38, 155, 181, 227, 289, 341, 349
– hypoxia   159, 221, 223–224, 228
– life   35, 158, 193, 200, 252
– respiration   101, 115, 120, 151, 160, 165,
    186, 203, 239, 251, 254, 261, 267
– respiratory organs   25
Arachnids   150, 278–279, 288

Arapaima gigas   236, 242, 267
Arboreal   2
Archilochus colubris   318
Arenicola cristata   187, 192
– *marina*   60, 99, 105, 115, 153, 167, 172, 334
Arhynchite pugettensis   103, 201
Aroid plants   273
Artemia   99, 105, 110
Arterial blood   13, 24, 49, 97, 112, 123, 125,
    128, 138, 143, 176–177, 193, 196, 234
    242–243, 248, 269, 275, 308, 316, 318–319,
    340, 344, 356
Arterial PCO_2 49, 128, 143, 159, 175–176, 191,
    236–237, 239, 247, 321, 343–344, 357
Arterial PO_2 48, 102, 104, 113, 138, 143, 191,
    237, 242, 248, 250–251, 276, 279, 321 344,
    357
Arterio-venous shunt   342
Arterio-venous anastomosis   124
Arthropods   2, 35, 66, 87, 91, 184, 188–189,
    245, 265, 274, 276–277
Arum maculatum   273
Ascidians   97, 143, 184, 335
Ascites   80
Asphyxia   182, 274, 334
Aspidonotus spirifer   201
Aspirational breathing   104, 242
Astacus leptodactylus   99
Astylosternus robustus   92
Atelectasis   210, 332, 342
Athoracophoridae   247, 280
Atmosphere   2–3, 8–9, 26–27, 29–34, 36–40,
    44–46, 54, 56, 87, 90, 99, 106, 121, 152–156,
    165, 169, 170, 173–175, 227, 265–266, 318,
    345–346, 348, 353
Atria   16, 66, 96, 208
Austropotamobius pallipes   116, 221
Autamorphism   53, 130
Avian pulmonary system   16, 86, 127, 143, 207,
    304–305, 308–309, 314–316, 319, 320–323
Axolotl   193, 300

Bacteria   2–3, 30, 32, 40–41, 44–48, 130, 153,
    166–167, 171
Balaena mysticetus   83
Bar-headed goose (see *Anser indicus*)
Barometric pressure   9, 38, 156, 173–176, 353
Basommatophora   200
Bats   38, 56, 69, 71, 77, 79, 86, 101, 149,
    309–314, 317, 320, 323, 349
Benthonic species   2
Bernoulli-Venturi effect   287, 348
Bicarbonate (HCO_3^-)   13, 48–50, 126, 155–156,
    231, 233–234, 236, 266, 274–275 277, 357
Bichir   20, 164, 219, 222, 273, 332
Bifurcation   304, 306
Bifurcation points   87
Bimodal breathing   68, 102, 194, 217, 242–245,
    250, 253–254, 275, 334
– respiration   254

493